Advances in Computer, Information, and Systems Sciences, and Engineering

Advances in Computer, Information, and Systems Sciences, and Engineering

Proceedings of IETA 2005, TeNe 2005, EIAE 2005

Edited by

Khaled Elleithy
School of Engineering, University of Bridgeport, USA

Tarek Sobh
School of Engineering, University of Bridgeport, USA

Ausif Mahmood
School of Engineering, University of Bridgeport, USA

Magued Iskander
Polytechnic University, USA

Mohammad Karim
Old Dominion University, USA

Springer

A C.I.P. Catalogue record for this book is available from the Library of Congress.

ISBN-10 1-4020-5260-X (HB)
ISBN-13 978-1-4020-5260-6 (HB)
ISBN-10 1-4020-5261-8 (e-book)
ISBN-13 978-1-4020-5261-3 (e-book)

Published by Springer,
P.O. Box 17, 3300 AA Dordrecht, The Netherlands.

www.springer.com

Printed on acid-free paper

Contents

Acknowledgements

The International Joint Conferences on Computer, Information, and Systems Sciences, and Engineering (CISSE 2005) would not have been possible to conduct without the work, efforts and dedication of many individuals and organizations.

The editors would like to acknowledge the technical co-sponsorship provided by the University of Bridgeport and the Institute of Electrical and Electronics Engineers (IEEE). We would like to express our gratitude to Prof. Toshio Fukuda, the Chair of the International Advisory Committee; Prof. James Ritchie, the Conference Chair of the International Conference on Engineering Education, Instructional Technology, Assessment and E-Learning (EIAE 2005); Prof. Bill Taylor, the EIAE 2005 Technical Program Co-Chair; Prof. Amr El Abbadi, the Conference Chair of the International Conference on Telecommunications and Networking (TeNe 2005); and Prof. Keyanoush Sadeghipour, the Conference Chair of the International Conference on Industrial Electronics, Technology and Automation (IETA 2005). The efforts of the CISSE Webmaster, Mr. Andrew Rosca, have been instrumental in the success of the conference. The work of Mr. Tudor Rosca in managing and administering the conference on-line presentation system has been crucial in conducting the world's first real-time on-line high caliber research conference. We also wish to recognize the roles played by Ms. Susan Kristie and Mr. Sarosh Patel, our administrative and technical support team.

Finally, and most importantly, we would like to express our thanks to our colleagues, the reviewers and technical committee members who did an exceptional job in reviewing the submitted manuscripts. In particular, we would like to acknowledge the contributions of Abhilasha Tibrewal, Atef Al Najjar, Edwin Yu, Meetu Walia, Nerik Yakubov, Noel Kriftcher, Sookram Sobhan, Vikram Kapilla, Nariman Sepehri, Angel Pobil, Bruno Siciliano, Elsayed orady, Alessandro Giua, John Billingsley, Junling (Joyce) Hu, Mohamed Kamel, Navarun Gupta, Sadiq M. Sait, Saeid Nahavandi, Abdelshakour Abuzneid, Anatoly Sachenko, Habib Youssef, Hesham El-Sayed, JoAnne Holliday, Nirwan Ansari and Torleiv Maseng.

Preface
Advances in Computer, Information, and Systems Sciences, and Engineering

This book includes the proceedings of the International Conference on Industrial Electronics, Technology & Automation (IETA'05), the International Conference on Telecommunications and Networking (TeNe'05), and the International Conference on Engineering Education, Instructional Technology, Assessment, and E-learning (EIAE'05). The proceedings are a set of rigorously reviewed world-class manuscripts addressing and detailing state-of-the-art research projects in the areas of Industrial Electronics, Technology & Automation, Telecommunications and Networking, Engineering Education, Instructional Technology, Assessment, and E-learning.

IETA'05, TeNe'05, and EIAE'05 were part of the International Joint Conferences on Computer, Information, and Systems Sciences, and Engineering (CISSE'05) (www.cisse2005.org), the World's first Engineering/Computing and Systems Research E-Conference.

CISSE 2005 was the first high-caliber Research Conference in the world to be completely conducted online in real-time via the internet. CISSE 2005 received 255 research paper submissions and the final program included 140 accepted papers, from more than 45 countries. The concept and format of CISSE 2005 were very exciting and ground-breaking. The PowerPoint presentations, final paper manuscripts and time schedule for live presentations over the web had been available for 3 weeks prior to the start of the conference for all registrants, so they could choose the presentations they want to attend and think about questions that they might want to ask. The live audio presentations were also recorded and were part of the permanent CISSE archive, which also included all power point presentations and papers.

IETA'05, TeNe'05, EIAE'05 provided a virtual forum for presentation and discussion of the state-of the-art research in Industrial Electronics, Technology & Automation (IETA'05), Telecommunications and Networking (TeNe'05), and Engineering Education, Instructional Technology, Assessment, and E-learning (EIAE'05). The virtual conferences were conducted through the Internet using web-conferencing tools, made available by the conference. Authors presented their PowerPoint, audio or video presentations using web-conferencing tools without the need for travel. The Conferences sessions were broadcasted to all the conference participants, where session participants were able to interact with the presenter during the presentation and (or) during the Q&A slot that followed the presentation. These international conferences were held entirely on-line. The accepted and presented papers were made available after the conference both on a CD and as a book publication by Springer.

The IETA'05, TeNe'05, and EIAE'05 conferences audio rooms provided superb audio even over low speed internet connections, the ability to display PowerPoint presentations, and cross-platform compatibility (the conferencing software runs on Windows, Mac, and any other operating system that supports Java). In addition, the conferencing system allowed for an unlimited number of participants, which in turn granted us the opportunity to allow all IETA'05, TeNe'05, and EIAE'05 participants to attend all presentations, as opposed to limiting the number of available seats for each session.

This volume of the conference's proceedings includes 74 papers that were presented in the three conferences. The papers cover an interesting range of topics in the area of industrial electronics, technology & automation such as signal modeling, distributed remote control, precise diagnosis, wireless sensor networks, resource management in cellular networks, and wavelet thresholding.

Furthermore, in the area of telecommunications and networking the papers cover a broad range of research issues such as voice over IP, design of pervasive architectures, ad-hoc wireless networks, network design and simulation, reliability of telecommunications laws and regulation, ATM networks, and enhancing QoS support in mobile ad-hoc networks.

Finally, in the area of engineering education, instructional technology, assessment, and E-learning, the included papers cover a range of interesting topics such as computer aided learning technique in engineering education, web-based teaching collaborative platforms, engineering education, interdisciplinary project-based learning, intelligent learning environment, relationship between pedagogy, ethics & technology, and visual modeling.

We hope that you will find the selected papers interesting and covering the state-of-the-art advances in the area of industrial electronics, technology & automation telecommunications and networking, engineering education, instructional technology, assessment, and E-learning. We are looking forward to your participation in CISSE'06 (www.cisse2006.org).

Editors

Prof. Tarek Sobh
Vice Provost for Graduate Studies & Research
Dean, School of Engineering
University of Bridgeport

Prof. Khaled Elleithy
Associate Dean, School of Engineering
Dept. of Computer Science and Engineering
University of Bridgeport

Prof. Magued Iskander
Professor
Department of Civil and Environmental Engineering
Polytechnic University

Prof. Mohammad Karim
Vice-President for Research
Old Dominion University

Prof. Ausif Mahmood
Professor
Dept. of Computer Science & Engineering
University of Bridgeport

A Method for Enabling Proactive Fault Monitoring in High-End Computer Servers

A.V. Usynin
Nuclear Engineering Department
University of Tennessee, Knoxville
Knoxville, TN 37996 USA

A.M. Urmanov
Physical Sciences Center
Sun Microsystems
San Diego, CA 92122 USA

***Abstract* – This paper describes a new method for monitoring electronic components in computer servers. Conventional monitoring techniques implemented in mid-range and high-end computer systems rely on low-speed low-resolution internal instrumentation that produces low quality measurements of critical system parameters. The poor quality measurements lead to degraded performance of diagnostic and prognostic algorithms. The developed method is based on the capability of an auto-associative memory to restore an associated response given an imprecise key vector. The method is performed in two stages. In the collection stage, high-quality signatures representing observable parameters of known good and faulty electronic components are collected using high-speed high-resolution instrumentation. The collected signatures are then stored in the auto-associative memory matrix. In the monitoring stage, low-quality measurements acquired by the low-speed, low-resolution internal instrumentation (an A/D converter accessible via a system bus) are fed into the auto-associative memory. The auto-associative memory restores the high quality signatures. The restored high-quality signatures are then used by diagnostic algorithms for fault detection and isolation and by prognostic algorithms for the remaining useful life estimation.**

I. INTRODUCTION

In recent years a great deal of attention has been directed toward the development and implementation of Prognostic and Health Management (PHM) solutions for mission-critical systems. A condition and health-monitoring predictive system is the key enabler for a PHM engine that encompasses on-line automated monitoring, data analysis, diagnostic routines and advanced prognostic applications predicting remaining useful life. The benefits of PHM have been recognized in many military and industrial applications.

Whisnant and Gross [1] use telemetry time series signals collected from enterprise-level computer servers to proactively monitor for a signal degradation that is an indicator of an upcoming failure for most types of failure mechanisms found in high-end computer systems.

Roemer and Kacprzynski [2] investigate diagnostic and prognostic technologies applied to gas turbine engines for performance anomaly detection and diagnosis. Ref. [2] considers the probabilistic fault identification process to diagnose particular fault patterns. Pattern recognition is performed using an unsupervised neural network for clustering fault types and a standard backpropagation neural network to classify particular fault patterns.

Price et al. [3] perform a study of integrated vehicle health monitoring (IVHM) systems that, in the longer term, will provide a basis for the development of self-repairing structures. In the shorter term, IVHM systems reduce maintenance and inspection requirements. The proposed algorithms enable a self-organized identification of damage. The detailed description of those algorithms can be found in [4]. Bengtsson et al. [5] discuss the technical components of a condition-based maintenance (CBM) approach. The study gives an illustration of how a CBM system can be designed using Case-Based Reasoning and Sound Analysis.

Vilalta et al. [6] perform a study of predictive algorithms developed for long- and short-term prediction of system events and performance variables observed in large computer systems. The short-term prediction of system events, e.g. a network router failure, is approached by using data-mining techniques to reveal data patterns frequently occurring before these events. The presented paper contributes to the efforts to extend the benefits of PHM toward computer servers and other electronic components.

Condition and health-monitoring predictive systems for computer servers running mission- or business-critical applications utilize proactive fault detection, dynamic reconfiguration of critical components and associated subsystems. A condition and health-monitoring system that can miss inactive fault indications cannot be efficiently used in proactive and preventive service offerings. Therefore, there is a need for a proactive fault signature detection method that is: (1) capable of detecting subtle and/or severely distorted fault signatures observed via low-speed, low-resolution internal instrumentation found in modern computer servers and (2) able to keep false alarm rate at acceptable levels.

Conventional monitoring approaches realized in most modern complex computer systems make use of threshold values defined for critical quantities. A threshold-based monitoring is performed by taking measurements of critical quantities at pre-determined intervals. The measured values are then compared against some pre-determined thresholds. If a particular quantity overshoots the corresponding threshold a faulty condition is then alarmed.

The disadvantage of such an approach is that the threshold values are usually set too high or too low to avoid an excessive number of false alarms. This results in that a threshold-based monitoring system can miss some subtle

K. Elleithy et al. (eds.), Advances in Computer, Information, and Systems Sciences, and Engineering, 1–4.

indications of latent faults that may have already occurred in the system but not yet produced a failure.

Another noteworthy aspect is that the signal sampling rate, in some cases, may not be sufficient to catch a complete signature of the failure precursor so that the fault remains undiscovered. One evident solution to this problem is to increase the sampling rate of the internal instrumentation e.g. to hundreds of kHz or faster sampling rates. However, due to bandwidth limitations of the system bus, this solution is not practical for enterprise-level servers. A modern high-end computer system contains hundreds different electronic components that need to be monitored. The bandwidth of the existing internal instrumentation cannot provide the desired level of resolution at which the failure precursors can be registered to evaluate the degree of degradation for all monitored components.

In addition to the mentioned drawbacks, a conventional monitoring system can miss fault indications because the captured signatures are distorted due to severely quantized output from analog/digital (A/D) chips widely used in most computer platforms today.

In this paper we introduce a novel technique for detection of faults in electronic components that may lead to failures in midrange and high-end computer systems. The proposed technique improves over existing methods by (1) taking advantage of an auto-associative memory based pattern recognition system for restoring severely distorted fault signatures, (2) providing online estimation of subtle failure precursor characteristics. As additional advantages, the method is readily applicable to existing low-speed, low-resolution internal instrumentation found in modern computer servers, and it is easily adaptable to newly discovered failure precursor patterns on existing and future computer systems.

II. METHODOLOGY

The electronic component fault signature detection method described in this paper employs an associative memory based pattern recognition system. The associative memory scheme operates as follows. There is a set of memory cells or patterns having an organization in which any pattern is accessed by a specific input key. This associative mapping is opposed to using of an explicit address. When a partially corrupted input key is fed into the associative memory, the stored pattern is retrieved as if the input key were undamaged.

This paper considers one class of associative memory: auto-associative memory. The term "auto-associativity" means that the patterns stored in memory are used as keys. The auto-associative memory retrieves the pattern that most resembles the input key, i.e. associating patterns with themselves.

A number of practical applications employing the auto-associativity paradigm has been proposed in recent years. Hines and Uhrig [7] applied an auto-associative neural network (AANN) to online sensor calibration monitoring. Gross et al. [8] used an auto-associative multivariate state estimation technique (MSET) for model-based fault detection. Krell et al. [9] presented an Associative Restoration algorithm combined with a Fuzzy Image Enhancement technique. The restoration algorithm makes use of an artificial neural network (ANN)-based auto-associative memory. The ANN auto-associative memory enables the inclusion of a-priori knowledge given by a high-quality x-ray image into an in-treatment image, which, usually, is of very poor quality due to the imaging physics. Cherkassky et al. [10] performed a comprehensive review of associative memory-based approaches to robust data retrieval.

This study considers a typical linear auto-associative memory scheme, which is given by the following expression

$$\hat{\mathbf{y}} = \mathbf{M}\mathbf{x} \tag{1}$$

where $\mathbf{x}$ is an input (key) vector ($n \times 1$), $\hat{\mathbf{y}}$ is a response vector ($k \times 1$), $\mathbf{M}$ is a memory matrix storing m associations. In the case of an auto-associative memory, the dimensions of a key vector equal to those of a response vector ($k = n$).

The pivotal element of an associative memory is the matrix $\mathbf{M}$. A number of different associative memory matrix schemes have been described in published literature [11-14]. The most popular schemes are Generalized Inverse (GI) and Correlation Matrix Memory (CMM). The GI memory matrix is given by the following expression.

$$\mathbf{M_G} = \mathbf{Y}\mathbf{X}^{+} \tag{2}$$

where $\mathbf{X}$ is a ($n \times m$) matrix of input (key) vectors, $\mathbf{Y}$ is a ($k \times m$) matrix of response vectors. $\mathbf{X}^{+}$ denotes the generalized inverse of $\mathbf{X}$ [15]. The expression of an auto-associative memory scheme is given by

$$\hat{\mathbf{y}} = \mathbf{Y}\mathbf{X}^{+}\,\mathbf{x} = \mathbf{Y}(\mathbf{X}^{T}\mathbf{X})^{-1}\,\mathbf{X}^{T}\mathbf{x} \tag{3}$$

The important assumption made in (3) is that the number of associations stored in the memory matrix is less than or equal to the dimension of the input vectors, i.e. $n \geq m$.

In the case of CMM, the memory matrix is constructed according to the following expression:

$$\mathbf{M_{CMM}} = \mathbf{Y}\mathbf{X}^{T} \tag{4}$$

The expression of the auto-associative scheme is given by:

$$\hat{\mathbf{y}} = \mathbf{Y}\mathbf{X}^{T}\mathbf{x} \tag{5}$$

A detailed description and performance evaluation of GI and CMM associative memories can be found in [16]. The GI associative memory is found to be preferable to the CMM method except when the number of stored associations is close to the key vector dimension.

The key idea behind the proposed method is to use the auto-associative memory to recall a high resolution signature of the failure precursor registered by low-speed-low-resolution internal instrumentation. The failure precursors caught by the internal instrument are incomplete or distorted images of those stored in the auto-associative memory. Having recalled the high resolution signature, one is able to assess the severity of the degradation indicated by the caught failure precursor.

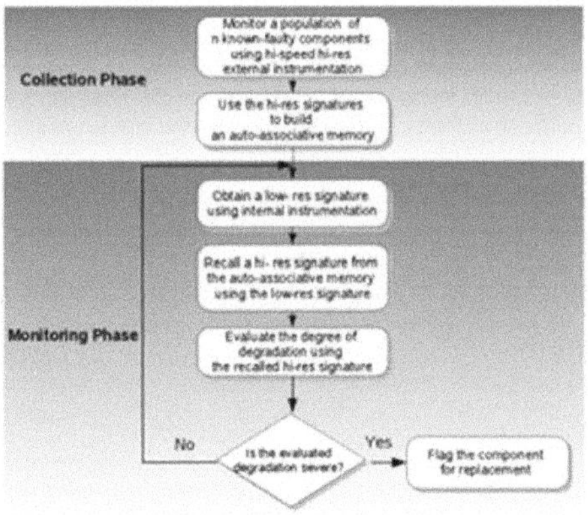

Fig. 1. A flowchart illustrating the use of the proposed method.

The algorithm of the method is illustrated in Fig. 1. In the collection phase, some number of the components known to be faulty are monitored with high speed, high resolution external instrumentation to capture high resolution signatures of failure precursors. The collected high resolution signatures constitute the auto-associative memory matrix. Having built the auto-associative memory matrix one initiates the monitoring phase using standard low speed, low resolution internal instrumentation. Each low resolution signature is considered to be a corrupted key vector that is input to the auto-associative memory. The memory then recalls a high-resolution signature of the failure precursor. The recalled high-resolution signature provides a detailed representation of the characteristics important for evaluation of degradation. A decision is then made as to whether the monitored component remains in an operational mode or should be replaced. The following section provides an example of practical application of the proposed method.

III. EXPERIMENTAL RESULTS

The proposed method is illustrated using the following example of detecting faulty power supplies found in high-end servers. A series of experiments has revealed that degradation in the power supplies is characterized by transients in the output voltage that have a characteristic pattern: a droop followed by a spike. The internal instrumentation (an A/D converter accessible via an i2c bus), due to certain limitations, can provide only few observations whose values are different from the nominal voltage level as shown in the upper pane of Fig. 2. Possessing the incomplete information it is difficult to evaluate the characteristics, such as amplitude and duration, of the transient. These characteristics are to be used to evaluate the severity of the damage found in a particular power supply.

Using an auto-associative memory based pattern recognition system one is able to reconstruct (recall) the complete signature of the registered transient, which is depicted in the lower pane of Fig. 2. To obtain such a representation the auto-associative memory should contain a database of high resolution signatures of possible transients. Collection of the high-resolution signatures is performed using external high-speed-high-resolution instrumentation on the power supplies known to be faulty.

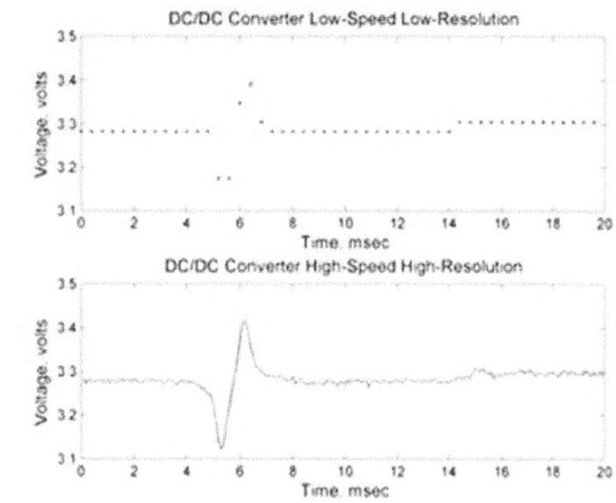

Fig. 2. Low resolution (the upper pane) and high resolution (the lower pane) signatures of the transient observed in the power supply.

Detection of such problems is of importance since the faults in power supplies can induce failures in higher-level components of a computer system. Fig. 3 shows an example representing a sequence of events that cause an overall system failure assuming a faulty power supply provides power to memory on the system board.

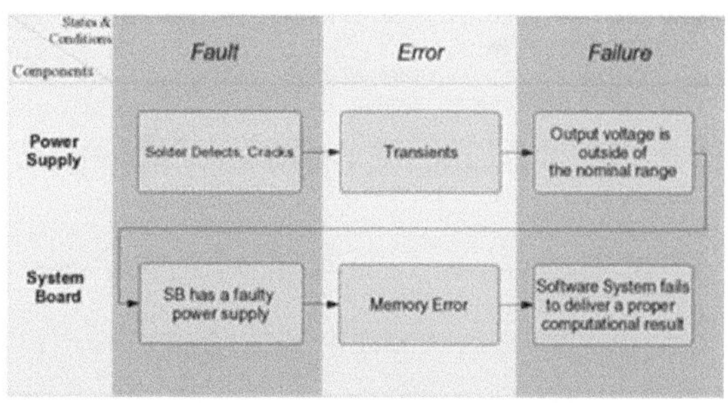

Fig. 3 A fault-error-failure chain illustrating how a faulty power supply induces an overall system failure.

A fault, such as a solder crack inside the power supply can introduce an error state. To be specific, the error state is a transient occurring at the time when the power supply is supposed to be in the normal operational mode, i.e. producing a steady nominal voltage. This error state causes a failure of the power supply in sense that it is not delivering its designated service while the transient is taking place in the output voltage. This power supply failure is considered to be a fault in terms of the system board having the power supply as a component. The fault occurred in the system board creates an error state of the system board. For instance, it can be a corrupted content of a memory cell located at the faulty system board. The corrupted memory cell is a potential source of the overall system failure in sense that the computer system fails to deliver a correct computation result.

IV. CONCLUSION

The proposed method provides a novel technique to detect faults in electronic components that may cause local and/or global hardware failures in midrange and high-end computer systems. The proposed detection technique has the following advantages over existing detection methods:

- Takes advantage of an auto-associative memory based pattern recognition system
- Provides online estimation of the failure precursor characteristics
- Readily applicable to existing systems with low-resolution low-speed internal instrumentation
- Easily adaptable to newly discovered failure precursor patterns

Utilization of the proposed method results in earlier detection of component degradation and better avoidance of system failures, which is crucial for achieving higher availability of computer systems. An example is presented that illustrates the use of the method in identification of faulty power supplies.

REFERENCES

[1] K. Whisnant, K. Gross and N. Lingurovska, "Proactive Fault Monitoring in Enterprise Servers", *Proceedings of the 2005 International Multiconference in Computer Science & Computer Engineering*, Las Vegas, NV, June 2005

[2] M.J. Roemer and G.J. Kacprzynski, "Advanced Diagnostics and Prognostics for Gas Turbine Engine Risk Assessment", *Aerospace Conference Proceedings, 2000 IEEE*, Vol. 6: 345-353, 2000

[3] D.C. Price, et al., "An Integrated Health Monitoring System for an Ageless Aerospace Vehicle", Published in *Structural Health Monitoring 2003: From Diagnostics & Prognostics to Structural Health Management*, ed. Fu-Kuo Chang, DEStech Publications (Lancaster, PA), pp. 310-318. 2003

[4] H.G. Lovatt, G. T. Poulton, D. C. Price, M. Prokopenko, P. Valencia, P. Wang. 2003. "Self-Organizing Impact Boundaries in Ageless Aerospace Vehicles". *Proceedings of the 2nd International Conference on Autonomous Agents and Multi-Agent Systems (AAMAS '03),* Melbourne, Australia, July 2003

[5] Bengtsson, M., Olsson, E., Funk, P., Jackson, M., "Technical Design of Condition Based Maintenance System: A Case Study using Sound Analysis and Case-Based Reasoning", *Proceedings of the 8th Congress of Maintenance and Reliability Conference*, Knoxville, TN,USA, 2004

[6] R. Vilalta, C. V. Apte, J. L. Hellerstein, S. Ma, and S. M. Weiss. "Predictive Algorithms in the Management of Computer Systems", *IBM Systems Journal*, Vol. 41., No. 3, 2002

[7] J. W. Hines and R. E. Uhrig, "Use of autoassociative neural networks for signal validation", *Journal of Intelligent and Robotic Systems*, Kluwer Academic Press, 1997

[8] K.C. Gross et al., "Application of a Model-based fault detection system to nuclear plant signals", *Proc. Of the Intl. Conf. On Intelligent System Application to Power Systems*, Seoul, Korea, pp. 60-65, 1997

[9] G. Krell, H.R. Tizhoosh, T. Lilienblum, C.J. Moore and B. Michaelis, "Enhancement and Associative Restoration of Electronic Portal Images in Radiotherapy", *Proceedings of the 10th IEEE Symposium on Computer-Based Medical Systems (CBMS'97)*, p.104, 1997

[10] V. Cherkassky, N. Vassilas and G.L.Brodt, "Conventional and Associative Memory-based Spelling Checkers", *Tools for Artificial Intelligence, Proceedings of the 2nd International IEEE Conference on*, p. 138-144, 1990

[11] T. Kohonen, "Correlation memory matrix", *IEEE Trans. Comput.*, vol. 21, pp. 353-359, Apr. 1972

[12] T. Kohonen, *Self-Organization and Associative Memory*, New York: Springer-Verlag, 1984

[13] K. Nakano, "Associatron: a model of associative memory", *IEEE Trans. Syst. Man. Cyber.*, vol. 2, pp. 380-388, July 1972

[14] K. Murakami, and T. Aibara, "Construction of a distributed associative memory on the basis of Bayes discriminant rule", *IEEE Trans. Pattern Analysis and Machine Intelligence.*, vol. 3, pp. 210-214, Mar. 1981

[15] R. Penrose, "A Generalized Inverse for Matrices", *Proc. Cambridge Phil. Soc.*, 51, pp. 406-413, 1955

[16] K.J. Raghunath and V. Cherkassky, "Noise Performance of Linear Associative Memories", *IEEE Trans. On Pattern Analysis and Machine Intelligence.*, vol. 16, no. 7, pp. 757-765, July 1994

Selection of Mill Cutter and Cutting Parameters through an Expert System

L. Rubio and M. De la Sen

Abstract— **This paper discusses the selection of tools in milling operations. An expert system hinged on numerical methods has been developed to carry out this research. The knowledge base is given by limitations in process variables, which let us define the allowable cutting parameter space. The mentioned process variables cause instabilities due to tool-work-piece interaction, knowing as chatter vibration, and the power available in the spindle motor. Then, a tool cost model is contrived. It is used to choose the suitable cutting tool, among a known set of candidate available cutters. Once the cutting tool is selected, the cutting parameters are calculated. For this purpose, other two cost functions are designed, which are depended on the time and frequency domains response properties. An example is presented to illustrate the method.**

I. Introduction

Machining, in particular milling operations, is a broad term used to define the process of removing material from a work-piece. Furthermore, the milling operation process is required, nowadays, to increase its productivity, reducing cost and improving the final product [1].

This paper bring forward the concept of selecting an appropriate mill cutter, among a known set of candidate cutters, and obtaining the adequate cutting parameters for milling operations through an expert system.

There are several versatile approaches for tool and/or cutting parameter selection based on the design of expert systems on manufacturing environments. Wong and Hamouda [2] developed an on-line fuzzy expert system. The system inputs, the tool type, the work-piece material hardness and the depth of cut and control the cutting parameters at the machine, as output. Cemar Cakir et al. [3] explained an expert system based on experience rules for die and mold operations. In that paper, the geometry and material of the work-piece, tool material and condition and operation type are considered as inputs. Then, the system provides recommendations about tool type, tool specifications, work-holding method, type of milling operation, direction of feed and offset values. Vidal et al. [4] focused on the problem of choosing the manufacturing route in metal removal process. They select the cutting parameters by optimizing the cost of the operation taking into account several factors, such as, material, geometry, roughness, machine and tool. Carpenter and Maropoulus [5] designed a system, which provides reliable tool selection and cutting data for a range of milling operations. The method employs rule based decision logic and multiple regression techniques for a wide range of materials.

Here, the developed expert system consists of the relative compliance between the tool and the work-piece and it is predicted with analytical methods. Moreover, milling process simulations have been developed in the time and frequency domains, which are, then used in the expert system definition.

Firstly, the knowledge base is explained. Basically, it defines the allowable cutting parameters, which are known as cutting parameter space, for a given tool-work-piece configuration. It is based on the chatter vibrations avoidance, which limits the productivity of the process, and on a spindle power limitation criterion.

On the other hand, a novel tool cost function is designed. It sub-optimizes a combination of several effects of interest, namely, the spindle power consumption, the material remove rate (MRR), as well as, a stability criterion against possible perturbations in the spindle speed variable (ΔN_s).

The MRR is a parameter which measures the process effectiveness. It is required to be as large as possible. But, if MRR increases beyond certain limits, chatter vibrations are appreciated and the process becomes unstable [6]. Other variable which limits the process effectiveness is the power available in the spindle motor (P_t) [7]. A third parameter, ΔN_s, taking part into the cost function is considered to ensure a well-posed behavior of the system if a perturbation in the spindle speed happened.

In conclusion, the proposed cost function is a measure of how the milling process is being carried out at certain operation conditions. The larger the cost function is, the worst operation will be. Thus, the cutter which minimizes the designed cost function is selected.

Once, the cutting tool is selected, the cutting parameters are obtained for a given tool configuration. A second cost function is designed. It is composed by the above mentioned cost function for the input cutting space parameter and the selected tool gives the programmed cutting parameters.

Then, the expert system considers tool characteristics, related tool-work-piece material parameters and milling operation as inputs and outputs the selected tool among the candidates and robust programmed cutting parameters.

L. Rubio and M. De la Sen are with the Instituto de Investigación y Desarrollo de Procesos, Facultad de Ciencia y Tecnología, Campus de Leioa, Universidad del País Vasco, Apartado 644, Bilbao, Spain (e-mail: {webrurol,webdepam}@lg.ehu.es).

K. Elleithy et al. (eds.), Advances in Computer, Information, and Systems Sciences, and Engineering, 5–12.

II. System Description

A model, which represents the dynamic compliance between the tool and work-piece in milling processes, has been developed. In this case, it is predicted with analytical methods. The model assumes the cutter to have two orthogonal degrees of freedom and the work-piece to be rigid. The feed is along the x-axes (fig.1). The milling cutter has n_t teeth, which are equally spaced. The dynamics of the system is given by differential equations [8, 9]:

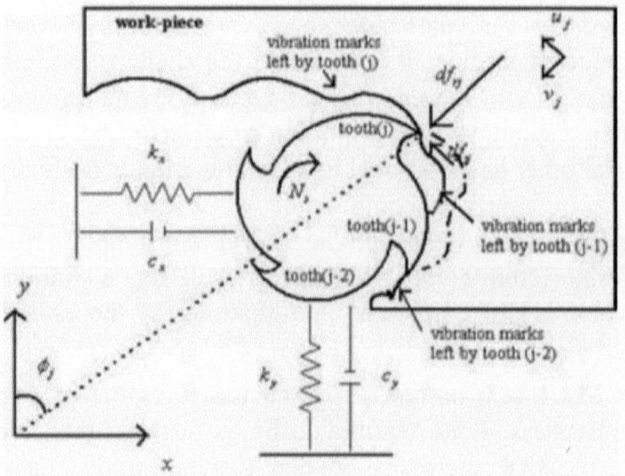

Fig. 1. End part of the milling system: tool and work-piece.

$$m_x \cdot \ddot{x}+c_x \cdot \dot{x}+k_x \cdot x=\sum_{j=0}^{n_t} f_{xj}(t)=f_x(t)$$
$$m_y \cdot \ddot{y}+c_y \cdot \dot{y}+k_y \cdot y=\sum_{j=0}^{n_t} f_{yj}(t)=f_y(t) \qquad (1,2)$$

where m_i, c_i and k_i are the mass, damping and stiffness of the tool, f_{xj} and f_{yj} are the components of the cutting force that is applied by the jth tooth, which are obtained by projecting f into the two orthogonal axis.

A simple model of the cutting forces will be discussed here which express the tangential cutting force to be proportional with the instantaneous chip thickness.

Despite this simplicity, this model captures the essence of the process. Hence,

$$f_t = k_t \cdot b \cdot h \qquad (3)$$

where k_t is the specific cutting force parameter, b is the axial depth of cut and h is the instantaneous chip thickness. In addition, the radial force may also be expressed in terms of the tangential force as,

$$f_r = k_r \cdot f_t \qquad (4)$$

where k_r is a proportional constant. This cutting force model has been used by several authors [6].

The most critical variable in (3) is the chip thickness because it changes not only with the geometry of cutting tool and cutting parameters, but also with the uneven surface left by the previous passes of the cutting tool. Hence, after determining the chip thickness for an uncut fresh surface, this thickness must be compared with the undulations left by the cutting tool during previous passes at the same position to obtain the instantaneous thickness of the material left to be removed. This process is known as *regenerative mechanism* [6].

The chip thickness is measured in the radial direction, with the coordinate transformation,

$$v_j = -x \cdot \sin\phi_j - y \cdot \cos\phi_j \qquad (5)$$

where ϕ_j is the instantaneous angular immersion of tooth j measured clockwise from the normal Y axis (fig.1).

$$h(\phi_j)=\left[s_t \cdot \sin\phi_j + \left(v_j^o - v_j\right)\right] \cdot g(\phi_j) \qquad (6)$$

where $g(\phi_j)$ is a unit step function which determines whether the tooth is in or out of cut, s_t is the feed rate per tooth, T is the tooth period and, if the spindle rotates at $N_s(rad/s)$, the immersion angle varies as $\phi_j(t)=N_s \cdot t$, and $\phi_j(t)=0$ if the j-tooth is not engaged with the part [6].

The 4th order Runge-Kutta method is employed to solve the differential equations (1) and (2) in the time domain [8, 9].

The stability lobes are obtained from the equations (3) and (4). This basic study has been yet developed by the authors in previous studies, and there exists an extensive bibliographic to develop the system stability in the cutting parameters space (see for instance [1, 6, 9]). In the next section, some mentions to the mathematical formulae of the stability lobes will be made. The main purpose of the present paper is not to develop analytically the stability border line but to take it into account in the design of the expert system. Therefore, the stability border line is supposed known. For a good understanding of the expert system, that will be presented in the section 3, above mentioned references can be used.

Figure 2 shows the lobes char and the analytical time and frequency domain response for a tool 2 system, which characteristics can be seen in section 5. The chatter stability lobes make up a spindle speed (frequency) dependent dividing line between stable (down part line) and unstable (up part line) depth of cut for a certain width of cut. The left place responses correspond to a stable state and the right place ones correspond to unstable state. Stable state corresponded figures present a delimit time response, and the tooth passing frequency and its harmonics, frequency response. Unstable state corresponded figures present a not delimit time response, and the chatter frequency is appreciated.

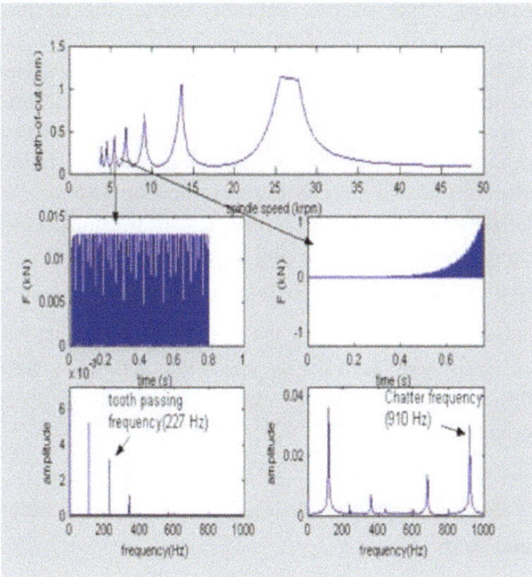

Figure 2: The milling system representation, stability chars, force time response and force frequency response.

The expert system, which is explained in the next secciōn, is obtained from the analytical prediction of the tool-work-piece system behavior presented above..

III. Expert System

The main objective of the expert system is to obtain a mill cutter, among the available ones, which have an operating point or adequate cutting parameters, with maximum productivity (MRR), robustness stability against spindle speed perturbations and minimum power consumption.

For this purpose, the allowable cutting space parameter, spindle speed, feed rate and axial depth of cut for a constant radial depth of cut, are considered. The regenerative chatter instability and the power available in the spindle motor are taken into account. Then, a novel cost function is schemed. It is inversely proportional to MRR and a parameter determinate as stability against spindle speed perturbation and proportional to power consumption. Each term of the cost function has a proportional factor to have terms of approximately similar magnitudes. Also, a weight factor which measures the importance of each term is incorporated. The weighting factors are intended to be programmed by the machine operator.

A. Milling process determination and preliminary rules

In order to evaluate the system performance, a suitable tool and performance indices are needed. Milling processes, basically, consists of two phases roughing and finishing the surface. The main difference between these operations is to decide the most appropriate performance index for a given tool. The quality and geometric profile of the cutting surface is of paramount importance in milling finishing operation, whereas roughing -milling consists on removing a large amount of material from a blank.

This paper deals with roughing milling operation. The rate at which the material is removed is called material removing rate (MRR). This parameter measures the productivity of machining processes. In milling operations, MRR is defined as the multiplication between axial and radial depth of cut, and feed per tooth. MRR upper limit is given by chatter vibrations and power deliver by the spindle motor. For certain combinations of cutting parameters, such as spindle speed, axial depth of cut and feed per tooth, either chatter vibrations are sensed, or the power available by the spindle motor is insufficient. Then, these parameters bound the roughing-milling productivity.

For those reasons, the input cutting parameter space is given by the cutting parameters as a first approximation, below the line at the stable stability lobes char while the power consumption is less than the power available by the spindle motor.

But, due to uncertainties in the model, the lobes are constructed, not by replacing pure imaginary roots into the characteristic equation, but adding a positive real number to them. Furthermore, to have a robust system, it has been taken into account a confine in a programmed maximum depth of cut.

Then, the following algorithmic methodologies are used, which are called ***preliminary rules***:

Rule1: Stability margin setting to ensure that the system plays in a stable region, despite the system model uncertainties.

Rule 1.1: For calculating secure stability lobes char, a small stability margin is selected, i.e, it is supposed that the chatter vibrations happen at $\delta + i \cdot \omega_c$ instead of at $i \cdot \omega_c$. The reason is that the stability border line is calculated from a linear approximation. Then, $i \cdot \omega_c$ is replaced by $\delta + i \cdot \omega_c, \delta > 0$, when the stability border line is calculated. This rule is applied to the equation (13).

Rule 1.2: For improving the robustness of the system, a margin at the final expression for chatter free axial depth of cut has been taken into account, equation (18), i.e, $b_{\lim} = \alpha \cdot b_{\lim}, 0 < \alpha < 1$. This rule lets a better control capacity in the spindle speed. On the other hand, a better MRR selection is lost because of the above design simplifying process.

Rule2: For searching the allowable input space parameter, the set of spindle speed, Ns, axial depth of cut, b and feed rate, s_t the following rules are applied.

Rule2.*1*: Calculate the boundary points, spindle speed and axial depth of cut pairs, which compose the line between stable and unstable zones, satisfying Rule 1. This rule is obtained by plotting the stability lobes char, which gives the line between stable and unstable zones

Rule2.2: Calculate the admissible input space, $Q := (N_s, b, s_t)$. The boundaries spindle speed and axial depth of cut, gives the maximum spindle speed and axial depth of cut pairs without chatter vibrations (rule 2.1). The time domain simulation is used to obtain the applied force by the milling machine. As it will be then seen in the next section, the spindle power is force-dependent, which is spindle speed, axial depth of cut and feed rate dependent. Then, for a given spindle motor power available, the admissible input *cutting parameter space* is obtained.

B. Tool selection

In this section, an approach for tool selection is suggested. For this purpose, a tool cost model function is designed. The designed tool cost model is used to select the appropriate tool between the candidates though the optimization Rules, explained below.

Then, the study requires a given set of candidates milling cutters. Each one is characterized by the following properties:

$R_i = (\omega_{nxi}, \omega_{nyi}, \xi_{xi}, \xi_{yi}, k_{xi}, k_{yi}, n_{ti}, D_i, \beta_i)$ where $(\omega_{xi}, \omega_{yi}) \in W$ is the tool natural frequency, $(\xi_{xi}, \xi_{yi}) \in \xi$ is the tool damping ratio, $(k_{xi}, k_{yi}) \in K$ is the tool static stiffness, n_{ti} is the tool number of teeth, D_i is the tool diameter and β_i is the tool helix angle. $R_i \in T$, $i = 1,2,..,N$, where N is the number of tools and T is the set of tools available to the designer. W is the set of tools' natural frequencies, conformed by the pairs $(\omega_{nx}, \omega_{ny})$ for each tool, ξ is the set of tools' damping ratio, conformed by the pairs (ξ_x, ξ_y) for each tool and tools' static stiffness is conformed by (k_x, k_y) for each tool.

C. Tool cost model definition

To carry out the selection of a suitable tool, a novel tool cost function has been conceived. The tool cost model for a single milling process can be calculated using the equation (20).

$$C(P_t, MRR, \Delta N_s; R, c_1, c_2) = c_1 \cdot NF_1 \cdot P_t + c_2 \cdot {NF_2}/{MRR} + c_3 \cdot {NF_3}/{\Delta N_s} \tag{20}$$

with $\sum_{i=1}^{3} c_i = 1$, $R \in T$, where $P_t = V \cdot \sum_{j=1}^{n_t} f_{tj}(\phi_j)$, $MRR = a \cdot b \cdot s_t$, ΔN_s takes its definition given below, and $q \equiv (N_s, b, s_t) \in Q$. Standardizing factors, NF_i, are defined as follow, $NF_1 = P_{tAv}^{-1}$, where P_{tAv} is the power available in the spindle motor, $NF_2 = MRR_{\max}$, where $MRR_{\max}$ is the maximum MRR with the chatter vibrations and spindle power restrictions calculated among all the candidate cutters and $NF_3 = \Delta N_{s,\max}$ where $\Delta N_{s,\max}$ is the maximum measured value of this variable among the candidate cutters.

The tool cost function is designed to be MRR, power consumption, and a range against possible perturbations in tool rotational motion, dependent and inversely proportional to MRR and a range against possible perturbations and directly to power consumption.

These parameters have the following definitions:

<u>Material or Metal -Removing Rate</u> (MRR)

$\mathrm{MRR} = a \cdot b \cdot s_t$, being a the radial depth of cut, the axial depth of cut and s_t the linear feed rate. The MRR is a parameter, which compares, the efficiency of the milling process. A larger MRR improves the process productivity.

<u>Cutting power draw from the spindle motor</u> (P_t)

The cutting power, P_t, drawn from the spindle motor is found from,

$$P_t = V \cdot \sum_{j=1}^{n_t} f_{tj}(\phi_j) \tag{21}$$

being $V = \pi \cdot D \cdot N_s$ the cutting speed and N_s the spindle speed. The tangential cutting force is given by:

$$f_{tj}(\phi_j) = K_t \cdot b \cdot h(\phi_j) \tag{22}$$

where b is the axial depth of cut, K_t is the cutting force coefficient, which are material dependent and is evaluated from experiments, and $h(\phi_j)$ is the chip thickness variation, which is feed rate s_t (mm/rev-tooth) dependent.

<u>Spindle speed security change</u> (ΔN_s)

An additional term, spindle speed security change, is added to the cost function model to be sure that chatter vibrations are avoided. The spindle speed security change, ΔN_s, measure the nearest spindle speed at which chatter vibrations happen to the supposed spindle speed it will be operated. This fact allows having an error margin due to possible perturbations in this variable.

To calculate analytically, ΔN_s, the following algorithmic methodologies are carried out. They are divided in two cases:

<u>Case I</u>: $k = 0$, this case corresponds to pairs, spindle speed, axial depth of cut, situated below the first lobe of the stability chars. Then, there is no lobe in the right part of the point as it can be shown in figure 3. Suppose that (N_{sI}, b_I) is the point which ΔN_s has to be calculated:

a) If $b_{\min,cri} > b$, $\Delta N_s = abs(N_{s,\min,cri} - N_{s,I})$.

b) If $b_{\min,cri} < b$, $\Delta N_s = abs(N_{s,cri}(b_I) - N_{sI})$.

$b_{\min,cri}$ is the minimum value of the axial depth of cut corresponding to the border line, $N_{s,\min,cri}$ is its corresponding spindle speed, $N_{s,cri}(b_I)$ is the left-projection of the point (N_{sI}, b_I) into the nearest lobe.

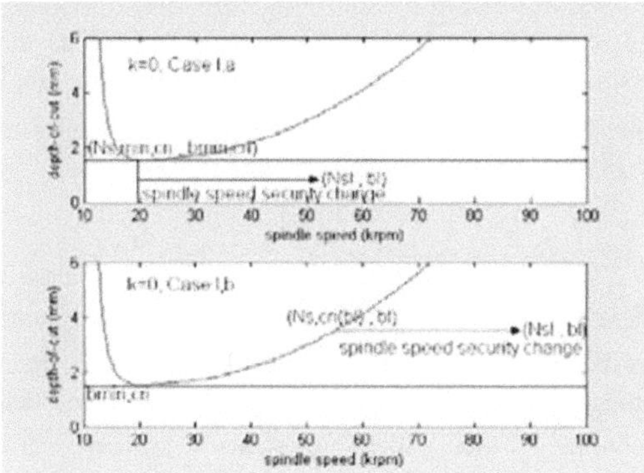

Fig. 3: Spindle speed security change, (ΔN_s), case I .

Case II: $k \neq 0$, in this case, the point ,which ΔN_s has to be calculated, is situated between two lobes in the stable region. Suppose that (N_{sII}, b_{II}) is the mentioned point, then $\exists k$ such that $N_{s,\min,cri}(k) < N_{sII} < N_{s,\min,cri}(k+1)$, where k is the lobe number, $k = 0,1..S-1$, and S is the number of printed lobes , and $N_{s,\min,cri}(k)$ is the spindle speed corresponding to the axial depth of cut minimum value on the border line, $b_{\min,cri}(k)$, for the k-lobe. Then:

a) If $b_{\min,cri}(k) > b < b_{\min,cri}(k+1)$

$$\Delta N_s = \min\begin{pmatrix} abs(N_{s,\min,cri}(k) - N_s), \\ abs(N_{s,\min,cri}(k+1) - N_s) \end{pmatrix}$$

b) If $b_{\min,cri}(k) < b > b_{\min,cri}(k+1)$

$$\Delta N_s = \min\begin{pmatrix} abs(N_{s,cri}(k) - N_s), \\ abs(N_{s,cri}(k+1) - N_s) \end{pmatrix}$$

where $N_{s,cri}(k)$ is the left-projection of the point (N_{sII}, b_{II}) into the k-lobe, and $N_{s,cri}(k+1)$ is the right-projection into the k+1-lobe. The case under consideration is graphically represented in figure 4.

Note that, the expert system does not take into account the process damping non-linearity, then, other cases to calculate ΔN_s have not been taken into consideration since $b_{\min}(k) = b_{\min}(k+1), \forall k$.

Furthermore, the other possible cases in the ΔN_s calculation are not considered, since they are unstable states cases. On the other hand, the calculated ΔN_s have been done taking into account Rules 1.1 and 1.2.

Standardization factors, NF_i, are also added to the cost function to have terms with the same magnitude. Moreover, they make to have a relative term between all the candidates cutters involved. On the other hand, these terms ensure that the cost function will be comparable among the different cutters.

The, values are the weights of the cost function terms. They $c_i, i = 1,2,3$ measure the importance of the cost function terms. The below optimization Rule 3 give a pattern to program the parameters c_i .

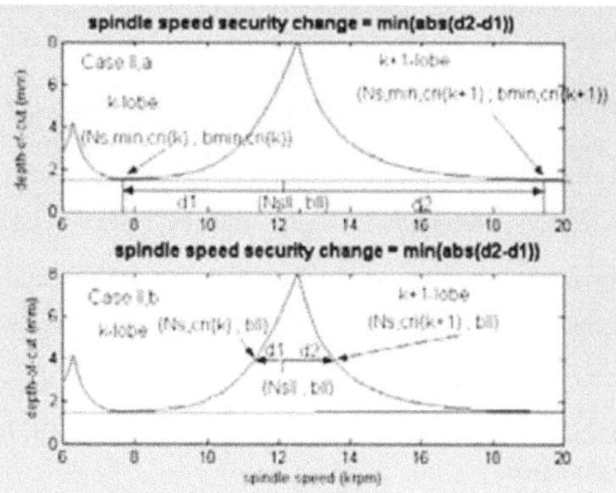

Fig. 4: Spindle speed security change, (ΔN_s), general case.

D. Optimization rules

The above defined tool cost function is used to select the appropriate tool and cutting parameters, through the following optimization rules.

Rule 3: Weight factors selection

The weight factors are intended to be programmed by the machine operator. An extended explanation of their meaning and their adequate selection is given in this section To select suitable values of c_i, $i = 1,..,3$, their meaning has to be perceived. The c_1-value measures the importance of the spindle speed consumption. A larger c_1 parameter is the more important to the spindle power consumption in the cost model function. The c_2 measures the machine productivity if the c_2 is near to one high productivity is required and if it is near to zero the productivity has no importance. The same reasoning is applied to the c_3, which measures the stability against possible perturbations in the spindle speed variable.

It has to be taken into account that the expert system, ensures that the spindle power consumption is always going to be smaller than the power available in the spindle motor, through Rule 1. Also, that the cutting parameter space has no sensed chatter vibrations through Rule 2.

Then, a possible criterion leading to a process with acceptable productivity, which is the main objective of the milling processes, c_2 about 0.75, and the other two constants will add 0.25, suitable values are $c_1 = 0.1$ and $c_2 = 0.15$.

Rule 4: Tool selection criterion
A simple tool selection criterion for cutter selection has been developed. For a given values of c_1, c_2, c_3, and a given tool characteristics, the cost function value is obtained for all the admissible input cutting parameter space. The minimum value of the cost function is stored. The procedure is repeated for all the available cutters. Comparing the minimum value of the cost function for all available or candidate cutters, the corresponding cutter to the minimum value of the minimum value of the cost function is the selected tool.
The selection criterion is, mathematically, expressed as:
Compute,

$$C\left(P_{tj}(q_j), MRR_j(q_j), \Delta N_{sj}(q_j), R_i, c_1, c_2\right); \qquad (23)$$

for each $R_i \in T$, $i \in \overline{N}$, and $\overline{N}$ is the set of candidate tools and $\forall q_j \equiv \left(N_{sj}, b_j, s_{tj}\right)$ where $j \in \overline{N}_p = \{1,.., N_p\}$ is a discrete sub-space of the cutting parameters space where the cost function (20) is calculated.

For obtaining the selected tool, ST, compute

$$ST = \arg\min_{i \in \overline{N}} \left\{C\left(P_t(q_j), MRR(q_j), \Delta N_s(q_j), R_i, c_1, c_2\right)\right\}$$

(24)
with $ST \in T$, obtaining the appropriate tool according to the criterion.
Following the rules, the expert system provides an appropriate cutter among the candidates.
Note that the objectives of the expert system are to obtain a tool which has an operating point or adequate cutting parameters where the MRR will be higher than the others available tools, with a prescribed stability robustness and without consumption more power than the available in the spindle motor. Hence, the tool cost model is designed so as to minimize it. Furthermore, the power consumption will be minimum, and the MRR and stability robustness will be maximum, for a given values of c_1, c_2, and c_3. The $c_i, i = 1,2,3$, are designed by the machine operator.
Rule 5: Cutting parameter selection
To obtain the cutting parameters a simple criterion, which consist of calculate the cutting parameters which corresponds to the minimum value of the cost function above defined for a certain values of c_1, c_2, c_3. But, here, a new approach thought an auxiliary cost function is going to be applied. In this case, once the tool has been selected, another novel and complete cost function is designed in order to obtain the best cutting parameters. It is composed by the above defined cost functions and other two, which are time and frequency domain responses related. Then, the first new cost function studies the temporal behavior of the input cutting parameters, and the second one its frequency response. The resultant cost function is used to obtain the cutting parameters for the selected tool.

E. Temporal response cost model definition

The temporal response cost model is defined as the maximum overshot (M_p) and the settling time (t_s) dependent function. Those characteristics are typical in the study of the time domain response of a system.

$$C_t\left(T_{tool}, Q_j, c_{1t}, c_{2t}\right) = c_{1t} \cdot {t_s}/{t_{s,\max}} + c_{2t} \cdot {M_s}/{M_{s,\max}}$$

where $t_{s,\max}$ and $M_{s,\max}$ are the maximum settling time and maximum overshot between the allowable input cutting space parameter, T_{tool} is the selected tool according with the previous section and $\sum_{i=1}^{2} c_{it} = 1, c_{it} \geq 0$.

a) Frequency response cost model definition

The frequency response cost model is dependent on the relation between the first and the second harmonic frequencies through the function, R_{12h}, and the relation between the first harmonic frequency and the chatter frequency, R_{1ch}. That is:

$$C_f\left(T_{tool}, Q_j, c_{1f}, c_{2f}\right) = c_{1f} \cdot {R_{12h}}/{R_{12h\max}} + c_{2f} \cdot {R_{12h}}/{R_{1ch\max}}$$

where $R_{12h\max}$ and $R_{12ch\max}$ are the maximum of those parameters between the allowable input space cutting parameter, T_{tool} is the selecting tool according the previous section, and $\sum_{i=1}^{2} c_{if} = 1, c_{if} \geq 0$.

b) Total cost-function model

The total cost function is, then, composed by the defined above three cost functions, the tool cost model, the temporal response cost model and the frequency response cost model.
For this case, the cutting parameters are calculated following the algorithm:

$$C_{resul\tan t}\left(T_{tool}, Q_j, c_{1r}, c_{2r}, c_{3r}\right) = c_{1r} \cdot C\left(T_{tool}, Q_j, c_1, c_2, c_3\right) + c_{2r} \cdot C_t\left(T_{tool}, Q_j, c_{1t}, c_{2t}\right) + c_{3r} \cdot C_f\left(T_{tool}, Q_j, c_{1f}, c_{2f}\right)$$

where $\sum_{i=1}^{3} c_{ir} = 1$, $c_{ir} \geq 0$, and T_{tool} is the selected tool.

Compute, $C_{\text{resultant}}\left(T_{tool}, Q_j, c_{1r}, c_{2r}, c_{3r}\right); \forall q_{jtool} \in Q_{tool}$, satisfying the rule 2.2.

Compute,

$$q^{**} = \arg\left\{ \min_{\forall q_j}(C_{resul\tan t}(T_{tool}, q_j, c_{1r}, c_{2r}, c_{3r})) \right\} \text{ and}$$

obtain the input cutting parameters for the selected tool.

Rule6: Resultant cost function weight factors selection.

To select the values of c_{ir} it has been taken into account the fact that the most important term in $C_{resul\tan t}$ is C by practical reasons. It is because C_t and C_r are corrected terms. For this reason, it should be taken the c_{1r} about 0.8, and c_{2r} and c_{3r} about 0.1 each one. The time and frequency domains weighting factors, $c_{1t}, c_{2t}, c_{1f}, c_{2f}$ are assumed to have the same value or be very similar.

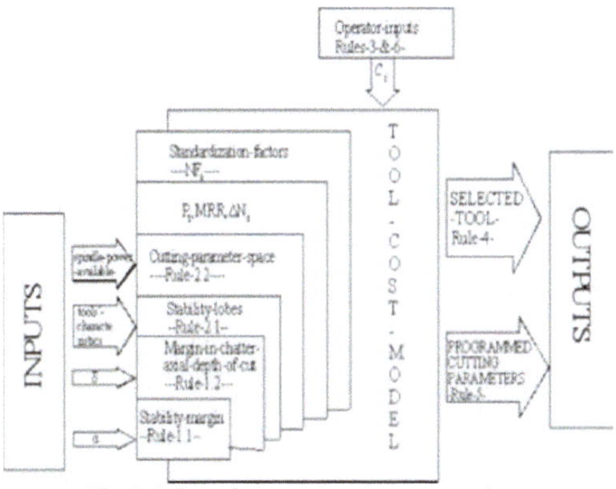

Fig. 5: Schematic expert system representation.

Finally, figure 5 shows a scheme of the expert system. The developed expert system takes the α and δ constants, the tools′ modal parameters such as its natural frequency, damping ratio, tool static stiffness, the number of teeth, the radius of the tool, the helix angle, and the cutting constants for the work material and cutter (tools′ characteristics), the spindle power available and the cost function weight factors, as inputs and outputs the appropriate tool among the candidates and robust programmed cutting parameters.

IV. EXAMPLE

For the validation of this method, the above study has been applied for two practical straight cutters and a full-immersion up-milling operation. The example considers the tools to have the following characteristics, according with the section III.B notation,

$R_1 = (603,666,3.9,3.5,5.59,5.715,3,30,0)$, and

$R_2 = (900.03,911.65,1.39,1.38,0.879,0.971,2,12.7,0)$.

The natural frequency is measured in hertz, the tool damping is in %, the tool stiffness is in $KN \cdot mm^{-1}$ and the diameter of the tool is in mm. The work-piece is a rigid aluminum block whose specific cutting energy is chosen to be $K_{1,2} = 600 KN \cdot mm^{-2}$ and the proportionally factor is taken to be $k_{r1} = 0.3$, for the tool one, and $k_{r2} = 0.07$ for the other one. Other design expert system parameters are, the stability margin factor, $\delta = 0.05$ and the stability margin factor for the axial depth of cut, $\alpha = 0.95$.

The analytical test for mill cutter selection was conducted using spindle speeds with increments of 1000 rpm, axial cutting depth started with its minimum value in the stability border line divided by ten, and it is increased in steps of this same size, for a given spindle speed. The operation constraint on the maximum feed per tooth is 0.55 mm and the step integration is selected to be 0.05. The spindle power availability is 745.3 w.

The resultant tool is that leading to the minimum tool cost function value.

In figure 6, it is shown the values of tool cost function as c_1-parameter varies, the c_3-value has been taken as a constant $c_3 = 0.075$ and the c_2 follow the rule $c_2 = 1 - c_1 - c_2$.

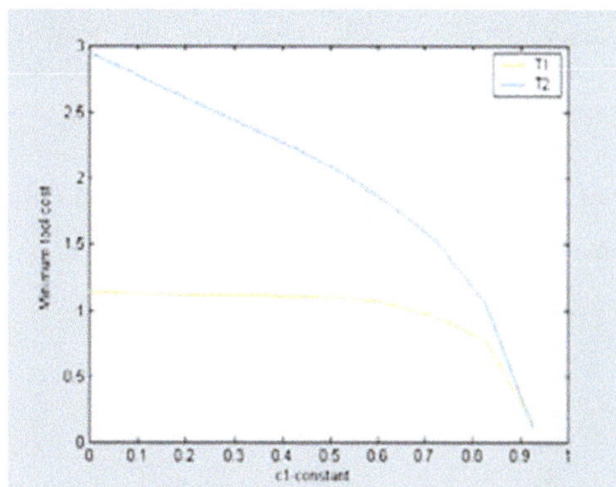

Fig. 6: Minimum tool cost function vs. c_1 varies, with $c_3 = 0.075$.

This study has been performed to illustrate the influence of the c_i parameters in the tool cost function. It is observed the tool R_1 has a better behavior respect to the tool R_2 for all possible value of c_1 and c_2, with $c_3 = 0.075$. Analyses with other values of c_1, c_2 and c_3 have been carried out and the results are similar, and the tool R_1 has a better behavior.

Then, a more general analysis shows in figure 7, in which the minimum value of the tool cost function for all possible combinations of c_1, c_2, c_3, with the restriction $c_1 + c_2 + c_3 = 1$ is displayed.

The analysis has revealed that the first tool has a better behavior than the second one for all combinations of the c_i parameters. Thus the output of the expert system is the first tool.

For the cutting parameters selection, two steps have been done. First, the cutting parameter corresponding to the minimum of the tool cost function for the selected tool for values of $c_1 = 0.2$, $c_2 = 0.725$, $c_3 = 0.075$ is obtained. These values are $q^* = (5800, 0.4924, 0.2722)$.

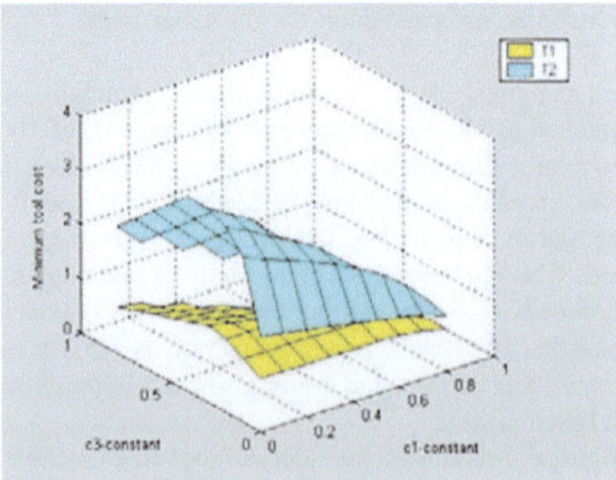

Fig. 7: Minimum tool cost function versus c_1, c_2, c_3 varies

It can be a well-done first approximation. For a more appropriate solution, taking into account the time and frequency domain system responses, the total cost function, C_{result} has been calculated for the allowable cutting space parameter. The minimum value is saved for $c_{1t} = c_{2t} = 0.5$, $c_{1f} = c_{2f} = 0.5$, and $c_{1r} = 0.8$, $c_{2r} = 0.1$, $c_{3r} = 0.1$ the resulted programmed cutting parameters are $q^{**} = (5680, 0.457, 0.265)$.

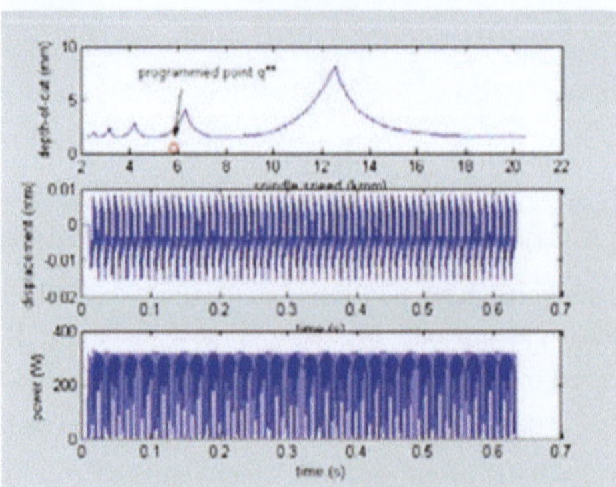

Fig. 8: Situation of the point q** in the stability diagram and tool displacement and power consumption time domain responses for the selected tool.

Figure 8 shows the situation for the stability lobes of the programmed point q^{**}, the tool displacement and the power consumption. It is observed that the point is robustly stable and the power consumption is less than the power availability in the spindle motor, while the MRR measure becomes acceptable.

This method can also be applied to any number of selected tools generating in automatic task the best one to be used in the system. Moreover, the method can be used to schedule the relative compliance between the available tools and the used work-pieces materials. On the other hand, the expert system can be used to optimize the manufacturing process, in the sense of planning the adequate sequence of work-pieces to be manufactured for each tool in order to minimize the changes of tools. Finally, apart from being a cheep method, the expert system could be easily used by an inexpert human operator.

V. CONCLUSION

An efficient approach for mill cutter selection has been developed through the design of an expert system. Such an expert system is instructed with the characteristics of the available candidates tools, as well as with the stability margin and constrains of operations, such as, power availability and robust. Furthermore, a tool cost model function, built from the expert systems preliminary rules, is proposed to evaluate the possible achievable performance by each candidate tool in the milling process. This performance index is then used to select an appropriate tool and cutting parameters for the operation which lead to the maximum productivity, while respecting tool stability and power consumptions margins though optimization rules. A simulation example, which shows the behavior of the system, is presented.

ACKNOWLEDGMENT

The Authors are very grateful to MCYT by its partial support through grant 2003-00164 and to the UPV/EHU through Project 9/UPV 00I06.I06-15263/2003.

REFERENCES

[1] S. Y. Liang, R. L. Hecker and R. G: Landers, "Machining Process Monitoring and Control: The State of the Art," in *Journal of Manufacturing Science and Engineering*, vol. 126, 2004, pp. 297–310.

[2] S. V. Wong and A. M. S. Hamouda, "The development of an online knowledge-based expert system for machinability data selection", Knowledge-Based Systems, Vol.16, pp. 215-219, 2003.

[3] M. C. Cakir, O. Irfan and K. Cavdar, "An expert system for die and mold making operations", *Robotics and Computer-Integrated Manufacturing,* Vol.21, pp. 175-183, 2005.

[4] A. Vidal, M. Alberti, J. Ciurana and M. Casadesús, "A decision support system for optimizing the selection of parameters when planning milling operations", *International Journal of Machine Tools and Manufacture,* Vol.45, pp. 201-210, 2005.

[5] I. D. Carpenter and P. G. Maropoulos, "A flexible tool selection decision support system for milling operations", *Journal of Materials Processing Technology,* Vol.107, pp. 143-152, 2000.

[6] Y. Altintas, *Manufacturing Automation*, Cambridge University Press, 2000.

[7] O. Maeda, Y. Cao and Y. Altintas, "Expert spindle design system", *International Journal of Machine Tools and Manufacture,* Vol.45, pp. 537-548, 2005.

[8] H. Li and X. Li, "Modeling and simulation of chatter in milling using a predictive force model", *International Journal of Machine Tools and Manufacture,* Vol.40, pp. 2047-2071, 2000.

[9] L.Rubio and M. De la Sen, "Analytical procedure for chatter in milling", Rediscover 2004, June 14-16, 2004, Cavtat, Croatia.

Selection of Mill Cutter and Cutting Parameters through an Expert System

L. Rubio and M. De la Sen

***Abstract*— This paper discusses the selection of tools in milling operations. An expert system hinged on numerical methods has been developed to carry out this research. The knowledge base is given by limitations in process variables, which let us define the allowable cutting parameter space. The mentioned process variables cause instabilities due to tool-work-piece interaction, knowing as chatter vibration, and the power available in the spindle motor. Then, a tool cost model is contrived. It is used to choose the suitable cutting tool, among a known set of candidate available cutters. Once the cutting tool is selected, the cutting parameters are calculated. For this purpose, other two cost functions are designed, which are depended on the time and frequency domains response properties. An example is presented to illustrate the method.**

I. INTRODUCTION

Machining, in particular milling operations, is a broad term used to define the process of removing material from a work-piece. Furthermore, the milling operation process is required, nowadays, to increase its productivity, reducing cost and improving the final product [1].

This paper bring forward the concept of selecting an appropriate mill cutter, among a known set of candidate cutters, and obtaining the adequate cutting parameters for milling operations through an expert system.

There are several versatile approaches for tool and/or cutting parameter selection based on the design of expert systems on manufacturing environments. Wong and Hamouda [2] developed an on-line fuzzy expert system. The system inputs, the tool type, the work-piece material hardness and the depth of cut and control the cutting parameters at the machine, as output. Cemar Cakir et al. [3] explained an expert system based on experience rules for die and mold operations. In that paper, the geometry and material of the work-piece, tool material and condition and operation type are considered as inputs. Then, the system provides recommendations about tool type, tool specifications, work-holding method, type of milling operation, direction of feed and offset values. Vidal et al. [4] focused on the problem of choosing the manufacturing route in metal removal process. They select the cutting parameters by optimizing the cost of the operation taking into account several factors, such as, material, geometry, roughness, machine and tool. Carpenter and Maropoulus [5] designed a system, which provides reliable tool selection and cutting data for a range of milling operations. The method employs rule based decision logic and multiple regression techniques for a wide range of materials.

Here, the developed expert system consists of the relative compliance between the tool and the work-piece and it is predicted with analytical methods. Moreover, milling process simulations have been developed in the time and frequency domains, which are, then used in the expert system definition.

Firstly, the knowledge base is explained. Basically, it defines the allowable cutting parameters, which are known as cutting parameter space, for a given tool-work-piece configuration. It is based on the chatter vibrations avoidance, which limits the productivity of the process, and on a spindle power limitation criterion.

On the other hand, a novel tool cost function is designed. It sub-optimizes a combination of several effects of interest, namely, the spindle power consumption, the material remove rate (MRR), as well as, a stability criterion against possible perturbations in the spindle speed variable (ΔN_S).

The MRR is a parameter which measures the process effectiveness. It is required to be as large as possible. But, if MRR increases beyond certain limits, chatter vibrations are appreciated and the process becomes unstable [6]. Other variable which limits the process effectiveness is the power available in the spindle motor (P_t) [7]. A third parameter, ΔN_S, taking part into the cost function is considered to ensure a well-posed behavior of the system if a perturbation in the spindle speed happened.

In conclusion, the proposed cost function is a measure of how the milling process is being carried out at certain operation conditions. The larger the cost function is, the worst operation will be. Thus, the cutter which minimizes the designed cost function is selected.

Once, the cutting tool is selected, the cutting parameters are obtained for a given tool configuration. A second cost function is designed. It is composed by the above mentioned cost function for the input cutting space parameter and the selected tool gives the programmed cutting parameters.

Then, the expert system considers tool characteristics, related tool-work-piece material parameters and milling operation as inputs and outputs the selected tool among the candidates and robust programmed cutting parameters.

L. Rubio and M. De la Sen are with the Instituto de Investigación y Desarrollo de Procesos, Facultad de Ciencia y Tecnología, Campus de Leioa, Universidad del País Vasco, Apartado 644, Bilbao, Spain (e-mail: {webrurol,webdepam}@lg.ehu.es).

K. Elleithy et al. (eds.), Advances in Computer, Information, and Systems Sciences, and Engineering, 5–12.

Section IV, assuming that both have been modelled in the normal form. It is shown that the structure remains invariant in the connection, and all structural properties of the connected system are thus easily determined from the basic theory of canonical forms. Internal stability is especially considered, and it is found to be the key concept to consider when determining the stability of connected converters. Implications to converter analysis are treated in Section V and conclusions are given in Section VI.

II. SMALL SIGNAL MODELS AND NORMAL FORMS

It is well-known that a dc-dc converter can be modeled by a linear small signal model quite accurately up to about half of the operating frequency [1], [2]. Such models for the basic converter types, buck, boost, buck-boost, flyback etc., converters are well reported in the literature and much used in industry in the design of special power supplies in different application areas.

Based on the small-signal model of the converter (or any two-port system in general) a *canonical representation* or *normal form* can be drawn [5]. In Fig. 1 this form is presented in the general unterminated case, where the system has not been connected to any particular source or load. Fig. 2 represents a model of a converter given in the canonical form, which is connected to a voltage source and electric load.

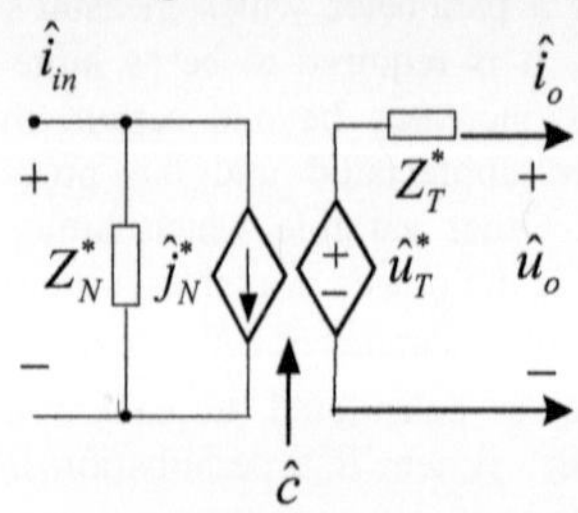

Fig. 1. Canonical unterminated system model

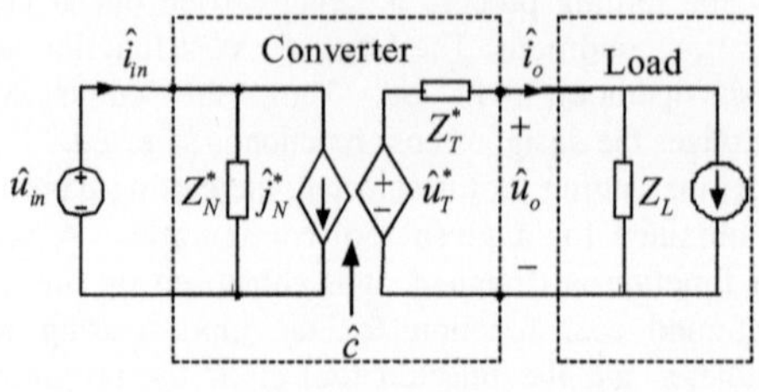

Fig. 2. Canonical converter model connected to a load

The canonical form (Fig. 1) can be described by the equations

$$\begin{cases} \hat{i}_{in} = \dfrac{1}{Z_N^*}\hat{u}_{in} + \underbrace{T_{ji-o}^*\hat{i}_o + G_{ci}^*\hat{c}}_{j_N^*} \\ \hat{u}_o = \underbrace{G_{i-o}^*\hat{u}_{in} + G_{co}^*\hat{c}}_{u_T^*} - Z_T^*\hat{i}_o \end{cases} \tag{1}$$

where (and later in the paper also) the Laplace variable s has been dropped for brevity. In the equation Z_N^* and Z_T^* are the input and output impedances, T_{ji-o}^* is the current susceptibility, G_{i-o}^* is the input output transfer function, and G_{ci}^*, G_{co}^* are the control to output transfer functions. The *input variables* are voltage $\hat{u}_{in}$, current $\hat{i}_o$ and the control variable $\hat{c}$. The *output variables* are current $\hat{i}_{in}$ and voltage $\hat{u}_o$. The *open loop canonical model* is given by

$$\begin{bmatrix} \hat{i}_{in} \\ \hat{u}_o \end{bmatrix} = \begin{bmatrix} 1/Z_N^* & T_{ji-o}^* & G_{ci}^* \\ G_{i-o}^* & -Z_T^* & G_{co}^* \end{bmatrix} \begin{bmatrix} \hat{u}_{in} \\ \hat{i}_o \\ \hat{c} \end{bmatrix} \tag{2}$$

and the *closed loop canonical model* by

$$\begin{bmatrix} \hat{i}_{in} \\ \hat{u}_o \end{bmatrix} = \begin{bmatrix} 1/Z_{Nc}^* & T_{ji-c}^* \\ G_{i-c}^* & -Z_{Tc}^* \end{bmatrix} \begin{bmatrix} \hat{u}_{in} \\ \hat{i}_o \end{bmatrix} \tag{3}$$

where the subscripts 'o' and 'c' denote open and closed loop functions, respectively. The "hat" (e.g. in $\hat{u}_{in}$) means that the variable represents a deviation from the averaged value, i.e. the dynamic variation of the variable around the nominal value.

From the system theoretical point of view these models represent multi-input multi-output (MIMO) systems, which can be described by the diagrams in Fig. 3.

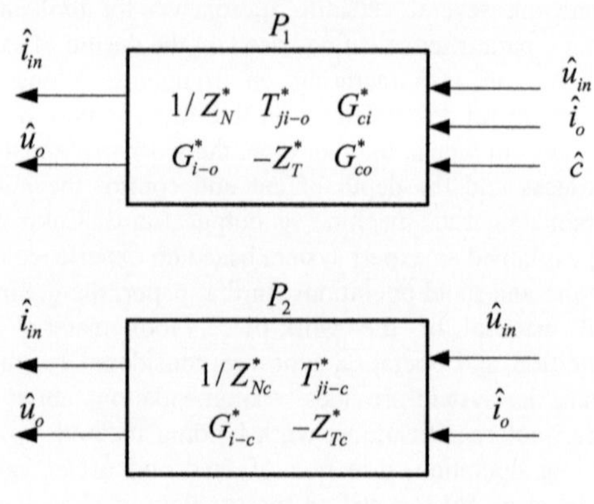

Fig. 3. Canonical forms in a system representation

The closed-loop equations can be derived applying the equations of the controller in the open-loop model. For example, in voltage-mode control the feedback law is (assume the regulator problem)

$$\hat{c} = -G_c^*\hat{u}_o \tag{4}$$

which, when applied in (2) gives (3) with

$$\frac{1}{Z_{Nc}^*} = \frac{1}{Z_N^*} - \frac{G_{i-o}^* G_{ci}^* G_c^*}{1 + G_{co}^* G_c^*}$$

$$T^*_{ji-c} = T^*_{ji-o} + \frac{Z^*_T G^*_{ci} G^*_c}{1 + G^*_{co} G^*_c} \tag{5}$$

$$G^*_{i-c} = \frac{G^*_{i-o}}{1 + G^*_{co} G^*_c}$$

$$Z^*_{Tc} = \frac{Z^*_T}{1 + G^*_{co} G^*_c}$$

The interesting point to note is that closing the loop does not change the canonical stricture (normal form) of the model (Fig. 1), which remains thus invariant. The same holds for all such controls, where the control law can be written by using the variables in the small-signal representation of the system model. As explained later a similar invariability holds when connecting canonical forms in series (or in parallel), which gives a clear indication to study the characteristics of the normal forms, because the results obtained are then general.

III. STRUCTURAL PROPERTIES

In classical control theory the three concepts *stability*, *controllability* and *observability* are the main tools for analyzing the system and for constructing a proper controller for it. This theory for linear systems is well developed indeed, and therefore the approach is well-suited for power systems modelled by linear small-signal models.

To investigate the above canonical models is therefore a key issue. Usually, two main approaches can be chosen. The system can be modelled by an input-output point of view, which gives raise to the use of *transfer function matrix models* or *algebraic polynomial theory*. On the other hand, internal descriptions by using the state-space approach have been widely used during the development of different aspects and methods of control theory in general. For literature of these two approaches, see [7], [8], [9], [10], [11], for example. Some system analytic tools in the study of stability problems in power electronics' applications can be found in [12], [13].

The system is called *bounded input - bounded output stable* (BIBO stable), if for any bounded input a bounded output results. The system is called *stable* if any solution of the related differential equations (representing the internal behaviour of the system) deviates only `little' from the equilibrium point, when disturbed from it initially. The system is called *asymptotically stable* if it is stable and also converges to the equilibrium. (Note that the above two definitions have been stated informally.)

The stability of the system can be analyzed by calculating the *poles* of the transfer function (matrix) or the eigenvalues of the system matrix of its *realization* (state-space representation). The poles are exactly the eigenvalues mentioned, if a *minimal realization* of the system transfer function matrix is used. If the realization is not minimal, the system described by it lacks some controllability and/or observability properties. That implies problems for example in the construction of state feedback controllers and observers. Asymptotic stability implies BIBO stability, but the converse holds only in the case that the system is both controllable and observable. Basically this means that all *pole-zero cancellations* of the transfer function matrix have been done, and a minimal realization of the system model is used. But note that unstable (RHP, right half plane) cancellations are forbidden, because they cause problems to the internal stability of the system.

The above facts are basic in the control engineering theory. For a reference, see any good textbook of classical control theory, e.g. [9], [11].

From the canonical system representations the poles can be seen directly from the denominators of the individual transfer functions. But their *multiplicities* cannot be seen directly, and this would be important in order to determine the *order* of the system (number of state variables in the minimal realization). The canonical model can represent a quite complex system, for example a controlled converter, or a system composed of several power stages in series, and therefore it is important to be able to see the complexity of the model. Also, as it will be seen later, it is of crucial importance to determine, whether there are pole-zero cancellations, especially unstable ones, when connecting subsystems together.

The poles with multiplicities can be calculated directly from the transfer function matrix by an algorithm developed by MacFarlane and Karcanias, see [9]. A similar algorithm exists also to determining the *zeros* of the system, which are not related to stability but to system performance instead. In fact, nonminimum phase zeros (RHP zeros) cause fundamental limitations to closed-loop system performance, which cannot be evaded by any means. More information on the fundamental limitations can be found in [11] and an example from the controller design of a boost-type DC/DC-converter (possessing a RHP zero) in [14].

IV. CONNECTIONS OF SYSTEMS

The connection of two dynamical systems is of major importance, because most complex systems are built from simple modules by connecting these in a desired topology. An example is shown in Fig. 4, where a DC-DC-converter is connected to an input voltage source and an input filter at the input side and to a general load at the output side. These components (filter, converter, load) can be represented by models in the canonical form, and therefore the whole system can be modelled by two series connections of such modules.

Another and a more general example is depicted in Fig. 5, where an AC/DC converter drives a DC Bus, which supplies several DC/DC converters connected in parallel and provided with input filters. The whole system can be analyzed by constructing unit models of each component in normal form, and then connecting these in series and in parallel.

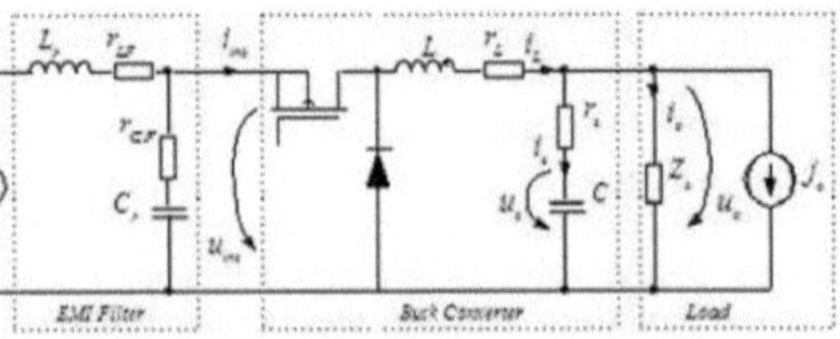

Fig. 4. Buck converter with EMI filter and load

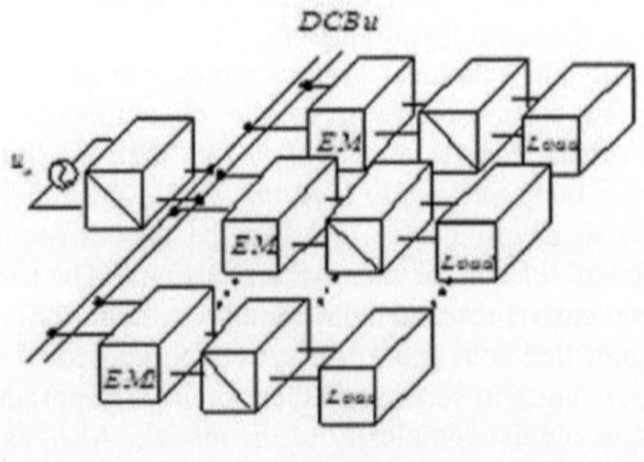

Fig. 5. A distributed power system

In Fig. 6 a series model of two normal forms is presented.

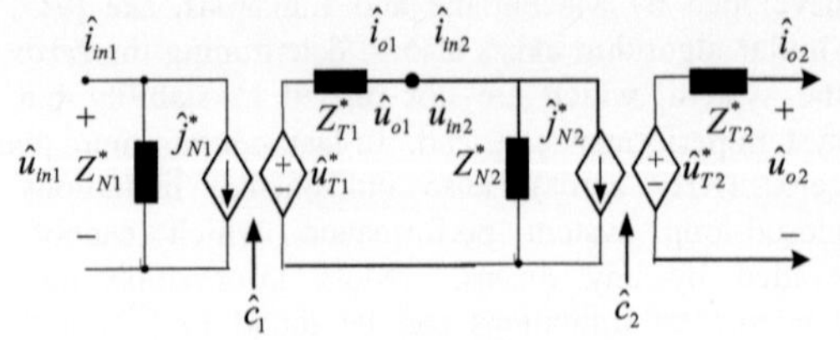

Fig. 6. Series connection of two normal forms

For proper modelling the *interconnection constraints* must be taken into account, here $\hat{u}_{01} = \hat{u}_{in2}$, $\hat{i}_{01} = \hat{i}_{in2}$.

A system theoretic diagram of the series connection is shown in Fig. 7 where the *input variables* are w and $\overline{w}$, *output variables* z and $\overline{z}$, and *internal variables* u and y. The correspondence to the true electrical variables is $w = \hat{u}_{in1}$, $\overline{w} = \hat{i}_{02}$, $z = \hat{i}_{in1}$, $\overline{z} = \hat{u}_{02}$, $u = \hat{i}_{01}$, $y = \hat{u}_{01}$.

The first input variable to system K is $y = \hat{u}_{01} = \hat{u}_{in2}$ and the first output variable of the same system $u = \hat{i}_{in2} = \hat{i}_{01}$; note the interconnection constraints.

The two subsystems can be modelled by

$$\begin{cases} z = p_{11}w + p_{12}u \\ y = p_{21}w + p_{22}u \end{cases} \tag{6}$$

$$\begin{cases} u = k_{11}y + k_{12}\overline{w} \\ \overline{z} = k_{21}y + k_{22}\overline{w} \end{cases} \tag{7}$$

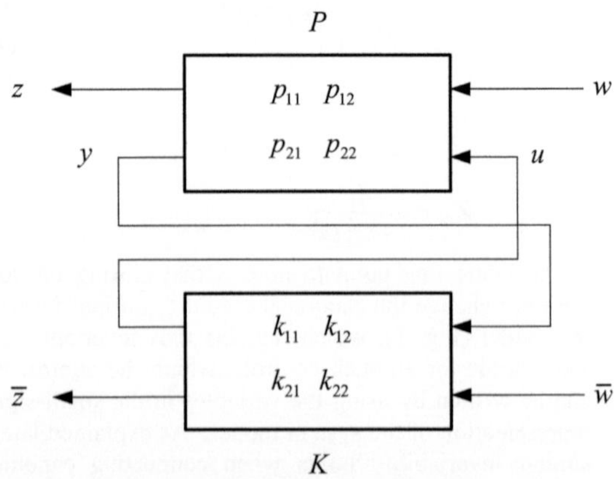

Fig. 7. Two systems connected in series

From these equations it is easy to solve the *intermediate equations*

$$\begin{bmatrix} y \\ u \end{bmatrix} = \begin{bmatrix} (I - p_{22}k_{11})^{-1} p_{21} & p_{22}(I - k_{11}p_{22})^{-1} k_{12} \\ k_{11}(I - p_{22}k_{11})^{-1} p_{21} & (I - k_{11}p_{22})^{-1} k_{12} \end{bmatrix} \cdot \begin{bmatrix} w \\ \overline{w} \end{bmatrix} \tag{8}$$

and the *input-output equations*

$$\begin{bmatrix} z \\ \overline{z} \end{bmatrix} = \begin{bmatrix} A & B \\ C & D \end{bmatrix} \begin{bmatrix} w \\ \overline{w} \end{bmatrix} \tag{9}$$

with

$$A = p_{11} + p_{12}k_{11}(I - p_{22}k_{11})_{-1} p_{21}$$
$$B = p_{12}(I - k_{11}p_{22})^{-1} k_{12}$$
$$C = k_{21}(I - p_{22}k_{11})^{-1} p_{21}$$
$$D = k_{22} + k_{21}p_{22}(I - k_{11}p_{22})^{-1} k_{12}$$

Note that the input-output equations can also be written as

$$\begin{bmatrix} z \\ \overline{z} \end{bmatrix} = (P * K) \begin{bmatrix} w \\ \overline{w} \end{bmatrix} \tag{10}$$

where

$$P = \begin{bmatrix} p_{11} & p_{12} \\ p_{21} & p_{22} \end{bmatrix}; \qquad K = \begin{bmatrix} k_{11} & k_{12} \\ k_{21} & k_{22} \end{bmatrix}$$

as usual and the asterisk * denotes the (Redheffer's) star product [9]. Note that in the above equations the identity matrix I has been used in some places to emphasize that the form of equations is valid in the case of multivariable signals also.

As explained, the input-output model shown in Fig. 8

is given by $N = P * K$, but it is interesting to note that the intermediate equations (8) are even more informative.

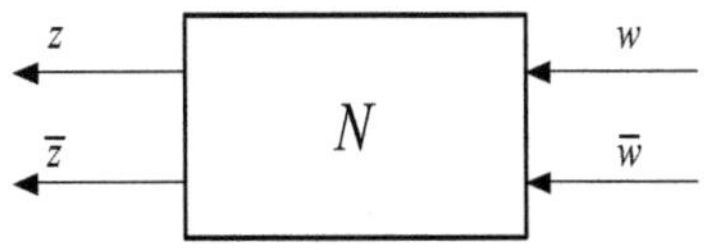

Fig. 8. Input-output model

To see this, consider the system in Fig. 9 and set $d_y = p_{21}w, d_u = k_{12}\overline{w}$.

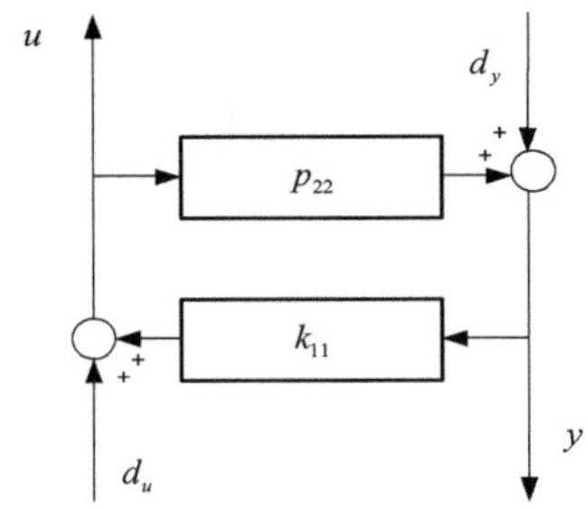

Fig. 9. Internal system topology

Short calculations show that the equations describing the system are exactly the intermediate equations (8). Moreover, the general configuration in the figure is used to study the *internal stability* of the system, see e.g. [9] [10]. Basically, internal stability means that if any bounded signal enters the system (from anywhere), no unbounded signals result in the system. This is a more severe condition than the ordinary input-output (BIBO) stability, because it also reveals possible instabilities in the hidden modes. Internal stability is in close relationship to asymptotic stability, but the latter is related to *realization* (state-space representation), not to the transfer function representation. For practical design properties, internal stability must exist in order for the system to perform properly.

Since the theorems of internal stability are well-known, it now becomes obvious that they are equivalent to series connections of canonical forms also. The following can be stated.

The series connection of two subsystems represented in canonical forms is internally stable if

1. The transfer functions p_{21} and k_{12} are stable (so that the signals d_y and d_u are bounded).

2. The transfer functions

$$\left(I - k_{11}p_{22}\right)^{-1}$$
$$k_{11}\left(I - p_{22}k_{11}\right)^{-1}$$
$$\left(I - p_{22}k_{11}\right)^{-1}$$
$$p_{22}\left(I - k_{11}p_{22}\right)^{-1}$$

are stable.

Moreover, if no *right half plane* (RHP) pole-zero cancellations occur in the transfer functions $k_{11}p_{22}$ and $p_{22}k_{11}$ the system is internally stable, if one of the above four transfer functions is stable. (Note that the functions $k_{11}p_{22}$ and $p_{22}k_{11}$ are the same in the single input – single output (SISO) case, but the result is stated here in the general MIMO form.)

From the basic relationship

$$\begin{cases} z = p_{11}w + p_{12}u \\ \overline{z} = k_{21}y + k_{22}\overline{w} \end{cases} \tag{11}$$

it is easily seen that if the system is internally stable and the transfer functions p_{11}, p_{12}, k_{21} and k_{22} are stable, the system is BIBO stable.

V. IMPLICATIONS TO CONVERTER ANALYSIS

Consider a DC/DC converter (or any power system, which can be modelled by a small-signal model in the canonical form) connected to a generalized load as in Fig. 2. Since the load system can easily be modelled in the canonical form also, the configuration according to figures 6 and 7 can be used with

$$P = \begin{bmatrix} p_{11} & p_{12} \\ p_{21} & p_{22} \end{bmatrix} = \begin{bmatrix} 1/Z^*_{Na1} & T^*_{ji-a1} \\ G^*_{i-a1} & -Z^*_{Ta1} \end{bmatrix} \tag{12}$$

and

$$K = \begin{bmatrix} k_{11} & k_{12} \\ k_{21} & k_{22} \end{bmatrix} = \begin{bmatrix} 1/Z_L & 1 \\ 1 & 0 \end{bmatrix} \tag{13}$$

The intermediate equations are then

$$\begin{bmatrix} \hat{u}_{o1} \\ \hat{i}_{o1} \end{bmatrix} = \begin{bmatrix} \dfrac{G^*_{i-a1}}{1+\dfrac{Z^*_{Ta1}}{Z_L}} & -\dfrac{Z^*_{Ta1}}{1+\dfrac{Z^*_{Ta1}}{Z_L}} \\ \dfrac{G^*_{i-a1}/Z_L}{1+\dfrac{Z^*_{Ta1}}{Z_L}} & \dfrac{1}{1+\dfrac{Z^*_{Ta1}}{Z_L}} \end{bmatrix} \begin{bmatrix} \hat{u}_{in} \\ \hat{j}_o \end{bmatrix} \tag{14}$$

where $\hat{j}_0$ means the output current of the load. The subscripts '$a1$' have been used to denote that those variables belong to the first power stage, which can be either open loop (o) or closed loop (c). The input-output equations are, correspondingly

$$\begin{bmatrix}\hat{i}_{in}\\ \hat{u}_o\end{bmatrix}=\begin{bmatrix}\frac{1}{Z^*_{Na1}}+\frac{T^*_{ji-a1}G^*_{i-a1}/Z_L}{1+\frac{Z^*_{Ta1}}{Z_L}} & \frac{T^*_{ji-a1}}{1+\frac{Z^*_{Ta1}}{Z_L}}\\ \frac{G^*_{i-a1}}{1+\frac{Z^*_{Ta1}}{Z_L}} & -\frac{Z^*_{Ta1}}{1+\frac{Z^*_{Ta1}}{Z_L}}\end{bmatrix}\begin{bmatrix}\hat{u}_{in}\\ \hat{j}_o\end{bmatrix} \quad (15)$$

It is assumed that the transfer functions Z^*_{Na1}, T^*_{ji-a1}, G^*_{i-a1}, Z^*_{Ta1} and Z_L are stable, but not necessarily minimum-phase. In transfer matrices *P* and *K* this means that p_{11} and k_{11} can be unstable.
If the interconnected (terminated) system is internally stable, then it is BIBO-stable provided that T^*_{ji-a1} is stable and Z^*_{Na1} is minimum phase. Surprisingly, the load impedance Z_L does not have any effect on BIBO stability. But let us consider the term

$$p_{22}k_{11}=-Z^*_{Ta1}\cdot\frac{1}{Z_L}=-\frac{Z^*_{Ta1}}{Z_L} \quad (16)$$

which is known as the *minor loop gain* in the power electronics' literature, see e.g. [1], [3], [4]. If no unstable (RHP) pole-zero cancellations occur in the minor loop gain, internal stability is determined by studying one of the four critical transfer functions.

The load impedance Z_L is usually stable, but it can be non-minimum phase. An unstable cancellation can happen, if Z^*_{Ta1} is non-minimum phase. If Z_L is unstable, a cancellation can take place if Z^*_{Ta1} is unstable. If there are no unstable cancellations, it is enough to check the stability of

$$\left(I-k_{11}p_{22}\right)^{-1}k_{12}=\frac{1}{1+\frac{Z^*_{Ta1}}{Z_L}} \quad (17)$$

to deduce internal stability. The frequency function (minor loop gain) Z^*_{Ta1}/Z_L then determines, whether the system is internally stable or not.

VI. CONCLUSIONS

A systematic and system theoretic way to analyze the properties of connected converter systems has been developed in the paper. The formulation of the basic unit models in a two port model was found efficient, because that particular model form was shown to remain invariant under open loop and closed-loop control and in the series connection of such models. A one to one relationship to system models used in the study of internal stability was established, thus giving a powerful tool for the design engineers to work with in practical converter design problems.

REFERENCES

[1] R. D. Middlebrook. "Input filter considerations in design and applications of switching requlators" *IEEE IAS Proceedings*, 1976, pp. 91-107.

[2] R. Middlebrook, "Small-signal modeling of pulse-width modulated switched-mode power converters". *Proceedings of the IEEE*, vol. 76, No 4, April 1988, pp. 343-354.

[3] R. D .Middlebrook, "Design techniques for preventing input-filter oscillations in switched-mode regulators", *Powercon Proceedings*, 1978, pp. A3.1 – A3.16.

[4] T. Suntio, I. Gadoura, and K. Zenger, "Input filter interactions in peak-current-mode-controlled buck converter operating in cicm", *IEEE Transactions on Industrial Electronics*, vol. 49, pp. 76-86, 2002.

[5] T. Suntio and I. Gadoura, "Use of unterminated modeling technique in analysis of input filter interactions in telecom DPS systems", *Proceedings of the IEEE INTELEC'02 Conference*, 2002, pp. 560-565.

[6] R. W. Erickson, *Fundamentals of PowerElectronics*, Chapman & Hall, 1997.

[7] J. Maciejowski, *Multivariable Feedback Design*, Addison-Wesley, 1989.

[8] H. Blomberg and R. Ylinen, *Algabraic Theory for Multivariable Linear Systems*, Academic Press, 1983.

[9] S. Skogestad and I. Postlethwaite, *Multivariable Feedback Control, Analysis and Design*, New York: John Wiley and Sons, 1996.

[10] K. Zhou and J. Doyle, *Essentials of Robust Control*, New Jersey: Prentice-Hall, 1998.

[11] T. Glad and L. Ljung, *Control Theory, Multivariable and Nonlinear Methods*, New York: Taylor and Francis, 2000.

[12] A. Altowati, K. Zenger, and T. Suntio, "Dynamic analysis of a buck converter with input filter via polynomial representation approach", in *Proceedings of the Nordic Workshop on Power and Industrial Electronics*, Trondheim, Norway, 14-16 June 2004.

[13] A. Hentunen, K. Zenger, and T. Suntio, "A systematic approach to analyze the stability of distributed power supply systems", in *Proceedings of the Nordic Workshop on Power and Industrial Electronics*, Trondheim, Norway, 14-16 June 2004.

[14] K. Zenger, T. Suntio and I. Gadoura, "Fundamental limitations in the control performance of switched-mode power supplies", in *Proceedings of the 2002 Nordic Workshop on Power and Industrial Electronics*, Stockholm, Sweden, 12-14 August 2002.

Texture Segmentation Using Kernel-based Techniques

Yu-Long Qiao, Zhe-Ming Lu and Sheng-He Sun
P. O. Box 339, Department of Automatic Test and Control, Harbin Institute of Technology, Harbin, 150001, China
qiaoyulong@dsp.hit.edu.cn

***Abstract-* Texture segmentation is an important component in texture analysis. Gabor wavelet shows potential capacity in describing the texture. Kernel-based methods have demonstrated excellent performances in a variety of pattern recognition problems. This paper extracts texture features in the Gabor wavelet transform domain and segments the computed feature image with kernel-based techniques, Spectral Clustering Algorithm (SCA) and Support Vector Machine (SVM). Due to the severe computational complexity of kernel-based techniques, we split the main segmentation process into two steps. SCA is firstly applied to the sampled feature image. Then the clustering result serves as training samples and is used to train SVM. The initial segmentation result is obtained by labeling the feature image with the trained SVM. A median filter is employed to improve the segmentation result. Four mosaic texture images and a natural scene are used to test our new algorithm.**

I. INTRODUCTION

Texture is one of the important characteristics that exist in many natural images. It also plays an important role in human visual perception and provides information for recognition and interpretation. Texture segmentation techniques have been found useful in the analysis and interpretation of document images, medical images and remote sensed images.

Texture segmentation is the partitioning of a texture image consisting of several regions with different textures into connected regions of similar texture. The texture segmentation algorithm often involves the subprocesses of texture feature extraction, feature selection or reduction if the number of features is too large, followed by a segmentation algorithm.

The first step of the texture segmentation is texture feature extraction. Tuceryan and Jain [1] identified four major categories of methods to extract texture features: statistical, geometrical, model-based, and signal processing based methods. Initially, texture analysis was based on first-order and second-order statistics, in which the co-occurrence matrix is one of the most important methods. Afterwards, Markov random field [2], fractal [3], wold decomposition [4] and various transforms were successfully applied to this field. However, a common weakness of the early methods is that the texture is analyzed only at a single scale. Law [5] convoluted some designed masks with a texture and calculated energy statistics to describe the texture. His work introduced the concept of multichannel processing. In the field of signal processing, there are some interesting spatial/frequency analysis techniques, such as the Gabor transform, Wigner distribution and wavelet transform, provide multiresolution analytical tools. Gabor wavelet (filter) shows its potential in the texture segmentation [6, 7].

The main purpose of texture feature extraction is to map differences in spatial structures, either stochastic or geometric, into differences in gray values. Segmentation methods then analyze the feature space in order to extract homogeneous regions. Many different approaches have been proposed, which are often classified as region-based, boundary-based, or as a hybrid of the two. In a region-based approach, one tries to identify regions of the image that have a uniform texture. Pixels or small local regions are merged based on the similarity of some texture properties. The regions with different textures are then considered to be segmented regions. The boundary-based approaches are based upon the detection of differences in texture in adjacent regions. Thus boundaries are detected where there are differences in texture.

A typical system for texture segmentation is illustrated in Fig.1. Jain and Farrokhnia [6] extracted the texture feature in the Gabor transform domain. Then the features are subject to a squared-error clustering algorithm to obtain a segmentation of the texture image. Unser [8] employed the wavelet frame to extract the texture feature. Acharyya and Kundu [9] proposed an unsupervised texture segmentation scheme based on M-band wavelet transform. Both techniques adopted *K*-means clustering algorithm to segment the texture image.

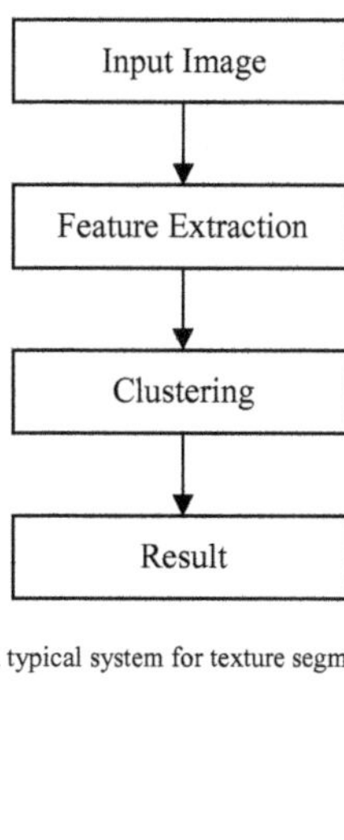

Fig. 1. A typical system for texture segmentation.

K. Elleithy et al. (eds.), Advances in Computer, Information, and Systems Sciences, and Engineering, 19–23.

In this paper, we present a texture segmentation scheme based on Gabor wavelet, Spectral Clustering Algorithm (SCA) [10] and Support Vector Machine (SVM) [11]. The main steps of the proposed algorithm are shown in Fig. 2. For an input texture image, we first compute the multiresolution texture feature in the Gabor wavelet domain, which includes Gabor wavelet transform, nonlinearity, smoothing and normalizing nonlinearity. Taking the computation complexity into account, we apply the kernel-based clustering algorithm SCA to the sampled feature image. Then SVM is trained with the clustering result and is used as a classifier to label pixels of the feature image. The segmented texture image is obtained after a proper post-processing. Our preliminary study on document image segmentation [12] employs the similar scheme. The detail and generalized texture segmentation algorithm is described in Section 2. Section 3 presents and discusses the experimental results on four mosaic textures and a natural scene.

II. TEXTURE SEGMENTATION ALGORITHM

A. Feature Extraction

A.1. Gabor Wavelet

A two-dimensional Gabor elementary function $g(x,y)$ and its Fourier transform $G(u,v)$ can be written as:

$$g(x,y) = h(x,y)\exp(2\pi j W x) = \left(\frac{1}{2\pi\sigma_x\sigma_y}\right)\bullet\exp\left[-\frac{1}{2}\left(\frac{x^2}{\sigma_x^2}+\frac{y^2}{\sigma_y^2}\right)+2\pi j W x\right] \quad (1)$$

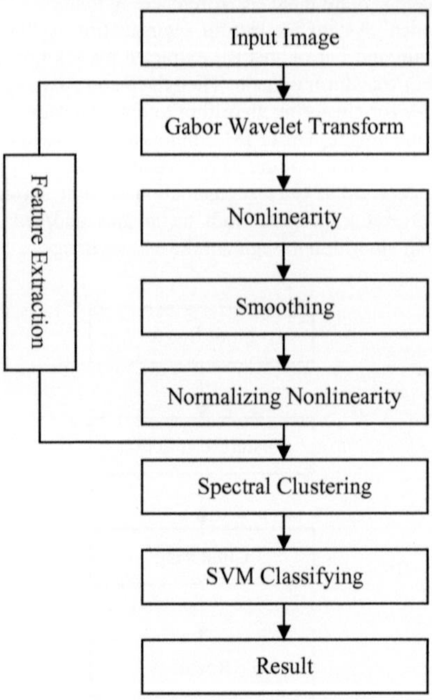

Fig. 2. Texture segmentation algorithm.

$$G(u,v) = H(u-W,v) = \exp\left\{-\frac{1}{2}\left[\frac{(u-W)^2}{\sigma_u^2}+\frac{v^2}{\sigma_v^2}\right]\right\} \quad (2)$$

where $h(x,y)$ is a Gaussian function. W denotes the center frequency of $G(u, v)$ along u axis, $\sigma_u = 1/2\pi\sigma_x$ and $\sigma_v = 1/2\pi\sigma_y$. σ_x and σ_y characterize the spatial extent along x and y axes, respectively, while σ_u and σ_v characterize the bandwidth along u and v axes, respectively. A class of self-similar functions, called Gabor wavelets, is now considered. Let $g(x,y)$ be the mother Gabor wavelet, then a self-similar function dictionary can be obtained by proper dilations and rotations of $g(x,y)$ as follows:

$$g_{mn}(x,y) = a^{-m} g(x',y'),\quad a>1,\quad m,n\in \mathbf{Z}$$
$$\begin{pmatrix} x' \\ y' \end{pmatrix} = a^{-m}\begin{pmatrix} \cos\theta & \sin\theta \\ -\sin\theta & \cos\theta \end{pmatrix}\begin{pmatrix} x \\ y \end{pmatrix} \quad (3)$$

where $\theta = n\pi/N$ and N is the total number of orientations. The scale factor a^{-m} in (3) is meant to ensure that the energy is independent of m.

An efficient Gabor filter bank design method is introduced in [12] to reduce the redundant information in the filtered images caused by the nonorthogonality of Gabor wavelets. Let U_l and U_h denote the lower and upper center frequencies of interest, respectively. Let N be the number of orientations and S be the number of scales in the multiresolution decomposition. Then the design strategy results in the following formulas for computing the wavelet parameters σ_u and σ_v.

$$a = (U_h/U_l)^{\frac{1}{S-1}},\quad \sigma_u = \frac{(a-1)U_h}{(a+1)\sqrt{2\ln 2}},$$
$$\sigma_v = \tan\left(\frac{\pi}{2N}\right)\left[U_h - (2\ln 2)\frac{\sigma_u^2}{U_h}\right]\left[2\ln 2 - \frac{(2\ln 2)^2\sigma_u^2}{U_h^2}\right]^{-\frac{1}{2}} \quad (4)$$

Where $W = U_h$ and $m = 0,1,\ldots,S-1$. In order to eliminate the sensitivity of the filter response to absolute intensity values, $G(0,0)$ in (2) are set to zero.

For the texture image, the Gabor wavelets with lower radial frequencies are not very useful, because they capture the feature too coarse to explain the texture. Thus, we use Gabor wavelets designed with N=6, S=3, $U_l = \sqrt{2}/8$ and $U_h = \sqrt{2}/2$ in the following experiment.

A.2. Computing Feature Image

An important step in texture segmentation is to develop a set of texture features that can successfully describe the texture so as to discriminate different textures. We employ the following procedures to compute features from each filtered image. First, each filtered image is subjected to the following bounded nonlinear transformation

$$\Phi(x) = \tanh(\alpha x) = \frac{1-e^{-2\alpha|x|}}{1+e^{-2\alpha|x|}} \quad (5)$$

Where α is a constant, we adopt an empirical value of $\alpha = 0.25$ as Jain and Farrokhnia [5]. The nonlinear transformation is followed with a Gaussian low-pass (smoothing) filter $h(x,y)$ with $\sigma = \sigma_x = \sigma_y$. Where σ determines the passband width of the smoothing window. Formally, the above procedures can be denoted by

$$F_k(p,q) = \frac{1}{R} \sum_{i=-\lfloor r/2 \rfloor}^{\lfloor r/2 \rfloor} \sum_{j=-\lfloor r/2 \rfloor}^{\lfloor r/2 \rfloor} L(p+i,q+j)\Phi(GC_k(p+i,q+j)) \quad (6)$$

Where $L(\cdot,\cdot)$ is the discrete Gaussian low-pass (smoothing) filter. GC_k denotes the *k*th filtered image. *r* is the window size and $R = r \times r$. The size *r* of the smoothing window is an important parameter. The larger the window size, the more reliable are the texture features found. On the other hand, the smaller the widow size, the more accurate are the boundaries between feature clusters localized. The reason is that the average blurs the boundaries between texture regions. Another important observation is that Gaussian weighted windows, rather than unweighted windows, are likely to result in more accurate localization of texture boundaries.

The second nonlinearity illustrated in Fig. 2 is not commonly used. Unser [13] proposed and tested several combinations of the first and second nonlinearities for texture segmentation. He concluded that squaring in combination with a logarithmic normalizing nonlinearity was the best combination. However, in our experiments, we find it is worse than the combination of nonliearity form (5) and the logarithmic normalizing nonlinearity.

B. Kernel-based Methods

In the last years, a number of powerful kernel-based methods have been proposed [11]. They provide an elegant way of dealing with nonlinear algorithms by reducing them to linear ones in some feature space *FS* nonlinearly related to the input space *IS*. Using kernel instead of a dot product in *IS* corresponds to mapping the data into a possibly high dimensional dot product space by a (usually nonlinear) map $\Psi: IS \to FS$ and taking the dot product there, i.e.,

$$k(\cdot,\cdot) = (\Psi(\cdot), \Psi(\cdot)) \quad (7)$$

where k is a kernel function, for example, linear kernel, polynomial kernel, Gaussian radial basis function kernel and sigmoid kernel [11]. By virtue of this property, we shall call Ψ a feature map associated with *k*. Any linear algorithm which can be carried out in terms of dot products can be made nonlinear by substituting a priori chosen kernel, e.g., kernel-based clustering, support vector machine and kernel principal component analysis.

The primary disadvantage of the kernel-based method is its severe computational burden. For an image of size 512×512, there are in total 262144 feature vectors (262144 pixels in the feature image). If we directly use kernel-based techniques, it needs to process a matrix of size 262144×262144. The severe computational complexity indicates that it is unreasonable to only use the kernel-based clustering algorithm to segment the texture image. Motivated by the work [6, 15], we firstly apply a clustering algorithm SCA to the sample feature image. Then, the clustering result is used to train a kernel-based classifier SVM. The trained SVM is employed to label pixels of the feature image and thus we obtain the initial segmentation result.

B.1. Spectral Clustering

Ng, Jordan and Weiss [10] presented an efficient spectral clustering algorithm SCA that shows surprisingly good performance on a number of challenging clustering problems. In this paper, we adopt it to segment the texture image. The detail steps of the clustering algorithm can be found in [10]. From the first step of this algorithm, we can see that it is also a kernel-based method because the affinity matrix is constructed by using Gaussian radial basis function kernel.

$$k(F^1, F^2) = \exp\left(-\left\|F^1 - F^2\right\|^2 / 2\sigma^2\right) \quad (8)$$

F^1, F^2 are two pixels in the feature image.

Taking the computational burden of SCA into account, we apply SCA to the sampled feature image. The sampling method is that we take a pixel from a subblock of the feature image. In general, the size of the subblock is adjusted according to that of the feature image so that the size of the sampled feature image is less than 50×50. In our following experiments, we will study the effect of the window size on the segmentation result.

After this step, a class label is assigned to each pixel of the sampled feature image. The labeled pixels cooperating with their class label serve as the training samples to train a SVM. Then the trained SVM is used to label pixels of the feature image so as to obtain the initial texture image segmentation. In spite of the good performance of SCA, there are still some pixels belonging to different groups have been clustered into the same cluster. To furthest reduce the infection of the misclustered pixels, we choose the largest connectivity region with the same class label as the training samples. This method makes use of the spatial relation of pixels and thus improves the final segmentation result.

B.2. Support Vector Machine (SVM)

Support vector machines construct a maximum margin hyperplane in the feature space *FS*. In the input space, this corresponds to a nonlinear decision boundary of the form

$$f(x) = \operatorname{sgn}\left(\sum_{i=1}^{l} \alpha_i k(x, x_i) + b\right) \quad (9)$$

where x_i are the training samples. l is the number of those samples. k is a kernel function. The kernel (8) is also employed in this step. The training samples with $\alpha_i \neq 0$ are called SV's. In many applications, most of α_i turn out to be zero. From (9), it can be seen that SVMs use the support vectors to determine the decision boundary, which have exploited the "margin" between classes. SVM classifier has often been found to provide higher classification accuracies than other widely used pattern recognition techniques and been used in various fields. An important advantage of the SVMs is that it is based on the principle of structural risk minimization, rather than on empirical risk minimization as do many other methods. Thus, SVMs aim at minimizing a bound on the generalization error, instead of simply minimizing the training error. LIBSVM developed by Chang and Lin [16] is one of the best softwares for SVMs, from which readers may find some interesting matters.

C. Post Processing

After finishing the classification with SVM, there may exist misclassified pixels usually take the form of noisy speckles in the segmentation image. A median filter of size 5×5 is iteratively applied to the initial segmentation image to improve the final segmentation result. In our experiment, the maximal iteration number is 10.

III. EXPERIMENTAL RESULTS

For fully evaluating our texture segmentation algorithm, we test it on four mosaic textures and a natural image. The mosaic textures, Fig. 3 (a), Fig. 4 (a) and Fig. 5 (a), are created by collaging subimages of several Brodatz textures [17].

Fig. 3 (a) shows a test D19-D57 texture pair. Fig. 3 (b) is the corresponding segmentation result. The segmentation accurate rate (SAR) is 98.3%. The segmentation errors mostly exist around the boundary between the texture regions. This segmentation result is satisfactory because the boundary is difficult to be determined even through careful identification.

Fig. 4 (a) is a 256×256 mosaic texture image. The corresponding segmentation result is presented in Fig. 4 (b). The SAR is 98.8%. The most segmentation errors occur around the boundary between D24 and D77.

Fig. 5 shows a 512×512 test image and its corresponding segmentation result. The SAR is 98.4%. There are some small missegmented regions in the D9 texture region.

Fig. 6 (a) is a 256×256 mosaic texture image used in [18]. Jain and Karu reported the SARs are 95.5%, 94.8% and 96.7% with Gabor filters (using the segmentation method [6]), Laws' masks and Learned masks, respectively, while the SAR of our new algorithm is 97.6%.

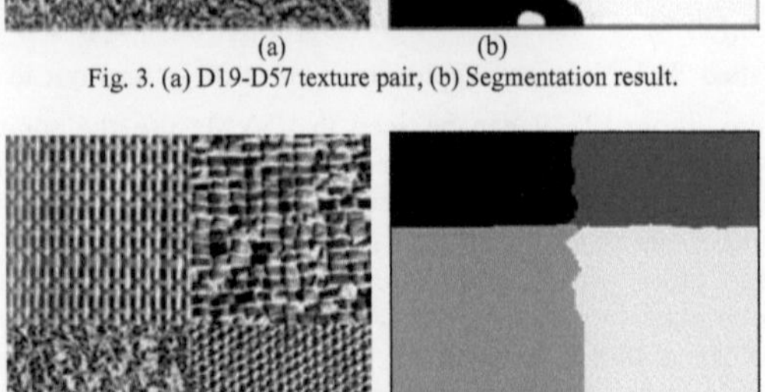

(a) (b)
Fig. 3. (a) D19-D57 texture pair, (b) Segmentation result.

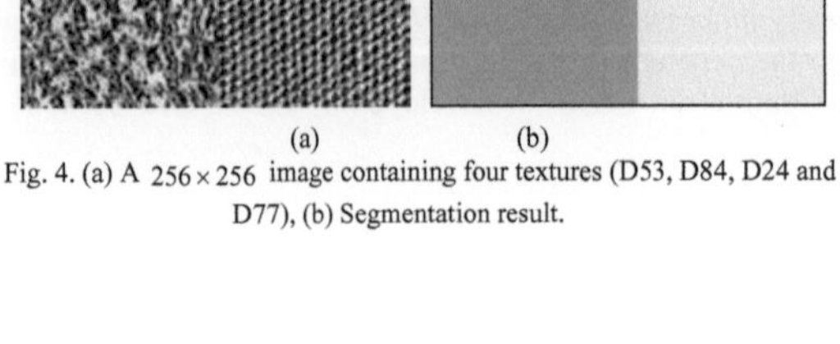

(a) (b)
Fig. 4. (a) A 256×256 image containing four textures (D53, D84, D24 and D77), (b) Segmentation result.

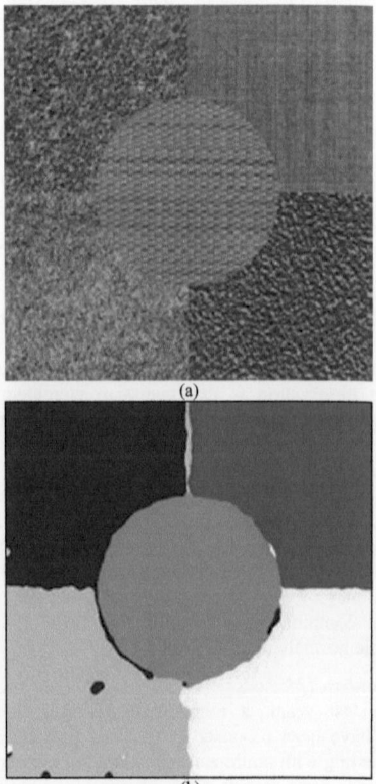

(b)
Fig. 5. (a) A 512×512 image containing five textures (D29, D16, D55, D9 and D57), (b) Segmentation result.

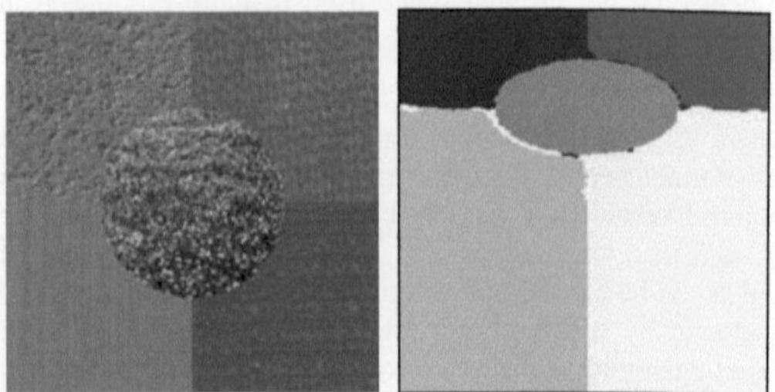

(a) (b)
Fig. 6. (a) A 256×256 image used in [18], (b) Segmentation result.

To further analyze the new scheme, we take a 256×256 natural scene shown in Fig. 7 (a) as the test image. The image is segmented into three regions, sky, tree and water. The result (Fig. 7 (b)) is very satisfactory.

To reduce the computational complexity of SCA, we sample the feature image. The sampling method is that we take a pixel from a subblock of the feature image. An interesting question is whether different subblock sizes result in very different final segmentation results. The SARs of Fig. 6 (a) with different subblock sizes are listed in TABLE I. From this table, we can see that the subblock size does not

Fig. 7. (a) A 256×256 natural scene image, (b) Segmentation result.

TABLE I
SARs WITH DIFFERENT SAMPLING WINDOW SIZES

Window Size	SAR (%)
5	97.60
7	97.18
9	97.27
11	97.17
13	97.18
15	97.16
17	97.07

cause large fluctuation of the segmentation result. When the window size is 17×17, the selected pixels are 0.44% of the total pixels. This sampling operation can reduce not only the computational complexity of SCA, but also that of the SVM, and thus speed up the segmentation.

IV. CONCLUSIONS

This paper proposes a texture segmentation algorithm. It makes full use of Gabor wavelet and kernel-based techniques (Spectral Clustering Algorithm and Support Vector Machine). Taking the computational burden of the kernel-based techniques into account, we firstly apply the clustering algorithm SCA to the sample feature image extracted from Gabor filtered images. Then, the clustering result is used to train a SVM. The segmentation is obtained by assigning a class label to each pixel of the feature image with the trained SVM. The segmentation results of four mosaic texture images and a natural scene sufficiently demonstrate the effectiveness of the new algorithm. Further discussion is made about the effect of different sampling window sizes on the final segmentation. The result suggests the sampling window size only causes a little fluctuation of the resulting segmentation. All experiments show the potential capacity of our proposed texture segmentation scheme.

REFERENCES

[1] M. Tuceryan and A. K. Jain, "Texture Analysis," In Handbook of pattern recognition and computer vision (2ed), World Scientific Publishing Co., 1998.
[2] G. C. Cross and A. K. Jain, "Markov random field texture models," IEEE Trans. Pattern Anal. Machine Intell., vol. 5, no. 1, pp. 25-39, 1983.
[3] A. P. Pentland, "Fractal-based description of natural scenes," IEEE Trans. Pattern Anal. Machine Intell., vol. 6, no. 6, pp. 661-674, 1994.
[4] F. Liu, R and W. Picard, "Periodicity, directionality, and randomness: Wold features for image modeling and retrieval," IEEE Trans. Pattern Anal. Machine Intell., vol. 18, no. 7, pp. 722-733, 1996.
[5] K. I. Law, "Texture Image Segmentation," Ph.D. dissertation, Univ. of Southern California, 1980.
[6] A. K. Jain and F. Farrokhnia, "Unsupervised texture segmentation using gabor filters," Pattern Recognition, vol. 24, no. 12, pp. 1167-1186, 1991.
[7] D. Dunn, W. E. Higgins and J. Wakeley, "Texture segmentation using 2-D gabor elementary functions," IEEE Trans. Pattern Anal. Machine Intell., vol. 16, no. 2, pp. 130-149, 1994.
[8] M. Unser, "Texture classification and segmentation using wavelet frames," IEEE Trans. Image Processing, vol. 4, no. 11, pp. 1549-1560, 1995.
[9] M Acharyya and M. K. Kundu, "An adaptive approach to unsupervised texture segmentation using M-band wavelet transform," Signal Processing, vol. 81, pp. 1337-1356, 2001.
[10] A. Y. Ng, M. I. Jordan and Y. Weiss, "On spectral clustering: analysis and an algorithm," In Advances in Neural Information Processing Systems, vol. 14, pp. 849-856, 2001.
[11] K. R. Müller, S. Mika, G. Rätsch, K. Tsuda and B. Schölkopf, "An introduction to kernel based learning algorithm," IEEE Trans. Neural Networks, vol. 12, no. 2, pp. 181-201, 2001.
[12] Y. L. Qiao, Z. M. Lu and S. H. Sun, "Document image segmentation using Gabor wavelet and kernel-based methods," submitted to 7th IAPR workshop on Document Analysis Systems.
[13] B. S. Manjunath and W. Y. Ma, "Texture feature for browsing and retrieval of image data," IEEE Trans. Pattern Anal. Machine Intell., vol. 18, no. 8, pp. 837-842, 1996.
[14] M. Unser and M. Eden, "Nonlinear operators for improving texture segmentation based on feature extracted by spatial filtering. IEEE Trans. Systems, Man, Cybernetics, vol. 20, pp. 804-815, 1990.
[15] M. Tuceryan, "Moment based texture segmentation," Pattern Recognition Letter, vol. 15, pp. 659-668, 1994.
[16] C.-C. Chang and C.-J. Lin, "LIBSVM: a library for support vector machines," Software available at http://www.csie.ntu.edu.tw/~cjlin/libsvm, 2001.
[17] P. Brodatz, "Textures: A Photographic Album for Artists & Designers. New York: Dover, 1966.
[18] A. K. Jain and K. Karu, "Learning texture discrimination masks," IEEE Trans. Pattern Anal. Machine Intell., vol. 18, no. 2, pp. 195-205, 1996.

A Blur Reducing Adaptive Filter for the Removal of Mixed Noise in Images.

S.Saraswathi janaki , D.Ebenezer

College of Engineering, Anna university,Chennai-600 025,India.

Abstract:-**In this paper, an algorithm has been developed to remove additive mixed noise in images with edge preservation. The noise characteristics may vary in the same application from one image to another. In these environments, nonlinear general filters will not perform well and adaptive non-linear filters are best suited. The algorithm based on local statistics such as signal variance and noise Variance is considered. It depends on minimum mean square estimation of the corrupted signal. The signal variance and noise variance are calculated by moving signal window and moving noise window respectively. This offers optimal adaptive filtering in the homogeneous regions as well as in the edges. The adaptive alpha trimmed mean filters with a threshold value have been proposed to preserve edges. The performance of the filter in the presence of different types of noise are evaluated and compared with general mean, median and alpha trimmed mean filters. The image enhancement factor has been calculated as the performance-measuring factor. The Lena image has been considered to carryout the subjective and objective analysis of the proposed filter.**

I. INTRODUCTION

The nonlinear filters are optimized for signals having specific statistical characteristics. Images are non-stationery processes and their statistics vary in the various regions[1]. The filter suggested by I.Pitas and A.N.Venetsanopoulos detects edges[2].If an image is corrupted by additive mixture of impulse and Gaussian noise, it could be eliminated by an adaptive filter based on local statistics proposed by Bernstein[3]. An adaptive version of the alpha trimmed mean filter has been proposed by A.Restrpo, A.C.Bovik[4] which is suitable for short tailed and medium tailed distributions. The preservation of edges with the simultaneous removal of impulse noise was given by Srinivasan[5]The noise characteristics may also vary from one application to another. Therefore adaptive is natural choice in this case.

Moreover edges of an image provide more information and often visual perception is based on them. Therefore, any image filter should preserve edges. An adaptive algorithm based on local statistics with threshold is proposed. The proposed algorithm is based on the variance of the image corrupted by a combination of impulse and Gaussian noise. The filter has been implemented in three different stages thereby eliminating impulses in first stage followed by edge detection and preservation in the second satge. The third stage has been implemented to reduce the blurring effect and to eliminate Gaussian noise. The statistics of the signal and noise are taken into consideration in order to maintain the originality of the signal.

The proposed adaptive nonlinear filter is a simple and optimal filter for the removal of noise without damaging the edges.

II. NON LINEAR FILTER FOR DIFFERENT NOISE MODELS.

In many impulse noise models for images, corrupted pixels are replaced by a pixel whose value is equal to the maximum or minimum value in the allowable dynamic range. The input signal s (k, l) is corrupted by additive noise n (k, l). The corrupted signal is given by

$$x(k, l) = s(k, l) + n(k, l)$$

n(k, l) is a mixture of Gaussian noise of various variance and impulse noise of various densities. If the image is corrupted by Gaussian noise alone, it could be removed by replacing the center pixel with the mean of the neighborhood pixels in a particular window. If the images are corrupted by impulse noise, the noise is eliminated by the replacement of center pixel with median of the neighborhood pixels .There exists a trade off between the elimination of noise and preservation of edges[6]. The general alpha trimmed mean filters smear edges, due to averaging operation. The adaptive alpha trimmed mean filters are proposed which can detect and preserve edges.

K. Elleithy et al. (eds.), Advances in Computer, Information, and Systems Sciences, and Engineering, 25–29.

III. FILTERING FRAMEWORK.

Fig.1 shows the filtering scheme that has been carried in three stages.

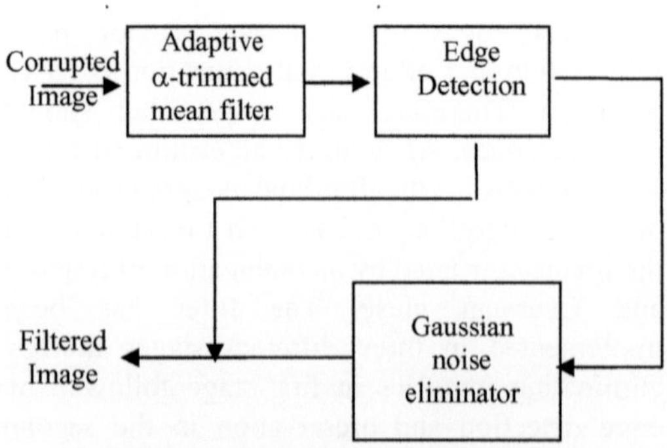

Fig. 1 Block diagram of the filtering scheme

A.. Stage 1:Adaptive alpha trimmed mean filter

The first stage of the filter incorporates adaptive alpha trimmed mean filter .

The value of alpha is determined by the variance of the corrupted image. The steps involved in the detection and elimination of impulses are as follows.

Step1: Let x(i,j) be the pixel values considered inside the window N*N. The mean , median and alpha trimmed mean of the pixels inside the window of size N*N are given by

$$\text{Mean}[x(i,j)] = \frac{1}{N*N}\sum_{m=1}^{N}\sum_{l=1}^{N} X(m,l)$$

Median [x(i,j)] = center pixel in the sorted set of pixels inside the window

Alpha trimmed mean[x(i,j)]=

$$\frac{1}{(1-2\alpha)(N*N)}\sum_{i=\alpha N}^{N-\alpha N}\sum_{j=\alpha N}^{N-\alpha N} x(i,j)$$

Step2: Mean of (N*N) and (N*N)-2 has been calculated

Step3: If the absolute difference between the mean of (N*N) and (N*N)-2 is lesser than the threshold 't', then all the (N*N) pixels are considered for the next stage. If it is greater than 't', then it is considered as the presence of impulses.

Step4: Next the difference between the mean of the (N*N)-2 pixels and the (N*N)-4 pixels is determined. This process is continued till the difference is lesser than the threshold, and once that condition is satisfied, the set of pixels, which were considered for finding the mean of the larger number of pixels among the two are carried on to the next stage.

Thus stage 1 is used to detect and remove impulse noise. Experimentally the best value of the threshold 't' has been calculated as 0.02

B Stage 2: Preservation of Edges

The stage2 is used to preserve the edges of an image. It is known that the variation between consecutive pixels is very high, if there is an edge. Hence, the variance of a set of pixels containing an edge will be high. The edges are preserved as follows.

Step1: The variance of the set of pixels obtained from the previous stage is calculated and is compared with a threshold value 'th'. If it is less than the 'th', it can be considered as the absence of edge and are passed on to the third stage.

Step 2: It is known that the set of pixels obtained from the first stage is free from impulse noise. Hence, the center pixel is compared with each of the pixels of the set obtained from the first stage. If it equals any one of the pixels in the set, it is considered that the center pixel is not corrupted by impulse noise. Hence it is left unaltered, and is retained in the resultant image. Otherwise, the set of pixels from the first stage is carried on to the third stage. It must be noted that during edge preservation, the Gaussian noise is not removed

The threshold value 'th' has been determined experimentally as 0.003 for less amount of Gaussian noise, where the edge preservation is more important.. For high amount of Gaussian noise, noise elimination is more important. Therefore a high value of th = 0.3 is chosen. Thus this stage preserves edges and retains the center pixel if it is an edge.

C. Stage 3: Reduction of Blur

The arithmetic mean filter averages the image pixels to reduce the Gaussian noise. In this process the neighborhood pixels get blurred. Reduction of blurring & elimination of Gaussian noise has been considered in stage 3.

Step 1: The variance of the pixels that are obtained from the previous stage is calculated and is compared with a threshold value.

Step 2: If the variance is less than the threshold, then the mean of the pixels inside the window is used to replace the center pixel of the window. Otherwise less number of pixels are involved in the mean operation. This process is continued till the condition is satisfied, and then the mean of the final set of pixels is calculated to replace the center pixel large values of Gaussian noise.

A larger threshold value ensures greater removal of Gaussian noise at the expense of more blurring of the image [7]. Here, again, the threshold value must be chosen as a large number for large Gaussian noise, as removal of the noise becomes more important when compared to blurring introduced in such a case.

It has been determined experimentally that a low value of 'pt' is offering best results for the removal of less amount of Gaussian noise .The value of pt is equal to 0.0001 for a noise variance of 100 & is 0.08 for Gaussian noise of variance 200. The image can be made smoother if the value of the threshold 'pt' is increased [8].In this way the third stage of the filter eliminates Gaussian noise and reduces the blurring effect.

IV. IMAGE ENHANCEMENT FACTOR

The performance of the filter has been studied by computing Image enhancement factor (IEF). It is the ratio of mean square error before filtering to the mean square error after filtering.

V. RESULTS OF PROPOSED ADAPTIVE ALPHA TRIMMED MEAN FILTER

The original image has been corrupted by the addition of white Gaussian noise of different variance and impulse noise of different percentages. The IEF of the proposed blur reducing adaptive filter is compared with the IEF of mean, median and alpha trimmed mean filters.

The IEF for the proposed new adaptive filter is found to be greater than mean, median filters, which shows the significant removal of noise. The IEF has been calculated for various filters with window size of 3X3, and are given in Table 1. and Table 2.

The impulses are eliminated, with the preservation of edges. Objective analyses of the filter performance shows that the mixture of white Gaussian noise and impulses are eliminated and edges are preserved with the reduction in blur. Fig2.a to Fig 2.g shows the results of the proposed adaptive alpha trimmed filter for various amounts of noise with the window size of 3x3.Fig 2.a shows the original test image. Fig.2.b is the image corrupted with the mixture of Gaussian noise of variance 100 and Salt and pepper noise of an amount of 4%. and the corresponding filtered result is shown in Fig 2.c.Fig 2.d is the image corrupted by the mixture of salt and pepper of 5% and Gaussian noise of variance 100 and its corresponding filtered image is shown in Fig 2.e.Fig 2.f represents the image corrupted by the mixture of salt and pepper noise of an amount of 10% and Gaussian noise is of variance 200. The corresponding filtered image is depicted in Fig 2.g.. Fig 3. shows the comparison of the results of the proposed blur reducing adaptive alpha trimmed mean filter with that of the results of the general mean, median and alpha trimmed mean filters. These results are shown for different amount of impulse noise. The comparison results for various amount of Gaussian noise is shown in Fig 4The comparison results are made for various amount of mixed noise. It is found that IEF for the proposed filter increases as the amount of impulse noise increases and is greater than the IEF of other filters. The IEF decreases as the corrupted amount of Gaussian Noise increases.

TABLE 1
IEF FOR MIXED NOISE (GAUSSIAN NOISE IS OF VARIANCE 100)

Percentage of impulse noise	Ief of Mean filter	Ief of median Filter	Ief of AATMF	Ief of blur reducing AATMF with edge preservation
1	2.0672	3.4177	4.3226	4.4
2	2.8857	5.613	7.005	7.1006
3	3.1698	7.3347	9.0963	9.1542
4	4.1295	9.6906	11.4696	11.5057
5	4.6002	11.6218	13.5606	13.93
10	8.3	15.7	16.213	17.10

TABLE 2
IEF FOR MIXED NOISE (IMPULSE NOISE IS OF 5%)

Variance of Gaussian Noise	Ief of Mean filter	Ief of median Filter	Ief of AATMF	Ief of blur reducing AATMF with edge preservation
50	4.0189	9.9345	13.0171	13.2102
100	4.0609	9.4506	11.2052	11.5057
150	4.2432	8.7348	10.565	10.8101
200	4.2798	8.5007	9.8991	10.0512
250	4.3711	8.3211	9.6988	9.4073
300	4.5481	8.1451	8.7943	8.748
350	4.6815	8.0925	8.6074	8.4509
400	4.7644	7.915	8.4681	8.1257

Fig 2.a

Fig 2.b

Fig 2.c

Fig 2.d

Fig 2.e

Fig 2.f

Fig 2.g

Fig 2.a to Fig 2.g Results of blur reduced adaptive alpha trimmed mean filter for various amount of noise.

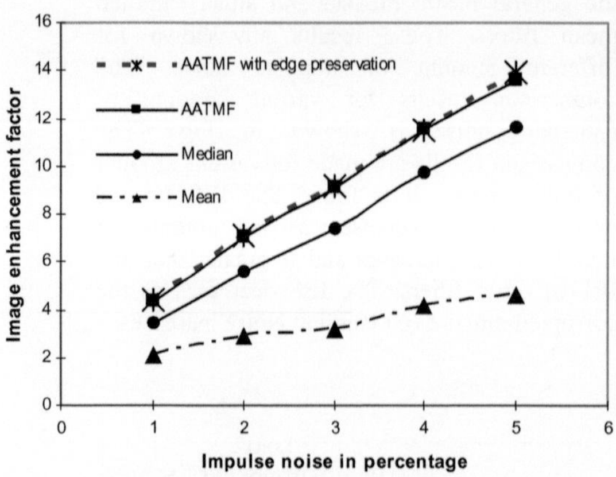

Fig.3 Comparison of IEF of AATMF with edge preservation with that of IEF of other filters for various amount of impulse noise

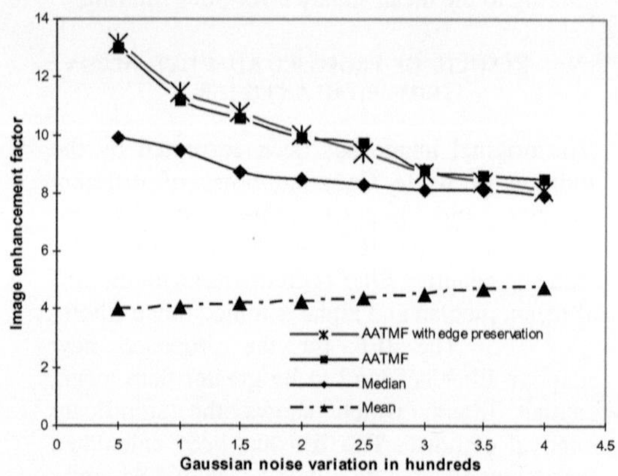

Fig.4 Comparison of IEF of AATMF with edge preservation with that of IEF of other filters for various amount of Gaussian noise

VI. CONCLUSION

A simple and novel adaptive alpha trimmed mean filter has been proposed. The subjective and Objective analysis of the filter has been made. The image enhancement factor is found to be greater than the mean, median and alpha trimmed mean filters. The filter preserves the edges, which provides information. Since the value of alpha adapts itself, this filter is optimized for any type of noise and any type of image. The Lena image is considered as the test image. Better noise elimination has been obtained while preserving the edges The proposed filter has been implemented in three stages using a threshold value to detect impulses and its removal, threshold value for the detection of edges and to reduce the blurring in images. The operation of this filter involves less complexity as it is compared with other filters. The performance of the proposed adaptive alpha trimmed mean filters are found to be better than mean, median, adaptive alpha trimmed mean filters without threshold.

REFERENCES

[1] I.Pitas and A. N. Venetsanopoulos " Non linear digital filters principles and applications", *Kluwer Academic Publishers,* Bostan. 1990

[2] I. Pitas, A. N. Venetsanopoulos, "Non linear order statistic filters for image filtering and edge detection", IEEE ,Vol.10, 1986, pp 395-413

[3] R.Bernstein "Adaptive nonlinear filters for simultaneous removal of different kinds of noise in images", *IEEE Trans,* Vol.CAS-34, No.11, Nov 1987, pp.1275- 1291.

[4] A.Restrepo, A.C.Bovik, "Adaptive trimmed Mean filters for image restoration," *IEEE Trans ,* Vol.36, no.8, 1998, pp 1326-1337.

[5] "Some nonlinear filtering strategies for eliminating uniform, Gaussian and impulse noise in images with edge and fine detail preservation properties", *ph.d dissertation,* Srinivasan.E Anna University.2003

[6] Sun.X. Z and Venetsanopoulos. A . N., "Adaptive schemes for Noise filtering and edge detection by Use f local statistics" , *IEEE Trans* Vol No.35 No.1 pp 57-70.1998

[7] Remi Oten and Rui J. P. Figueirdo, "Adaptive Alpha-Trimmed Mean Filters Under Deviations From Assumed Noise Model" *IEEE Trans* Vol.13,No.5,May 2004.

[8] Saraswathi. S ., and D.Ebenezer 'Adaptive alpha trimmed median filter for the removal of mixed noise in images, Proceedings of the International conference TIMA 2001, MIT, Anna University, Chennai

Signal Modeling Using Singular Value Decomposition

Sobia Baig, Fazal-ur-Rehman
Faculty of Electronic Engineering, GIK Institute of Engineering Sciences & Technology
Topi 23460- Swabi, NWFP, Pakistan.
Email: sobia@giki.edu.pk, frahman@giki.edu.pk

Abstract

Developing a signal model for approximation of a signal, involves reducing the error between the original signal and that reconstructed from the signal model. This topic has been a research area in signal processing and various mathematical techniques have been applied to solve this problem. This paper aims at presenting a solution using a uniform and redundant transmultiplexer structure, with Singular Value Decomposition (SVD) technique.

I. INTRODUCTION

Signal modeling is the representation of an arbitrary signal by the best possible approximation from a class of bounded signals that belong to the l_2 norm space. Signals that are part of l_2 norm space are square summable signals. The criterion for finest approximation is also characterized by the l_2 norm. The approach here is to model an arbitrary bounded signal, by the output of a linear system, which in turn is driven by an arbitrary bounded signal.

Signal modeling finds application in various signal processing fields, such as multiresolution theory, sampling theory and signal denoising. Spline interpolation system is another example of signal modeling. In [1], author explores the possibility of using linear combinations of B-splines to obtain a stable signal representation. Aldroubi in [2] studies a general problem of oblique projections in discrete shift-invariant spaces of l_2 and gives error bounds on the approximation. A general signal model is proposed in [3], consisting of a bank of filters with different interpolation ratios and authors have shown that optimal prefilters in general require time varying systems. Authors in [4] find an optimal condition in order to minimize the mean squared error between a deterministic signal and its approximate representation. They refer to the choice of synthesis filters as an energy compaction problem.

In this paper, we focus on a multi-channel, redundant and a uniform transmultiplexer system that is used to develop a system model, using SVD technique. This signal model utilizes the concept of filter banks, with a uniform interpolation and decimation ratio. SVD is a powerful and widely used technique of matrix computation and has found use in a variety of applications including least squares problem, solution of linear equations, data compression, noise filtering, and similar other signal processing fields [5]. In this paper SVD is employed to decompose a circulant matrix in order to determine the constituent analysis and synthesis filter banks of a perfect reconstruction transmultiplexer. Redundancy can be added to a transmultiplexer by either utilizing the concept of cyclic prefix or by zero-padding. In the former case, we get a square circulant matrix and in the later we have a rectangular matrix, which is then analyzed using SVD. We employ the zero-padding method here. Mathematical analysis for the perfect reconstruction of redundant transmultiplexer based signal model is also presented. Section II presents the signal modeling problem and its solution, while the role played by a redundant transmultiplexer in signal regeneration is discussed in Section III and SVD technique based solution is presented in Section IV. Uniform structure for a signal model is presented with an example in Section V, followed by conclusions drawn in Section VI.

II. SIGNAL MODEL PROBLEM AND SOLUTION

Given an arbitrary signal $x(n)$, the problem of signal approximation is finding $y(n)$ so that it can be the best possible estimate of $x(n)$. We begin by suggesting a linear system, as a mathematical signal model. This is shown in Fig. 1, which represents a single channel signal model. Here $c(n)$ represents the coefficients of an arbitrary channel oversampled by a factor of P, and passed through a filter $F(z)$, to get the output $y(n)$. Instead of finding a sequence $c(n)$, we assume that these coefficients $c(n)$ are known and now the signal modeling problem is, how to process the output signal $y(n)$, so that we can regenerate the input coefficients $c(n)$, in order to approximate $x(n)$. Solution to this problem, as suggested in [6], pertains to the theory of biorthogonal partners.

An important application of biorthogonal partners is the reconstruction of signal that has been oversampled by an integer amount, say P. We process the signal $y(n)$ and filter it through $G(z)$ giving $x(n)$, which gives us the reconstruction of original input signal, as shown in Fig. 2.

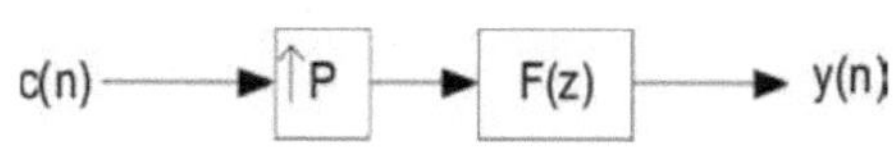

Fig. 1.Single channel signal model

K. Elleithy et al. (eds.), Advances in Computer, Information, and Systems Sciences, and Engineering, 31–35.

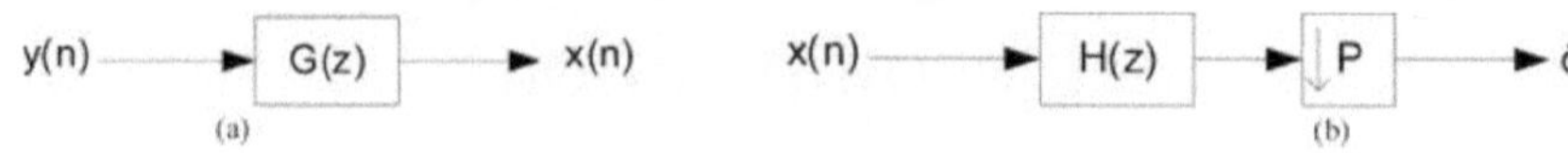

Fig. 2.(a)Filtering of signal *y(n)* by *G(z)* (b) Signal Reconstruction using Biorthogonal partner *H(z)*

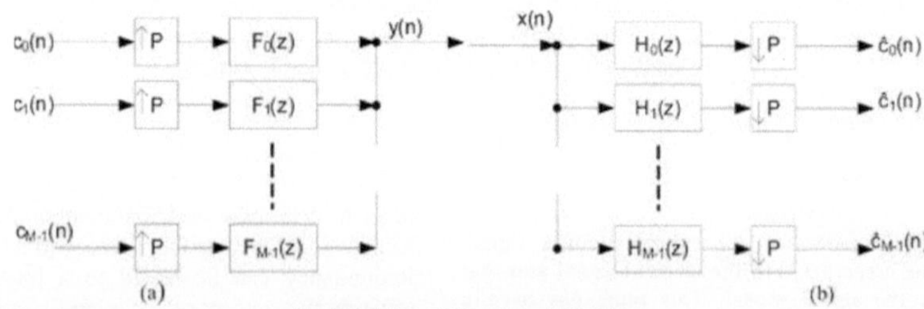

Fig. 3 Solution of Signal Model Problem

If we can find a filter *H(z)*, such that it is FIR and a biorthogonal partner of *F(z)*, then we can recover the coefficients *c(n)* as shown in Fig. 2(b). A possible solution is suggested here, as shown in Fig. 3. This signal model can be considered as a uniform and redundant channel model, as presented in Fig. 3(a) and the problem objective is to recover the coefficients, as shown in Fig. 3(b).

For M branches the system is made redundant with P interpolators, such that $P>M$ and the output signal *y(n)* is the sum of outputs from all branches, and in time-domain it is described by the following:

$$y(n) = \sum_{i=0}^{M-1} \sum_{k} c_k(i) f_k(n - iP) \qquad (1)$$

The solution to the problem of signal modeling is the application of a redundant transmultiplexer structure based on synthesis and analysis filter banks [7], which is described in the next section.

III. Redundant Transmultiplexer Structure

The transmultiplexer is presented in Fig. 4 with M branches, upsampled and downsampled by P, multiplexed into one filter *G(z),* so we have the interpolation and decimation factor $P>M$. This introduces redundancy in the filter banks and helps in perfect signal reconstruction. Also, with $P>M$, it is easier to implement the filter bank structure practically.

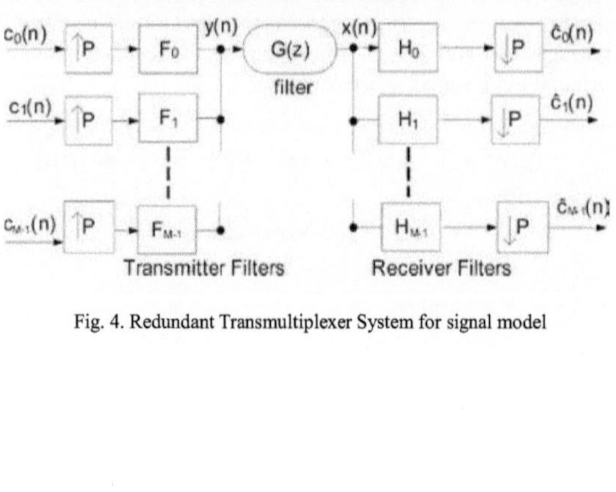

Fig. 4. Redundant Transmultiplexer System for signal model

For redundant transmultiplexer application, we take:

$$P = M + L \qquad (2)$$

where L is the order of filter *G(z)*. The redundancy in the transmultiplexer can be added using cyclic prefixing technique, as is done in case of Multicarrier Modulation schemes like Orthogonal Frequency Division Multiplexing (OFDM) or by zero-padding method [8].

B. Zero-Padding for Redundant Transmultiplexer

A redundant transmultiplexer structure is depicted in Fig. 6, with polyphase decomposition of synthesis and analysis filter banks, as proposed in [8]. The filters in polyphase form are given by:

$$\begin{aligned} H_k(z) &= \sum_{i=0}^{P-1} z^i E_{ki}(z^P) \\ F_m(z) &= \sum_{i=0}^{P-1} z^{-i} R_{im}(z^P) \end{aligned} \qquad (3)$$

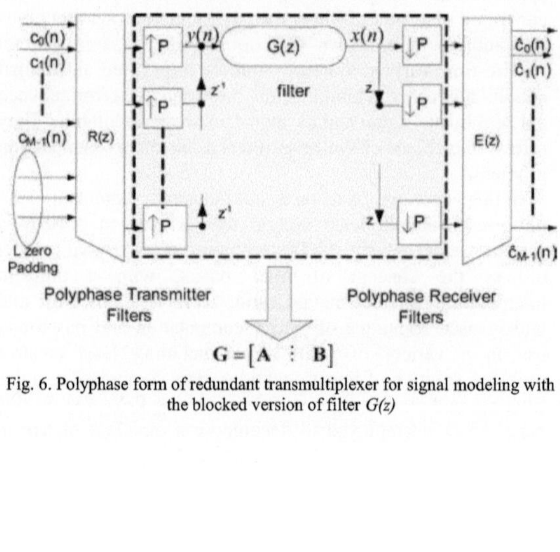

Fig. 6. Polyphase form of redundant transmultiplexer for signal modeling with the blocked version of filter *G(z)*

In Fig. 6, the box shows the blocked version of filter *G(z)*, which is a pseudocirculant matrix **G**, for $M=6$ and $L=3$, the first $P\text{-}L=3$ columns have constant entries the last $L=3$ columns have delay terms z^{-1}. For a filter of order *L*, we have:

$$G(z)=\sum_{n=0}^{L} g(n)z^{-n} \tag{4}$$

G is given by the following:

$$\mathbf{G}=\left[\begin{array}{ccc|ccc} & \mathbf{A} & & & \mathbf{B} & \\ g(0) & 0 & 0 & z^{-1}g(3) & z^{-1}g(2) & z^{-1}g(1) \\ g(1) & g(0) & 0 & 0 & z^{-1}g(3) & z^{-1}g(2) \\ g(2) & g(1) & g(0) & 0 & 0 & z^{-1}g(3) \\ g(3) & g(2) & g(1) & g(0) & 0 & 0 \\ 0 & g(3) & g(2) & g(1) & g(0) & 0 \\ 0 & 0 & g(3) & g(2) & g(1) & g(0) \end{array}\right]$$

The delay terms are eliminated by zero-padding, so that the synthesis filter bank is represented by a constant matrix **R**, as:

$$\mathbf{R}=\begin{bmatrix}\mathbf{R}_1\\ \mathbf{0}\end{bmatrix} \tag{5}$$

where $\mathbf{R}_1$ is a *MxM* matrix, called the precoder matrix and **R** is a *PxM* matrix, with *L* zero-padding, as shown in Fig. 6. And so the equivalent *PxM* filter matrix **A** is given by:

$$\mathbf{A}=\begin{bmatrix} g(0) & 0 & \dots & 0 \\ g(1) & g(0) & \dots & 0 \\ \vdots & \vdots & \ddots & \vdots \\ g(L) & & & \\ 0 & g(L) & & \\ & & \ddots & \\ 0 & 0 & \dots & g(L) \end{bmatrix}$$

For the redundant transmultiplexer, the reconstructed signal can be mathematically represented in terms of polyphase components **E** and **R** as:

$$\hat{\mathbf{c}}(\mathbf{n})=\mathbf{EAR}_1\mathbf{c}(\mathbf{n}) \tag{6}$$

The condition for perfect reconstruction of $\hat{\mathbf{c}}(\mathbf{n})$ is:

$$\mathbf{EAR}_1=\mathbf{I} \tag{7}$$

and this condition can be realized through SVD of matrix **A**.

IV. Singular Value Decomposition

SVD is applied to decompose the equivalent filter matrix **G,** so that we can determine the polyphase form of synthesis and analysis filter banks of the redundant transmultiplexer. In the preceding Section, we formulated a redundant transmultiplexer structure, by zero-padding, producing a pseudocirculant **G** matrix, which is then reduced to a tall fully-banded Toeplitz matrix **A**. In this case we can decompose **A** matrix using SVD.

A SVD of a rectangular *PxM* matrix is of the form:

$$\mathbf{A}=\mathbf{USV}^{\mathbf{T}} \tag{8}$$

where **U** is a *PxP* orthogonal matrix, **V** is a *MxM* orthogonal matrix and **S** is *PxM* diagonal matrix, with $\Sigma_{ij}=0$ if $i\neq j$. More specifically, we can write a tall matrix **A** using SVD, as:

$$\mathbf{A}=\mathbf{USV}^{\mathbf{T}}=[\mathbf{U}\quad \mathbf{U}^{\perp}]\begin{bmatrix}\mathbf{\Sigma} & 0\\ 0 & 0\end{bmatrix}\begin{bmatrix}\mathbf{V}\\ \mathbf{V}^{\perp}\end{bmatrix} \tag{9}$$

Thus the reconstructed signal $\hat{\mathbf{c}}(\mathbf{n})$ represented in "(6)", can be written as:

$$\hat{\mathbf{c}}(\mathbf{n})=\mathbf{EU\Sigma VR}_1\mathbf{c}(\mathbf{n})=\underbrace{\mathbf{EUK}}\ \underbrace{\mathbf{KVR}_1}\mathbf{c}(\mathbf{n}) \tag{10}$$

where $\mathbf{K}=\Sigma^{1/2}$ and now in order to satisfy "(7)" we can choose the matrices **E** and **R** as:

$$\mathbf{E}=\mathbf{K}^{-1}\mathbf{U}^{\mathbf{T}}\ \textit{and}\ \mathbf{R}_1=\mathbf{V}^{\mathbf{T}}\mathbf{K}^{-1} \tag{11}$$

Therefore the reconstructed signal $\hat{\mathbf{c}}(\mathbf{n})$ in "(12)" becomes:

$$\begin{aligned}\hat{\mathbf{c}}(\mathbf{n})&=\underbrace{\mathbf{K}^{-1}\mathbf{U}^{\mathbf{T}}\mathbf{UK}}\,\underbrace{\mathbf{KVV}^{\mathbf{T}}\mathbf{K}^{-1}}\mathbf{c}(\mathbf{n})\\ \hat{\mathbf{c}}(\mathbf{n})&=\mathbf{Ic}(\mathbf{n})\end{aligned} \tag{12}$$

V. Signal ModelSimulation And Results

In order to demonstrate the proposed multi-channel and uniform solution to the signal modeling problem, we utilize an example, as shown in Fig. 7(a). In this example we consider an arbitrary FIR filter and use the zero-padding method to formulate a tall, fully banded Toeplitz matrix **A** from **G**, which is a pseudocirculant matrix. Let the filter order $L=4$, then *G(z)* filter is given by:

$$G(z)=1-z^{-4} \tag{13}$$

We choose a multi-channel system, so $M=4$, and since $L=4$, we have $P=M+L=8$. It follows that after *L* zero-padding as shown in Fig. 7(b), the polyphase form of synthesis filter banks can be written as matrix **R**:

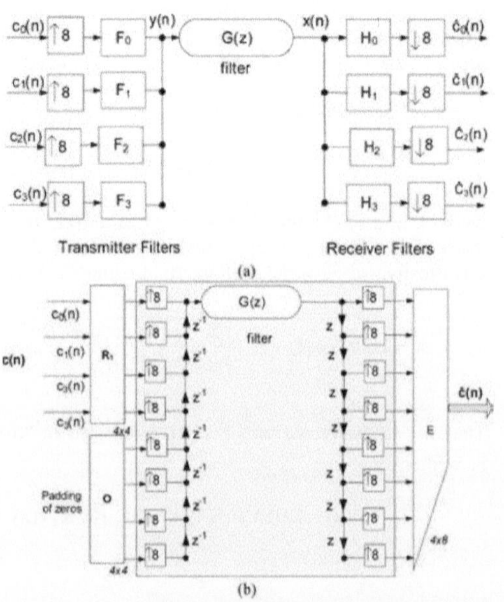

Fig. 7 (a). Uniform Signal Model (b). Polyphase Form of Transmitter & Receiver Filter Banks

$$\mathbf{R} = \begin{bmatrix} R_{00} & R_{01} & R_{02} & R_{03} \\ R_{10} & R_{11} & R_{12} & R_{13} \\ R_{20} & R_{21} & R_{22} & R_{23} \\ R_{30} & R_{31} & R_{32} & R_{33} \\ 0 & 0 & 0 & 0 \\ 0 & 0 & 0 & 0 \\ 0 & 0 & 0 & 0 \\ 0 & 0 & 0 & 0 \end{bmatrix}$$

and the transmitter filters as:

$$\begin{aligned} F_0 &= \sum_{i=0}^{3} z^{-i} R_{i0}(z^8) \\ F_1 &= \sum_{i=0}^{3} z^{-i} R_{i1}(z^8) \\ F_2 &= \sum_{i=0}^{3} z^{-i} R_{i2}(z^8) \\ F_3 &= \sum_{i=0}^{3} z^{-i} R_{i3}(z^8) \end{aligned} \tag{14}$$

The fully banded Toeplitz matrix **A** is an *8x4* matrix, and is given as:

$$\mathbf{A} = \begin{bmatrix} g_0 & 0 & 0 & 0 \\ g_1 & g_0 & 0 & 0 \\ g_2 & g_1 & g_0 & 0 \\ g_3 & g_2 & g_1 & g_0 \\ g_4 & g_3 & g_2 & g_1 \\ 0 & g_4 & g_3 & g_2 \\ 0 & 0 & g_4 & g_3 \\ 0 & 0 & 0 & g_4 \end{bmatrix} = \begin{bmatrix} 1 & 0 & 0 & 0 \\ 0 & 1 & 0 & 0 \\ 0 & 0 & 1 & 0 \\ 0 & 0 & 0 & 1 \\ -1 & 0 & 0 & 0 \\ 0 & -1 & 0 & 0 \\ 0 & 0 & -1 & 0 \\ 0 & 0 & 0 & -1 \end{bmatrix}$$

A uniform structure based on a redundant transmultiplexer as suggested in the above mentioned example is simulated using SIMULINK and Matlab. Evaluating SVD of **A** and following the procedure given in the previous Section, we are able to determine the constant polyphase matrices **E** and **R** and consequently reconstruct the original signal. Simulation results are shown in Fig. 8, which shows comparison of the original input signal with the reconstructed signal. There is a small error and we have almost perfect signal regeneration at the output of this uniform signal model.

VI. CONCLUSIONS

Signal approximation problem and it solution using multirate filter banks has been presented in this paper. A redundant and uniform filter bank based transmultiplexer is used to reconstruct a signal so that we can best approximate an arbitrary l_2 norm signal. Redundant transmultiplexer structure can be formed through zero-padding at the input of synthesis filter bank. A pseudo-circulant matrix is then decomposed using SVD. This powerful technique of matrix decomposition is utilized in order to find the polyphase matrices of synthesis and analysis filter banks in terms of constituents of the circulant matrix. In this way we can realize the perfect reconstruction of input signal and ensure best possible approximation of desired signal.

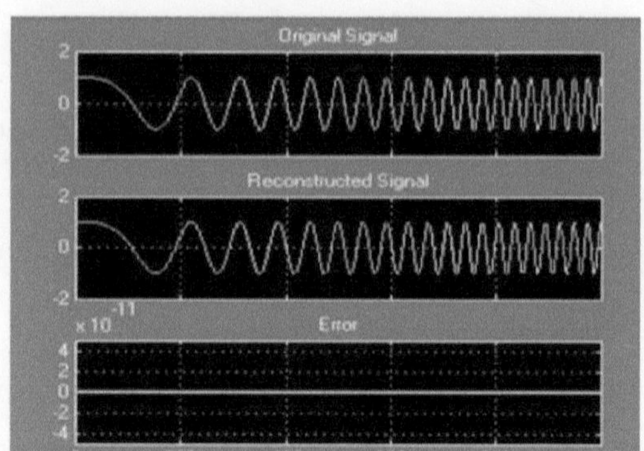

Fig.8. Matlab Simulation of Multi-channel & Uniform Signal Model

REFERENCES

[1] Manuel J. C. S. Reis, Paulo J. S. G. Ferreira and Salviano F. S. P. Soares."Linear Combinations of B-Splines as Generating Functions for Signal Approximation," Digital Signal Process., vol. 15, no. 3, pp. 226-236, May 2005.

[2] A. Aldroubi and M. Unser, "Oblique projections in discrete signal subspaces of l_2 and the wavelet transform," Proc. SPIE, vol. 2303, Wav. appl. in signal and image proc., II, San Diego, CA, 1994.

[3] B. Vrcelj and P. P. Vaidyanathan, "Least squares signal approximation using multirate systems: multichannel nonuniform case," Proc. 35th Asilomar Conference on Signals, Systems and Computers, Pacific Grove, CA, Nov. 2001.

[4] A. Tkacenko and P. P. Vaidyanathan, "On the least squares approximation model for overdecimated rational nonuniform filter banks and applications," Proc. IEEE Int. Conf. on Acoust., Speech, and Signal Process., Hong Kong, Apr. 2003.

[5] Applications of Singular Value Decomposition (SVD). Mathematics and Computers in Simulation, " Vol. 67, 15-31, 2004. (With G.I. Malaschonok) Special Issue: Applications of Computer Algebra in Science, Engineering, Simulation and Special Software).

[6] B. Vrcelj and P. P. Vaidyanathan,"Fractional biorthogonal partners and application in signal interpolation,"*Proc. IEEE Int. Symp. on Circuits and Systems, Scottsdale, AZ,* May 2002.

[7] Vaidyanathan, P. P., "Multirate Systems and Filter Banks" Prentice-Hall, Englewood Cliffs, NJ, 1993.

[8] P. P. Vaidyanathan and B. Vrcelj, "Transmultiplexers as precoders in modern digital communication: a tutorial review"', *Proc. IEEE Int. Symp. on Circuits and Systems*, Vancouver, May 2004.

An approach to distributed remote control based on middleware technology, MATLAB/Simulink - LabMap/LabNet framework

Cecil Bruce-Boye
Fachhochschule Lübeck
3 Stephensonstrasse
Lübeck, 23562 Germany

Rüdiger zum Beck
Fachhochschule Lübeck
3 Stephensonstrasse
Lübeck, 23562 Germany

Dmitry A. Kazakov
cbb software GmbH
1 Charlottenstrasse
Lübeck, 23560 Germany

***Abstract*- In this paper we present an example of a distributed control system implementing state observer controller. Distribution is based on middleware approach. As the middleware for the hardware and network abstraction we used the LabMap/LabNet. For the controller platform we took MATLAB/Simulink. The approach illustrated in this work offers an outlook for the cases where computational resources are limited and distributed among several hardware components, especially in view on embedded application area.**

I. INTRODUCTION

Many control processes are not centralized at one particular location. The input and output operations are performed on different knots of the system distributed both physically and logically. Similarly the control activities can be bound not only to the locally connected inputs and outputs but also to the remote ones. Further they can be distributed as well. A middleware stretching through the whole system provides an access to distributed data, integrating the software and hardware components of the system. LabMap [1] represents an example of such bus middleware used in automotive area. LabMap provides abstraction of the application level from the hardware specific and decoupling the hardware interface modules from the application level. Another important advantage of LabMap is a smooth integration of numerous components with their variety of software and hardware protocols, supporting a component based software design [2, 3].

MATLAB/Simulink [4] is one of the most wide spread tools used for design, simulation, testing and final production of control systems. Though MATLAB/Simulink supports interaction with hardware in the loop, this requires relatively expensive plug-in cards. Furthermore the choice of the hardware one can communicate with it very limited.

LabMap interface to MATLAB/Simulink drastically enlarges the scope of application of this comfortable and popular engineering framework. The interface allows using MATLAB/Simulink in simulation mode, yet controlling hardware in real time. This requires no real-time workshop. The simulation time of MATLAB/Simulink is mapped to the real time by the interface.

As a software bus LabMap is natively networking. This opens a wide perspective for research and construction of controlling system on the fly, without even having any embedded target system in hand. Of course these opportunities come at some cost. In particular one cannot rely any more on the hard real time behavior of the system. In this paper we will show that soft real time might be an option.

II. RELATED WORKS

The middleware architecture should respond to many, sometimes mutually exclusive, requirements. In the area of automation and control there exist numerous middleware products varying in their architecture.

OPC [7] is an initiative for open data connectivity. Similarly to LabMap OPC views data as variables. Though OPC is based on client-server architecture. Many hardware vendors provide OPC servers for their hardware. MATLAB OPC Toolbox 2 [8] offers an OPC client to Simulink, but it does not provide a server component essential for publishing the internal signals. A weakness of OPC is in relatively big overhead and its client-server architecture, which seems unsuitable for developing distributed controlling applications. Presently OPC is moving away from strict client-server approach by introducing OPC DX [11] for sever-to-server distribution of variables, but client-to-client communication is still impossible.

An alternative approach offers the MATLAB Distributed Computing Toolbox 2 [9], which though lacks Simulink support.

CORBA [10] is basically a software bus and does not suffer the problems of client-server approach, though it lacks description of timing constraints and is based on method invocation mechanism which does not well fit into MATLAB/Simulink framework.

III. MIDDLEWARE ARCHITECTURE

The middleware architecture should respond to many, sometimes, mutually exclusive requirements. In the area of automation and control there exist various middleware products varying in their architecture. The OPC [7] is an initiative for open data connectivity. Similarly to LabMap OPC views data

K. Elleithy et al. (eds.), Advances in Computer, Information, and Systems Sciences, and Engineering, 37–41.

as variables. The difference is though it has a client-server architecture. Though MATLAB OPC Toolbox 2 [8] offers an OPC client to Simulink, but it does not provide a server component essential for publishing the internal signals. In general a client-server architecture seems unsuitable for developing distributed controlling applications. An alternative approach offers the Distributed Computing Toolbox 2 [9], which though lacks Simulink support. CORBA [10] is basically a software bus and does not suffers the problems of client-server approach, though it lacks description of timing constraints and is based on method invocation mechanism which do not well fit to MATLAB/Simulink framework.

By an application the middleware LabMap is viewed as a set of variables. Each variable has a type, the current value and the current timestamp. No configuration is required. An application may access a value immediately after the system starts. Each variable responds to basic requests:

- *Get* reads the value of a variable in a safe way. It is guaranteed that the read value bits and the timestamp are consistent. The application is relieved from the burden of locking the value in presence of concurrent tasks accessing it. The application is unaware of the source of the value and the policy used to actualize or obtain the value
- *Set* writes the value of a variable.
- *Request* queries a new value of a variable from underlying hardware. The actions necessary to undertake is the responsibility of the driver. Variable request is asynchronous and the application is not blocked until input / output completion.
- *Send* initiates writing the value on the underlying hardware. Like in the case of request the application is unaware of the actions the hardware undertakes upon send.

The middleware performs most of its operations asynchronously to the process invoking the operation. An I/O operation on the hardware is the most important example. To synchronize processes the middleware offers a variety of synchronization mechanisms:

- *Wait for I/O completion*. An application may enter time-limited waiting for completion of I/O involving a variable
- *Waiting for value change*. An application may enter time-limited waiting for the time moment the value of a variable is getting changed. This method is very often used for monitoring system state variables. The main advantage of this approach is that the application developer is relived from the burden of polling the variable value and may concentrate on the domain objectives.
- *Blackboard*. Sometimes it is necessary to trace all changes of a variable. For instance an application visualizing signal waveforms would like to trace the signal. A usual approach for catching value changes involves some kind of notification mechanism between the source of the value and the application. Such point-to-point bias is hard to implement without overstraining the system resources and a danger of resource leaks when an application ends abnormally. The software bus uses an alternative approach. All value changes are written onto the blackboard and remain for a certain time there. Any application may inspect the blackboard content in order to trace the changes of desired variables.

IV. NETWORK INTERFACE

The software bus provides abstraction mechanisms hiding the distributed nature of the bus from the applications and their components. Further the software bus itself abstracts networking as a hardware interface. LabMap provides a highly optimized implementation of the networking interface called

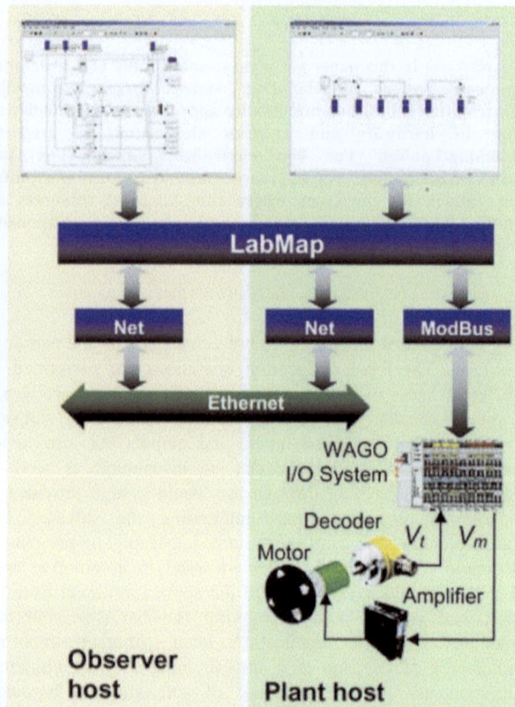

Fig 1. System setup

LabNet. LabNet corresponds to the presentation and application levels (6-7) within the OSI model. LabNet itself may function over different transport layers, but its most compelling use is based on Ethernet TCP/IP sockets.

What are the features of the networking interface LabNet, which make it attractive for distributed control?

- LabNet is fully transparent for applications using it.
- Data synchronization between distributed partitions is maintained in background, invisibly for applications.
- Middleware requests, such as I/O initialization etc are transported by LabNet and executed on the remote partitions

as if the hardware they control were directly connected to the local host.

- Measurement units are consistently handled, not compromising an ability to deal with different though compatible unit systems on different network hosts.
- An application may ensure that no other application accesses data for write. Further local and global access variables are supported. Local variables are seen by the applications running on the same network host where the variable is declared. Global variables are accessible from potentially any host. Access operations on global variables are also global.
- Any value is provided with a time stamp. A hardware driver is responsible for supplying correct time stamps. An application may figure out that the values of several are consistent by inspecting their time stamps.
- A peer-to-peer time synchronization mechanism is available to ensure consistency of the time stamps. Time stamps are transparently translated from one side clock reading to another. This feature can be used when there is no way to synchronize local clocks. At the same time any external time synchronization, such as NTP, can be used.
- Local variables can be browsed from remote hosts.

V. DISTRIBUTED STATE OBSERVER CONTROLLER SETUP

Fig 1. represents the test system we used for the controller. It consists of two hosts running MATLAB/Simulink. The hosts are connected via Ethernet. One of them has the WAGO I/O System [6] attached. It is used to acquire the revolution (via digital encoder HEDL 5540 A12) and set the voltage for the amplifier (4-Q-DC LSC 30/2) of an electric motor (DC Maxon A-max 32). The hosts run LabMap. This allows them to exchange data. LabMap natively provides time stamping for all values. The network interface of LabMap is shown as "Net" on the figure. It has a peer-to-peer time synchronization mechanism to ensure consistency of the time stamps. This does not require any synchronization of the partition's clocks, because the time stamps are translated from one clock reading to other transparently. WAGO I/O System is attached using the ModBus interface of LabMap.

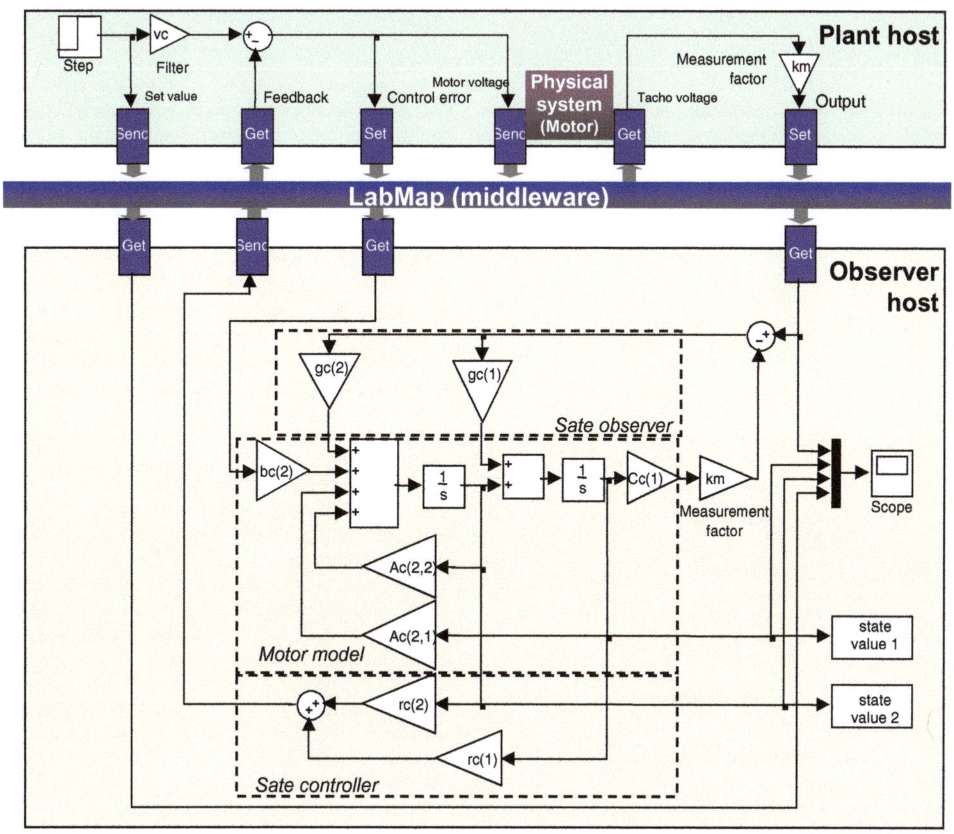

Fig. 2. MATLAB / Simulink model. Parts of the model running on different hosts shown on differently colored backgrounds

VI. State Observer and Controller

For the test system shown on fig 1. we chose a Luenberger state observer controller [5]. The plant part and the observer part are running on two separate network hosts. Each host is a conventional PC. Fig 2. shows the controller outline. The state observer *gc* and the state controller *rc* are calculated by the Ackermann formula. In order to ensure input – output accuracy the filter *vc* is calculated with respect to the state controller *rc*, shown in the following MATLAB m-file:

```
ObserverCalc.m
%calculation of state space controller and
%luenberger observer
%the system identificatin was carried out
%by the system identification
%software toolbox IDICON
num = [3.2637]      %result from system identification
den = [1 10.862 11.205]       %frequency domain
Ac=[0 1; -den(3) -den(2)];   %state spaces discription
bc=[0;num];
Cc=[1 0];

km=den(3)/num                 %measurment factor
cm=km*Cc;
p=[-10 -20];                  %pole placement observer
gc=(acker(Ac',Cc',p))';       %state observer
                              %gc=[19.138; -19.082]
prc = [-2 -2.5];              %pole placement controller
rc=acker(Ac,bc,prc);          %state controller
                              %rc=[-1.9013 -1.9493]
vc=inv(cm*inv(bc*rc-Ac)*bc);%filter vc=0.44622
```

Fig 3. shows the control revolution, output values and two state variables. As seen from the figure the control performance is satisfactory despite distribution of the controller over the network and consequent delays imposed by the data transfer of the network.

Our experiments with the system showed robust characteristic of the distributed controller against external disturbances.

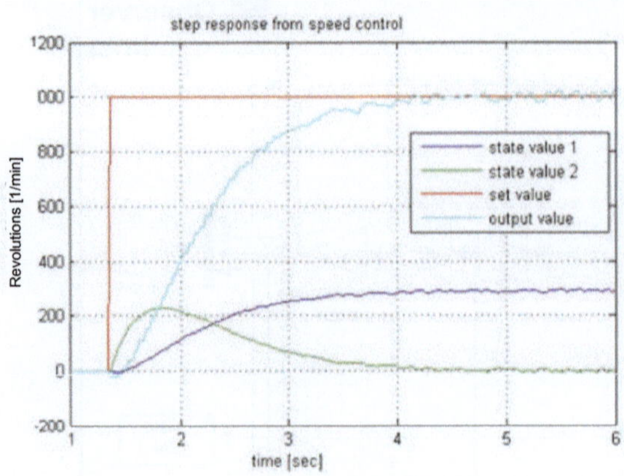

Fig. 3. State-controller behavior

In fig 4. the open-loop control and the close-loop control by the previously discussed state-contoller with observer is illustrated. In the first 5.5s the DC-motor is running in an open-loop mode without any feedback signal. Instead of running with the reference value of 1000 revolutions per minute the output is approximately 400 revolutions per minute. After switching the state controller into the circuit the close-loop performance is satisfactory, i.e. the output value matches the reference value.

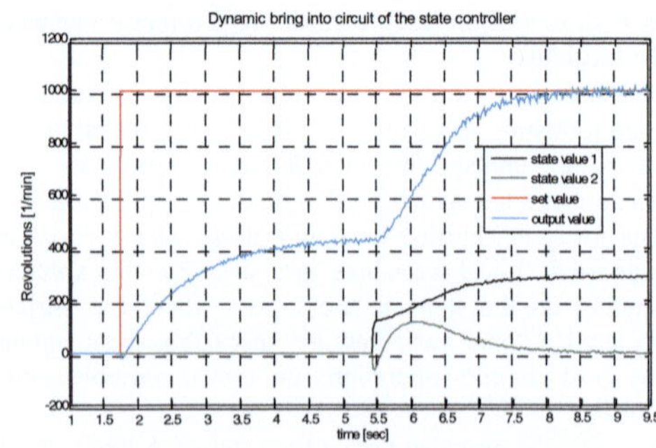

Fig. 4. Open-loop and close-loop performance

VII. Conclusion

The approach illustrated in this work shows feasibility of distributed control for the cases where computational resources could be limited and distributed among several hardware components. In such cases it is essential to be able to balance the resources to ensure meeting functional and non-functional requirements of control system, especially in embedded application area.

On the other side, the approach provides a framework for engineers accustomed to MATLAB/Simulink, who might wish to use the hardware unsupported directly by the MATLAB/Simulink or simulated by the software.

LabMap provides an interface to decouple the control application development from the hardware layer. A distributed application communicates with the hardware through and its partitions via variables and the middleware carries out the actual sending of the data through the appropriate protocol and interfaces for the given hardware.

In our future work we plan to investigate applicability of the approach for achieving transferability of the distributed controlling components.

The presented framework is succesfully applied for educational use in the control laboratories of the Universities of Applied Sciences in Luebeck and Bremen [12]. It allows to share laboratory resources between groups of students. Students can design their controller on their own observer hosts offline. After completing their task they gain access to the laboratory plant host through the middleware as shown in fig.1 to perform experiments.

References

[1] http://www.cbb-software.com/labmap.html

[2] C. Bruce- Boye, D.Kazakov, A. Fechner, "The hard and soft option LabMap the ultimate auto test platform", Testing Technology International. May 2001

[3] C.Bruce-Boye, D.Kazakov, "Distributed data acquisition and control via software bus", International Industrial Ethernet Development High Level Forum 2004 (IEHF 2004) in Peking, Automation Panorama No. 5

[4] M. Schetzen, V.K. Ingle, "Discrete systems laboratory in MATLAB," Thompson Engineering, 2000

[5] D.Luenberger, "An introduction to observers." IEEE Trans. Automatic Control, AC-16 1971

[6] http://www.wago.com

[7] F. Iwanitz, J. Lange "OPC - Fundamentals, Implementation and Application", 2002

[8] MATLAB OPC Toolbox 2 http://www.mathworks.com

[9] MATLAB Distributed Computing Toolbox 2, http://www.mathworks.com

[10] "The Common Object Request Brocker: Architecture and Specification", OMG Document 99-10

[11] "OPC DX 1.00 Specification 2003-03-11"

[12] H.-W. Philippsen „Einstieg in die Regelungstechnik", 2003, ISBN 3-446-22377-0

AN ANALOG COMPUTER TO SOLVE ANY FIRST ORDER DIFFERENTIAL EQUATION

T. ElAli, A. Sopeju, A. Fapohunda, O. Olorode
Department of Engineering
Benedict College
1600 Harden Street
Columbia, SC 29204

Abstract
A very simple analog computer was designed and tested to solve any first order constant-coefficients and linear differential equation. The analog computer was built using operational amplifiers, resistors and capacitors. Using the Pspice simulator, various input types were tested across the input terminals of the analog computer.

I. INTRODUCTION

Our goal is to build a generic Operational amplifier circuit to solve a generic 1st order differential equation with any input.
Consider the generic differential equation to be solved

$$\frac{dy(t)}{dt} + ay(t) = bx(t) \tag{1}$$

where x(t) is the forcing function (the input to the system represented by this differential equation) and y(t) is the solution (the output of the same system). The variables a and b are some real constant numbers. [1]
In the last equation, (assuming zero initial conditions) if we solve for $\frac{dy(t)}{dt}$ then integrate both sides of the equation we will get

$$y(t) = -a\int y(t)dt - b\int x(t)dt \tag{2}$$

In Equation (2) we have two terms to be integrated where one is negative and the other is positive.

II. METHODS

Consider the Operational amplifier circuit shown in Figure 1. The input-output relationship is given as

$$y(t) = -A\frac{1}{RC}\int x1(t)dt - B\frac{1}{RC}\int x2(t)dt \tag{3}$$

In Figure 1, the output y(t) is the integral of the input arriving at the negative terminal of the Operational amplifier. Thus the negative of the derivative of y(t) is located at the negative terminal of the Operational amplifier. [2]
If we set RC=1 in equation (3) we will have

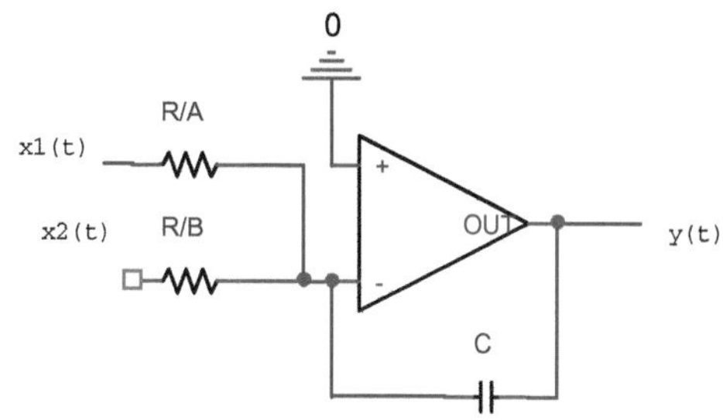

Fig. 1. Operational Amplifier Circuit

$$y(t) = -A\int x1(t)dt - B\int x2(t)dt \tag{4}$$

One final step before we attempt to implement Equation (4), the solution of a generic 1st order linear constant coefficient differential equation.
Now consider the circuit given in Figure 2.
The input-output relationship is

$$y(t) = -\frac{Rf}{R}x(t) \tag{5}$$

You also can see that if Rf = R then we have pure inversion (unity gain). The circuit containing an inverter and an integrator connected in series can solve the above differential equation. Figure 3 is a typical example of such a circuit.

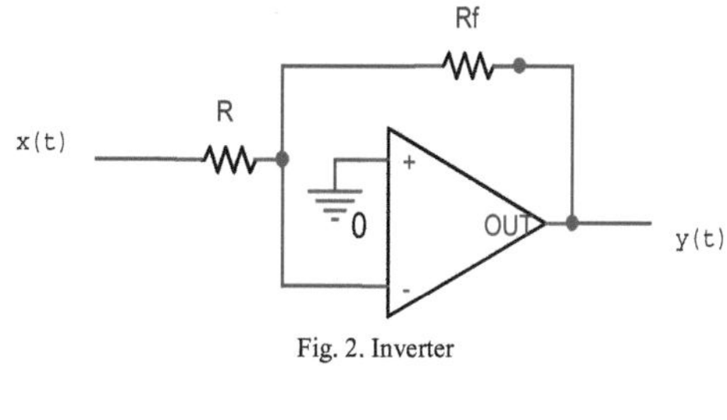

Fig. 2. Inverter

K. Elleithy et al. (eds.), Advances in Computer, Information, and Systems Sciences, and Engineering, 43–45.

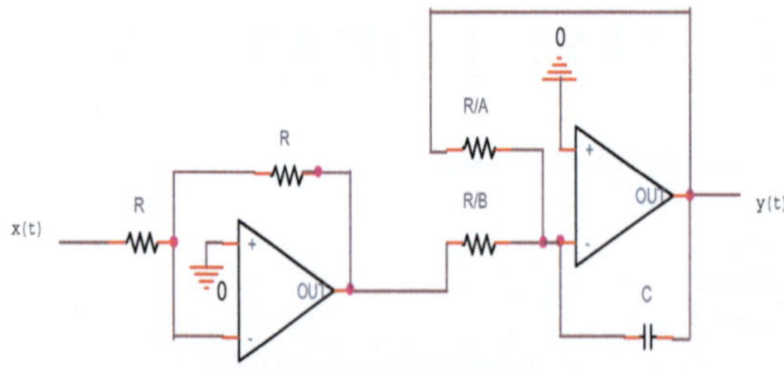

Fig. 3. A Circuit to Solve Equation 6

The circuit in Figure 3 would solve any first order differential equation of the form

$$y'(t) + Ay(t) = Bx(t) \tag{6}$$

In building a circuit to solve the given differential equation we can consider various types of input functions, which we could refer to as the forcing function of the equation. In this, we would apply 3 different forcing functions and then observe the response of this circuit to the various input functions. We will consider the following differential equation

$$y'(t) + 3y(t) = x(t) \tag{7}$$

CASE I

If a delta function is applied across the circuits input terminals in Figure 3, then the differential equation becomes

$$y'(t) + 3y = \delta(t) \tag{8}$$

In solving this differential equation, we use the Laplace's transformations.

Firstly, recall that by Laplace's transform it is necessary to first try to convert the given time functions and this conversion would yield the following results in the Laplace domain

$$\psi(s) = [1/(s+3)] \tag{9}$$

Converting equation (9) to the time domain would yield the result as shown below

$$y(t) = Ae^{-3t} \qquad t \geq 0 \tag{10}$$

The response from the output terminal would die off very fast due to the presence of the negative powered exponential. [4].

CASE II

If a sinusoidal function is applied across the circuits input terminals in Figure 3, then the differential equation becomes

$$y'(t) + 3y = \sin(\omega t) \tag{11}$$

Converting the given time functions would yield the following results in the Laplace domain

$$\psi(s) = [\omega/((s^2+\omega^2)(s+3))] \tag{12}$$

Breaking down equation (12) into partial fractions yield the following

$$\psi(s) = A/(s+3) + [(B+Cs)/(s^2+\omega^2)] \qquad t \geq 0 \tag{13}$$

Converting equation (13) to the time domain would yield the result as shown below

$$y(t) = Ae^{-3t} + [\frac{B}{\omega}\sin(\omega t)] + [C\cos(\omega t)] \qquad t \geq 0 \tag{14}$$

The response from the output terminal would have three components. One of the components dies while the others experience sustained oscillations.

CASE III

If a step function is applied across the circuits input terminals in Figure 3, then the equation of the differential equation becomes

$$y'(t) + 3y = u(t) \tag{15}$$

Converting the given time functions would yield the following results in the Laplace domain

$$\psi(s) = [1/s(s+3)] \tag{16}$$

Breaking down equation (16) into partial fractions yield the following

$$\psi(s) = [(A/s) + (B/(s+3))] \tag{17}$$

Converting equation (17) to the time domain would yield the result as shown below

$$y(t) = Au(t) + Be^{-3t} \qquad t \geq 0 \tag{18}$$

The response from the output terminal would have two components.

III EXPERIMENT AND SIMULATIONS

The circuit in Figure 4 was built to solve

$$y'(t) + 3y = \sin(\omega t) \tag{19}$$

The circuit in Figure 4 would solve the given differential equation with a sinusoidal function connected across it input terminals. [3]. Its outputs are as shown in Figures 4.1 and 4.2.

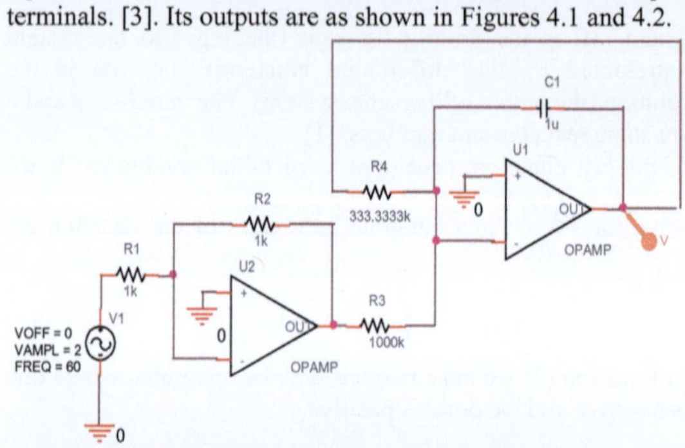

Fig. 4. Circuit to solve $y'(t) + 3y = \sin(\omega t)$

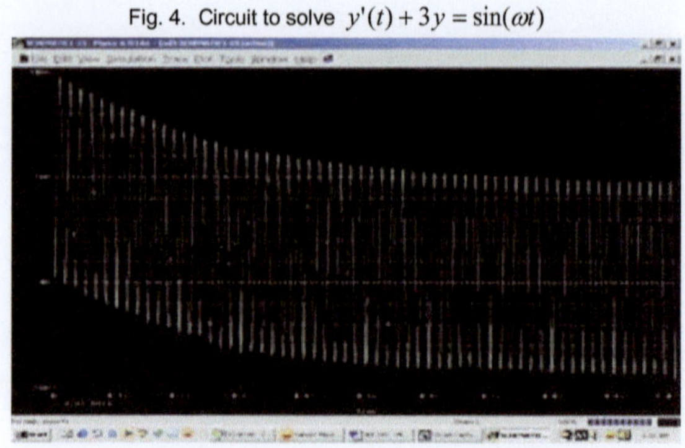

Fig. 4.1. Output when time is less than 10 seconds.

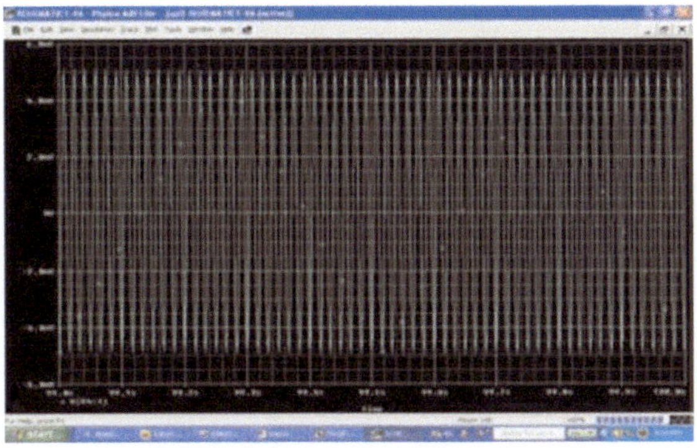

Fig. 4.2. Output after circuit has been on for a long tim

The circuit in Figure 5 was built to solve

$$y'(t) + 3y = u(t) \qquad (20)$$

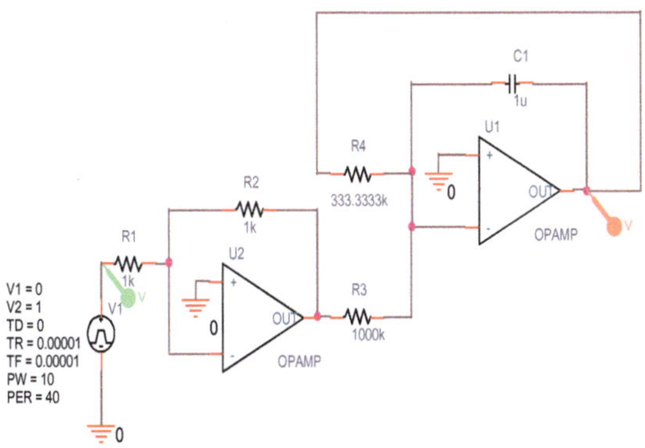

Fig. 5. Circuit to solve $y'(t) + 3y = u(t)$

The above circuit would solve the given differential equation with a step function connected across it input terminals. Its output is as shown below:

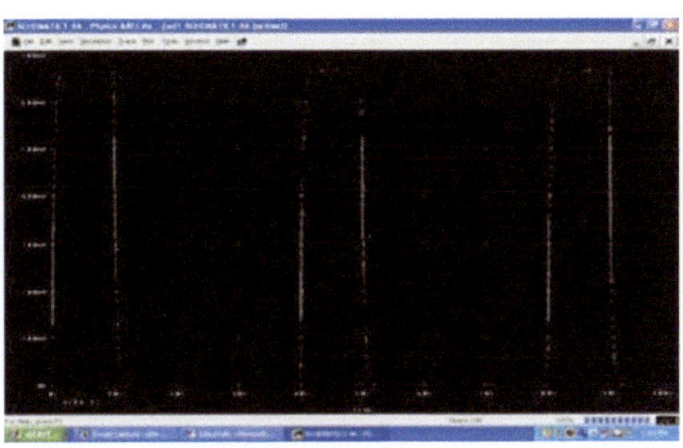

Figure 5.1. Output of circuit in Fig. 5

IV CONCLUSION

It would be discovered by looking at the graphs and also by comparing these results with what was gotten analytically that the circuits worked as desired. The differential equation was solved and its outputs were a solution to the given input. In case I where a delta function is applied across the input terminals of the circuit, due to limitations of the software it is almost impossible to simulate the response of the circuit to a delta function input. But from analysis it can be deduced that the output signal of the circuit would die off as time increases, this is due to the presence of the exponential function that has a negative power. In case II where a sinusoidal function is applied across the input terminals of the circuit, the analytical solution is obtained as in the following.

In Equation (14) with a frequency of 60Hz and amplitude of 2 we get

$$y(t) = [5.2e^{-3t} + [\frac{15.6}{120\pi}\sin(120\pi t)] - [5.2\cos(120\pi t)]]*10^{-3} \qquad t \geq 0 \qquad (20)$$

From the graphs, it would be discovered that although the output was a tiny fraction of the input, it remained a sinusoidal. This result is further established by mathematical analysis as shown above. This is due to the fact that after a couple of seconds the exponential part of the signal dies off, and the sinusoidal part persists. This persistent sinusoidal part would be a combination of both two different sinusoids lagging each other. In case III where a step function is applied across the input terminals of the circuit, this function is derived by connecting a pulsating voltage source with a very large period to the circuits input terminals. The output of this circuit would include an exponential part that settles after a couple of seconds and resembles a step function. It can also be noted that the output of the circuit is a fraction of the input. This can also be proven mathematically and its result is as shown below.

Solving equation (20) yields

$$y(t) = \frac{1}{3}(u(t) - e^{-3t}) \qquad t \geq 0 \qquad (21)$$

These circuits in the past played a very major role in the construction of special purpose computers, in recent times they do not have very wide application but are still been used in the building of certain internal components of some systems. The knowledge of these circuits also helped in the study and evolution of better and more effective circuits.

V REFERENCES

[1] Henry Edward, "Elementary Differential Equations", 4th edition, *Prentice Hall*, 2000.

[2] Robert Boylestad, "Electriconic Devices and Circuit Theory", 8th edition, *Prentice Hall*, 2002.

[3] J. W. Nilson, "Electric Circuits", 7th edition, *Prentice Hall*, 2005.

[4] J. W. Nilson, "Introduction to Pspice Manual using Orcad", 7th edition, *Prentice Hall*, 2005.

Product traceability integration within process for more precise diagnosis

P. Vellemans, B. Riera, P. Billaudel
CReSTIC – LAM
7 Boulevard J. Delautre
08000 Charleville-Mézières, France

To master the production it is necessary to be able to adapt the control of this one to the characteristics of a product and individually to ensure the traceability of this last. This way, the product can be considered as the pivot that ensures coherence between the levels of management and manufacture execution. Thus, it becomes necessary to control the production through the product, at the moment "t" and/or through time interval. We present in this article, the problems of production follow-up met within an existing system of production industrial; as well as the used methodology and tools which made it possible to rebuild, by means of systematic recordings, all the process history, to identify the useful elements with precision and to determine the element which acted, which was made or used, at which time, etc.

I. Introduction

Since the crisis of Bovine Spongiform Encephalopathy, the traceability term acquired its letters of media nobility. Currently the word is abundantly used in the agribusiness sector, but also in others domains like the automobile industry or New Technologies Information and Communication. Actually, we can say that the traceability is one of the essential concerns as regards quality [1], [2].

The first objective of the traceability is to be able to find, for a given product, the trace of all the stages of its manufacture and source of all its components and vice versa [4]. It also offers the advantage of being able to intervene upstream of the distribution, while making it possible; for example, to control the product quality until the origin of its raw materials. That permits a clear reduction in the costs of non quality intervening traditionally on the finished products. Moreover, that adds an additional layer to the structuring of the factors; this one is due to the sedentary requirements, the universalization and diversity of the production chains, to the offer complexity, the mass productions and the loss of proximity and finally to a more and more regulated environment. In addition to tracing the advance of a product (or a process), the traceability promises stakes, we can't get away from it; such as: Quality control, to ensure the consumer safety, to optimize the products recalls, the control of logistic flows and to respect it regulation.

II. Context of research

A Problematic

The study case with which we confronted is the production follow-up of an aluminum crosspiece support integrating a sand part which we will call, thereafter, "mold core" [4]. The first problem encountered by this production workshop is the loss of all traces when the mold core forwards by a stock. This hindrance is due to impossibility of marking mold core with an unspecified device ("physical": impossibility to engrave, note a number or another reference mark on sand). With that, another complication, caused by the operators is added storing the mold cores: those have each one their own logic of classification (the first produced mold core is not always the first used, for example). Consequently, it would be utopian to believe that one can find the origin of a mold core which made it possible to mould an aluminum crosspiece support.
Moreover, before the mold cores pair is not used, this one passes by a control operator point which makes it possible to check their conformity (thicknesses, forms, etc). Mold cores pairs are consequently eliminated from the production chain. This information loss is not being reflected in the data bases; this fact, the event is not recorded and consequently non known by the production management. Thus, theoretically, aluminum crosspieces are manufactured with mold cores which were removed production chain by control operator.

B. *The simulated system*

To solve the whole of the problems referred to above, we simulated the production system, following, Fig. 1:

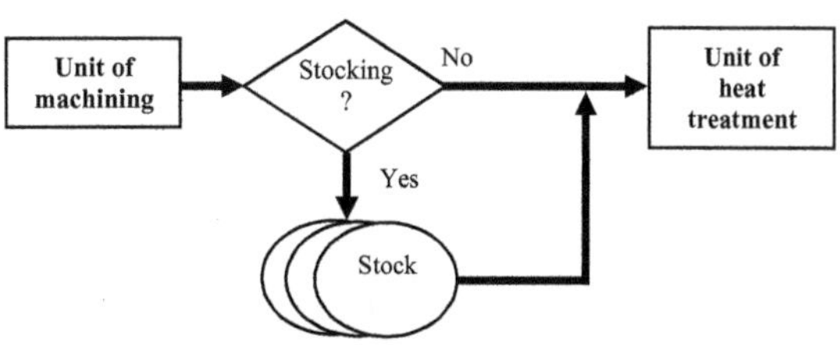

Fig. 1. Simulation of parts machined production and treated thermically.

This last is made up as follows:
- A unit of machining: drilling of a part, according to a defined diameter,
- A stock,
- A unit of heat treatment.

K. Elleithy et al. (eds.), Advances in Computer, Information, and Systems Sciences, and Engineering, 47–52.

Notes:

- Durations of machining and heat treatment are considered as constant,
- When the part diameter is lower than five centimeters, operator changes the tool. (There can be one tool change per batch; batches are defined latter.)

At the beginning of production chain, our part undergoes a machining and receives, at end of machining, a number (which we use only as witness), a time (instant of machining end) and a diameter (which varies with time). Then, this one is stored (or not). According to the storage parameters, the part is piled up either in "First In First Out" or in "Last In First Out" (we stopped with these two academic cases, which at all doesn't prevent the addition of other stacking kinds). Moreover, a time of storage can be allotted to him (this time makes it possible to represent time during which the mold cores remain in a stock). Lastly, the part undergoes a heat treatment: parts are treated by batch (a part number per batch is to be defined). Let us insist on the fact that the filling of the batches can be random or not. So, the parts come out from this treatment with three new parameters: the instant of heat treatment end (that we use only like witness), a pH and a treatment temperature (see Table I).

TABLE I

DATA BASE OF SIMULATED SYSTEM

Part number	Instant of machining end	Instant of heat treatment	Part diameter	pH	Temperature

We have just seen that the parts, constituting the batches, can be mixed in a random way, in order to be able to represent the total loss of traceability and by this fact, the inconsistency of the data within the system. Thus, the production parameters of a part correspond any more neither to the part number, nor at the instant of machining end. The most annoying consequence is that theoretically rejected parts will be considered as consumable and reciprocally. (We will see thereafter how the rejects are defined.)

To obtain an idea of the information loss evolution, we varied the number of parts constituting a batch (see Fig. 2). So, when the parts in the batches are ordered, we obtain, logically, a traceability of 100 %. On the other hand in the contrary case, we lose this traceability very quickly.

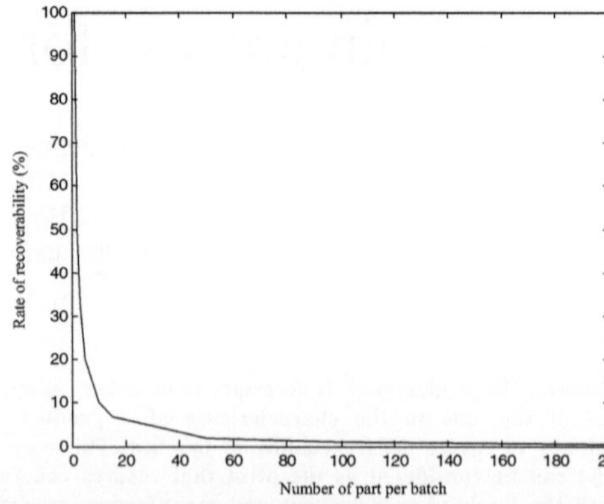

Fig. 2. "Recoverability" rate according to the part number constituting a batch.

As for the rejects designation, we defined the following criteria, a part is rejected if:

- The diameter is lower than 7 centimeters[1].
- The pH and the temperature, of the heat treatment, are not classified in the classes known as "good part". Two zones (1 & 2) were defined where the parts are considered as right (see Fig. 3), apart from these two zones the parts are rejected (for that we use a fuzzy K-nearest neighbors algorithm).[2]

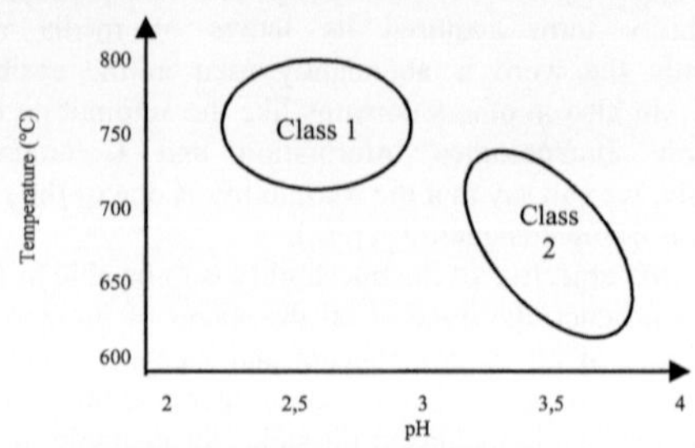

Fig. 3. "Good parts" classification.

III. Contribution

To summarize the context: we machine parts. Once forwarded by stock, the parts are treated thermically. This treatment is carried out by batch. The batches constituted of parts, are take in stock by random. So, our goal is to find the order -the traceability- of the parts once those treated.

A. Simulation: When wear is linear...

Initially, we simulate the part number to machining and we determine the instants to which the parts, turn down by operator, will have to be removed. Then, to lose the traceability

[1] The wear measurement can be taken only at exit of the heat treatment.

[2] Concerning the variations of the pH and the temperature, no model can be built, since these two parameters change in a random way.

of our parts, we make them forward by stock. Then, the last stage of our system consists in simulating the parts heat treatment; we treat the parts by batch (the size of the batches is defined by operator). Moreover, to add an additional degree to our loss traceability, we mix in a random way the parts constituting batch. If we visualize the historic of the parts diameters, we obtain the following representation, Fig. 4:

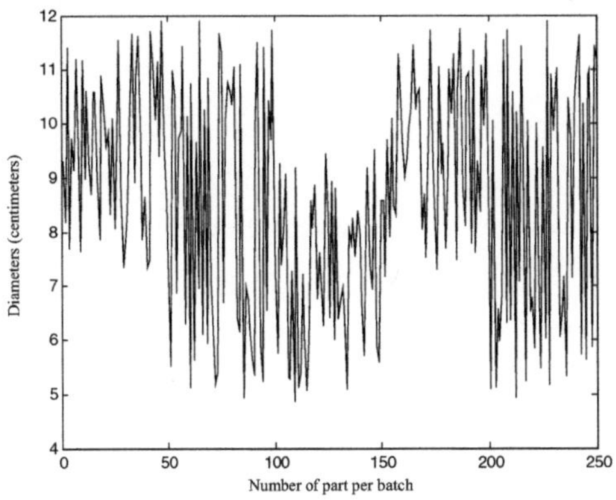

Fig. 4. Parts diameters mixed during the heat treatment.

So, once the mixed parts, we don't have any more information representative of machining (visually). The only information of which we lay out is the variation of the part diameter and more exactly the tool wear. We leave the assumption that wear grows in a linear way; what comes down to saying that the diameter of our part decrease in a linear way. From there, to find a more homogeneous representation, we classify the whole of the diameters of a batch (with the associated parameters) in a decreasing way (see Fig. 5); thus, we can partially obtain a machining representation according to time.

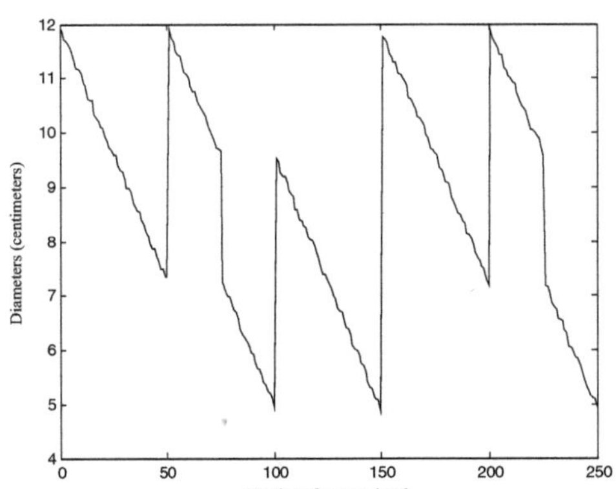

Fig. 5. Classification decreasing of the parts diameters according to the batches

Moreover, we have at our disposal other information which is the tool change instant. This information is considered as precise and reliable: an operator changes the tool and records the change instant (this last is done in an instantaneous way). Thus, before with a change of tool, the parts diameters are of low value; subsequently, the diameters will have a value close to the initial value. Knowing that, we can reclassify the parts diameters in a way even more coherent (see Fig. 6).

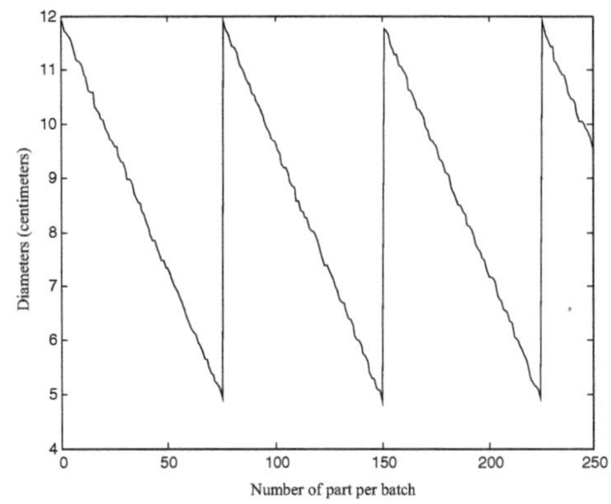

Fig. 6. Reclassification, according to the tools change.

From this rebuilding, we obtain a curve shape which approaches reality (see Fig. 1Fig. 7), but having a linear aspect (smoother).

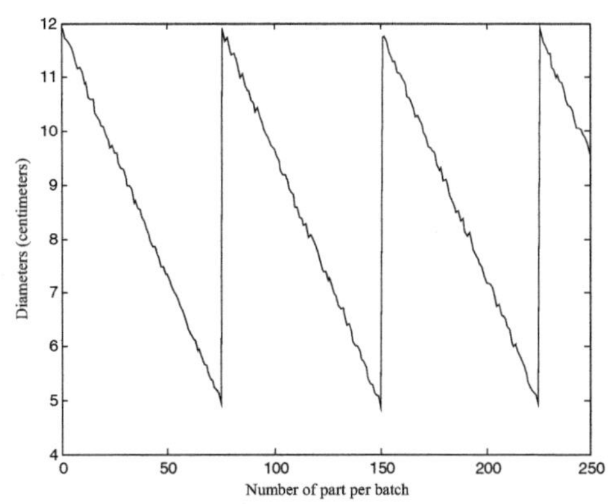

Fig. 7. "Real" evolution of the part diameter.

We compared the real curve with that rebuilt and we calculated the percentage of badly classified points; the results are visible to the Table II:

TABLE II

BADLY CLASSIFIED POINTS PERCENTAGE

Number of change tool	0	1	2	3	4	5
Percentage of wrong classified points (%)	80	41	31	21	15	9,5

We can notice that: more we change tool, less we have badly classified points. That being due to the fact, that we can better delimit the points thanks to the various changes of tools.

B. ... *When we don't know the wear form*

The results presented previously are obtained if we know the diameters variation through time. Now, we consider that we don't have any idea of the wear evolution. This in order to simulate that no information referring to the traceability can be used. Consequently, we need the knowledge of an expert. This last will have as task to model, by segments, the type of curve which one should obtain according to parameters' of the system. To allow rebuilding the diameters evolution according to the expert model, we use a linear regression method: least squares method. (This part is on the way of improvement.)

TABLE III

BADLY CLASSIFIED POINTS PERCENTAGE WITHOUT MODELS

Number of change tool	0	1	2	3	4	5
Percentage of wrong classified points (%)	89	85	83	81	76	70

In spite of a close rebuilding to the operator model, we can notice, on the Table III, that the percentage of badly classified points increases considerably. (For best rebuilding, this part is on the way of improvement.)

C. *Precision of the defective parts localization*

So, these histories rebuilding enable us to solve the problem of production follow-up and thus of our mold cores' traceability; consequently, at every instant, we have an idea of the mold cores position through the production and thus a thorough knowledge of our system. On the other hand, this traceability is only one model of the parts diameters evolution. Consequently, if we determine that the whole of the parts whose diameter is lower than a certain threshold (for our case, equal to 7), we cannot affirm that all the parts, whose diameter is lower than 7, are rejected parts and reciprocally.

So to affirm or cancel that a part is rejected, we used the fuzzy sets, defined in the [0, 1] interval by membership functions [5].

In practical applications, the trapezoidal membership functions provide generally a good representation of the subjective information [6]. A trapezoidal function is represented by a 4-tuples (e,f,g,h) of parameters whose meaning is explained in the Fig. 8, (g and h are non negative.)

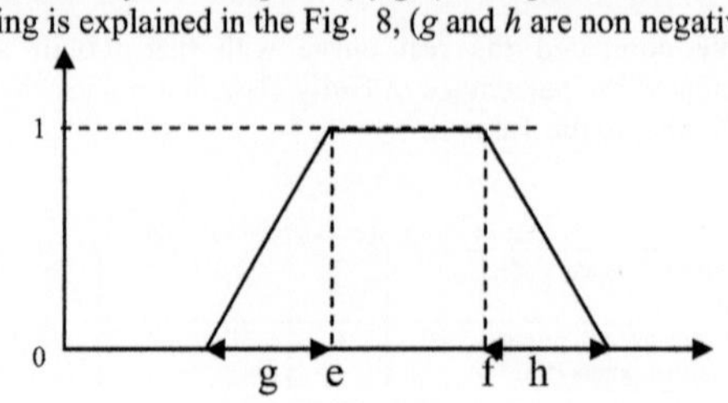

Fig. 8. R representation of a membership function by a trapezoidal.

The values of the 4-tuples parameters are defined according to the imprecision included in the information. If $g=h=0$ and $e=f$, then we have a precise non fuzzy information while if only $g=h=0$, the information is imprecise and non fuzzy. [7]

So, these memberships' functions make it possible to visualize the membership gradation of a point to a unit. To build our membership functions and thus to know position of parts consumable, defective (and those state is fuzzy), we conceived several scenarios. These memberships' functions we will help to quantify the membership degree of each part (=1: part is defective; =0: part is right) and also indicate us to which instant we are likely to obtain defective parts in a production.

Continuously with our simulated example, we determined four possible cases, Fig. 9:

- Case 1: n defective parts are in the batch "*i*",
- Case 2: n defective parts are in the batches "i" and "*i+1*",
- Case 3: n defective parts are in the batches "*i-1*" and "*i*",
- Case 4: n defective parts are in the batches "*i-1*", "*i*" and "*i+1*",

(n represents the number of defective parts, m the number of produced parts and i the produced batch.)

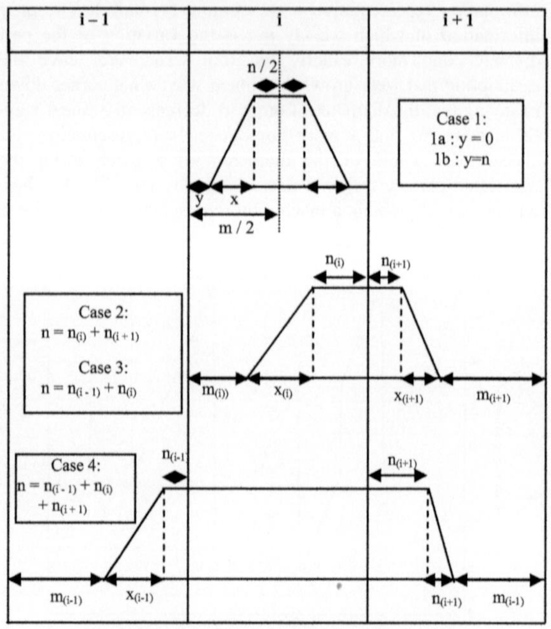

Fig. 9. Scenarios of the various memberships' functions.

Note: knowing that after a tool change, the produced parts are always correct, the membership' functions will have the following form (see Fig. 10):

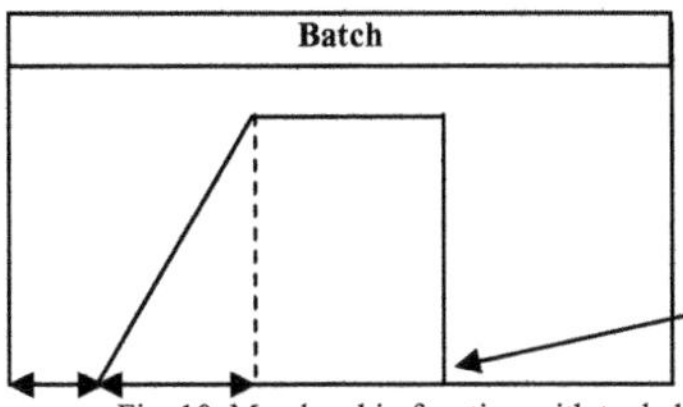

Fig. 10. Membership function with tool change.

Always according to our example, we obtain the general membership function following, Fig. 11:

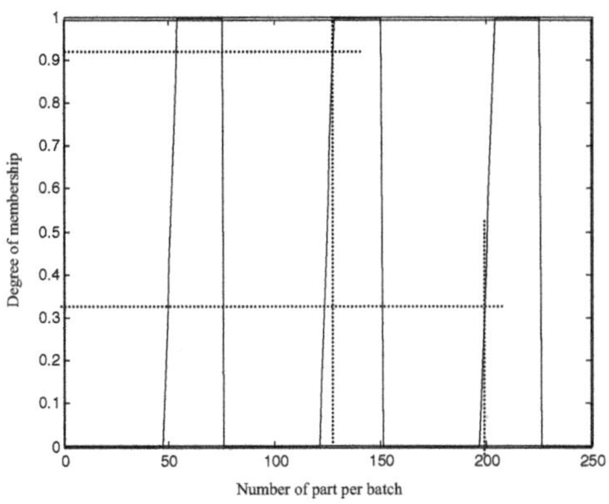

Fig. 11. Membership function

The membership function enables us to give the truth degree of the assertion: "this part is a reject". For example, we can say that the part "132" will be considered as reject, whereas the part "199" is highly likely to be a part whose diameter is right.

As we see it, the membership functions offer a great flexibility at the time of modeling. The disadvantage of fuzzy sets is that they represent primarily the imprecise information nature, uncertainty being represented in an implicit way and is accessible only by deduction from the various membership functions. A membership function is not a measurement of confidence; on the other hand, the possibilities' theory introduces such measurements on fuzzy sets [8].

The possibilities' theory, derived from the fuzzy sets, provides two measurements of confidence making it possible to represent knowledge on a field. They are measurements of "possibility" and "necessity". The first constitutes the most pessimistic measure (or most careful) and represents a degree preferably; whereas the necessity measurement translates the priority character of an event. (The results, on this part, are on the validation way.)

D. Localization of the removed parts

When machining is finished, an operator controls the parts, this one removes the parts which have defects (at this stage, the operator doesn't have any possibility of measuring the part diameter). These parts eliminations, at machining end, are not reflected in the data base. Thus, we cannot know if a part "α" is include in the production; if it isn't the case, we don't know the moment to which this one was removed from the production.

To cure this information loss and to determinate the instant when the part was removed, we use the operator observations; this one will have to indicate, Fig. 12:

- The instant to which the part was removed
- An interval of time, which makes it possible to encircle in a more flexible way the instant of suppression,
- The confidence degree which it has for information above.

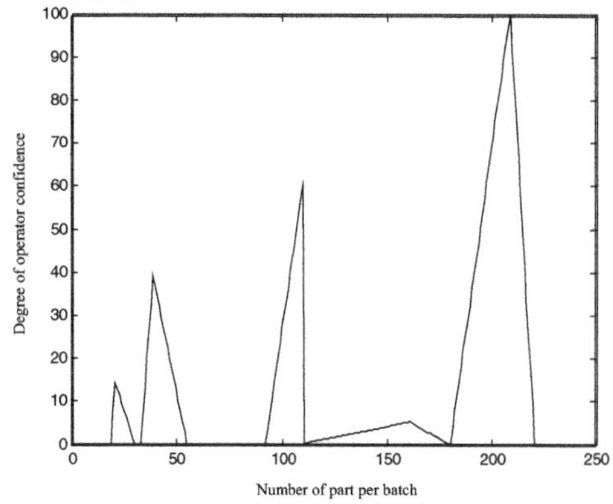

Fig. 12. Data relating to the parts losses.

Thanks to this last information, we will be able to have traceability even more effective and obtain better results (reduction of badly classified parts percentage) at the time of the various histories rebuilding.

IV. Progress Report

We were interested in this article to the follow-up of aluminum crosspiece support integrating a sand part. In this process, the product is considered as the pivot that ensures coherence between the levels of management and manufacture execution. Thus, it becomes necessary to control the production through the product, at the moment "t" and/or through time interval; so, we had to think of a system which would retrace the whole of the events occurring on this part.

The diversity of information relating to the products gives more and more problems of management and especially of the production follow-up. To compensate for this lack of legibility through a production, we will use the concept of traceability.

So, in the first time, we tried to rebuild the historic of the production for that we used the fuzzy sets and more exactly the memberships' functions associated with the possibilities' theory. These combined methods enable us to represent the imprecise and uncertainties of the data; and give us the possibility to determine the defective parts appearance, which enables us to obtain a more precise rejects' detection and consequently, thereafter, a better diagnosis.

In the second time, we are confronted to the loss of data. To solve this, we use the operators' knowledge; what enables us to add knowledge with our system, in order to obtain more reliable and more robust traceability.

The next stage (which is in the finalization step) is to provide to the operator a human adapted tool for a diagnosis aid, which will give him an overall vision of the system; over a time of production, for example, and/or the product evolution will show to him, at every moment "t", throughout the production chain.

References

[1] P. Vellemans, B. Riera, and P. Billaudel, "Human adapted system for traceability: a multi-methodological approach," *Third IEEE International Conference on Systems, Signals & Devices (SSD'05),* March 21-24, 2005, Susa Tunisia.

[2] A. Arana, B. Lasa, and L. Alfonso, "Meat traceability using DNA markers: application to the beef industry," *Meat Science*, 2002, Vol.61, PP. 367-37.

[3] ISO 8204, "Management de la qualité et assurance de la qualité," *Vocabulaire Concept, sélection et terminologie*, réédité en 2000, *Norme ISO 9000*, 1994.

[4] P. Vellemans, B. Riera, and P. Billaudel, "Human adapted system for analysis of the failure causes: a multi-method approach," *IFAC/IFIP/ IFORS/IEA Symposium, Analysis, Desig and Evaluation of Human-Machine Systems, Error and Safety*, 7-9 September 2004, Atlanta.

[5] L.A. Zadeh, "Probability Measures of Fuzzy Events," *J. Math. Anal. App.*, vol. 23, pp. 421-427, 1968. *IFAC/IFIP/ IFORS/IEA Symposium, Analysis, Desig and Evaluation of Human-Machine Systems, Error and Safety*, 7-9 September 2004, Atlanta.

[6] T. Ölmez, Z. Dokur, "Classification of heart sounds using an artificial neural network," *Pattern Recognition Letters,* Vol.24, PP 617-629.

[7] MS. Mouchaweh, B. Riera, P. Billaudel, "Fusion of objective and subjective information in multi-criteria decision-making"

[8] D. Dubois, H. Prade, "Théorie des possibilités, applications à la représentation des connaissances en informatique," Paris: Masson, 292p., 1988.

Voltage Control Measures by Using STATCON through PSS/E in WAPDA Power System

Dr. Muhammad Ahmad Choudhry, Kashif Naeem Bangash, Tahir Mahmood and Aamir Hanif
Department of Electrical Engineering
University of Engineering and Technology
Taxila, Pakistan

***Abstract*—Water and Power Development Authority (WAPDA) Pakistan has lack of reactive power reserves. The station which needs reactive power to boost the voltage during under voltage problem also requires some source to absorb reactive power to overcome over voltage problem.**

In this research work, the behavior of static Voltage control through STATCON has been reviewed and real data of WAPDA 500/220 kV power system regarding extreme cases of voltage control has been simulated on power system simulator PSS/E.

Different rating of simulated STATCON has been proposed at various locations to overcome voltage stability problem, increase steady state, dynamic state and transient stability.

Due to practical nature of the research work, it would be effectively utilized for resolving issues and bottlenecks related to WAPDA primary transmission line.

I. INTRODUCTION

In ac power systems, given the insignificant electrical storages, the electrical generation and load must balance at all times. To some extent, the electric system is self regulating. If generation is less than load, the voltage and frequency drop, and thereby the load, goes down to equal the generation minus the transmission losses. However, there is only a few percent margins for such regulation. If voltage is propped up with reactive power support, then the load will go up, and consequently frequency will keep dropping, and the system will collapse.

Real-power losses arise because aluminum and copper (the materials most often used for transmission lines) are not perfect conductors; they have resistance. The reactive-power nature of transmission lines is associated with the geometry of the conductors themselves (primarily the radius of the conductor) and the geometry of the conductor configuration (the distances between each conductor and ground and the distances among conductors).

The reactive-power behavior of transmission lines is complicated by their inductive and capacitive characteristics. At low line loadings, the capacitive effect dominates, and generators and transmission-related reactive equipment must absorb reactive power to maintain line voltages within their appropriate limits. On the other hand, at high line loadings, the inductive effect dominates, and generators, capacitors, and other reactive devices must produce reactive power. The balance point at which the inductive and capacitive effects cancel each other (what is called surge-impedance loading) is typically about 40% of the line's thermal capacity.

The power systems of today by and large, are mechanically controlled. There is a wide spread use of microelectronic, computers and high speed communication for control and protection of present transmission system; however, when operating signals are mechanical and there is a little high speed control. Another problem with mechanical devices is that control cannot be initiated frequently, because these mechanical devices tend to wear out very quickly compared to static devices. In effect, from the point of view both dynamic and steady state operation, the system is really uncontrolled. The possibility of controlling power flow in an electric system without generation rescheduling or topology changes can improve the power system performance.

The concept of FACTs (Flexible AC Transmission System) was first defined by Hingorani, N.G. in the 1988. Up to now, lots of advanced FACTS devices have been put forwarded due to rapid development of modern power electronic technology.

In WAPDA power system breaker switched shunt reactor and shunt capacitor are used with regard to voltage control requirements, STATCON is the most economic special means available for reactive power absorption and production respectively.

II. STATIC SYNCHRONOUS CONDENSER OR COMPENSATOR

Static synchronous condenser or compensator is a static synchronous generator operated as a shunt connected static VAR compensator whose capacitive or inductive output current can be controlled independent of the ac system voltage. It is also called STATCON/STATCOM which is one of the key FACTS controllers.

The STATCON is a solid-state shunt device that generates or absorbs reactive power and is one member of a family of devices known as flexible AC transmission system (FACTS) devices. Actually STACON is a new type of SVC based on inverter technology and GTO thyristors. STATCON is similar to the SVC in response, speed, control, capabilities, and the use of power electronics. Rather than using conventional capacitors and inductors combined with fast switches, however, the STACON uses power electronics synthesize the reactive power output.

Consequently the aim of this approach is to produce a variable reactive shunt impedance that can be adjusted (continuously or in a step like manner) to meet the compensation requirements of the transmission network. These (ac to dc or dc to ac) converters are operated as a current or voltage sources and they produce reactive

K. Elleithy et al. (eds.), Advances in Computer, Information, and Systems Sciences, and Engineering, 53–59.

power essentially without reactive power storage components by circulating alternating current among the phases of the ac system. Functionally from the standpoint of reactive power generation, their operation is similar to that of ideal synchronous machines whose reactive power output is varied by excitation control. Because of these similarities with rotating synchronous generators, they are termed Static Synchronous Generators (SSGs). When an SSG is operated without an energy source, and with appropriate controls to function as a shunt connected reactive compensator, it is termed, analogously to the rotating synchronous Compensator (Condenser) or STATCOM/ STATCON.

STATCOM comprises of a large number of gate controlled semi conductor power switches (GTO thyristors). The gating commands for these devices are generated by internal converter control in response to the demands for reactive and real power reference signal(s). The reference signals are provided by the external or system control, from operator instructions and system variables, which determines the functional operation of the STATCOM.

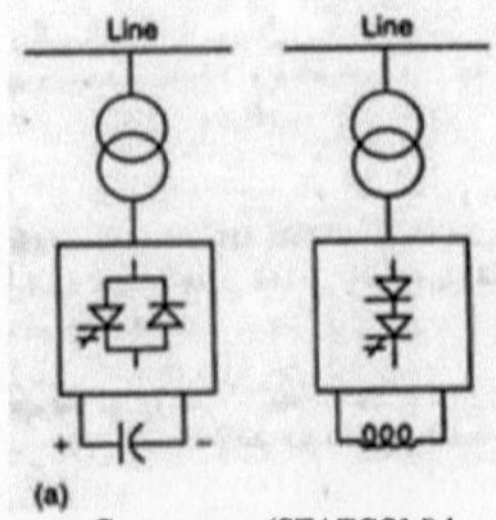

Fig. 1. Static Synchronous Compensator (STATCOM) based on voltage sourced and current sourced converters.

III. VOLTAGE CONTROL MEASURES BY USING STATCON THROUGH PSS/E IN WAPDA POWER SYSTEM

In normal practice it is much more practical and economic to size the power system according to the maximum demand for real power and to manage the reactive power by means of compensators and other equipment which can be deployed more flexibly than generating units and which make no contribution to fault levels, but in WAPDA power system the same generating units are mostly used for reactive power managements.

Followings are the voltage problems at 500/220 kV WAPDA network that will cause system collapse if remedial measures are not taken. In order to avoid forced load shedding at the end of each problem, different ratings of STATCON are suggested at optimal locations and their impact on voltage profile is shown.

1. Low Voltage Problem

1.1. Low Water Indent during period of CANAL CLOSURE

During the month of canal closure most of power comes from South to meet the load demand of North region. Due to low water indent machines of Tarbela & Mangla remain off the bar for most of time.

In evening and morning peak hours, we take the machines of Tarbela and Mangla on bar while after peak hours we start desynchronizing the machines of Tarbela and Mangla keeping in view the system stability. During these days we keep the voltage of Hubco high and transfer the MVAR on 500 kV lines towards North region.

Outage of Rousch

Rousch powerhouse is connected to WAPDA grid system through 500 kV transmission line it play a vital role by controlling system voltage due to its peculiar location. In the absence of this power house the reactive power reserve affects considerably which results in voltage problem and under such conditions tripping of any 500 kV transmission line or by generating unit tripping it may cause severe voltage depression in North which may further lead to partial / complete system collapse.

Total system load in Morning peak hours is 6750 MW and total generation by power houses at that time is listed in Table 1.1.

TABLE 1.1

Sr. No	Power house	Active power (MW)	Reactive power (MVAR)
1	Tarbela	290	+520
2	Mangla	200	+265
3	Warsak	43.2	+40
4	S Hydel	56.2	+6.7
5	Jamshoro	518.6	+75.9
6	Kotri	70	-17.1
7	Lakhra	30.3	-9.2
8	Guddu	942	+53
9	Mzaffargarh	1024	+338
10	Kapco	558	+285
11	Hubco	1150	-226
12	KEL	120	+50
13	Lalpir	200	+180
14	Pakgen	340	-13
15	SEPCOL	107	+50
16	HCPC Quetta	101	+60
17	UCH	280	+34
18	Fauji Kabirwala	126.2	+100
19	Saba	125	+70
20	Japan	107	+60
21	Liberty	159.9	+100
22	Chashnupp	200	+136

Voltage profile at different stations during outage of Rousch is shown in Table 1.2. Due to low water, indent machines of Tarbela and Mangla are desynchronized shortly after peak hours. During this time voltages are so depressed that there is a chance of system collapse. To avoid the system collapse we forcefully shed the load.

TABLE 1.2

Sr. No	Station	Voltage (kV)
1	Sheikh Muhammadi	506 / 217
2	Daud Khail	226
3	Chushnupp	231
4	Tarbela	505 / 223 / 208
5	Mardan	206 / 129
6	Warsak	136
5	New Rewat	496 / 208
6	Mangla	202
7	Gakhar	194
6	Sialkot	193
7	Kala Shah Kaku	209
8	Sheikhupura	496 / 214
10	Gatti	499 / 220
11	Sarfaraznagar	208
12	Yousafwala	212
11	Rousch	495
12	New Multan	529 / 226
13	Muzaffargarh	537 / 233
14	Guddu	533 / 234
15	Dadu	538 / 234
16	Jamshoro	533 / 227
17	Hubco	535

For system stability the load shall be shed in following areas.

TABLE 1.3

Sr. No.	Station	Load shedding (MW)
1	Sheikh Muhammadi	50
2	Mardan	100
3	Gakhar	100

After load shedding, voltages at different stations will improve. Voltage profile after load shedding is shown in Table 1.4.

TABLE 1.4

Sr. No	Station	Voltage (kV)
1	Sheikh Muhammadi	501 / 216
2	Daud Khail	226
3	Chushnupp	232
4	Tarbela	497 / 217 / 206
5	NewRewat	489 / 204
6	ISPR	201
7	Burhan	201
8	Mardan	204
9	Sheikhupura	492 / 212
10	Gatti	495 / 219
11	Rousch	515
12	New Multan	527 / 226
13	Muzaffargarh	536 / 233
14	Guddu	539 / 232
15	Dadu	538 / 234
16	Jamshoro	533 / 234
17	Hubco	535

In order to avoid load shedding and for system security different ratings of simulated STATCON are installed at different stations shown below. These STATCON will supply MVAR to the system, which will help in stabilizing the system voltages.

Simulated STATCON at following stations are recommended.

TABLE 1.5

Sr. No	Station	MVAR
1	220 kV Mardan	208
2	220 kV Gakhar	218
3	500 kV Sheikhupura	59

After installing STATCON at different stations voltages will improve without implementing the load shedding. Voltages at different stations after installation of STATCON are shown in Table 1.6.

TABLE 1.6

Sr. No	Station	Voltage (kV)
1	Sheikh Muhammadi	508 / 218
2	Daud Khail	226
3	Chushnupp	231
4	Tarbela	507 / 227 / 221
5	Mardan	220 / 136
5	New Rewat	500 / 220
6	ISPR	217
7	Burhan	216
8	Mardan	198
9	Sheikhupura	500 / 223
10	Bund Road	221
11	Kala Shah Kaku	218
12	Sialkot	214
13	Ghakkar	220
14	Ravi	218
15	Gatti	501 / 222
16	Rousch	519
17	New Multan	529 / 226
18	Muzaffargarh	537 / 233
19	Guddu	537/ 231
20	Dadu	537 / 233
21	Jamshoro	533 / 234
22	Hubco	535

Outage of Chashnupp

Chashma Nuclear power house also finds crucial location in WAPDA network in controlling system voltages of North. Its maximum reactive power sharing is 150 MVAR on lagging side and 60 MVAR on leading side. The outage of Chashnupp will affect the voltages to a greater extent particularly when Tarbela is sharing its minimum generation during low water indents. Under these conditions load shedding is the only option left for the system engineers to restore the system stability and keep the voltage at a desired level. In the base case as mentioned above if Chashnupp power plant trips, following load shedding is to be carried out.

TABLE 1.7

Sr. No.	Station	Loadshedding (MW)
1	Sheikh Muhammadi	203
2	Mardan	20

The voltage profile will be as under given in Table 1.8.

TABLE 1.8

Sr. No	Station	Voltage (kV)
1	Sheikh Muhammadi	508 / 217
2	Daud Khail	206
3	Chushnupp	207
4	Tarbela	509 / 226 / 210
5	Mardan	209
6	NewRewat	500 / 210
7	Sheikhupura	500 / 216
8	Gatti	502 / 222
9	Rousch	520
10	New Multan	531
11	Muzaffargarh	539
12	Guddu	540 / 232
13	Dadu	539 / 234
14	Jamshoro	534 / 234
15	Hubco	535

The above load shedding will compensate the outage of Chashnupp; however the above load shedding could be avoided if STATCON are installed at some particular stations for regulation of voltages in case of tripping of Chashnupp power plant. Following STATCON has been simulated to improve the system voltages and the system stability as well.

TABLE 1.9

Sr. No	Station	MVAR
1	220 kV DaudKhail	130
2	500kV NewRewat	108
3	500kV Sheikhupura	60.7

The voltage profile will be as mentioned in Table 1.10.

TABLE 1.10

Sr. No	Station	Voltage (kV)
1	Sheikh Muhammadi	504 / 215
2	Daud Khail	220
3	Chnupp	220
4	Tarbela	506 / 223 / 220
5	Mardan	206
6	NewRewat	500 / 220
7	Mangla	211
8	Gakhar	203
9	Sheikhupura	500 / 219
10	Gatti	50 / 221
11	Rousch	517
12	New Multan	531
13	Muzaffargarh	536
14	Guddu	538
15	Dadu	537
16	Jamshoro	533 / 234
17	Hubco	535

Transmission Line outages

During the month of canal closure most of power flow from south to north region. 500 kV Gatti – Rousch circuit play a role as an interface between South and North region. The outage of 500 kV Gatti – Rousch circuit will effect the system voltages when Tarbela share its minimum generation during low water indents. Under such circumstances load shedding is the only option left for system engineer to restore the system stability and keep the voltages at a desired level. Following load shedding is to be carried out.

TABLE 1.11

Sr. No.	Station	Loadshedding (MW)
1	Jaranwala Road	196
2	Ravi	100

The voltage profile will be as under given in table 1.12.

TABLE 1.12

Sr. No	Station	Voltage (kV)
1	Sheikh Muhammadi	504 / 216
2	Daud Khail	226
3	Chushnupp	231
4	Tarbela	502 / 222 / 207
5	New Rewat	493 / 207
6	Sheikhupura	494 / 214
7	Ravi	211
8	Gatti	494 / 220
9	Jaranwala road	220
10	Rousch	533
11	New Multan	531 / 226
12	Muzaffargarh	538 / 233
13	Guddu	535/ 234
14	Dadu	539 / 237
15	Jamshoro	534 / 234
16	Hubco	535

The above loadshedding will compensate the outage of 500 kV Rousch – Gatti circuit. The above loadshedding could be avoided if STATCON are installed at some particular stations for regulation of voltages in case of tripping of 500 kV Rousch – Gatti circuit.

Simulated STATCON at following stations are recommended.

TABLE 1.13

Sr. No	Station	MVAR
1	500 kV New Rewat	37
2	500 kV Sheikhupura	114
3	500 kV Gatti	233

After installing STATCON, voltages will raise and there will be a no need of load shedding. The voltage profile will be as mentioned in table 1.14.

TABLE 1.14

Sr. No	Station	Voltage (kV)
1	Sheikh Muhammadi	509 / 218
2	Daud Khail	226
3	Chushnupp	231
4	Tarbela	508 / 224 / 214
5	New Rewat	500 / 213
6	Burhan	209
7	ISPR	210
8	Lahore	500 / 216
9	Gatti	500 / 220
10	Rousch	532
11	New Multan	530 / 226
12	Muzaffargarh	537 / 233
13	Guddu	534 / 232
14	Dadu	536 / 233
15	Jamshoro	532 / 234
16	Hubco	535

Summer peak Base Case

During summer season in peak hours we usually keep the voltages high at both the tail ends i.e. Tarbela & Hubco to meet the MVAR demand of load centre i.e. Lahore and Gatti. During daytime IPPs surrounding Lahore are desynchronized for economic dispatch but usually we take out of merit generation from IPPs just for voltage stability to share MVARs. In afternoon usually we take expensive IPP like KEL & SEPCOL on bar because voltages of area like Gakhar, Lahore Bund road, Sialkot, Sarfaraznagar are depressed. In Evening peak hours usually total system load rises up to 10305 ~ 10600 MW.

Total system loads during this case are 10492 MW and Total generation by power houses for this case is shown in Table 1.15.

TABLE 1.15

Sr. No	Power house	Active power (MW)	Reactive power (MVAR)
1	Tarbela	2429	+772
2	Mangla	1100	+772
3	Warsak	125	+68
4	Chashma	78	+3.3
5	S Hydel	65	+20
6	Jamshoro	500	+170
7	Kotri	81	+25
8	Lakhra	35	+8.2
9	Guddu	1090	-175
10	Mzaffargarh	900	-3.3
11	NGPS Multan	106	+50
12	GTPS Faisalabad	118	+70
13	SPS Faisalabad	75	+50
14	Kapco	1135	+470
15	Hubco	600	-205
16	KEL	114	+50
17	Pakgen	344	+170
18	SEPCOL	90	+50
19	HCPC Quetta	117	+90
20	UCH	270	+129
21	Rousch	275	-9.5
22	Fauji Kabirwala	146	+60
23	Saba	123	+54
24	Japan	108	+60
25	Liberty	185	+80
26	Chashnupp	283	+149

Voltage profile during evening peak hours is shown in Table 1.16.

TABLE 1.16

Sr. No	Station	Voltage (kV)
1	Sheikh Muhammadi	530 / 224
2	Daud Khail	227
3	Chushnupp	233
4	Tarbela	536 / 233
5	Mardan	213
6	NewRewat	524 / 216
7	ISPR	210
8	Burhan	225
9	Sheikhupura	511 / 218
10	New Kot Lakhpat	207
11	Gatti	515 / 222
12	Rousch	530
13	New Multan	535 / 227
14	Muzaffargarh	538 / 231
15	Guddu	537 / 238
16	Dadu	538 / 238
17	Jamshoro	533 / 231
18	Hubco	535

Outage of 500 kV Tarbela – New Rewat circuit

During peak summer months Tarbela imparts almost 3700 MW which is dispersed to various load centres through 500 & 220kV network. At this time 500 kV Tarbela – New Rewat circuit become heavily loaded as most of power flow on this line to meet the demand of Lahore area. The power ranges from 1100 MW to 1200 MW. The outage of this circuit may catastrophic effect on the system stability from both overloading and voltage levels point of view and the chances of system collapse cannot be ignored.

In the above base case about 900 MW load was flowing on this circuit and its outage causes depression of system voltages which leads to the partial system collapse. The system could however be protected by implementing automatic loadshedding of about 100 MW in New Rewat area. The voltage profile will be as under given in Table 1.17.

TABLE 1.17

Sr. No	Station	Voltage (kV)
1	Sheikh Muhammadi	530 / 224
2	Daud Khail	227
3	Chushnupp	233
4	Tarbela	536 / 235
5	Mardan	215
6	NewRewat	488 / 204
7	ISPR	198
8	Burhan	228
9	Sheikhupura	491 / 210
10	Ravi	206
11	Gatti	503 / 217
12	Rousch	528
13	NMULTAN	530 / 226
14	Muzaffargarh	534 / 231
15	Guddu	534 / 232
16	Dadu	537 / 237
17	Jamshoro	532 / 231
18	Hubco	535

The other option is to install STATCON at 500 kV grid station New Rewat & Sheikhupura. The detail of STATCON simulated is as under along with post effect on voltages at various stations as mentioned in table 1.18.

TABLE 1.18

Sr. No	Station	MVAR
1	500 kV New Rewat	222
2	500 kV Sheikhupura	85

The voltage profile will be as mentioned in Table 1.19.

TABLE 1.19

Sr. No	Station	Voltage (kV)
1	Sheikh Muhammadi	530 / 224
2	Daud Khail	227
3	Chushnupp	233
4	Tarbela	536 / 235
5	Mardan	215
6	New Rewat	500 / 219
7	ISPR	212
8	Burhan	228
9	Sheikhupura	500 / 214
10	Bund Road	213
11	Gatti	507 / 219
12	Rousch	530
13	New Multan	532 / 226
14	Muzaffargarh	534 / 231
15	Guddu	535 / 232
16	Dadu	537 / 237
17	Jamshoro	532 / 231
18	Hubco	535

2. *Over Voltage problem*

2.1. *Winter Base case*

During winter season WAPDA experiences over voltage problems particularly during night when system load dropped considerably. Since very few hydel units are keep on bar due to restricted water indents and expensive thermal power houses are also taken off the bar on account of economic dispatch therefore the reactive power capability of the overall system is reduced to a greater extent. All the E.H.V transmission lines become lightly loaded thus reactive power is generated by these lines which causes voltage rise in the system. All efforts are made to cope with these over voltages but due to limited reactive power reserves sometime the voltages become out of control. Under such conditions the only option left is to open the lightly loaded E.H.V lines.

WAPDA has faced severe over voltage problem on Eid day dated 17/12 at 1500 hrs when system load decreased to 4112 MW. Total power generation at that time was as under given in Table 2.1.

TABLE 2.1

Sr. No	Power house	Active power (MW)	Reactive power (MVAR)
1	Tarbela	371	+44
2	Mangla	303	-27
3	Warsak	55	+24
4	Chashma	182	+80
5	Small Hydel	50	-5.1
6	Jamshoro	175	-40
7	Kotri	61	+20
8	Lakhra	34	+20
9	Guddu	811	-56
10	Muzaffargarh	380	-75.2
11	GTPS Faisalabad	56	-55.6
12	Kapco	490	+192
13	HCPC Quetta	120	+60
14	UCH	550	-38.3
15	Rousch	76	-82
16	Fauji Kabirwala	45	+40
17	Liberty	167	-2
18	Chashnupp	257	+88

For voltage stability following circuits had to open in various stages.

Openings

500 kV Hubco – Jamshoro CCT # 2
500 kV Jamshoro – Dadu CCT # 2
500 kV Dadu – Guddu CCT # 2
500 kV Guddu – New Multan CCT # 2
500 kV Gatti – Tarbela CCT # 2
220 kV KDA – JMSHR CCT # 2

After openings of above lightly loaded high voltage lines, we were able to control the voltages, the voltages at various stations were as given in table 2.2.

TABLE 2.2

Sr. No	Station	Voltage (kV)
1	Sheikh Muhammadi	504 / 215
2	Daud Khail	220
3	Chushnupp	224
4	Tarbela	505 / 224
5	NewRewat	509 / 218
6	Sheikhupura	515 / 225
8	Gatti	512 / 231
9	Rousch	520
10	New Multan	523 / 231
11	Guddu	519 / 222
12	UCH	224
13	Sibbi	221
14	Quetta Industry	219/135
15	HCPC	136
16	Dadu	535 / 233
17	Jamshoro	533 / 236 / 134
18	Hala Road	235
19	Hubco	543

During this situation if the unit of Jamshoro had tripped than voltages of Hubco could raises to 545 kV further causing tripping of the only energized 500 kV circuit i.e. Hubco – Jamshoro.

If during this situation, let us see what happen if for security of system we do not open all these circuits. Voltage profile will be as shown in table 2.3.

TABLE 2.3

Sr. No	Station	Voltage (kV)
1	Sheikhupura	524
2	Gatti	522
3	New Multan	541
4	Guddu	554
5	Dadu	585
6	Jamshoro	592
7	Hubco	603

As evident from the above the voltages at Guddu, Dadu, Jamshoro and Hubco stations are beyond the allowing limits which may adversely affect the system stability. The opening of high voltage lines for voltage control could be avoided if STATCON are installed at some particular stations for regulation of voltages.

Following STATCON have been simulated which helps to control the system voltage and system stability as well.

TABLE 2.4

Sr. No	Station	MVAR
5	500 kV Sheikhupura	95
1	500 kV New Multan	277
2	500 kV Dadu	196
3	500 kV Jamshoro	133

The voltages are as mentioned in Table 2.5 below.

TABLE 2.5

Sr. No	Station	Voltage (kV)
1	Tarbela	505
2	NewRewat	505
3	Sheikhupura	500 / 218
4	Gatti	499 / 226
5	Rousch	499
6	New Multan	500 / 219
7	Guddu	499 / 217
8	Dadu	500 / 217
9	Jamshoro	500 / 221
10	Hubco	509

STATCON at Jamshoro will help to increase the reactive power reserves as 220 kV KESC (Karachi Electric Supply Corporation) finds a peculiar place. It is connected to WAPDA power system through two 220 kV KDA – Jamshoro circuits. Usually it gets reactive power demand from 220 kV Jamshoro. Whenever these ccts trip during time when Jamshoro power house is absorbing maximum voltages of Jamshoro and Hala Road could raise to 239 kV and 244 kV respectively because there is no nearby reactive power absorbing source.

IV. DISCUSSIONS AND CONCLUSIONS

The voltage control problems have been investigated in different prospective for the WAPDA power system. Methods used for voltage control in WAPDA primary system is contrary to basic rule i.e. voltage is a local problem not global. WAPDA power system has lack of reactive power generation plus absorption reserves. The primary system is basically designed to meet the load demand. For this purpose, power houses have been located at optimal locations. The same power houses are primarily used to control the voltages. As a result, whenever power houses are OFF the bar for economic dispatch, scheduled maintenance outage and forced outages cause severe voltage problems. There has been no such solution evolved so far to curb the voltage collapse problem at the WAPDA power system.

In this research with the aid of power system simulator, various rating of static condenser has been simulated at various locations to overcome the problem. Static Condenser (STATCON) has the ability to synthesize the need of reactive power demand thus keeping the voltages within permissible range. It also increases steady state, dynamic state and transient stability. This device creates reactive power margins on generators & transformers. The system is so stable that it can face any major disturbance thus saving the billions of revenue to WAPDA.

ACKNOWLEDGMENT

The authors greatly fully acknowledge the useful discussions with various technical persons of Water and Power Development Authority Pakistan.

REFERENCES

[1] N. P. Padhy and M. A. Abdel Moamen, "Power Flow Control and Solutions with Multiple and Multi-type FACTS Devices", Electric Power System Research, 74 (2005) 341-351.

[2] R. Natesan and G. Radman, "Effects of STATCOM, SSSC and UPFC on Voltage Stability", Proceedings of the Thirty-Sixth Southeastern Symposium on System Theory, 2004, pp. 546-550.

[3] N. G. Hingorani & L. G. Understanding, "Concepts and Technology of Flexible AC Transmission Systems" Wiley-IEEE Press (December 10, 1999)

[4] R. P. Sadikovic et al, "Effect of FACTS devices on steady state voltage stability", University of Tuzla, Bosnia.

[5] K. M. Son, "Coordination of an SVC with a ULTC Reserving Compensation Margin for Emergency Control" IEEE Trans. on Power delivery, Vol. 15, No. 4 Oct 2000.

[6] C. R. Fuerte-Esquivel, et al "A Comprehensive Newton-Raphson UPFC Model for the Quadratic Power Flow Solution of Practical Power Networks" IEEE Transactions on power systems, VOL., 15, NO 1 February 2000

[7] M. Noroozian, et al "Use of UPFC for Optimal Power Flow Control" IEEE Transactions on Power delivery, Vol. 12, No. 4, October 1997.

Decentralized Kalman Filter in Wireless Sensor Networks – Case Studies

Vesa Hasu, Heikki Koivo
Control Engineering Laboratory, Helsinki University of Technology
vesa.hasu@hut.fi, heikki.koivo@hut.fi

Abstract

Application of data fusion techniques in new fields brings always up new sets of practical problems. This paper studies the decentralized Kalman filter (DKF) in out-of-sequence (OOSM) and clusterized topology problems, which rise in application to wireless sensor networks. The OOSM problem is closely related to the uncertainty of wireless links. Data may be randomly delayed or completely lost. The possible clusterized topology of large wireless sensor networks raises also problems for DKF. Namely, how much information should be kept in memory, if the sensor node moves from one network cluster to another. The above problems and solutions are discussed and examined in case simulations.

1. Introduction

Wireless sensor networks (*e.g.* [5]) are an up and coming technology for the future of automation systems. The main concepts in the use of wireless sensors include firstly exploitation of wireless radio links and secondly massive number of sensors due to cheap components without wiring. Hence the use of a large number of sensors becomes more economically feasible. The increased number of sensors adds also the need for filtering and data fusion in networks.

In addition to wide range of possibilities, the new application field brings new problems for old tools. The scope of this work is to examine and discuss the decentralized Kalman filter (DKF) [1] in the two specified problem field closely related to wireless sensor networks. DKF filter is decentralized version of Kalman filter (KF) [1], [9] to estimate the state of the system. DKF is based on the information filter form of KF.

The first problem rises from the out-of-sequence-measurements (OOSM). The wireless radio link is more likely to cause random delays and completely lost data packets than the wired network. The OOSM problem is well documented for Kalman filter, see *e.g.* Nettleton and Durrant-Whyte [2]. This paper compares a truncated back-propagation of measurements (suggested in [2]) with heuristic and sub-optimal approaches, which do not require as much system memory and computational performance from the wireless network nodes. The communication overhead is fairly large in DKF, all fusion centers communicate to all other centers, and keeping it as low level as possible is a desirable in order to keep the filter applicable in wireless framework.

The second DKF problem in wireless sensor networks is the filter performance in a large and clusterized wireless sensor network. If the sensor network has a flood of nodes, the network must be divided into clusters in order to reduce the required communication. The communication between clusters happens through so-called cluster heads, a sort of cluster beacons [4], [6], [7]. The emphasis in here is laid especially on problems caused by changes of clusters that DKF suffers from. This case is raised in with moving nodes or variable communication capabilities. The change of cluster within the network brings up the problem of filter initialization and error covariance memory after the change of clusters.

DKF requires distribution of state error information and covariance error information of all nodes throughout the network. The biggest problems in the clusterized data fusion are whether it is necessary to distribute error covariance information between clusters or not, and if information is not distributed throughout the system, what kind of memory of estimate error covariance should be applied.

This paper is organized as follows. Section 2 introduces the mathematical foundation of DKF. The OOSM and clusterization problems are discussed with simulations in Sections 3 and 4, respectively. Conclusions are presented in Section 5.

2. Decentralized Kalman Filter

As the name indicates, DKF is a decentralized version of the normal centralized KF. One of the main characteristics of DKF is that it is mathematically identical filter to the centralized KF. Hence DKF is the

K. Elleithy et al. (eds.), Advances in Computer, Information, and Systems Sciences, and Engineering, 61–68.

optimal linear filter with the assumption of independent and white noises. The following presentation of DKF follows the presentation of Grime and Durrant-Whyte in [1].

In order to obtain decentralized version of KF, the system model must be partitioned. The partitioning of the centralized system model is done by unstacking the observations vector **z** into m subvectors of dimension m_i, depending on the number of observations in each observation node. The similar procedure is done for the observations matrix and measurement noise vector. In mathematical terms, observation, observation weight matrix and observation noise partitionings are

$$\mathbf{z}(k) = \left[\mathbf{z}_1^T(k),\ldots,\mathbf{z}_m^T(k)\right]^T,$$

$$\mathbf{H}(k) = \left[\mathbf{H}_1^T(k),\ldots,\mathbf{H}_m^T(k)\right]^T,$$

and

$$\mathbf{v}(k) = \left[\mathbf{v}_1^T(k),\ldots,\mathbf{v}_m^T(k)\right]^T.$$

DKF is defined for systems, which local models for sensing node i look like

$$\mathbf{x}(k+1) = \mathbf{F}(k)\mathbf{x}(k) + \mathbf{G}(k)\mathbf{w}(k), \quad (1)$$

$$\mathbf{z}_i(k) = \mathbf{H}_i(k)\mathbf{x}(k) + \mathbf{v}_i(k), \quad (2)$$

where $\mathbf{F}(k)$ is the state transition matrix between time steps k and $k + 1$, **G** is the control distribution matrix, **x** is the state vector, and **w** is the process noise. Note that (1) defines the system dynamics, and it is the same for all nodes.

Let's consider an information filter form of KF. This alternative formulation is based on defining the information state vector $\hat{\mathbf{y}}_i(k\,|\,l) \triangleq \mathbf{P}_i^{-1}(k\,|\,l)\hat{\mathbf{x}}(k\,|\,l)$ and the information matrix $\mathbf{Y}_i(k\,|\,l) \triangleq \mathbf{P}_i^{-1}(k\,|\,l)$, where $\mathbf{P}_i(k|l)$ is the error covariance of the state estimate in the node i of time k given the information up to time l. Process and observation noises **w** and **v** are assumed to be uncorrelated white sequences with covariance matrices **Q** and **R**, respectively.

The advantage of information filter approach is its applicability for the decentralized data fusion. According to [1], the DKF equations for the local update of node i are

$$\mathbf{Y}_i(k+1\,|\,k) = \left(\mathbf{F}_i(k+1)\mathbf{Y}_i^{-1}(k\,|\,k)\mathbf{F}_i^T(k+1) + \mathbf{G}_i(k+1)\mathbf{Q}_i(k+1)\mathbf{G}_i^T(k+1)\right)^{-1}, \quad (3)$$

$$\hat{\mathbf{y}}_i(k+1\,|\,k) = \mathbf{Y}_i(k+1\,|\,k)\mathbf{F}_i(k)\mathbf{Y}_i^{-1}(k\,|\,k)\hat{\mathbf{y}}_i(k\,|\,k), \quad (4)$$

$$\mathbf{I}_i(k+1) = \mathbf{H}_i^T(k+1)\mathbf{R}_i^{-1}(k+1)\mathbf{H}_i(k+1), \quad (5)$$

$$\mathbf{i}_i(k+1) = \mathbf{H}_i^T(k+1)\mathbf{R}_i^{-1}(k+1)\mathbf{z}_i(k+1) \quad (6)$$

$$\tilde{\mathbf{Y}}_i(k+1\,|\,k+1) = \mathbf{Y}_i(k+1\,|\,k) + \mathbf{I}_i(k+1), \quad (7)$$

$$\tilde{\mathbf{y}}_i(k+1|k+1) = \hat{\mathbf{y}}_i(k+1|k) + \mathbf{i}_i(k+1) \quad (8)$$

where the tilde refers to local estimates, **I** is called the contributed information matrix, which is also known as *variance error information*, and **i** is called the information state contribution, which is sometimes referred as *state error information*.

For global update, all DKF nodes communicate their information state contribution **i** and their contributed information matrix **I** for all of the other nodes. The global estimates of the information state vector and the information matrix are given by

$$\hat{\mathbf{y}}_i(k+1\,|\,k+1) = \hat{\mathbf{y}}_i(k+1\,|\,k) + \sum_{j=1}^{m}\mathbf{i}_j(k), \quad (9)$$

and

$$\mathbf{Y}_i(k+1\,|\,k+1) = \mathbf{Y}_i(k+1\,|\,k) + \sum_{j=1}^{m}\mathbf{I}_j(k), \quad (10)$$

respectively. Naturally, by definitions, the state estimate $\hat{\mathbf{x}}_i(k+1\,|\,k+1)$ can be extracted by

$$\hat{\mathbf{x}}_i(k+1|k+1) = \mathbf{Y}_i^{-1}(k+1|k+1)\hat{\mathbf{y}}_i(k+1|k+1). \quad (11)$$

The global estimates of state and state error covariance given by (11) are mathematically equal to the estimate of the centralized KF.

3. Out-of-Sequence-Measurements Problem in Decentralized Kalman Filter and Simple Solutions

The out-of-sequence-measurements (OOSM) mean that measurements are arriving late and/or in the incorrect order to the data fusion center. OOSM problem is covered well in the literature for the centralized KF, see *e.g.* [8]. The algorithms dealing with OOSM in information filter formation of KF can be easily extracted to DKF [2].

An algorithm for taking delayed measurements exactly into account is introduced in [2]. In there the asequent information is propagated through matrix iterations. For example, if coming information, $\mathbf{i}(k)$ and $\mathbf{I}(k)$, is delayed with one step, their propagated versions, $\mathbf{i}^+$ and $\mathbf{I}^+$, are added to estimates given by (9) and (10). Definitions for the propagated information are given by (12) and (13) in the bottom of the next page, where $\mathbf{M}^Y(k) = \mathbf{F}^{-T}(k)\mathbf{Y}(k\,|\,k)\mathbf{F}^{-1}(k)$, $\mathbf{M}^I(k) = \mathbf{F}^{-T}(k)\mathbf{I}(k)\mathbf{F}^{-1}(k)$, $\mathbf{M}(k) = \mathbf{M}^Y(k) + \mathbf{M}^I(k)$, and

$\mathbf{\Sigma}(k) = \mathbf{G}^T(k)\mathbf{M}^Y(k)\mathbf{G}(k) + \mathbf{Q}^{-1}(k)$. If the delay of the arrived information is longer than one step, (12) and (13) are iterated with the information of step $k+1$ with $\mathbf{i}^+(k+1)$ and $\mathbf{I}^+(k+1)$ being substituted for $\mathbf{i}$ and $\mathbf{I}$.

Since the above approach for exact propagation uses substantially computational capacity and memory, checking the sub-optimal OOSM algorithms is worthwhile. Simple, sub-optimal strategies for DKF in OOSM problem include the following.

1. Use only the information gotten in the current time-step, and ignore the delayed information.
2. Use the latest information gotten from each node, and ignore the possible delays.
3. Use the iterative propagation of old measurements introduced in [2], but only if the new information is at most n steps delayed. In here n is a small positive integer.

The first approach is the simplest way to handle the old measurements, and it does not require any memory for the old measurements. The second strategy seems suitable especially for time-invariant systems, since the contribution information matrices $\mathbf{I}_j$ are constant for each node j. On the other hand, it can result to relying on quite old measurements, which are not even close to the correct ones anymore. This strategy requires memory on the latest measurements of each node.

The third approach is a sort of truncated version of the propagation of old data using (12) and (13). It requires memory of a few KF iteration matrices over the propagation window, which is assumed to be small. In [2], all statistics of propagation ($\mathbf{M}^Y$, $\mathbf{M}^I$ and $\mathbf{\Sigma}$) are assumed to be stored in memory. The suggested short propagation window corresponds to the case, where much delayed measurements are handled as they were missing.

All of these approaches have one downside in common compared to the normal operation of DKF: nodes do not have a common global estimate anymore. The OOSM problem leads to the loss of the global update synchronization, and every node ends up having just an own upgraded local estimate.

3.1. Simulation Model for 3-D Tracking

The above OOSM solutions are tested in a three-dimensional tracking of velocity and acceleration based on a 3-D position measurement. If the sampling time is denoted by ΔT, the corresponding state-space model is

$$\mathbf{x}(k+1)=\begin{bmatrix}1 & \Delta T & 0 & 0 & 0 & 0\\ 0 & 1 & 0 & 0 & 0 & 0\\ 0 & 0 & 1 & \Delta T & 0 & 0\\ 0 & 0 & 0 & 1 & 0 & 0\\ 0 & 0 & 0 & 0 & 1 & \Delta T\\ 0 & 0 & 0 & 0 & 0 & 1\end{bmatrix}\mathbf{x}(k)+\begin{bmatrix}\Delta T^2/2 & 0 & 0\\ \Delta T & 0 & 0\\ 0 & \Delta T^2/2 & 0\\ 0 & \Delta T & 0\\ 0 & 0 & \Delta T^2/2\\ 0 & 0 & \Delta T\end{bmatrix}\mathbf{v}(k),$$

$$\mathbf{z}(k)=\begin{bmatrix}1 & 0 & 0 & 0 & 0 & 0\\ 0 & 0 & 1 & 0 & 0 & 0\\ 0 & 0 & 0 & 0 & 1 & 0\end{bmatrix}\mathbf{x}(k)+\mathbf{w}(k),$$

where $\mathbf{x}$ is the state vector including position and velocity in all three dimensions

$$\mathbf{x}(k)=\left[x_1(k)\ \ \dot{x}_1(k)\ \ x_2(k)\ \ \dot{x}_2(k)\ \ x_3(k)\ \ \dot{x}_3(k)\right]^T,$$

$\mathbf{z}$ is the measured position in each dimension, $\mathbf{v}$ is a Gaussian system noise affecting to acceleration of all three dimensions, and $\mathbf{w}$ is a Gaussian measurement noise affecting to all measurements.

The 3-D tracking system is simulated for an insight of performance of the sub-optimal OOSM procedures. The system has four nodes with measurement and estimation. The measurement for each nodes own sensor comes always in time, but the other three updates may arrive with delay. Sampling time ΔT is set to 0.1, and simulations last 200 iterations.

The delay in the communication between the DKF nodes is simulated by Poisson-distributed random numbers. As an illustration, the probability density functions of Poisson distribution with parameter values from 0.5 to 3 are drawn in Fig. 1. The parameter λ is equal to the mean of the distribution. The sub-optimal approaches for DKF to handle OOSM problem, which were discussed in above, are compared to each other. Initial state is chosen to be in origin, and known perfectly to all nodes.

The sub-optimal approaches to OOSM are considered as presented in Chapter 3. The third approach is applied with a memory of two previous statistics. This means that measurements with at most two step delays can be applied in the back propagation of the state estimation update.

$$\begin{aligned}\mathbf{I}^+(k+1) &= \mathbf{M}^I(k)+\mathbf{M}^Y(k)\mathbf{G}(k)\left(\mathbf{G}^T(k)\mathbf{M}^Y(k)\mathbf{G}(k)+\mathbf{Q}^{-1}(k)\right)\mathbf{G}^T(k)\mathbf{M}^Y(k)\\ &\quad -\mathbf{M}(k)\mathbf{G}(k)\left(\mathbf{G}^T(k)\mathbf{M}(k)\mathbf{G}(k)+\mathbf{Q}^{-1}(k)\right)^{-1}\left(\mathbf{M}(k)\mathbf{G}(k)\right)^T\end{aligned} \tag{12}$$

$$\begin{aligned}\mathbf{i}^+(k+1) &= \mathbf{F}^{-T}(k)\mathbf{i}(k)+\left(\mathbf{M}^Y(k)\mathbf{G}(k)\mathbf{\Sigma}^{-1}(k)\mathbf{G}^T(k)\right)\mathbf{F}^{-T}(k)\hat{\mathbf{y}}(k\mid k)\\ &\quad -\left(\mathbf{M}(k)\mathbf{G}(k)\left(\mathbf{\Sigma}(k)+\mathbf{G}^T(k)\mathbf{M}^I(k)\mathbf{G}(k)\right)^{-1}\mathbf{G}^T(k)\right)\mathbf{F}^{-T}(k)\left(\hat{\mathbf{y}}(k\mid k)+\mathbf{i}(k)\right)\end{aligned} \tag{13}$$

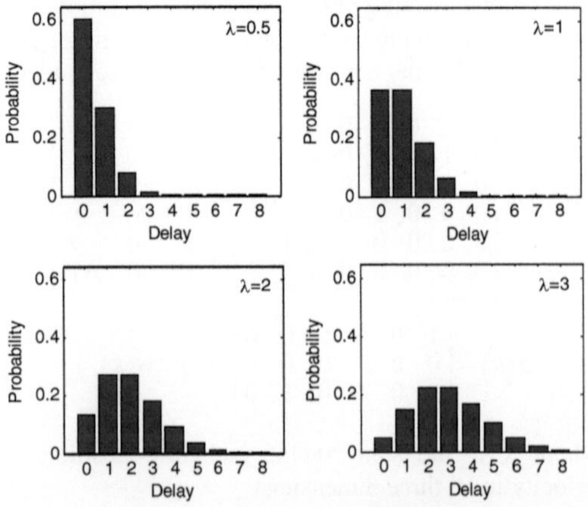

Fig. 1: The delay probability densities.

3.2. Simulation Results and Conclusions

The mean square error (MSE) results of simulations with different delay distributions are collected to Tables 1, 2, 3, 4. Due to the completely different scales of position and velocity (see Figs 2, and 3), the comparing of MSEs of different methods are divided to position and velocity MSEs. Q and R refer to the variances of model and measurement noises, respectively. Note that each simulation is just one realization of the system. Therefore only the methods should be compared, not the state estimation performance differences between the simulation cases.

The results in Tables 1-4 indicate that the two-step truncated iterative propagation has substantially better performance compared to heuristic ones. Additionally, the performance difference decreases, while the average delay increases. This can be interpreted as a sign that the two-step iteration propagation window is becoming too short.

4. Clustered Sensor Network

The communication in a wireless sensor network can be more efficient, if messages to be communicated are collected to a cluster head [4]. This cluster head communicates the fused information further. Hence the data fusion should be done in clusters of network nodes. The receiver of this communication can be either in other parts of same the network or in other networks, such as a fixed network.

In KF data fusion, the clusterization of sensor network can lead to a few different configurations. First of all, clusters can communicate all their

Table 1: MSEs of DKF, while the delay distribution is $P(0.5)$.

	Ignore old	Newest available	Iterative propagation
	Distance MSEs		
$Q = 0.01, R = 1$	0.1399	0.1483	0.0977
$Q = 0.01, R = 0.1$	0.0647	0.0740	0.0133
$Q = 0.04, R = 0.1$	0.0179	0.0163	0.0159
	Velocity MSEs		
$Q = 0.01, R = 1$	0.0671	0.0875	0.0056
$Q = 0.01, R = 0.1$	0.5815	0.8267	0.0028
$Q = 0.04, R = 0.1$	0.1162	0.1705	0.0059

Table 2: MSEs of DKF, while the delay distribution is $P(1)$.

	Ignore old	Newest available	Iterative propagation
	Distance MSEs		
$Q = 0.01, R = 1$	0.2708	0.2506	0.0765
$Q = 0.01, R = 0.1$	0.0376	0.0276	0.0308
$Q = 0.04, R = 0.1$	0.0139	0.0200	0.0079
	Velocity MSEs		
$Q = 0.01, R = 1$	0.1843	0.2221	0.0073
$Q = 0.01, R = 0.1$	0.0249	0.0423	0.0028
$Q = 0.04, R = 0.1$	0.0815	0.1493	0.0060

Table 3: MSEs of DKF, while the delay distribution is $P(2)$.

	Ignore old	Newest available	Iterative propagation
	Distance MSEs		
$Q = 0.01, R = 1$	0.2205	0.2924	0.1043
$Q = 0.01, R = 0.1$	0.0224	0.0165	0.0098
$Q = 0.04, R = 0.1$	0.0230	0.0289	0.0077
	Velocity MSEs		
$Q = 0.01, R = 1$	0.0460	0.1418	0.0071
$Q = 0.01, R = 0.1$	0.0523	0.1400	0.0042
$Q = 0.04, R = 0.1$	0.2535	0.8042	0.0055

Table 4: MSEs of DKF, while the delay distribution is $P(3)$.

	Ignore old	Newest available	Iterative propagation
	Distance MSEs		
$Q = 0.01, R = 1$	0.4807	0.7435	0.4761
$Q = 0.01, R = 0.1$	0.0541	0.0630	0.0218
$Q = 0.04, R = 0.1$	0.0353	0.0224	0.0357
	Velocity MSEs		
$Q = 0.01, R = 1$	0.0098	0.0348	0.0099
$Q = 0.01, R = 0.1$	0.1303	0.4243	0.0033
$Q = 0.04, R = 0.1$	0.0368	0.1188	0.0055

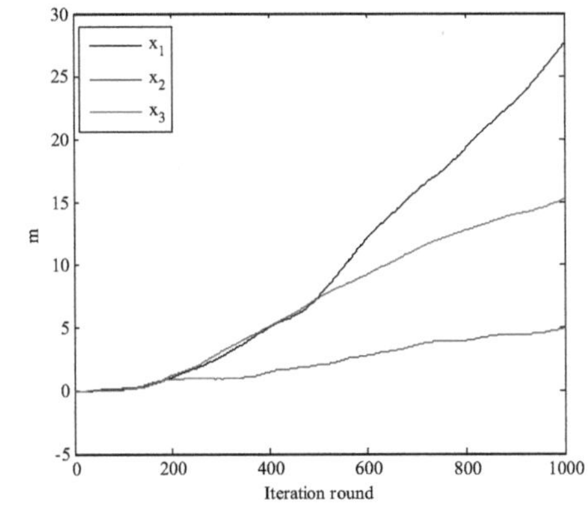

Fig. 2: An example of position in different dimensions.

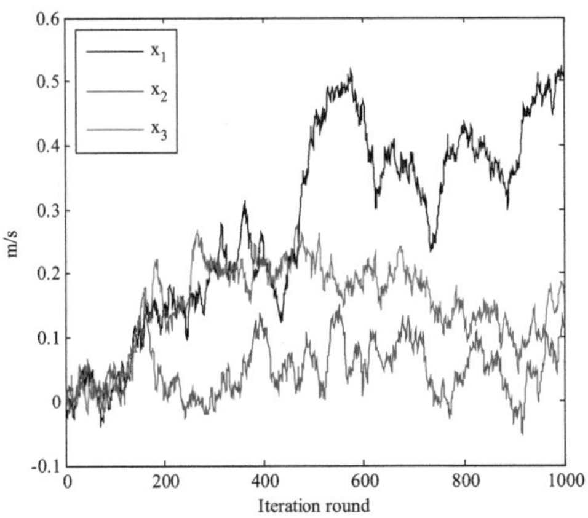

Fig. 3: An example of velocity in different dimensions.

information to all clusters. This corresponds to the regular global DKF and it leads to results mathematically equivalent to the normal centralized KF. The second possibility is that each cluster fuses its own data locally, thus producing only local estimates. The local level KF inside a cluster can be done either in ad-hoc -topology by DKF or in star-topology by centralized KF.

The global KF option, which uses full communication, has no problem in a changing cluster configuration. The full information is distributed throughout the wireless sensor network, and changes in clusters are not affecting the performance of data fusion.

If the KF data fusion is done locally, the changing clusters become a problem. The reason is that the structure of state and covariance error information matrices changes, and error covariance matrices cannot be updated accurately. The key point in here is how much of error covariance should be in memory during a change of clusters.

4.1. Error Covariance Memory Possibilities for Clusterized Kalman Filter

There are a few possible levels for error covariance matrix memory during a change of data fusion cluster for KF systems. Obviously, the best option to prevent cluster change problems is to use global filter, in which all nodes are included, and keep the covariance matrices of the whole system in memory. However, the global fusion is not decentralized, which may be desirable in some applications, and it causes too heavy burden for a sensor network communication, if the system is not small scale. The tradeoff between the use of communication resources and filter performance does not necessarily support using global filter.

In a genuinely clusterized system, it is possible to remember either error covariance matrix for sensor nodes making the same cluster transition or just the error variance of own node. One potential option is to reset error covariance memory in every cluster change. Unfortunately, this leads to very large errors in transition stage of the filtering, and therefore it can not be recommended.

The memory of error covariance between nodes making the same cluster transition requires first memory of the whole error covariance matrix of the previous cluster, and then transmission of the covariance matrix to the new cluster head. The cluster head must detect nodes, which make the same cluster change and adjust the cluster error covariance matrix accordingly.

The error covariance memory types and communication requirements during cluster changes are collected in Table 5.

Table 5: Specifications of node error memories in a system with n nodes in cluster (when applicable).

Type	Memory	Communication during a cluster change
Global fusion	Full system error covariance matrix in central fusion node	-
Memory of error covariance between nodes making the same cluster transition	Memory of cluster $n \times n$ error covariance matrix in each node	Detection of transition nodes and communication of their error covariance matrix to the cluster head
Memory of own covariance	Memory of own error covariance in each node	Communication of own error covariance to the cluster head

4.2. Cluster Change Simulation

The scope of the following simulations is to examine the performance loss due to data fusion cluster changes, and compare the memory possibilities during cluster changes. Simulations are made by using DKF in moving measurement nodes in a spatio-temporal field, which is in this case nominated as a temperature field. The temperature is simulated in an area of 10 x 10 km. Without going into details, the heuristic spatio-temporal model is structured to include a mean value, a spatial part, a temporal part, and a spatio-temporal part. An example temperature profile is drawn in Fig. 4.

The measurement network includes 50 moving nodes, which are divided to six clusters. The measurement nodes are set to move inside the area in constant speed to a randomly chosen direction. Simulations include two cases: high and low mobility with speeds restricted to 60 km/h and 20 km/h, respectively. If the movement is taking the node outside the area, the direction is changed.

The clusterization of nodes is made according to position by K-means clustering [3]. The initial clusterization is made with 100 iterations, and the following clusterizations are made with 10 iterations from the previous clusterization. An example of node locations and a clusterization is shown in Fig. 5. Numbers next to nodes correspond to numbers of allocated clusters.

Temperature measurements are filtered by taking location-based weighted averages over clusters. The weight of node i in the estimation of temperature in node j, c_{ji}, is determined by firstly distance weighting and secondly scaled. Mathematically this corresponds to

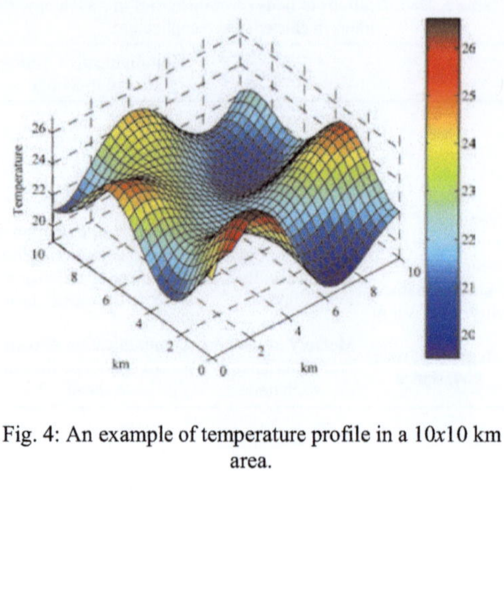

Fig. 4: An example of temperature profile in a 10x10 km area.

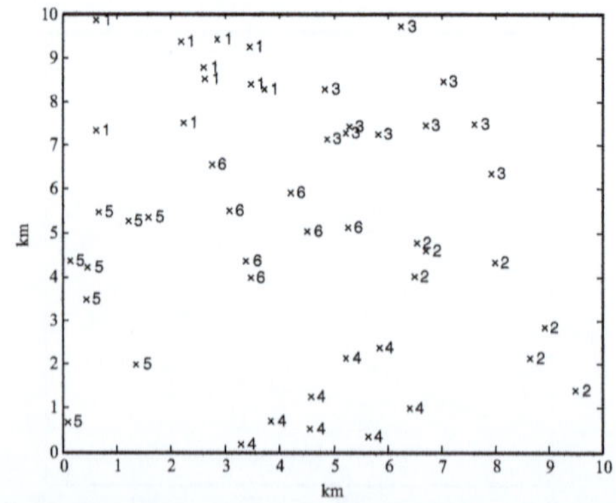

Fig. 5: An example of node locations and clustering. Nodes are marked with x, and numbers correspond to clusters.

$$\tilde{c}_{ji} = e^{-d_{ji}^2}, \tag{14}$$

$$c_{ji} = \frac{c_{ji}}{\sum_{k=1}^{N} \tilde{c}_{jk}}, \tag{15}$$

where d_{ji} is the distance between i and j, and N is the number of nodes in the cluster. The above weights were applied in KF as elements of matrix **C** in a cluster-wide state-space model

$$\begin{aligned} \mathbf{T}(k+1) &= \mathbf{C}(k)\mathbf{T}(k) + \mathbf{w}(k), \\ \mathbf{T}_m(k) &= \mathbf{T}(k) + \mathbf{v}(k), \end{aligned} \tag{16}$$

where **T** is the node temperature vector and $\mathbf{T}_m$ is the measured temperature vector. The model (16) corresponds to the local model (1) and (2) in DKF.

Initial filtered temperature is set to 18 degrees. Process and measurement noise covariance matrices (**Q** and **R**) are set to identity matrices. Note that the process noise covariance matrix **Q** is not accurate, since the simulated temperature obeys predetermined model unknown to the filter.

4.3. Simulation Results and Conclusions

For convenience, let's use abbreviate full system KF, clustered KF with memory of error covariance matrix between the nodes making same cluster change, and clustered KF with memory of node error variance with FKF, CKF1, and CKF2, respectively. Simulation results are divided into two cases: high and low mobility, which refer to speeds of the nodes. Evenly distributed speeds of mobiles are restricted to 60 and 20 km/h in high and low mobility cases, respectively.

In the cases of high and low mobility, temperature mean-square-errors (MSEs) over the four simulated repetitions with 1200 iterations and 50 nodes are given in Table 6 and Table 7, respectively. For normalization, MSEs of CKF1 and CKF2 scaled with MSEs of FKF are referred as relative MSEs, and they are drawn in Figures 6 and 7. Relative MSEs are shown as functions of cluster changes in Fig. 8.

Table 6: MSEs of different versions of KF in clustered data fusion network in high mobility cases.

Number of clusters	Cluster changes	No filtering	FKF	CKF1	CKF2
2	1129.75	0.9968	0.3659	0.3965	0.4088
3	1559.75	0.9938	0.3639	0.4097	0.4236
4	1944.25	0.9967	0.3656	0.4233	0.4380
5	2546.25	1.0016	0.3687	0.4323	0.4473
6	2906	1.0020	0.3685	0.4356	0.4502
8	3611.5	0.9981	0.3654	0.4430	0.4588
10	4117.75	1.0018	0.3699	0.4523	0.4645
12	4654	1.0025	0.3686	0.4537	0.4678
14	4903.5	0.9997	0.3682	0.4553	0.4686

Table 7: MSEs of different versions of KF in clustered data fusion network in low mobility cases.

Number of clusters	Cluster changes	No filtering	FKF	CKF1	CKF2
2	681.5	1.0036	0.3604	0.3876	0.3959
3	1012.75	0.9962	0.3585	0.4018	0.4111
4	1144.25	0.9990	0.3615	0.4137	0.4222
5	1535	0.9967	0.3608	0.4177	0.4275
6	1751.5	1.0017	0.3612	0.4256	0.4347
8	2199.25	1.0026	0.3592	0.4293	0.4391
10	2521.5	0.9974	0.3613	0.4348	0.4438
12	2799.5	0.9945	0.3576	0.4368	0.4450
14	2957.5	1.0020	0.3603	0.4414	0.4498

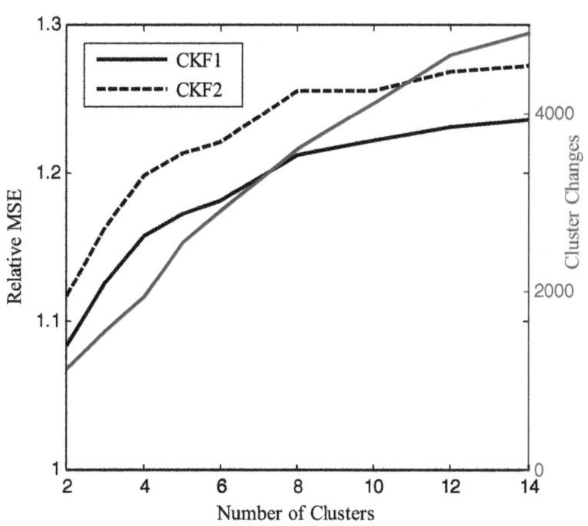

Fig. 6: Relative MSEs of CKF1 and CKF2 with cluster changes as a function of clusters in the high node mobility case.

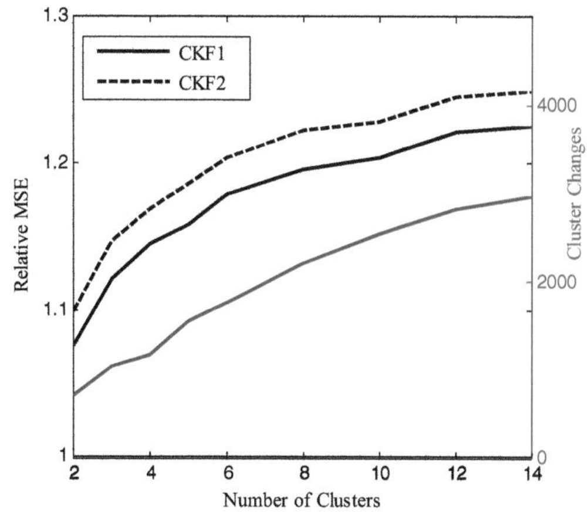

Fig. 7: Relative MSEs of CKF1 and CKF2 with number of cluster changes as a function of number of clusters in the case of low node mobility.

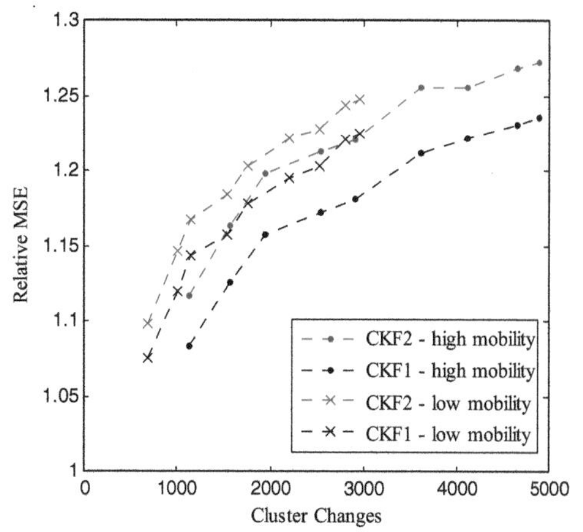

Fig. 8: Relative MSEs as functions of number of cluster changes.

Based on the results, a few conclusions can be drawn. The main result is that CKF2 has indeed some performance loss in compared to CKF1. Nevertheless, this performance loss is not considerably large – about 5% at the most. If the level of performance loss is compared to differences in the signaling and memory requirements, CKF2 seems to have more desirable qualities in regards to application to wireless sensor networks. Additionally, results indicate that the performance loss of clustered KFs, CKF1 and CKF2, compared to full system Kalman filter, FKF, is dependent on number of clusters (see Figures 6 and 7) and number of cluster changes (see Figures 6, 7 and 8).

5. Conclusions

MSE results of Tables 1-4 indicate that iterative propagation has superior performance to the trivial solutions to handle the OOSM problem, even if the propagation window is truncated. However, Table 4 suggests that the performance difference reduces while the average delay grows larger than the maximum number of back propagated steps.

MSEs of the 'ignore old data' and 'use newest data' techniques vary much more than the ones of the data back propagation. This suggests the robustness of the truncated analytic approach. Besides, applying the newest available data may not be robust approach for the time-varying systems. The difference of old variance error information may come too large.

The simulations suggest that applying the iterative propagation of OOSM in KF is worthwhile, even if the back propagation window is fairly small. On the other hand, applying (12) and (13) requires a quite lot of computational and memory capacity, which can be restricting for limited capacity nodes.

Clustered data fusion architectures lead to a new set of problems. For KF, the problem of error covariance update after node changes from one cluster to another is apparent. Here is discussed a few possibilities to cope with this problem.

The simplest mode in error covariance memory, CKF2, is noted to have slightly worse performance in simulation cases as the more complex ones, but much lighter communication and memory requirements make it the most attractive alternative for wireless sensor networks.

Both of the examined problems have the same characteristic – the need for more accurate handling of OOSM and cluster change problems are dependent on the filter accuracy requirements. If the communication is the restricting bottleneck in the system, it is obvious that the lighter protocols, such as presented in here, must be used. If the wireless network size is not large and the filter must operate accurately, the use of accurate techniques is justified.

References

[1] S. Grime, H.F. Durrant-Whyte: "Data Fusion in Decentralized Sensor Networks", Control Eng. Practice, Elsevier Science, 2 (5), pp. 849-863, 1994.

[2] E.W. Nettleton, H.F. Durrant-Whyte: "Delayed and Asequent Data in Decentralised Sensing Networks", Proc. SPIE Conf. #4571, Oct./Nov. 2001.

[3] Seber, G.A.F.: "Multivariate Observations", Wiley, New York, 1984.

[4] X. Du, F. Lin, "Improving Routing in Sensor Networks with Heterogeneous Sensor Nodes", *Proc. 61st IEEE Veh. Tech. Conf. 2005 Spring*, Stockholm, Sweden, 2005.

[5] K. Römer, F. Mattern, "The Design Space Of Wireless Sensor Networks", *IEEE Wireless Comm.*, vol. 11, no. 6, pp. 54-61, Dec. 2004.

[6] T. Pham, E.J. Kim, M. Moh, "On data aggregation quality and energy efficiency of wireless sensor network protocols - extended summary", *Proc. First Int. Conf. Broadband Networks*, pp. 730-732, 2004.

[7] S. Bandyopadhyay, E.J. Coyle, "An energy efficient hierarchical clustering algorithm for wireless sensor networks", *Proc. 22nd Ann. Joint Conf. of the IEEE Comp. and Comm. Soc., INFOCOM 2003*, vol. 3, pp. 1713-1723, 2003.

[8] Y. Bar-Shalom, "Update with Out-of-Sequence Measurements in Tracking: Exact Solution", *IEEE Trans. Aerospace and Elect. Systems*, vol. 38, no. 3, pp. 769-778, 2002.

[9] A. Gelb (ed.), *Applied Optimal Estimation*, The MIT Press, Cambridge, MA, USA, 15th ed., 1999.

NUMERICAL MODELING OF GMAW ARC

J. Hu*, H.L. Tsai and P.C. Wang
*Author for correspondence
Department of Mechanical Engineering,
University of Bridgeport,
Bridgeport, CT 06604, USA

ABSTRACT

A comprehensive model has been developed to simulate the transient, coupled transport phenomena occurring during a gas metal arc welding process. This includes the arc plasma; melting of the electrode; droplet formation, detachment, transfer, and impingement onto the workpiece; and weld pool fluid flow and dynamics. The fluid flow and heat transfer in both the arc and the metal were simulated and coupled through the boundary conditions at the arc-metal interface at each time step. The detached droplet in the arc and the deformed weld pool surface were found to cause significant changes in the distributions of arc temperature and arc pressure, which are usually assumed to have Gaussian distributions at the workpiece surface. The comprehensive model could provide more realistic boundary conditions to calculate the heat transfer and fluid flow both in the plasma and the metal. The predicted arc plasma distribution, droplet flight trajectory, droplet acceleration and final weld bead shape compared favorably with the published experimental results. This paper was to present the heat transfer and fluid flow in the arc plasma.

I. INTRODUCTION

Gas metal arc welding (GMAW) is an arc welding process that uses a plasma arc between a continuous, consumable filler-metal electrode and the weld pool. GMA welding is one of the most important and popular welding technologies. Very complicated transport phenomena, including the arc plasma, electrode melting, and weld pool dynamics, occur during the GMA welding process. The trial-and-error procedures have been used in the industry to identify key welding parameters and to develop the GMA welding technologies. However, this weld-and-cut method, is not only very expensive and time-consuming, but also cannot achieve the fundamental understanding on how the transport phenomena affects weld quality, such as weld penetration, weld bead shape, and the formation of porosity.

Many models have been developed to model the heat transfer and fluid flow in the arc plasma for both GTAW [1-18] and GMAW [19-24]. Mckelliget and Szekely [1], Choo et al. [2] and Goodarzi et al. [3] have simulated the arc column by assuming the current density distribution at the cathode surface in GTAW. Fan et al. [4-5] used fixed temperature boundary condition at the cathode tip to calculate the arc column in GTAW. Zhu et al. [6] developed a unified model to simulate the arc column, the cathode and the cathode sheath in GTAW. Lowke et al. [7-8] simplified the unified model to treat the electrode in a special way at the cathode surface to account for electrode effects [7] or neglect the electrode sheath [8]. The simplified models [7-8] reduced the computation time to 1% of the original unified model and gave fair results in agreement with experimental results when 0.005-0.01 cm mesh size was chosen around the cathode tip. These simplified models have been used and further developed by many researchers [9-18] to calculate the heat transfer and fluid flow in the arc column.

Both GTAW and GMAW have a plasma arc struck between an electrode and a workpiece. Even though the GTAW has an inert tungsten cathode as the electrode and the electrode of GMAW is a melting metal and usually set as the anode, the GTAW arc model can be adopted to model the GMAW arc. Jonsson [19] adopted the GTAW arc model of Mckelliget and Szekely [1] to calculate the arc column by assuming a current density distribution at the cathode spot. Zhu et al. [20] calculated the anode temperature profile by incorporating the simplified arc model of Lowke et al. [8] into a one-dimensional conduction model of the moving electrode in GMAW. The heat input to the electrode was estimated from the arc plasma, and the 'molten' metal was discarded when its temperature reached the melting point. Haidar and Lowke [21] and Haidar [22] extended the simplified arc model of Lowke et al. [8] to simulate the droplet formation in GMAW. They were the first to simulate the dynamic interaction of the arc plasma and the droplet. Haidar [13, 23, 24] further developed this GMAW model to take into account the sheath effect at the anode surface. However, the droplet was eliminated immediately when it was detached from the electrode tip. The weld pool dynamics was also neglected and the workpiece was treated as a flat plate. The fluid flow in the weld pool was not calculated and only conduction was considered. Zhu et al. [25] have developed a comprehensive model to simulate the arc column, droplet formation, detachment, transfer and impingement onto the workpiece and the weld pool dynamics. However, the simulated arc plasma poorly matched the experimental [26-30] results and the simulation results from aforementioned arc models [1-24].

Many mathematical models [31-43] have been proposed to simulate the transport phenomena during the GMA welding process without considering the arc plasma. They were limited to a portion of the welding process, for example, weld pool dynamics, and/or involved many simplifications, such as a Gaussian distribution of the arc pressure.

In this paper, a comprehensive model has been developed to simulate the transient, coupled transport phenomena

K. Elleithy et al. (eds.), Advances in Computer, Information, and Systems Sciences, and Engineering, 69–74.

occurring during a gas metal arc welding process. This includes the arc plasma; melting of the electrode; droplet formation, detachment, transfer, and impingement onto the workpiece; and weld pool fluid flow and dynamics.

II. MATHEMATICAL MODLELS

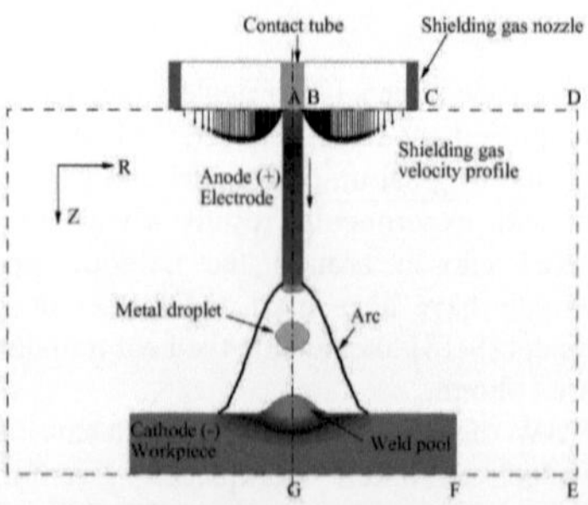

Fig. 1. A schematic representation of a GMAW system including the electrode, the arc, and the weld pool (not to scale).

Fig. 1 is a schematic representation of a two-dimensional axisymmetric GMAW system, with the computational domain marked by ABCDEFGA. There are three phases inside the computational domain: a solid phase, a liquid phase and a gas phase. The solid phase includes the unmelted electrode and part of the workpiece, while the liquid phase includes the melted electrode, falling droplet, and part of the workpiece. The gas phase includes the partially ionized arc plasma and shielding gas. Between the liquid zone and solid zone, there is a small zone called mushy zone where the solid and liquid metal coexist. A continuum formulation [44] was used to handle the metal domain consisting of the solid phase, liquid phase and mushy zone. Latent heat during melting and solidification was considered using the enthalpy method. As the properties of gas are far different from those of metal, two computational domains are used for computational robust and efficiency. One computational domain is used to calculate the heat transfer and fluid flow in the gas phase and another is used for metal, which includes both solid phase and liquid phase. The heat transfer and fluid flow in both computational domains are coupled with the electromagnetic field in both domains.

The differential equations governing the conservation of mass, momentum, and energy based on the continuum formulation given by Diao and Tsai [44] are employed in the present study, and the current continuity equation is used to calculate the current density distribution. The equations are given below:

Mass continuity

$$\nabla \cdot (\rho V) = 0 \tag{1}$$

Momentum

$$\frac{\partial}{\partial t}(\rho u) + \nabla \cdot (\rho V u) = \nabla \cdot \left(\mu_l \frac{\rho}{\rho_l} \nabla u \right) - \frac{\partial p}{\partial r} - \frac{\mu_l}{K} \frac{\rho}{\rho_l}(u - u_s)$$

$$- \frac{C \rho^2}{K^{1/2} \rho_l} |u - u_s| (u - u_s) - \nabla \cdot (\rho f_s f_l V_r u_r) - J_z \times B_\theta \tag{2}$$

$$\frac{\partial}{\partial t}(\rho v) + \nabla \cdot (\rho V v) = \nabla \cdot \left(\mu_l \frac{\rho}{\rho_l} \nabla v \right) - \frac{\partial p}{\partial z} - \frac{\mu_l}{K} \frac{\rho}{\rho_l}(v - v_s)$$

$$- \frac{C \rho^2}{K^{1/2} \rho_l} |v - v_s| (v - v_s) - \nabla \cdot (\rho f_s f_l V_r v_r) + \rho g \beta_T (T - T_0) + J_r \times B_\theta \tag{3}$$

Energy

$$\frac{\partial}{\partial t}(\rho h) + \nabla \cdot (\rho V h) = \nabla \cdot \left(\frac{k}{c_s} \nabla h \right) + \nabla \cdot \left(\frac{k}{c_s} \nabla (h_s - h) \right) - \nabla \cdot$$

$$\left(\rho (V - V_s)(h_l - h) \right) - \Delta H \frac{\partial f_l}{\partial t} + \frac{J_r^2 + J_z^2}{\sigma_e} - S_R + \frac{5 k_b}{e} \left(\frac{j_r}{c_s} \frac{\partial h}{\partial r} + \frac{j_z}{c_s} \frac{\partial h}{\partial z} \right) \tag{4}$$

Current continuity

$$\nabla^2 \phi = \frac{1}{r} \frac{\partial}{\partial r} \left(r \frac{\partial \phi}{\partial r} \right) + \frac{\partial^2 \phi}{\partial z^2} = 0 \tag{5}$$

Ohm's law

$$J_r = -\sigma_e \frac{\partial \phi}{\partial r}, \quad J_z = -\sigma_e \frac{\partial \phi}{\partial z} \tag{6}$$

Maxwell's equation

$$B_\theta = \frac{\mu_0}{r} \int_0^r J_z r dr \tag{7}$$

In Eqs. (1)-(4), u and v are the velocities in the r and z directions, respectively. $V_r (= V_l - V_s)$ is the relative velocity vector between the liquid phase and the solid phase in the mushy zone. The subscripts s and l refer to the solid and liquid phases, respectively, and the subscript 0 represents the initial condition. p is the pressure; T is the temperature; h is the enthalpy; ϕ is the electrical potential; ρ is the density; μ is the viscosity; k is the thermal conductivity; g is the gravitational acceleration; β_T is the thermal expansion coefficient; c is the specific heat; σ_e is the electrical conductivity; J_r and J_z are current densities, in the respective r and z directions; B_θ is the self-induced electromagnetic field; S_R is the radiation heat loss; μ_0 is the magnetic permeability; k_b is the Stefan-Boltzmann constant; and e is the electronic charge.

The third and fourth terms on the right-hand side of Eqs. (2) and (3) represent the respective first- and second-order drag forces for the flow in the mushy zone. The fifth term on the right-hand side of Eqs. (2) and (3) represents an interaction between the solid and the liquid phases. The second term on the right-hand side of Eq. (4) represents the net Fourier diffusion flux. While the third term represents the energy flux associated with the relative phase motion, and the forth term is used to consider the latent heat of fusion. All the terms mentioned in this paragraph are zero, except in the mushy zone. When Eqs. (2)-(4) are used to calculate the arc plasma, these terms associated with the mushy zone are set to zero and all the thermal physical properties are replaced by those of the arc plasma.

The second-to-last term on the right-hand side of Eq. (3) is the thermal expansion term. The last term of Eq. (2) and Eq. (3) is the electromagnetic force term. The last three terms in Eq. (4) are Ohmic heating, radiation loss, and electron enthalpy flow, respectively.

Only half of the entire physical domain is calculated due to the cylindrical symmetry along the centerline AG. The wire feed rate is incorporated through a boundary condition on axial velocity along AB. The imposed shielding gas flow is set through a boundary condition on axial velocity along BC. A constant mass flow boundary condition is used for the open boundaries CD and DE. The temperature boundaries along AD, DE, and EG are determined by the ambient condition, which is set as room temperature. Uniform current density is specified along AB. The voltage is set to zero at the bottom of the workpiece FG.

Two computational domains are used to calculate arc plasma and metal. For the arc plasma domain, the metal domain was treated as inner obstacles, while the plasma temperature, velocity, and pressure distributions were calculated. For the metal domain, a volume-of-fluid (VOF) method [45] was used to handle the free surfaces for the droplet and the surface of the weld pool. Additional body force source terms are added to the momentum transport equations at the metal free surface to consider the effects of surface tension, Marangoni shear stress, arc plasma shear stress and arc pressure. Additional source terms [46] are added to the energy equation for the special treatment of heat transfer near the anode sheath and the cathode sheath.

The current distribution is greatly influenced by the temperature in the arc column and the shape of the metal domain, but it is only slightly influenced by the temperature distribution in the metal domain as the electrical conductivity of metal varies slightly with temperature. Therefore, the current continuity equation and its associated boundary conditions are solved in the entire domain, while other primary variables, including p, u, v, and T, are calculated separately in the metal domain and the arc domain. The current continuity equation is iterated with the transport equations in the arc domain to obtain the current density distribution for both the arc domain and the metal domain. Iterations are required to assure convergence of each domain and then the boundary conditions are calculated from each domain for the coupling between the two domains.

For the metal domain, the method developed by Torrey et al. [45] was used to solve p, u, v and T. This method is Eulerian and allows for an arbitrary number of segments of free surface with any reasonable shape. The basic procedure for advancing the solution through one time step, δt, consists of three steps. First, at the beginning of the time step, explicit approximations to the momentum equations (2) and (3) are used to find provisional values of the new time velocities. Second, an iterative procedure is used to solve for the advanced time pressure and velocity fields that satisfy Eq. (1) to within a convergence criterion at the new time. Third, the energy equation is solved.

For the arc plasma domain, a fully implicit formulation is used for the time-dependent terms, and the combined convection/diffusion coefficients are evaluated using an upwind scheme. The SIMPLE algorithm [47] is applied to solve the momentum and continuity equations to obtain the velocity field. At each time step, the current continuity equation is solved first, based on the updated parameters. Current density and electromagnetic force are then calculated for the momentum and energy equations. The momentum equations and the continuity equation are then solved in the iteration process to obtain pressure and velocity. The energy equation is solved to get the new temperature distribution. Next, the temperature-dependent parameters are updated, and the program goes back to the first step to calculate the current continuity equation. This process is repeated for each time step until the convergence criteria are satisfied.

III. Results And Discussion

The electrode is mild steel with a 0.16 cm diameter. The workpiece is also a mild steel disk with a 3 cm diameter and a 0.5 cm thickness. The current is set to be constant at 220 A and the welding time is 1 s. The imposed argon shielding gas flows out of a gas nozzle with a 1.91 cm inner diameter at a rate of 24 l/min. The contact tube is set flush with the bottom of the gas nozzle and is 2.54 cm above the workpiece. The initial arc length is set as 0.8 cm. The wire feed rate is 4.5 cm/s.

In this section, the arc characteristics during a welding process are presented. In order to increase the readability of flow direction, only a quarter of the grid nodes were used in Fig. 2 and 3. The shape of the electrode and workpiece are marked with thick lines.

Fig. 2 (a)-(d) shows the distributions of temperature, pressure, plasma velocity, and current at different instants, from t = 100 ms to t = 400 ms. From the temperature contours in the plasma at t = 100 ms in Fig. 2 (a), it can be seen that the arc has a bell-shaped envelope, which covers the droplet and expands as it approaches the workpiece. The maximum temperature of the plasma is found to be 19300 K on the axis near the bottom of the droplet. The corresponding velocity distribution at t = 100 ms in Fig. 2 (c) shows a strong downward arc plasma flow underneath the droplet. The maximum axial velocity in the arc column is found to be 230 m/s on the axis. The corresponding arc pressure contours at t = 100 ms are shown in Fig. 2 (b), which shows two high pressure regions. One is underneath the droplet with a maximum of 800 Pa above the ambient pressure, and the other is near the cathode with a maximum of 600 Pa above the ambient pressure. The high temperature arc column and high speed arc plasma jet are formed by the high current flow in the arc column, which is drawn in Fig. 2 (d). The high current flow in the arc column provides heat to keep the arc column hot, and in turn, the high temperature arc column maintains a highly conductive path for the current to flow through. From Fig. 2 (d), it can be seen that current tends to diverge in the arc column after it flows out of electrode tip. The divergence in current generates an inward and downward force underneath the electrode tip, which has a pinch effect on arc plasma. The pinch effect of the electromagnetic force in the arc column, which is shown in Fig. 3, draws the arc plasma and the surrounding shielding gas to flow inward at the electrode tip and then downward along the axial direction to the workpiece. The inward arc plasma flow forms a high arc pressure zone underneath the electrode tip. When the downward arc plasma flow reaches the wrokpiece at the bottom, the downward momentum was retarded and a high pressure zone forms at the workpiece.

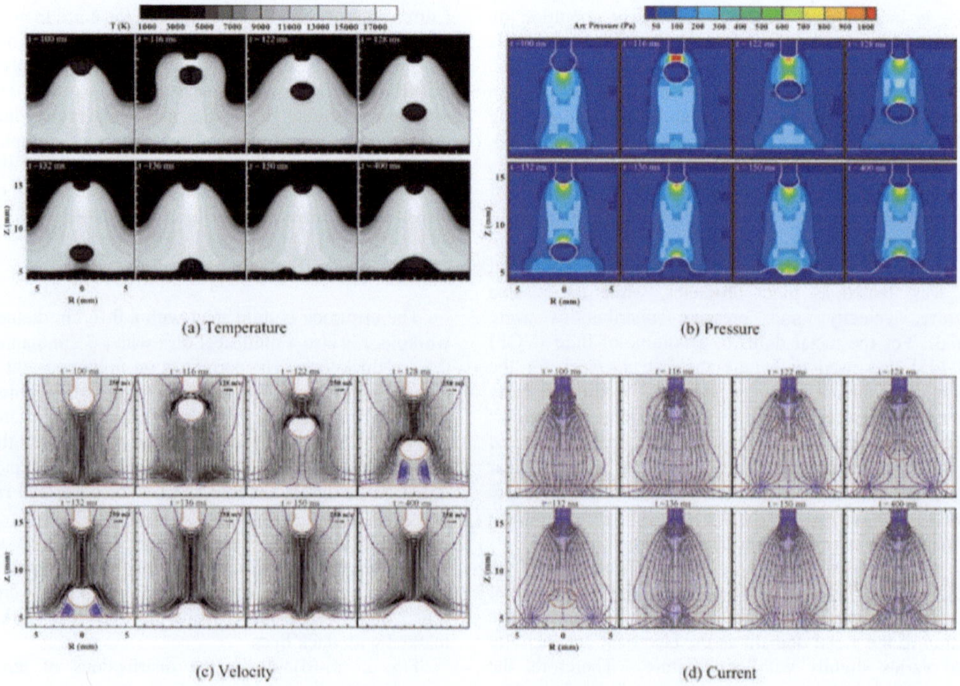

Fig. 2. Distributions of temperature, pressure, velocity and current in the arc plasma.

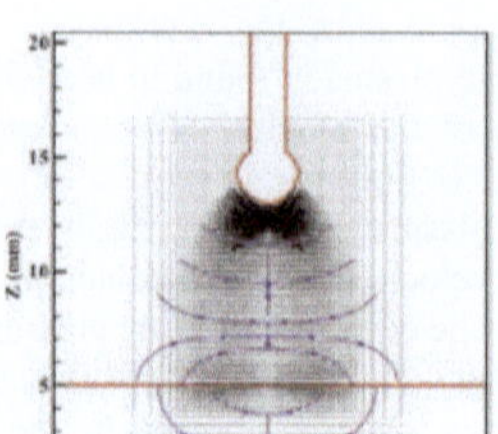

Fig. 3. Electromagnetic force at t = 100 ms.

After the droplet is detached from the electrode, new arc plasma is struck between the electrode tip and the top surface of the detached droplet. During the process of the detached droplet being transferred to the workpiece, the existence of the moving droplet greatly distorts the arc shape. From the temperature contour and velocity distribution from t = 116 ms to t = 132 ms in Fig. 2 (a) and (c), it can be seen that arc plasma flows around the moving droplet. The high temperature and high velocity arc column is limited to the region between the electrode tip and the top of the detached droplet. The arc plasma temperature underneath the moving droplet decreases continuously after the droplet is detached from the electrode. The high velocity plasma flow induced by the pinch effect of the electromagnetic force underneath the moving droplet subsides and vortices form underneath the droplet when it moves down to the workpiece. These phenomena are supported by the experimental results of Jones et al. [27-30], which has observed that arc plasma tended to flow around the detached droplet.

From the current distribution in Fig. 2 (d), it can be seen that current also flows around the detached droplet. Only a small amount of current flows through the detached droplet, except at t = 116 ms, when the droplet has just been detached and the temperature underneath the droplet is still relatively high. When the detached droplet moves farther away from the electrode tip in the cases of t = 122 ms to 132 ms, more current flows around the detached droplet. The arc plasma temperature quickly decreases when current flow decreases underneath the detached droplet due to the high radiation loss and low capacity of the plasma. The lower plasma temperature underneath the detached droplet further reduces the current flow in the plasma beneath the droplet, and hence the plasma temperature continues to drop. At the surface of the workpiece, the current bypassed around the detached droplet tends to converge at a place other than the spot directly underneath the droplet.

As it is shown in Fig. 2 (b), the existence of the detached droplet also dramatically changes the arc pressure distribution underneath the droplet. The high arc pressure, which was under the droplet before it was detached, decreases rapidly. The pressure difference between the upper and lower surfaces of the droplet helps to push the detached droplet down to the workpiece.

The current distribution in the weld pool is greatly influenced by the shape of the weld pool surface. The temperature, arc plasma velocity, current and arc pressure distributions from t = 136 ms to t = 400 ms in Fig. 2 show the influence of the weld pool shape on the arc plasma. The current tends to converge on the projected area at the workpiece, which may be at the workpiece center as in the cases of both t = 136 ms and t = 400 ms or not at the center as that of t = 150 ms. The temperature distribution and the arc pressure distribution at the deformed weld pool surface from t = 136 ms to t = 400 ms in Fig. 2, also show a different pattern from those at the flat weld pool surface.

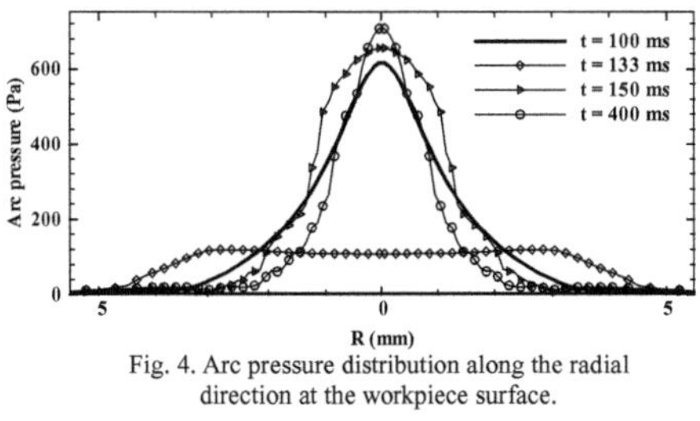

Fig. 4. Arc pressure distribution along the radial direction at the workpiece surface.

In the previous models of simulating the weld pool dynamics, the arc pressure distribution at the center of the workpiece surface was assumed as a Gaussian distribution with a fixed amplitude and distribution radius. However, the arc pressure distribution at the workpiece surface changes dramatically during the welding process, as shown in Fig. 4. Thus, it shows that the assumed Gaussian distribution of the arc pressure cannot reflect the real arc pressure distribution at the weld pool surface. Similarly, the current distribution and heat flux cannot be assumed as Gaussian distributions with fixed amplitude and fixed distribution radius. Thus, a comprehensive model that simulates the coupling of the arc and metal domain is needed to provide better boundary conditions at the metal surface for both domains.

IV. CONCLUSION

A comprehensive model has been used to simulate the transport phenomena occurring during a gas metal arc welding process. An interactive coupling between the arc plasma; the melting of the electrode; the droplet generation, detachment, transfer, and impingement onto the workpiece; and weld pool dynamics were considered. The heat transfer and fluid flow in the arc column were studied based on the transient distributions of current, temperature, velocity, and pressure in the arc plasma, droplet, and weld pool calculated in the comprehensive model. The moving droplet stuck between the electrode tip and the workpiece and the deformed weld pool were found to distort the arc flow and affect the current, temperature, velocity, and pressure distribution in the arc column. The assumed Gaussian distributions of the arc pressure, current and heat flux at the weld pool surface in the traditional models were shown not to be representative of the real distributions in the welding process. The coupled model can provide more realistic boundary conditions to the metal domain and thus can more accurately predict the heat transfer and fluid flow phenomena during the electrode melting, droplet generation, droplet transfer, weld pool dynamics and weld formation.

REFERENCES

[1] J. Mckelliget and J. Szekely, "Heat Transfer and Fluid Flow in the Welding Arc," Metall. Trans. 17A, 1986, pp. 1139-1148.

[2] R.T.C. Choo, J. Szekely and R.C. Westhoff, "On the calculation of the free surface temperature of gas-tungsten-arc weld pools from first principles: Part I. Modeling the welding arc," Metall. Trans. 23B, 1992, pp. 357-369.

[3] M. Goodarzi, R. Choo and J.M. Toguri, "The Effect of the Cathode Tip Angle on the GTAW Arc and Weld Pool: I. Mathematical Model of the Arc," J. Phys. D: Appl. Phys. 30, 1997, pp. 2744-2756.

[4] H.G. Fan, S-J Na and Y.W. Shi, "Mathematical model of arc in pulsed current gas tungsten arc welding," J. Phys. D: Appl. Phys. 30, 1998, pp. 94-102.

[5] H.G. Fan, Y.W. Shi, "Numerical simulation of the arc pressure in gas tungsten arc welding," Journal of Material Processing Tech. 61, 1996, pp. 302-308.

[6] P. Zhu, J.J. Jowke, R. Morrow, "A Unified Theory of Free Burning Arcs, Cathode Sheaths and Cathodes," J. Phys. D: Appl. Phys. 25, 1992, pp. 1221-1230.

[7] J.J. Lowke, R. Morrow and J. Haidar, "A Simplified Unified Theory of Arcs and Their Electrodes," J. Phys. D: Appl. Phys. 30, 1997, pp. 2033-2042.

[8] J.J. Lowke, P. Kovitya and H.P. Schmidt, "Theory of Free-Burning Arc Columns Including the Influence of the Cathode," J. Phys. D: Appl. Phys. 25, 1992, pp. 1600-1606.

[9] P. Zhu, J.J. Jowke, R. Morrow and J. Haidar, "Prediction of Anode Temperatures of Free Burning Arcs," J. Phys. D: Appl. Phys. 28, 1995, pp. 1369-1376.

[10] J. Haidar, "Departures from Local Thermodynamic Equilibrium in High-Current Free Burning Arcs in Argon," J. Phys. D: Appl. Phys. 30, 1997, pp. 2737-2743.

[11] L. Sansonnens, J. Haidar, J.J. Lowke, "Prediction of Properties of Free Burning Arcs Including Effects of Ambipolar Diffusion," J. Phys. D: Appl. Phys. 33, 2000, pp. 148-157.

[12] M. Tanaka, H. Terasaki, M. Ushio and J.J. Lowke, "A Unified Numerical Modeling of Stationary Tungsten-Inert Gas Welding Process," Metall. Mater. Trans. 33A, 2002, pp. 2002-2043.

[13] J. Haidar, "A Theoretical Model for Gas Metal Arc Welding and Gas Tungsten Arc Welding. I.," J. Appl. Phys. 84 (7), 1998, pp. 3518-3529.

[14] R.J. Ducharme, P.D. Kapadia, J. Dowden I.M. Richardson, and M.F. Thornton, "A Mathematical Model of TIG Electric Arcs Operating in the Hyperbaric Range," J. Phys. D: Appl. Phys. 29, 1996, pp. 2650-2658.

[15] J. Menart, J. Heberlein and E. Pfender, "Theoretical Radiative Transport Results for a Free-Burning Arc Using a Line-by-Line Technique," J. Phys. D: Appl. Phys. 32, 1999, pp. 55-63.

[16] J. Menart, S. Malik and L. Lin, "Coupled Radiative, Flow and Temperature-Field Analysis of a Free-Burning Arc," J. Phys. D: Appl. Phys. 33, 2000, pp. 257-269.

[17] H.P. Schmidt and G. Speckhofer, "Experimental and Theoretical Investigation of High-Pressure Arcs-Part I: the Cylindrical Arc Column (Two-Dimensional Modeling)," IEEE Trans. on Plasma Science, 24, 1996, pp. 1229-1238.

[18] G. Speckhofer and H.P. Schmidt, "Experimental and Theoretical Investigation of High-Pressure Arcs-Part II: the Magnetically Deflected Arc (Three-Dimensional Modeling), IEEE Trans. on Plasma Science, 24, 1996, pp. 1239-1248.

[19] P.G Jonsson, R.C.Westhoff, and J. Szekely, "Arc Characteristics in Gas-Metal Arc Welding of Aluminum Using Argon as the Shielding Gas," J. Appl. Phys. 74, 1993, pp. 5997-6006.

[20] P. Zhu, M. Rados, and S.W. Simpson, "A Theoretical Study of Gas Metal Arc Welding System," Plasma Sources Sci. Technol. 4, 1995, pp. 495-500.

[21] J. Haidar, and J. Lowke, "Predictions of Metal Droplet Formation in Arc Welding," J. Appl. Phys. D: Appl. Phys. 29, 1996, pp. 2951-2960.

[22] J. Haidar, "An Analysis of the Formation of Metal Droplets in Arc Welding," J. Phys. D: Appl. Phys. 31, 1998, pp. 1233-1244.

[23] J. Haidar, "Prediction of Metal Droplet Formation in Gas Metal Arc Welding II," J. Appl. Phys. 84 (7), 1998, pp. 3530-3540.
[24] J. Haidar, "An Analysis of Heat Transfer and Fume Production in Gas Metal Arc Welding. III.," J. Appl. Phys. 85 (7), 1998, pp. 3448-3459.
[25] F.L. Zhu, H.L. Tsai, S.P. Marin and P.C. Wang, "A Comprehensive Model on the Transport Phenomena during Gas Metal Arc Welding Process," Progress in Computational Fluid Dynamics, Vol. 4, No. 2, 2004, pp. 99-117.
[26] L. A. Jones, T.W. Eagar and J.H. Lang, "Images of Steel Electrode in Ar-2%O_2 Shielding during Constant Current Gas Metal Arc Welding," Welding Journal, Welding Research Supplement 1998, pp. 135-141s.
[27] L.A. Jones, T.W. Eagar and J.H. Lang, "Magnetic Forces Acting on Molten Drops in Gas Metal Arc Welding," J. Phys. D: Appl. Phys. 31, 1998, pp. 93-106.
[28] L.A. Jones, T.W. Eagar and J.H. Lang, "A Dynamic Model of Drops Detaching from a Gas Metal Arc Welding Electrode," J. Phys. D: Appl. Phys. 31, 1998, pp. 107-123.
[29] H.G. Fan, and R. Kovacevic, "Droplet Formation, Detachment, and Impingement on the Molten Pool in Gas Metal Arc Welding," Metall. Trans. 30B, 1999, pp. 791-801.
[30] Fan, H.G. and Kovacevic, R., 1998, "Dynamic Analysis of Globular Metal Transfer in Gas Metal Arc Welding - a Comparison of Numerical and Experimental Results," J. Phys. D: Appl. Phys. 31, pp. 2929-2941
[31] S.K. Choi,, C.D. Yoo, and Y-S Kim, "The Dynamic Analysis of Metal Transfer in Pulsed Current Gas Metal Arc Welding," J. Phys. D: Appl. Phys. 31, 1998, pp. 207-215.
[32] S.K. Choi, C.D. Yoo, and Y-S Kim, "Dynamic Simulation of Metal Transfer in GMAW, Part 1: Globular and Spray Transfer Modes," Welding J., 1998, pp. 38-44s.
[33] G. Wang, P.G. Huang, and Y.M. Zhang, "Numerical Analysis of Metal Transfer in Gas Metal Arc Welding," Metall. Trans. 34B, 2003, pp. 345-353.
[34] F. Wang, W.K Hou, S.J. Hu, E. Kannatey-Asibu, W.W. Schultz and P.C. Wang, "Modelling and Analysis of Metal Transfer in Gas Metal Arc Welding," J. Phys. D: Appl. Phys. 36, 2003, pp. 1143-1152.
[35] K.C. Tsao and C.S. Wu, "Fluid Flow and Heat Transfer in GMA Weld Pools", Weld. J., 1998, pp. 70s-75s.
[36] S. Kumar and S.C. Bhaduri, "Three-Dimensional Finite Element Modeling of Gas Metal-Arc Welding," Metall. Trans. 25B, 1994, pp. 435-441.
[37] S.-D. Kim and S.-J. Na, "A Study on the Effect of Constant Tube-to-Workpiece Distance on Weld Pool Shape in Gas Metal Arc Welding," Welding J. 74(5), 1995, pp. 141-152s.
[38] C.-H. Kim, W. Zhang and T. DebRoy, "Modeling of Temperature Field and Solidified Surface Profile during Gas-Metal Arc Fillet Welding," J. Appl. Phys. 94, 2003, pp. 2667-2679.
[39] S. Ohring and H.J. Lugt, "Numerical Simulation of a Time-Dependent 3-D GMA Weld Pool due to a Moving Arc" Welding J., 1999, pp. 416-424s.
[40] Z. Cao, Z. Yang and X.L. Chen., "Three-Dimensional Simulation of Transient GMA Weld Pool with Free Surface," Welding J., 2004, pp. 169-176s.
[41] Y. Wang and H.L. Tsai, "Impingement of Filler Droplets and Weld Pool Dynamics during Gas Metal Arc Welding Process," Int. J. Heat and Mass Transfer 44, 2001, pp. 2067-2080.
[42] Y. Wang, Q. Shi and H.L. Tsai, "Modeling of the Effects of Surface-Active Elements on Flow Patterns and Weld Penetration," Metall. Trans. 32B, 2001, pp. 145-161.
[43] Y. Wang and H.L. Tsai, "Effects of Surface Active Elements on Weld Pool Fluid Flow and Weld Penetration in Gas Metal Arc Welding," Metall. Trans. 32B, 2001, pp. 501-515.
[44] Diao, Q.Z. and Tsai, H.L., 1993, "Modeling of Solute Redistribution in the Mushy Zone during Solidification of Aluminum-Copper Alloys", Metall. Trans., **24A**, pp. 963-973.
[45] Torrey, M.D., Cloutman, L.D., Mjolsness, R.C., and Hirt, C.W., 1985, "NASA-VOF2D: A Computer Program for Incompressible Flows with Free Surfaces," LA-10612-MS, Los Alamos National Laboratory.
[46] J. Hu, "Heat and Mass Transfer in the Gas Metal Arc Welding Process," Ph.D. Dissertation, 2005, University of Missouri-Rolla, Rolla, Missouri, USA.
[47] Patanka, S.V., 1980, "Numerical Heat Transfer and Fluid Flow", New York: McGraw-Hill.

A Self-repairable MEMS Comb Accelerometer

Xingguo Xiong
Department of Electrical and Computer Engineering
University of Bridgeport, Bridgeport, CT 06604, USA
Email: xxiong@bridgeport.edu,
ylw@cse.cuhk.edu.hk*, wjone@ececs.uc.edu**

Yu-Liang Wu*, Wen-Ben Jone**
Department of Computer Science and Engineering*
The Chinese University of Hongkong, Shattin, Hong Kong
Department of ECECS, University of Cincinnati**
Cincinnati OH 45221, USA

***Abstract*—In this paper, a built-in self-repair technique for the MEMS comb accelerometer device is proposed. The main device of the comb accelerometer consists of *n* identical modules, and *m* modules are introduced as the redundancy. If any of the working module in the main device is found faulty during a built-in self-test (BIST), the control circuit will replace it with a good redundant module. In this way, the faulty device can be self-repaired through redundancy. The implementation of dual-mode BIST on the BISR module is discussed. The sensitivity loss due to device modularization can be well compensated by different design alternatives. The yield model for MEMS redundancy repair is developed. The simulation results show that the BISR (built-in self-repair) design leads to effective yield increase compared to non-BISR design, especially for a moderate non-BISR yield. The yield as well as the reliability of the accelerometer can be improved due to the redundancy repair.**

Keywords: Microelectromechanical System (MEMS), Built-in self-test (BIST), Built-in self-repair (BISR), comb accelerometer, yield analysis.

I. INTRODUCTION

As a newly developed discipline, Micro Electro Mechanical System (MEMS) has achieved exciting progress during last decades [1]. As more and more MEMS devices are used, *fault-tolerant* MEMS design is extremely crucial with the following three reasons. *First*, with increasing applications of MEMS to safety-critical fields, such as aerospace, automobile and medical applications, MEMS reliability is becoming a very important issue. Especially, many MEMS devices have movable parts and their repeated movements (vibrations, etc.) may lead to different kinds of structural material fatigues. Thus, even if a MEMS device is tested as fault-free, it still may fail after serving for a certain lifetime. Such a failure during in-field usage is a potential threat especially for safety-critical applications. *Second*, due to the involvement of multiple fields in MEMS design and fabrication, in contrast to the well-developed VLSI technology, MEMS fabrication is vulnerable to more defect sources. *Third*, there is increased tendency that MEMS is going to be integrated into system-on-chip (SoC) designs using a standard CMOS process [2]. That is, MEMS devices will be fabricated on the same chip with digital, analog, memory, and FPGA circuit technologies. Fault-tolerant design for traditional CMOS circuits have been proposed and many available techniques are existent [3][4]. However, fault-tolerant design for MEMS devices has never been studied. It will be uneconomical to get rid of the entire SoC chip, if there exist minor MEMS defects. Thus, it is emergent to find a solution to have a defective MEMS device fix itself, whenever the test process (in-field or manufacturing) finds defects existent. By implanting the built-in self-repair (BISR) feature into MEMS devices, the reliability as well as yield rate can be greatly improved.

Efforts on trimming the device geometry parameters with certain physical or chemical processes have been reported [5][6]. These can be treated as somewhat the "*hard repair*" of MEMS devices. However, "hard repair" requires an extremely precise control over the process to avoid any over-trimming or over-milling. Further, the repairing process may have to be performed for each individual device separately with different adjustment, because each defective device may have its own geometry deviation. This leads to extremely high cost for the device repairing process through trimming, milling or etching, so it is not suitable for batch fabrication processes. Moreover, the defects that can be repaired are very limited, because only defects involving deviation in geometry parameters can be dealt with.

In this paper, a self-repairable MEMS comb accelerometer based on modular redundancy is proposed. Due to the modularized design, a faulty module can be virtually separated from the main device and replaced with another redundant module, though the defective module is still physically connected to the device. Thus, a defective MEMS device can repair itself given the number of faulty modules is smaller than that of redundant ones. The entire BISR MEMS device is partitioned into $m+n$ identical modules. Among them, n modules serve as the main device, while the other m modules serve as redundant modules. As the prerequisite of BISR, the dual-mode built-in self-test (BIST) of each BISR module is also introduced. The sensitivity loss due to modularized design can be compensated by revising the design parameters such as shrinking the beam width, enlarge the mass width, etc. If the design parameters are fixed, the sensitivity can still be effectively compensated by electrostatic force. The yield model for MEMS redundancy repair is developed. Simulation results demonstrate effective yield increase due to the redundancy repair of the MEMS comb accelerometer.

K. Elleithy et al. (eds.), Advances in Computer, Information, and Systems Sciences, and Engineering, 75–82.

An optimized design of the MEMS BISR accelerometer is suggested and a comparison with the non-BISR device is proposed.

II. NON-BISR MEMS COMB ACCELEROMETER

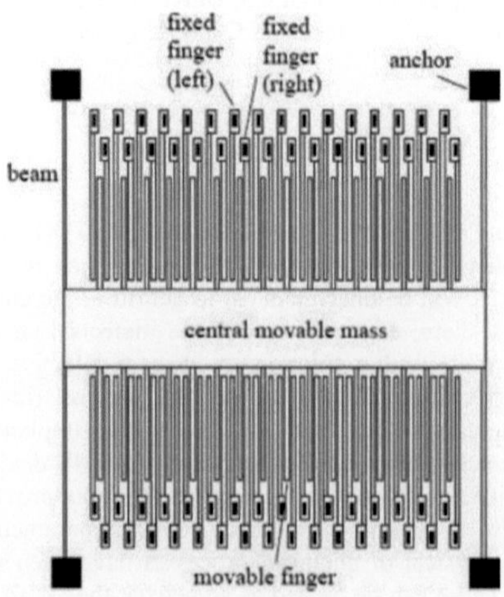

Fig. 1. The general design of MEMS comb accelerometer

A typical surface-micromachined comb accelerometer [7] is shown in Figure 1. The central movable mass is connected to the four anchors through four flexible beams. The movable fingers extrude from both sides of the central mass, and can move together with it. There is a pair of fixed fingers around the left and right sides of each movable finger. Each movable finger and its left and right fixed fingers constitute a differential capacitance pair c_1 and c_2 separately as shown in Fig. 2. In the static state, each movable finger stays in the middle position between the left and right fixed fingers, and the capacitance gaps of both c_1 and c_2 are equal to d_0. Assume there are n_f pairs of finger groups in the MEMS device, and let $C_1(C_2)$ represent the sum of all $c_1(c_2)$ capacitances. We have

$$C_1 = C_2 = \frac{n_f \varepsilon_0 (L_f - \Delta)h}{d_0}$$

where n_f is the total number of differential capacitance groups, ε_0 is the dielectric constant of air, L_f is the length of each movable finger, Δ is the non-overlapped length at the root of each movable finger, and h is the thickness of the device.

Assume the mass of both the central movable mass and all the movable fingers as M. If there is an acceleration a in perpendicular to the beams while in parallel to the device plane, the central mass will experience an inertial force $-M{\cdot}a$. This will result in a certain amount of beam deflection along the direction of the inertial force, hence the equivalent amount of displacement of the central mass and the movable fingers. Thus, each capacitance gap will be changed accordingly which leads to the change of corresponding capacitances (Fig. 2).

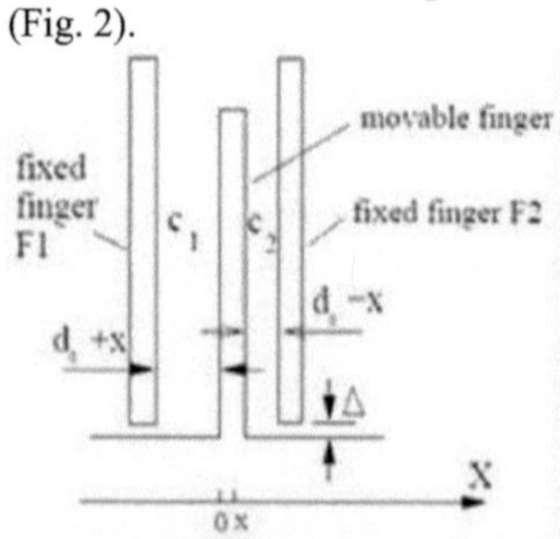

Fig. 2. The schematic diagram of differential capacitance.

As shown in Figure 2, the inertial force results in a deflection of the beams and a certain displacement x of movable fingers along the X direction. Given $x<<d_0$, we have C_1 and C_2 changed to [1]

$$C_1 = \frac{n_f \varepsilon_0 (L_f - \Delta)h}{(d_0 + x)} \approx \frac{n_f \varepsilon_0 (L_f - \Delta)h}{d_0}\left(1 - \frac{x}{d_0}\right),$$

$$C_2 = \frac{n_f \varepsilon_0 (L_f - \Delta)h}{(d_0 - x)} \approx \frac{n_f \varepsilon_0 (L_f - \Delta)h}{d_0}\left(1 + \frac{x}{d_0}\right).$$

In order to sense the displacement x of the movable plate M due to the inertial force, modulation voltages V_{mp} and V_{mn} are applied to left and right fixed fingers (F1 and F2) separately

$$V_{F1} = V_{mp} = V_0 sqr(\omega t),$$

$$V_{F2} = V_{mn} = -V_0 sqr(\omega t).$$

where V_0 represents the modulation voltage amplitude, ω denotes the frequency of the modulation voltage, and t gives the time for operation. According to the charge conservation law, the charge in capacitances C_1 and C_2 must be equal, so we have

$$C_1(V_{F1} - V_M) = C_2(V_M - V_{F2}).$$

where V_M is the voltage level sensed by the movable plate M. Solving the above equations, we have

$$V_M = (x / d_0)V_0 sqr(\omega t).$$

It can be observed from this result that under the above modulation voltage biasing, the central movable plate M acts just as a voltage divider between the top and bottom fixed plates $F1$ and $F2$ respectively. By measuring the voltage level on central movable electrode V_M, we can find the displacement x of the central movable plate M, which in turn is directly proportional to the experienced acceleration. Thus, we can derive the value of the applied acceleration along the sensitive direction (X direction). This is the working principle for most differential capacitive MEMS devices.

III. BISR MEMS COMB ACCELEROMETER DESIGN

Due to the tiny size (in the range of microns) of MEMS accelerometers, its overall capacitance is generally below

1pF, and the capacitance change in working mode is in range of fF. In order for the tiny capacitance to be detected by the signal sensing circuit, it is desirable to enlarge the device capacitance. Thus, a MEMS comb accelerometer device generally contains large number of repeated comb finger groups in a very compact manner. For example, an ADXL50 accelerometer contains 42 differential comb finger groups [7], while an ADXL150 accelerometer contains 54 differential comb finger groups [8]. A bulk-micromachined comb accelerometer in [9] contains 80 differential comb fingers groups. Such a highly dense comb structure with many long and narrow capacitance gaps is extremely vulnerable to various defects such as particle contamination, stiction [10] Taking the ADXL50 accelerometer as an example, the length of each movable finger is 120μm, while the capacitance gap between each pair of fixed and movable fingers is only 1.3μm. If a conductive particle with diameter larger than 1.3μm falls into any of the 84 capacitance gaps, it will lead to a short-circuit of the device capacitance and result in a failure of the entire device. Thus, a large number of finger groups unavoidably lead to the decrease in yield as well as reliability.

Fig. 3. Modularized comb accelerometer structure

Yield and reliability have been two major concerns for MEMS commercialization. In order to enhance the yield and reliability of MEMS comb accelerometer, a self-repairable MEMS comb accelerometer design is proposed, as shown in Figure 3. Here, the device consists of six identical modules, and each module has its own beams, mass and finger structures (fixed and movable). By assumption, four modules are connected together as the main device, while the remaining two modules serve as redundancy. The movable parts of each module are physically connected to those of adjacent modules through the common anchors, and signals sensed by all movable fingers in the device are connected to the sensing circuit directly. However, the fixed fingers of each module are connected to the modulation signal circuit through switches made of analog MUXes. By turning on or off these switches, we can determine whether a module works as part of the main device or the redundant device. For example, if the modulation signals of module M_c are turned off, then the movable fingers of M_c cannot sense any signal. Thus, M_c is electronically disconnected from the MEMS device (though it is still connected to the entire MEMS structure physically), and is not involved in the MEMS function.

During the BISR mode, the device will first perform the BIST process for each individual module. The dual-mode BIST technology presented in [11] will be used here to ensure a thorough test. The BIST result of each module is compared with the pre-stored good module response to judge whether the module is good or faulty. This information is fed to the BISR control circuit as input for self repairing. If a module is tested as faulty, the control circuit will permanently exclude the module from the main device and replace it with a good redundant module (if there is any). Thus, after repairing, the main device can still be ensured to work properly.

In the following discussion, we call the MEMS comb accelerometer with (without) the BISR feature as a BISR (non-BISR) accelerometer. In order for a fair comparison, we assume that the total number of finger groups of the non-BISR accelerometer should equal to that of the main device in the BISR accelerometer. In this way, the area overhead of the modularized BISR accelerometer includes the area for two redundant modules, plus the area for extra beams and gaps between modules in the BISR accelerometer, as well as the area for control circuits.

IV. BUILT-IN SELF-TEST OF THE ACCELEROMETER

In order to implement the built-in self-repair of the accelerometer, each module need to perform a built-in self-test (BIST) on itself. The BIST results will be fed to the control circuit. The control circuit will construct main device from available good modules. If any module in the main device is tested to be faulty during in-field usage, the BISR control circuit will separate it out from the main device and replace it with a good redundancy. The BIST acts as the prerequisite for the BISR technique. A dual-mode BIST method [11] will be used here for the self-repairable accelerometer. The dual-mode BIST technique partitions the fixed instead of the movable capacitance plates, so that both sensitivity and symmetry BIST can be implemented on each module of the MEMS comb accelerometer. Since each of the sensitivity and symmetry BIST has its own fault coverage, a combination of both modes ensures a better fault coverage. The capacitance partition of a BISR module for dual-mode BIST implementation is shown in Figure 4. In order for simplification, only eight finger groups are shown in the module. In the figure, M1~M8 are movable fingers, M_s is the sensing mass, D1~D8 are driving fingers, and S1~S8 are sensing fingers.

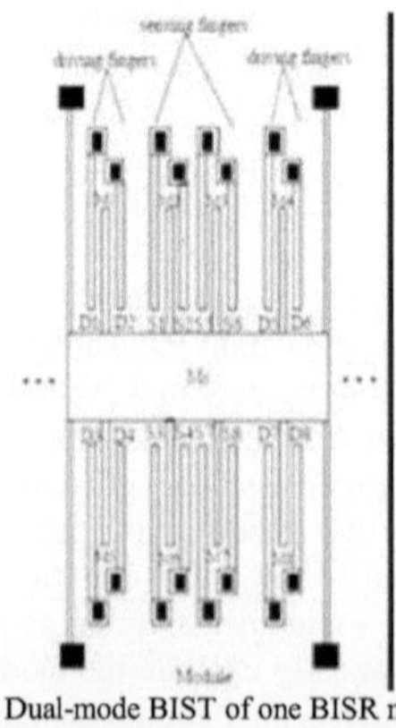

Fig. 4. Dual-mode BIST of one BISR module.

During normal operation, TE=0, modulation voltage V_{mp} is applied to {S1, S3, S5, S7, D1, D3, D5, D7}, and V_{mn} is applied to {S2, S4, S6, S8, D2, D4, D6, D8}. The voltage level in the movable fingers V_{Ms} is measured as the output voltage to determine the acceleration. When TE=1 and TS=0, the device works in the sensitivity test mode. A certain test driving voltage V_d is applied to {D1, D3, D5, D7} to activate the device with electrostatic force. The modulation voltage V_{mp} is applied to {S1, S3, S5, S7}, while V_{mn} is applied to {S2, S4, S6, S8}. The output voltage on movable mass Ms is measured for the device sensitivity. This value is compared with the expected good device value within a certain tolerance level to find whether the device is faulty. When TE=1 and TS=1, the device is in the symmetry test mode. Test driving voltage V_d is applied to {D1, D3, D5, D7}, modulation voltage V_{mp} is applied to {S1, S5}, while V_{mn} is applied to {S3, S7}. The sensing circuit checks whether the output voltage on movable fingers is a constant zero to detect any asymmetry caused by local defects. If there is a non-zero voltage on the movable electrode Ms, then it indicates there are local defects which alter the symmetry of the device. The voltage biasing scheme for the comb accelerometer in the normal and both BIST modes is shown in Table I. Since each of the sensitivity BIST and symmetry BIST has its own fault coverage, a combination of both mode ensures better fault coverage.

TABLE I. VOLTAGE BIASING FOR COMB ACCELEROMETER.

Voltage biasing	Normal operation	Sensitivity BIST	Symmetry BIST
V_d	-	D1,D3,D5,D7	D1,D3,D5,D7
V_{nom}	-	D2,D4,D6,D8 M1,M4,M5,M8	D2,D4,D6,D8 M1,M4,M5,M8 S2,S4,S6,S8
V_{mp}	S1,S3,S5,S7 D1,D3,D5,D7	S1,S3,S5,S7	S1,S5
V_{mn}	S2,S4,S6,S8 D2,D4,D6,D8	S2,S4,S6,S8	S3,S7

To implement the BIST technique, a control circuit is needed to switch the device among the normal operation mode and both BIST modes. Such a control circuit is not complex and only contains some switches made of analog Muxes. The control circuit design for the BISR MEMS accelerometer is shown in Figure 5. In the control circuit, totally only six Muxes are needed: three 3-to-1 Muxes and three 2-to-1 Muxes. The differential capacitance detection circuit for BIST modes can be shared with that of the normal operation mode. Thus, the circuit overhead for the BIST technique implementation is small.

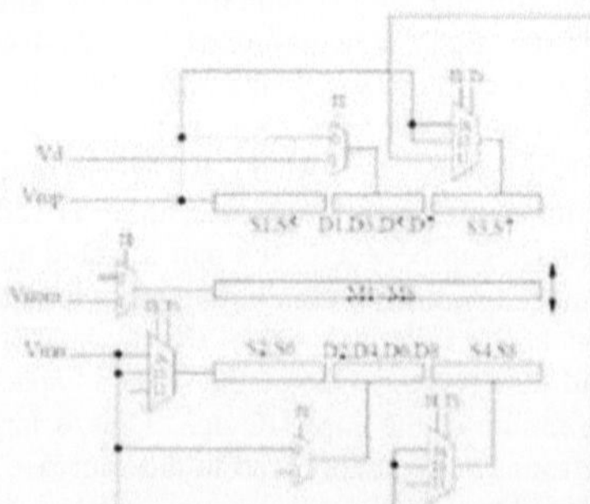

Fig. 5. The control circuit for dual mode BIST of one BISR module.

V. PERFORMANCE ANALYSIS

A. Sensitivity Analysis

A MEMS comb accelerometer actually can be simplified by a spring-mass model. The width and length of each tether (seismic mass) are represented by $W_b(W_m)$ and $L_b(L_m)$ separately, while the width and length of each movable finger are denoted by W_f and L_f respectively. There are totally N_f number of movable fingers, and the thickness of the device is h. Assume the density and the Young's modulus of poly-Si as ρ and E respectively. The sensing mass M_s of the accelerometer, which includes the seismic mass and all the movable fingers attached to it, can be expressed as follows:

$$M_s = \rho h(W_m L_m + N_f W_f L_f)$$

The total spring constant k_{tot} of all four beams is:

$$k_{tot} = 4 \cdot k_s = \frac{4EhW_b^3}{L_b^3}$$

The displacement sensitivity S_d of the device, which is defined as the displacement of movable fingers per unit gravity acceleration (g) along the sensitive direction, can be expressed as:

$$S_d = \frac{M_s g}{k_{tot}} = \frac{\rho g(W_m L_m + N_f W_f L_f)L_b^3}{4EW_b^3}$$

Assume all other parameters are fixed, the relationships between the device displacement sensitivity S_d and the beam width W_b, central mass width W_m are shown in Figure 6.

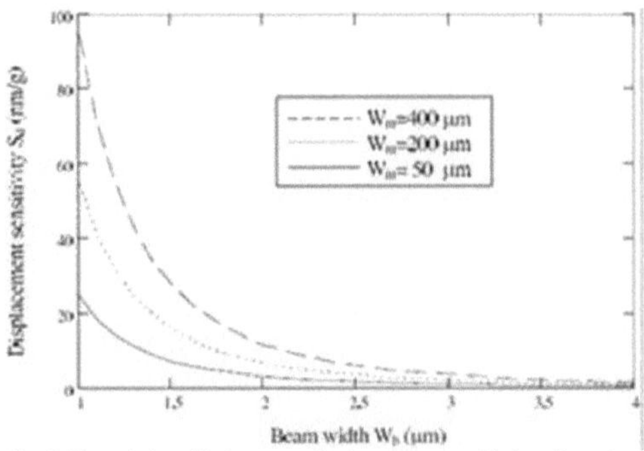

Fig. 6. The relationship between displacement sensitivity of accelerometer and beam/mass width.

We denote the displacement sensitivity of the BISR (non-BISR) accelerometer by S_{db} (S_{dnb}). In the BISR device, we partition all N_f comb finger groups into n modules in the main device, and add m number of modules as redundancy. That is, each BISR module contains $N_f = n$ comb finger groups. Correspondingly, the sensing mass of each BISR module also falls to $1/n$ of the non-BISR version. If we keep the same geometry parameters (i.e., width W_b, length L_b and thickness h) for the beams of both BISR and non-BISR designs, we have:

$$S_{db} \approx \frac{(M_s / n)g}{k_{tot}} = \frac{1}{n} S_{dnb}$$

where M_s and k_{tot} are the sense mass and spring constant of the non-BISR accelerometer.

For the BISR accelerometer design, the sensitivity loss due to the modularized design can be recovered back by various methods, such as reducing the beam width of each module, using folded beams instead of straight beams, or enlarging the width Wm of the central mass, etc. Among them, shrinking the beam width W_b is the most efficient way for sensitivity compensation without changing the device area. Assume the beam width of the non-BISR accelerometer as W_{b0}, for the case of $n=4$, if the beam width of the BISR accelerometer is reduced to $0.63W_{b0}$, the sensitivity of the BISR accelerometer will be fully compensated back to the non-BISR value. The designer can also compensate the sensitivity loss of the BISR accelerometer by the combination of the above methods according to the specific design requirements.

Even if all the device parameters (such as beam width and mass width, etc) are fixed, the sensitivity of an accelerometer in the open-loop mode can still be conveniently enhanced by electrostatic force with an appropriate DC biasing voltage [12]. The electrostatic force acts as a spring with a negative spring constant. This will help reduce the effective spring constant of the accelerometer and increase its sensitivity. Since this electrostatic force actually deflects the mass, it is equivalent to the effect of an inertial force acting on the mass. Thus, the electrostatic force is a powerful tool to enhance the sensitivity. A small DC biasing voltage (several volts) can increase the sensitivity to infinity, and this offers a great flexibility in the sensitivity recovery. Thus, if electrostatic force is used for sensitivity compensation, the beam width and mass width can be kept the same as those of the non-BISR device.

B. Frequency Analysis

Using the simplified spring-mass model discussed above for the comb accelerometer, we have the resonant frequency f_{nsr} of the non-BISR design given by the following equation [7]:

$$f_{nsr} = \frac{1}{2\pi}\sqrt{\frac{k_{tot}}{M_s}}$$

where M_s and k_{tot} are defined in the last section.

Assume the dimension of each beam in the BISR design remains the same as that of the non-BISR design, but the comb finger groups are divided into n identical modules. The resonant frequency f_{sr} of the BISR accelerometer can thus be given by

$$f_{sr} = \frac{1}{2\pi}\sqrt{\frac{k_{tot}}{(1/n)M_s}} = \sqrt{n} f_{nsr}$$

Further, the relationship between resonant frequency and displacement sensitivity can be given by:

$$f_{nsr} = \frac{\sqrt{g}}{2\pi\sqrt{S_d}}$$

Consequently, if the displacement sensitivity of the BISR accelerometer is compensated to the same as that of the non-BISR design (e.g., by adjusting the k_{tot} value), the resonant frequency will also remain the same as that of the non-BISR device.

VI. YIELD ANALYSIS OF BISR COMB ACCELEROMETER

A. Yield Model for MEMS Redundancy Repair

Yield model for VLSI redundancy repair has been well developed [13][14]. However, yield model for MEMS redundancy repair has not been available. MEMS has its unique properties which are different from VLSI. Thus a yield model for MEMS redundancy repair must be developed. Assume a set of defects can occur to N number of locations in a MEMS device, i.e., there are N possible defects in the MEMS device. Further, assume every defect occurs independently of each other, and the probability for each defect to occur is equal and defined as q. Thus, based on the defect distribution discussed in [13], the probability $P(X = x)$ that x number of indistinguishable randomly distributed defects occurring to the MEMS device can be expressed as a Poisson distribution:

$$P(X = x) = \frac{\lambda^x}{x!} e^{-\lambda}$$

The simple Poisson distribution is too pessimistic for yield estimation because the defect clustering effect is not considered. Hence, the compound Poisson distribution is more popular by considering the normalized distribution of a chip defect density clustering factor. The next problem is how the average defect λ distributes. In this work, we assume that the defect distribution function F(λ) for λ is a gamma function given by the following equation [13]:

$$F(\lambda)=\frac{1}{\Gamma(k)(Ab)^k}(\lambda)^{k-1}e^{-\lambda/(Ab)}$$

where A is the device area, b is a defect density coefficient, λ is the average defect, and *k* is the clustering parameter. Further, the average defect density D_0 is given by:

$$D_0=b\cdot k$$

The probability that x defects occur in a MEMS device with area A can be given by the following equation [13]:

$$P(x,A)=\int_0^\infty \frac{\lambda^x}{x!}e^{-\lambda}F(\lambda)d\lambda$$

$$=\frac{1}{x!\Gamma(k)(Ab)^k}\cdot\int_0^\infty(\lambda)^{x+k-1}e^{-(1+1/Ab)\lambda}d\lambda$$

$$=\frac{\Gamma(x+k)(Ab)^x}{x!\Gamma(k)(1+Ab)^{x+k}}$$

$$=\binom{x+k-1}{x}\left(\frac{1}{1+Ab}\right)^k\left(\frac{Ab}{1+Ab}\right)^x$$

For a non-BISR MEMS accelerometer, the yield Y_0 is the probability that no defect occurs (i.e., X = 0), which can be expressed as:

$$Y_0=P(0,A_0)=\left(\frac{1}{1+A_0b}\right)^k$$

where A_0 is the area of the non-BISR accelerometer. From this equation, we see that the yield drops as the device area increase. This is reasonable because the larger the device area is, the more likely it may suffer from some defects.

Assume there are totally *i* possible defects in the BISR MEMS accelerometer, *n* number of modules in the main device, and m number of redundant modules as shown in Figure 7. The yield of the BISR MEMS accelerometer after redundancy repairing, denoted as Y_r, equals the probability that none of the *i* defects occurs, plus the probability that some among the *i* defects do occur but they all fall into no more than m number of modules. The former is the case where all the *(n+m)* number of modules are healthy, while the latter is the case in which some modules are faulty, but the device can still be self-repaired into a good one through redundancy. Each of both cases will be investigated in the following discussions.

Assume the main device of the BISR accelerometer has the same number of finger groups as the corresponding non-BISR design. Let A_0 denotes the area of the non-BISR accelerometer, A_b denotes the area of beams in each module of the BISR accelerometer (and also the non-BISR modules/

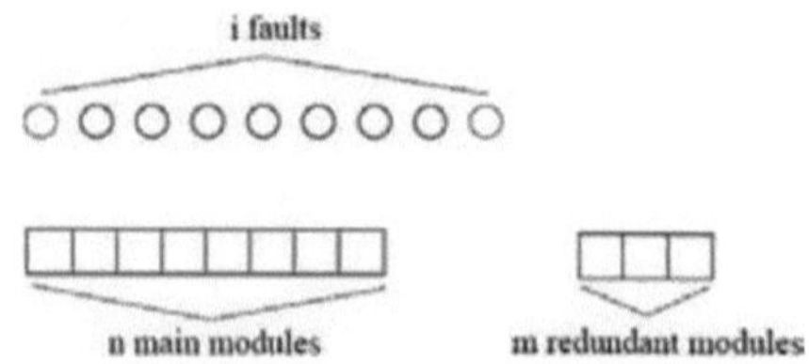

Fig. 7. Fault distribution among the modules of BISR accelerometer.

devices set by the design rules. In our simulation, we select $A_b=0.12A_0$. The area A_r of the BISR accelerometer is given by:

$$A_r=\frac{(n+m)(A_0-A_b)}{n}+(n+m)A_b$$

$$=\frac{(n+m)[A_0+(n-1)A_b]}{n}$$

The probability that none of the i number of defects occurs can be expressed as:

$$P(0,A_r)=\left(\frac{1}{1+A_r\cdot b}\right)^k$$

For the second case in which some defects do occur but the device can still be self-repaired through redundancy, it can be further divided into the following two sub-cases:

- The total number *(i)* of defects is smaller than or equal to the number *(m)* of redundant modules, i. e., $i\leq m$. For this sub-case, regardless of whatever distribution for the defects, the BISR comb accelerometer can always be repaired into a good device.
- The total number *(i)* of defects is larger than the number of the redundant modules m; however, all the defects fall into m number of modules or less. For this sub-case, the BISR accelerometer can still be repaired into a good device.

Both the above two sub-cases must be counted into the yield of the BISR accelerometer. The first sub-case can be easily solved, while the second sub-case requires an extensive analysis. For the first sub-case, the probability P1 that i number ($i\leq m$) of defects occur in the BISR MEMS accelerometer can be expressed as:

$$P_1=\sum_{x=1}^{x=m}P(x,A_r)$$

For the second sub-case *(i>m)*, the faulty device can be repaired into a good device only if all the *i* number of defects fall into m number of modules or less. First, we examine the probability that *i* defects are distributed into *j* *(j≤m)* MEMS modules and each of the *j* modules contains at least one defect. So, we distribute one defect to each of the *j* modules to ensure that each of the *j* modules contains at least one defect. Thus, there are *i-j* defects remaining to be distributed to the j modules with any number of defects (maybe 0) for this distribution. As we know, there are $\binom{n_0+r-1}{r}$ ways to distribute *r* identical balls into n_0 distinct cells with any number of balls per cell. Here, we have $r=i-j$ and $n_0=j$ and the total number of ways of distribution is $\binom{i-1}{j-1}$. Finally, the probability that the BISR accelerometer (with n modules in the main device, and m modules in the redundancy device) can be repaired when a certain i number ($i>m$, sub-case 2) of defects occur is:

$$R(m,n,i) = \frac{\sum_{j=1}^{m} \binom{m+n}{j} \cdot \binom{i-1}{i-j}}{\sum_{j=1}^{\min(i,m+n)} \binom{m+n}{j} \cdot \binom{i-1}{i-j}}$$

Thus, the probability P2 that more than m number of defects occur to a MEMS device, but it still can be self-repaired into a good device can be expressed as follows.

$$P_2 = \sum_{x=m+1}^{\infty} P(x, A_r) \cdot R(m,n,x)$$

$$= \sum_{x=m+1}^{\infty} P(x, A_r) \cdot \frac{\sum_{j=1}^{m} \binom{m+n}{j} \cdot \binom{x-1}{x-j}}{\sum_{j=1}^{\min(x,m+n)} \binom{m+n}{j} \cdot \binom{x-1}{x-j}}$$

Hence, the yield for the BISR accelerometer after redundancy repair is:

$$Y_r(m,n) = P(0, A_r) + P_1 + P_2$$

$$= \sum_{x=0}^{x=m} P(x, A_r) + \sum_{x=m+1}^{\infty} P(x, A_r) \cdot \frac{\sum_{j=1}^{m} \binom{m+n}{j} \cdot \binom{x-1}{x-j}}{\sum_{j=1}^{\min(x,m+n)} \binom{m+n}{j} \cdot \binom{x-1}{x-j}}$$

The yield increase $IY(m,n)$ of the BISR accelerometer, when compared with the corresponding non-BISR accelerometer, can be given by:

$$IY(m,n) = Y_r(m,n) - Y_0$$

$$= \sum_{x=0}^{x=m} P(x, A_r) + \sum_{x=m+1}^{\infty} P(x, A_r) \cdot \frac{\sum_{j=1}^{m} \binom{m+n}{j} \cdot \binom{x-1}{x-j}}{\sum_{j=1}^{\min(x,m+n)} \binom{m+n}{j} \cdot \binom{x-1}{x-j}} - P(0, A_0)$$

We can see that under this theory, if we set $m=0$ and $n=1$, then $Y_r(0,1)$ comes back to $P(0,A_0)$ which is exactly the yield of the non-BISR device. Thus, the general case of $Y_r(m,n)$ includes the yield of the non-BISR device as a special case.

Based upon the MEMS yield model for redundancy repair, we can derive the relationship between the yield increase and the non-BISR device yield for different m and n numbers. Figure 8 shows the simulation result for $n = 4$ and $m=2,4,6,8$ separately. From this figure, we can observe that the BISR device always gives a positive yield increase regardless of the non-BISR yield. This demonstrates the effectiveness of the redundancy repair technique for MEMS devices. If the non-BISR yield is too low (approaches *0*) or too high (approaches *1*), the yield increase by redundancy repair is not significant. This is a reasonable result. For a very low non-BISR yield (approaches *0*), the defect density is extremely high and there are too many faulty modules in the main device. Compared with so many defective modules in the main device, the redundancy is relatively deficient to repair all of them. Thus, the yield increase by redundancy BISR is not significant. For a very high non-BISR yield (approaches *1*), the main device itself is highly likely to be fault-free, and hence repair is not necessary.

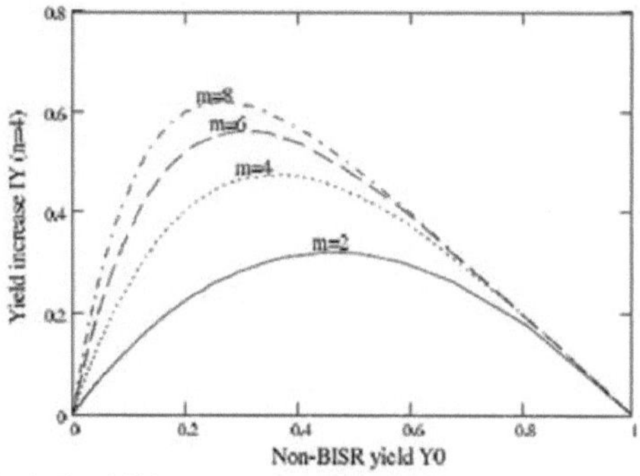

Fig. 8. The yield increase vs. non-BISR yield for different m numbers

In order to verify the effectiveness of redundancy repair for moderate initial yield, we randomly select $A_0 = 0.24mm^2$, $b=1.8/mm^2$, and $k=1$, the yield of the non-BISR accelerometer is 0.698. For the BISR accelerometer with $m=2$ and $n=4$, the yield becomes 0.947 which is an increase of *35.7%* (when compared with the non-BISR yield *0.698*). This demonstrates an effective improvement on the yield for a comb accelerometer through redundancy repair.

VII. DESIGN AND SIMULATION OF BISR ACCELEROMETER

The geometry parameters of the BISR comb accelerometer with $m=2$ and $n=4$ are listed in Table II, using a set of design rules comparable to ADXL accelerometers [7]. For comparison, a none-BISR accelerometer with the same number of capacitance groups (as that at the main device of the BISR accelerometer) is also designed. The geometry parameters of the non-BISR accelerometer are also listed in the same table. The simulation results for the performance of both BISR and none-BISR accelerometers are shown in Table III. From Table III, we can see that, by narrowing the beam width from *3μm* (non-BISR beam width) to *2μm* (BISR beam width), the sensitivity loss of the BISR accelerometer due to device modularization can be fully compensated. The BISR device demonstrates a displacement sensitivity of *9.3nm/g*, which is even larger than that of the non-BISR device (*8.1nm/g*). This shows that the sensitivity loss can be effectively compensated by shrinking the beam width of the comb accelerometer.

TABLE II Design of BISR/non-BISR accelerometers

Design Parameters	BISR device	Non-BISR device
device area (μm^2)	1500×900	980×900
thickness t (μm)	6	6
no. of capacitance groups	20×6	80
capacitance gap d_0 (μm)	2	2
beam width W_b (μm)	2	3
beam length $L_b(\mu m)$	300	300
mass width $W_m(\mu m)$	300	200
mass length $L_m(\mu m)$	220×6	880
finger width W_f (μm)	4	4
finger length L_f (μm)	200	200

TABLE III Simulation results of the accelerometers

Performance	BISR main device	non-BISR device
sensing mass M_s (μg)	1.15×4=4.6	3.4
capacitance C_0 (pF)	0.1×4=0.4	0.4
sensitivity S_d (nm/g)	9.3	8.1
sensitivity $S_c(fF/g)$	0.96×4=3.84	3.3
spring constant k_m (N/m)	1.21×4=4.84	4.08
frequency f_0 (kHz)	5.17	5.55

VIII. CONCLUSIONS

In this paper, a self-repairable MEMS comb accelerometer device design is proposed. The strategy of the BISR device is to partition the large number of comb fingers into several modules. Thus each module contains less number of comb fingers, the risk of being faulty is reduced. Further, redundant modules are introduced. If any of the working module is found to be faulty during the BIST, it will be replaced with a good redundant module. In this way, the device can be self-repaired into good one. The modularized design leads to a sensitivity loss. However, it can be compensated by adjusting the design parameters, such as shrinking the beam width, enlarge the mass width, etc. Even if all the design parameters are already fixed, a powerful electrostatic force compensation method can still be used to fully recover the device sensitivity without changing the device parameters. The built-in self-test of each BISR module is discussed. In order to quantitatively evaluate the effectiveness of the BISR technique on the yield increase, a yield model for MEMS redundancy repair is developed. The simulation results based on the yield model demonstrate an effective yield increase due to the redundancy repair. The yield increase is especially effective for moderate initial yield. However, if the initial yield is too large or too small, the yield increase due to redundancy repair is not significant. For demonstration purpose, for a moderate initial yield of 0.626, the yield after repair is 92.5%. A yield increase of 33.5% can be achieved due to redundancy repair. Compared to the "hard" repair with physical or chemical process, the redundancy repair has the advantage of low cost, suitable for in-field usage, etc. Furthermore, it is virtually applicable to any local defect within one module. This is because if any of the modules is faulty due to local defect, it will be separated out from the main device and replaced with a good redundant module. However, the proposed BISR technique cannot be applied to global defects. For global defects, all the modules are faulty and no good module is available. Thus, there will be no good redundant module available for repair.

Due to the fabrication variation, the performance of the MEMS accelerometer modules cannot be perfectly the same. During the BISR, when the faulty module is replaced with good redundancy, the tiny performance difference between the switched modules must be compensated. Our future research is to develop a built-in self-calibration (BISC) technique to ensure a smooth switch between the defective and healthy modules, so that the device performance will maintain the same after redundancy repair.

ACKNOWLEDGEMENT

The authors are thankful to Dr. Tarek Sobh, Dr. Lawrence Hmurcik and Dr. Stephen Grodzinsky in University of Bridgeport for their valuable discussions in the paper.

REFERENCES

[1] G. T. A. Kovacs, *Micromachined Transducers Sourcebook*, WCB/McGraw-Hill, New York, 1998.
[2] URL: http://public.itrs.net.
[3] K. Chakraborty and P. Mazumder, *Fault-Tolerance and Reliability Techniques for High-Density RAMs*, Prentice Hall, Upper Saddle River, NJ 07458, 2002.
[4] B. Johnson, *Design and Analysis of Fault Tolerant Digital Systems*, Addison Wesley, 1989.
[5] J. Ramirez-Angulo, and R. L. Geiger, "New Laser-trimmed Film Resistor Structures for Very High Stability Requirements", *IEEE Transactions on Electron Devices*, Vol. 35, Issue 4, pp. 516-518, Apr. 1988.
[6] K. Tanaka, Y. Mochida, M. Sugimoto, K. Moriya, T. Hasegawa, K. Atsuchi, and K. Ohwada, "A Micromachined Vibrating Gyroscope", *Sensors and Actuators A: Physical*, Vol. 50, pp. 111-115, 1995.
[7] W. Kuehnel, and S. Sherman, "A Surface Micromachined Silicon Accelerometer with On-chip Detection Circuitry," *Sensors and Actuators A: Physical*, Vol. 45, pp. 7-16, 1994.
[8] "ADXL150/ADXL250: _5g to _50g, Low Noise, Low Power,Single/Dual Axis iMEMS Accelerometers", *ADXL150/ADXL250 datasheet*, URL: http://www.analog.com.
[9] X. Xiong, D. Lu, and W. Wang, "A Bulk-micromachined Comb Accelerometer with Floating Interconnects", *Proceedings of 48th IEEE InternationalMidwest Symposium on Circuits and Systems (MWSCAS'05)*, Cincinnati, Ohio, Aug 7-10, 2005, accepted to be published.
[10] N. Deb, and R. D. Blanton, "Analysis of Failure Sources in Surfacemicromachined MEMS", *Proc. of International Test Conference*, pp. 739-749, Oct. 2000.
[11] X. Xiong, Y. Wu, W. Jone, "A Dual-Mode Built-In Self-Test Technique for Capacitive MEMS Devices", *Proc. of the 22nd IEEE VLSI Test Symposium (VTS'04)*, pp. 148-153, May 25-29, 2004.
[12] B. Li, D. Lu, and W. Wang, "Open-loop Operating Mode of Micromachined Capacitive Accelerometer", *Sensors and Actuators A: Physical*, Vol. 79, pp. 219-223, 2000.
[13] J. Hirase, "Yield Increase of VLSI after Redundancy-repairing," *Proc. of 10th Asian Test Symposium*, pp.353-358, Nov. 2001.
[14] T. L. Michalka, R. C. Varshney, and J. D. Meindl, "A Discussion of Yield Modeling with Defect Clustering, Circuit Repair, and Circuit Redundancy", *IEEE Transactions on Semiconductor Manufacturing*, Vol. 3, Issue 3, pp.116-127, Aug. 1990.

Preservation of stability and passivity in irrational transfer functions

Guillermo Fernández-Anaya, José Álvarez-Ramírez., and José-Job Flores-Godoy, *Member, IEEE*

***Abstract*—In this work it is presented a generalization on the properties preserved on $\mathcal{RH}_\infty$ functions to irrational $\mathcal{H}_\infty$ functions. Particularly, we are interested in passive systems with distributed parameters and delay linear systems. Using substitutions of the variable s by passive functions we show that stability, stabilizability, strong coprime factorizations and passivity are preserved.**

I. Introduction

As it is pointed out in [1], the concept of positive realness of a transfer function plays a central role in Stability Theory. The definition of rational Positive Real (PR) functions arose in the context of Circuit Theory and Network Synthesis [1], [2], [3]. This concept is one of the oldest in systems, circuits and control theory. In fact, the driving point impedance or admittance of a passive network is rational and positive real. On the other hand, if the network is dissipative (due to the presence of resistors), the driving point impedance of the network is a Strictly Positive Real (SPR) transfer function. Conceptually, we say that SPR systems are systems that dissipate energy. Similarly, PR systems, also called passive systems, are systems that do not generate energy. These systems are relevant to the analysis of absolute stability of nonlinear Lur'e systems, [3], [4], [5]; and to the design of adaptive schemes where there is a well known fact that for some model reference adaptive control, a sufficient condition for convergence of several recursive algorithms of adaptation is the SPR property of a suitable family of transfer functions [6], [7], [8].

In previous papers, [9], [10], [11], [12], results for finite dimension linear time-invariant systems were presented. The common thread among them is the used of substitutions of strictly positive real functions with zero relative degree (SPR0), or more generally, positive real functions with zero relative degree (PR0), for the s variable in real rational functions. It has been shown that this type of substitutions preserve stability, passivity, $\mathcal{RH}_\infty$-norm and coprime factorizations among other properties.

In this note we are interested to present an extension of some of the previous results for more general systems, i.e., passive systems with irrational $\mathcal{H}_\infty$ functions. Particularly, some systems that exhibit this type of transfer functions are linear systems with distributed parameters and delayed linear systems.

G. Fernández-Anaya is with the Departamento de Física y Matemáticas, Universidad Iberoamericana. Prol. Paseo de la Reforma 880, México, D. F. 01210, MÉXICO (e-mail: guillermo.fernandez@ uia.mx).

J. Álvarez-Ramírez is with the Dept. IPH, Universidad Autónoma Metropolitana Iztapalapa, México, D. F., MÉXICO

J.-J. Flores-Godoy is with the Departamento de Física y Matemáticas, Universidad Iberoamericana (e-mail: job.flores@uia.mx).

II. Motivation

Passivity is an important property in several science fields. In particular for distributed systems we present two examples.

A. Passive linear distributed systems

The dynamics of a lossless line is give by the wave equation

$$\frac{\partial^2 v(x,t)}{\partial t^2} = c^2 \frac{\partial^2 v(x,t)}{\partial x^2}$$

where $v(x,t)$ is the voltage along the line and c the wave propagation speed. Using Kirchhoff's voltage and current law for an element of length dx of transmission line, it is possible to describe the transmission line dynamical behavior. The transmission line is modeled by a distributed serial inductance Ldx and a parallel capacitance. Therefore the dynamics are described by

$$\frac{\partial v(x,t)}{\partial t} = -L\frac{\partial i(x,t)}{\partial t}$$

$$\frac{\partial i(x,t)}{\partial t} = -C\frac{\partial v(x,t)}{\partial t}$$

where the current is denoted by $i(x,t)$ and the voltage is denoted by $v(x,t)$, with x the coordinate along the line and t the time. Applying the Laplace transform with respect to the time variable we have

$$\frac{\partial v(x,t)}{\partial t} = -Lsi(x,s)$$

$$\frac{\partial i(x,t)}{\partial t} = -Csv(x,s)$$

introducing new variables

$$a(x,s) = v(x,s) + z_0 i(x,s)$$

$$b(x,s) = v(x,s) - z_0 i(x,s)$$

where $z_0 = \sqrt{\frac{L}{C}}$ is the characteristic impedance of the transmission line, and let $c = \frac{1}{\sqrt{LC}}$ be the wave propagation speed. Then, the new variables satisfy

K. Elleithy et al. (eds.), Advances in Computer, Information, and Systems Sciences, and Engineering, 83–87.

$$\frac{\partial a(x,s)}{\partial x} = -\frac{s}{c}a(x,s)$$
$$\frac{\partial b(x,s)}{\partial x} = \frac{s}{c}a(x,s)$$

and solving the ordinary differential equations for the new variables we obtain the following

$$a(x,s) = a(0,s)\exp\left(-\frac{x}{c}s\right)$$

$$b(x,s) = b(l,s)\exp\left(-\frac{l-x}{c}s\right)$$

The voltage v and current i are found from

$$v(x,s) = \frac{a(x,s)+b(x,s)}{2}$$
$$i(x,s) = \frac{a(x,s)+b(x,s)}{2z_0}$$

With the boundary conditions $v_2 = v(l,s)$ and $i_2 = i(0,s)$, we have the two-port description of an infinite lossless transmission line is given by

$$\begin{bmatrix} v_2(s) \\ i_2(s) \end{bmatrix} = \begin{bmatrix} \cosh(Ts) & -z_0\sinh(Ts) \\ -\frac{1}{z_0}\sinh(Ts) & \cosh(Ts) \end{bmatrix}$$

where s is the Laplace transform variable, $T = \frac{l}{c}$ is the wave propagation time along the line from $x = 0$ to $x = l$.

If a passive load impedance $z_L(s)$ is connected at $x = l$ such that $v_2(s) = z_L(s)i_2(s)$, then the transfer function from $i_1(s)$ to $v_1(s)$ is found to be

$$\frac{v_1(s)}{i_1(s)} = z_0\frac{z_L(s)\cosh(Ts) + z_0\sinh(Ts)}{z_0\cosh(Ts) + z_L(s)\sinh(Ts)}$$

Noting that $z_L(s)$ is a PR function, then $\frac{v_1(s)}{i_1(s)}$ is a PR function (see [4]).

1) The case when the load is a short circuit, which means that $v_2 = 0$ and $z_L(s) = 0$. Then

$$h_{1_1}(s) = \frac{v_1(s)}{i_1(s)} = z_0\tanh(Ts)$$

is a PR function. Notice that $\tanh(Ts) \notin \mathcal{H}_\infty$.

2) Also, we have the case when the load is an open circuit, so that $i_2 = 0$ and $z_L(s) = \infty$. Then

$$h_{1_2}(s) = \frac{v_1(s)}{i_1(s)} = z_0\coth(Ts)$$

is a PR function. Again, note that $\coth(Ts) \notin \mathcal{H}_\infty$.

B. Linear delay systems

In the case of passive linear delay systems, we present two examples.

Example 1 ([13], [14]) Let the system

$$\dot{x}(t) = -a_1x(t) - a_2x(t-\tau) + u(t)$$
$$y(t) = x(t)$$

under an appropriate initial condition defined on the interval $[-\tau, 0]$, $x \in \mathbb{R}$, $\tau > 0$, and assuming $a_1 > 0$, $a_1 + a_2 > 0$, $|a_1| > |a_2|$ the transfer function

$$h_1(s) = \frac{Y(s)}{U(s)} = \frac{1}{s + a_1 + a_2e^{-\tau s}}$$

is PR. For instance, $a_1 = 2$ and $a_1 = 1$, in this case $h_1(s) \in \mathcal{H}_\infty$.

Example 2 ([13], [15]) Motivated by the analysis of combustion dynamics, the following example of a second order system is given

$$\dot{x}(t) = \begin{bmatrix} -a_1 & k \\ -k & -a_2 \end{bmatrix}x(t) + \begin{bmatrix} 0 & 0 \\ -c & 0 \end{bmatrix}x(t-\tau) + \begin{bmatrix} 0 \\ 1 \end{bmatrix}u(t)$$

where the initial condition is defined on the interval $[-\tau, 0]$, $x \in \mathbb{R}$, $\tau > 0$. By Lemma 1 in [13], the transfer function

$$h_2(s) = \frac{Y(s)}{U(s)} = \frac{s + a_1}{(s+a_1)(s+a_2) + k^2 + kce^{-\tau s}}$$

is passive when $a_1 > 0$, $a_2 > 0$ and $a_2^2 > c^2$, i.e., for instance, let $a_1 = 1$, $a_2 = 2$, $k = 1$ and $c = 1$, then the transfer function $h_2(s)$ is PR.

Remark 1 It is important to notice that passive systems (PR functions) do not need to belong to $\mathcal{H}_\infty$.

III. PRELIMINARIES

In this section, we give the notation and the basic definitions that will be used throughout the presentation.

Let $\mathbb{R}$ be the field of real numbers, $\mathbb{C}$ the complex plane, $\operatorname{Im}\mathbb{C}$ the imaginary axis, $\mathbb{C}^+$ the open right-half-plane of the complex plane, $\overline{\mathbb{C}}^+ \equiv \mathbb{C}^+ \cup \operatorname{Im}\mathbb{C}$.

Definition 1 ([16])

1) $\mathcal{L}_\infty$ is the Banach space of scalar-valued functions that are (essentially[1]) bounded on $\operatorname{Im}\mathbb{C}$, with the norm

$$\|f(s)\|_\infty \equiv \operatorname*{ess\,sup}_{\omega\in\mathbb{R}}|f(j\omega)|$$

2) $\mathcal{H}_\infty$ is the closed subspace of $\mathcal{L}_\infty$ with functions that are analytic and bounded in $\mathbb{C}^+$. These functions are bounded in the sense that the norm

$$\|f(s)\|_\infty \equiv \sup_{s\in\mathbb{C}^+}|f(s)| = \sup_{\omega\in\mathbb{R}}|f(j\omega)|$$

is finite.

Notice that based on the sub-multiplicative property of

[1] The measure of the set for which function is unbounded is zero [17].

the $\mathcal{H}_\infty$-norm

$$\|fg\|_\infty \le \|f\|_\infty \|g\|_\infty,$$

and that the product of analytic functions in $\mathbb{C}^+$ is an analytic function in $\mathbb{C}^+$, $\mathcal{H}_\infty$ is in effect a commutative Banach algebra, due to its vectorial space structure, i.e., a commutative complex algebra and a Banach space.

Definition 2 ([18])

1) A complex algebra is a vector space $\mathcal{V}$ over $\mathbb{C}$ in which an associative and distributive multiplication is defined.
2) A commutative Banach *-algebra is a commutative Banach algebra $\mathcal{A}$ together with an "involution"

$$a \to a^* \text{ for } a \in \mathcal{A}$$

satisfying:

a) * is conjugate linear, i.e.,

$$(\alpha a)^* = \overline{\alpha} a^* \text{for } \alpha \in \mathbb{C},\ a \in \mathcal{A}$$

b) $a^{**} = a$, for every $a \in \mathcal{A}$;

c) $(ab)^* = b^* a^*$, for any $a, b \in \mathcal{A}$;

d) $\|a^*\| = \|a\|$, for every $a \in \mathcal{A}$.

By (2d), we see that $^*: \mathcal{A} \to \mathcal{A}$ is continuous.

3) A $\mathbb{C}^*$-algebra $\mathcal{A}$ is a Banach *-algebra for which $\|a^* a\| = \|a\|^2$, for all $a \in \mathcal{A}$.
4) A $\mathbb{C}^*$-algebra $\mathcal{A}$ is unital, if $\mathcal{A}$ has an identity element $e \in \mathcal{A}$ such that $\|e\| = 1$.
5) The self-adjoint elements of a unital $\mathbb{C}^*$-algebra $\mathcal{A}$ are the elements that satisfy $a^* = a$.

Notice that the subset of all self-adjoint elements of a unital $\mathbb{C}^*$-algebra is itself a unital $\mathbb{C}^*$-algebra. An example of a commutative unital $\mathbb{C}^*$-algebra is $\mathcal{H}_\infty$.

Definition 3 ([5], [4])

1) A function $p(s)$ of complex variable $s = \sigma + j\omega$, with $j = \sqrt{-1}$, is positive real (PR) or passive function if
   a) $p(s)$ is real for s real,
   b) $p(s)$ is analytic for $s \in \mathbb{C}^+$.
   c) $\mathrm{Re}[p(s)] \ge 0$ for all $s \in \mathbb{C}^+$.
2) A function $p(s)$ is strictly positive real (SPR) or strictly passive if $p(s-\varepsilon)$ is PR for some $\varepsilon > 0$.
3) A function $p(s)$ is bounded real (BR) if $p(s)$ is analytic in $\mathbb{C}^+$ and $\|p(s)\|_\infty \le 1$.

Definition 4

1) A function $p(s)$ is stable, if it is analytic in $\overline{\mathbb{C}}^+$.
2) A function $p(s)$ is critically stable, if it is analytic in $\mathbb{C}^+$.

IV. Main Results

In this section the main results are presented. We assume that all PR functions used in this section are single-valued and not constants.

Proposition 1

1) If $p(s)$ is a critically stable function and $q(s)$ is a PR function, then $p(q(s))$ is a critically stable function.
2) If $p(s)$ is a stable function and $q(s)$ is a PR function, then $p(q(s))$ is a stable function.

Proposition 2

1) The composition of PR functions, is a PR function.
2) If $p(s)$ is a PR function and $q(s)$ is a SPR function, then $p(q(s))$ is a SPR function.
3) The composition of SPR functions, is a SPR function.

Notice that the set of all the PR functions, form a monoid with composition of functions as binary operation in this set. Furthermore, the set of all the SPR functions form a groupoid with the same operation.

Definition 5 ([[18]]) Let $\mathcal{A}$ and $\mathcal{B}$ be unital $\mathbb{C}^*$-algebras. A homomorphism $\varphi: \mathcal{A} \to \mathcal{B}$ is a linear map satisfying

1) $\varphi(ab) = \varphi(a)\varphi(b)$ for all $a, b \in \mathcal{A}$
2) $\varphi(a^*) = (\varphi(a))^*$ for all $a \in \mathcal{A}$;
3) $\varphi(1_\mathcal{A}) = 1_\mathcal{B}$.

Theorem 1 The function

$$\xi_q : \mathcal{H}_\infty \to \mathcal{H}_\infty$$

defined by

$$\xi_q[p(s)] \equiv p(q(s))$$

for any $p(s) \in \mathcal{H}_\infty$ and for a fixed single-valued PR function $q(s)$ is an endomorphism of the commutative unital $\mathbb{C}^*$-algebra $\mathcal{H}_\infty$ such that:

1) ξ_q is continuous;
2) $\|\xi_q[p(s)]\|_\infty \le \|p(s)\|_\infty$ for all $p(s) \in \mathcal{H}_\infty$;
3) let $\mathrm{Ker}\xi_q = \{p(s) \in H_\infty | \xi_q[p(s) = 0]\}$. If $\mathrm{Ker}\xi_q$ is trivial for some single-valued PR function $q(s)$, then $\|\xi_q[p(s)]\|_\infty = \|p(s)\|_\infty$ for all $p(s) \in \mathcal{H}_\infty$.
4) $\xi_q[\mathcal{H}_\infty] \equiv \mathrm{Im}\,\xi_q$ is a commutative unital $\mathbb{C}^*$-algebra,

furthermore, when the Kerξ_q is trivial for some single-valued PR function $q(s)$, then $\xi_q[\mathcal{H}_\infty]$ and $\mathcal{H}_\infty$ are isomorphic $\mathbb{C}^*$-algebras.

This result has interesting consequences in the synthesis of positive real $\mathcal{H}_2$ controllers [19], Synthesis of robust strictly positive real systems [8], [20], [21], in adaptive control [6], in filtering theory [7], absolutely stability [3], for suboptimal control in $\mathcal{H}_\infty$ theory [16], and others fields.

Corollary 1 Let $p(s)$ be a function in $\mathcal{H}_\infty$ and $\xi_q:\mathcal{H}_\infty \to \mathcal{H}_\infty$ be defined as in Theorem 1, and suppose that

$$h(s) = \frac{p(s)-1}{p(s)+1}$$

where $p(s) \neq -1$ for all $s \in \mathbb{C}^+$. Then if $h(s)$ is a BR function, then $\xi_q[h(s)]$ is a BR function.

Now based on Corona Theorem it is possible to state the preservation of strongly coprime factorization using the homomorphism ξ_q as defined in Theorem 1.

Theorem 2 (Corona Theorem [22], [23]) Suppose that $f_1(s),\ldots,f_n(s) \in \mathcal{H}_\infty$. There exists a $\delta > 0$ such that

$$\inf_{s\in\mathbb{C}^+}\left(|f_1(s)| + \cdots + |f_n(s)|\right) > \delta$$

if and only if there should exist $g_1(s),\ldots,g_n(s) \in \mathcal{H}_\infty$ satisfying the Corona Equation:

$$g_1(s)f_1(s) + \cdots + g_n(s)f_n(s) = 1, \quad s \in \mathbb{C}^+$$

Notice that $\mathcal{H}_\infty$ is a Hermite ring (see [23], [24]), but it is not a Bezout domain [25].

Definition 6 [26], [27]) A strongly coprime factorization for $h(s)$ over $\mathcal{H}_\infty$ is a pair $(n(s),d(s))$ such that $n(s),d(s) \in \mathcal{H}_\infty$ and there exist $x(s),y(s) \in \mathcal{H}_\infty$ with the following properties

$$h(s) = n(s)d^{-1}(s)$$

and

$$n(s)x(s) + d(s)y(s) = 1$$

Notice that the pair $(n(s),d(s))$ is strongly coprime if and only if

$$\inf_{s\in\mathbb{C}^+}\left(|n(s)| + |d(s)|\right) > 0$$

This is a consequence of the Theorem 5 and the Theorem 2 in [27]. Also, note that if $h(s)$ has a strongly coprime factorization, then $h(s)$ is stabilizable (see [27]).

Corollary 2

1) Let $\xi_q:\mathcal{H}_\infty \to \mathcal{H}_\infty$ be defined as Theorem 1. If $(n(s),d(s))$ is a strongly coprime factorization of $h(s)$ over $\mathcal{H}_\infty$, then $(\xi_q[n(s)], \xi_q[d(s)])$ is a strongly coprime factorization of $\xi_q[h(s)]$ over $\mathcal{H}_\infty$.
2) If $h(s)$ is stabilizable by a controller $c(s) = x(s)y^{-1}(s)$, then $\xi_q[h(s)]$ is stabilizable by a controller $\xi_q[c(s)] = \xi_q[x(s)]\xi_q[y^{-1}(s)]$.

Also, there exists a similar result for the preservation of the set of all stabilizing controllers of $h(s)$.

V. Some Examples for Passivity

Using the Proposition 2, item (1), we have the following academic examples to show the preservation of passivity.

Based on Subsection II-A, consider the following function:

$$f_1(s) = \tanh(Ts)$$

which is a PR function. On the other hand we know that

$$q(s) = \frac{as+b}{cs+d}$$

is a PR function when $a,b,c,d > 0$ and $ad - bc \neq 0$. Therefore, the function

$$f_2(s) = f_1(q(s)) = \tanh\left(T\frac{as+b}{cs+d}\right)$$

is a PR function by Proposition 2. Similarly, the same conclusion can be drawn for the hyperbolic cotangent function.

Consider now the PR function from Example 1 and the PR functions $q_1(s) = as + b$; $a,b > 0$ and $q_2(s) = \frac{1}{s}$, then the functions

$$f_3(s) = h_1(q_1(s)) = \frac{1}{as + b + a_1 + a_2e^{-\tau(as+b)}}$$

$$f_4(s) = h_1(q_2(s)) = \frac{s}{1 + a_1s + a_2se^{-\tau\frac{1}{s}}}$$

are PR functions by Proposition 2. Notice that $f_3(s)$ may be written as

$$f_3(s) = \frac{\frac{1}{a}}{s + \bar{a}_1 + \bar{a}_2e^{-\beta s}}$$

where $\bar{a}_1 = \frac{b+a_1}{a}$, $\bar{a}_2 = \frac{a_2e^{-\tau\beta}}{a}$ and $\beta = \tau a$. Observe that $\bar{a}_1 > 0$, $\bar{a}_1 + \bar{a}_2 > 0$, $|\bar{a}_1| > |\bar{a}_2|$ satisfy the requirements needed for a function of this type to be a PR function.

For the case of the function from Example 2, consider the PR function $q_1(s) = as + b$ where $a,b > 0$, then the function

$$f_5(s) = h_2(s)$$
$$= \frac{as + b + a_1}{(as + b + a_1)(as + b + a_2) + k^2 + kce^{-\tau(as+b)}}$$
$$= \frac{1}{a}\frac{s + \hat{a}_1}{(s + \hat{a}_1)(s + \hat{a}_2) + \hat{k}^2 + \hat{k}\hat{c}e^{-\beta s}}$$

where $\hat{a}_1 = \frac{b + a_1}{a}$, $\hat{a}_2 = \frac{b + a_2}{a}$, $\hat{k} = \frac{k}{a}$ and $\hat{c} = \frac{ce^{-\tau b}}{a}$, is a PR function by Proposition 2. It is also clear that the following inequalities are satisfy: $\hat{a}_1 > 0$, $\hat{a}_2 > 0$, $\hat{a}_2 > \hat{c}^2$, as is requested in the Example 2.

VI. CONCLUSIONS

In this work, we presented a generalization of results on preservation of properties for SISO irrational $\mathcal{H}_\infty$ functions such as stability, stabilizability, passivity and coprime factorization. These properties are preserved using the substitution of the variable s in the transfer functions, by a passive transfer function $q(s)$ this substitution is an endomorphism of the $\mathbb{C}^*$-algebra $\mathcal{H}_\infty$ which preserves the $\mathcal{H}_\infty$-norm. Also, it is mentioned that passive and strictly passive functions are closed under function composition.

REFERENCES

[1] K. S. Narendra and A. M. Annaswamy, *Stable Adaptive Systems*. Englewood Cliffs, NJ, USA: Prentice-Hall, 1989.

[2] B. D. O. Anderson and S. Vongpanitlerd, *Network Analysis and Synthesis. A modern Systems Approach* Englewood Cliffs, NJ, USA: Prentice Hall, 1972.

[3] H. K. Khalil, *Nonlinear System*. Englewood-Cliffs, NJ, USA: Prentice-Hall International Editions, 1996.

[4] R. Lozano, B. Brogliato, B. Maschke, and O. Egeland, *Dissipative Systems Analysis and Control: Theory and Applications*. London, UK: Springer-Verlag, 2000.

[5] W. M. Haddad and D. S. Bernstein, "Explicit construction of quadratic Lyapunov functions for the small gain, positivity, circle, and Popov theorems and their application to robust stability Part I: Continuous-time theory," *Int. J. Robust and Nonlinear Control*, vol. 3, pp. 313-339, 1993.

[6] B. D. O. Anderson, R. Bitmead, C. Johnson, P. Kokotović, R. Kosut, I. Mareels, L. Praly, and B. Riedle, *Stability of Adaptive Systems---Passivity and Averaging Analysis*. Cambridge, USA: MIT Press, 1982.

[7] G. C. Goodwin and K. S. Sin, *Adaptive Filtering, Prediction and Control*. Englewood-Cliffs, NJ, USA: Prentice-Hall International Editions, 1984.

[8] B. D. O. Anderson, S. Dasgupta, P. Khargonekar, K. J., Krauss, and M. Mansour, "Robust strict positive realness: characterization and construction," *IEEE Trans. Circuits Syst. I*, vol. 37, pp. 869-876, 1990.

[9] G. Fernández-Anaya, "Preservation of SPR functions and stabilization by substitutions in SISO plants," *IEEE Trans. Automat. Contr.*, vol. 44, no. 11, pp. 2171-2174, 1999.

[10] G. Fernández-Anaya, J. C. Martínez-García, and V. Kučera, "$\mathcal{H}_\infty$-robustness properties preservation in SISO systems when applying SPR substitutions," *International Journal of Control*, vol. 76, pp. 728-740, 2003.

[11] G. Fernández-Anaya, J. C. Martínez, V. Kučera, and D. Aguilar-George, "MIMO systems properties preservation under SPR substitutions," *IEEE Trans. Circuits Syst. II*, vol. 51, pp. 222-227, 2004.

[12] J. C. Martínez-García, G. Fernández-Anaya, V. Kučera, and J. J. Flores-Godoy, "Control-oriented properties preservation in linear systems when applying PR0 substitutions," in *Proceedings of the 16th IFAC World Congress*, Prague, Czech Republic, July 2005.

[13] S. I. Niculescu and R. Lozano, "On the passivity of linear delay systems," *IEEE Trans. Automat. Contr.*, vol. 46, pp. 460-464, 2001.

[14] L. Lefèvre, J. Lefèvre, and G. Dauphin-Tanguy, "Passivity of linear delay systems," in *Proceedings of the 6th IEEE Mediterranean Conf. Control Automation*, Alghero, Italy, 1998.

[15] M. Fleifil, J. P. Hathout, A. M. Annaswamy, and A. F. Ghoniem, "Reduced order modeling of heat release dynamics and active control of time-delay instability," in *Proceedings of the 38th AIAA Aerospace Sciences Meeting*, Reno, NV, 2000, in AIAA 2000-0708.

[16] K. Zhou, J. C. Doyle, and K. Glover, *Robust and Optimal Control*. Upper Saddle River, NJ, USA: Prentice-Hall, Inc., Simon & Schuster, 1995.

[17] G. B. Folland, *Real analysis: Modern techniques and their applications*. New York, NY: John Wiley & Sons, 1984.

[18] I. F. Wilde, *$\mathbb{C}^*$-algebras---Notes*, D. Maskell and I. Robinson, Eds. London, UK: Haven Books, 2001.

[19] J. C. Geromel and P. B. Gapsky, "Synthesis of positive real $\mathcal{H}_2$ controllers," *IEEE Trans. Automat. Contr.*, vol. 42, pp. 988-992, 1997.

[20] A. T. G. Bianchini and A. Vicino, "Synthesis of robust strictly positive real systems with ℓ_2 parametric uncertainty, *IEEE Trans Circuits Syst. I*, vol. 48, pp. 438- 450, 2001.

[21] C. H. Huang, P. A. Ioannou, J. Maroulas, and M. G. Safanov, "Design of strictly positive real systems using constant output feedback," *IEEE Trans. Automat. Contr.*, vol. 44, pp. 569-572, 1999.

[22] R. Krzysztof, "Corona theorem and isometries," *Opuscula Mathematica*, vol. 24, no. 1, pp. 124-131, 2004.

[23] M. Vidyasagar, *Control System Synthesis: a Factorization Approach*. Cambridge, Massachusetts: MIT Press, 1985.

[24] N. K. Nikol'ski, *Treatise on the Shift Operator: Spectral Function Theory*. Springer, 1986.

[25] M. von Renteln, "Hauptideale und aussere funktionen im ring {H1}," *Archiv der Mathematik*, vol. 28, pp. 519-524, 1977.

[26] J. J. Loiseau, M. Cardelli, and X. Dusser, "Neutral-type time--delay systems that are not formally stable are not bibo stabilizable," *IMA J. Mathematical Control and Information*, vol. 19, pp. 217-227, 200.

[27] M. C. Smith, "On stabilization and the existence of coprime factorizations," *IEEE Trans. Automat. Contr.*, vol. 34, pp. 1005-1007, 1989.

Resource Management in Cellular Networks

Adeel Akram, Shahbaz Pervez, M. Nadeem Majeed, M. Zafrullah
University of Engineering and Technology Taxila, Pakistan
Email: {shahbaz, adeel}@uettaxila.edu.pk
nadeemajeedch@yahoo.com
profzafrullah@uettaxila.edu.pk

Abstract

Now a days cellular phone companies are facing a lot of problem regarding resource management. Resources for establishing cellular networks are very expensive, Antennas are one these resource. Cellular companies are facing problem regarding proper management of these antennas. In this paper we address the issue of antenna placement problem (APP) to reduce the number of antennas in a specific area. We propose a self tuning algorithm that identifies the location of antennas to be placed in the desired coverage area using minimum number of antennas.
Digital Elevation Maps for the proposed coverage area are processed to create a candidate location matrix where antennas can be placed. This matrix provides a mean to isolate areas where antenna deployment is restricted due to Legal or other concerns.
This paper utilizes omni directional antennas and provides x, y coordinates of the proposed location in the area of interest.

Keywords:-Antenna Placement Problem (APP), Digital Elevation Models (DEM), optimal cellular Coverage

1. Introduction

Setting up Cellular network in a specific area not only requires exclusive frequency assignment but also require optimal placement of antennas. To provide full area coverage using minimum number of antennas is itself a challenging task. Within the selected geographical location where the Cellular network is to be installed, engineers require the minimum number of antennas to be installed, their type and location of installation.

This paper provides a solution to the *Antenna Placement Problem (APP)* and proposes algorithm to optimize the coverage of an area. This paper does not take into consideration the *Frequency Assignment Problem (FAP)* that minimizes electromagnetic interference due to multiple use of same frequencies in different parts of the network. Although FAP has been intensively investigated, little work has been done for APP.

One reason is that the APP is much more difficult to model, and even a simple model - treating only coverage of the working area - can be shown to be NP-Hard.

Furthermore APP is a multi-objective and constrained problem with all its known difficulties.

The objectives are to minimize number of antennas used and to have practically feasible location of placement to provide full area coverage.

The development of an appropriate model is one of the most difficult tasks in solving complex real world problems. The model for the APP used in this project is the result of several loops of design, evaluation and redesign. The quality of solutions of the APP is assessed by two constraints, which deal with coverage and resource optimization. They address economical and technical aspects of the project.

Digital Elevation Map (DEM) gives Latitude, Longitude and Elevation information of the area where the Cellular networks is to be deployed.

The complete area is divided into smaller squares to simplify the problem. We can provide deterministic coverage by placing an omni directional antenna in each square, provided the size of the square is less than the range of the antenna.

K. Elleithy et al. (eds.), Advances in Computer, Information, and Systems Sciences, and Engineering, 89–92.

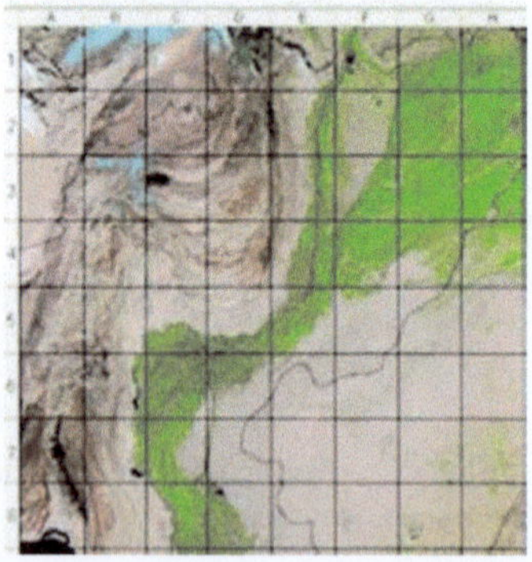

Figure 1. Digital Elevation Map of deployment site

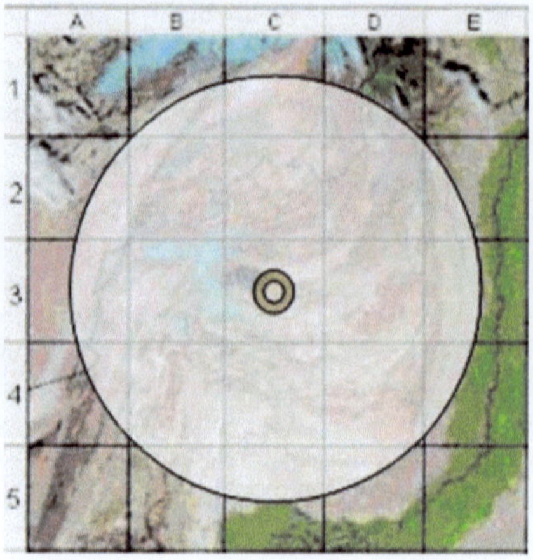

Figure 2. Omni Directional Antenna in Cell "C3"

Computational geometry and graph theoretic techniques, specifically the Voronoi diagram can also be applied to find the best and worst case coverage as explained in [1]

2. Constraints & Objectives

A *feasible* solution has to fulfill the following two constraints:

- **Antenna Constraint (AC):** Minimize the number of Antennas,
- **Coverage Constraint (CC):** Points in the covered area receives signal above a certain threshold.

The algorithm assumes that the omni-directional antenna covers eight adjacent grid squares around it. The size of these squares depends on the coverage radius of the antenna used.

3. Antenna Location selection Scheme

We have implemented an antenna placement scheme by considering the area of interest as a grid map. The scheme provides information about the locations of the antennae in terms of the x and y coordinates of the grid points. Since the optimal solution for this problem is NP-Hard, a heuristic algorithm for large maps is also required.

Our algorithm first tries to get all the possible points in the whole area where antenna can be placed. Once we have these potential areas, it is possible to use the branch and bound method, to go through all the possible positions. While we're going through all the different possibilities, we keep track of the lowest number of antennas needed.

After the first possible antenna placement is found, the lowest number of antennas for this candidate location is calculated, we use this value as a bound for all future calculations. For example, if the first possible point returns the lowest numbers of antennas to be 3, then the branch should not have to go down any further than three.

This bounding technique will save a large percentage of calculations. This method covers all possibilities and returns the smallest number of antenna coverage.

Following figure shows a 4x5 area with 7 possible antenna locations indicated by the • sign.

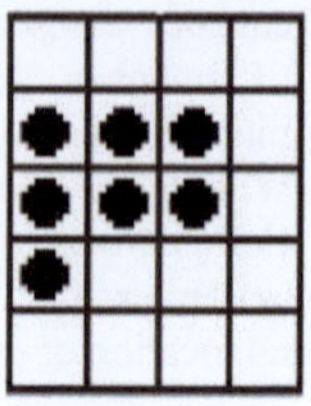

Figure 3. Candidate Locations for Placing Antenna

There are seven possible locations to start. Choose the first location and recursively take the result and choose the remaining locations. Continue this process recursively until all the desired coverage areas are covered. For this, we assume that an antenna will cover the surrounding 8 areas including the area it occupies, and the desired coverage area happens to be antenna placement candidate zone.

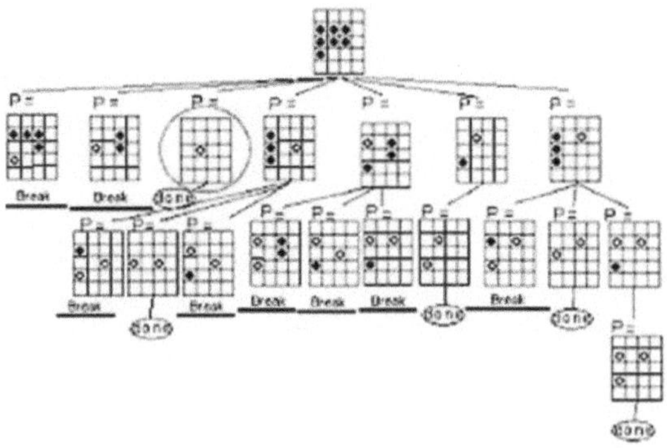

Figure 4. Search space of antenna placer algorithm

The optimal algorithm first follows the right branch for the above search space graph. The first branch node from the root along the right branch shows the first antenna was placed at the location (1,1) and it covers four of the desired coverage areas.

The ◆ sign of the four covered locations are removed from the map. The map shows that after the first antenna is placed, there are only three locations on the left yet to be covered.

The second branch node from the root along the right branch shows the second antenna was placed at the location (3,1) and it covers two of the remaining three uncovered locations.

The leaf node of the right branch shows the third antenna was placed in location (3,3) and all desired coverage area are covered. The number of antennae used for the current solution is 3 and it was used to bound and screen the other branches in the search space.

The second right branch puts second antenna at location (3,2) and with just two antennae, this solution covers all desired coverage areas. From this point on, any branch with antennae more than two will "break" from the search, thus reducing the search iterations of the algorithm.

The algorithm continues to search other possibilities of antenna placement and keeps reducing the search depth to the bound set by the number of antennas used to provide complete coverage of candidate area.

The encircled branch node shows that the algorithm finds an optimal location for placement of antenna where the desired region of interest is covered using only one antenna. From this point onwards on all other branches the algorithm will cut off after the first antenna is placed.

This way the complete area of interest is covered using minimum number of antennas.

4. Future Work

The result of this algorithm served as a base line data for comparing different heuristic algorithms and see how far their computation results deviated from the optimal solutions.

Heuristics values such as density of candidate location matrix can also contribute in reduction of number of iterations of search algorithm.

This algorithm can further be modified to include the radiation pattern of antenna to be deployed and its coverage radius to re-calculate the scale of grid squares.
If directional/sector antennas are used, the effect of other factors such as fresnel zones, multipath fading and environment losses can also be considered.

In Future the algorithm may also be improved to provide height of antenna to be placed in addition to its x, y coordinates.

5. References

1. Mischa Schwartz, "Network Management and Control Issues in Multimedia Wireless Networks," IEEE Personal Communications, Vol. 2, No. 3, June 1995, pp. 8-16.
2. Leung, Kin K., "Traffic Models for Wireless Communications Networks," IEEE JSAC, Vol. 12, No. 8, Oct. 1994.
3. S. Hurley, Planning effective cellular mobile radio networks, IEEE Transactions on Vehicular Technology 51(2) (2002) 243-253.
4. Rudolf Mathar , Thomas Niessen, Optimum positioning of base stations for cellular radio networks, Wireless Networks, v.6 n.6, p.421-428, Dec. 2000
5. Meguerdichian, Farinaz, "Coverage Problems in Wireless Ad-hoc Sensor Networks", Infocom '01
6. C.-H. E. Chow, "NETSIM: A network simulation system for the design and analysis of network survival techniques," CASI proposal, November, 1993.

7. Web Page: National Geologic Map Database, USGS, National Cooperative Geologic Mapping Program, National Geologic Mapping Act of 1992.
8. A. Ganz, C.M. Krishna, D. Tang and Z.J. Haas, On optimal design of multitier wireless cellular systems, IEEE Communications Magazine (February 1997) 88-93.
9. J.K. Han, B.S. Park, Y.S. Choi and H.K. Park, Genetic approach with a new representation base station placement in mobile communications, in: Proc. of the 54th IEEE Conf. on Vehicular Technology, Vol. 4 (2001) pp. 2703-2707.
10. M. Hata, Empirical formula for propogation loss in land mobile radio services, IEEE Transactions on Vehicular Technology 29(3) (1980) 317-325.
11. I. Howitt and S.-Y. Ham, Base station location optimization, in: Proc. of the IEEE VTC'99 Conf., Vol. 4 (1999) pp. 2067-2071.
12. Jerry D. Gibson, "The Mobile Communications Handbook", CRC Press (Published in Cooperation with IEEE Press), 1996.
13. Antenna Handbook, MCRP 6-22D, US Marine Corps 1992.
14. X. Huang, U. Behr and W. Wiesbeck, Automatic cell planning for a low-cost and spectrum efficient wireless network, in: Proc. of Global Telecommunications Conf. (GLOBECOM), Vol. 1 (2000) pp. 276-282.
15. A. Molina, G.E. Athanasiadou and A.R. Nix, The automatic location of base-stations for optimised cellular coverage: A new combinatorial approach, in: Proc. of the IEEE VTC'99 Conf. (1999) pp. 606-610.

Real-time Vehicle Detection Using Information of Shadows Underneath Vehicles

Yoichiro Iwasaki and Hisato Itoyama
Department of Information Engineering, Graduate School of Engineering, Kyushu Tokai University
9-1-1, Toroku, Kumamoto 862-8652, Japan

***Abstract-* We propose a vehicle detection method for traffic flow images obtained from a video camera set up on a low place such as the roadside, sidewalk, etc. The method uses shadows underneath vehicles as the information to detect vehicles. The method distinguishes the size of each vehicle according to the distance between both ends of front- and rear-tires, and discriminates the lanes on which vehicles exist. The method has functions to create and update automatically a background image, and to estimate and update automatically a threshold value to binarize background subtraction images in order to enhance the vehicle detection accuracy. The proposed algorithm can realize a high-speed processing without complicated calculations, and real-time vehicle detection by using a general-purpose personal computer. Experimental results show that the vehicle detection accuracy is 98.9%.**

I. INTRODUCTION

In vision-based traffic monitoring, it is desirable to set up a camera on a high place in order to secure a measurement area more widely, and to prevent overlapping of vehicles. In many of previous studies for vision-based traffic monitoring, images obtained from a camera set up on a high place are used [1][2][3]. However, it is restricted to secure the high places such as buildings, pedestrian bridges, etc. The construction of a pole to attach a camera is needed if such a high place does not exist near the measurement area.

This paper proposes a real-time vehicle detection method for traffic flow images obtained from a video camera set up on a low place such as the roadside, sidewalk, etc. By using the proposed method, a measurement area can be selected more freely, and traffic flow data such as traffic volumes, space headways, etc. can be collected easily.

In vehicle detection, various algorithms have been proposed. Some previous work is shown below.

The work of [1] used spatio-temporal analysis in daytime images, and morphological analysis of headlight pairs in night images. Yoneyama *et al.* [2] proposed a method to eliminate moving cast shadows for robust vehicle extraction. The work of [3] was based on edge detection and template matching. A vehicle detection method based on edge detection in a tunnel was proposed [4]. The work of [5] was intended for image data from a single camera placed in a moving vehicle, and used a combination of three clues: shadow, entropy, and horizontal symmetry.

The proposed method uses only shadows underneath vehicles as the information to detect vehicles, and discriminates the lanes on which vehicles exist. The shadows underneath vehicles exist not only at fine weather but also at other various weather conditions since the bottom of a vehicle is extremely close to the surface of a road. We confirmed that the shadows underneath vehicles exist under cloudy, rainy, twilight, and nighttime illuminated by streetlights conditions from observations.

The proposed algorithm offers a high-speed processing without complicated calculations compared with other vehicle detection method [6] targeted to vehicular side views, and real-time vehicle detection by using a general-purpose personal computer.

II. PROPOSED VEHICLE DETECTION ALGORITHM

A. Detecting Locations of Vehicles and Discriminating their Sizes

An input image is a gray scale image, the size of the image is 640x480 pixels, and a pixel has 256 gray levels. The height of the installation of the video camera is about 1.7m, and the measurement is targeted to a general street of multilane.

A background subtraction is done by (1), the obtained background subtraction image is binarized, and the binary image in which vehicles and shadows underneath the vehicles exist only is obtained. In the binarization, the regions of vehicles and shadows underneath the vehicles are converted to white pixels, and the background regions are converted to black ones.

$$f_{\mathrm{sub}}(i,j,t)=\begin{cases} g(i,j,t)-f(i,j,t): g(i,j,t)-f(i,j,t)>0 \\ 0: \text{otherwise} \end{cases} \tag{1}$$

where $f_{\mathrm{sub}}(i, j, t)$ is the background subtraction image, $g(i, j, t)$ is the background image which contains no vehicle, $f(i, j, t)$ is the input image, i and j are x and y coordinates, and t is time.

We will discuss a method of creating and updating a background image, and a method of estimating and updating a threshold value to binarize images in Sections *B* and *C*, respectively.

A vertical projection is done by searching white pixels along the vertical columns in the binary image. If there is one

K. Elleithy et al. (eds.), Advances in Computer, Information, and Systems Sciences, and Engineering, 93–97.

white pixel at least in a vertical column, the black pixel at the same row on the lowest horizontal line in the measurement area is changed into white. The white box in Fig. 1 shows the measurement area. Although the lowest horizontal line in Fig. 1 is drawn in white, the white pixels do not exist in actual images. If eight pixels or more are white in fifteen continuous pixels on the lowest horizontal line, corresponding fifteen continuous pixels are connected. Fig. 2 shows a result of the vertical projection. The location and width of the lower horizontal line in the rectangle in Fig. 2 is obtained by the processing described above. The height of the rectangle is decided from the height of the measurement area. By this processing, the vehicle and shadow regions are grasped roughly.

Next, the connected components of the shadows underneath the vehicles on/over the lower horizontal line of a rectangle are searched. The shadow region underneath each vehicle is distinguished by the difference between the vertical positions of shadows, and each pair of left and right edges of the shadow is identified. Fig. 3 shows a result of identifying left and right edges of the shadows. The "S" and "E" in Fig. 3 indicate the left and right edges, respectively. Moreover, a horizontal projection for the rectangle region just over each shadow is done by summing the number of black pixels along the horizontal rows. Each boundary position between a vehicle and the shadow underneath the vehicle can be identified from the horizontal projection result because the frequency of the black pixels increases extremely at the boundary position. Fig. 4 shows a distribution of black pixels, and the arrow shows the boundary position.

A tire, which touches a road, is black, and the gray-level values of the tire and those of the shadow underneath the vehicle are very similar. The regions of tires become the white pixels that are the same as the regions of shadows when a binary image is created. Therefore, the region of the shadow and the tires of a vehicle can be scanned as one connected component, and then the positions of tires can be identified. As a result, the length of a wheelbase plus a wheel diameter can be obtained, and the length is used as the information to distinguish the size of each vehicle. Fig. 5 shows a detection result. In Fig. 5, the sideways arrow shows the boundary position, and the two upward arrows show the both ends of front- and rear-tires.

There is a difference in the vertical position of each lane in the images of vehicular side views. Therefore, if the vertical positions of the vehicles can be measured, the lanes on which vehicles exist are specified. The proposed algorithm discriminates the lanes on which vehicles exist from the measured boundary positions.

Even if the positions of front- and rear-tires cannot be obtained, the vehicle location can be detected from the location of the shadow underneath the vehicle.

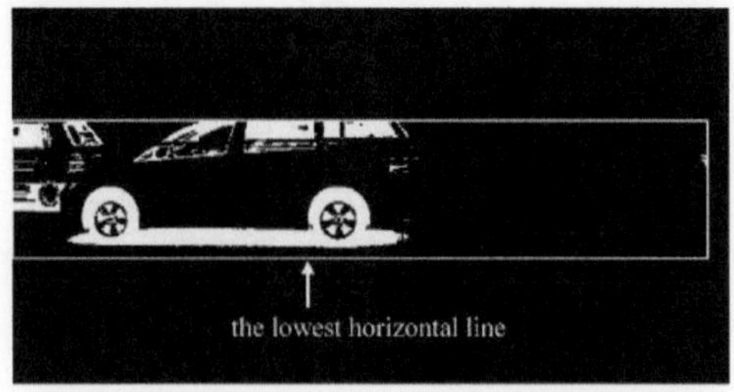

Fig. 1. The measurement area and the lowest horizontal line.

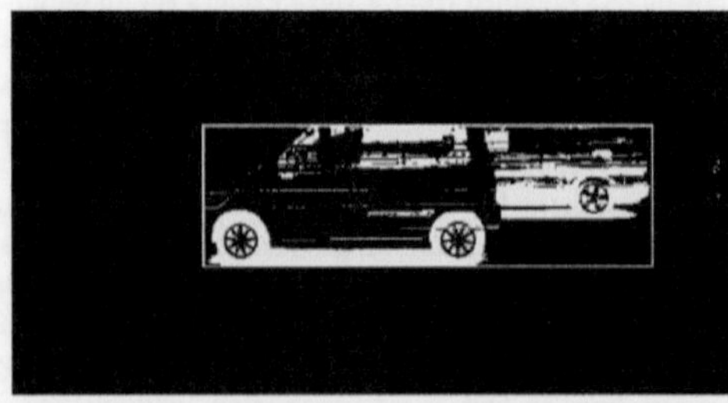

Fig. 2. A region of vehicles and their shadows.

Fig. 3. Left and right edges of shadows.

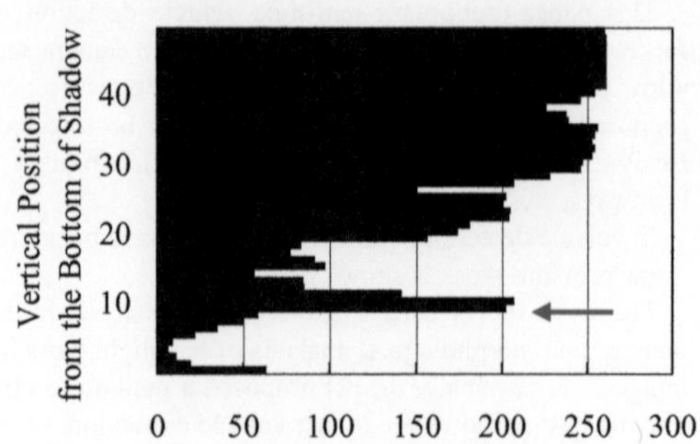

Fig. 4. A distribution of black pixels.

Fig. 5. Positions of tires and the boundary between a vehicle and its shadow.

B. Creating and Updating a Background Image

Since gray-level values in the background region change from time to time due to environmental factors, it is important to update the background image in a short time interval in order to enhance the vehicle detection accuracy.

An initial background image is created in advance by use of a background image renewal processing method proposed in [7] from 50 continuous image flames. In the process of creating the initial background image, moving objects are eliminated automatically.

First, we will explain the background image renewal processing method by using (2), (3), and Fig. 6, which are proposed in [7].

$$h(i,j,t) = f(i,j,t) - g(i,j,t-T) \tag{2}$$

$$g(i,j,t) = g(i,j,t-T) + \text{Table}\{h(i,j,t)\} \tag{3}$$

where $g(i, j, t\text{-}T)$ is the previous background image with the time interval T, which equals input intervals of image, and Table$\{h(i, j, t)\}$ is the response function shown in Fig. 6.

If $h(i, j, t)$ is within the range from $-H$ to H, the pixel position is regarded as the background region, then the gray-level value at the position in the previous background image is updated rapidly by use of Table$\{h(i, j, t)\}$. On the other hand, when $h(i, j, t)$ is under $-H$ or over H, the gray-level value is also changed and updated in a long period of time. Therefore, the response function is effective for the change of gray-level values not only due to environmental factors but also due to the noises, garbage on the road, movement of a camera, etc. which are not caused by the environmental factors. The speed of these updates depends on the shape of the response function in Fig. 6. The method proposed in [7] takes the stop vehicles into the background region in a long time. Therefore, the stop vehicles cannot be detected. So we propose a method of adding a new algorithm to the method in [7]. Next, we will explain the new algorithm.

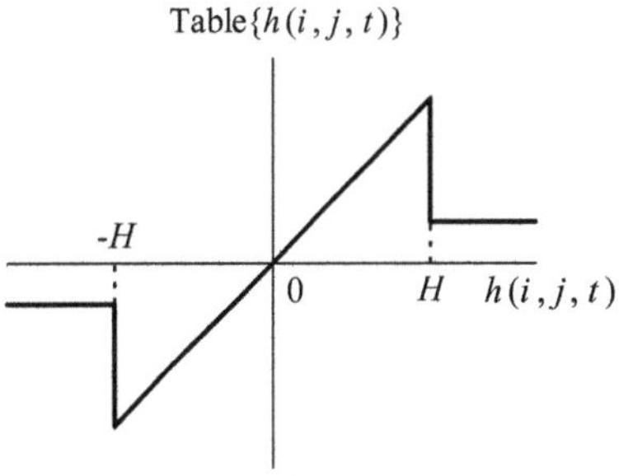

Fig. 6. Response function proposed in [7].

The measurement area is divided into 16x4 blocks. The number of pixels whose values are greater or equal to a threshold value in each block is measured in a background absolute subtraction image given by

$$f_{\text{abs\_sub}}(i,j,t) = |f(i,j,t) - g(i,j,t)| \ , \tag{4}$$

where $f_{\text{abs\_sub}}(i, j, t)$ is the background absolute subtraction image.

In this process, absolute values are adapted as the background subtraction factor in contrast with (1). If the number of pixels is greater or equal to a threshold number, it is judged that there is a stop vehicle, then the block is not updated. By only the variation of gray-level values in a very small area such as a pixel, the existence of a stop vehicle cannot be judged. By comparing the many pixels in a block, the existence of a stop vehicle can be identified.

In addition, when a block is not updated continuously over 30 flames in the past, the surrounding eight blocks are examined. If the seven or more blocks of the surrounding eight blocks are updated, it is assumed that the judgment of not updating is wrong, then the block is updated immediately. Because the outermost blocks in the measurement area do not have the surrounding eight blocks, it is assumed that the insufficient blocks are update always. Fig. 7 shows an example of updating a background image using proposed algorithm. Fig. 7 shows that the blocks of the vehicles and their shadows are not updated, and the gray and white blocks are updated. The white is a block for which the update is begun because the miss judgment turned out.

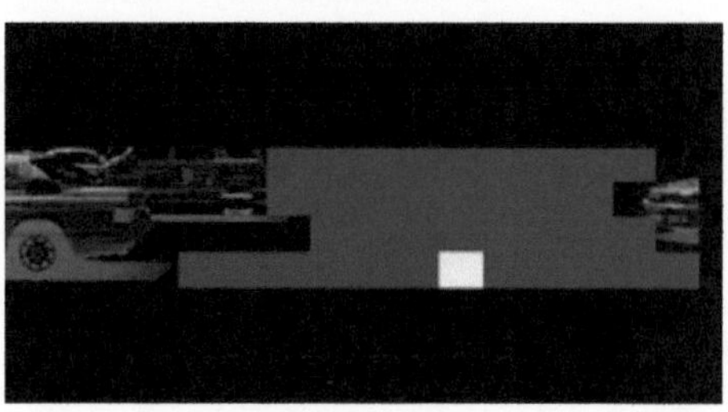

Fig. 7. Updating a background image.

The proposed algorithm for updating a background image can reduce the influences of environmental factors, noises, garbage on the road, movement of a camera, etc.

C. Estimating and Updating a Threshold Value to Binarize Images

Fig. 8 shows an example of histogram of a background subtraction image given by (1). We specified the range of gray-level values of shadows underneath vehicles from the characteristics of histograms of many background subtraction images. It is desirable to set the minimum in the range to the threshold value in order to binalize. Then, we propose an automatic estimation method for the threshold value.

The curve of the histogram is traced from the high gray-level value side, and the difference of the frequency that the arrow in Fig. 8 indicates is detected. We found that the value that is subtracted 35 from the gray-level value at the difference of the frequency approximates to the best threshold value from the statistical results of the traffic image processing. We confirmed that this method could be used in two weather conditions, fine and cloudy, in which there is a clear difference in the range of gray-level values. In order to estimate a reliable threshold value, the proposed method collects seven flames of input images in which a vehicle exists at least, obtains seven candidates for the threshold value, and selects the median of the seven values as the initial threshold value.

In addition, in each image processing, the threshold value is updated at input intervals. In the process, the fluctuations of gray-level values in the updated blocks in background image are reflected to the threshold value. As a result, the best threshold value can be maintained at any time.

III. EXPERIMENTAL RESULTS

We evaluated the validity of the proposed method by using traffic flow images captured from a camera position on the National Root 57 in Kumamoto City, Japan. Fig. 9 shows some vehicle detection results. The white boxes in Fig. 9 show the vehicles whose front- and rear-tires are detected, and the black boxes show the vehicles detected from the location of their shadows because both positions of tires cannot be measured.

The traffic flow images contain 454 vehicles, and 449 vehicles among those are detected. Therefore, the vehicle detection accuracy is 98.9%. The vehicles whose shadows cannot be seen because the vehicles are behind other vehicles are not included in the 454 vehicles.

The personal computer used in the experiments has a Pentium IV 3.73GHz CPU, a video capture board, and a 2GB RAM. The system is developed by Visual C++ under Windows XP. Vehicle detection is achieved more than nine frames per second even for heavy traffic scenes.

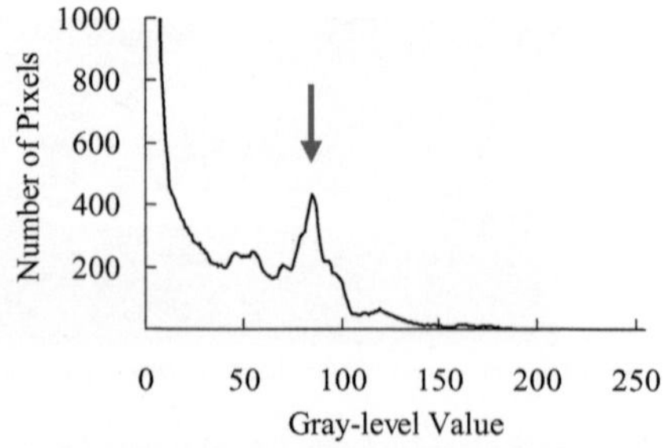

Fig. 8. Histogram of a background subtraction image.

Fig. 9(a)

Fig. 9(b)

Fig. 9. Some vehicle detection results.

Fig. 9(c)

Fig. 9(d)

Fig. 9(e)

Fig. 9. (Continued.)

IV. CONCLUSIONS

We presented a real-time vehicle detection method that uses the shadows underneath vehicles as the information to detect vehicles. The method distinguishes the size of each vehicle according to the distance between both ends of front- and rear-tires, and discriminates the lanes on which vehicles exist. The method contains a new algorithm for automatic renewal of the background image and the threshold value to binarize image in order to enhance the vehicle detection accuracy.

In the traffic monitoring targeted to vehicular side views, we confirmed that the shadows underneath vehicles can be used as robust information to detect vehicles.

REFERENCES

[1] R. Cucchiara, M. Piccardi, and P. Mello, "Image Analysis and Rule-based Reasoning for a Traffic Monitoring System," *Proc. IEEE/IEEJ/JSAI Int. Conf. Intelligent Transportation Systems*, pp.758-763, October 1999.

[2] A. Yoneyama, C. H. Yeh, and C. -C. J. Kuo, "Moving Cast Shadow Elimination for Robust Vehicle Extraction based on 2D Joint Vehicle/Shadow Models," *Proc. IEEE Conf. Advanced Video and Signal Based Surveillance*, pp. 229-236, July 2003.

[3] K. Uchimura, and K. Matsushima, "Traffic Flow Measurement Considering Occlusion," *Trans. IEE Japan*, vol. 122-C, no. 12, pp. 2120-2127, December 2002 (in Japanese).

[4] H. Kuboyama, and S. Ozawa, "Measurement of Heavy Traffic in a Tunnel from Image Sequences," *IEICE Trans.*, vol. J85-D-II, no.2, pp. 210-218, February 2002 (in Japanese).

[5] T. K. ten Kate, M. B. van Leewen, S. E. Moro-Ellenberger, B. J. F. Driessen, A. H. G. Versluis, and F. C. A. Groen, "Mid-range and Distant Vehicle Detection with a Mobile Camera," *Proc. 2004 IEEE Intelligent Vehicles Symposium*, pp. 72-77, June 2004.

[6] S. Imai, Y. Imai, and M. Iwahashi, "Traffic Monitoring System based on Minutia Matching," *Proc. 2005 IEICE General Conf.*, AS-2-2, March 2005 (in Japanese).

[7] T. Tanizaki, K. Ueda, K. Ikegaya, and I. Horiba, "A Detection of Wet Condition on Road Using a Background Image Renewal Processing," *IEICE Trans.*, vol. J80-D-II, no.9, pp. 2270-2277, September 1997 (in Japanese).

Invariant Control of Wastewater Aeration

A. A. Sniders, A. D. Laizans
Institute of Agricultural Energetics
Latvia University of Agriculture
5, J.Cakstes boulv.
Jelgava, LV3001, Latvia

***Abstract* – This paper considers the possibility of wastewater biological treatment process improvement by optimization of the wastewater aeration control system. Therefore, the mathematical modeling and simulation of this process is made. The control object is wastewater aeration tank with one input impact to control – the air blower capacity, one measured output parameter of the process – the dissolved oxygen concentration and the main perturbance as a load impact – the oxygen consumption. The block diagram for transient process simulation using the "Matlab" sub-program "Simulink" is compiled, allowing to start practical design of optimized air blowers control system. The simulations prove that the optimal solution for the transient process improvement in wastewater aeration tank, taking in account fluctuating load, is invariant control with perfect compensation of the load impact by self tuning of PID & Preemptive control circuits.**

I. INTRODUCTION

In accordance with the demands of the wastewater biological treatment technology the wastewater aeration process must be permanent and uninterruptible. The actual problems and tasks on the subject of development of the wastewater treatment technology, efficiency and automatic control have been widely searched and discussed last time [1, 2, 3, 4, 5, 6, 7]. This work surveys only one theme devoted to improvement of the wastewater aeration using an invariant control principle.

The optimal regulation of air blowers capacity in order to get maximum efficiency is the major and most important activity on the way to increase overall economical feasibility of communal wastewater biological treatment [6, 8, 9].

The steady-state and dynamic equations and characteristics of the wastewater aeration tank and aeration process control system have been compiled on the purpose to study, analyze and make recommendations for practical design of air blowers control system [8, 9, 10, 11]. For mathematical modeling of steady-state characteristics and the transient process in a wastewater aeration tank as an ecological object the ordinal and partial differential equations are used [12].

The analyses of the expressions for static gains prove that the wastewater aeration tank is a non-stationary object with variable steady-state parameters [11], but the dynamic transient process may be described by one-dimension model because of even air distribution all around the aeration tank surface, if membrane type air diffusers are used [8,9, 13].

Therefore the aeration tank of the biological wastewater treatment system is considered as the first order control object with one entrance impact – the air blower capacity L_{ab} (m^3/min), one output parameter to control – dissolved oxygen concentration C (g/m^3), and the main perturbance as a load – oxygen consumption q (g/min). The control object parameters for modeling are as follows (the data from real municipal wastewater biological treatment plant in Latvia): aeration tank capacity $Q = 500$ m^3/dn; set dissolved oxygen concentration $C_o = 2$ g/m^3; variable load $q = (40 - 100)$ g/min; variable air blower capacity $L_{ab} = (6 - 15)$ m^3/min.

The static gains and time constants of the oxygen concentration control system component parts are calculated on the basis of their technical parameters.

Laplace transforms, principles of process superposition and perturbance compensation from the automatic control theory are used to make the mathematical model and compose the simulation block-diagram of this system [14, 15, 16, 17].

The reference [17] helps to do the design of the structure and calculations of differentiation filter parameters for Preemptive control circuit of the wastewater aeration control system.

Transient process simulation and optimization are realized using "Matlab" sub-program "Simulink" and methods for operation with "Simulink" libraries [16, 18].

The main task of this study is to design the optimized wastewater aeration control system in order to improve the transient process of dissolved oxygen concentration in the wastewater aeration tank using Preemptive control circuit for oxygen consumption affect compensation on the wastewater aeration process [2,19].

II. WASTEWATER AERATION SYSTEM WITH INVARIANT CONTROL

The aeration control system is made as a two parameter control system with a closed loop PID control circuit and an open loop Preemptive control circuit (Fig. 1.).

The simulation block-diagram is compiled using standard blocks from "Simulink" libraries.

The block-diagram consists of a feedback control circuit with "Oxygen transmitter", "Input" Step signal generator, "PID control " circuit, "Frequency converter" with output signal limitation ± 50 Hz, "Air blower" and "Aeration tank" as a control object and an open loop "Preemptive control" circuit which consists of "Load transmitter" with gain 0.15 V/g/min, "Differentiation filter" with gain 0.5 and time constants 15 and 20 minutes and output "Signal limitation" ± 10 V.

The model of "Oxygen transmitter" consists of the "Transport Delay" block with delay time 0.5 min and first

K. Elleithy et al. (eds.), Advances in Computer, Information, and Systems Sciences, and Engineering, 99–102.

order inertial block with gain 2.0 V/(g/m^3) and time constant 1.2 min.

The block-diagram of "PID Control" is composed according to the algorithm installed in modern controllers [7, 9]and consists of proportional, integral, derivative and filter circuits.

The load physical affect on the system output is simulated by the "Load impact" block with gain 0.12 (g/m^3)/(g/min) and time constant 20 min.

Three components of the load are formed by signal generators – initially stable constant load, then steady growing component is added, and finally random component of the load is activated. To obtain the simulation response curves, the "Scopes" 1, 2, 3, 4 are set up.

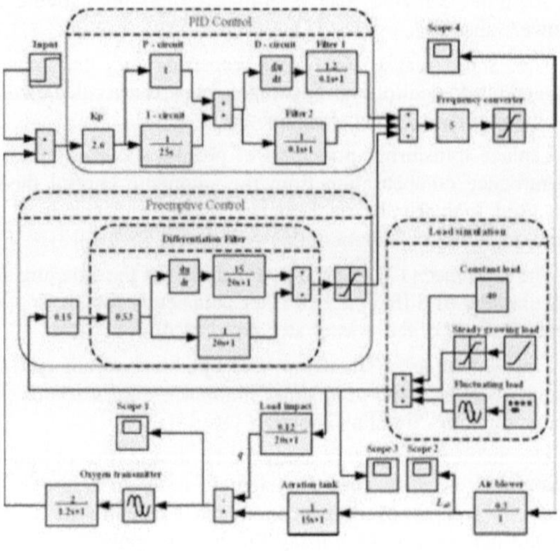

Fig.1.Simulation block diagram of the invariant PID & Preemptive control system using "Simulink".

The process variable – dissolved oxygen concentration $C\ (t)$ is measured by the "Oxygen transmitter", but perturbation $q\ (t)$ - by the "Load transmitter" .

The control system is based on feedforward from $q\ (t)$ as a variable load (oxygen consumption), and feedback from the process output $C\ (t)$ (oxygen concentration). As shown in Fig. 1., perturbance $q\ (t)$ has an influence to $C\ (t)$ via two circuits. The first circuit is through the aeration tank as a physical control object, the second one is through the frequency converter, and thou – directly to the air blower.

The main task of the control system is to ensure perfect compensation the physical influence $q\ (t)$ on $C\ (t)$ by use of the "Preemptive control" circuit which operates under the invariant control principle. If the action is done correctly, the controlled variable $C\ (t)$ should remain undisturbed even if $q\ (t)$ changes.

III. MATHEMATICAL MODELING OF THE INVARIANT CONTROL SYSTEM

On the basis of the simulation block diagram (Fig. 1.) we can compile an operator equation for the open loop system with PID & Preemptive control circuits:

$$C(s)=U_c(s)\cdot\left[G_c(s)\cdot K_{fc}\cdot K_{ab}\cdot G_a(s)\right]-\\ -q(s)\cdot\left[G_q(s)-G_f(s)\cdot K_{fc}\cdot K_{ab}\cdot G_a(s)\right] \quad (1)$$

where s – Laplace variable, s^{-1};
$C(s)$ – dissolved oxygen concentration in the aeration tank, g/m^3;
$U_c(s)$ – input voltage of PID controller, V;
$G_c(s)$ = 2.6(1 + 1/25 s)(1.2 s+ 1)/(0.1 s + 1) – transfer function of a real PID controller;
K_{fc} = 5 Hz/V – gain of the frequency converter;
K_{ab} = 0.3 (m^3/min)/Hz – gain of the air blower;
$G_a(s)$ = 1.0/(15s + 1) – transfer function of the aeration tank;
$q(s)$ – dissolved oxygen consumption as a load, g/min;
$G_q(s)$ = 0.12/(20s + 1) – transfer function of the load;
$G_f(s)$ – transfer function of the preemptive control circuit.

In conformity with the invariant control principle we can get the expression of the transfer function for the Preemptive control circuit. From formula (1) we can formulate such a condition. If

$$G_q(s)-G_f(s)\cdot K_{fc}\cdot K_{ab}\cdot G_a(s)=0\,, \quad (2)$$

then the load $q(s)$ does not affect the transient process in the aeration tank.

From formula (2) and the simulation block diagram (Fig. 1.), we get the expression for the Preemptive control circuit transfer function:

$$G_f(s)=\frac{G_q(s)}{K_{fc}\cdot K_{ab}\cdot G_a(s)}=\\ =\frac{0.12\cdot(15s+1)}{5\cdot 0.30\cdot 1\cdot(20s+1)}=\frac{0.08\cdot(15s+1)}{(20s+1)} \quad (3)$$

To obtain an operator equation for the closed loop feedback control system with the Preemptive control circuit we have to solve the following system of equations:

$$\begin{cases} U_c(s)=U_i(s)-U_o(s) \\ U_o(s)=C(s)-G_o(s) \\ C(s)=U_c(s)\left[G_c(s)\cdot K_{fc}\cdot K_{ab}\cdot G_a(s)\right]- \\ -q(s)\cdot\left[G_q(s)-G_f(s)\cdot K_{fc}\cdot K_{ab}\cdot G_a(s)\right] \end{cases} \quad (4)$$

Where $U_i(s)$ – input voltage of the system, V;

$U_o(s)$ – output voltage of the oxygen transmitter, V;
$G_o(s) = K_o/(\tau_o \cdot s+1) \cdot \exp(-s \cdot \tau)$ – transfer function of the oxygen transmitter;
$K_o = 2$ V/(g/m^3) – gain of the oxygen transmitter;
τ_o = 1.2 min – time constant of the oxygen transmitter;
τ = 0.5 min – transport delay of the oxygen transmitter.

After solution of the equation system (4) we obtain an operator equation which describes the transient process in a closed loop PID & Preemptive control system (5):

$$C(s) = (U_i(s) \cdot [G_c(s) \cdot K_{fc} \cdot K_{ab} \cdot G_a(s)] - -q(s) \cdot [G_q(s) - G_f(s) \cdot K_{fc} \cdot K_{ab} \cdot G_a(s)]) / /(1 + G_o(s) \cdot G_c(s) \cdot K_{fc} \cdot K_{ab} \cdot G_a(s)) \quad (5)$$

IV. SIMULATION PROCEDURE AND THE RESULTS

Load values used during simulation of the wastewater aeration system are presented in Fig. 2.

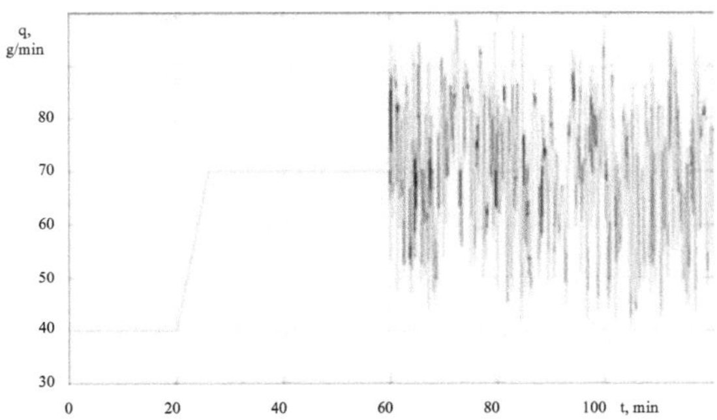

Fig. 2. Simulated oxygen consumption *q(t)* as a load with constant, steady growing and random components

Simulation of the wastewater aeration system is started with the constant load q = 40 g/min formed by the signal generator "Constant load".

After 20 minutes the steady growing load with slope 5 g/min/min is added to the constant load 40 g/min. "Load limitation" block limits the load growth maximum to 70 g/min.

After 60 minutes from the beginning the random fluctuating load component with amplitude ± 30 g/min and frequency 0.05 min^{-1} formed by the generator "Fluctuating load" is added. Therefore, the final load is 70 ± 30 g/min.

Simulated response of oxygen concentration *C(t)* in the feedback wastewater aeration control system with PID controller and constant oxygen consumption q = 40 g/min

shows that the transient process has maximal overshoot $\sigma_{max} \leq 8\%$ and transient response time $t_r \approx 7.5$ min (Fig. 3.).

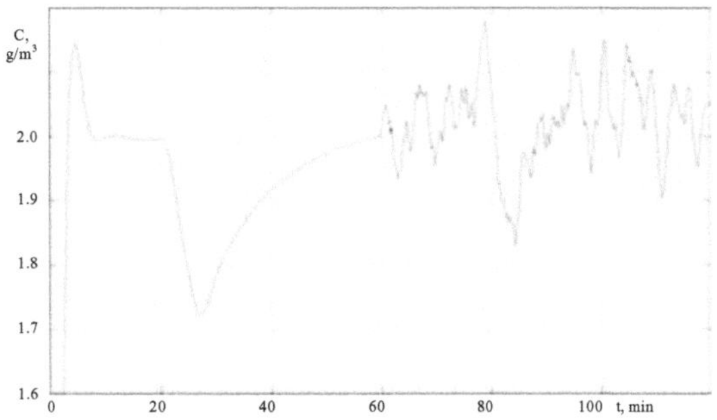

Fig. 3. Simulated response of oxygen concentration *C(t)* in the PID control system with variable load.

Optimal parameters of PID controller are calculated using Ziegler – Nichol's criteria and optimized during simulation. They are: gain of proportional circuit – 2.6; integral time constant – 11 min; derivative time constant – 1.2 min; filter time constant – 0.1 min.

We can make a conclusion that the control system without preemptive control circuit is sufficiently stable and fast in operation under constant load.

Under steady growing load with slope 5 g/min/min, the response curve has a great deviation – about 14 % (Fig. 3.), which slowly decreases to zero during 40 minutes. Under random fluctuating load the maximum amplitude of controlled variable oscillations is ± 8 %.

Simulation testifies that the feedback control system operation quality is insufficient under variable load (Fig. 3.).

If the parameters of the control object and the load exchange, the process stability and quality disimproves. Therefore we should introduce an adaptive controller with auto tuning.

Simulation results using PID and Preemptive control are presented in Fig. 4.

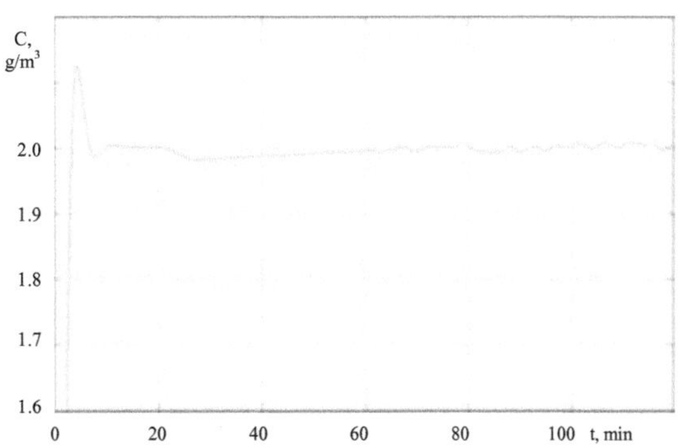

Fig. 4. Simulated response of oxygen concentration *C(t)* in the invariant PID & Preemptive control system with variable load.

The simulated response of oxygen concentration *C(t)* in the wastewater aeration control system with PID & Preemptive control circuits, the parameters of which are correctly calculated using the invariant control principle (Formula 3), shows that the process parameter *C(t)* remains practically constant under any type of loads applied – constant, steady changing or randomly fluctuating.

If the parameters of the Preemptive control circuit deviate from the optimal value on ± 8 %, deviation of the process parameter is only (1.5 ÷ 2) % (Fig. 4.).

The proportional gain, derivative time constant and filter time constant of PID controller are the same as for only feedback control, but the integral time constant changes from 11 to 25 minutes. If the parameters of the control object and load change, the parameters of PID control and Preemptive control circuits should be optimized. So an adaptive controller must be used.

CONCLUSIONS

Adaptive feedback PID control system with a Preemptive compensating circuit, the parameters of which are correctly calculated according to the invariant control principle, cancel out the affect of perturbance *q(t)* on the output variable *C(t)*.

Simulation shows that for feedback control system with PID Controller and optimally tuned parameters the maximum deviation of the control variable *C(t)* reaches 14 % under linear growing load *q(t)* with slope 5 g/min/min, but under random fluctuating load with amplitude ± 30 g/min and frequency 0.05 min^{-1} it is ± 8% - showing overall unacceptable quality of control (Fig. 3.).

Introduction in practice the invariant PID & Preemptive control system with sub-optimally tuned parameters of perturbance *q(t)* compensation circuit decreases the output variable *C(t)* deviations from ± 8 % (Fig. 3.) to ± (1.5 – 2) % (Fig. 4.), and through this makes the process of wastewater aeration more stable and more efficient.

REFERENCES

[1] G. Olsson, M. Nielsen, Y. Zuan, A. Lunggard – Jensen, *Instrumentation, Control and Automation in Wastewater Systems,* London: IWA Publishing, 2005, 264 p.

[2] P. Ingildsen, U. Jeppsson, G. Olsson," Dissolved oxygen controller based on on-line measurements of ammonia combining feed-forward and feedback", *Wat. Sci. Tech.,* Vol. 45, No. 4-5, 2002, pp. 453-460.

[3] C. Rosen, M. Larsson, U. Jeppsson, Z. Yuan, "A framework for extreme-event control in wastewater treatment", *Wat. Sci. Tech.* ,Vol. 45, No. 4-5, 2002, pp. 299-308.

[4] C. Rosen, G. Olsson, "Disturbance detection in wastewater treatment plants", *Wat. Sci. Tech.* , Vol. 32, No. 2, 1998, pp. 197-205.

[5] P. Ingildsen, G. Olsson, Z. Yuan ," A Hedging Point Strategy – Balancing Effluent Quality, Economy and Robustness in the Control of Wastewater Plants", *Proceedings ICA 2001 Vol. 2* , Malmo, Sweden, pp. 449-456, June 3-7, 2001.

[6] J. Ferrer, "Energy saving in the aeration process by fuzzy logic control", *Wat. Sci. Tech.,* Vol.38, No. 3, 1998, pp. 209 – 217.

[7] T. Zipper, N. Fleishmann, R. Haberl, "Development of a new system for control and optimization of small wastewater treatment plants using oxidation – reduction potencial", *Wat. Sci. Tech.,* Vol.38, No. 3, 1998, pp. 307– 314.

[8] A. Sniders, "Transient process modeling in wastewater aeration unit" , *Proceedings of the International Conference "Environment & Technology & Resources"*, Rezekne High School, pp. 268-274, June, 2003 (in Latvian).

[9] A. Sniders, "Simulation and energy-saving control in wastewater aeration system" , *Proceedings of the 3rd International Conference "Ecology and Agricultural Machinery"*, Saint- Petersburg, vol. II, pp. 294-302, June, 2002 (in Russian).

[10] A. Sniders, U. Skrastins,"Identification of the aeration tank as the object of oxygen transfer under static regime", *Proceedings of the Latvia Univ. of Agric.*, Jelgava, vol. 2 (279), 1995, pp. 79-85 (in Latvian).

[11] A. Sniders, "Static indices of wastewater pneumatic aeration", *Proceedings of the Latvia Univ. of Agric.* Jelgava, vol. (8), 1997, pp. 43-47 (in Latvian).

[12] S. E. Jorgensen, *Fundamentals of Ecological Modeling,* 2nd ed. Amsterdam: Elsevier Science B.V.,1994, 663 p.

[13] Fine Bubble Aeration, I TT Flygt AB, 1998, 7p.

[14] Carlos A. Smith, Armando B. Corripio, *Principles and Practice of Automatic Process Control,* 2nd ed. New York: John Willey & Sons, 1997, 768 p.

[15] A. Cropt , R. Davison , M. Hargraves *Engineering Mathematics. A Foundation for Electronics, Electrical, Communications and Systems Engineers* ,3rd ed. Marlow: Plarson Education Ltd., 2001, 969 p.

[16] Finn Haugen, *Modeling and Control of Dynamic Systems,* Skien, Norway: Control Consult, 1997, 242 p.

[17] J. Topcheyev, *Atlas for Automatic Control Systems Design,* Moscow: Engineering Industry, 1989, 752 p.(in Russian).

[18] A. Gultyayev, *Simulation in Windows Environment. Practical Appliance* ,Saint-Petersburg: Korona Print, 2001, 400 p. (in Russian).

[19] A. Sniders,"Feedback & Feedforward control of wastewater aeration", *Proceedings of the International Conference "Advanced technologies for energy producing and effective utilization"*, Jelgava, pp. 124-129, June 2004.

Denoising of Infrared Images by Wavelet Thresholding

Dietmar Wippig, Bernd Klauer and Hans Christoph Zeidler
Department of Electrical Engineering
Helmut-Schmidt-University
22043 Hamburg, Germany
{dietmar.wippig, bernd.klauer, h.ch.zeidler}@hsu-hamburg.de

Abstract - **Infrared imaging devices become more important for civil and military navigation. Noisy images are often a problem especially at poor visibility. Therefore denoising could improve the image quality by wavelet thresholding. Different popular threshold estimation methods are compared with regard to the hard-, soft-, firm- and non-negative garrote thresholding function. Experimental results show that the BayesShrink thresholding estimator applied on the non-negative garrote delivers the best results.**

***Keywords*: Image Denoising, Wavelet Thresholding, Dyadic Wavelet Transform, Infrared Images**

I. Introduction

Infrared imaging devices support monitoring of navigation and ship recognition systems. Together with radar they are often the only means to visualize the environment at night or poor visibility. Especially at poor visibility noise degrades the infrared images significantly. Thus the target of this paper is to choose a method to denoise these infrared images properly and fast for streaming images.

Fig. 1: Example of an infrared ship image

Wavelet denosing has become more and more popular for image denoising, because additive noise can be removed while preserving important features of the image. As shown by Donoho and Johnstone [1] [2] [3] [4] [5] thresholding the coefficients is the most important task in wavelet denoising. Therefore, different algorithms for computing the threshold and thresholding functions were investigated for denoising infrared images.

In this work we use the discrete dyadic wavelet transform proposed by Mallat and Zhong [6] to denoise the infrared images. The noise distribution of the infrared images is unknown, so we assume Gaussian noise. Under this assumption orthogonal wavelet transforms have the advantage that the variance distributions of the wavelet coefficients are identical at each level [7].

Our paper is structured as follows. Section 2 gives a short description of the discrete dyadic wavelet transform. Section 3 introduces the concept of wavelet denoising. Section 4 describes the threshold estimation methods to be compared. Experimental results are given in section 5. Finally the conclusion is given in section 6.

II. Discrete Dyadic Wavelet Transform

The dyadic wavelet transform of $f(x)$ at scale 2^j is

$$W_{2^j} f(x) = f * \varphi_{2^j}(x),$$

where $\varphi_{2^j}(x)$ is the dilation of wavelet $\varphi(x)$ by a factor 2^j.

The wavelet can be designed to display local maxima for sharp variations of f, which is important for analyzing the properties of signals and images. This can be done by choosing a wavelet that is the first-order derivative of a smoothing function $\theta(x)$. A smoothing function $\theta(x)$ is any function whose integral is equal to 1 and which converges to zero at infinity.

In order to have a fast filterbank implementation in discrete time Mallat and Zhong [6] chose $\theta(x)$ and correspondingly $\varphi(x)$ using the filters $H(\omega)$ and $G(\omega)$

$$H(\omega) = e^{i\omega/2}(\cos(\omega/2))^{2n+1}$$
$$G(\omega) = 4ie^{i\omega/2}\sin(\omega/2)$$

to derive

$$\theta(\omega) = \left(\frac{\sin(\omega/4)}{\omega/4}\right)^{2n+2}$$

$$\varphi(\omega) = i\omega\left(\frac{\sin(\omega/4)}{\omega/4}\right)^{2n+2}$$

The corresponding function in the time domain of $\varphi(\omega)$ is a quadratic spline with compact support and the time-domain function of $\theta(\omega)$ is a cubic spline. For our experiments we chose $n = 1$ as proposed in [6].

K. Elleithy et al. (eds.), Advances in Computer, Information, and Systems Sciences, and Engineering, 103–108.

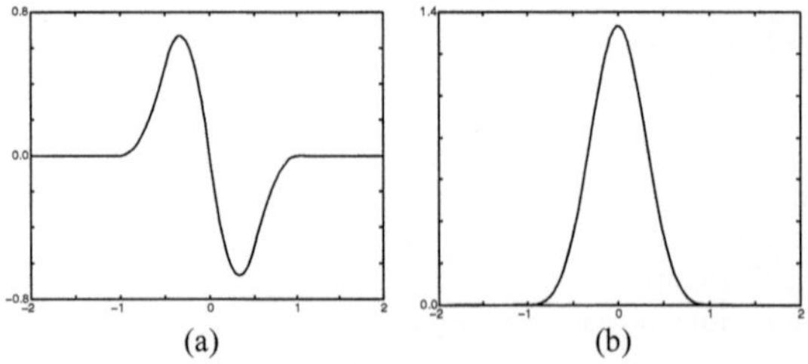

Fig. 2: (a) Quadratic spline wavelet $\varphi(x)$ with $n=1$, (b) Corresponding cubic spline smoothing function $\theta(x)$

The function $f(x)$ can be reconstructed from its dyadic wavelet transform $W_{2^j}f(x)$

$$f(x)=\sum_{j=-\infty}^{+\infty} W_{2^j}f*\chi_{2^j}(x),$$

where $\chi_{2^j}(x)$ is the dilation of the reconstructing wavelet $\chi(x)$ by a factor 2^j.

For the reconstructing filter bank a filter $K(\omega)$ is defined

$$K(\omega)=\frac{1-|H(\omega)|^2}{G(\omega)}$$

In two dimensions the wavelet transform of a function $f(x,y)$ at scale 2^j has the two components:

$$W_{2^j}^1 f(x)=f*\varphi_{2^j}^1(x)$$
$$W_{2^j}^2 f(x)=f*\varphi_{2^j}^2(x)$$

The function $f(x)$ can be reconstructed from its dyadic wavelet transform:

$$f(x,y)=\sum_{j=-\infty}^{+\infty}(W_{2^j}^1 f*\chi_{2^j}^1(x,y))+(W_{2^j}^2 f*\chi_{2^j}^2(x,y))$$

Using the quadratic spline wavelt from above another filter $L(\omega)$ is defined for the reconstruction filter bank:

$$L(\omega)=\frac{1+|H(\omega)|^2}{G(\omega)}$$

The filter bank implementation of the discrete dyadic wavelet transform is shown in figure 3.

For computing the filter bank in spatial domain the implementation described in [8] is used.

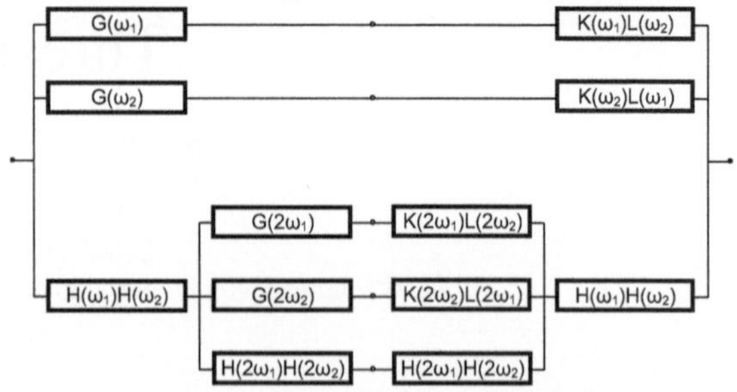

Fig. 3 Filter bank implementation of a two dimensional discrete dyadic wavelet transform decomposition (left) and reconstruction (right) for two sales.

III. Wavelet Denoising

Wavelet denoising decomposes the image into wavelet series, extracts important wavelet coefficients through thresholding algorithms, and then reconstructs the image with the thresholded wavelet coefficients. The optimal manipulation of the wavelet coefficients and therefore the noise estimation of the wavelet coefficients are the most challenging issues in this context. Most approaches assume that the images are corrupted by independent and identically distributed (i.i.d.) zero mean, white Gaussian noise ε with standard deviation σ. Generally a noisy image can be modeled by

$$g(x,y)=f(x,y)+\varepsilon(x,y).$$

The noise of the corresponding coefficients $W_{2j}^1 f(x)$ and $W_2^j f(x)$ may be considered Gaussian distributed and can be estimated by the robust median estimator [1]:

$$\hat{\sigma}_{2^j}^1=\frac{1}{0.6745}Median\left|W_{2^j}^1 f\right|$$
$$\hat{\sigma}_{2^j}^2=\frac{1}{0.6745}Median\left|W_{2^j}^2 f\right|$$

Some approaches only estimate the variance in the diagonal subband of the first scale. For a better adaptivity we always calculate the local variance for every subband and scale.

In this work we consider the four different thresholding functions: hard-, soft-, firm- [10] and non-negative garrote [11]. These functions have in common that depending on an estimated threshold λ small wavelet coefficients should be set to zero. These coefficients represent the noise. The other coefficients should not be reduced too much, because they contain the image details. The threshold functions T are defined as follows:

$$T_{hard}(x,\lambda)=\begin{cases} x & , |x|\geq \lambda \\ 0 & , |x|<\lambda \end{cases}$$

$$T_{soft}(x,\lambda)=\begin{cases} x-\lambda & , x\geq \lambda \\ 0 & , |x|<\lambda \\ x+\lambda & , x\leq \lambda \end{cases}$$

$$T_{firm}(x,\lambda_1,\lambda_2)=\begin{cases} x & , |x|>\lambda_2 \\ \operatorname{sgn}(x)\frac{\lambda_2(|x|-\lambda_1)}{\lambda_2-\lambda_1} & , \lambda_1<|x|\leq \lambda_2 \\ 0 & , |x|\leq \lambda_1 \end{cases}$$

$$T_{nng}(x,\lambda)=\begin{cases} x-\lambda^2/x & , |x|>\lambda \\ 0 & , |x|\leq \lambda \end{cases}$$

Wavelet denoising can be summarized as follows:

1. Compute the wavelet transform of the image g up to J scales.
2. Calculate the threshold λ and threshold the coefficients by T.
3. Reconstruct the image from the thresholded wavelet coefficients.

IV. THRESHOLD ESTIMATION METHODS

Donoho and Johnstone propose a universal threshold [2], which only depends on the number of image pixels N and the estimated standard deviation $\hat{\sigma}$. They called it VisuShrink, because of the "noise-free" character of the reconstructed images.

$$\lambda_{VisuShrink}=\hat{\sigma}\cdot\sqrt{2\cdot l\,n(N)}$$

In [2] Donoho and Johnstone propose a minimax threshold called RiskShrink. The minimax threshold λ^* is defined as the value of λ, which minimizes the quantity

$$\Lambda_N^*=\inf_{\lambda}\sup_{\theta}\left\{\frac{R_\lambda(\theta)}{N^{-1}+R_{oracle}(\theta)}\right\}.$$

For diagonal linear projection the oracle risk is

$$R_{oracle}(\theta)=\min(\theta^2,1).$$

The L_2 risk functions can be found in the appendix.

SureShrink [3] is a subband-adaptive threshold method which calculates a threshold based upon Stein's unbiased risk estimator (SURE). The threshold is selected to minimize the L_2 risk when applying the thresholding function. The SURE estimator estimates the MSE depending on the thresholding function [3][11]. For an ordered N-dimensional vector X with $X\sim N(0,1)$ the SURE estimator is:

$$SURE^{soft}(\lambda,X)=N-2\cdot\#\{i:|X_i|\leq\lambda\}+\sum_{i=1}^{N}\min|X_i|,\lambda)^2$$

$$SURE^{nng}(\lambda,X)=1+\sum_{i=1}^{\#\{i:|X_i|\leq\lambda\}}X_i^2-2\cdot\#\{i:|X_i|\leq\lambda\}+(\lambda^4+2\lambda^2)\sum_{i=\#\{i:|X_i|\leq\lambda\}+1}^{N}X_i^{-2}$$

where X is the set of wavelet coefficients in a subband and $\#\{i:|X_i|\leq\lambda\}$ is the number of coefficients set to zero.

To find the optimal threshold we have to minimize the SURE estimator:

$$\lambda_{SureShrink}=\arg\min_{\lambda}(SURE(\lambda,X))$$

A subband-adaptive threshold can be calculated with the BayesShrink method [12] for soft thresholding:

$$\lambda_{BayesShrink}=\frac{\hat{\sigma}^2}{\hat{\sigma}_X}$$

where

$$\hat{\sigma}x=\sqrt{\max\left(\hat{\sigma}_Y^2-\hat{\sigma}^2,0\right)}$$

and

$$\hat{\sigma}_Y^2=\frac{1}{N}\sum_{i=1}^{n}\sum_{j=1}^{m}g_{i,j}^2$$

If $\hat{\sigma}_X$ is zero, all coefficients are set to zero by taking $\lambda=\max(|g_{ij}|)$.

Another method to calculate a threshold by minimizing the MSE is the generalized cross validation (GCV) [13] which does not need to estimate the noise variance. The threshold that minimizes the GCV function is calculated for soft thresholding:

$$GCV(\lambda)=\frac{\frac{1}{N}\cdot\|g-T(g,\lambda)\|}{\left[\frac{N_0}{N}\right]^2}$$

$$\lambda_{GCVShrink}=\arg\min_{\lambda}(GCV(\lambda,X))$$

N is the number of wavelet coefficients g of the subband and N_0 is the number of coefficients g set to zero by the soft thresholding function with parameter λ. The GCV function of an example is shown in figure 4.

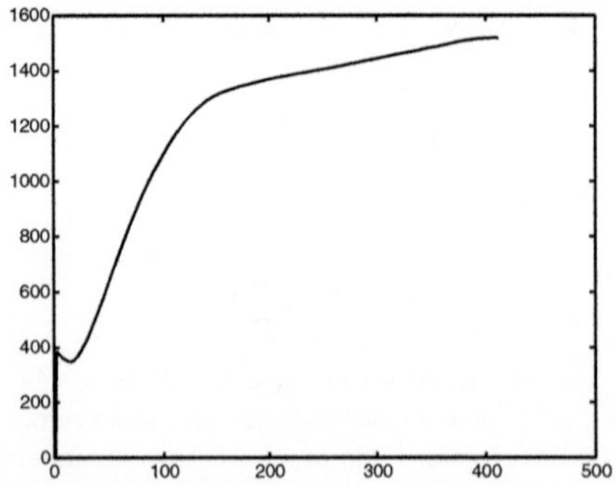

Fig. 4: GCV Function example.

Further wavelet thresholding approaches [14][15][16] consider the neighborhood characteristics of the wavelet coefficients. NeighShrink [14] takes the square sum of the wavelet coefficients within a centered neighbourhood window $B_{i,j}$ around each wavelet coefficient $g_{i,j}$ and the VisuShrink threshold λ to calculate a shrinkaging factor as:

$$\beta_{i,j} = \left(1 - \frac{\lambda^2_{VisuShrink}}{S^2_{i,j}}\right)_+ .$$

The factor β can be expressed in a soft threshold $\lambda(i,j)$:

$$\lambda_{NeighShrink}(i,j) = |g_{i,j}| \cdot \frac{\lambda^2_{VisuShrink}}{S^2_{i,j}}$$

where

$$S_i^2 = \sum_{(k,l)\in B_{i,j}} g_{k,l}^2 .$$

The best windows size is found by 3x3 [14].

In [15] another threshold based on the VisuShrink threshold λ is proposed according to the means of the absolute values of the wavelet coefficients within a centered window $B_{i,j}$. The threshold is defined for the hard and soft thresholding function by

$$\lambda_{NeighShrink2}(i,j) = \frac{\left(M_{i,j} - \min_{i,j} m_{i,j}\right) \cdot \lambda_{VisuShrink}}{\left(m_{i,j} - \min_{i,j} m_{i,j}\right)}$$

where

$$m_{i,j} = \frac{1}{N} \sum_{(k,l)\in B_{i,j}} |g_{k,l}|$$

$$M_{i,j} = \frac{1}{N} \sum_{i=1}^{n} \sum_{j=1}^{m} m_{i,j} .$$

[16] proposes a spatially adaptive wavelet thresholding using the same threshold estimator like the BayesShrink method, but considering the local characteristics in $\hat{\sigma}_X$. They take the means of the eight nearest neighbors of a coefficient $g_{i,j}$ as the context of the coefficient $z_{i,j}$. Thus the variance of $g_{i,j}$ can be estimated from other coefficients with contexts in an interval around $z_{i,j}$ The L closest points above and below $z_{i,j}$ are taken building the set of coefficients $g_{i,j}$ in the moving window $B_{i,j}$. To ensure that $B_{i,j}$ contains enough points estimating the variance $L = \max(50, 0.02^* N)$ is choosen [17].

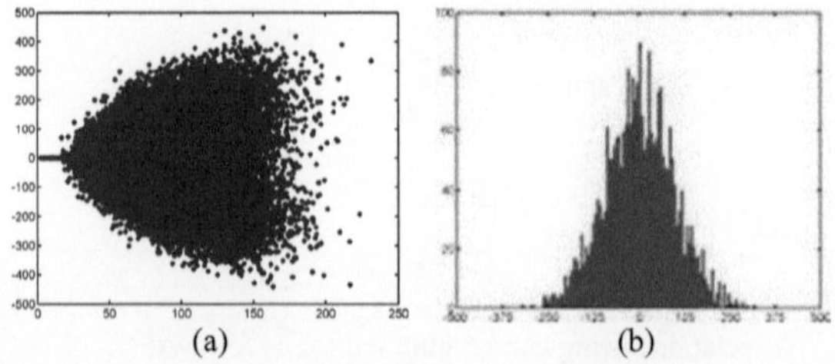

Fig. 5: (a) Plot of $z_{i,j}, g_{i,j}$, (b) Histogram of $g_{107,184}$ in $B_{107,184}$ which includes the points corresponding to $z_{i,j}$ [43.5,53.5].

The local threshold is then:

$$\lambda_{AdaptShrink}(i,k) = \frac{\hat{\sigma}}{\hat{\sigma}_X(i,j)}$$

where

$$\hat{\sigma}_X(i,j) = \max\left(\frac{1}{2L+1} \sum_{(k,l)\in B_{i,j}} g_{k,l}^2 - \sigma^2, 0\right).$$

[18] and [19] adapt the local variance estimation of Wiener filtering to the wavelet domain by a minimum mean-square error (MMSE) estimator:

$$\hat{g}(i,j) = \frac{\sigma^2_{i,j}}{\sigma^2_{i,j} + \hat{\sigma}^2} g_{i,j}$$

The local variances $\sigma^2_{i,j}$ in a local neighborhood $B_{i,j}$ centered at $g_{i,j}$ are computed using an maximum likelihood (ML) estimator:

$$\sigma^2_{i,j} = \left(\frac{1}{N} \sum_{(k,l)\in B_{i,j}} g_{i,j}^2 - \hat{\sigma}^2\right)_+$$

The LAWMLShrink [18] shrinkage factor can be transformed in the following soft threshold:

$$\lambda_{LAWMLShrink}(i,j) = |g_{i,j}| \cdot \frac{\hat{\sigma}^2}{S^2_{i,j}}$$

where

$$S^2_{i,j} = \sum_{(k,l)\in B_{i,j}} g_{k,l}^2 .$$

While in [18] the window size has to be choosen, in [19] the optimal window size is selected by a Bootstrap method taking the minimum local variance for each location. This leads to taking the smallest local square sum for soft thresholding.

BivariateShrink [20] is a locally adaptive denoising algorithm using the bivariate shrinkage function by a maximum a posteriori (MAP) estimator:

$$\hat{g}(i,j)=\frac{\left(\sqrt{g_{i,j}^2+\tilde{g}_{i,j}^2}-\frac{\sqrt{3}\hat{\sigma}^2}{\hat{\sigma}_{i,j}}\right)_+}{\sqrt{g_{i,j}^2+\tilde{g}_{i,j}^2}}\cdot g_{i,j}$$

where $\tilde{g}_{i,j}$ is the corresponding coefficient to $g_{i,j}$ in the coarser scale.

The local standard deviation is estimated in a local neighborhood $B_{i,j}$ centered at $g_{i,j}$ by:

$$\hat{\sigma}_{i,j}=\sqrt{\left(\hat{\sigma}_y^2-\hat{\sigma}^2\right)_+}$$

where

$$\hat{\sigma}_y^2=\frac{1}{M}\sum_{(k,l)\in B_{i,j}} g_{k,l}^2 .$$

Adapted to the soft thresholding function the bivariate shrinkage function leads to the threshold:

$$\lambda_{BivariateShrink}(i,j)=\frac{\frac{\sqrt{3}\hat{\sigma}^2}{\hat{\sigma}_{i,j}}}{\sqrt{g_{i,j}^2+\tilde{g}_{i,j}^2}}\cdot\left|g_{i,j}\right|$$

We consider more multiscale approaches as [21],[22] and [23] which are more sophisticated but have also a higher computational demand. Due to their high complexity we will not consider these approaches.

V. RESULTS

Having only noisy infrared images we start our experiments on the well-known image Lena. We choose the 512x512 8-bit gray-level images from [24] and as quality measure we use the peak signal-to-noise ratio (PSNR) which is defined in [20] as

$$PSNR=20\log_{10}\left(\frac{256}{RMS}\right)$$

where

$$RMS=\sqrt{\frac{1}{N}\sum_{i=1}^{n}\sum_{j=1}^{m}(f_{i,j}-g_{i,j})^2}$$

Adding Gaussian noise with standard deviation 10 we have a PSNR value of 18.25 before applying wavelet denoising in 3 scales. Each thresholding function is used with all thresholds estimated by the estimation functions. As for all estimation functions, except RiskShrink the thresholds for firm thresholding cannot be calculated, we take the computed threshold as λ_1 and the threshold calculated by VisuShrink as λ_2. Except for RiskShrink and SureShrink we also take the computed threshold for the non-negative garrote function. As it can be seen in the appended table I, hard thresholding has for the most estimation functions higher PSNR values than the other thresholding functions, but always lower visual quality. For the other thresholding functions BayesShrink and GcvShrink show the best numerical results. So we take them for our further experiments on the infrared images.

We found out that for both thresholding estimation functions the non-negative garrote gives the best results. On some test images GcvShrink has slightly better numerical and visual results than BayesShrink. But BayesShrink is more robust and therefore delivers a better mean quality (Fig. 6).

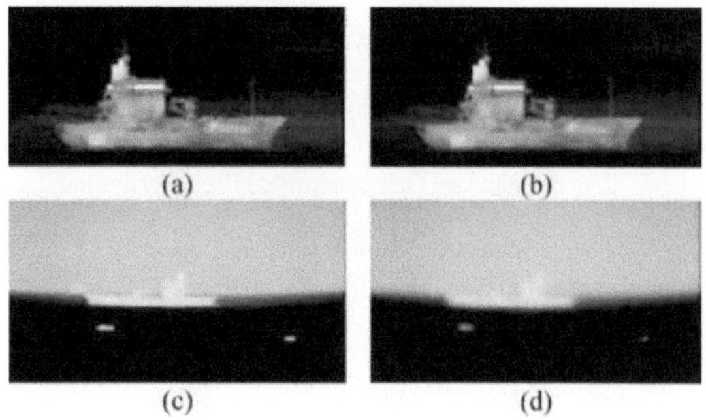

Fig. 6: Example results (a) and (c) with BayesShrink (b) and (d) with GcvShrink

VI. CONCLUSION

The target of our investigations was to improve the visual quality of infrared images. To achieve our goal we compared different estimation functions for threshold values and typical threshold functions. We have shown that the BayesShrink threshold estimator together with the non-negative garrote function delivers best results for our image set.

Our future work will be focused on the real-time applicability of the algorithms for streaming images and on proving our results on a bigger image set.

APPENDICES

I. L_2 RISK FUNCTIONS

The L_2 risk functions of the thresholding functions are defined in [9][10][11]as follows:

$$R_\lambda^{hard}(\theta)=1+\left(\theta^2-1\right)\left(\Phi(\lambda-\theta)-\Phi(-\lambda-\theta)\right)+(\lambda+\theta)\phi(\lambda+\theta)+(\lambda-\theta)\phi(\lambda-\theta)$$

$$R_\lambda^{soft}(\theta)=1+\lambda^2+\left(\theta^2-\lambda^2-1\right)\left(\Phi(\lambda-\theta)-\Phi(-\lambda-\theta)\right)-(\lambda-\theta)\phi(\lambda+\theta)-(\lambda+\theta)\phi(\lambda-\theta)$$

$$R_{\lambda_1\lambda_2}^{firm}(\theta) = \theta^2\left(\Phi(\lambda_1-\theta)-\Phi(-\lambda_1-\theta)\right) +1-\Phi(\lambda_2-\theta)+\Phi(-\lambda_2-\theta) +\Psi\left(\lambda_1-\theta,\lambda_2-\theta,r_2,r_1(\theta-\lambda_2)\right) +\Psi\left(\lambda_1+\theta,\lambda_2+\theta,r_2,-r_1(\theta+\lambda_2)\right) +(\lambda_2-\theta)\phi(\lambda_2-\theta)+(\lambda_2+\theta)\phi(\lambda_2+\theta)$$

$$R_\lambda^{nng}(\theta) = R_\lambda^{hard}(\theta)+2\theta\lambda^2 A_\lambda(\theta)+\lambda^4 B_\lambda(\theta) -2\lambda^2\left\{1-\Phi(\lambda-\theta)+\Phi(-\lambda-\theta)\right\}$$

where Φ is the standard Gaussian probability distribution function

ϕ is the standard Gaussian probability density function

$$r_1 = \frac{\lambda_1}{(\lambda_2-\lambda_1)},\ r_2 = \frac{\lambda_2}{\lambda_2-\lambda_1},\ with\, \lambda_1 < \lambda_2$$

$$\Psi(a,b,c,d) = (c^2+d^2)\left(\Phi(b)-\Phi(a)\right) + c\phi(a)(ac+2d)-c\phi(b)(bc+2d)$$

$$A_\lambda(\theta) = \int_\lambda^\infty \left\{\frac{\phi(x-\theta)-\phi(x+\theta)}{x}\right\}dx$$

$$B_\lambda(\theta) = \int_\lambda^\infty \left\{\frac{\phi(x-\theta)+\phi(x+\theta)}{x^2}\right\}dx$$

II. PSNR Values Of Lena Image With PSNR 18.25

Function	Hard	Soft	Firm	Garrote
VisuShrink	24.142	22.927	24.142	23.227
RiskShrink	24.302	23.498	24.141	24.057
SureShrink	27.131	25.797	26.631	26.221
BayesShrink	27.122	26.169	26.814	26.889
GcvShrink	27.058	26.193	26.839	26.916
NeighShrink	27.146	26.054	26.606	26.698
NeighShrink2	26.320	25.184	25.907	25.851
AdaptShrink	25.395	26.040	26.330	26.368
LAWMLShrink	21.176	25.928	26.168	25.869
LAWMLShrink2	23.093	25.767	26.370	26.313
BivariateShrink	25.968	25.289	25.845	25.853

Table 1:
PSNR Values of Lena image with PSNR 18.25

References

[1] D.L. Donoho: "Wavelet shrinkage and W.V.D.: A 10-minute tour" *Proceedings International Conference on Wavelets and Applications*, Toulouse, France , 1992

[2] D.L. Donoho and I.M. Johnstone: "Ideal spatial adaption via wavelet shrinkage" *Biometrica*, Vol. 81, pp. 425-455, 1994

[3] D.L. Donoho and I.M. Johnstone: "Adapting to unknown smoothness via wavelet shrinkage" *Journal of American Statistical Association*, Vol. 90, No. 432, pp. 1200-1224, 1995

[4] D.L. Donoho and I.M. Johnstone: "Wavelet shrinkage: Asymptopia?" *Journal of the Royal Statistical Society Series B 57*, pp. 301-369, 1995

[5] D.L. Donoho: "De-Noising by Soft-Thresholding" *IEEE Transactions on Information Theory*, Vol. 41, No. 3, pp. 613-627, 1995

[6] S. Mallat and S. Zhong: "Characterization of Signals from Multiscale Edges" *IEEE Transactions on Pattern Analysis and Machine Intelligence*, Vol. 14, No. 7, pp. 710-732, 1992

[7] S. Mallat and W.-L. Hwang: "Singularity Detection and Processing with Wavelets" *IEEE Transactions on Information Theory*, Vol. 38, No. 2, pp. 617-643, 1992

[8] I. Koren and A. Laine: "A Discrete Dyadic Wavelet Transform for Multidimensional Feature Analysis" *Time-Frequency and Wavelet Transform in Biomedical Engineering*, M. Akay (Editor), IEEE Press, New York, 1997

[9] A.G. Bruce, H.Y. Gao: "Understanding WaveShrink: Variance and Bias Estimation" *Research Report*, No. 36, StatSci Division of MathSoft, 1996

[10] A.G. Bruce, H.Y. Gao: "WaveShrink with Firm Shrinkage" *Research Report*, No. 39, StatSci Division of MathSoft, 1996

[11] H.Y. Gao: "Wavelet shrinkage denoising using the non-negative garrote" *Journal of Computational and Graphical Statistics*, Vol. 7, No. 4, pp. 469-488, 1998

[12] S.G. Chang, B. Yu, M. Vetterli: "Adaptive Wavelet Thresholding for Image Denoising ans Compression" *IEEE Transaction on Image Processing*, Vol. 9, No. 9, pp.
1532-1546, 2000

[13] M. Jansen: "Wavelet Thresholding and Noise Reduction" *PHD Thesis*, Department of Computer Science, KULeuven (Leuven, Belgium), 2000

[14] G.Y. Chen, T.D. Bui, A. Krzyzak: "Image denoising using neighbouring wavelet coefficients" *Proceedings of ICASSP 2004*, Vol. 2, pp. 917-920, 2004

[15] W. Shengqian, Z. Yuanhua, Z. Daowen: "Adaptive shrinkage de-noising using neighbourhood characteristic" *Electronics Letters*, Vol. 38, No. 11, pp.502-503, 2002

[16] S.G. Chang, B. Yu, M. Vetterli: "Spatially adaptive wavelet thresholding with context modeling for image denoising" *Proceedings of International Conference on Image Processing*, Vol. 1, pp.535-539, 1998

[17] S.G. Chang, B. Yu, M. Vetterli: "Spatially Adaptive Wavelet Thresholding with Context Modeling for Image Denoising" *IEEE Transactions on Image Processing*, Vol. 9, No. 9, pp.1522-1531, 2000

[18] M.K. Mihacak, I. Kozintsev, K. Ramchandran: "Low-Complexity Image Denoising Based on Statistical Modeling of Wavelet Coefficients" *IEEE Signal Processing Letters*, Vol. 6, No. 12, pp. 300-303, 1999

[19] M.K. Mihacak, I. Kozintsev, K. Ramchandran: "Spatially adaptive statistical Modeling of wavelet image Coefficients and its application to denoising" *Proceedings of ICASSP*, Vol. 6, pp. 3253-3256, 1999

[20] L. Sendur, I.W. Selesnick: "Bivariate Shrinkage with local variance estimation" *IEEE Signal Processing Letters*, Vol. 9, No. 12, pp. 438-441, 2002

[21] Y. Xu, J.B. Weaver, D.M. Healy, J. Lu: "Wavelet Transform Domain Filters: A Spatially Selective Noise Filtration Technique" *IEEE Transaction on Image Processing*, Vol. 3, No. 6, pp. 747-758, 1994

[22] T.-C. Hsung, D. P.-K. Lun, W.-C. Siu: "Denoising by Singularity Detection" *IEEE Transaction on Signal Processing*, Vol. 47, No. 11, pp. 3139-3144, 1999

[23] J. Scharcanski, C.R. Jung, R.T. Clarke: "Adaptive Image Denoising using Scale and Space Consistency" *IEEE Transaction on Image Processing*, Vol. 11, No. 9, pp. 1092-1101, 2002

[24] Mike Wakin: "Standard Test Images" *Webpage*, http://www.ece.rice.edu/~wakin/images , 2005

Intelligent Technologies for the Conformity Assessment in the Chain of Agricultural Production

G.A. Moskvin
Latvia University of Agriculture, e-mail: logicor@cs.llu.lv
E.G. Spakovica
Latvia University of Agriculture, e-mail: evelina@aldems.lv

Abstract. **In this article are discussed main results of new intelligent technology for the conformity assessment in application area of agricultural production. In the chain of agricultural production the agricultural product's properties, which have to be fixed by means of measuring devices, can be so different that there can be no "essence". Therefore on the basis of modeling of "intellect of consumer " a new intelligent technology and compact low- cost electronic device "artificial tongue" Logicor-AT has been developed, in order to make sure the conformity assessment and protection of consumers interests and their rights.**

Key words: **intelligent technologies, conformity assessment, consumers' protection.**

I. INTRODUCTION

Intelligent techniques for measuring human sensory response to food texture have been undertaken since 1980s to study relations between physiological and sensory testing of perception. Since the half of the eighties the technological mimic of the main functions of human olfaction became possible. Since that, an increasing number of researchers have dedicated their efforts to improve the original idea pursuing the fabrication of electronic tongue.

Practical applications, in a wide number of cases, appeared in the literature, and in the nineties some companies have introduced the electronic tongue technology to the market. Recently in food industry and in agriculture for quality control of agricultural products are even more often applied sensors. Much research was done in order to find new and more diverse sensors, and to date there are several companies offering ready-to-use electronic tongue. Traditional chemical and biological sensors are highly specific for certain compounds. Large, expensive gas chromatographs, for instance, which separate organic chemicals based on molecular weight through coiled glass tubes, must be attachedto at least four different detectors to identify complicated compounds. Technologies today are not easily adaptable to the field and theones that are happen to be both bulky and expensive.

Historically the first instrument with artificial intellect in Latvia has been built in Agricultural University at Jelgava 15 years ago under supervision of Prof. Gennady A. Moskvin. It was an artificial tongue-device based on couple of electrodes and signal generating recognition parts (Fig.1).

Fig. 1. Artificial tongue "Logicor-AT" - intelligent device for consumer's protection

For ages, the human tongue has been an important tool in assessing the quality of many products, food and agricultural products being good examples. While all other parts of production processes, including as in food industry, were getting more and more automated, there was still no "objective" instruments for using the "subjective" information confined in the taste of products. This changed in 1988, when Gennady A. Moskvin introduced the new concept of an electronic tongue. The "Artificial Tongue" (AT) ES and AI device is an electronic instrument, which consists of systems for data acquisition and data analysis. Analysts expect that as regulations, pertaining to food and agricultural products testing, continue to be adopted, the shift toward rapid nonconformity assessment methods will continue. Normally such standards become effective because the majority of producers agree to them. Seldom they are related to safety, but

K. Elleithy et al. (eds.), Advances in Computer, Information, and Systems Sciences, and Engineering, 109–114.

often to a characteristic quality which industry considers as to be useful to establish credibility for the market. These standards are commonly referred to as "commodity standards" or "standards of identity".

II. ESTIMATION OF AGRICULTURAL PRODUCT'S TESTING INDUSTRY

The purpose of a quality control program is to acquire reliable information on all the attributes of a product, which affect its quality. The methods used to measure quality can be subjective, as in taste tests or they can be objective, such as physical, chemical or microscopic analysis. Subjective methods are based on the opinions of the examiners and because they require the use of our various senses, they are often called sensory analysis. Quality management systems (QMS) force operators to document which and how processes are done, in order to prove though records and audit that the process, however described, is consistent. QMS do not require specific or high quality standards, but the meeting of required standards. QMS are also a convenient framework to establish safety standards. The first types are legal standards - those which are commonly established by national governments and generally relate to safety. These standards are often mandatory and represent minimum standards of quality. Their spoilt major purpose is to ensure that products are not adulterated or do not carry dangerous contamination. These might involve undesirable microorganisms, insects, pesticides or potentially toxic additives. They may even consider processing conditions to ensure that foods are not contaminated or excessively damaged. Few of us would argue the importance of standards genuinely related to safety of food product.

In the most generic sense, quality refers to the combination of characteristics that are critical in establishing a product's consumer acceptability. In food industry, this is usually an integrated measure of taste, purity, flavor, texture, color, appearance and workmanship. In a highly-competitive market, another criteria of quality can be 'value' or a consumer's perception of the worth of the product based upon the funds available for the consumer. This is true for all stages of quality's traceability– from environment to home.

The measurement and evaluation of quality is a complicated affair. Most organizations employ professional technicians to carry out this task, but this has not always been the case.

In the past, many companies assumed that the quality of their raw materials could be guaranteed simply by paying the highest prices. However, this did not prove to be very reliable and almost all firms now use various analytical methods for quality determinations. For a short time in agriculture even more often biosensors are applied for quality control of agricultural products. The biosensors industry is new but growing. The market is comprised of four segments- medical, environmental, food, and military. Ninety percent of sales come from glucose-detecting biosensors for medical applications. The market is generating a need for pathogen detecting biosensors across all segments [1].

Quality control ensures that raw materials meet set standards, processing methods perform as designed, finished products meet company standards, and consumer confidence in the company remains high. Problems arise when there is disagreement on what is actually required to ensure safety. Another important area of standardization relates to the information presented to the consumer. In this case it is not the product itself, but rather its description that must conform to a particular standard.

Much effort has been devoted to harmonizing labeling information and very large market segments do have common requirements. There may be some disputes arising out of a culturally-based philosophy regarding the role of food in the diet. Some societies traditionally confer great health benefits to certain foods while others may not. This may lead to health claims that are allowed in one country and not in another. Industry standards are sometimes established by an organized industry association in order to establish a reliable identity for a particular product [2].

The processed food sector accounts for the largest number of tests, with over 52.2 million performed annually. This represents over 36% of total tests performed and is likely driven by the larger number of processing plants, which is 38% of all plants. Industry analysts expect that as regulations pertaining to pathogen testing continue to be adopted, the shift toward rapid-screening methods will continue.

The overall food product testing industry is growing steadily. For example, the US food industry performed around 144.3 million microbiological tests. The dairy sector has the highest testing rate per processing plant, averaging over 630 tests per plant per week. The beef and poultry sectors perform the least number of tests per plant averaging 369 tests per plant per week. As a result, the beef and poultry sectors account for only 22.3% of all testing in the industry. The fruit and vegetable sector is currently the smallest of the four sectors accounting for only 9.7% of testing. There is growing recognition of toxins as health risks, especially in grains and fish/seafood, which are two fast-growing food categories because consumers perceive them as healthful. Sales in the US for pathogen, pesticide and GMO products combined used by food processors are projected to increase from $149.5 million in 2000 to $239.4 million in 2005 at an AAGR (average annual growth rate) of 9.9%. The pathogen specific testing market is expected to grow for all segments at a compounded annual growth rate (CAGR) of 4.5% with a total market value of $563 million by 2003 [1].

III. APLICATION OF "CHERNOFF FACES" IMAGE RECOGNITION METHOD FOR THE ASSESSMENT OF CONFORMITY

The concept concerns the roles of marketing ethics in transactions between producers, marketers and poor consumers.

Therefore we describe our research results looking representing some problems and obstacles faced by poor consumers. For protection of their interests and rights we have developed technical decisions by use of new "watch-fractal" method for quality assessment of agricultural products, raw materials and goods, which was developed on the basis of "Chernoff faces" image recognition method (Fig.2).

Subsequently, the real "organisms" of AI systems can be expressed by means of the accepted conditioned standard of the perceptual model (experts knowledge). Besides, the most significant "biological" features remain. It can be said also in other words: real organisms are "the projections" of the initial organism, designed by experts , are models on in reality existing organic reason forms.

Internal motivation in self-organizing of biological systems, reasonable by chaos theory, can be used for the control of their quality.

Internal intention of bionic systems from chaos by self motivation to order, to self harmonized structure, is an important internal engine of process in search of an optimum of the biological system. Such biological system continuously forms individual metric coordinates "space- time", selected space - measure and required "speed" of time for the conformity control, obtaining needed useful information for the modification and its processing by means of application of "artificial" or "natural" intellectual tools.

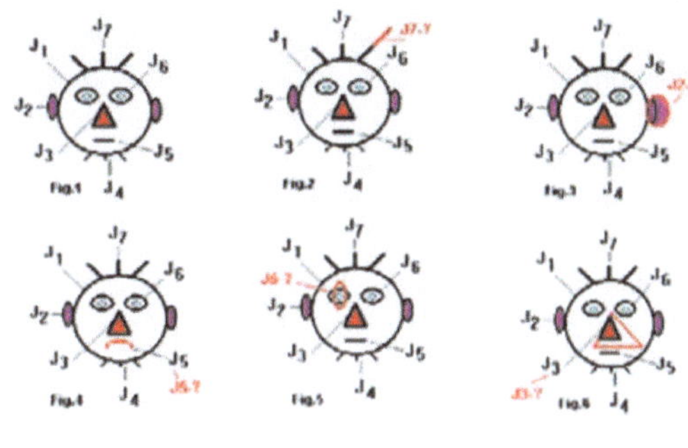

Fig.2. "Chernoff faces" image recognition method for the conformity assessment.

Such motivation makes the identification process all-embracing, core-aimed. This process directed at the core of the matter never is local but global. Harmony degree of a external test influence on the researched biological medium, can be estimated by methods of functional systems according to preservation of afferent principle. In bionic systems such coherent approach in strategy of measurement and assessment of useful medium's properties allows to allocate such set of key test signals, which does not contradict the keeping of afferent principles of the maintenance [3,4].

Bionic approach in the modeling of intelligent AT measuring systems allows to examine two types of intelligent control models. For the elaboration of the first type models it is sufficient to study in isolation only "inner" parameters and processes of the object under exploration without taking into consideration the impact of outer medium factors and, in relation with it, "behaviour" changes of the structure intended for synthesis. The modeling of this type can be useful for a preliminary metric image identification and conformation of the object under exploration. The further use of the model depends only on the success of the acquired model's theoretic and technical continuation [5].

"Behaviour" factor analysis of the object of interest has to be considered at the basis of the second type of bionic models. Further on "reference" functions of the influence factors are determined and feedback algorithm is synthesized in the form of ANN "self- learning" programs. The basic contours technical realization of the models are formed in complete agreement with the existing notions, data, and levels of knowledge about the investigation process or object together with the exploration task and aim. Technological and informative revolutions in all production spheres, especially in technology branches comprising computing and research aver determining the application of local (divided intellect) systems and further development of local microcomputers. Subsequently, the leading parts in automation belong to the intellectualization of measuring devices [7-9].

Many artificial tongue usually consist of three functional knots: a primary sensing element (measuring transducer), processor and a registering device. Sensing elements usually have electric exit signals and further processing measuring-information is completed by using different electrical schemes, mainly, of an analogous type. For functional opportunities, preciseness and signal stability are important ,therefore, the processing of digital data has significant preferences.

The presence of a microcomputer in the measuring channel allows the use of special testing programs carrying out identification experiments of a measurable medium with the help of intelligent complex means to make use of definite physical effects [6],[11]. This device uses electrical currents to produce recognizable patterns on a graph that are different for different compounds. Device uses electrical currents to produce recognizable patterns on a graph that are different for different compounds. The human tongue functions in a similar way because the sense of taste is no more than the recognition of electrical signal patterns for foodstuffs and other substances that have previously been encountered and remembered. Taste is learned, he added. Automatic identification at the critical control points should be achieved for all stages of food production starting with the obtaining of raw materials or the production of component parts up to their marketing (Fig.3). Therefore, at the start of the precise, safe, operative and objective technological process a information flow has to be established throughout all the production stages [9-11].

The solution of this problem is hindered by the lack of such measuring devices. The suppliers information about testing, regulation and control of technological processes with systematical, energetical, constructive, informative, exploitative and metrological parameters could be joined with such a measuring devices in real conditions. Just in such a way the world tendency towards the "intellectualization" of measuring devices and sensors can be explained.

Possibilities of the computing technique allow to process operatively great amount of information by definite algorithm. The promotes not only an optimal collection of measuring data from suppliers, but also a correction of their mistakes and an interpretation of the measuring result in accordance with the required information for ,users producers and consumers.

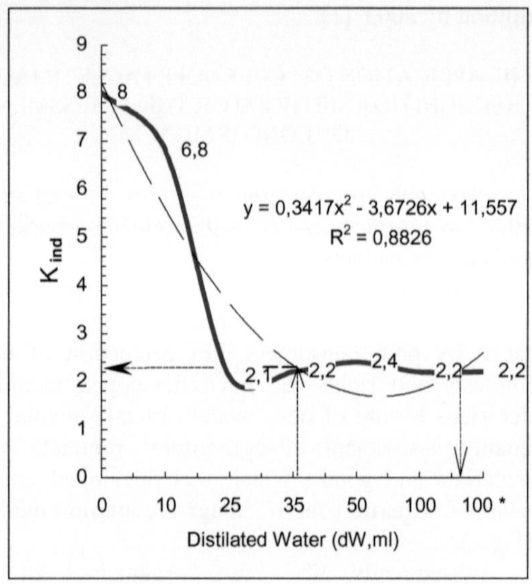

Fig. 3. Determination optimal K_{ind} for the standard apple with add to sample 35 ml distillated water, $K_{ind} = f$ (dW); A_s = 1,0…5,0 dB; F_{ind} = 7,0 kHz. dW =100* after waiting 40 min.

In industrial production there are no analogous for such food product properties as stochastic and not uniform flow of materials and informative resources, significant changes of their properties and quality in time, or presence of inertia in the communications with a bionic system. The above said does not allow to apply the traditional methods and means on the control of technological processes and on food production processes. The situation is worsened by the low technical level of the existing suppliers and devices used in agriculture and their level of preciseness and credibility. Theoretic investigations prove that sensors lag behind the development of food and other technologies. Therefore all over the world intensive financing is observed in the field of technical progress.

IV. RESULTS

The overall quality of a fruit is not a linear combination of all measurable quality parameters. This presents major problems as to how these measurements should be combined to quality indices and grading decisions. The quality of fruit is a combination of numerous parameters such as: firmness, acidity, aroma, color, color uniformity, bruises, scars, cuts, presence of soil, size, shape or insects diseases. The main parameters are specific to the individual fruit. Thus, the concept of this work is to develop a system that can classify fruits based upon several parameters (vision, taste, firmness, smell and weight) by using multi-sensor data acquisition. Despite the numerous techniques developed for non-destructive evaluation of quality, for example of fruits and vegetables, quality sorting is still primarily based on manual decisions and manual work (Fig.3-5).

Techniques and criteria for training sets for the classifier were developed in such a way that only about 155 dates was needed to achieve good conformity classification resulting in 87% of correct classifications for objects that were tested at different dates. A classifier that was trained can achieve 77% accuracy in the same classification. The main results of preliminary experimental research prove that quality and conformity control of agricultural products and raw materials can be determined by fractal conformity methods by using the intelligent artificial tongue "Logicor-AT". sensor. Research has focused on the development of classification algorithms. The multi - sensor system utilizes an imaging system, an impact sensor, a sensor of electroconductivity (constant and alternating current), an electronic chronometer for determination of relaxation time (T_{rel}), an ultrasonic sensor, a gauge for measuring electrical resistance (R_a), a force gauge and an “artificial tongue” measurement device.

The peculiarities, conditions and specifics of food production require elaborate simple, safe, inexpensive and precise electronic conformity control devices. The elaboration of such devices is the decisive factor in operation of the conformity control systems for the quality of food and other products.

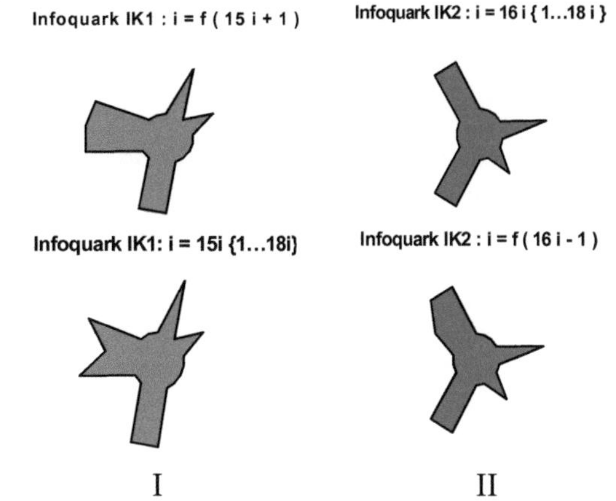

Fig. 4. Conformity control by fractal geometrical images (I – II - fractal images of info quarks IK1, K2)

Intelligent sensors can essentially perfect the whole control system due to increasing preciseness and a rational processing of signals received from the sensory element. An existing problem of measuring devices and sensors is the problem of a precise control of production processes.

The problem of consumer’s provision with qualitative food products is the efficiency problem of any production. Therefore it is topical to design new generation measuring devices with the application of fractal methods.

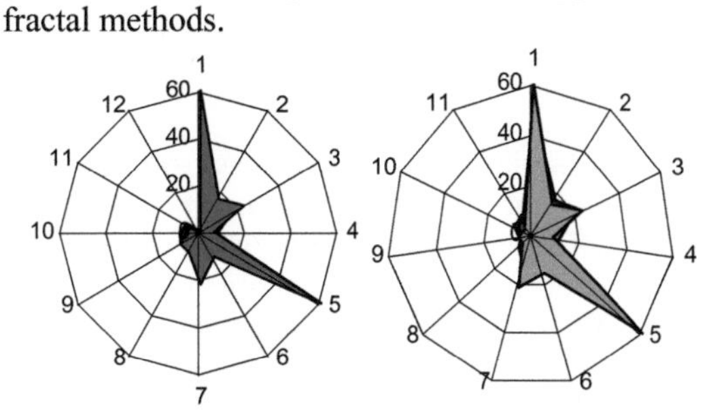

Fig. 5. Conformity control by watch – fractal method

One direction of the developments can be the elaboration of low-cost electronic rapid conformity control devices with "Artificial Intelligence" (AI) elements, which continue the development of microprocessor technology. Such intellectual suppliers have the "artificial tongue" sensible sensor elements in form of measurement transducers with digital or analogous electrical or other exit signal. These devices can effectively work under the changing operation regime of equipment, as well as adopt themselves to specific agricultural and food technology processes under real exploitation conditions.

The compatibility of measuring processes and functions in the "compensating stage" can be taken over by the cognition subject with its intellectual apparatus. In the elaboration of intellectual measuring systems it has to be understood that such system can be open to man's (expert's, specialist's) intellect, knowledge, practical experience (also not formalized and not systemized) and even to intuition [5].

These devices are already used in laboratories and in business [10],[11]. Preliminary interrogations of consumers show a positive relation of consumers and businessmen to the application of an intelligent device "artificial tongue" for conformity control in areas of agricultural production and business.

In general, it allows to improve scientific knowledge's on the basis of informative service and quality conformity control programs and guarantee legal protection of interests and rights of each consumer. The research on "consumer intellect" models are carried out through the synthesis of the non-traditional "watch -fractals" conformity method with the conformity control of agricultural products by using low- cost, rapid control intelligent instruments.

V. CONCLUSIONS

1. In agriculture are even more often applied biosensors for quality control of agricultural products. The biosensors industry is growing.

2. In generic sense, quality refers to the combination of product characteristics that are critical for consumer acceptance. In the food industry, this is usually an integrated measure of taste, purity, flavor, texture, color, appearance and handling. In a highly-competitive market, another criteria of quality can be 'value' or the consumer's perception of the worth of the product based upon the funds available for it.

3. On the basis of results from experimental researches and modeling methods "of the intellect of the consumer " a new "watch-fractal" method and compact, low-cost electronic AI device the "artificial tongue", for conformity and quality control of agricultural products, goods and raw materials was developed. These intelligent technology and devices are already in use in laboratories and in business. Preliminary interrogations of consumers show positive relation of consumers and businessmen to new possibilities for the protection of their rights and interests.

ACKNOWLEDGEMENT

This project was funded by two grants of Latvian Council of Science – LZP 05-1594 and LZP 05-1595.

REFERENCES

[1] Stephen M. Radke, Evangelyn C. Alocilja, (2002), Market Analysis of Biosensors for Food Safety, ASAE Paper 027025;

[2] Charles R. Hurburgh (2003), Food quality and international trade, Resource, Engineering & Technology for a Sustainable World, Vol. 10 No. 4 ,April 2003, (ISSN 1076-3333);

[3] Moskvin G.A. Artificial Intelligence Measuring, Automatic Control and Expert Systems in Agriculture: in 3-rd IFAC/CIGR Workshop on Artificial Intelligence in Agriculture, Makuhari, Japan, April 24-26,1998, Preprints, p.176-181.

[4] Moskvin G.A. Fractal and Perceptual Images in Info-Ergonomics: in 1-st IFAC Workshop on Control Applications and Ergonomics in Agriculture, Athens, Greece, June 15-17,1998, p.255-261.

[5] Moskvin G.A. Intellectualized automatic measuring, dosing and accounting systems. Latvia University of Agriculture, Summary of habilitation. Jelgava, 91p.

[6] Moskvin G., Spakovica E. New Method and Low - Cost Intelligent Instruments for the Fraud Detection and Conformity Control of Agricultural Products. 2002 ASAE Annual Meeting and CIGR WORLD Congress . July 29-July 31, Hyatt Regency, Chicago, IL, USA, ASAE Paper Number 023077.

[7] Moskvin G. United States Patent 5,016,569

[8] Moskvin G. United States Patent 4,989,445

[9] Moskvin G., Development of Intelligent Systems and Technologies in Agriculture. Proceedings of International Scientific Conference "Motor Vehicle, Logistics, Alternative Fuels". Jelgava, April 24, 2003, p. 165-172.

[10] Spakovica E. Interests of the Consumers and their Protection. Jelgava, Latvia University of Agriculture, 138 p.

[11] Moskvin G., Spakovica E. Application of Artificial Intelligence for Quality Control of Agricultural Production. Proceedings of International Scientific Conference "Motor Vehicle, Logistics, Alternative Fuels". LLU, Jelgava, April 24, 2003, p.158-164.

IEC 61499 in Factory Automation

K. Thramboulidis
Electrical & Computer Engineering
University of Patras
26500, Patras, Greece
thrambo@ee.upatras.gr

Abstract-- **The International Electro-technical Commission (IEC) has adopted the function block (FB) concept to define the IEC 61499 standard for the development of distributed industrial control applications. However, even though many researchers have been working during the last years to exploit this standard in factory automation, it is clear that the standard has a long way towards its adoption by the industry. Most practitioners are unfamiliar with the semantics of this standard and even more modifications and extensions are required to the model in order to be effectively used in the context of a process that will addresses the whole life cycle of factory automation systems. This paper surveys research results reported so far about the IEC 61499 model and attempts to highlight the inefficiencies of this paradigm to support the whole development process of distributed control applications as far as the software engineering point of view is considered. Open problems and future challenges are discussed as well.**

Index terms—IEC 61499, Function Block, Factory Automation, Engineering Support System, distributed control application.

I. Introduction

The Function Block (FB) is a well-known and widely used construct by control engineers. It was first introduced by the IEC1131 standard on programming languages for programmable logic controllers. However, languages defined by IEC1131 as well as vendor's proprietary tools, proved inefficient to address the increased demand for a more flexible development process in the control and automation domain. To address this demand, the International Electro-technical Commission (IEC) has defined the basic concepts for the design of distributed industrial-process measurement and control systems (IPMCSs). The IEC 61499 standard [1] extended the FB concept of IEC1131 to share many of the well defined and already widely acknowledged benefits of concepts introduced by object technology. This standard describes also a methodology that utilizes the FB as the main building block and defines the way that FBs can be used to define robust, re-usable software components that constitute complex IPMCSs. Complete control applications, can be built from networks of FBs that are formed by interconnecting their inputs and outputs.

Even though many researchers have been working during the last years to exploit the IEC 61499 in the development process of distributed IPMCSs, it is clear that the standard has a long way towards its adoption by the industry. Christensen [2] states that "several aspects of this standard are unfamiliar to most practitioners of control systems engineering, especially the ideas of distributed applications, event driven execution control and service interface function blocks ..." He argues that the use of design patterns can simplify the job of becoming familiar with the application of these new concepts. Thramboulidis [3] argues that framework technology can also play a significant role to this direction. Stromman et al. in their research to increase the understanding of the use of the IEC 61499 by industrial practitioners claim that "is no obvious way to define good guidelines" [4].

This paper surveys research results reported so far about the IEC 61499 model and attempts to highlight the inefficiencies of this paradigm to support the whole development process of distributed control applications as well as to discuss open problems at least as far as the software engineering point of view is considered. We argue that modifications and extensions are required to the model in order to be effectively used in industry by a process that will addresses the whole life cycle of factory automation systems. The IEC model does not address, for example, requirements elicitation and partially addresses the design phase. Furthermore a lot of implementation issues are left for the vendors that will provide IEC-compliant tools and products, but this may become a source of incompatibility problems. The most important limitation is that the FB construct and the diagrams proposed by the standard i.e. the Function Block Network and the Execution Control Chart (ECC), prove insufficient to capture the different aspects of a control application. The FB network is defined as an aggregation of interconnected FB instances. Its semantics as defined by the standard are not enough to capture the structure and behavior of control applications. Even more there is no: a) methodology that will guide the control engineer to define the FB network that models the control application at design time, and b) guidance on how to identify the FB types that are required to compose a control application. The use of already defined FB types is a significant starting point; however, a lot of other FB types should be defined to represent specific functionality of the control application as well as to capture the application logic, with the control engineer having no guidance to this direction.

Further issues that have to be addressed for the IEC 61499 model to be effectively used in the development process of factory automation systems include:

- location transparency in FB design diagrams,

K. Elleithy et al. (eds.), Advances in Computer, Information, and Systems Sciences, and Engineering, 115–123.

- handling Quality of Service (QoS) characteristics in design level,
- better integration with reliable communication infrastructures,
- validation of models,
- execution environments and implementation models,
- integration with legacy hardware/software (h/s) systems.

In [5] is stated that "Although many researchers are already working on different aspects of the IEC proposal … the absence of tools and products that are compliant with this approach is evident. The Function Block Development Kit (FBDK) by Rockwell Automation and the CORFU-FBDK are the only known tools supporting this approach." The situation one year later is somehow better. The interest from industry [6][7][8] and academia is growing. The standard was adopted by OOONEIDA [9], an IMS [10] initiative for an open knowledge economy for intelligent distributed automation, and MIM [11], a new paradigm for the development of the new generation Mechatronic systems.

The remainder of the paper is organized as follows. In the next section a brief introduction to the FB construct is given. In section 3, issues related with the specification of requirements of control applications are examined. In section 4, the design phase of IEC 16499 is discussed as well as the relation of the new standard with the huge amount of legacy systems. In section 5, the process of moving from FB design specifications to executable code is considered. The Engineering Support System (ESS) as a means of automating the development process is discussed in section 6, and finally the paper is concluded in the last section.

II. The IEC 61499 Standard

IEC 61499 defines the application model as "a function block network, whose nodes are function blocks or subapplications and their parameters and whose branches are data and event connections." The FB, the main building block of IPMCS applications, is defined as "a functional unit of software, comprising an individual, named copy of the data structure specified by the function block type, which persists from one invocation of the function block to the next" [1].

The FB consists of a head and a body; the head is connected to the event flows and the body to the data flows. An FB can be simple enough such as the one shown in fig.1(a) that identifies temperature alarms or complex, such as the one that controls part of a production line.

The functionality of the FB is provided by means of algorithms, which process inputs and internal data and generate output data. The sequencing of algorithm invocations is defined in the FB type specification using a variant of statecharts called Execution Control Chart (ECC). An ECC consists of EC states, EC transitions and EC actions as shown in fig. 1(b). An EC state may have zero or more associated EC actions, except from the initial state that shall have no associated EC actions. An EC action may have an associated algorithm and an event that will be issued after the execution of the algorithm. EC transitions are directed links that represent the transition of the FB instance from one state to another. An EC transition is fired when the associated Boolean expression becomes true.

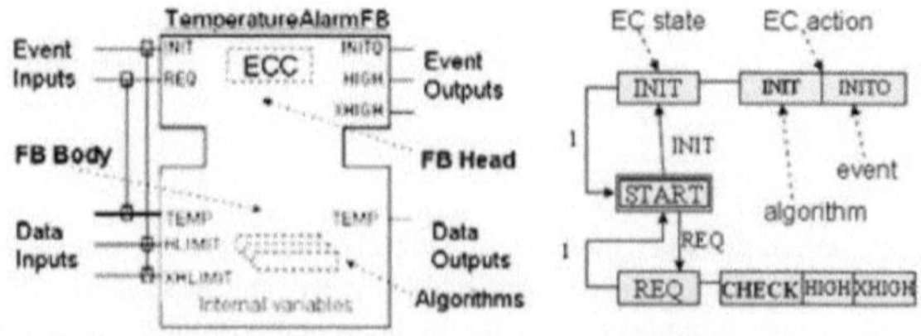

(a) Graphical representation (b) ECC
Fig. 1. IEC 61499 Function Block type.

III. The Analysis Phase

IEC 61499 defines the FB network as the top level model of the application and assumes that this is the first model created by the developer. It does not define the way that requirements will be captured and formalized neither the way that these requirements will be transformed to design specifications. These issues have to be addressed by an effective development process. The need for an analysis phase is mandatory.

A. Requirements elicitation

The Unified Modeling Language (UML), the new industry standard in software and system development, provides the constructs and diagrams that are required for requirements elicitation. A detailed discussion on the use of the UML in control and automation is given in [5]. In the same paper a hybrid approach that integrates UML with the FB construct and exploits the model driven development paradigm is described. Component interaction diagrams are utilized to realize use cases. UML diagrams are next transformed to FB design specs and finally to C++ or Java executable code. A simple system, namely the Teabag Boxing system, is used to demonstrate the applicability of the proposed hybrid approach from UML analysis models through FB design models to the implementation model of the application.

The use of UML in the development process of IPMCSs has been considered by many researchers [12][13][14]. However, these approaches adopt a direct use of UML, in contrast to the one presented in [15] which provides specific UML constructs to exploit the already existing skills of control engineers acquired from the wide use of the FB concept in industry. A framework-based approach is adopted in [15] to increase re-usability not only in implementation space but also in design space artifacts and design decisions. The first of the two considered alternatives utilizes UML's extensibility mechanisms, while the second is based on the meta-model pattern. A set of constructs was defined to assist the control engineer in the development process. The concept of the (Industrial Process Terminator) IPT-stereotype is introduced to represent any mechanical component of the controlled industrial process that acts as source or sink of data or control information for the controlling system. Later on, the concept of Industrial Process Parameter (IPP) was introduced in [5] to model every parameter of the IPT that is either monitored or controlled by the controlling system. The FunctionBlock-stereotype (FB-stereotype), the Data-Dependency-stereotype (DD-stereotype), and the Control-Dependency-stereotype (CD-stereotype) are introduced and their semantics are defined. A UML model of the above concepts was defined to give a formal base for their use.

B. From requirements to design specifications

The use of UML for requirements elicitation imposes the need for an integration of UML with the FB approach. A transformational approach is adopted in [16][17] and a process is defined for the transformation of requirements expressed in the form of use cases, to design specifications expressed in the form of FB diagrams. The proposed process, presented in the form of workflows, utilizes a set of Transformation Rules to smoothly pass from the UML analysis model to the FB-based design model. Specific rules to transform class diagrams and interaction diagrams to equivalent FB network diagrams are defined. CORFU ESS that automates the transformation process is presented in [17] and a detailed presentation of the transformation rules along with a case study based on the steam boiler control application [56] is presented in [18].

IV. The Design Phase

Lewis [19] argues that even though the IEC 61499 represents an important step towards a unified design architecture it provides just one of the five design views required for distributed control systems. He also claims that the others views will emerge as designers start to face the challenge of building large distributed systems.

The semantics of the FB Network and the ECC that are the only diagrams proposed by IEC61499 are not enough to capture the structure and behavior of control applications. There is a need for location transparency in FB design diagrams, for handling QoS characteristics in design level, for validation of the design models. In the following, design alternatives in the context of IEC61499 are discussed and an attempt is made to address the above topics.

A. IEC 61499 design alternatives

Christensen [2] describes the traditional model-view-controller (MVC) pattern that was developed to address separation of concerns in user interfaces for object-oriented systems. He claims that this pattern can be adapted for use in the modeling, simulation and testing of IPMCs in the IEC61499 context. In the same paper an engineering methodology that has been found to be successful in accomplishing system design and simulation using this pattern is briefly described.

Stromman *et al.* [4] describe an interesting working course in which academic and industrial users of the IEC61499 standard were assigned to automate a pallet lift in limited time using an embedded controller. They briefly describe the alternative solutions that were given by automation designers using the IEC61499 model and state that one of their primary objectives was to "identify possible reconfiguration and reuse strategies with IEC61499 as well as context-related factors that can be used to evaluate the alternatives." They argue that the evaluation of design approaches is highly context-dependent since parameters such as the background of designers, the existence of legacy software, the possible use of subcontractors, the organizations' policy regarding open systems and the business goals determine the best design approach. However, even though they state that one major problem is the gathering of user requirements the experiment did not gave significant emphasis on the way that these requirements will be converted to design specs. As was already mentioned, we believe that the absence of a well defined analysis phase that covers the semantic gap between user requirements and design specs is the source of many problems in the development process. We estimate that the same experiment carried out with software engineering analysis methods will result in more interesting findings. Adopting for example the approach presented in [5] a solution quite similar with the second one presented in [4] is obtained. The use case technique and classical component composition rules guide the designer to this solution that has nothing to do with the term mechatronical approach used to refer to this solution. In any case: a) more research is required to the direction of fully understanding the use of IEC 61499 by industrial practitioners and producing a coherent methodology that will guide the control engineer to best exploit the benefits of the IEC model, and b) new skills are required for the proper use of the IEC61499 FB construct since its semantics are quite different and more complicated than the semantics of the already used constructs (see IEC61131 languages).

Thramboulidis *et al.* [20] propose two extensions to the semantics of the FB design model that are related with the ECC and intend to improve the expressiveness of the design model and the efficiency of the implementation one. The first one concerns the introduction of the concept of transitory state, while the second provides for algorithms the possibility to generate output events. The proposed extensions were adopted in [21] to demonstrate that they simplify the translation process of FB design diagrams to executable code.

Panjaitan *et al.* [22] propose a task scheduling approach as a probable solution to the re-configurability requirement of IEC61499 applications. Scheduling is proposed as a means of describing the process sequences of a system in an effective and re-configurable way. UML sequence diagrams and Gantt charts are investigated for the planning of task schedules.

Cengic *et al.* [23] describe a method that utilizes the IEC 61499 FB model to provide a means for the supervisory control theory to be applied in the development of distributed control applications. They claim that the Supremica tool that implements the method significantly reduces application development time.

Hussain and Frey [24] state that even though the IEC61499 evolved mainly to provide a generic distributed modeling platform that will simplify the modeling of distributed systems it will also "diminish the differences between business and industrial systems." In the same paper the authors consider the new generation of controllers, i.e. the network enabled controllers, to demonstrate the applicability of the IEC 61499 FB model. A simple application executed on a Netmaster microcontroller is described as well.

The urgent need for extra notation to express real-time constrains and QoS characteristics on the design model is stated in [21].

B. Location Transparency

Service Interface FBs (SIFBs) are introduced by 61499 to provide interfaces to the underlying communication subsystem. In this way the standard defines how data and event connections of the FB diagram should be implemented. All the design alternatives except the one proposed by CORFU adopt the use of SIFBs in the design level. CORFU proposes a better approach that provides distribution flexibility and favors location transparency. It defines a set of services that have to be provided by the execution environment and used by the ESS and the devices to automatically setup and implement both the event and data connections. This approach that was adopted in the Archimedes execution environments: a) simplifies the FB design diagrams, b) de-couples them from the physical architecture, and c) results in a more flexible reconfiguration process that is required during the operational phase. To interface with the controlled mechanical process a special kind of FBs, called Mechanical Process Interface FBs (MPIFBs), quite similar with SIFBs, is introduced. It is expected that IEC-compliant sensors and actuators will encapsulate the required MPIFBs to provide IEC interfaces. This means that there is no need for SIFBs in the design phase; application designers will represent in the FB design-model monitored and controlled parameters of the mechanical process utilizing the (Mechanical Process Parameter) MPP construct.

C. Management and event FBs

IEC61499 suggests the use of Management function blocks to load, create and initiate FB's execution. However, management operations are outside of the problem domain and should not be defined by the control engineer [3]. They should be provided by the engineering infrastructure on which the control application should be deployed. Even more, these operations are not modeled in the best way using the FB paradigm. Instead, more robust paradigms such as the OO and the component based one can be utilized for their development. This approach adopted in the Archimedes execution environments resulted in a more robust implementation allowing the functionality of the device to be expanded to cover run-time reconfiguration requirements [25].

Another modification proposed in [26] is related to the event function blocks, a special kind of FBs defined by 61499 to allow a wide range of different event scenarios to be modeled. According to this modification a) the FB network editor of the ESS provides for the design time, a palette with event icons that are used by the designer to import in the FB design-model different event scenarios, and b) the device should provide an event-API that is used by the ESS to automatically configure the device so as to provide the required behavior during the operation phase. This API was defined and it is argued that its use simplifies the function block design diagram and improves the performance of the resulting application.

D. Distribution of the control application

The control application that has been defined in the form of network of interconnected FB instances can be executed on a device, but usually it will be executed on a network of interconnected nodes (devices). This means that the control engineer has to address the FB allocation problem as well as the FB execution scheduling problem for the FBs that are allocated to the same node.

Designers usually address the execution scheduling problem by capturing the solution to the FB design diagram. However, such an approach destroys location transparency of the design diagram and reduces reusability in the design level. A more flexible solution is obtained utilizing the task scheduling services provided by processing nodes.

Feng Xia *et al.* [27] propose a FB allocation scheme and a FB execution scheduling scheme that takes into account time, precedence and hardware constraints and targets minimized system response time. The FB network is modeled as a task graph that shows precedence constraints among FBs. They adopt FB clustering to eliminate large communication costs and propose a parallel pipeline based algorithm for the scheduling of FBs.

Thramboulidis *et al.* [28] describe a flexible deployment process in the context of the proposed model driven development approach that exploits the CORBA component model. A CCM-based implementation model and the corresponding execution environment in the context of Archimedes system platform are presented and a case study demonstrates the applicability of the proposed approach.

Prayati *et al.* [29] claim that address the problem by the proposed FBALL algorithm. However, the presented material (methodology, algorithm description and case study) contains so many inconsistencies and merits that raises many questions for its correctness [30].

E. Verification of the design model

Many researchers are already working on the verification of IEC61499 based applications and significant results were reported so far. However, it seems that a long way is required to get a mature verification process supported by the corresponding tool that will be integrated in the development environment. Vyatkin *et al.* [31] propose the use of the Net Condition/Event Systems modeling formalism. Stanica *et al.* [32] utilize timed automata to verify the execution of FB network. RT-UML as a means for the verification of the design model is considered in [20]. Khalgui *et al.* [33] highlight limitations in the 61499 model and propose new semantics in the model to allow its validation. Hagge *et al.* [34] argue that the 61499 model does not prevent design errors causing data inconsistencies, based on unconsidered dependencies between data and event connections. They propose a new FB modeling language, based on Petri Nets, that models events and data combined as colored tokens to get a verifiable model.

F. Integrating legacy systems

For the migration of existing hardware and software to IEC-compliant systems two approaches should be highlighted. The first one utilizes the IEC FB model to address the interoperability problem between existing field bus segments of different types and the second attempts to utilize the IEC FB model on existing hardware scan based devices. Another quite interesting approach considers the re-use of existing programs that were developed using IEC61131 function blocks [4].

The IEC61499 FB model can be exploited to address the interoperability problem that arises between different fielbus segments. This model is adopted in [35] to address the problem of interconnecting different fieldbus segments of various types. The architecture of an interworking unit that is required to interconnect the different fieldbuses is described. The concepts of local fieldbus proxy, remote fieldbus proxy as well as the FB proxy are introduced to simplify the development process and obtain a flexible architecture for the interworking unit. The concepts of virtual field bus and virtual field device were also introduced in [3] to facilitate the migration of no IEC-compliant field buses and devices with IEC-compliant products and tools. The structure of both artifacts is given along with paradigms of their use. The concepts of Event Connection Table and Data Connection Table are introduced in [35] to get a modular interworking unit that would also support the run-time reconfiguration of the control application. A wrapper layer is proposed on top of the fieldbus API to get an IEC-compliant interface for each specific fieldbus. The applicability of this architecture is demonstrated in [36] and [37] where the case of two fieldbuses of Profibus type is considered and a case study is conducted. A wrapper was developed to abstract Profibus to an IEC 61499 compliant virtual field bus and an implementation of the interworking unit using RTLinux is presented. Rtnet, an RTLinux protocol stack was adopted and extended to provide TCP connectivity to fulfill the requirements imposed by the configuration management of field buses.

Scan-based controllers constitute a significant part in the existing infrastructure in industrial systems. The deployment of IEC61499 FB based control applications on scan-based controllers is examined in [38]. The authors argue that an event based FB design diagram can be transformed to a scan based control application and executed on legacy scan based devices. An application generator is utilized to automate this translation process assuming that each FB instance is translated to a Java program with its own thread of execution.

V. Moving from Design Specs to Executable Code

A. Device's execution environment

The generation of executable code from FB design specifications is still an open issue that preserves the effective use of the IEC61499 model in practice. Even though many researchers are working in this direction [20][39][40][41] the only environments that support at the time of writing this paper the execution of FB based design specifications are the FBDK by Rockwell Automation [42] and the Archimedes System Platform [11][57] (see table I).

IEC environments	**Analysis phase**	**Design phase**	**Deployment**	**Execution environment**	**implementation model language**
FBDK	not supported	FBDK	manual	FBRT (J2ME)	Java
CORFU Framework	supported through ROSE	CORFU FBDK	semi-automatic	RTAI-AXE RTSJ-AXE CCM-AXE	Java, C++
Archimedes System Platform	supported through CORFU ESS	Archimedes ESS	automatic	Real-time Linux, Real-time Java, CORBA component model	Java, C++ Real-time Java

TABLE I
PUBLICLY AVAILABLE IEC61499 ENVIRONMENTS

FBRT [43] that is the first publicly available execution environment for FB design models, utilizes Java but does not support timeliness neither run-time re-configuration.

Archimedes System Platform (http://seg.ee.upatras.gr /MIM) follows a more flexible approach than the one adopted by FBDK. It exploits the model based development paradigm to provide: a) an environment that favors the construction of implementation models for IEC61499 compliant designs, and b) the infrastructure required to meet the real-time constraints and support run-time re-configurability of such applications. This approach allows for the concurrent use of different execution environments for the execution of the same control application. RTOS, Java virtual machine and CORBA component model are examples of general-purpose execution environments already examined in the context of Archimedes. An implementation model framework was constructed for each case to capture the design decisions made during the mapping process of FB design model constructs to the corresponding implementation space constructs. The implementation model framework along with the corresponding execution environment and the model-to-model transformers are provided in the form of a package. The RTSJ-AXE package [25], which is based on the real-time Java specification mainly focus on run-time reconfiguration. The RTAI-AXE package [44], which is based on RTAI a real-time Linux variant, mainly focus on performance.

Brennan *et al.* [39] propose a FB-based model to support configuration and reconfiguration of DCSs and discuss its implementation on real-time Java. The same authors even though describe an operating system that allows the execution of function block based control applications, they do not provide any prototype implementation or performance metrics and they do not describe the used implementation meta-model.

Ferrarini and Veber [40] present and analyze a set of implementation approaches presenting advantages and disadvantages for each solution. However, it is not clear how the influence of the garbage collector is taken into account in the presented performance metrics neither the fact that Windows XP executes the Java virtual machine in a round robin scheme.

Zoilt *et al.* [41] discuss and compare different implementation and scheduling approaches towards a real-time execution of IEC61499 applications. They also present an approach for scheduling real-time constrained applications and denote their intention to implement this model.

B. Inter-device communication

The modification to the standard proposed for location transparency is better implemented adopting the extension that considers low-level communication services. Such an approach is adopted in [3] where the Industrial Process Control Protocol (IPCP) is defined to provide the services required to load, initiate, configure and re-configure FBs. These services are required both for the real-time interaction between FBs allocated in different devices, and for the no real-time interaction of the ESS and the HMI modules with the devices. The presented in [3] approach was adopted in the first prototype implementation of the execution environment of an IEC-compliant device that is presented in [21]. The result was to simplify the function block design model and enhance the performance of the resulting implementation one. For this prototype implementation of the IPCP protocol, NDDS, a RTPS [45] implementation, was selected first, as well as the real time CORBA middleware that provides QoS required by real-time applications. It is argued that the publish-subscribe model that was adopted, preserves compliance with the IEC model, satisfies the QoS required by distributed control applications and allows a flexible deployment, re-deployment and re-configuration of control applications, even during run-time.

The possibility of using commercially available standard-based real-time middleware for managing the inter device communication links is also examined by Sierla *et al.* [46] with the argument that these links usually have unique properties with respect to communication mechanisms and real-time QoS characteristics. The requirements for process control are described and used to derive test cases for middleware products. In the test arrangement they describe, continuous data distribution, event notifications and scalability were tested. They argue that RTPS supports all the key communication mechanisms required, claims sub-millisecond real-time performance and supports the automatic and dynamic reconfiguration of the application at run-time. The presented results suggest that the performance of the NDDS implementation that was selected for detailed testing is sufficient for modern automation systems and that, the reliability of data transfer is satisfactory even under peak load.

The applicability of Object Oriented middleware in FB based applications is examined in [47]. The DOME middleware is introduced and results of its validation on relevant automation architectures as well as quality measurements are given.

VI. AUTOMATING THE DEVELOPMENT PROCESS

It is clear that for the above infrastructure to be effectively utilized during the development process a tool that will guide and help the engineer during the development process is required. The big advantage of the IEC61499 model is that it favors the definition of such a tool not as a monolithic tool but as an aggregation of user selected IEC-compliant tools where each one offers specific services required to support the steps of the development process. This approach was adopted in [48] and the basic requirements of an integrated ESS that should utilize mature software engineering practices such as object-orientation are defined. CORFU ESS and Archimedes ESS are examples to this direction. Aendenroomer et al. [49] describe also an approach to better tailor IEC-compliant publicly available development platforms such as FBDK and CORFU, for rapid application generation. An XSL (eXtensible Stylesheet Language) model is defined and a parser was developed to generate C/C++ code from the XML representation of the control application. The authors claim that the presented approach allows the generation of code in any language even hardware specification languages such as VHDL or Verilog. As a proof-of-concept a simple PID tank-level controller is presented.

The need for an architecture that should: a) facilitate the development process of IEC-complinat ESSs, and b) favor the automation of the application of the IEC FB model in the development process of IPMCSs, was evident during the early steps of the CORFU framework.

A. Engineering Support Systems

The first prototype implementation of the CORFU ESS was presented in [50]. This ESS adopting the extensions and modification already presented in [26] provides a more functional tool compared with FBDK which only addresses the design phase. The CORFU ESS to be close with the latest trends in the development of CASE tools, is composed of a well-known general-purpose CASE tool and a custom Function Block development tool namely CORFU FBDK.

IBM's Rose, was selected for the first prototype implementation of the CORFU ESS mainly due to the extension mechanisms that supports either in the form of scripting language, or as a COM automation server. Scripting language was utilized to extend Rose's toolbars in order to simplify the design of object interaction and class diagrams by allowing the control engineer to use application specific stereotypes such as the ones described in [15]. The main components of the CORFU FBDK are briefly presented in [50] along with implementation details.

Archimedes ESS [28] is an attempt to implement CORFU FBDK on the GME meta-case tool, to better exploit the benefits of the model driven development.

B. The 4-layer Architecture

A 4-layer architecture is presented in [35]. The first attempt to utilize the IEC61499 FB in the context of the 4-layer CORFU architecture is presented in [51] using the well known steam boiler system. The distribution of the control application to system layer components is done transparently through the automatic construction of XML configuration files for the underlying system layer components. These human readable XML representations are translated off-line in machine readable format, which is the one expressed by the proper update of ECT, DCT and required proxies. The required proxies (devices or FBs) are automatically created by the ESS in the interworking unit's LFP module. At the same time the related DCTs and ECTs entries are updated to transparently establish the inter segment connection between FB instances. This technique was adopted later in the TORERO project [52] and is referred as weaving.

The 4-layer architecture is enhanced in [26] to satisfy the requirements for the development of an IEC-compliant ESS. A detailed description of the enhanced 4-layer architecture, a description of its layers and the way that this architecture unifies the FB concept with UML can be found in [17]. In the same paper it is claimed that this architecture proved to be very significant for a) the identification of the key abstractions that the ESS must provide as building blocks of its various diagrams used during the modeling process of control systems and b) a number of significant extensions and modifications to the IEC-model that improve the development process [53]. These modifications have already been discussed in the previous sections since we still believe that these modifications are very important towards a more effective development process.

The requirement to model the system layer of the 4-layer architecture was considered very crucial for the effectiveness of the development process. The concept of the device proxy that is a software representative of the real world device is introduced to allow the distribution of the control application components to system layer devices. Partitioning, assignment, and scheduling as well as verification were identified as major importance actions that should have to be supported by the ESS [26].

C. The device model

The urgent need for a common device model is stated in [54]. This device model should allow an ESS to: a) assign functionality to the great number of different field devices that already exist in the market, and b) exploit FBs provided by intelligent field devices, which are expected to appear in the market in the near future. The requirements for the field device model imposed both by the development of the FB-oriented ESSs and the demand for device interoperability during the fieldbus operation phase, are also considered in [54]. A modular field device model is defined using UML with primary objective to hide complexities associated with field devices and allow the ESS to be able to provide the functionality required by the FB's distribution and assignment process to the system layer building blocks and mainly to field devices. The proposed model combined with the 4-layer CORFU architecture, greatly simplifies the development of open IEC-compliant ESSs. Information hiding of those ESS's actions that have no meaning in the application design phase but refer to the configuration of the underlying communication subsystem, was of primary concern. The functionality of the ESS that is related to field device specification is examined in detail to define an efficient device model. The same device model with some additions regarding timeliness is presented in [55] and [29].

A detailed design and implementation of an IEC-compliant device is described in [44][25]. A device model is also described in [52].

VII. CONCLUSIONS

Undoubtedly the IEC 61499 standard represents an important step for the exploitation of current software engineering practices in factory automation and mainly in low level control. However, the standard does not address the whole development process and the need for a reliable reference implementation as well as mature products and tools is the first step towards its adoption by the industry. Even though many researchers have already reported encouraging results working with different aspects of the standard, many issues are still open.

More over, since the semantics of the new standard are not known to control engineers, specific mature methodologies, frameworks and Engineering Support Systems are required. In any case it is expected that IEC-compliant products and tools will soon be available in the market and this will be the first step towards an open market in factory automation.

REFERENCES

[1] International Electro-technical Commission, (IEC), International Standard IEC61499, Function Blocks, Part 1 - Part 4, IEC Jan. 2005. (http://www.iec.ch/)

[2] J.H.Christensen, Design patterns for systems engineering in IEC 61499, Otto-von-Guericke-Universität Magdeburg, Germany, 22-23 March 2000, 63-71.

[3] Thramboulidis, K., "Development of Distributed Industrial Control Applications: The CORFU Framework", 4th IEEE International Workshop on Factory Comm. Systems, August 2002, Vasteras, Sweden.

[4] Stromman, M., S. Sierla, K., Koskinen, "Control Softawre Reuse Strategies with IEC 61499" 10th IEEE Int. Conf. on Emerging Technologies and Factory Automation, (ETFA'05), Catania, Italy, Sept. 2005.

[5] Thramboulidis, K., "Using UML in Control and Automation: A Model Driven Approach", 2nd IEEE Int. Conf. on Industrial Informatics, 24-26 June, Berlin, Germany, (INDIN'04).

[6] Western Reserve Controls, Inc., W2 Series IEC61499 Development Kit http://www.wrcakron.com/IEC61499.html

[7] ICS Triplex ISaGRAF, Commercially Available IEC 61499 Software, http://www.icstriplex.com/

[8] DACHSview, visual Programming of Real-Time Applications, Steinhoff Automation, http://www.steinhoff-automation.com/

[9] Vyatkin V., J., Christensen, J., Lastra, "OOONEIDA: An Open, Object-Oriented Knowledge Economy for Intelligent Distributed Automation," IEEE Trans. on Industrial Informatics, vol. 1, No. 1. February 2005.

[10] Intelligent Manufacturing Systems research and development (R&D) program http://www.ims.org/

[11] Thramboulidis, K., "Model Integrated Mechatronics – Towards a new Paradigm in the Development of Manufacturing Systems", IEEE Transactions on Industrial Informatics, vol. 1, No. 1. February 2005.

[12] K., Young, R. Piggin, P. Rachitrangsan, "An Object-Oriented Approach to an Agile Manufacturing Control System Design", Int. Journal of Advanced Manufacturing Technology, Vol. 17, Springer-verlag 2001.
[13] L. Bekker, C. Pereira, "On the suitability of the RT_UML profile for modeling distributed real-time systems", ETFA'03, Lisbon, Sept. 2003.
[14] H. Mosemann, F. Wahl, "Automatic Decomposition of Planned Assembly Sequences Into Skill Primitives", IEEE transactions on Robotics and Automation, Vol. 17, No. 5, October 2001.
[15] Thramboulidis, K., "Using UML for the Development of Distributed Industrial Process Measurement and Control Systems", IEEE Conference on Control Applications (CCA), September 2001, Mexico.
[16] Tranoris, C., K. Thramboulidis, "From Requirements to Function Block Diagrams: A new Approach for the design of industrial applications", *10th IEEE Mediterranean Conf. on Control and Automation*, (MED'02), Lisbon, Portugal 2002.
[17] Thramboulidis, K., and C. Tranoris, "Developing a CASE Tool for Distributed Control Applications", The International Journal of Advanced Manufacturing Technology, Volume 24, Number 1-2, July 2004, pages 24-31, Springer-Verlag.
[18] Tranoris, C., and K. Thramboulidis, "Integrating UML and the Function Block concept for the development of distributed control applications" 9th IEEE Int. conf. on Emerging Technologies and Factory Automation, Lisbon, Portugal, 16-19 Sept. 2003.
[19] Lewis, R., *Modeling control systems using IEC 61499*, IEE 2001.
[20] Thramboulidis, K., G. Doukas, A. Frantzis, "Towards an Implementation Model for FB-based Reconfigurable Distributed Control Applications", 7th IEEE Int. Symposium on Object-oriented Real-time Distributed Computing, Vienna, Austria 2004. (ISORC 04)
[21] Thramboulidis, K., G. Doukas, T. Psegianakis, "An IEC-Compliant Field Device Model for Distributed Control Applications", 2nd IEEE Int. Conf. on Industrial Informatics, 24-26 June, Berlin, Germany, 2004.
[22] Panjaitan, S., T. Hussain, G., Frey, "Development of re-configurable Distributed Controllers in 61499 based on Task Schedules described by UML diagrams or Gant charts", 3rd IEEE International Conference on Industrial Informatics, Perth, Australia, August 2005, (INDIN'05).
[23] Cengic, G.; Akesson, K.; Lennartson, B.; Chengyin Yuan; Ferreira, P., "Implementation of full synchronous composition using IEC 61499 function blocks", Automation Science and Engineering, 2005. IEEE International Conference on Aug. 1, 2005 Page(s):267 - 272
[24] Hussain, T., Frey, G., "Developing IEC 61499 Complinat Distributed Systems with Network Enabled Controllers", IEEE Conf. on Robotics, Automation and Mechatronics, Singapore, pp. 613-618, Dec. 2004.
[25] K. Thramboulidis, A. Zoupas, "Real-Time Java in Control and Automation: A Model Driven Development Approach", 10th IEEE Int. Conf. on Emerging Technologies and Factory Automation, (ETFA'05), Catania, Italy, September 2005.
[26] Thramboulidis, K., "Towards an Engineering Tool for Implementing Reusable Distributed Control Systems", ACM SIGSOFT Software Engineering Notes, Vol. 28 , Issue 5 (September 2003).
[27] Feng Xia; Zhi Wang; Youxian Sun, "Allocating IEC function blocks for parallel real-time distributed control system", IEEE International Conference on Control Applications, Vol. 1, 2004 pp. 254 – 259.
[28] Thramboulidis, K., D. Perdikis, S. Kantas, "Model Driven Development of Distributed Control Applications", (accepted) The International Journal of Advanced Manufacturing Technology.
[29] Prayati, A., C. Koulamas, S. Koubias, G. Papadopoulos, "A methodology for the development of distributed real-time control applications with focus on task allocation in heterogeneous systems", IEEE Trans. on Industrial Electronics, Dec. 2004 vol. 51 No 6.
[30] Thramboulidis, K., "Comments on "A methodology for the development of distributed real-time control applications …" " (submitted) IEEE Transactions on Industrial Electronics.
[31] Vyatkin, V., H.Hanisch, Formal-modelling and Verification in the Software Engineering Framework of IEC61499:a way to self-verifying systems, ETFA'01, Nice, pp 113-118, 15-18 October, 2001.
[32] M. Stanica, H. Guéguen. "*Using timed automata for the verification of IEC 61499 applications",* Workshop on Discrete Event Systems, WODES'04, Reims, France, Sept 22-24, 2004.
[33] Khalgui, M., Rebeuf, X.; Simonot-Lion, F., "A behavior model for IEC 61499 function blocks", *Third Workshop on Modelling of Objects, Components, and Agents, Aarhus, Denmark, October 8-11, 2004.*
[34] N. Hagge and B. Wagner "A New Function Block Modeling Language Based on Petri Nets for Automatic Code Generation", IEEE Trans. on Industrial Informatics, Vol. 1, No 4, pp. 226-237, Nov. 2005
[35] Thramboulidis, K., C. Tranoris, "An Architecture for the Development of Function Block Oriented Engineering Support Systems", IEEE International Conference on Computational Intelligence in Robotics and Automation (CIRA'01), Canada August 2001.
[36] Thramboulidis, K., P.Parthimos, G. Doukas, "Using RTLinux to Interconnect Field Buses: The Profibus Case Study", ICMEN - International Conference on Manufacturing Engineering, 3 - 4 October 2002, Thessaloniki, Greece.
[37] Tranoris, C.et al. "Using RT_Linux for the interconnection of industrial fielsbuses" ASME international, 1st National Conference on Recent Advances in Mechanical Engineering, Sept. 17-20, 2001 Patras, Thessaloniki, Greece.
[38] Lastra, J., A. Lobov, L. Godinho, "Closed Loop Control Using an IEC 61499 Application Generator for Scan-Based Controllers", 10th IEEE International Conference on Emerging Technologies and Factory Automation, (ETFA'05), Catania, Italy, Sept. 2005.
[39] Brennan, R, S. Olsen, M. Fletcher, D. Norrie, (2002), "Comparing two Approaches to Modelling Decentralized Manufacturing Control Systems with UML Capsules", 13th IEEE International Workshop on Database and Expert Systems Applications, Sept. 2-6, 2002, France.
[40] Ferrarini, L., C. Veber, "implementation approaches for the execution model of IEC 61499 applications", 2nd IEEE International Conference on Industrial Informatics, 24-26 June, Berlin, Germany, (INDIN'04).
[41] Zoilt, A., G. Grabmair, F. Auinger, C. Sunder, "Executing real-time constraint Control Applications modeled in IEC61499 with respect to Dynamic Reconfiguration", 3rd IEEE International Conference on Industrial Informatics, Perth, Australia, August 2005, (INDIN'05).
[42] FBDK (Function Block Development Kit), Rockwell Automation, http://www.holobloc.com.
[43] FBRT (Function Block Run-time Toolkit), Rockwell Automation, http://www.holobloc.com
[44] Doukas, G., K. Thramboulidis, "A Real-Time Linux Execution Environment for Function-Block Based Distributed Control Applications",3rd IEEE *Int. Conf.*on Industrial Informatics, Perth, Australia, August 2005, (INDIN'05).
[45] IE/PAS 62030, Real-Time Publish Subscribe (RTPS) Wire Protocol Specification, Version 1.0, 2004.
[46] Sierla, S., J. Peltola, K. Koskinen, « Real-Time Middleware for the Requirements of Distributed Process Control», 3rd IEEE Int. Conference on Industrial Informatics, Perth, Australia, August 2005, (INDIN'05).
[47] Riedl, M., C. Diedrich, F. Naumann, "Function block Applications Based on an Object Oriented Middleware", 3rd IEEE Int. Conference on Industrial Informatics, Perth, Australia, August 2005, (INDIN'05).
[48] Thramboulidis, K., "Towards a UML based Engineering Support System", 9th IEEE Mediterranean Conference on Control and Automation, MED'01, Croatia 2001.
[49] Aendenroomer, A., H. He, K. V. Ling, K. A. Sreenidhi, M. A. Mullamitha and K. M. Goh, "Closed-loop Modeling and Rapid Application Generation, using IEC 61499 Function Blocks and XML", 3rd Int. Conf. on Reconfigurable Manufacturing, 10-12 May 2005.
[50] Tranoris, C., and K. Thramboulidis, "An IEC-compliant Engineering Tool for Distributed Control Applications", 11th Med. Conference on Control and Automation - MED'03, Rhodes, Greece 2003.
[51] Thramboulidis, K., C. Tranoris, "A Function Block Based Approach for the Development of Distributed IPMCS Applications", 10th IEEE International Conference on Advanced Robotics (ICAR 2001), August 22-25, 2001, Budapest, Hungary.
[52] C., Schwab, M., Tangermann, L., Ferrarini, "*Wed based methodology for Engineering and maintenance of Distributed Control Systems: The TORERO Approach",* INDIN'05, Perth, Australia, August 2005.

[53] Thramboulidis, K., "An Architecture to Extend the IEC61499 Model for Distributed Control Applications", 7th *International Conference on Automation Technology*, (Automation 2003), May 8-11, Taiwan 2003.
[54] Thramboulidis, K., A. Prayati, "Field Device Specification for the Development of Function Block Oriented Engineering Support Systems", International Conference on Emerging Technologies and Factory Automation, (ETFA 2001), French Riviera 2001.
[55] Prayati, A., Koubias, S., Papadopoulos, G., "Real-Time aspects in the development of function block oriented engineering support systems", 4th IEEE WFCS, , August 2002, Vasteras, Sweden
[56] J.- R. Abrial, "Steam-boiler control specification problem", August 10, 1994., http://www.informatik.unikiel.de/ ~procos/dag9523.
[57] Archimedes System Platform web page http://seg.ee.upatras.gr/MIM/archimedes.htm

Experimental Investigation on Transverse Vibration Characteristic of Laminate Square Plates by ESPI

Soňa Rusnáková, Juraj Slabeycius, Pavel Koštial, Vladimír Rusnák*
Faculty of Industrial Technologies, University of Alexander Dubček in Trenčín,
Department of Physical Engineering of Materials, I. Krasku 491/30, 020 01 Púchov,
Slovak Republic
* Form s.r.o., Horní Lideč, 756 12 Střelná, Czech Republic

***Abstract*- This work provides non-contact optical technique to investigate the transverse vibration characteristics of laminate square plates in resonance. Most of the works on vibration analysis of plates published in the literature are analytical and numerical and very few experimental results are available. The edges of these laminated plates are free. Electronic Speckle Pattern Interferometry (ESPI) can obtain both resonant frequencies and mode shapes of vibrating plates simultaneously. Excellent quality of the interferometric fringe patterns for the mode shapes is obtained. Seven different composite plates with the same geometric configuration are investigated in this study.**

I. Introduction

Plates are structural elements of great importance and are used extensively in all fields of engineering applications such as aerospace and electronic industry. There have been extensive studies on the vibration of classical plates for various shapes, boundaries, and loading conditions for nearly two centuries. The analysis methods for vibrations of plates can be classified into three types, which are analytical [1], numerical [2,3] and experimental [4].

The analytical method is restricted to problems with one pair of opposite simply supported edges. For the plates with other boundary conditions, the exact solutions generally cannot be obtained by this method. As regards the numerical method, it can solve very complicated plate problems, but a large number of data needed to be processed and the enlarged computing time for increased accuracy is the serious drawbacks. The experimental methods commonly used for vibration analysis are the modal analysis and the spectrum analyser; they are point-wise measurement techniques and generally are used in conjunction with certain types of accelerators and shakers. These point-wise techniques have suffered from irregular mode shape estimation as a result of lengthy data acquisition period. Hence, they are very time-consuming for vibration mode shapes extraction. Therefore, the need for experimental techniques with the characteristics of much faster and full-field configuration is the first motivation for the study presented herein.

II. Vibration characteristics of laminate square plates

Conventional structures have many additional sources of energy dissipation, such as bolted and riveted joints, lubricated bearings, and so on. In space applications, because of the absence of a surrounding fluid or gas, aerodynamic damping is essentially zero, thus removing an important source of energy dissipation, especially in thin-sheet structures. However, when using composite materials, it is usually necessary to use adhesively bonded joints, because bolts and rivets tent to pull out. This seriously reduces structural damping, which makes material damping far more important. This reduces situation can be alleviated in fibre-reinforced materials by making a suitable choice of components to that the damping derives essentially from the matrix and fibre-matrix interface. It is therefore more important to understand the mechanisms of damping in composites and to appreciate their significance than is the case for metallic materials. The main sources of internal damping in a composite material arise from microplastic or viscoelastic phenomena associated with the matrix and from relative slipping at the interface between the matrix and the reinforcement. Thus, excluding the contribution from any cracks and debonds, the internal damping of the composite will be influenced by the properties and relative proportions of matrix and reinforcement in the composite it is usually represented by the (volume fraction of the reinforcement (Vf), the size of the inclusions, the orientation of the reinforcing material to the loading axis, the surface treatment of the reinforcement and the loading and environmental factors, such as amplitude, frequency and temperature.

Mode shapes characterize behavior of the plates under vibration loading. The loading is performed by loudspeaker in our case. "Fig. 1 ", characterizes development of several mode shapes of composite samples in relation with their dimension coefficient. It is well known that resonant frequency the mode shapes of the composite plates increases with their thickness, respectively decreases with their main dimensions.

III. Determination of vibration modes by ESPI

The schematic layout of ESPI optical system, as shown in "Fig. 2 ", is employ to perform the out-of-plane vibration measurement of the resonant frequencies and mode shape for composite plate.

Composite plates with all edges free are used to have the ideal boundary conditions for experimental simulation. The resonant frequency and correspondent mode shape for the

K. Elleithy et al. (eds.), Advances in Computer, Information, and Systems Sciences, and Engineering, 125–128.

vibrating plate are determined experimentally using the no contacting optical method ESPI. A He-Ne laser with wavelength is 632,8 nm is used as the coherent light source.

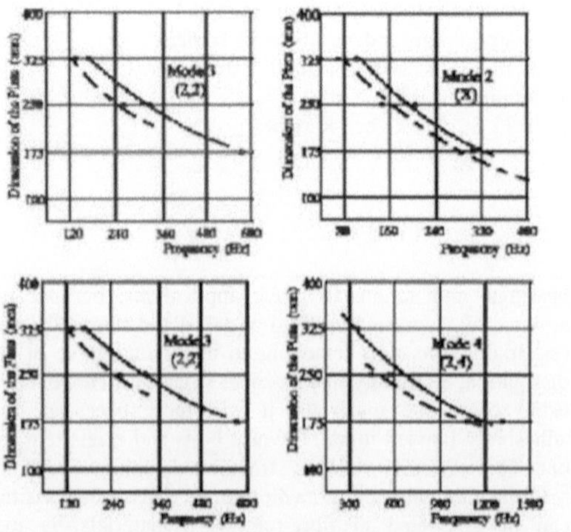

Fig. 1 Variation of the frequency of several mode shapes with dimensions of the plate

. . . .Carbon fibre composite – 8 layers (0°) Volume fraction of fibre=0,5

- - - Aramid fibre composite - 8 layers (0°), Volume fraction of fibre=0,5

The out-of-plane set-up can be briefly described as follows: A laser light beam is split into two. One of the beams, the object beam, is used to illuminate the object. A video camera is then used to view the illuminated object. The other beam, which is called the reference beam, is directed in such a way that it intersects the line of view between the object and video camera. At that point, a partial mirror is used to deflect the reference beam into the video camera causing it to combine with the light reflected off the object. Due to the monochromatic properties of the laser light, the object and reference beam interfere to produces a unique speckle pattern. The speckle pattern is recorded by the video camera and digitised in a computer.

Out of plane displacement is cancelled out as both beam path lengths are altered by the same amount.

To perform an ESPI inspection process, a speckle image of the unstressed object is first captured and stored in the computer. The object is then stressed causing the object's surface to displace. This causes the beam path length of the light reflected off the object surface to alter, which in turn causes the unique speckle pattern to change. When compared with the original stored speckle image, a final image containing the familiar zebra-like fringe patterns is produced.

The resonant frequencies and corresponding mode shapes can be determined at the same time using the ESPI optical system.

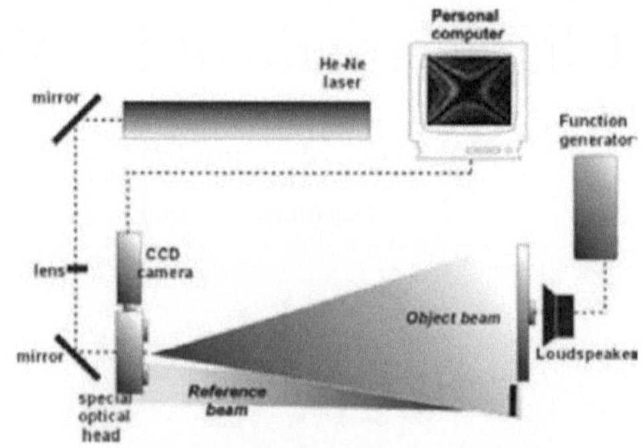

Fig. 2 The experimental setup for ESPI method

IV. DETERMINATION OF DYNAMIC PROPERTIES BY ESPI

The results, which provide the relations between measured resonant frequency and dynamic material parameters, are very good known in the literature [5].

Isotropic Square Plate

At the beginning have to be determined the X – and O – mode shape of isotropic square plate. Poisson's ratio N can be obtained by

$$N = \frac{1}{0{,}72}\frac{\left(\frac{f_0}{f_x}\right)^2 - 1}{\left(\frac{f_0}{f_x}\right)^2 + 1} \tag{1}$$

where f_0, f_x are resonant frequency of O and X mode shapes. The next step is to calculate the fundamental frequency $f_{2,0}$ by

$$f_{2,0} = f_{0,2} = \frac{f_0 + f_x}{2} \tag{2}$$

and to specify the Young's modulus

$$E_0 = \left(\frac{2}{3}\right)^4 f_{2,0}{}^2 \frac{48dl^4}{\pi^2 h^2}\left(1 - N^2\right) \tag{3}$$

where "l" is the side length, "h" is the thickness of the square plate and "d" is the material density.

Orthotropic square plate

For the orthotropic square plate it is possible to determine the resonant frequencies of higher modes from the fundamental mode frequencies $f_{0,2}$, $f_{2,0}$ and $f_{1,1}$. Then we can calculate the five unknown constitutive parameters: Young's modulus in longitudinal direction E_x, Young's modulus in the transverse direction E_y, *in–plane* shear modulus of laminate G_{xy}, Poisson's ratio in the longitudinal direction N_x, Poisson's ratio in the transverse direction N_y.

The stiffness parameter can be obtain by:

$$D_x = \left(\frac{2}{3}\right)^4 f_{2,0}^2 \frac{48dl^4}{\pi^2 h} \tag{4}$$

TABLE 1
DIMENSION OF THE TESTED PLATES

Nr.	Description of materials	Density d $[kg/m^3]$	l $[m]$	h $[m]$	f_x $[Hz]$	f_0 $[Hz]$	$f_{1,1}$ $[Hz]$	$f_{0,2}$ $[Hz]$
1	izoftalic gelcoat+izoftalic polyester resin / 4 layers glass fabric	1872	0,175	0,003	208	267	-	-
2	izoftalic gelcoat+izoftalic polyester resin/ 3 layers Kevlar fabric	1420	0,175	0,003	252	288	-	-
3	izoftalic gelcoat + vinylester resin / 1 layer glass matt+7 layers glass fabric	1700	0,175	0,004	380	336	-	-
4	izoftalic gelcoat+vinylester resin / 8 layers Kevlar fabric	1150	0,175	0,006	639	488	-	-
5	epoxy resin / 4 layers carbon fabric	1150	0,175	0,005	399	570	-	-
6	epoxy resin / 4 layers carbon fabric	1300	0,175	0,003	286	378	-	-
7	Izoftalic polyester resin / 6 layers glass unidirectional fabric	1600	0,175	0,005	-	-	230	355

$$D_y = \left(\frac{2}{3}\right)^4 f_{0,2}^2 \frac{48dl^4}{\pi^2 h} \tag{5}$$

$$D_{xy} = f_{1,1}^2 \frac{48dl^4}{2\pi^2 h} \tag{6}$$

$$\frac{D_x}{D_y} = \frac{E_x}{E_y} = \frac{N_x}{N_y} = A \tag{7}$$

The Young's modulus E_x, E_y can be calculated by using

$$E_x = D_x(1 - N_x, N_y) \tag{8}$$

$$E_y = D_y(1 - N_x N_y) \tag{9}$$

Poisson's ratio N_x it is possible determined by using

$$N_x = \sqrt{A\left(1 - \frac{E_x}{E_y}\right)} \tag{10}$$

$$N_y = \frac{N_x}{A} \tag{11}$$

$$G_{xy} = \frac{1}{2}\left[D_{xy} - N_y D_x\right] \tag{12}$$

V. EXPERIMENTAL RESULTS

Seven laminate composite plates (isotropic and orthotropic) are used in this study for experimental investigations. The dimension of the tested plates are shown in the "Tab. 1", where h is the thickness of investigated plates, l is the length of the plates, f_x is the resonant frequency of X – mode shape, f_o is resonant frequency of O – mode shape, $f_{1,1}$ is the resonant frequency of mode shape *(1,1)*, $f_{0,2}$ is the resonant frequency of mode shape *(0,2)*.

Fibre reinforced composite plates with all edges free are used to have the ideal boundary conditions for experimental simulation. However, in reality the free suspension of a plate can only be approximated. Therefore a compromise has to be found of the free boundary condition on the one side and avoidance of the rigid body motion on the other side. Earlier experiments showed that there is no significant difference of resonant frequencies between supporting the plate on thin needles and suspending the plate with soft wires.

In generally we can say that this value strongly depends on type of resin, type of hardener of epoxy resin and type of production technology.

The accuracy of results strongly depends on determination of X - mode and O - mode shapes.

These results indicate that is possible to measure the stiffness parameters of various composite materials.

Fibre reinforced composites may have unknown non – isotropic properties due to the manufacturing process.

Dynamic properties of composite materials, especially polyester/glass fibre composite are known as the materials, where Poisson's ratio and Young's modulus depends on content of glass fibre, kind of synthetic resin, orientation of fibre, method of production etc.

"Tab. 2" describes results of dynamical properties of laminate composite plates obtained by ESPI.

We used isopthalic resin (Nr. of specimens 1,2) or vinylester resin (Nr. of specimens 3,4) with various types of reinforcements (glass fibre, woven glass roving, woven Kevlar roving,) for production of specimens (Nr. of specimens 1,2,3,4). We determined small deviations of Young's modulus between specimens 1,2,3 and the values given in the literature [6]. When we use epoxy resin with woven Kevlar fabric (Nr. of specimen 5), with woven carbon fabric (Nr. of specimen 6), – we can see bigger deviations between Young's modulus: the table value of Young's modulus for carbon/epoxy composites is 132 GPa, respectively 76 GPa for Kevlar/epoxy composite.

To obtain the stiffness parameters of such materials, ESPI method is developed. Our experimental method is based on the evaluation of resonant frequencies of the mode shapes of square laminate plate. To avoid influences of the measurement setup, the mode shapes and resonant frequencies are determined using ESPI, which provides a non-destructive and non – contacting measurement.

The relations of IV section provide determination Young's modulus E, Poisson's ratio N, Shear modulus G in the case of isotropic materials and Young's modulus in longitudinal direction E_x, Young's modulus in the transverse direction E_y, the major Poisson's ratio N_{xy}, shear modulus G_{xy} in the case of orthotropic materials.

"Fig. 3", shows the first four mode shapes for isotropic laminate plate. Those mode shapes are very important because on the basic resonant frequencies those mode shapes we can obtain very important dynamic properties of laminate composites. The zero – order fringe, which is the brightest fringe on experimental image, represents the nodal lines of the vibrating plate at resonant frequencies. The rest of the fringes are constant amplitudes of displacement.

Because of the symmetry of the geometric configuration of the isotropic square plate and boundary conditions, two possible mode shapes may exist at the same resonant frequency. If the plate is excited at this frequency, a linear combination of these modes will be obtained in experimental observation, and it is possible to define uniquely the mode shape of the vibration.

The mode shapes of laminate composite plates are influenced by material properties, boundary conditions geometry, and lamination arrangement. The last factor provides the designers with more flexibility optimally to synthesize the characteristics (i.e., the stiffness, the fundamental frequencies, and the corresponding mode shapes) of the designed structures.

The mode shapes of laminated plates have some particular characteristics that are different from isotropic plates. This is noticed in the case of an orthotropl5ic square plate with stacking sequence $[0]_{16}$, where the (1,3), (2,3), (1,4), and the (2,4) modes are found to be associated with frequencies that are lower than the (3,1) mode. This feature can be explained by comparing the structural properties, boundary conditions, and dimensions of the plate in the *x* and *y* directions. For the orthotropic plate, the overall bending stiffness with respect to the y direction is weaker than in the x direction. Therefore, the plate tends to form nodal lines in the *y* direction first. In contrast, an isotropic square plate has the same frequency for both the (3,1) and (1,3) modes. [7]

VI. Conclusions

Optical techniques have certain advantages for vibration analysis and ESPI is applied to many vibration problems.

TABLE 2
DYNAMIC PROPERTIES OF INVESTIGATED COMPOSITE PLATE

Nr.	Description of materials	*E* [GPa]	*G* [GPa]	*N*	N_x	N_y	*Gxy* [GPa]
1	izoftalic gelcoat+izoftalic polyester resin / 3 layers glass fabric	9,5	4,1	0,34	-	-	-
2	izoftalic gelcoat+izoftalic polyester resin / 3 layers Kevlar fabric	10	4,6	0,18	-	-	-
3	izoftalic gelcoat + vinylester resin / 1 layer glass matt/ layers glass fabric	11,6	5,3	0,17	-	-	-
4	izoftalic gelcoat+vinylester resin / 3 layers Kevlar fabric	16,7	6,9	0,36	-	-	-
5	epoxy resin / 4 layers kevlar fabric	25,8	10,4	0,47	-	-	-
6	epoxy resin / 4 layers carbon fabric	37	15,6	0,36			
7	Izoftalic polyester resin / 6 layers glass unidirectional fabric	Ex 4,9	Ey 1,1	-	0,25	0,48	0,37

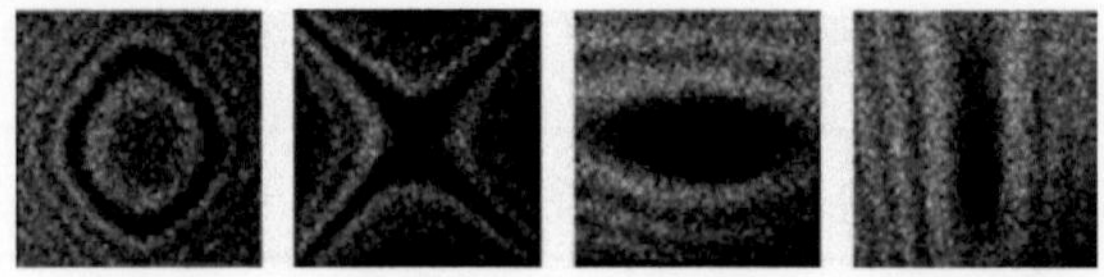

O - mode shape X – mode shape (0,2) mode shape (2,0) mode shape

Fig. 3 The mode shapes for isotropic laminate plate obtained from ESPI

It has been shown that the optical ESPI method has the advantages of non-contact and full field measurement, submicron sensitivity, validity of both static deformation and dynamic vibration, and direct digital image output.

Moreover, no equipment, transducer or sensor, needs to be attached to the plate in order to record the vibration. This method can be applied to many cases within a range of displacements between tens of nanometers and tens of micrometers. Because ESPI uses video recording and display, it works in real time to measure dynamic displacement, which enables implementation of this technique for vibration measurement.

The vibration behaviour of isotropic plates is well understood theoretically with computational results and data stretching back many years. However, there are only very few experimental results available in the literature, especially for the full-.field measurement of vibration mode shapes.

Composite industry is very popular today, with big volume of different and unexplored materials. As well, we can say that each composite product is original. We are determining the type of materials only in productions process. For this reason, ESPI (for its non-destructive characters) can be very useful tool for basic materials properties (Young's modulus, Poisson's ratio, Shear modulus) identification in composite industry.

References

[1] I.E. Harik, X. Liu, N. Balakrishnan, "Analytic Solution to Free Vibration of Rectangular Plates", *Journal of Sound and Vibration*, vol.153, 1992, pp. 51– 62.
[2] W.Q. Chen, R.Q. Xu, H.J. Ding, "On Free Vibration of a Piezoelectric Composite Rectangular Plate", *Journal of Sound and Vibration*, vol.218, 1998, pp. 741–748.
[3] T. Sakiyama, H. Matsuda, "Free Vibration of Rectangular Mindlin Plate with Mixed Boundary Conditions", *Journal of Sound and Vibration,* vol. 113,1987, pp. 208–214.
[4] K.H. Low, G.B. Chai, T.M. Lim, S.C. Sue, "Comparisons of Experimental and Theoretical Frequencies for Rectangular Plates with Various Boundary Conditions and Added Masses", *International Journal of Mechanical Sciences,* vol. 40, pp. 1998, pp. 1119–1131.
[5] G. W. Caldersmith, "Vibration of Orthotropic Rectangular Plates", *Acoustica*, vol. 56, 1984, pp. 144-152.
[6] T. J. Reinhart, et al, "*Materials-Handbooks, manuals*", I. ASM INTERNATIONAL, ISBN 0-87170-279-7,1987.
[7] C. – C. Ma, C. – C. Lin, "Experimental Investigation of Vibrating Laminated Composite Plates by Optical Interferometry Method". *AIAA Journal,* vol. 39, 2001, pp. 491-497.

FEM Modeling of Electromechanical Impedance for the Analysis of Smart Damping Treatments

C. P. Providakis
M. E. Voutetaki
M. E. Stauroulaki
Department of Applied Science
Technical University of Crete
Chania, GREECE

D.- P. N. Kontoni
Department of Civil Engineering
Technological Educational Institute of Patras
Patras, GREECE

***Abstract*-The present paper investigates the numerical modeling of the electro-mechanical impedance for smart plate structures partially treated with active constrained layer damping treatments. The example used is a smart cantilever plate structure containing a viscoelastic material (VEM) layer sandwiched between a piezoelectric constrained layer and the host vibrating plate. Comparisons are made between active constrained layer and active damping only and based on the resonance frequency amplitudes of the electrical admittance numerically evaluated at the surface of the piezoelectric model of the vibrating structures.**

I. INTRODUCTION

There has been an extensive use of passive and active surface treatment strategies for damping the vibration of structures since the 80s. Important among these strategies is the active constrained layer damping (ACLD) treatment [1-4]. It is now well established that the active constrained layer damping configuration (ACLD) can be designed and is more effective to reduce the lower frequency modes. The idea of combining active materials such as piezoelectric materials, with viscoelastic materials (VEM) to produce an active constrained layer has been attractive since the added passive damping provides a fail-safe mechanism and improves the stability and robustness of the system. In the contrary, it is well known that VEM reduces the transmissibility of the control force of the active layer and therefore, several different types, of ACLD treatments should be investigated in detail to perform more effective controllability and in the same time more vibration reduction when compared to purely active damping systems. Further, the modeling and analysis of active constrained layer damping (ACLD) system represent a high level of sophistication and complexity from the structural analysis viewpoint.

The finite element method has become a powerful and versatile tool in designing smart structures containing piezoelectric sensors and actuators. Numerous theories and modeling techniques relating to the finite element method have been proposed for the analysis of adaptive piezoelectric structures [5-10].

Any active material system with integrated induced strain actuators can be generalized as a 1-D simple-electromechanical system as shown in Fig. 1. The entire electro-mechanical system may be electrically represented by an electrical impedance which is affected by the dynamic characteristics of the host vibrating structure. In other words, the electromechanical admittance (inverse of electrical impedance) is modified by structural damping and/or stiffness. Any change in the electrical impedance (admittance) signature is considered as an indication of a change in structural damping and/or stiffness. Therefore, the electromechanical impedance (EMI) of the electrical modeling describes very accurately the dynamic behavior of the smart structural systems.

The purpose of the present work is to provide a numerical technique for the design of vibrating smart structures. It focuses on studying the effect of different active constrained layer damping treatments on the system damping of a smart plate. A cantilever plate with partial active constrained layer damping treatment is modeled by using the finite element method. Finally simulation results are presented for the damping reduction of the first modes of vibration.

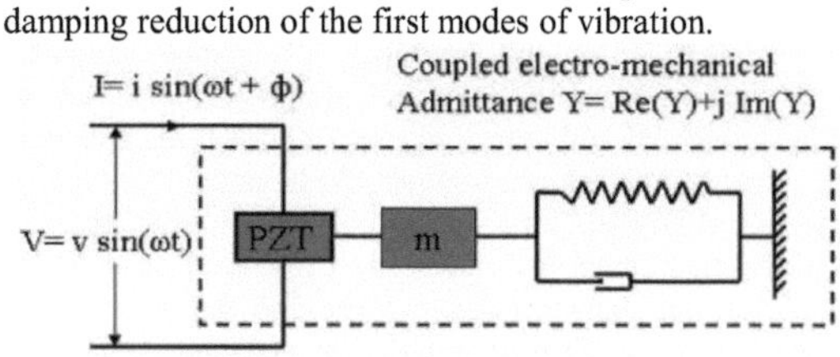

Fig. 1. A schematic description of a PZT actuated dynamic system

K. Elleithy et al. (eds.), Advances in Computer, Information, and Systems Sciences, and Engineering, 129–133.

II. FEM MODELING AND FORMULATION

A finite element formulation for dynamic analysis of piezoelectric material has been presented in many papers [11, 12] and can be in written partition form as:

$$[M_{uu}]\ddot{u} + [K_{uu}]u + [K_{u\varphi}]\Phi = \{F\} \qquad (1)$$

$$[K_{u\varphi}^{T}]u + [K_{\varphi\varphi}]\Phi = \{Q\}$$

where $[M_{uu}] = \int \rho N_u^T N_u dV$, kinematically constant matrix

$[K_{uu}] = \int B_u^T C^E B_u dV$, elastic stiffness matrix

$[K_{u\varphi}] = \int B_u^T h^T B_\phi dV$, piezoelectric coupling matrix

$[K_{\varphi\varphi}] = - \int B_\phi^T b^S B_\phi dV$, dielectric stiffness matrix

$\{F\} = \int_v N_b^T f_b dV + \int_{A_1} N_{A_1} f_S dA_1 + N_a^T f_c$,

mechanical force

$\{Q\} = - \int_v N_{A_2}^T q_S dA_2 - N_S^T q_c$, electrical charge

Where fb, is the body force, fs, is the surface force, fc, is the concentrated force, qs is the surface charge, qc is the point charge, A_1 is the area where mechanical forces are applied and A_2 is the area where electrical charges are applied.

If the time harmonic case is considered the following matrix equations can be obtained from the finite element formulation:

$$\left(-\omega^2\begin{bmatrix}[M_{uu}] & 0\\ 0 & 0\end{bmatrix} + \begin{bmatrix}[K_{uu}] & [K_{u\phi}]\\ [K_{u\phi}]^T & [K_{\phi\phi}]\end{bmatrix}\right)\begin{Bmatrix}\{u\}\\ \{\Phi\}\end{Bmatrix} = \begin{Bmatrix}\{K\}\\ \{Q\}\end{Bmatrix} \qquad (2)$$

Where ω is the angular frequency. The response then can be obtained in the frequency domain for the displacements {u} and potential {Φ}, for a given charge of force excitation at a given angular frequency. It should be noted that all these vectors must now be considered as complex amplitudes.

Constrained active layer damping consists of three layers: the host plate, the viscoelastic layer and the piezoelectric constraining layer. The viscoelastic layer, being much softer than the host plate and constrained layer, has considerable transverse shear deformations. In order to model this difference, the viscoelastic layer needs to deform differently from the structure or constrained layer. In this paper, in order to account for the different deformation of each layer the viscoelastic, the host structure and the constraining layer is modeled separately with different finite elements.

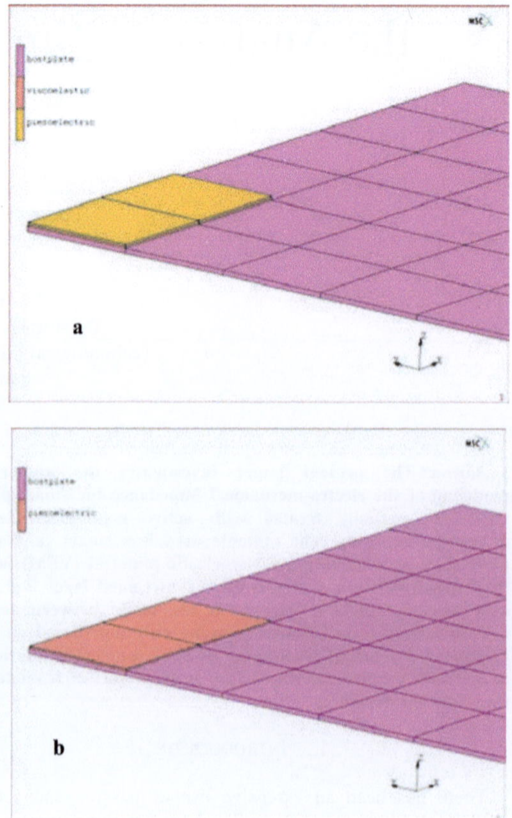

Fig. 2. a) ACLD treatment; b) AD treatment

The idea of moving a piezoelectric actuator from direct contact with the host structure to placing a viscoelastic layer between the host structure and the piezoelectric actuator is questionable. The actuator is more effective in applying force when it is acting directly at the host structure rather than through a very soft damping material as being the viscoelastic one. However an active constraining layer can significally enhance the damping effectiveness via active control. In comparing different types of damping treatments, the active constrained layer (ACLD) and the pure active layer (AD) using piezoelectric film actuators are designed along with the host structure in this paper to demonstrate the potential of each type of treatments (Fig. 2). In order to perform the comparison studies of different damping configurations coupled electromechanical harmonic analyses are performed by using the finite element program MSC.MARC2005 [13] enhanced by its piezoelectric analysis capabilities.

III. THE ELECTROMECHANICAL IMPEDANCE FEATURE

During the last few years the Electro-Mechanical Impedance (EMI) technique has demonstrated its potential for cost-effective structural monitoring for a wide range of engineering structures [14-18]. The key point of the impedance-based qualitative monitoring technique is the electromechanical coupling of the piezoelectric PZT actuator/sensor patch. The PZT patches are made of "piezoelectric" materials, which generate surface charges in response to applied mechanical stresses and conversely undergo mechanical deformation in response to applied electric fields. Hence, when a PZT patch surface bonded to the monitored structure is electrically excited by means of an impedance analyzer it produces deformations in the surrounded local area of the host structure. The response of this area is transfered back to the PZT patch in the form of admittance function (admittance is the reciprocal of impedance), comprising the conductance (the real part) and the susceptance (the imaginary part).

The acquisition of the coupled electromechanical impedance can be performed by providing a constant voltage (1 V rms) to the PZT patch over a pre-set frequency range. The magnitude and phase of the steady state current draw (after transient decay) of the PZT is recorded in real and imaginary admittance. The variation in the electrical admittance over the pre-selected frequency is due to two factors: (1) the electrically capacitive nature of the PZT, and (2) the interaction between the PZT and the structure to which it is bounded. The variations in the capacitive nature show up as a gradual change in the impedance curve, whereas the structural interaction corresponds to sharp peaks that represent part of structural resonance. Therefore, any change in the dynamic behavior of the host structure like the damping characteristics will result in an electrical admittance change.

Following the above description when a unit voltage is applied to the piezoelectric layer of the finite element model of the active constrained damping layer the admittance could be calculated. The application of the unit voltage is simulated by considering a positive and a negative electrode located at the upper and lower, respectively, surface of the PZT patch. In the model these electrodes are made by tying the potential degree of freedom of all the nodes belonging to an electrode to one node. In this way, the admittance can be easily calculated. The admittance is the ratio of the resulting current and the applied voltage. The resulting current I is related to the total reaction charge on the piezoelectric surface as:

$$I = i\omega \sum_{i} Q_i \tag{3}$$

Where ΣQi is the sum of the reaction charges resulted from the finite element analysis on all nodes belonging to the appropriate piezoelectric electrode. This summation is automatically done by the finite element program since all the nodes of an electrode are tied to one node. Conductance is the real part of the resulted admittance and suscuptance is the imaginary part. These electrical "signatures" of the structure (in the frequency domain), contain vital information concerning the phenomenological nature of the structural parameters i.e., the stiffness, the damping and the mass. The structural dynamic behavior corresponds to a unique pattern of the sharp peaks generated above the baseline electrical capacitive admittance in the admittance – frequency plots.

IV. NUMERICAL EXAMPLE

The purpose of this section is to numerically evaluate and to compare the damping resulting from an active constrained damping layer (ACLD) system and a purely active damping system (AD). A thin aluminum cantilever plate, which is partially treated with a viscoelastic (VEM) material which also has attached to its surface a piezoelectric actuator, is studied. The VEM layer is under the actuator for the ACLD as can be seen in figure 2 and considered as perfectly bonded to each other. This is a relative simple structure and it is assumed that there is no structural damping so as to show clearly the effects of active restrained and purely active damping. It is also assumed that the piezoelectric elements are bonded perfectly to the host plate and to the VEM for the case of purely active (AD) and active constrained layer damping (ACLD) model, respectively. The nodes at the interface of the PZT and VEM (for the ACLD case) and at the interface of the PZT and host plate (for the AD case), respectively, are connected to the common ground to make a closed circuit for the potential. In both ACLD and AD plate cases there is not control loop that regulates the piezoelectric actuator and the structure response. Two 8-node piezoelectric solid elements (Fig. 3) are used to model the PZT devices each of which have four degrees of freedom, the first three are for the x-, y-, and z- displacements, and the fourth is for the electric potential. The material parameters for the PZT devices are ρ_a=7500 Kgr/m^3, E_a=3.6x10^{11}N/m^2 and v_a=0.3.

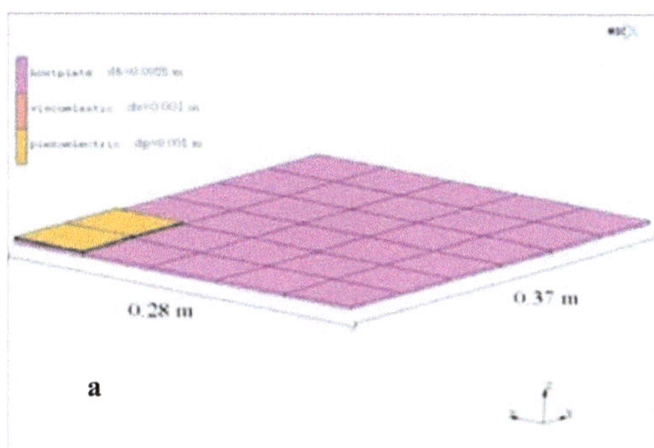

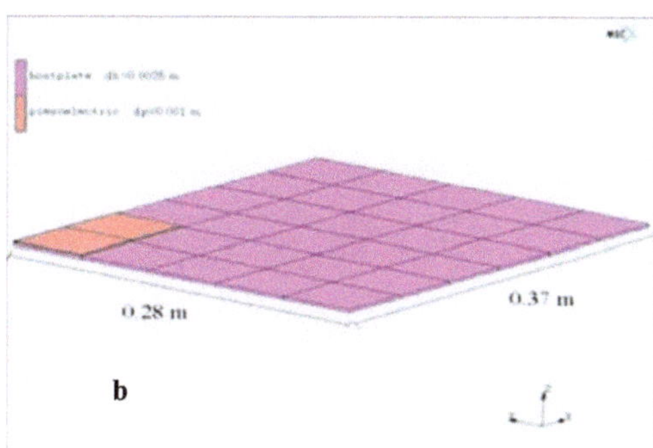

Fig. 3. a) ACLD finite element model; b) AD finite element model

To model the VEM layer, two 8-node solid finite elements (Fig. 3) are assumed with a viscoelastic relax function for the shear modulus given by $G(t) = 100 + 9900e^{(t/0.4170316)}$, while $E_v = 2\text{x}10^7$ N/m^2 and $v_v = 0.3$. The host aluminum plate is modeled by 8-node solid elements (Fig. 3) with boundary conditions: one short side of the host plate is clamped and the others are free. The material properties of the host plate are $\rho_p = 2700$ Kgr/m^3, $E_p = 9\text{x}10^{10}$ N/m^2 and $v_p = 0.3$. The dimensions of the host plate are 0.35 m x 0.28 m while the thickness of piezoelectric, viscoelastic and host plate layer are 0.001 m, 0.001 m and 0.0025 m, respectively.

Fig. 4 represents the conductance and one peak may be observed at 19.18 Hz. For the case of ACLD plate which coincide with the first natural frequency of the cantilever ACLD plate under study.

Figure 5 shows the susceptance curve. We notice that, in both figures, the AD patch directly attached to the host plate without VEM stiffness the structure and the first mode natural frequency shifts near to 19.42 Hz. The resonance conductance and susceptance aptitude decreases significantly as shown in figures 4 and 5. This reduction outperforms the most effective damping actuation for the ACLD plate model as compared to the pure active AD model.

The same results are shown in figures 6, 7, 8 and 9 for the next two mode natural frequencies. The ACLD curve clearly depicts superior damping for all first modes, which, as expected, increases the integrity of the system and results in increased performance bandwidth for vibration damping.

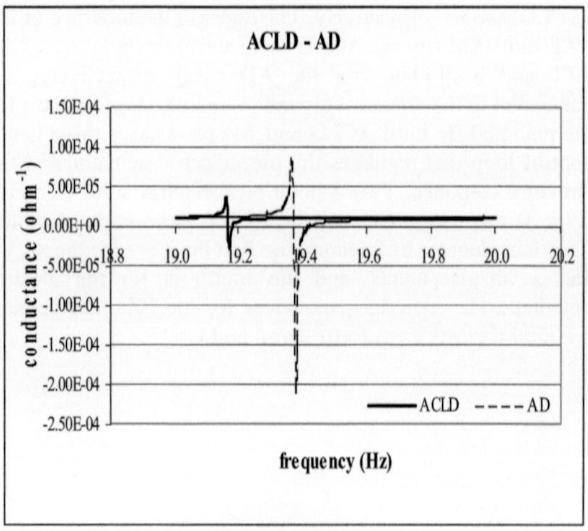

Fig. 4. Conductance signature for the 1st mode of ACLD, AD cases

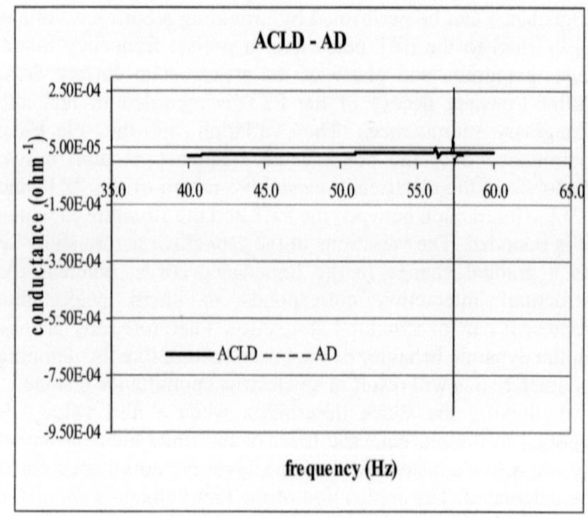

Fig. 6. Conductance signature for the 2nd mode of ACLD, AD cases

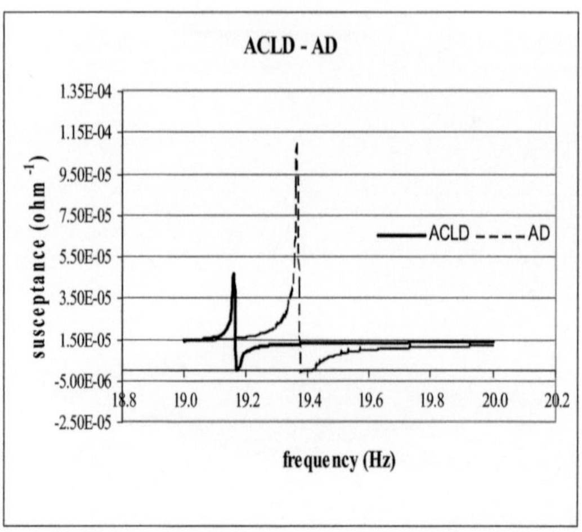

Fig. 5. Susceptance signature for the 1st mode of ACLD, AD cases

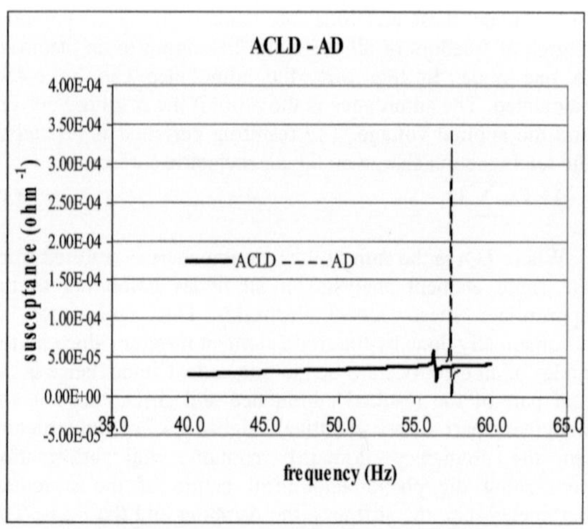

Fig. 7. Susceptance signature for the 2nd mode of ACLD, AD cases

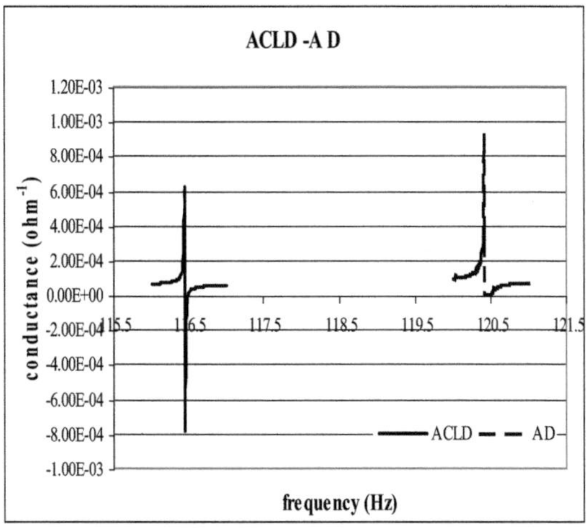

Fig. 8. Conductance signature for the 3rd mode of ACLD, AD cases

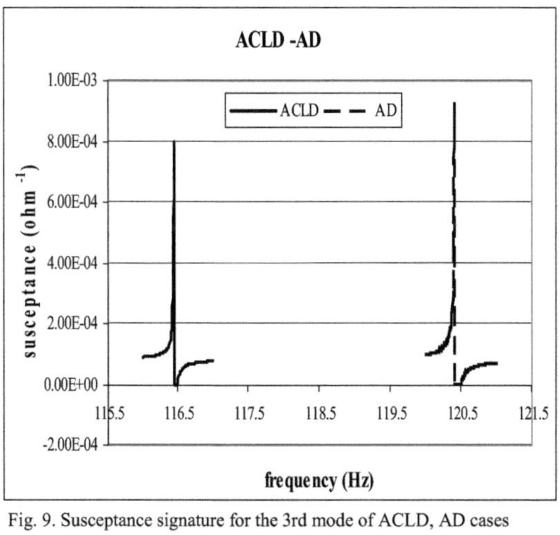

Fig. 9. Susceptance signature for the 3rd mode of ACLD, AD cases

The present paper has presented a numerical finite element investigation of different damping performance of smart structures employing VEMs and piezoelectric actuators.

Some comparison studies have been presented which examine the performance of the active constrained damping layer (ACLD) model relative to the purely active (AD) model. These studies enabled a complete description of the numerical evaluation of the electrical input admittance of the damping systems and a faithful reproduction of the dynamical characteristics of the systems.

It is clear from the results presented here for a simple cantilever plate structure that an ACLD outperforms the active system and provides the benefit of active/passive damping.

ACKNOWLEDGMENTS

The authors acknowledge financial support by "Archimedes - EPEAEK II Research Programme" co-funded by European Social Fund and National Resources.

REFERENCES

[1] I.Y. Shen, "Hybrid damping through intelligent constrained layer damping treatments", *ASME J. Vibr. Acoust.*, vol.116, pp.341-349, 1994.
[2] A. Baz, "Boundary control of beams using active constrained layer damping", *Journal of Vibration and Acoustics*, vol. 119, pp.116-172, 1997.
[3] L.C. Hau and E.H.K. Fung, "Effect of ACLD treatment confifuration on damping performance of a flexible beam", *Journal of Sound and Vibration*, vol. 269, pp.549-567, 2004.
[4] T. Liu, H. Hua and Z. Zhang, "Robust control of plate vibration via active constrained layer damping", *Thin-walled Structures*, vol. 42, pp. 427-448, 2004.
[5] W.C. Van Nostran, G. Knowles, and D. Iuman, "Finite element model for active constrained damping", *Proc. Conf. Smart Struct. Mater. SPIE*, vol.2173, pp.269-281, 1994.
[6] V.V. Varadan, Y-lt. Lim, and V.K. Varadan, "Closed loop finite element modeling of active/passive damping in structural vibration control", *Smart Mater. Struct.* vol.5, pp. 685-694, 1996.
[7] Y_H. Lim, V.V. Varadan, and V.K. Varadan, "Closed loop finite element modeling of active structural damping in the time domain", *Smart Mater. Struct.* vol. 8, pp390-400, 1999.
[8] J. Kim, V.V. Varadan, and V.K. Varadan, "Finite element modeling of a smart cantilever plate and comparison with experiments", *Smart Mater. Struct.* vol. 5, pp.165-170, 1996.
[9] Q. Wang, W.H Duan and S.T. Quek, "Repair of notched beam under dynamic load using piezoelectric patch", *Intern. Journal of Mechanical Sciences*, vol. 46, pp.1517-1533, 2004.
[10] A. Littlefield, J. Fair-weather and K. Craig, "Use of FEA derived impedances to design active structures", *J. Intelligent Material System and Structures*, vol. 13, pp.377-388, 2002
[11] H.S. Tzou, and C.I. Tseng, "Distributed piezoelectric sensor/actuator design for dynamic measurement control of distributed parameter systems : a piezoelectric finite element approach", *J Sound Vib.*, vol. 138, pp.17-34, 1990.
[12] S.S. Rao, and M. Sunar, , "Analysis of distributed thermopiezoelectric sensors and actuators in advanced intelligent structures", *AIAA J.*, vol.31, pp.1280-1286, 1993.
[13] MSC. MARC 2005, Users Guide, MSC. Software, CA, USA, 2004
[14] Sun, F.P., Chaudhvy, Z. Rogers, C.A.., Majmn, M and Liang, C., "Automatel real-time structure health monitoring via signature pattern recognition", In Proceedings, *SPIE conference on Smart Structures and Materials*, San Diego, CA. vol. 7, pp. 559-605, 1995.
[15] J.W. Ayres, F. Lalande, Z. Chandhvy, and L.A. Rogers, Qualitative impedance-based health monitorind of civil infrastructures", *Smart Materials Structures*, vol. 7, pp.599-605, 1998.
[16] C. K. Soh, K.K.H. Tseng, S. Bhalla, and A. Gupta, "Performance of smart piezoceramic patches in helth monitoring of a RC Bridge", *Smart Mater. Struct.*, vol. 9, pp.533-542, 2000.
[17] K. K. Tseng and L. Wang, "Structural damage identification for thin plates using smart piezoelectric transducers", *Comp. Meth. Appl. Mech. Engng.*, vol. 194, pp. 3192-3209, 2005.
[18] S. Bhalla, C. K. Soh, and Z. Liu, "Wave propagation approach for NDE using surface bonded piezoceramics", *NDT & E Intrn.*, vol. 38, pp. 143-150, 2005.

Spectral Characteristics of Quantum Associative Memories

G.G. Rigatos
Unit of Industrial Automation
Industrial Systems Institute
26504, Rion Patras, Greece
email: grigat@isi.gr

S.G. Tzafestas
Department of Electrical and Computer Eng.
National Tech. Univ. of Athens
15773, Zografou Athens, Greece
email:tzafesta@softlab.ece.ntua.gr

***Abstract*- Quantum associative memories are derived from the Hopfield memory model under the assumption that the elements of the correlation weight matrix W are stochastic variables. The probability density function of each weight is given as a solution of Schrödinger's diffusion equation. Spectral analysis of quantum associative memories follows previous studies on the wavelets' power spectra. Spectral analysis shows that the energy stored in quantum associative memories is distributed in discrete levels.**

I. INTRODUCTION

It is assumed that each weight w_{ij} of the associative memory (Fig. 1) is a stochastic variable and that its probability density function $\psi(w_{ij},t)$ is given by the solution of Schrödinger's diffusion equation [1]. The solution of Schrödinger's equation can be written as a generalized Fourier series, i.e. $\psi(w_{ij},t)=\sum_{k=0}^{\infty} c_k \psi_k(w_{ij},t)$, where $\psi_k(w_{ij},t)$ are orthogonal basis functions. The coefficients c_k are an indication of the probability the value of $\psi(w_{ij})$ at time instant t to be described by the eigenfunction $\psi(w_{ij},t)$. This results into the mean value $<w_{ij}>=\sum_{k=1}^{\infty} c_k^2 a_k$, where c_k^2 denotes the probability for the weight to be described by the eigenfunction ψ_k , and α_k is the eigenvalue associated with ψ_k . If the probability c_k^2 is substituted by a fuzzy membership function μ_{A_k} , and $\mu_{A_k}(w_{ij})$ is selected so as to satisfy the strong partition condition $\sum_{k=1}^{\infty} \mu_{A_k}(w_{ij})=1$ then it holds that $<w_{ij}>=\mu_{A_k} a_k$ [2],[3].

Taking the weights w_{ij} of the $N \times N$ weight matrix W to be stochastic variables means that W can be decomposed into superimposed associative memories $W=\sum_{i=1}^{2^N} \mu_i \bar{W}_i$ where the nonnegative coefficients μ_i indicate the contribution of each local associative memory $\bar{W}_i$ to the total outcome [4]. Thus the number of attractors is increased by a factor of 2^N .

Spectral analysis of quantum associative memories follows previous studies on the wavelets' power spectra [5]-[7]. Spectral analysis of quantum associative memories shows the following:

• The Gaussian basis functions of the weights w_{ij} of a quantum associative memory express the discrete levels in which the energy of w_{ij} is distributed. The smaller the spread σ of the basis functions is, the larger becomes the spectral (energy) content that can be captured therein. Narrow spread of the Gaussians in w_{ij} , results in wide range of frequencies in the Fourier transform $F(w_{ij})$.

• The energy levels of the weights w_{ij}'s are discrete. Indeed, the eigensolution of Schrödinger's equation is of the form $\psi_k(w_{ij},t)=X_k(w_{ij})e^{-i\hbar\omega t}$. It holds that $HX_k(w_{ij})=E_k X_k(w_{ij}),\ k=0.1,2,\ldots$ where H is the Hamiltonian of the system and $E=\hbar\omega$ is the corresponding energy eigenstate.

• The stochastic weights w_{ij} satisfy the principle of uncertainty. The dispersion $D(g)$ of the Gaussians in $<w_{ij}>$ and the dispersion $D(G)$ of the associated Fourier transform results in $D(g)D(G)=\frac{1}{4}$. This is analogous to quantum mechanics uncertainty principle, i.e. $\Delta x \Delta p \geq \hbar$.

The structure of the paper is as follows: In Section II, the probability density function of the weights of quantum

K. Elleithy et al. (eds.), Advances in Computer, Information, and Systems Sciences, and Engineering, 135–142.

associative memories is calculated as a solution of Schrödinger's equation. The equivalence between the weight w_{ij} and a quantum particle is demonstrated. In Section III the spectral decomposition of the Morlet wavelets is presented and Heisenberg uncertainty boxes are introduced. In Section IV spectral analysis of the weights w_{ij} of the quantum associative memory is performed with the use of Fourier transform. Finally, in Section V, concluding remarks are stated.

II. QUANTUM ASSOCIATIVE MEMORIES

Previous studies on quantum associative memories can be found in [8]-[10]. In quantum associative memories the probability density function of element w_{ij} in the weight matrix W is given as a solution of Schrödinger's differential equation. It is shown that the weight matrix W can be decomposed into a superposition of matrices $\bar{W}_i$, thus increasing the number of attractors by a factor of 2^N . The eigenvalues of matrices $\bar{W}_i$ denote the levels of energy quantization.

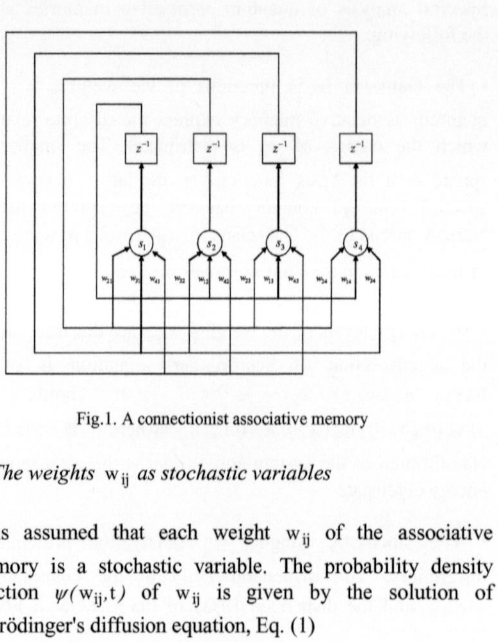

Fig.1. A connectionist associative memory

A. The weights w_{ij} as stochastic variables

It is assumed that each weight w_{ij} of the associative memory is a stochastic variable. The probability density function $\psi(w_{ij},t)$ of w_{ij} is given by the solution of Schrödinger's diffusion equation, Eq. (1)

$$i\frac{\partial\psi(w_{ij},t)}{\partial t}=-\frac{\hbar^2}{2m}\nabla^2\psi(w_{ij},t)+V(w_{ij})\Rightarrow \qquad i\frac{\partial\psi(w_{ij},t)}{\partial t}=H\psi(w_{ij},t) \tag{1}$$

where i is the complex variable, $\hbar$ is Planck's constant , V denotes the potential and H is the Hamiltonian. An attempt to associate the weights of neural networks with diffusion equations can be found in [11]. The solution $\psi(w_{ij},t)$ of Eq. (1) equation can be written as generalized Fourier series [1]:

$$\psi(w_{ij},t)=\sum_{k=0}^{\infty}c_k\psi_k(w_{ij},t) \tag{2}$$

where $\psi_k(w_{ij},t)$ are orthogonal basis functions. The coefficients c_k are an indication of the probability the value of w_{ij} at time instant t to be described by the eigenfunction $\psi_k(w_{ij},t)$. The probability amplitude that the weight (particle) w_{ij} lies between w_{ij} and $w_{ij}+\Delta w_{ij}$ is $P(w_{ij})dw_{ij}=|\psi(w_{ij})|^2\,dw_{ij}$. The aggregate probability should equal 1 , i.e. $\int_{-\infty}^{+\infty}|\psi(w_{ij})|^2\,dw_{ij}=1$. Thus, the mean value of the weight w_{ij} is then given by:

$$<w_{ij}>=\int_{-\infty}^{+\infty}P(w_{ij})w_{ij}dw_{ij}=\int_{-\infty}^{+\infty}\psi^* w_{ij}\psi dw_{ij} \tag{3}$$

where ψ^* is the Hermitian of the wavefunction $\psi(w_{ij},t)$. Since $\psi(w_{ij},t)$ has been analyzed in a generalized Fourier series, the coefficients c_k can be calculated using the property of orthonormality, i.e.

$$\psi(w_{ij},t)=\sum_{k=1}^{\infty}c_k\psi_k(w_{ij},t)\Rightarrow \int_{-\infty}^{+\infty}\psi(w_{ij},t)\psi_k(w_{ij},t)dw_{ij}=c_k$$
$$\Rightarrow c_k=\int_{-\infty}^{+\infty}\psi_k^*(w_{ij},t)\psi(w_{ij},t)dw_{ij} \tag{4}$$

The eigenvalues and the eigenfunctions of the quantum position operator $x=w_{ij}$ are defined as

$$x\psi_k=a_k\psi_k \tag{5}$$

which implies a projection of position $x=w_{ij}$ on the axis ψ_k, where ψ_k is the eigenfunction and a_k is the associated eigenvalue. Thus, the mean value of the weight is given by

$$<w_{ij}>=\int_{-\infty}^{+\infty}\psi^* w_{ij}\psi dw_{ij}\Rightarrow <w_{ij}>=\int_{-\infty}^{+\infty}\sum_{k=1}^{\infty}c_k\psi_k^* w_{ij}\sum_{k=1}^{\infty}c_k\psi_k dw_{ij}\Rightarrow$$
$$<w_{ij}>=\int_{-\infty}^{+\infty}\psi^*\sum_{k=1}^{\infty}\alpha_k c_k\psi_k dw_{ij}\Rightarrow <w_{ij}>=\sum_{k=1}^{\infty}c_k\alpha_k\int_{-\infty}^{+\infty}\psi_k^*\psi dw_{ij}\Rightarrow$$
$$<w_{ij}>=\sum_{k=1}^{\infty}c_k^2\alpha_k \tag{6}$$

where c_k^2 denotes the probability the weight to be described by the basis function ψ_k. When the weight (position of the particle) w_{ij} is described by the basis function ψ_k then the value of w_{ij} which results from a measurement is the associated eigenvalue a_k. This is the so-called filtering problem, which means that if one tries to measure a system being initially at condition $\psi = \sum_{k=1}^{\infty} c_k \psi_k$, then the result of the measurement is to change ψ to the eigenfunction ψ_k. The only measurable value will be the eigenvalue a_k. The eigenvalue a_k is selected with probability $P = c_k^2$.

B. Equivalence between weight w_{ij} and a quantum particle

If the probability c_k^2 is substituted by the fuzzy membership $\mu_{A_k} = e^{-(w_{ij}-k\cdot c)^2/2\sigma^2}$, and $\mu_{A_k}(w_{ij})$ is selected so as to satisfy the strong partition condition

$$\sum_{k=1}^{\infty} \mu_{A_k}(w_{ij}) = 1 \qquad (7)$$

then it holds

$$< w_{ij} >= \sum_{k=1}^{\infty} e^{-\frac{1}{2}\frac{(w_{ij}-k\cdot c)^2}{\sigma^2}} \alpha_k \qquad (8)$$

From Eq. (6) and Eq. (8) the following are deduced [2], [3]:

1. The Gaussian $e^{-\frac{1}{2}\frac{(w_{ij}-k\cdot c)^2}{\sigma^2}}$ corresponds to c_k^2, i.e. it is a measure of the probability the weight w_{ij} to belong to the basis function ψ_k.
2. The eigenvalues α_k correspond to the centers of the fuzzy sets A_k (Fig. 2).
3. The basis functions A_k correspond to the orthonormal eigenfunctions ψ_k of the solution $\psi(w_{ij},t)$ of Schröndiger equation.
4. The assumption that the Gaussians $\mu_{A_k}(w_{ij})$ satisfy condition of Eq. (7) is equivalent to the requirement that the aggregate probability $\psi(w_{ij},t)$ to find the weight (quantum particle) between the positions w_{ij} and $w_{ij}+\Delta w_{ij}$ equals unity, i.e. $\int_{-\infty}^{+\infty} |\psi(w_{ij},t)|^2 \, dw_{ij} = 1$.
5. The measurement in a quantum system gives as output only the eigenvalue α_k which is associated with the eigenfunction ψ_k, with probability $P = \mu_{A_k}(w_{ij})$.
6. Both the particle's wave function $\psi(w_{ij},t) = \sum_{k=1}^{\infty} c_k \psi_k(w_{ij},t)$ and the mean position of the quantum particle $< w_{ij} >= \sum_{k=1}^{\infty} \mu_{A_k}(w_{ij}) \alpha_k$ can be represented as vectors in a Hilbert's space, where the basis functions will be taken to be ψ_k and A_k respectively.

Comparing Eq. (8) to Eq. (6) one can observe the isomorphism:

$$|\psi_k(w_{ij})|^2 \rightarrow \mu_{A_k}(w_{ij}) = e^{-\frac{1}{2}\frac{(w_{ij}-k\cdot c)^2}{\sigma^2}} \qquad (9)$$
$$\alpha_k = k \cdot c$$

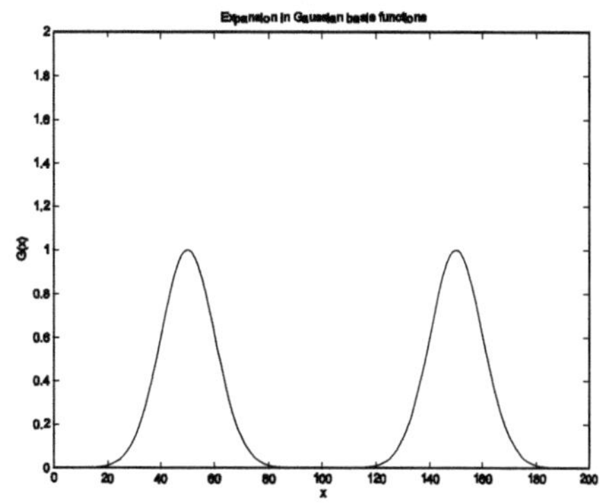

Fig.2 Gaussians as a measure of uncertainty of the weights' values

C. Compatibility with quantum mechanics postulates

In quantum associative memories the weights follow Schrödinger's differential equation, given in Eq. (1). This implies that the exact value of the weights is not explicitly known. This parameter is described by a probability density function $|\psi(w_{ij},t)|^2$ which stems from the solution of Schrödinger's differential equation.
It has been shown that quantum associative memories, satisfy basic postulates of quantum mechanics, i.e. [12]: (i) quantum learning of w_{ij}'s with the use of unitary operators (ii) existence of the variables w_{ij} in a superposition of states, (iii) evolution between the eigenvector subspaces of the weight matrix W with the use of unitary operators.
Taking the weights w_{ij} of the weight matrix W to be stochastic variables means that W can be decomposed into a superposition of associative memories (Fig. 3). The associative memory W equals a weighted averaging of the

individual weight matrices $\bar{W}_i$, i.e. $W=\sum_{i=1}^{2^N}\mu_i\bar{W}_i$ where coefficients μ_i indicate the contribution of each local associative memory $\bar{W}_i$ to the aggregate outcome (Fig. 3).

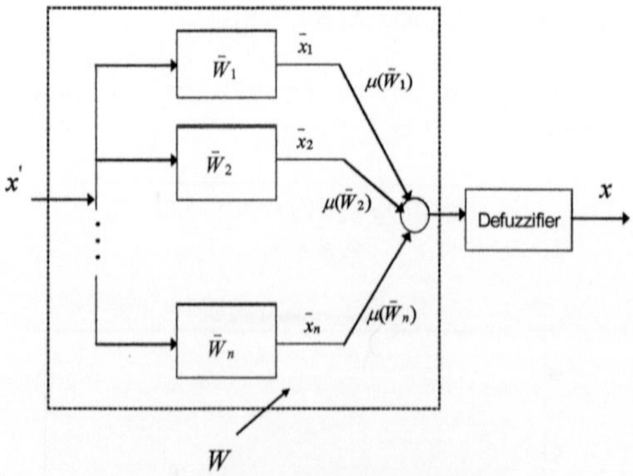

Fig. 3 Superposition of Hopfield weight matrices

III. Spectral Analysis of Wavelets

The study of the power spectra of wavelets will be used as the basis of the spectral analysis of quantum associative memories.

A. The Morlet wavelet

The Morlet wavelet is the most commonly used complex wavelet and is given by [5]

$$\psi(x)=\pi^{-\frac{1}{4}}\left(e^{i2\pi f_0 x}-e^{\frac{-(2\pi f_0)^2}{2}}\right)e^{-\frac{x^2}{2\sigma^2}} \tag{10}$$

where f_0 is the central frequency of the mother wavelet. The second term in the bracket is known as the correction term as it corrects for the non-zero mean of the complex sinusoid of the first term. In practice, it becomes negligible, for values of $f_0 >> 0$ and can be ignored. In that case the Morlet wavelet can be written in a simpler form as

$$\psi(x)=\pi^{-\frac{1}{4}}e^{i2\pi f_0 x}e^{-\frac{x^2}{2}} \tag{11}$$

This wavelet is simply a complex wave within a Gaussian envelope. The complex sinusoidal waveform is contained in the term $e^{i2\pi f_0 x}=cos(2\pi f_0 x)+i\,sin(2\pi f_0 x)$. The real and the imaginary part of the Morlet wavelet for various central frequencies are depicted in Fig. 4.

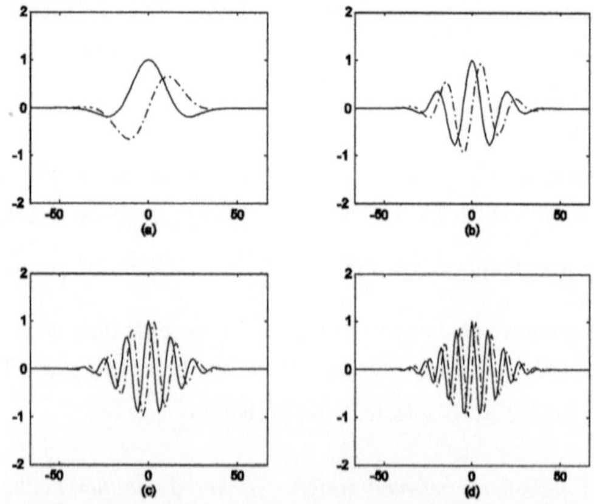

Fig. 4 Real (continuous) and imaginary (dashed) part of Morlet

It can be seen that the real and the imaginary part of the wavelet differ in phase by a quarter period. The $\pi^{1/4}$ term is a normalization factor which ensures that the wavelet has unit energy.

B. Spectral decomposition and Heisenberg boxes

The Fourier transform of the Morlet wavelet is given by

$$\hat{\psi}(f)=\pi^{\frac{1}{4}}\sqrt{2}e^{-\frac{1}{2}(2\pi f-2\pi f_0)^2} \tag{12}$$

which has the form of a Gaussian function displaced along the frequency axis by f_0. The energy spectrum (the squared magnitude of the Fourier transform) is given by

$$|\hat{\psi}(f)|^2=2\pi^{\frac{1}{2}}e^{-(2\pi f-2\pi f_0)^2} \tag{13}$$

which is a Gaussian centered at f_0. The integral of Eq. (13) gives the energy of the Morlet wavelet. The energy spectrum of the Morlet wavelet depicted in Fig. 4(c), for different values of the variance σ^2 is given in Fig. 5.

The central frequency f_0 is the frequency of the complex sinusoid and its value determines the number of significant sinusoidal waveforms contained within the envelope (i.e. those which are not very close to zero amplitude). The dilated and translated Morlet wavelet $\psi(\frac{x-b}{a})$ is

$$\psi\left(\frac{x-b}{a}\right)=\pi^{-\frac{1}{4}}e^{i2\pi f_0\left(\frac{x-b}{a}\right)}e^{-\frac{1}{2}\left(\frac{x-b}{a}\right)^2} \tag{14}$$

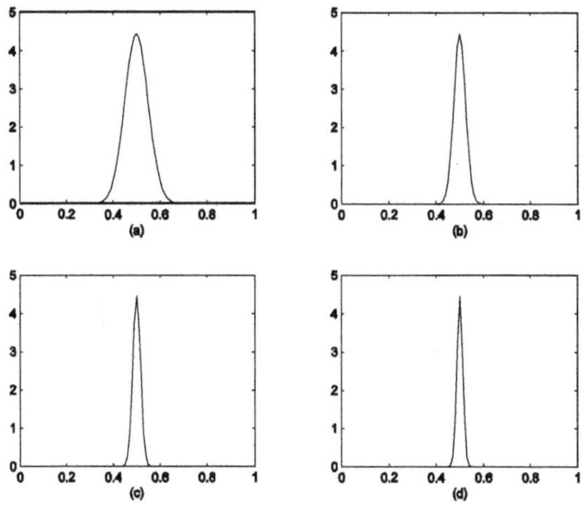

Fig. 5 Energy spectrum of the Morlet wavelet of Fig. 4(c) for central frequency $f_0 = 0.5$Hz and different variances: (a) σ^2, (b) $4\sigma^2$, (c) $8\sigma^2$, (d) $16\sigma^2$

The Heisenberg boxes in the time-frequency plane for a wavelet at different scales, are shown in Fig. 6. To evaluate frequency composition a sample of a long region of the signal is required. If instead, a small region of the signal is measured with accuracy, then it becomes very difficult to determine the frequency content of the signal in that region. That is, the more accurate the temporal measurement (smaller σ_x) is, the less accurate the spectral measurement (larger σ_f) becomes, and vice-versa [6].

The Morlet central frequency f_0 sets the location of the Heisenberg box in the time-frequency plane. Thus, when comparing Heisenberg-boxes centered at the same location in the time-frequency plane, lower values of f_0 correspond to Heisenberg boxes that are wider in frequency and narrower in time than boxes corresponding to higher f_0 s. Thus Morlet wavelets with lower f_0 s correspond to time-frequency decompositions that are more temporal than spectral, than their higher central frequency counterparts.
If the time-length of the wavelets remains the same, then no matter the change of the central frequency f_0 the associated Heisenberg boxes will have the same dimensions. This is depicted in Fig. 7.

Finally, in Fig. 8 are shown the Heisenberg boxes in the time-frequency plane for a number of wavelets with three different spectral frequencies (low, medium and high). The confining Gaussian windows are all of the same dimensions in time. Therefore, altering the central frequency of the wavelet shifts the Heisenberg box up and down the time-frequency plane without altering the its dimensions.

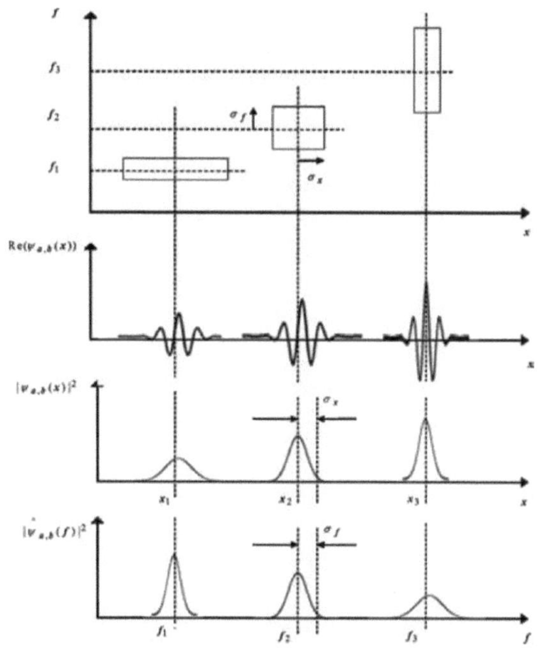

Fig. 6 Heisenberg boxes in the time-frequency plane for a wavelet at various scales. f_c is the standard deviation of the spectrum around the origin; σ^2 is the standard deviation of the spectrum around the mean spectral components f_1, f_2 and f_3.

IV. Spectral Analysis of the Weights w_{ij}

Spectral analysis of quantum associative memories is carried out following previous studies on wavelets power spectra [5]-[7]. Spectral analysis in quantum associative memories shows that: (i) the weights w_{ij} satisfy the principle of uncertainty, (ii) the energy of the weight matrices $\bar{W}_i$ is quantized in discrete levels, which are indicated by the associated eigenvalues.

A. Spectral content of the weights w_{ij}

Theorem 1: The Gaussian basis functions of the weights w_{ij} of a quantum associative memory express the discrete levels in which the energy of w_{ij} is distributed. The smaller the spread σ of the basis functions is, the larger becomes the spectral (energy) content that can be captured therein.

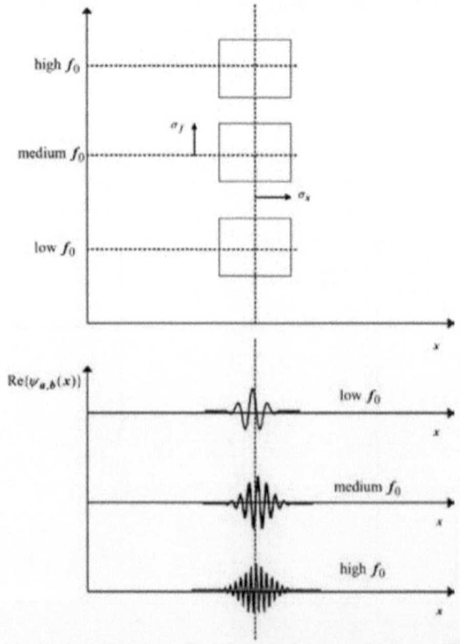

Fig. 7 Heisenberg boxes for wavelet of the same time length. Changes of the central frequency f_0 do not affect the size of the Heisenberg boxes.

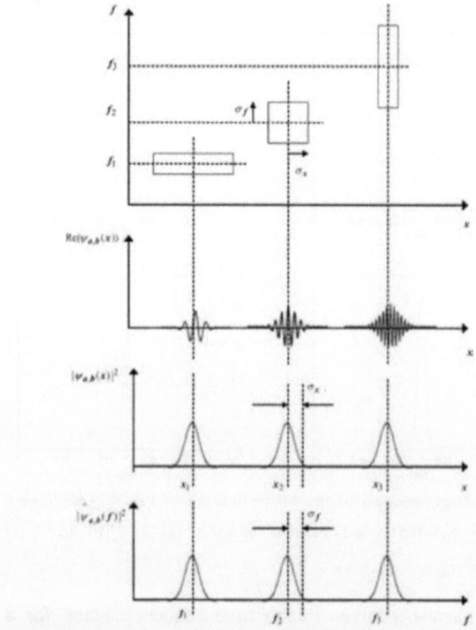

Fig. 8 Heisenberg boxes in the time-frequency plane for a number of wavelets with three different spectral frequencies set to low, medium and high value.

Proof: Consider the Fourier transform $G(f)$ of the Gaussian $g(x)=I_0 e^{-ax^2}$. It holds that

$$G(f)=\int_{-\infty}^{+\infty} g(x)e^{-ifx}dx \Rightarrow G(f)=\int_{-\infty}^{+\infty} I_0 e^{-ax^2}e^{-ifx}dx \Rightarrow$$

$$G(f)=I_0\sqrt{\frac{\pi}{a}}e^{-\frac{f^2}{4a}}$$

Consequently for $<w_{ij}>=\sum_{k=1}^{\infty} e^{-(w_{ij}-k\cdot c)^2/2\sigma^2}(k\cdot c)$ holds

$$G(f)=\sum_{k=1}^{\infty}(k\cdot c)G(e^{-\frac{1}{2}\frac{(w_{ij}-k\cdot c)^2}{\sigma^2}}) \Rightarrow$$

$$G\{\sum_{k=1}^{\infty} e^{-\frac{1}{2}\frac{(w_{ij}-k\cdot c)^2}{\sigma^2}}\}=\sum_{k=1}^{\infty}(k\cdot c)e^{ifkc}\sqrt{2\pi}e^{-\frac{1}{2}\sigma^2 f^2}$$

where c is the distance between the centers of two adjacent fuzzy basis functions. If only the real part of $G(f)$ is kept then one obtains

$$G\{\sum_{k=1}^{\infty} e^{-\frac{1}{2}\frac{(w_{ij}-k\cdot c)^2}{\sigma^2}}\}=\sum_{k=1}^{\infty}(k\cdot c)cos(fkc)\sqrt{2\pi}e^{-\frac{1}{2}\sigma^2 f^2} \quad (16)$$

The diagram of the Fourier transform $G(f)$ for $k=1$ and several values of the variance σ^2, is depicted in Fig. 9.

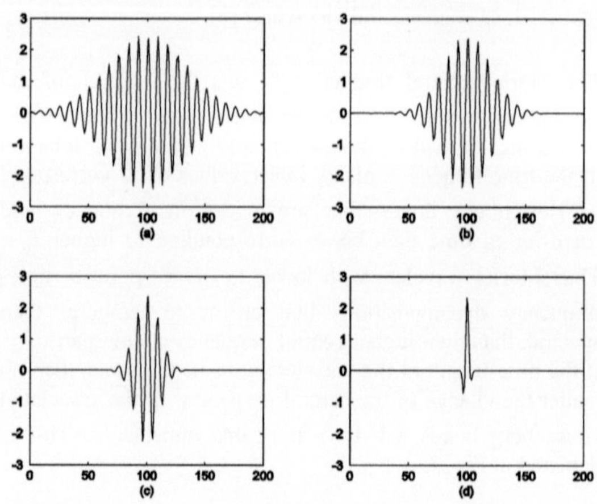

Fig. 9 Fourier transform of the weight $<w_{ij}>$ of a quantum associative memory, assuming the following variances of the Gaussians (a) $3\sigma^2$, (b) $5\sigma^2$, (c) $10\sigma^2$, (d) $25\sigma^2$

If the weight w_{ij} can be decomposed in the time domain into shifted Gaussians of the same variance σ^2. then, in the frequency domain, w_{ij} will be analyzed in a superposition of filters, which have a shape similar to the one given in Fig. 10.

B. Discrete energy levels in weights w_{ij}'s

1) Energy eigenstates in the weights w_{ij}'s

Schrödinger's equation given in Eq. (1) is studied again i.e.

$$i\hbar\frac{\partial\psi(w_{ij},t)}{\partial t}=-\frac{\hbar^2}{2m}\nabla^2\psi(w_{ij},t)+V\psi(w_{ij},t) \quad (17)$$

where V is the potential. Using the principle of variables separation $\psi(w_{ij},t)=X(w_{ij})T(t)$ one obtains

$$i\hbar\frac{1}{T(t)}\frac{\partial T(t)}{\partial t}=-\frac{1}{X(w_{ij})}\frac{\hbar^2}{2m}\frac{\partial^2 X(w_{ij})}{\partial w_{ij}^2}+ + V(w_{ij})=\mu \quad (18)$$

Setting $\mu=\hbar\omega$, where ω denotes the frequency, gives

$$i\hbar\frac{1}{T(t)}\frac{\partial T(t)}{\partial t}=\hbar\omega \quad (19)$$

$$\frac{1}{X(w_{ij})}\frac{\hbar^2}{2m}\frac{\partial^2 X(w_{ij})}{\partial w_{ij}^2}+V(x)=-\hbar\omega \quad (20)$$

Eq. (19) gives $T(t)=e^{-i\hbar\omega t}$, which results into $\psi(w_{ij},t)=X(w_{ij})e^{-i\hbar\omega t}$. An energy eigenstate is a state of the network with clearly defined energy, i.e. $E=\hbar\omega$. From Eq. (20) one gets [13]

$$\begin{aligned}&\frac{\hbar^2}{2m}\frac{\partial^2 X(w_{ij})}{\partial w_{ij}^2}-V(w_{ij})X(w_{ij})=-\hbar\omega X(w_{ij})\Rightarrow\\&[-\frac{\hbar^2}{2m}\frac{\partial^2}{\partial w_{ij}^2}+V(w_{ij})]X(w_{ij})=-\hbar\omega X(w_{ij})\Rightarrow\\&HX(w_{ij})=EX(w_{ij})\end{aligned} \quad (21)$$

where $H=[-\frac{\hbar^2}{2m}\frac{\partial^2}{\partial w_{ij}^2}+V(w_{ij})]$ is the Hamiltonian of the system and $E=\hbar\omega$ is the energy eigenstate. Eq. (21) is the equation of the eigenvalues of the Hamiltonian operator H. The application of the operator H to the spatial component $X(w_{ij})$ of the solution of Eq. (18) gives again $X(w_{ij})$ multiplied by the eigenvalue of the energy E. Therefore, according to Eq.(18), the possible energy values of the weight w_{ij} are the eigenvalues of the Hamiltonian operator H and are discrete (quantized).

The solutions of Eq. (21) are the eigenfunctions $X_k(w_{ij})$, $k=0,1,2,...$ and the corresponding energy eigenvalues are E_k.

$$HX_k(w_{ij})=E_kX_k(w_{ij}) \quad (22)$$

Thus, it is anticipated that the energy stored in weights w_{ij}'s will be quantized in discrete levels.

2) Weights w_{ij} and the principle of uncertainty

The significance of Eq. (21) is that the product between the information in the space domain $g(w_{ij})$ and the information in the frequency domain $G(s)$ cannot be smaller than a constant. In the case of a Gaussian function $g(w_{ij})=e^{-(w_{ij}^2/2\sigma^2)}$ with Fourier transform $G(s)=e^{-s^2\sigma^2/2}$ it can be observed that: (i) If σ is small then $g(w_{ij})$ has a pick at $w_{ij}=0$ while $G(s)$ tends to become flat. (ii) If σ is large then $g(w_{ij})$ is flat at $w_{ij}=0$ while $G(s)$ makes a peak at $s=0$.

This becomes more clear if the dispersion of a function $g(w_{ij})$ round $w_{ij}=0$ is used [14]

$$D(g)=\frac{\int_{-\infty}^{+\infty}w_{ij}^2/g(w_{ij})/^2\,dw_{ij}}{\int_{-\infty}^{+\infty}/g(w_{ij})/^2\,dw_{ij}} \quad (23)$$

In case that $g(w_{ij})=e^{-(w_{ij}^2/2\sigma^2)}$ the dispersion of functions $g(w_{ij})$ and of its Fourier transform $G(s)$ becomes

$$\begin{aligned}D(g)&=\frac{\int_{-\infty}^{+\infty}w_{ij}^2e^{-w_{ij}^2/2\sigma^2}dw_{ij}}{\int_{-\infty}^{+\infty}e^{-w_{ij}^2/2\sigma^2}dw_{ij}}=\frac{1}{2}\sigma^2\\D(G)&=\frac{\int_{-\infty}^{+\infty}s^2e^{-s^2\sigma^2/2}ds}{\int_{-\infty}^{+\infty}e^{-s^2\sigma^2/2}ds}=\frac{1}{2\sigma^2}\end{aligned} \quad (24)$$

which results into the uncertainty principle for the weights of quantum associative memories

$$D(g)D(G)=\frac{1}{4} \quad (25)$$

Eq. (25) means that the accuracy in the calculation of the weight w_{ij} is associated with the accuracy in the calculation of its spectral content. When the spread the Gaussians of the stochastic weights is large (small) then their spectral content is poor (rich). Eq. (25) is an analogous of the quantum mechanics uncertainty principle, i.e. $\Delta x \Delta p \geq \hbar$, where Δx is the uncertainty in the measurement of particle's position, Δp is the uncertainty in the measurement of the particle's momentum and $\hbar$ is Planck's constant.

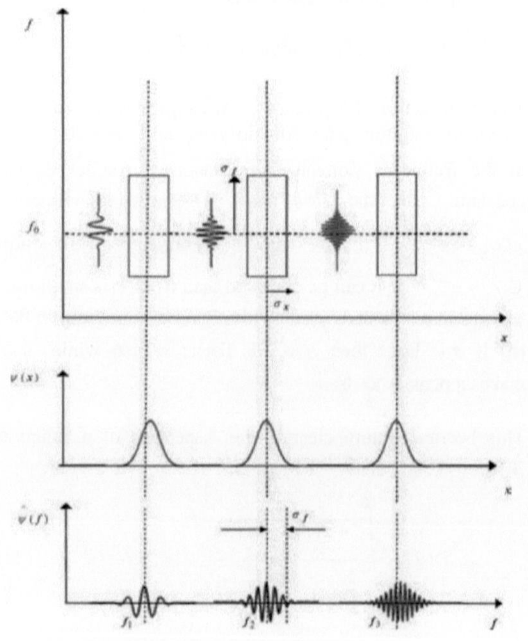

Fig. 10 Heisenberg boxes in the space-frequency plane for a number of Gaussian basis functions shifted on the time axis.

V. Conclusions

Quantum associative memories stem from neural associative memories if the weights w_{ij} are considered to be stochastic variables. In quantum associative memories the probability density function of element w_{ij} in the weight matrix W is given as a solution of Schrödinger's differential equation. It was also proved that the weight matrix W can be decomposed into a superposition of matrices $\bar{W}_i$, thus increasing the number of attractors by a factor of 2^N.

Spectral analysis of quantum associative memories follows previous studies on the wavelets power spectra. Spectral analysis of quantum associative memories has shown the following:

- The Gaussian basis functions of the weights w_{ij} of a quantum associative memory express the discrete levels in which the energy of w_{ij} is distributed. The smaller the spread σ of the basis functions is, the larger becomes the spectral (energy) content that can be captured therein. Narrow spread of the Gaussians in $< w_{ij} >$, results in wide range of frequencies of the Fourier transformed pulse.

- The energy levels of the weights w_{ij} 's are discrete. Indeed, the eigensolution of Schrödinger's equation is of the form $\psi_k(w_{ij},t) = X_k(w_{ij})e^{-i\hbar\omega t}$. It holds that $HX_k(w_{ij}) = E_k X_k(w_{ij})$, where H is the Hamiltonian of the system and $E = \hbar\omega$ is the corresponding energy eigenstate, $k = 0,1,2,...$

- The stochastic weights w_{ij} satisfy the principle of uncertainty. The dispersion $D(g)$ of the Gaussians in $< w_{ij} >$ and the dispersion $D(G)$ of the associated Fourier transform results in $D(g)D(G) = \frac{1}{4}$. This is analogous to quantum mechanics uncertainty principle, i.e. $\Delta x \Delta p \geq \hbar$.

References

[1] C. Cohen-Tannoudji, B. Diu and F. Laloë, *Mécanique Quantique I*, Hermann, 1998.

[2] G.G. Rigatos and S.G. Tzafestas, Parallelization of a fuzzy control algorithm using quantum computation, *IEEE Transactions on Fuzzy Systems*, vol. 10, no. 4, pp. 451-460, 2002.

[3] G.G. Rigatos and S.G. Tzafestas, Fuzzy learning compatible with quantum mechanics postulates, *Computational Intelligence and Natural Computation, CINC '03*, North Carolina, 2003.

[4] B. Kosko, *Neural networks and fuzzy systems : A dynamical systems approach to machine intelligence*, Prentice Hall, 1992.

[5] P.A. Addison, *The Illustrated Wavelet Transform Handbook*, Institute of Physics Publishing, 2002.

[6] S. Mallat, *A Wavelet Tour of Signal Processing*, Academic Press, 1998.

[7] I. Debauchies, The wavelet transform, time-frequency localization and signal processing, *IEEE Transactions on Information Theory*, vol. 36, pp. 961-1005, 1990.

[8] D. Ventura and T. Martinez, Quantum Associative Memory, *Information Sciences*, Elsevier, vol. 124, no. 1-4, pp. 273-296, 2000.

[9] M. Perus, Multi-level Synergetic Computation in Brain, *Nonlinear Phenomena in Complex Systems*, vol. 4, no. 2, pp. 157-193, 2001.

[10] G. Resconi and A. J. Van der Waal, Morphogenic neural networks encode abstract rules by data, *Information Sciences*, 2000, pp. 249-273.

[11] S. Haykin, *Neural Networks: A Comprehensive Foundation*, McMillan, 1994.

[12] M.A. Nielsen and A. Chuang, *Quantum Computation and Quantum Information*, Cambridge, 2000.

[13] W.A. Strauss, *Partial Differential Equation: An Introduction*, J. Wiley, 1992.

[14] M.A. Pinsky, *Partial Differential Equations and Boundary Value Problems with Applications*, McGraw-Hill, 1991.

Virtual Navigation System for the disabled by Motor Imagery

Dongjun Suh, Hyun Sang Cho, Jayoung Goo, Kyung S. Park, Minsoo Hahn
Digital Media Laboratory, Information and Communications University, 517-10 Dogok-dong, Gangnam-gu, Seoul, 135-854, Korea
{linuxer, haemosu, bucsu, park, mshahn}@icu.ac.kr

Abstract —**This paper shows a motor imaginary based Brain-Computer Interface (BCI) virtual navigation system. The virtual navigation system is designed for motion disabled people to navigate and control between virtual environment as well as real environment. This system consists of the brain-computer interface using motor imagery brain signals, the communication module, the brain signal analyzer and the virtual world linked with a physical miniature model of real home environment. A preliminary user evaluation showed that our BCI-testbed produced a reasonable classification rate to be used in virtual reality and real world. In the near future, we will integrate this system in a ubiquitous computing environment.**

Keywords — **Virtual Reality, BCI, Motor Imagery, Virtual Control, Virtual Navigation, ActiveHome**

I. INTRODUCTION

There have been growing interests in treating psychological and physical disorders using virtual reality. This area of research is called clinical virtual reality [2]. Some examples are treatment for anxiety disorders, stress disorder, pain distraction and physical rehabilitation. In this paper, we present a virtual navigation system that enables motion disabled people to control the real world environment using a motor imaginary based brain-computer interface.

Motion disable people have a lot of psychological disorder such as depression caused by their uncontrollable body. They can not avoid charging a burden on their family members. We believe that virtual reality can provide psychological compensations to motion disabled people by immersion so that they can act as a regular interactive family. However, most motion disabled people have difficulty in interacting with virtual environment using the current VR input devices.

Recently, Brain-Computer Interface (BCI) research has been evolved tremendously. BCI provides control capabilities to people with motor disabilities. There are many BCI based approach to control virtual environment. Such examples include VEP, SSVEP, P300 and motor imagery [9]. The motor imagery based BCI has some advantages for motion disabled people to use it in the virtual environment because virtual reality provides the sense of presence [3, 7].

The basic mechanism of motor imagery based BCI is that human imagination of a movement (such as left hand movement) results in a similar neural change on the brain as a real execution of the same movement does [3]. Hence, motor imagery brain signal allows virtual users to navigate in a virtual environment by giving left-right direction information. Left hand motor imagery makes μ-rhythm (8~12 Hz) decrease in the sensory motor region of right hemisphere [3, 7, 9].

This work describes a motor imagery-based BCI used as an input feedback to virtual environment, suggesting a new chance for diagnosis and therapeutic environments. This paper presents the motivation, design, implementation, and evaluation of our BCI based virtual control system for the motion disabled people to control real home environment.

II. RELATED WORK

Over the past decade, there have been many virtual reality applications developed for treating psychological and physical disorders [2, 4]. For example, SIREN, a virtual reality driving simulator developed by Iowa University, assists motion disorders (such as Alzheimer's disease or Parkinson's disease) to improve their mental abilities. Similarly, a virtual reality driving simulator developed by Lee [4] serves for motion disabled people to help their driving ability. Also, the Rutgers ankle system [10] is a virtual reality simulation developed for ankle rehabilitation.

Recently, BCI research is also targeted for motion disabled people's rehabilitation. Gao in Tsinghua university developed BCI based environment controller to help motion disabled people control home appliances [1]. This system uses VEP (Visual Evoked Potential) and SSVEP (Steady State Visual Evoked Potential) brain signals occurred by LED stimulator. However, it only gives a single two dimensional feedback on the monitor. It does not provide immersive user interaction. Very recently, Leeb in Graz University has implemented the ideas of virtual navigation using motor imagery BCI [3, 7]. This system uses Head-Mounted Display and hence it easily creates cyber sickness. In addition, this system is operated in only virtual environment.

In our motor imagery-based BCI system, users can control navigation and manipulation in a virtual environment and also trigger the controls in the real home

K. Elleithy et al. (eds.), Advances in Computer, Information, and Systems Sciences, and Engineering, 143–148.

environment. Motor imagery refers to human imagination of movements that result in the changes of brain signals as a real execution of physical movement. A basic phenomenon is an event-related de-synchronization (ERD) appeared on the contra lateral hemisphere during the unilateral hand movement imagery. The power-value of the sensorimotor μ-rhythm (EEG 8~12Hz) on C3 left side hemisphere decreases after human imagine the right hand motor imagery and vice versa.

III. BCI CLASSIFICATION TEST

First, we conducted a small BCI classification test. The goal was to evaluate how our motor imagery based BCI can be applicable to control a virtual environment. We used a Laxtha Inc.'s 8 channel EEG measurement device.

A. *Experiment Method*

Fig. 1 shows an electrode placement for EEG data gathering. First, we have tried various brain parts to find the appropriate electrode placements for better motor imagery measurements. We found that the position (C3, C4) has shown the best result for motor imaginary classification. Hence, we attached the electrodes linearly from C3 to Cz and from Cz to C4 to capture the gradation of amplitude decrease from C3 and C4. This placement is followed with international 10/20 system of electrode placement.

There are human sensory motor cortex in C3 and C4. In this region, human hand movement imaginations affect the amplitude of the specific range (μ-rhythm 8~12Hz) of EEG signals. Generally, the cross part of hemisphere linked cross human body. The results showed that the imagination of left hand movement decreases the power value of right part of hemisphere. In this sense, we can use this principle of 'motor imagery' in virtual navigation.

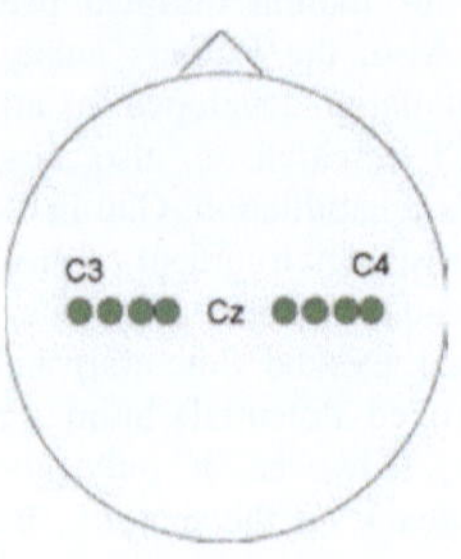

Fig. 1. Electrode placement for EEG data gathering: measuring placements are distributed from C3 to Cz and from C4 to Cz

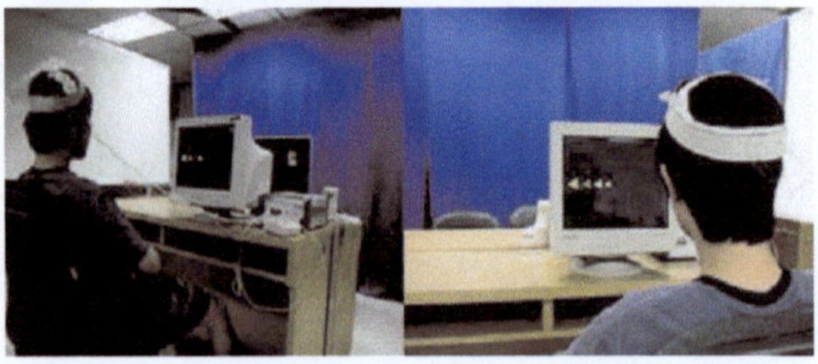

Fig. 2. Experiment setting for motor imagery

Fig. 2 shows the experiment setting for the BCI classification test. Four graduate students participated in this experiment. Their age is ranged from 26 to 36 years old. They have moderate experienced with virtual reality systems such as CAVE and passive-stereoscopic VR wall display. The main objective of this experiment was to develop a classification method to detect different (such as, left or right) motor imagery. Each subject was asked to perform three different tasks: looking at a fixation cross on the monitor without motor imagery, motor imagery of left hand movement, and motor imagery of right hand movement. For each task, the subject was given visual cues of direction and fixation. Fixation time is needed to remove previous motor imagery. EEG during the fixation time can be reference signal to classify left or right direction.

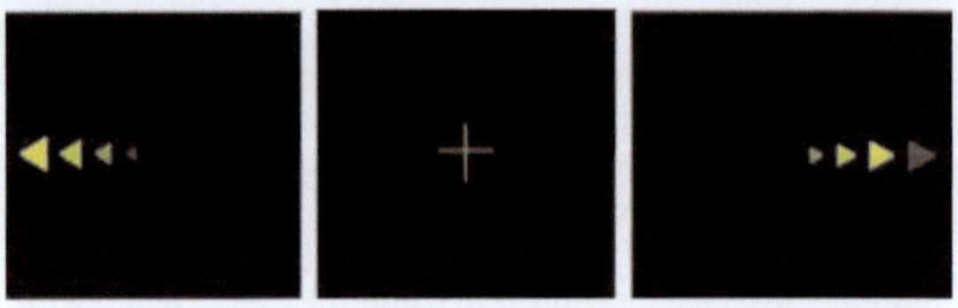

Fig. 3. Visual cues of the direction in session: blinking left arrow , middle fixation cross and right arrow

Fig. 3 shows the visual cues of direction, the blinking left arrow is displayed to draw a subject to imagine left motor imagery. The experimental procedure is following: first the fixation time (5 sec), followed by the left motor imagery (5 sec), then the fixation time (5 sec), and finally the right motor imagery (5 sec). This process repeated twenty-five times, and the 30-seconds rest period was offered after every five trials. During the fixation time, the subject concentrates on the fixation cross on the monitor. The fixation time was needed to remove any remaining from previous motor imagery. The brain signals during the fixation time can be a reference point to classify left or right direction.

B. Experiment Results

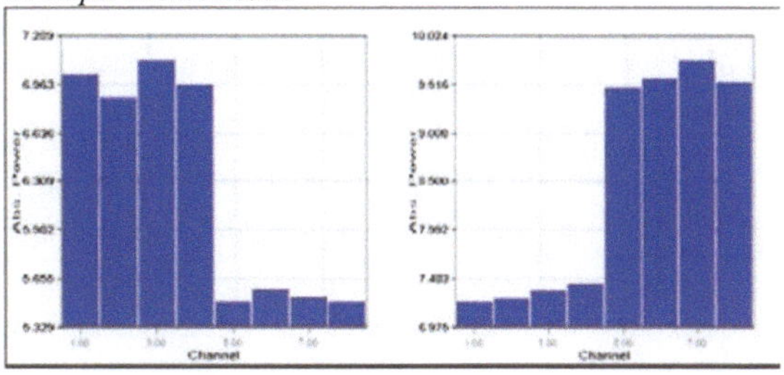

Fig. 4. Subject A's session 1 power spectral analysis result

Fig. 4 shows one subject's power spectrum value of μ-rhythm. This is a well-classified result shown as left or right hand motor imagery. The left graph shows a distinct decrease of μ-rhythm amplitude on the right hemisphere (EEG ch#5 to ch#8) during the subject's left motor imagery. Likewise, the right graph shows a clear decrease of μ-rhythm amplitude on the left hemisphere (EEG ch#1 to ch#4) during the subject's right motor imagery. However, we noticed that the results have been strongly affected by subject's body and mental conditions. For example, the subject's mental fatigue led to many failures in the classification since he could not concentrate on motor imagery. Due to this fact, the overall classification rate of our BCI module was approximately 70%. Other factors contributed to low classification rate were inappropriate electro rode placements and environmental distractions such as noise.

C. Classification method

We have tried various classification methods in this motor imagery BCI classification test to find left and right directions. This test is how to find appropriate index which can separate left and right motor imagery. We adopted nearby value compared method that means average values of around C3 (left motor imagery) and around C4 (right motor imagery) through many experiments. Fig.5 explains classification methods more specifically.

In the first trial, we tried to find exact C3 and C4 points which can decide direction. Many experiments were conducted but we failed to find adequate C3 and C4 points (as shown in the left image of Fig.5). In the second trial, we tried different method to get useful data in classification tests as shown in the right image of Fig. 5. We adopted average values of around C3 and C4 that were grouped points together. This trial is better than previous one, however classification rate was down as time goes by. So, we also could not determine second trial as a classification method.

Motor imagery signal is emitted from sensory motor around parietal lobe position [3, 7, 9]. In this part, we could get new idea to attach electrode arrangement. According to the parietal lobe position, we rearranged electrodes placement to acquire more useful motor imagery signal. That is, in-line attached trial as seen in the above experiment method (Fig. 1) was tested as a classification method. After several trials using this method, we could get very useful index value to separate left and right direction.

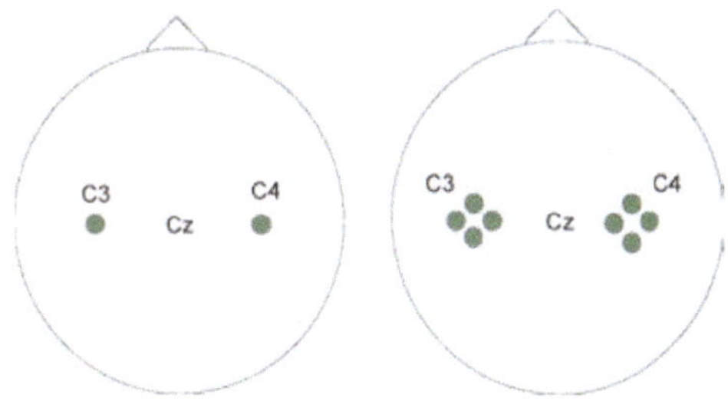

Fig. 5. Different electrode placement test for classification method

IV. BCI BASED VIRTUAL CONTROL TESTBED

We developed a virtual environment (as shown in Fig. 6) using Ygdrasil [16]. It runs on the ICU digital media lab's VR table display (Red hat 9.0 linux machine, 2560 x 1024 screen resolutions). Ygdrasil is a framework for creating networked virtual environments. It is focused on building the behaviors of virtual objects from re-usable components, and sharing the state of an environment through a distributed scene graph mechanism. Ygdrasil has been used to construct a variety of artistic and education virtual reality applications. Fig. 6 shows the virtual environment on the left image and a physical miniature mockup model representing a home environment on the right image.

Fig. 6. The left image shows virtual environment; the right image shows a miniature model of real home

The virtual environment is composed of the UDP communication module between the BCI interface and the virtual environment, the serial communication module for sending control signals to the real environment, the user navigation modules, and the static object placement. When the virtual environment receives a message 'R', a user navigates to the right. When it receives a message 'L', a user turns to the left. When it receives a message 'U', a user moves forward in the virtual environment. In addition, when a user gets close to the table or stand lamp in the virtual environment, the light will be on. When a user gets away from the table or the stand light, the light is turned off. By the serial communication interface, the user's action in the virtual environment triggers to send event message to the real world miniature model. Therefore, when a user turns the light on, the light in the real world miniature model will be on as well.

V. System Architecture

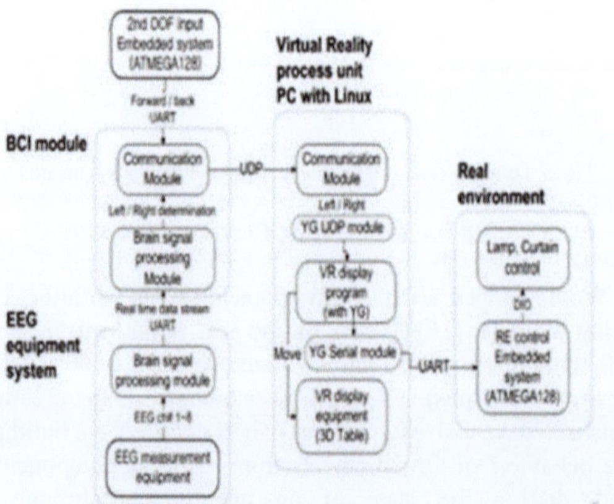

Fig. 7. Hardware architecture of the BCI virtual control testbed

Fig. 7 shows the system architecture of the BCI based virtual control testbed. The testbed consists of three modules: the user control process, the virtual environment processing unit and the real environment control. The user control processing consists of three module: physical control signal input process (i.e. ATMEGA128 processor with 2 signal input buttons to go forward and lock/unlock to make direct decision; in this case, physical input means input signal from physical body), the Laxtha LXSM-D1WD9 8 channel EEG measurement equipment, and the brain signal analysis unit.

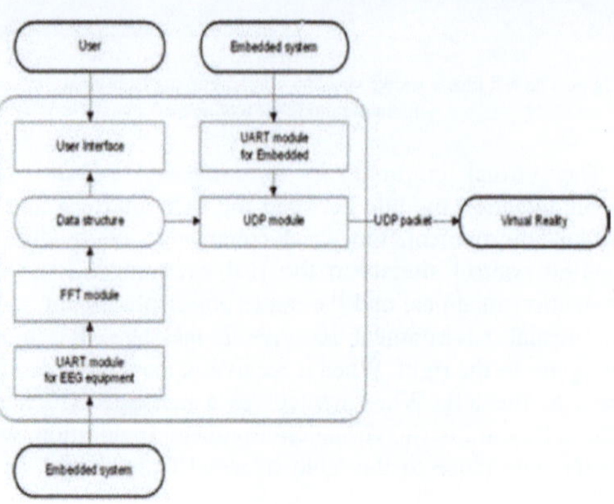

Fig. 8. Software structure for the BCI module

As shown in Fig. 8, the signal analysis unit has 2 UART and 1 UDP communication connection. The 2 UART connections get the data from users: one for physical signals from ATMEGA 128 embedded system and the other for EEG data collection. The EEG data is gathered through the UART module and Laxtha's dynamic linked library (Dll) interface. The data gathering rate is every 0.25 second (256 Hz sampling). We have gotten the 256 data every one second and process them by FFT numerical calculation to extract μ-rhythm (8~12 Hz). For this extraction, we have developed MFC Extension Dll for the FFT analysis.

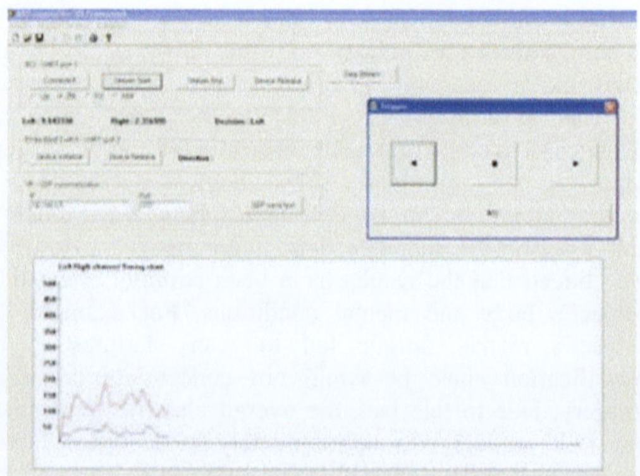

Fig. 9. User's direction trajectory analysis interface

As shown in Fig. 9, our system provides a graphical analysis of brain signals visualization and interface for each channel. It has 8 channels capacity of brain signals. The average of ch#1~4 and ch#5~8 are compared to make a decision whether it is left or right. In Fig. 9, the red curve represents an average amplitude value of the left part channels (i.e., ch#1~4) and the blue curve represents an average amplitude value of right part channel (ch#5~8). Fig. 9 shows the user orienting to the left direction since the red curve is dominant over the blue curve.

Our system also provide a physical foot pedal interface (shown in Fig. 10) for users to toggle between start and stop BCI to better support user navigation in the virtual environment. This was intended to prevent the direction decision error (although it is less than 10%) that may cause practical inconvenience. We experienced that users seemed to feel cyber sickness or fatigue induced by unwilling wander in the virtual environment when only BCI was used. Our system usability has been dramatically improved after introducing this lock/unlock user interface. In our system, a user can first select the desired moving direction using BCI signals. Once the direction is selected, he/she can lock the direction using the foot pedal interface, then move forward.

He/she can also unlock the direction to precede a direction change.

Cyber sickness (also called simulator sickness) is one of the important issues in virtual reality system [17, 18, 19, 20, 21]. It is a side effect of high fidelity visual simulators and virtual interfaces. Cyber sickness resembles more familiar disease of motion sickness, including, but not limited to pallor, dizziness, headache, and nausea [22].

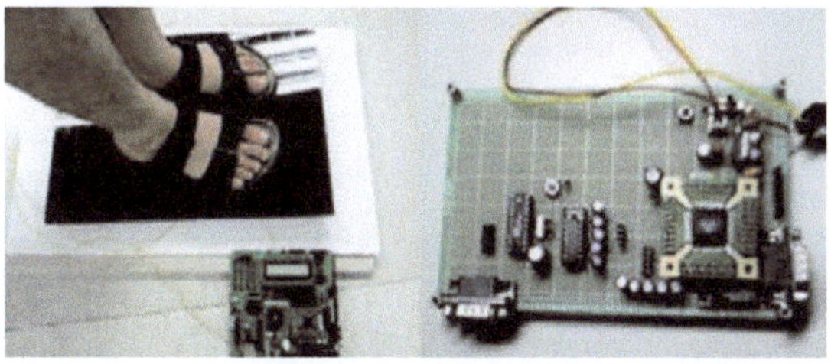

Fig. 10. Foot Pedal Controller for Locking, Unlocking and ATMEGA 128 Embedded System

Our first prototype of BCI-based virtual control system (without lock/unlocking mechanism) caused a severe cyber sickness problem during user navigation in the virtual environment. If the user could not exactly select his/her desired direction, the user tended to get lost of controlling navigation and in some cases the user could not even stop rotating in the space. Cyber sickness can be generated by watching moving virtual world scenes. The effect of navigation speed that triggers the level of cyber sickness is interesting topics in virtual reality research [23, 24].

Therefore, we developed the locking-unlocking interface additionally to solve these problems. From our experience, when we provided the locking-unlocking foot pedal interface shown in Fig. 10, the cyber sickness problem seemed to be almost disappeared even if the user was exposed to the virtual environment for a long time. Moreover, we could get more accurate user navigation controls in the virtual environment with the provision of the fixed direction method. However, motion disabled people may not be able to use the foot pedal interface because of paralyzed their body. In the future, we will develop other multi-modal user interfaces that do not require physical body involvement, for example the interface using physiological signals from brain or whole body.

VI. User Test

Fig. 11 shows a user test on a VR table display. The VR table display uses two DLP projectors and passive-stereoscopic polarization filters and 3D movie glasses to give users 3D effects. A participant accomplished four tasks: navigation, on/off a stand lamp, table light and draw up/down a curtain. The user could select left or right directions by user's motor imagery thought and he could also move forward or stop and lock or unlock the direction using foot pedal interface. Surprisingly, our BCI system has shown much better classification rate when we tested on the VR display as compared to the simple monitor imagery classification test in the section III. We believe that rich three dimensional visual feedbacks in the virtual environment have more impacted on classifying motor imagery.

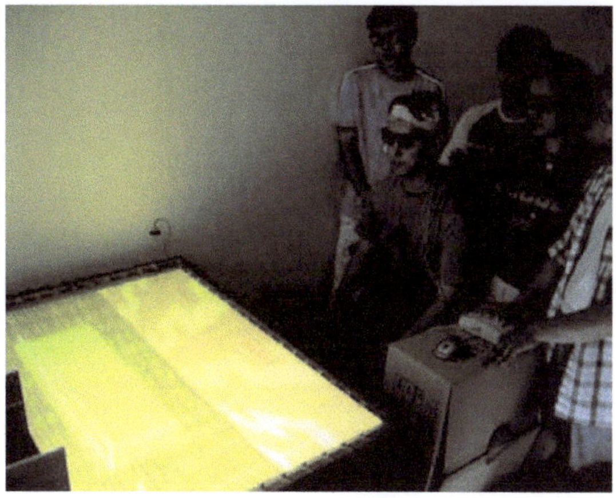

Fig. 11. User test in ICU digital media lab's VR table display (stereoscopic type)

VII. Conclusions And Future Works

We proposed a new BCI-based virtual control testbed for motion disabled people. We intended to create a testbed that enable motion disabled people to control home environment such as navigation and trigger actions. In a user test, the result shows that our motor imagery based BCI virtual environment give users' good feedback of navigation and interaction with objects in home environment. To specify, users can rotate their position into right and left directions by motor imagery. Therefore, virtual control strategy can give possibility of a new style ubiquitous computing combined by BCI based virtual reality system. Moreover, this system can be applied with not only motion disabled people but also normal people. In the near future, we are planning on applying our system to motion disabled people. In addition, we will continue to enhance our BCI classification interface and improve this interface to manage not only the classification of left or right but also of up and down. Furthermore, we will investigate a multi-modal approach consisted of brain signals and human eye signals like EOG. Finally, we will integrate our testbed in the ActiveHome, ubiquitous computing environment in ICU Digital Media Lab. We believe our BCI based virtual control testbed introduce a new type of ubiquitous computing research.

ACKNOWLEDGMENT

The research was made possible by funding from the ICU Digital Media Lab. First of all, I would like to say many thanks to our subjects. I would also like to give sincere thanks to other DML members, who gave us perceptive criticism, kind encouragement, and willing assistance that helped bring the project to a successful conclusion. Also, we gratitude Dr. Bin He (a neural engineer and a fellow member of IEEE) for giving us useful information about motor imagery based BCI.

This research was supported by the MIC(Ministry of Information and Communication), Korea, under the Digital Media Lab. support program supervised by the IITA(Institute of Information Technology Assessment).

REFERENCES

[1] X. Gao, D. Xu, M. Cheng, and S. Gao, "A BCI-Based Environmental Controller for the Motion-Disabled," *IEEE Trans. Neural Syst. Rehab. Eng.*, vol. 11, no. 2, June 2003

[2] L. Hodges et al., "Treating Psychological and Physical Disorders with VR," *IEEE Computer Graphics and Applications*, vol. 21, November 2001, pp.25-33

[3] R. Leeb, R. Scherer, F. Lee, H. Bischof, and G. Pfurtscheller, "Navigation in Virtual Environments through Motor Imagery," *Proc. 9th Computer Vision Winter Workshop*, 2004, pp. 99-108

[4] J. Lee, "Virtual Reality and Psychology," *Korean Jounal of Psychology 2004,* vol 23, no. 2, 87-104

[5] J. Bayliss, "Use of the Evoked Potential P3 Component for Control in a Virtual Apartment," *IEEE Trans. Neural Syst. Rehab. Eng.,* vol. 11, no. 2, June 2003

[6] J. Bayliss and D. Ballard, "A Virtual Reality Testbed for Brain–Computer Interface Research," *IEEE Trans. Neural Syst. Rehab. Eng.,* vol. 8, no. 2, June 2000

[7] R. Leeb, and G. Pfurtscheller, "Walking through a Virtual City by Thought," *Proc. 26th Annual International Conference of the IEEE EMBS San Francisco,* CA, USA • September 2004

[8] B. Kamousi, Z. Liu, and B. He, "Classification of Motor Imagery Tasks for Brain-Computer Interface Applications by Means of Two Equivalent Dipoles Analysis," *IEEE Trans. Neural Syst. Rehab. Eng.,* vol. 13, no. 2, June 2005

[9] G. Pfurtscheller, C. Neuper, "Motor Imagery and Direct Brain–Computer Communication," *Proc. IEEE,* vol. 89, no. 7, July 2001

[10] S. Adamovich et al., "A Virtual Reality Based Exercise System for Hand Rehabilitation Post-Stroke: Transfer to Function," *Proc. 26th Annual International Conference of the IEEE EMBS San Francisco*, CA, USA • September 2004

[11] G. Schalk, D. McFarland, T. Hinterberger, N. Birbaumer, and J. Wolpaw, "BCI2000: A General-Purpose Brain-Computer Interface (BCI) System," *IEEE Trans. Biomed. Eng.*, vol. 51, no.6, June 2004

[12] G. Townsend, B. Graimann, and G. Pfurtscheller, "Continuous EEG Classification During Motor Imagery—Simulation of an Asynchronous BCI," *IEEE Trans. Neural Syst. Rehab. Eng.*, vol. 12, no. 2, June 2004

[13] T. Wang, J. Deng, B. He "Classification of Motor Imagery EEG Patterns and Their Topographic Representation," *Proc. 26th Annual International Conference of the IEEE EMBS San Francisco*, CA, USA • September 2004

[14] S. Mason, G. Birch, "A General Framework for Brain–Computer Interface Design," *IEEE Trans. Neural Syst. Rehab. Eng.*, vol. 11, no. 1, March 2003

[15] J Pineda, D. Silverman, A. Vankov, J. Hestenes, "Learning to Control Brain Rhythms: Making A Brain-Computer Interface Possible," *IEEE Trans. Neural Syst. Rehab. Eng.*, vol. 11, no. 2, June 2003

[16] YGdrasil documentation http://www.evl.uic.edu/yg/overview.html

[17] K. Fujita, "Influence of attention and predictive visual cue on motion perception and sickness in immersive virtual environment," *Proc. 26th Annual International Conference of IEEE EMBS San Francisco,* CA, USA•'September 2004

[18] J. Joseph, "A Discussion of Cybersickness in Virtual Environments," *SIGCHI Bulletin*, vol.32, no1, January 2000

[19] R. Mourant, T. Thattacheny, "Simulator Sickness in a Virtual Environments Driving Simulator," *Proc. 44th Human Factors and Ergonomics Society*, San Diego, California, July 2000

[20] J. Lin, H. Abi-Rached, M. Lahav, "Virtual Guiding Avatar: An Effective Procedure to Reduce Simulator Sickness in Virtual Environments," *Proc. SIGCHI Conference on Human factors in computing systems,* Vienna, Austria, April 2004, pp.719-726

[21] M. Mollenhauer, R. Romano, "The evaluation of a motion base driving simulator in a cave at tacom," *Proc. 24th Army Science Conference*, December 2004

[22] B.Jeager, R. Mourant, "Comparison of simulator sickness using static and dynamic walking simulations," *Proc. 45th Human Factors and Ergonomics Society*, Minneapolis, Minnesota, October 2001, pp.1896-1900

[23] R. Kennedy, N. Lane, K. Berbaum, M. Lilienthal, "Effects of navigation speed on motion sickness caused by an immersive virtual Environment," *Proc. 45th Human Factors and Ergonomics Society,* Minneapolis, Minnesota, October 2001, pp. 452-61

[24] J. Wu, Y. Lei, B. Chen, M. Ouhyoung, "A 3D Tracking Experiment on Latency and Its Compensation Methods in Virtual Environments," *The 8th Annual Symposium on UIST '95*, ACM Press, 1995, pp. 41-49

Density Function Based Medical Image Clustering Analysis and Research

SONG Yuqing[1] XIE Conghua [1,2] ZHU Yuquan[1] LI Cunhua[3] CHEN Jianmei[1]

(1 School of Computer Science and Engineering, Jiangsu University, Zhenjiang, China 212013)

(2 Department of Computer Science, Changshu Institute of Technology, Changshu, China 215500)

(3 Department of Computer Science, Huaihai Institute of Technology, Lianyungang, China 222005)

Abstract- **Image clustering analysis is one of the core techniques for image indexing, classification, identification and segmentation for image processing. On the basis of investigation on the laws of gray distribution of medical image, density function construction based medical image clustering analysis method is designed. For the special kind of data object -- medical images, firstly its density function is constructed; secondly, a method of density function construction based on medical image clustering analysis and its implementation algorithm, that is, hill climbing, is offered; At last, abdomen medical images are used to do experiments according to those two algorithms. Experiment results show that medical image clustering on the ground of density function construction achieves good effects and can clearly express the content and semanteme of medical images.**

I. Introduction*

Clustering analysis has wide applications in the fields of pattern recognition and image processing, especially in image identification and segmentation. Some of those researches have achieved great successes. In 1979 Coleman and Andrews[1] used clustering analysis method to segment image. Stewart uses fuzzy clustering analysis to recognize and classify radar objects[2]. Current applications of clustering analysis in this field are focused on improving traditional clustering analysis algorithms and methods for better quality and processing speed of segmentation and classification.

Density based clustering method is to cluster data objects according to the concepts of density. It produces clusters by density of data object or some kind of density function. DBSCAN[3], DBCLASD[4], DENCLUE[5] and so on, are typical density based clustering methods. From the end of 50s to the beginning of 60s in last century, Rosenblatt[6] and Parzen[7] offered famous density estimation method, which doesn't need prior knowledge of data distribution and any presupposition but can accurately express laws of data distribution. Data distribution mode of density estimation provides direct mean for definition and discovery of clustering.

The clinical value of medical images is to show the projections of different tissues and organs of body on medical images and distinguish normal projections from abnormal ones. Expressive forms of projections are pixels gray of digital image and law of gray distribution. Investigation on the distribution law of different gray density has special clinical value. Currently, making use of clustering analysis method to identify and segment medical images is on germinal stages.

On the ground of deep investigation on gray distribution for medical images, we present a new clustering analysis method for medical image processing -- density function construction based medical image clustering analysis method. Researches show that this method can express medical image content very well and has obvious clustering effect; therefore this method is fit for medical image clustering analysis. This novel

* Supported by the National Natural Science Foundation of P.R.China under Grant No. 60572112 and the Social Development Foundation of Zhenjiang, Jiangsu province of P.R.China under Grant No. SH2003014.

K. Elleithy et al. (eds.), Advances in Computer, Information, and Systems Sciences, and Engineering, 149–155.

method has great theoretical and practical values for medical image indexing, classification, identification and segmentation.

II Constructing Density Function For Medical Image Data

Kernel density estimation problem can be described as following: let ξ be continuous random variant for distribution feature, its distribution function and density function are $F(X)$ and $f(X) = dF(X)/dX$ respectively. Suppose $\{X_i\}_{i=1,2...,n}$ is the independent samples set of random variant ξ, we will find the approximate estimation $\hat{f}(X)$ of distribution function $f(X)$.

Definition 1 (Kernel function) Let dataset $D = \{X_1,..., X_n\} \subset F^d$ be a set of independent similar distribution random variants, $f(X)$ be distribution density function of variants, kernel density estimation $\hat{f}_\sigma(X)$ of density function $f(X)$ is defined as:

$$\hat{f}_\sigma(X) = \frac{1}{n\sigma}\sum_{i=1}^{N} K(\frac{X_i - X}{\sigma}), \quad X \in R \tag{1}$$

Where K is kernel function, σ is a given positive number and usually called window width or smoothness parameter[8].

DENCLUE (DENsity based CLUstEring) defines data point density function as the sum of its neighbor influence functions by Hinneburg et al[5] and its main idea is in following definitions.

Definition 2 (Influence function and density function) Given dataset $D = \{X_1,..., X_n\} \subset F^d$, the influence function of a data object X_0 D is a function $f_B^{X_0} : F^d \to R_0^+$, such that:

$$f_B^{X_0}(X) = f_B(X_0, X), \quad X \in F^d \tag{2}$$

Density function in space F^d generated in D is defined as:

$$f_B^D(X) = \sum_{i=1}^{N} f_B^{X_i}(X), \quad X \in F^d \tag{3}$$

Through comparison, gauss function is selected as kernel function [9][10], and then the density function form is:

$$f_{Gauss}^D = \frac{1}{N\sigma^d (2\pi)^{d/2}}\sum_{i=1}^{N} exp(-\frac{1}{2\sigma}(X - X_i)^T (X - X_i)) \tag{4}$$

To improve computation speed, only consider the influence functions of data points nearby to compute density function of a data point, neglect small influence functions of data far away points, then approximate density function is defined as:

$$\hat{f}_{Gauss}^D(X) = \sum_{X_i \in near(X)} e^{\frac{d(X,X_j)^2}{2\sigma^2}} \tag{5}$$

Medical images have a lot of noises; therefore the data point distribution law is bad. If density of data point is only the count of neighborhood data points, then its reliability is poor. We take advantage of gauss kernel function to define density in order to reduce the impact of noise data. Medial images are two-dimensional plane; each data point has properties of coordinate (x, y) and gray. In this paper, *Near* function can be defined as : $Near(X) = \{X^* \mid d(X, X^*) < k\sigma, k \in Z\}$, Experiment results show that, when $k = 4$, it is enough to express the impact of neighbor data. All in all, density function for medial image data is constructed as:

$$\hat{f}_{Gauss}^D(X) = \sum_{X_1 \in \{X^* | d(X,X^*) < 4\sigma\}} e^{\frac{-d(X(x,y,gray), X_1(x,y,gray))^2}{2\sigma^2}} \tag{6}$$

In order to implement medical image clustering, including center-defined cluster and arbitrary-shape cluster, definition of density attractors is needed.

Definition 3 (Density attractor) X^* in F^d is called local attractor, if and only if density function $\hat{f}_{Gauss}^D(X)$ has the local maximum at X^*. A point $X \in D$ is density-attracted to density attractor X^* if there exits a series of points $X = \{X_0, X_1 ... X_k\}$, $d(X_k, X^*) \le \varepsilon$ such that

$$X_i = X_{i-1} + \delta.\frac{\nabla f(X_{i-1})}{\| f(X_{i-1}) \|}, i = 1,2,...k. \tag{7}$$

, where $\nabla f_{Gauss}^D(X) = \sum_{i=1}^{N}(X - X_i) f_{Gauss}^{X_i}(X)$ is the gradient of density function at point X.

We define the density function value of data points satisfying with $\hat{f}_{Gauss}^{D}(X) \geq \xi$ as image clustering feature.

III Medical Image Clustering Analysis Based On Density Function Construction

According to above-mentioned discussions, clustering features of medical images can be extracted in 3 steps.

(1)Read all pixels of medical image one by one and compute density function value $\hat{f}_{Gauss}^{D}(X)$ of each point X. In practice, for high efficient accessing of image pixel information, tree structure index (K-D tree) is used.

(2) Processing all clustering feature data points that have greater density value than threshold.

Firstly, according to definition 3, hill-climbing strategy is used to find density attractor of each pixel data. Hill-climbing algorithm begins to climb at would-be-clustered data points with the guide of gradient direction by cursive searching and the end point of hill climbing way is attractor of all data points along the way. Then compute density function value $\hat{f}_{Gauss}^{D}(X)$ of the end point to judge them as cluster or noise data. If $\hat{f}_{Gauss}^{D}(X) \geq \xi$, all the data points on the way are marked as feature data of this cluster and write coordinate (x, y) and density function value $\hat{f}_{Gauss}^{D}(X)$ into cluster database; or all the data points along the way are marked as non-clustering feature points.

(3) Processing all clustering feature points. In order to clearly express the validity of medical image clustering analysis on the ground of density function construction, we replace each clustering feature point discovered by this algorithm as the interpolation of pixels in original image for displaying the clustering image.

Above steps describe the whole procedures of medical image clustering analysis and displaying of clustering image. The formalizations of image clustering analysis algorithm and hill climbing algorithm implementing the procedure are as following:

Algorithm clustering analysis for medical image
Input: Image G, Threshold σ, ξ
Output: Sets of cluster and its feature of G
Scan all pixels of G and set all as unmarked;
For each unmarked point p in G, do
Compute p 's density value $\hat{f}_{Gauss}^{D}(X)$;
Compute gradient at p and search for attractor p^* of p by hill_climbing scheme directed by gradient at p.
If $(\hat{f}_{Gauss}^{D}(X^*) > \xi)$
mark p, p^* and all points on the way of searching to cluster attracted by p^* ,write x, y and $\hat{f}_{Gauss}^{D}(X)$ to cluster feature database,
else
Mark all point from p to p^* as no-cluster.
For each marked point in G do
Replace related pixel attribute in G by mark pixel
Rebuild image G'.

HILL_CLIMBING:
For $X \in D$
Compute $\hat{f}_{Gauss}^{D}(X), \nabla f_{Gauss}^{D}(X)$;
if exist $Y \in D$ on direction $\nabla f_{Gauss}^{D}(X)$, $\hat{f}_{Gauss}^{D}(Y) > \hat{f}_{Gauss}^{D}(X)$ and fulfill formula (7) then
$X = Y$
else
if $\hat{f}_{Gauss}^{D}(X) \geq \xi$ then
link all data points visited to cluster.
else
mark all data points visited as no-cluster.

In clustering analysis algorithm, two thresholds σ, ξ control clustering feature effect, where σ controls clustering feature points, ξ controls storage of clustering feature points. When ξ keeps unchanged, σ gets smaller, comparison range become smaller, so computation complexity is lower, difference between clustering feature points and other pixel points is more ill-defined and less clustering feature points can be discovered. On the contrary, too big σ will make most of image area as the neighbor points of a point and get too many clustering feature points, which leads to smaller different degree of image clusters. On the other hand, when σ has a stable value, ξ has a bigger value, more number of attractors have smaller density value than threshold, so less of clustering feature points can be obtained, which leads to bigger different degree of image clusters and is in favor of image clustering analysis result.

Ⅳ EXPERIMENT AND RESULT

Experiments are done on image workstation in Image Research Institute of Jiangsu University. This workstation is specially designed for image processing by Dell Company.

Experiment data collect 40,000 abdomen CT images of 2,000 cases from Image Diagnostic Department of the Affiliated Hospital of Jiangsu University. After selection, we determine 1,000 normal images as experiment data for our research. The reason for choosing abdomen images is that abdomen part is the most complicated and difficult for diagnosis. So if problems of identification and diagnosis of abdomen images are solved, image diagnostic problems of any other parts of body can be solved easily.

In consideration of high rate of tumor pathogenesis of liver and kidney, our research is focused on those two of most important tissues and organs. Firstly, preprocess image data, standardize gray of abdomen images[11]; secondly, remove project gray data that is far off outside the arrange of liver or kidney respectively; lastly, use above algorithm to compute density function of image data points and implement data clustering analysis by hill climbing algorithm.

According to the experiment results of normal kidneys of image, clustering result is obviously good. Fig.1 and Fig.2 show the two kidneys in original image and kidneys extracted manually by expert respectively. Fig.3 and Fig.4 show the kidneys extracted by the method of this paper and clustering feature image of the kidneys respectively. All kidneys are pointed by arrow.

According to the experiment results of normal liver of image, clustering result is clear. Fig.5 and Fig.6 show the liver in original image and liver extracted manually by expert respectively. Fig.7 and Fig.8 show the liver extracted by the method of this paper and the liver clustering feature image respectively. All livers are pointed by arrow.

Experiment results show that density function based medical image clustering effect is obvious and that clustering analysis results can distinguish the projections of different tissues and organs of abdomen medial images. Therefore, density function based medical image clustering analysis can segment medical image contents according to medical semantic content effectively.

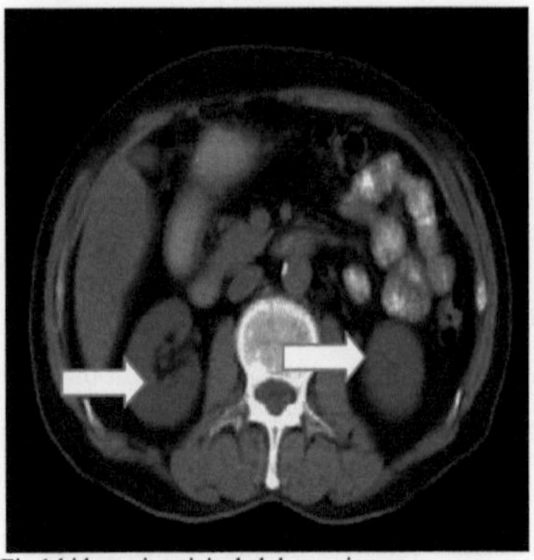

Fig.1 kidneys in original abdomen image

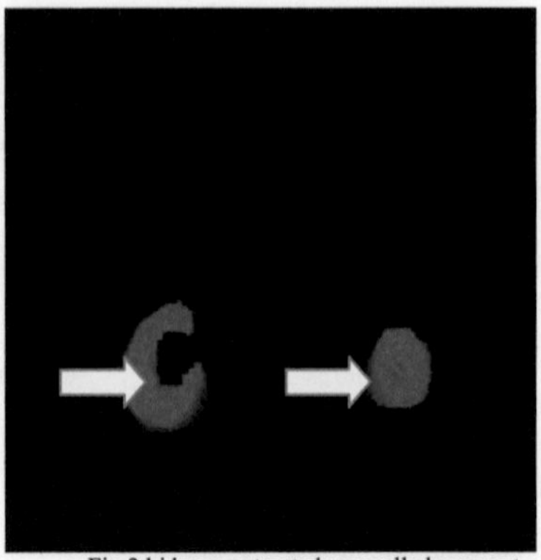

Fig.2 kidneys extracted manually by expert

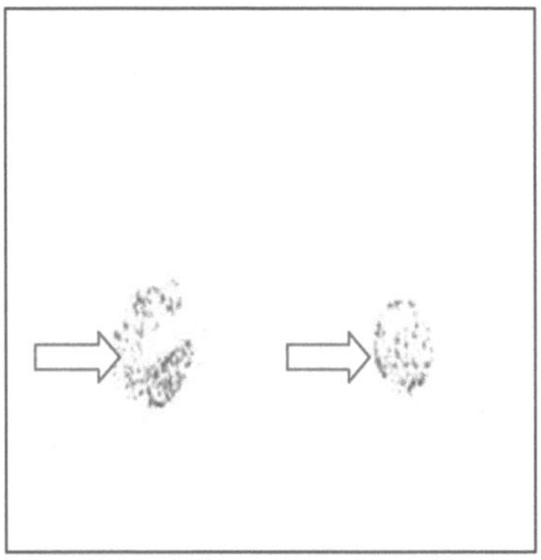

Fig.3 kidneys extracted by the method of this paper

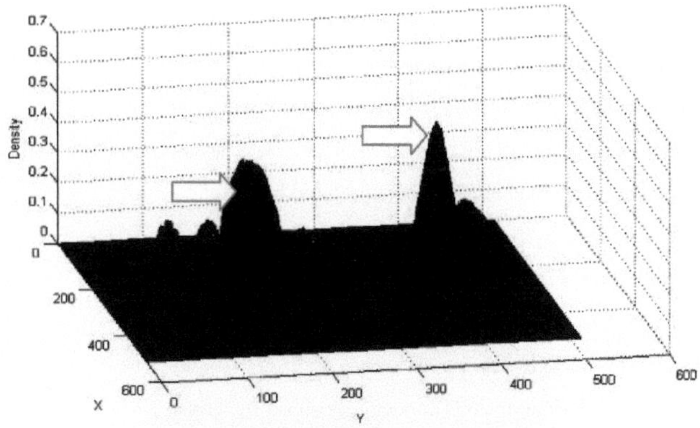

Fig 4. Density clustering image of kidneys

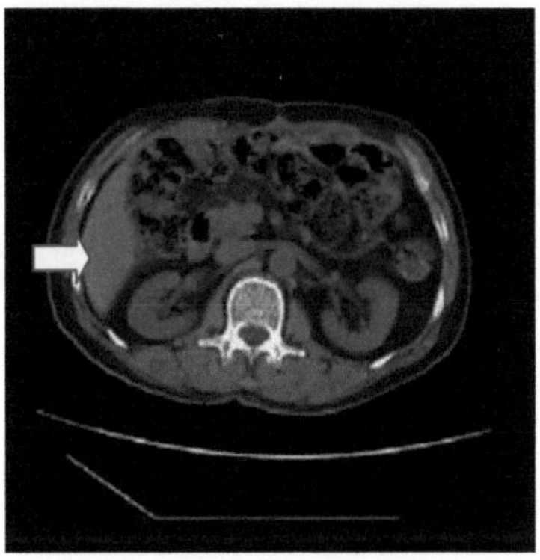

Fig.5 liver in original abdomen image

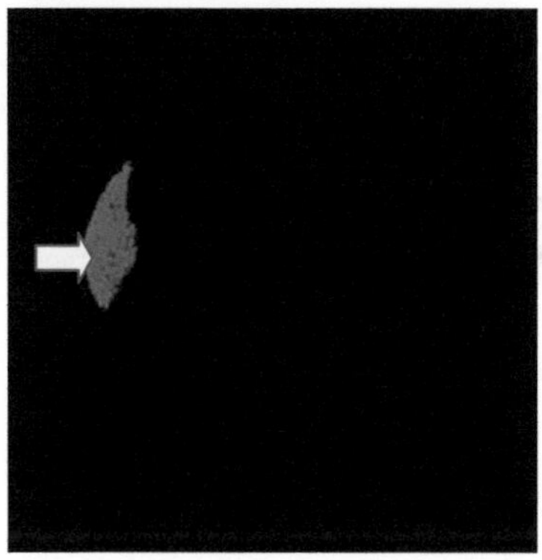

Fig.6 liver extracted manually by expert

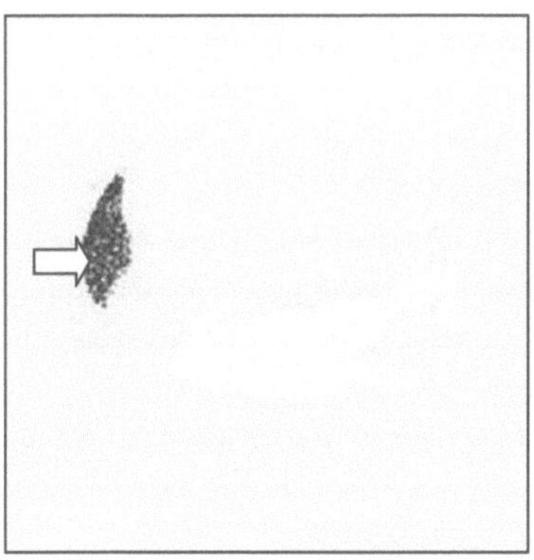

Fig.7 liver extracted by the method of this paper

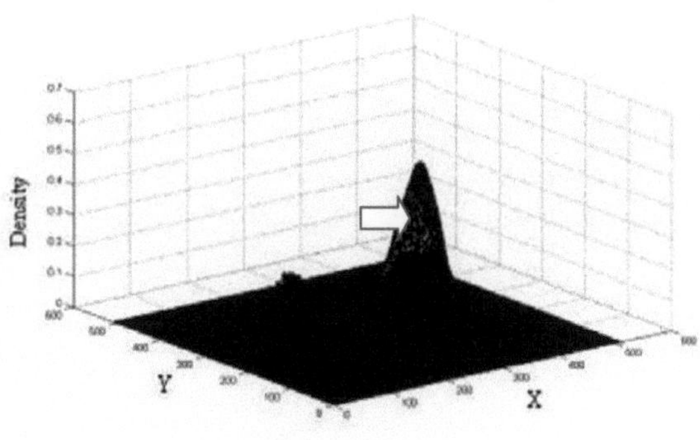

Fig 8. Density clustering image of liver

The number of pixels of image produced by error segmentation is an important weight index for medical image result. One method of evaluation so called probability of error (PE)[12] can evaluate the quality of image segmentation very well. When an image is composed of object and background, PE can be computed by formula:

$$PE = P(O) \times P(B \mid O) + P(B) \times P(O \mid B) \quad (8).$$

Where $P(O)$ and $P(B)$ represent the prior probability of object and background respectively, $P(B \mid O)$ is the probability of background pixels divided into object by wrong and $P(O \mid B)$ is the probability of object pixels divided into background by error. In order to evaluate our method, experts extract the kidneys and liver in images manually as standard images. Each image repeats to cluster 10 times with the same smooth parameter and different density threshold and PE is the average value of 10 probabilities. The maximum entropy based multi-thresholds segmentation algorithm (MEMTS) is used to compare with the method of this paper, each image repeats to cluster 10 times with different numbers of threshold and PE is the average value of 10 probabilities. The clustering results of 10 different kidneys and liver images are shown in Fig.9 and Fig.10 by these two methods. According to those two figures, the method of this paper has better results than MEMTS.

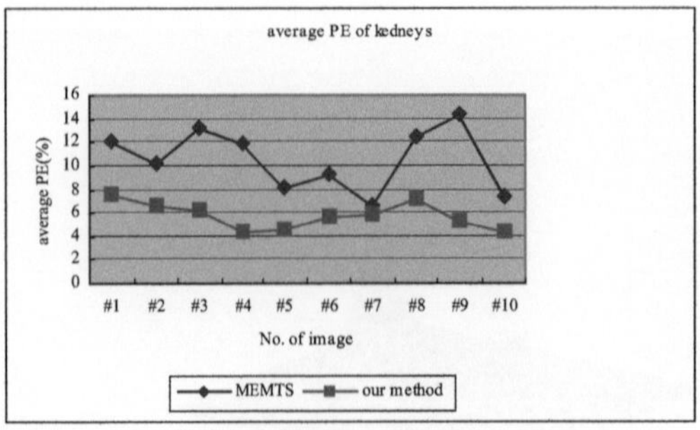

Fig9. Average PE of 10 kidneys images

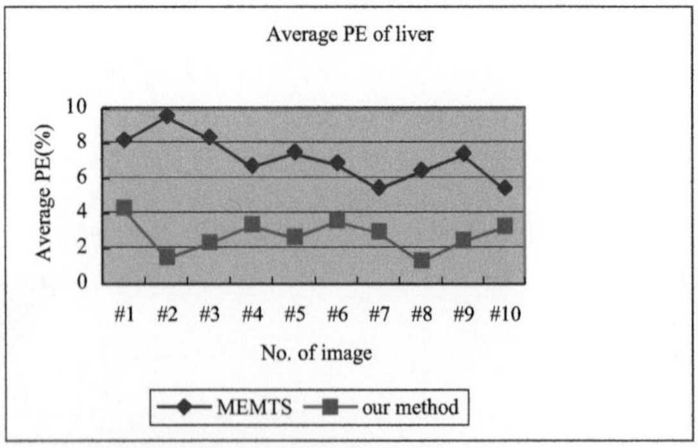

Fig10. Average PE of 10 liver images

V CONCLUSIONS

We work over the experiments on distribution of gray density of human abdomen medical images and get the anticipating results by constructing the density function for medial image data, which shows that density function has great impact on the investigation on medial images. With the development of medical image clustering analysis and local feature expression of medical image content, they will have important effect on CBIR, indexing, classification, identification and segmentation for medical images.

This article only has an introductory research for gray density distribution of medical image. There will be a lot of research fruits after systematical investigation on application of methods of approximate density function construction based medical image clustering analysis for abdomen image.

References

[1] Coleman GB, Andrews HC, "Image segmentation by clustering", Proceedings of the IEEE.1979, 7(5).pp. 773-785.

[2] Stewart C, Lu YC, Larson V, "Neural clustering approach for high resolution radar target classification" ,.Pattern Recognition.1994,7(4)pp.503-513.

[3] Ester M., Kriegel H. and Sander J. et al., "A density-based algorithm for discovering clusters in large spatial databases with noise", Proc. of the 2nd Int'l Conf., on Knowledge Discovery and Data Mining (KDD'96), Portland: AAAI Press, 1996, pp226-231.

[4] Xu X., Ester M. and Kriegel H.P. et al., "A distribution-based clustering algorithm for mining in large spatial databases", Proceedings of the 14th ICDE., Orlando, FL. 1998, pp.324-331.

[5] Hinneburg A. and Keim D., "An efficient approach to clustering in large multimedia databases with noise", Proc. of the 4th Int'l Conf. on

Knowledge Discovery and Data Mining (KDD'98), New York: AAAI Press, 1998, pp58-65

[6] Rosenblatt M., "Remarks on some nonparametric estimates of a density function", Annals of Mathematical Statistics 1956,27.pp.832-837.

[7] Parzen E, "On estimation of a probability density and mode", Annals of athematical Statistics, 1962,35.pp. 1065-1076.

[8] Silverman, B. W., "Density Estimation for Statistics and Data Analysis", Chapman and Hall, London, 1986

[9]Alexander Hinneburg and Daniel A.Keim,"A General Approach to Cluster ing in Large Databases with Noise", Knowledge and Information Systems ,2003, 5(4):387-415

[10]Li Cunhua, Sun Zhihui, Song Yuqing, "DENCLUE-M: Boost-ing DENCLUE algorithm by Mean Approximation on Grids", Proc. 4th Int. Conf. on Advances in Web-age Information Management, LNCS2762. New York: Springer Verlag, 2003,pp.202-213.

[11] Song Yuqing , "Some Key Techniques Research for Medial Image Data Mining", Nanjing, China, Southeast University doctor dissertation,2005: 15-17

[12] Lee SU, et al, "A comparative performance study of several global threshold techniques for segmentation ", CVGIP,1990,21:171~190.

Hierarchical Secret Sharing in Ad Hoc Networks through Birkhoff Interpolation

E. Ballico[1], G. Boato[2], C. Fontanari[1], and F. Granelli[2]
[1]Dept. of Mathematics, [2]Dept. of Information and Communication Technology
University of Trento
Via Sommarive 14, I-38050 Trento (Italy)
{ballico, fontanar}@science.unitn.it, {boato, granelli}@dit.unitn.it

Abstract–**Securing ad hoc networks represents a challenging issue, related to their very characteristics of decentralized architecture, low-complexity, and multiple hops communications. Even if several methods are available, this paper presents a novel approach to allow secret sharing of information at lower levels of the node protocol stack. In fact, secret sharing schemes provide a natural way of addressing security issues in ad hoc networks. To this aim, a flexible framework for secure end-to-end transmission of confidential information is proposed which exploits multipath source routing and hierarchical shares distribution. Such a goal is achieved by designing an ideal, perfect, and eventually verifiable secret sharing scheme based on Birkhoff polynomial interpolation and by establishing suitable hierarchies among independent paths.[1]**

I. Introduction

Nowadays, ad hoc networks represent a relevant research topic in the field of telecommunication networks. In an ad hoc network wireless hosts communicate with each other in the absence of a fixed infrastructure. They can be used in several applications, ranging from tactical operations, to establish quickly military communications during the deployment of forces in unknown and hostile terrain, to rescue missions, for communication in areas without adequate wireless coverage; from exhibitions or conferences or virtual classrooms, to sensor networks, for communication between intelligent sensors.

A wireless ad hoc network presents a larger spectrum of security problems than conventional wired and wireless networks, due to the broadcast nature of the transmission medium and vulnerability because nodes are often placed in a hostile or dangerous environment where they are not physically protected. As in any wireless environment, the nodes are easy to capture, compromise and hijack. An attacker can listen to and modify all the traffic on the wireless communication channel, and may attempt to masquerade as one of the participants. Due to the absence of any central support infrastructure, authentication based on public key cryptography and certification authorities may be difficult to accomplish.

There has been a flurry of research and development effort in the field of security in ad hoc networks, but results are still incomplete. Due to the severe hardware and energy constraints, selecting appropriate cryptographic primitives and security protocols in ad hoc networks is a problematic point. A usual approach for keeping sensitive data secret is to encrypt the data with a secret key known also by the receiver, ensuring in this way confidentiality. As already said, the general energy constraint in these networks creates limits also for security, due to consumption of processor power. Moreover, in asymmetric cryptographic algorithms (used for example in Encapsulating Security Payload to provide confidentiality in IPSec [1]) the length of keys which provide security is often too high for node's working memory. To achieve confidentiality means to prevent intermediate or non-trusted nodes from understanding the contents of packets. A lot of protocols offer solutions using cryptographic algorithms (see [2], [3], [4]).

The idea of the paper is to exploit the specific characteristics of an ad hoc network (multihop data delivery, absence of fixed infrastructure, decentralized architecture) in order to enforce security at the lower levels of the protocol stack (i.e., MAC layer). In particular, we propose a method to achieve end-to-end data protection against passive attacks in ad hoc networks where the nodes are not highly mobile. We focus on a method which permits to achieve confidentiality exploiting multiple paths between source and destination and taking into account different characteristics of the paths. As a consequence, the next paragraphs are focused on *secret sharing schemes*. A method which exploits multiple paths avoiding message retransmission is to transmit

[1] This research is partially funded by the T.A.S.C.A. project of I.N.d.A.M., supported by P.A.T. (Trento) and M.I.U.R. (Italy), and by the DIPLODOC project, funded by P.A.T. (Trento).

K. Elleithy et al. (eds.), Advances in Computer, Information, and Systems Sciences, and Engineering, 157–164.

redundant information through additional routes for error detection and correction [5]: part of the disjoint routes are used to transmit data and part for redundant information. In such a way if certain routes are compromised, the receiver is able to recover the message. In our method the transmission exploits m ultiple paths but improving the concept of redundant information. In the previous example confidentiality can be achieved only by adding encryption. We pretend to avoid the use of cryptographic algorithms for the already mentioned reasons. The key point of our scheme is that a non-trusted node intercepting a packet gets no information about the transmitted data. This is achieved with a secret sharing scheme. In [5], the usage of threshold schemes for key management is proposed. The same principle together with multipath routing is exploited, also in the case of data transmission, in [6]. A (k, n) threshold sharing scheme allows to divide a confidential message into n shares and requires the knowledge of at least k out of n shares to reconstruct the original content. Each share does not carry any meaningful partial plaintext of the original message and, if the number of shares available is less than k, a potential attacker can do no better than guessing, even with infinite computing time and power. The basic scheme, due to Shamir [7], relies on standard Lagrange polynomial interpolation and introduces a hierarchical approach by simply assigning a higher number of shares to higher level (more important, or reliable) participants. More recently, a refined hierarchical scheme was obtained by Tassa [8] from subtler properties of Birkhoff polynomial interpolation.

In order to exploit different characteristics of paths and in particular different trust levels, it is natural to apply a hierarchical model. For instance, the protocol SAR (Security Aware Ad Hoc Routing) proposes the Quality of Protection bit vector to classify routes [3]. Even though a hierarchical secret sharing scheme seems to be suitable also for an ad hoc network, nevertheless the previous approaches suffer from severe constraints for practical implementation. Namely, efficiency of Shamir's proposal is compromised by the systematical delay due to multiple assignments. On the other hand, Tassa's algorithm works only on large finite fields (see [8], Theorem 4), making it unsuitable for an ad hoc node with limited computational resources.

In order to overcome such difficulties, the present paper introduces an alternative scheme for secure information sharing in ad hoc networks applications. As in Tassa's approach, Birkhoff interpolation theory is applied, but with some crucial improvements. In particular, random allocation of participants enables to exploit stronger mathematical tools and drastically reduce the size of the base field. Furthermore, each participant receives only one share (overcoming the main drawback of Shamir's hierarchical scheme) and the secret is identified with a sequence of elements of the field (reflecting the natural structure of a message as a sequence of packets). As a consequence, both the delay and the overhead are significantly reduced.

The structure of the paper is the following. In Section II, after recalling the definition of hierarchical secret sharing, the proposed sharing scheme is introduced and a mathematical proof that it is ideal, perfect, and eventually verifiable is given. In Sections III and IV, a hierarchical multipath framework for ad hoc networks is described, providing both a general scheme and a sample application scenario. Finally, in Section V, some concluding remarks and outlines about future work on the topic are presented.

II. Description and Analysis of the Proposed Algorithm

The basic scheme proposed by Shamir [7] relies on standard Lagrange polynomial interpolation. To be explicit, Shamir's idea is simply to identify a secret $S \in R$ with some coefficient of a polynomial

$$p(x) = \sum_{i=0}^{k-1} a_i x^i$$

where for instance $a_0 = S$ and $a_1, \ldots, a_{k-1}$ are arbitrary real numbers. In order to distribute S among n participants, just fix n distinct real numbers $v_1, \ldots, v_n$ and assign to the j-th participant the share

$$p(v_j) = \sum_{i=0}^{k-1} a_i v_j^i$$

In order to reconstruct the secret, a subset of participants with associated real numbers $\{v_{i_1}, \ldots, v_{i_s}\}$, with $1 \le i_1 < i_2 < \ldots < i_s \le n$ has to solve the following linear system:

$$A \begin{pmatrix} a_0 \\ \vdots \\ a_{k-1} \end{pmatrix} = \begin{pmatrix} p(v_{i_1}) \\ \vdots \\ p(v_{i_s}) \end{pmatrix} \qquad (1)$$

where

$$A = \begin{pmatrix} 1 & v_{i_1} & \ldots & v_{i_1}^{k-1} \\ \vdots & & & \vdots \\ 1 & v_{i_s} & \ldots & v_{i_s}^{k-1} \end{pmatrix}$$

is a so-called *Vandermonde matrix*. It follows that the linear system (1) admits a unique solution if and only if $s \ge k$. In particular, at least k out of n shares are needed to reconstruct S, hence we obtain a (k, n) secret sharing scheme.

As pointed out by Shamir himself in [7], a hierarchical variant can be introduced by simply assigning a higher number of shares to higher level participants. More recently, a refined hierarchical scheme was obtained by Tassa [8] from subtler properties of Birkhoff polynomial interpolation.

Let U be a given set of n participants and fix a collection Γ of subsets of U, which is monotone in the sense that if $I \in \Gamma$ then any set containing I also belongs to Γ. A threshold secret sharing scheme with *access structure* Γ is a method of sharing a secret among the members of U, in such a way that only subsets in Γ can recover the secret, while all other subsets have no information about it. Assume that U is divided into levels, i.e. $U=\bigcup_{l=0}^{t}U_l$ with $U_i \cap U_j = \varnothing$ for every $i \neq j$. If $0 < k_0 < \ldots < k_t$ is a strictly increasing sequence of integers, then a $(k_0,\ldots,k_t;n)$ hierarchical threshold secret sharing scheme distributes to each participant a share of a given secret S, in such a way that $\Gamma = \{V \subset U : \#V \cap (\bigcup_{l=0}^{i} U_l) \geq k_i \quad \forall i = 0,\ldots,t\}$ where # states cardinality.

Roughly speaking, a subset of participants can reconstruct the secret if and only if it contains at least k_0 members of level 0, at least k_1 members of level 0 or level 1, at least k_2 members from levels 0, 1, and 2, and so on.

In order to construct a suitable $(k_0,\ldots,k_t;n)$ hierarchical threshold secret sharing scheme, it is natural to apply Birkhoff interpolation instead of Lagrange interpolation. The key point is that the Birkhoff scheme involves not only a polynomial, but also its (higher order) derivatives. To be formal, as in [11], p.124, let $E=(E_{i,j})$, $i=1,\ldots,m$; $j=0,\ldots,k-1$, be an mxk *interpolation matrix*, whose elements are zeros or ones, with exactly k ones. Let $X = x_1,\ldots,x_m, x_1 < x_2 \ldots < x_m$, be a set of m distinct *interpolation points*. For polynomials

$$p(x)=\sum_{i=0}^{k-1} a_i x^i$$

of degree $\leq k-1$ we consider the k interpolation equations

$$p^{(j)}(x_i) = B_{i,j}$$

for $E_{i,j}=1$, where $p^{(j)}$ denotes the *j*-th derivative of *p* and $B_{i,j}$ are given data. Here the unknowns are the k coefficients $a_0,\ldots,a_{k-1}$ of *p(x)*. However, it is easy to convince ourselves that a Birkhoff interpolation problem can admit infinitely many solutions even if the number of equations equals the number of unknowns. Indeed, think for a moment at the case in which $E_{i,0}=0$ for every $i=1,\ldots,m$. In such a case, the interpolation system involves only derivatives of the polynomial *p*, hence it keeps no track of the constant term a_0, which remains undetermined. More generally, elementary linear algebra considerations show that if a Birkhoff interpolation problem admits a unique solution then its associated interpolation matrix $E=(E_{i,j})$, $i=1,\ldots,k$; $j=0,\ldots,k-1$, has to satisfy the following *Pólya condition*

$$\#\{E_{i,j}=1 : j \leq h\} \geq h+1 \quad 0 \leq h \leq k-1$$

(see for instance p.126 of [11]).

The idea now is to exploit this necessary condition in order to ensure that only authorized subsets can reconstruct the secret. Intuitively speaking, an evaluation of the polynomial itself carries more information than an evaluation of any of its derivatives since it involves more coefficients; therefore it sounds reasonable to assign to a participant of higher level the evaluation of a lower order derivative.

For shares generation, we propose the following algorithm:

1. Select a finite field F with characteristic $p \geq k_t$ and cardinality $q \geq k_t(k_t-1)/2$. Identify the secret S with a sequence $(S_0,\ldots,S_z)$, with $0 \leq z \leq k_t - 1$ and $S_i \in F$ for every i.
2. Set $k = k_t$ and pick a polynomial
$$p(x)=\sum_{i=0}^{k-1} a_i x^i$$
where $a_i = S_i$ for every $0 \leq i \leq z$ and a_i arbitrary elements of F for $z+1 \leq i \leq k-1$.
3. Identify each participant of level l with a random element $u \in F$ and associate to u the share $p^{(k_{l-1})}(u)$, where $p^{(h)}$ denotes the h-th derivative of p and $k_{-1} := 0$.

We stress that for $z=0$ our scheme is *ideal*, i.e. the size of the shares is equal to the size of the secret; if $z \geq 1$ the information rate is even higher. We also point out that the above scheme becomes *verifiable* after the following additional procedure, inspired by [9]. The dealer selects an appropriate elliptic curve E over the field F in such a way that the discrete logarithm problem for E is hard (see for instance [10], § 5.2), chooses a point Q on E and broadcasts Q and $Q_i \triangleq a_i Q, 0 \leq i \leq k-1$. Any node receiving the share $p^{(h)}(u)$ can verify its integrity by simply checking

$$p^{(h)}(u)Q = \sum_{i=0}^{k-1} h! \binom{i}{h} u^{i-h} Q_i .$$

Fix now $V \triangleq \{u_1,\ldots,u_m\} \subset U$. Up to reordering we may assume that $u_i \in U_{l(i)}$ with $l(i) \leq l(j)$ for every $i \leq j$. Consider the square $k \times k$ matrix M_V whose i-th row is given by

$$\frac{d^{k_{l(i)-1}}}{dx^{k_{l(i)-1}}}(1,x,x^2,\ldots,x^{(k-1)})(u_i) \qquad (2)$$

In order to reconstruct the secret S, the combiner has to solve the following linear system:

$$M_V \begin{pmatrix} a_0 \\ \ldots \\ a_{k-1} \end{pmatrix} = \begin{pmatrix} p^{k_{l(1)-1}}(u_1) \\ \ldots \\ p^{k_{l(k)-1}}(u_k) \end{pmatrix} \qquad (3)$$

This amounts to the Birkhoff interpolation problem with associated interpolation matrix $E_V = (e_{i,j})_{i=1 j=0}^{k \; k-1}$ defined as follows:

$$e_{i,j} = \begin{cases} 1 & \text{if } j = k_{l(i)-1} \\ 0 & \text{otherwise} \end{cases}$$

It is easy to check by direct inspection that $V \in \Gamma$ if and only if E_V satisfies the Pólya condition. In particular, we deduce the following:

Theorem 1. If $V \notin \Gamma$ then $\det(M_V)$ is identically 0, hence V cannot reconstruct the secret.

Next, we introduce some standard terminology. A *shift* Λ of E is a translation of a one from a place (i,j) into the place $(i,j+1)$, assuming that this is possible, i.e. $j+1<k$ and that $e_{i,j+1}=0$; such operation produces a new matrix $\Lambda(E)$. We say that two rows of E have a *collision* if they have ones in the same column.

Define α to be the minimum number of shifts $\Lambda_1,\dots,\Lambda_\alpha$ such that $\Lambda_\alpha \cdot \Lambda_{\alpha-1} \cdot \ldots \cdot \Lambda_1(E)$ has no collisions in rows 1 and 2. With this notation we have the following:

Lemma 1. *Let $Ma=b$ be a Birkhoff interpolation problem with associated interpolation matrix $E=\left(e_{i,j}\right)_{i=1\,j=0}^{k\quad k-1}$ defined over a field F of characteristic $p=0$ or $p>\max\{k-1,\alpha!\}$. Then $\det(M)$ is not identically 0 if and only if E satisfies the Pólya condition.*

Proof: The case of characteristic 0 is Theorem 10.1 in [11]. In the other case, following verbatim the same proof we obtain (see equation (10.4)):

$$\frac{d^\alpha}{dx_1^\alpha}\det(M)=\varepsilon CD$$

where $\varepsilon=\pm 1$, D is not identically 0 and C is the number of different representations of $\Lambda_\alpha \cdot \Lambda_{\alpha-1} \cdot \ldots \cdot \Lambda_1(E)$ as a composition of shifts. Our assumption on the characteristic of F implies that εCD (hence $\det(M)$) is not identically 0 for every $C \le \alpha!$. In order to prove that this last inequality always holds, we may argue by induction on α. If $\alpha=1$, the result is trivial. For $\alpha \ge 2$, fix any composition Π of α shifts. We claim that in order to determine a representation $\Pi=\Lambda_\alpha \cdot \Lambda_{\alpha-1} \cdot \ldots \Lambda_1$ we have at most α choices for Λ_1. If it is true, then for every choice by induction we have at most $(\alpha-1)!$ choices to complete $\Lambda_\alpha \cdot \Lambda_{\alpha-1} \cdot \ldots \cdot \Lambda_1$, hence we obtain at most $\alpha!$ representations. Assume by contradiction to have at least $\alpha+1$ choices for Λ_1, say $e_{s_i,t_i} \to e_{s_i,t_i+1}$, $i=1,\dots,\alpha+1$. Let $N_i(G):=\#\left\{g_{s_i,j}=1 : j \le t_i\right\}$ for every $k \times k$ square matrix G with entries $g_{i,j} \in F$ and let $E' := \Lambda_\alpha \cdot \Lambda_{\alpha-1} \cdot \ldots \cdot \Lambda_1(E)$. Then $N_i(E') < N_i(E)$ for every $i=1,\dots,\alpha+1$ and $\Lambda_\alpha \cdot \Lambda_{\alpha-1} \cdot \ldots \cdot \Lambda_1$ would be composed by $\alpha+1$ shifts, contradiction. In order to apply the above general results in our context, first of all notice that E_V satisfies Pólya condition if and only if $V \in \Gamma$ (essentially by definition). As a consequence, we can prove that a subset of participants can reconstruct the secret if and only if it belongs to the access structure. In particular, for $z=0$ our scheme is *perfect*.

Theorem 2. In the notation above, if $V \in \Gamma$ then $\det(M_V)(u_1,\dots,u_k) \ne 0$, hence V can reconstruct the secret.

Proof: If $t=0$, then our scheme reduces to Shamir's one ([7]) and there is nothing to prove. If instead $t \ge 1$, up to reordering the rows of M_V we can assume that there are no collisions in rows 1 and 2, hence we have $\alpha=0$. By Lemma 1, $\det(M_V)$ does not vanish identically and by [12], Proposition 1.2, it is a homogeneous polynomial of total degree $k(k-3)/2$. Since $u_1,\dots,u_k$ are random, the conclusion follows from the next easy result.

Lemma 2. Let $p(x_1,\dots,x_k)$ be a not identically 0 homogeneous polynomial of degree δ defined over a finite field F_q with $q>\delta+k$. Then there exists $(u_1,\dots,u_k) \in \left(F_q\right)^k \setminus \cup_{i,j}\Delta_{i,j}$ such that $p(u_1,\dots,u_k) \ne 0$, where $\Delta_{i,j} := \{(x_1,\dots,x_k) \in \left(F_q\right)^k : x_i = x_j\}$

Proof: By induction on k, the case $k=0$ being obvious. Write

$$p(x_1,\dots,x_k,y)=p_\delta(x_1,\dots,x_k)y^\delta+\ldots+p_0(x_1,\dots,x_k)$$

and for a fixed $(\overline{x}_1,\dots,\overline{x}_k) \in \left(F_q\right)^k \setminus \cup_{i,j}\Delta_{i,j}$ define $f(y):=p(\overline{x}_1,\dots,\overline{x}_k,y)$. If $f(y)=0$ for every $y \in F_q \setminus \{\overline{x}_1,\dots,\overline{x}_k\}$ then f has $q-k \ge \delta+1$ zeros and f vanishes identically. In particular, we have $p_i(\overline{x}_1,\dots,\overline{x}_k)=0$ for every i and if this is true for every choice of $(\overline{x}_1,\dots,\overline{x}_k)$ in $\left(F_q\right)^k \setminus \cup_{i,j}\Delta_{i,j}$ then by inductive assumption each $p_i(x_1,\dots,x_k)$ is identically 0, contradiction.

III. Hierarchical Secret Sharing in Ad Hoc Networks

In an ad hoc network end-to-end security is a major issue. Indeed, we can assume that neighboring nodes can easily exchange security keys, in order to establish an authenticated and secure channel on a link basis (a usual assumption in this situation, see for instance [14] and [6]). Therefore, the problem to solve is related to grant secure transmissions on an end-to-end basis. Our approach aims at exploiting multiple independent paths from source to destination in such a way to spread along the ad hoc network the secret information. Depending on various factors, paths can be classified into several levels and identifying a suitable hierarchy of disjoint routes allows to apply the hierarchical secret sharing scheme presented in Section II in order to enforce security. However, the selection of a path hierarchy is a non trivial task in an ad hoc network. In the next paragraphs a general model is proposed, which is based on both global and local properties of the paths.

Let us identify a path from a source (dealer) A to a destination (combiner) B as an ordered sequence of nodes $x=(x_1,\dots,x_s)$, where $x_1=A$ and $x_s=B$; two paths $(x_1,\dots,x_s)$ and $(y_1,\dots,y_t)$ are *independent* or *node disjoint* if $\{x_1,\dots,x_s\}\cap\{y_1,\dots,y_t\}=\{A,B\}$. As it is well known, the problem of finding a maximum cardinality set of node disjoint paths in a network (thought as a digraph) can be solved in polynomial time (see for instance [13], Proposition 2.4 (i)). Choose a maximum cardinality set U of independent paths (or more generally any sufficiently big subset) and fix n:=#U.

We can assume that neighboring nodes can easily exchange security keys, in order to establish an authenticated and secure channel on a link basis (a usual assumption in this situation, see for instance [14] and [6]). Therefore, the problem to solve is related to grant secure transmissions on an end-to-end basis.

Let $p_{loss}(x)$ be the packet loss probability of the path x. Its value depends on several properties of the path (for instance, signal-to-noise ratio, number of hops, interference, etc.).

It is possible to define $p_{sniff}(x)$ as the probability for a packet traveling on x to be eavesdropped. As a function of x, p_{sniff} can be expressed as the algebraic sum of several contributions of different nature:

$$p_{sniff}(x)=\sum_j F_j(x)-\sum_{i,j} G_j(x_i)$$

normalized in such a way that $0\le p_{sniff}(x)\le 1$. The term F_j includes the potential flaws related to the whole path, such as the total number of hops, while G_j takes into account security robustness of each node, such as terminal's reliability, tamper resistant harware equippment, user's trustworthiness. It is clear that the functionals F_j contribute to p_{sniff} with positive sign (in particular, if the number of hops is higher, then a path is more likely to be subject to external attacks), while the functionals G_j are considered with the opposite sign.

The set of values assumed by p_{sniff} induces a natural hierarchy on the set of paths. Namely, if $p_{sniff}(U)=\{p_0,\dots,p_t\}$ with $0\le p_0<\dots<p_t\le 1$, then we can define:

$$U_l:=\{x\in U: p_{sniff}(x)=p_l\}.$$

Finally, we have to determine a strictly increasing sequence of thresholds k_i. It is clear that higher thresholds produce a safer scheme; on the other hand, by definition, in order to recover the secret, the receiver needs at least k_i shares from $\bigcup_{l=0}^{i} U_l$. Hence, it seems reasonable to fix k_i as the expected number of shares reaching the destination via paths of level i or less. More precisely, we define:

$$k_i:=\left\lfloor \sum_{\substack{x\in U_l\\ 0\le l\le i}} \left(1-p_{loss}(x)\right)\right\rfloor, \tag{4}$$

where $\lfloor r \rfloor$ denotes the biggest integer $\le r$.

Summarizing, the proposed scheme can be implemented as follows:

Source A:

1. finds a set U of node disjoint paths to destination B with a multipath routing process (see for instance [15] and [16]) and for each $x\in U$ collects all relevant properties of x (signal-to-noise ratio, number of hops, interference, terminal's reliability, user's trustworthiness...);
2. computes $p_{loss}(x)$ and $p_{sniff}(x)$ for every path $x\in U$;
3. defines $(k_0,\dots,k_t;n)$ with k_i as in (threshold) and $n:=\#U$;
4. selects a finite field F with characteristic $p\ge k_t$ and cardinality $q\ge k_t(k_t-1)/2$ and identifies the secret with an element $S\in F$;
5. picks a polynomial
$$p(x)=\sum_{i=0}^{k-1} a_i x^i$$
where $k=k_t$, $a_0=S$ and a_i arbitrary elements of F for $1\le i\le k-1$;
6. identifies each path $x\in U_l$ with a random element $u\in F$ and associates to u the share $p^{(k_{l-1})}(u)$, where $p^{(h)}$ denotes the h-th derivative of p and $k_{-1}:=0$;
7. transmits every share along an independent path through source routing, exploiting private communication between neighbouring nodes (as in [14], [6], it is possible to assume that neighbouring nodes in an ad hoc network can exchange encryption keys during link initialization in order to establish an authenticated channel on the physical link).

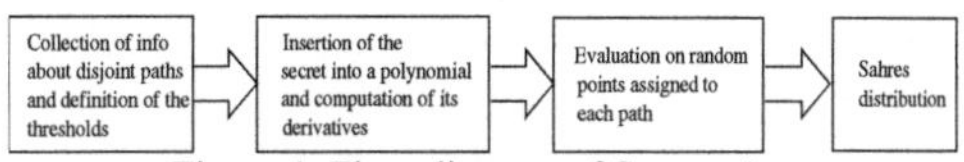

Figure 1. Flow diagram of Source A

Destination B:

1. receives $m\le n$ shares, say from the subset of paths $V:=\{u_1,\dots,u_m\}\subseteq U$;
2. constructs the matrix M_V as in (2);
3. recovers the secret by solving the linear system (system) in the indeterminates $a_0,\dots,a_{k-1}$.

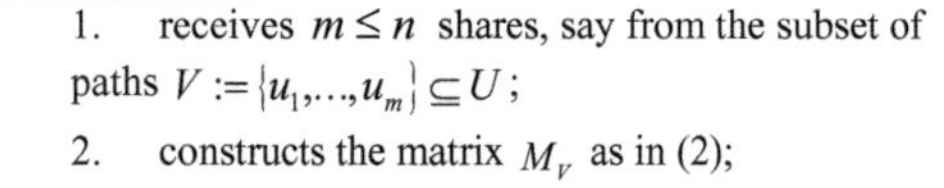

Figure 2. Flow diagram of Destination B

Notice that any intruder eavesdropping up to $k_i - 1$ shares from $\cup_{0 \le l \le i} U_l$ has no information about the confidential message sent over the network (according to Theorem 1).

IV. A Possible Application Scenario

This section provides a possible application scenario for the proposed hierarchical secret sharing scheme. Let us assume to have an ad hoc network of N nodes uniformly distributed in a planar region of area R, for example sensors disseminated in a terrain which need to communicate protecting data. In this situation, the problem of finding independent paths between a pair of nodes has been addressed in several papers. In [6] a modified Dijkstra algorithm is used; however, this solution assumes the knowledge of the graph associated with the network, which is not always available in such a network. Different approaches are proposed, like [15] (Selective Broadcast) and [16] (AODVM), just to quote some recent contributions. In particular, simulation results are available for $N = 50$ and $R = 1500 \times 500\, m^2$ ([15], § VI) and for $N = 250, 350, 500$ and $R = 2500^2\, m^2$ ([16], § IV). Here instead we are going to determine a suitable choice of parameters for the implementation of our scheme in this context, providing also a numerical example (see Table 1). We denote by $\rho = N/R$ the node spatial density and by r the average distance between any pair of neighbouring nodes. As motivated in [17], II.A, the relationship between ρ and r is approximately $r \approx 1/\sqrt{\rho}$. As a consequence, an estimate for number h_0 of hops in the shortest path between two nodes A and B having mutual distance d is given by $\lceil \sqrt{\rho}\, d \rceil$, where $\lceil s \rceil$ denotes the smallest integer $\ge s$. An arbitrary path x between A and B is a sequence of $h(x) \ge h_0$ hops, and it is possible to split the set U of independent paths between A and B into levels as follows:

$$V_i := \{x : h(x) = h_0 + i\}.$$

Moreover, if p is the probability for each node to be active, it is natural to set the packet loss probability increasing with the number of hops:

$$p_{loss}(x) = \left(1 - p^{h(x)}\right).$$

In order to implement the proposed scheme, we introduce the integer:

$$\mathrm{i}_0 := \min\left\{ i : \sum_{\substack{x \in V_l \\ 0 \le l \le i}} \left(1 - p_{loss}(x)\right) \ge 1 \right\} \qquad (5)$$

and we define a hierarchical structure based on length of paths:

$$U_0 := \bigcup_{0 \le i \le i_0} V_i$$

$$U_1 := \bigcup_{i > i_0} V_i.$$

Hence the set U naturally splits according to two priority levels into U_0, corresponding to paths with lower number of hops and therefore lower probability of eavesdropping, intrusion and capturing, and U_1, collecting all remaining ones. The discriminating number of hops is chosen as the minimum value compatible with the condition that the threshold k_0 defined in (4) is a positive integer. In this simplified scenario, we have $p_{sniff}(U_0) = \{p_0\}$ and $p_{sniff}(U_1) = \{p_1\}$ with $0 \le p_0 \le p_1 \le 1$, but the proposed scheme gives the freedom to establish the path hierarchy according to different parameters (for instance, high priority could be granted to paths with high performance: high signal-to-noise ratio, low delay,...), as well as a combination of the above.

In Figure 3 we report a set of 70 nodes uniformly distributed in a 1000×1000 meters terrain. We find 5 independent paths between the fixed source A and destination B. The number of hops in the shortest path x_3 is equal to 5, exactly as estimated above, and 3 sets of paths V_0, V_3, V_5 are defined. Now it is possible to compute p_{loss} for each path and calculate i_0 by applying (5). The path hierarchy is now completely determined (see Table 1). In order to quantify the security enforcement produced by our hierarchical approach, we compare the probabilities of reconstructing the secret after capturing r nodes in three different cases: Shamir's standard scheme (where just 3 nodes are needed among $x_1, \ldots, x_5$), Shamir's hierarchical scheme (where 2 shares are assigned to paths x_2, x_3, x_4 and just one to paths x_1, x_5, so at least 4 shares, being expected to reach the destination, are requested to recover the secret) and our hierarchical scheme (where at least 2 nodes are needed among x_2, x_3, x_4 and at least 3 among $x_1, \ldots, x_5$). We make the assumption that source A and destination B cannot be captured. The remaining $n = 68$ nodes, having the same probability of being captured, naturally split into 6 subsets, 5 corresponding to the independent paths $x_i, i = 1, \ldots, 5$ (respectively with cardinalities $n_1 = 9, n_2 = 7, n_3 = 4, n_4 = 7, n_5 = 9$) and the last are x_6 collecting external nodes (with cardinality $n_6 = n - n_1 - n_2 - n_3 - n_4 - n_5 = 32$). In this model, the probability of getting r_i nodes from the set x_i ($i = 1, \ldots, 6$) by capturing exactly $r = r_1 + r_2 + r_3 + r_4 + r_5 + r_6$ nodes is given by the formula

$$P = \frac{\prod_{i=1}^{6} \binom{n_i}{r_i}}{\binom{n}{r}}$$

Hence, by adding the contributions of all relevant cases, we obtain the probability of reconstructing the secret as a function of r as reported in Figure 4. It is apparent that Shamir's hierarchical scheme is unsuitable for this application, while our approach outperforms the standard non-hierarchical scheme. For instance, after capturing 10% of nodes, the probability of secret reconstruction is 32% for our method against 49% for Shamir's one.

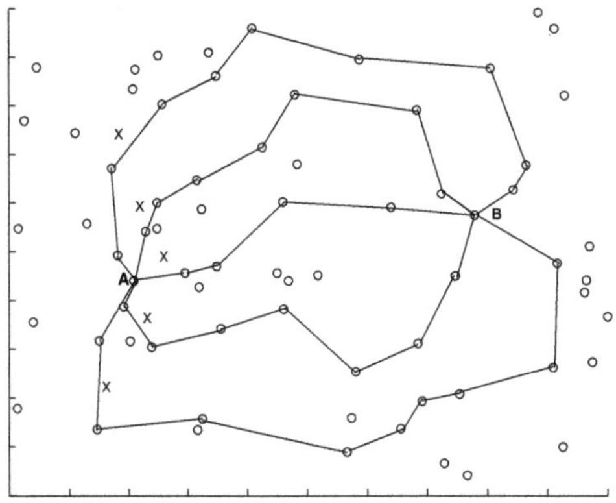

Figure 3. Example of uniform node distribution

Table 1
Values corresponding to Figure 3

$R = 1000^2\ m^2$
$N = 70$
$\rho = 7 \times 10^{-5}$
$r \approx 119,5\ m$
$h_0 = 5$
$V_0 = \{x_3\}$, $V_3 = \{x_2, x_4\}$, $V_5 = \{x_1, x_5\}$
$p = 0,95$
$p_{loss}(x_3) \approx 0,23$
$p_{loss}(x_2) = p_{loss}(x_4) \approx 0,34$
$p_{loss}(x_1) = p_{loss}(x_5) \approx 0,4$
$i_0 = 3$
$U_0 = V_0 \cup V_3$, $U_1 = V_5$
$k_0 = 2$, $k_1 = 3$

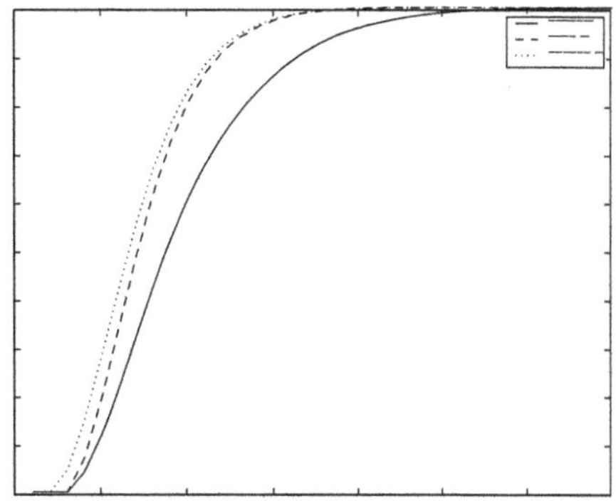

Figure 4. Probability of reconstructing the secret after capturing r=0,…,34 nodes in Shamir's standard scheme (shamir), Shamir's hierarchical scheme (h-shamir) and our hierarchical scheme (birkhoff)

V. CONCLUSIONS AND FUTURE WORK

The paper presents an innovative approach that exploits multi-path routes in order to define a flexible framework for end-to-end secure transmission in ad hoc networks by distributing the shares in a hierarchical way. This is achieved by designing an ideal, perfect, and eventually verifiable secret sharing scheme based on Birkhoff polynomial interpolation and by establishing suitable hierarchies among independent paths. An explicit algorithm is provided and a numerical example is presented in order to validate it.

Future work on the topic will deal with the implementation of the proposed scheme in a simulation environment.

REFERENCES

[1] RFC2401, Security Architecture for the Internet Protocol

[2] R. Molva and P. Michiardi, "Security in Ad Hoc Networks," Invited Paper in Personal Wireless Communication (PWC'03), September 2003, Venice, Italy.

[3] S. Yi, P. Naldurg, and R. Kravets, "Security-Aware Ad-Hoc Routing for Wireless Networks," Technical Report of the University of Illinois at Urbana-Champaign, 2001.

[4] L. Buttyan and J.-P. Hubaux, "Working Session on Security in Wireless Ad Hoc Networks," Mobile Computing and Communications Review, Vol. 6, No. 4, 2002.

[5] L. Zhou and Z. Haas, "Securing Ad Hoc Networks," IEEE Network 13(6) (1999), 24--30.

[6] W. Lou and Y. Fang, "Securing Data Delivery in Ad Hoc Networks," Proc. of the Ninth International Conference on Distributed Multimedia Systems, September 2003, Miami (USA), 599-604.

[7] A. Shamir, "How to share a secret," Communications of the ACM 22 (1979), 612-613.

[8] T. Tassa, "Hierarchical Threshold Secret Sharing," Proc. of the Theory of Cryptography Conference 2004, MIT, Cambridge MA, USA, February 2004, LNSC 2951, Springer-Verlag, 2004, 473-490.

[9] T. P. Pedersen, "Distributed Provers with Application to Undeniable Signatures," Proc. of Eurocrypt '91, Lecture Notes In Computer Science, LNCS 547, Berlin: Springer-Verlag, 1991, 221-238.

[10] N. Koblitz, A. Menezes, and S. Vanstone, "The State of Elliptic Curve Cryptography," Designs, Codes and Cryptography 19 (2000), 173-193.

[11] R. A. DeVore and G. G. Lorentz, "Constructive Approximation," Grundlehren der Mathematischen Wissenschaften 303, Springer-Verlag, Berlin, 1993.

[12] G. G. Lorentz, K. Jetter, and S. D. Riemenschneider, "Birkhoff interpolation," Encyclopedia of Mathematics and its Applications 19, Addison-Wesley Publishing Co., Reading, Mass., 1983.

[13] J. Cheriyan, "Randomized algorithms for problem in matching theory," SIAM J. Comput. Vol. 26, No. 6, December 1997, 1635-1655.

[14] M. Hietalahti, "Key Establishment in Ad Hoc Networks," Technical Report, Helsinki Univ. of Technology, Dept. of Computer Science and Engineering, 2001.

[15] K. Wu and J. Harms, "Multipath Routing for Mobile Ad Hoc Networks," IEEE ComSoc/KICS Journal of Communications and Networks, Special Issue on Innovations in Ad Hoc Mobile Pervasive Networks, Vol. 4, No. 1, March 2002, 48-58.

[16] Z. Ye, S. V. Krishanamurthy, and S. K. Tripathi, "A Framework for Reliable Routing in Mobile Ad Hoc Networks," IEEE INFOCOM 2003.

[17] G. Ferrari and O. K. Tonguz, "Minimum Number of Neighbors for Fully Connected Uniform Ad Hoc Wireless Networks," Proc. IEEE Intern. Conf. Commun. (ICC'04), Paris, France, June 2004, 4331-4335.

An Analysis Format for Client-Server Performance Using GEO & LEO Satellite Networks (Inmarsat vs. Globalstar)

Terry C. House
Nova Southeastern University
hterry@nova.edu

***Abstract*-Secure Role-Based Access Control (SRBAC) is an entry protocol implemented to protect sensitive database information. An individual's security clearance and role description will determine what privileges they are authorized. Prior research focused on a Mobile Secure Role Based Access Control (MS-Ro-BAC) Database, for wireless and wired access connectivity. "Ref. [4]." In order to implement a control procedure to protect sensitive database information, over wired or wireless communication channels, required a combination of new server and programming security implementations. Proprietary software was developed as a front end authentication when logging into the server and website databases; two levels of security decreased the chances of compromising the database system and design. "Ref [5]." An individual's username, password, role description and security clearance determines what privileges they are authorized and what data they have access to. The wireless Iridium data connection proved not to be the choice of satellite communication service to best support the MS-Ro-BAC database system in a global environment. Nevertheless, the wired connectivity proved to be robust and stable, the world wide connectivity and data transmission between the database activities performed consistently at high level of proficiency "Ref. [3]." The capabilities provided by the MS-Ro-BAC research would support the 'Global War on Terrorism' and increase the security of US forces and Department of Defense Personnel around the world.**

I. INTRODUCTION

A. The MIS-Ro-BAC Ideology

The "Mobile Secure Role Base Access Control Project" (MS-Ro-BAC) is a single unit system with the ability to instantly connect to other identical devices around the world through Mid Earth Orbit Satellite (MEO) communication. "Ref. [2]." Satellite technology was initially designed to support the National Oceanic and Atmospheric Administration (NOAA). This organization supported two systems of orbiting terrestrial bodies: Polar and Geosynchronous Orbits. The first satellite system known as Television and Infrared Observation Experimental Satellite (TIROS) was launched into space April 1, 1960. Later, the International Telecommunications Satellite Corporation (INTELAT) was founded by 11 nations to monitor the development and design of global satellite communications systems. Soon after, in April 1965 the first communication satellite, INTELAT-I (Early Bird) was launched into space. This system was capable of carrying 240 voice channels with a life span of 18 years; however, the satellite became inactive after 3 years "Ref. [8]." Today's satellite technology is efficient and productive on an international level.

A. Background

The most widely used satellite systems for communications are the Geosynchronous Orbit (GEO) and Low Earth Orbit (LEO) satellite constellation. During this investigation, the Inmarsat and Globalstar systems are the subjects of comparison and analysis for this study. The Globalstar satellite constellation system consists of 48 active satellites and 4 orbiting spares, at a height of 876 miles above the earth. This particular set of LEO satellites orbit the earth at 3.4 kms. per second; however, the Inmarsat-3 constellation system consists of only 5 satellites. "Ref. [8]." Each satellite individually covers 1/3 of the earth's surface at a height of 22,000 miles, while traveling at the speed of the earth's rotation. Therefore, any satellite placed in geosynchronous orbit would appear from the earth as a stationary object. Two of the most powerful data systems available today are Globalstar and Inmarsat. Each system boast data rates of 56 – 64 kilo bits per second (kbs.), while connected to their various data gateway terrestrial stations. Both devices will serve as a modem connection that is coupled with a laptop as the user interface. "Ref. [12]."

II. PROBLEM STATEMENT

The current WiFi approach used in the military, DoD and civilian sector does not incorporate a formal template and model to conduct GEO Satellite data transmission for client/server interaction using a Role Based Access Control (RBAC) interface. Such deficiencies create dangerous latency problems and security vulnerabilities when sending and receiving strategic information in a global setting. This is true

K. Elleithy et al. (eds.), Advances in Computer, Information, and Systems Sciences, and Engineering, 165–170.

particularly in the fight against terrorism, where a mobile global system may be the primary means of secure communications in a covert and austere environment. "Ref. [4], [7]."

III. CURRENT RESEARCH SIGNIFICANCE

A credible standardized framework to access and submit information via LEO & GEO Satellite-ATM TCP/IP networks in a global environment is the relevance of this research. The significant contribution to the field of information technology is the development and implementation of a proven methodology designed to send data in a global client to server, secure RBAC framework. Many global communication systems can benefit from architectural design templates for hardware and software requirements that maximize the data quality and throughput with minimum latency; therefore, reducing the data transmission time and enhancing the throughput of information between the client and server. "Ref. [3]."

IV. PREVIOUS IRIDIUM EXPERIMENTATION

A. Iridium Research Methodology

Research experiments have been conducted using the Iridium technology in an independently funded laboratory in Fayetteville, NC. The experiments involved active duty military Special Operations personnel from Fort Bragg NC. Such experimentation proved the Iridium data communication system is incapable of supporting the bandwidth and communication requirements of MS-Ro-BAC Network. "Ref. [8]."

B. Laboratory Configuration

Components of the research laboratory were: Various software products, 3 servers, 5 desktop nodes, 2 mobile laptop computers, 2 routers with wireless capability, private IP address, two hubs, DSL Internet capability, software, hardware for Inmarsat HSD and Globalstar Satellite connectivity. Each scenario incorporated in the research has become the standard template topology, of how to best use the technology to maximize the quality of service (QOS) over GEO & LEO satellite technology. "Ref. [7]." The researcher is a fulltime professor of computer science and has access to fully operable technology center if the primary location wasn't feasible or appropriate for the required experiments.

C. Industry Engineering Support

Design engineers of Iridium technology corporations, as well as US. Army Special Operations Communicators were instrumental in their support for this research. "Ref. [1]." The US Military is currently using Iridium SATCOM technology around the world, particularly in Iraq and Afghanistan; however, the data rate is approximately 9 kbs. and the quality of service is poor Professional interviews and real-world experience was very helpful when ensuring the experimentation and analysis process was rigorous and substantiated through the IT industry's current professionals and end users. "Ref. [8]."

D. Relational RBAC Role Inheritance

In Fig. 2. The CMDR0 is the base model design for implementing a RBAC system. CMDR1 and CMDR2 both encompass CMDR0's privileges; however, they both have independent features as well. The CMDR3 design adds the ability of inheritance to the overall model. This indicates roles can inherit permissions from other roles in the data base system as seen in "Fig. 3, 4," "Ref. [5], [9]."

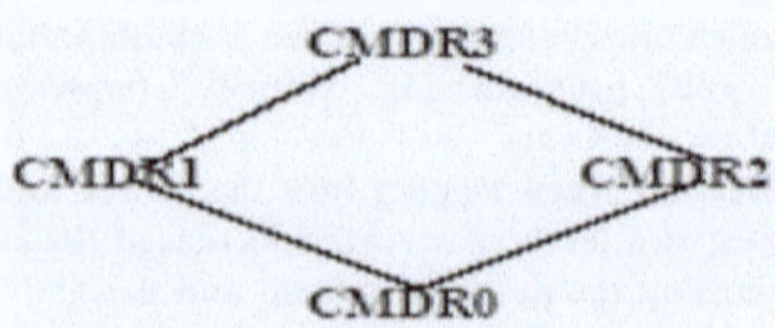

Fig. 1. Relationship model of hierarchy

E. Data Collection Process

Example Form for 10 Scenarios of: WIRED ACCESS DAY

	Sec.Min	quality	Errors	Attempts	Rating
Scenario 1	00.000	00.000	00.000	00.000	00.000
Scenario 10	00.000	00.000	00.000	00.000	00.000
AVERAGE	00.000	00.000	00.000	00.000	00.000
MODE	00.000	00.000	00.000	00.000	00.000
MEDIAN	00.000	00.000	00.000	00.000	00.000
RANGE	00.000	00.000	00.000	00.000	00.000

Fig. 2. Wireless Scenario Data Collection Form

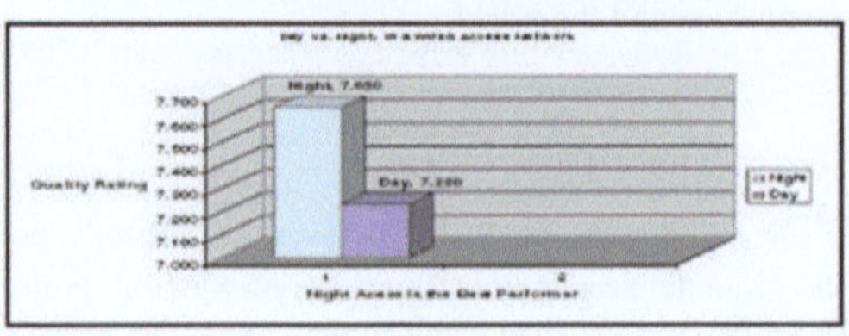

Fig. 3. *'Wireless night access,'* consistently connected at a higher QOS than day access, to the MS-Ro-BAC database network.

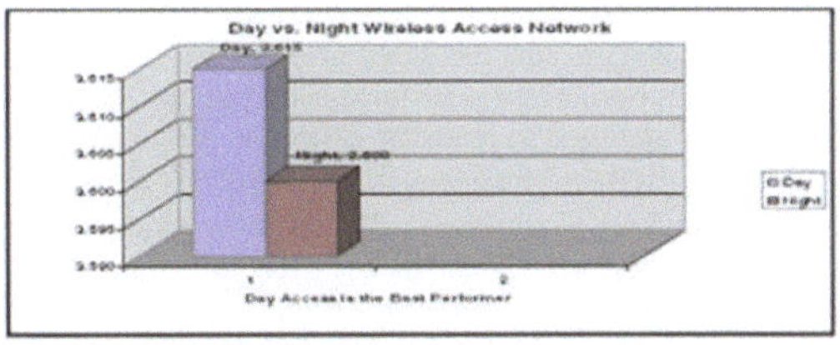

Fig. 4. *'Wired night* access,' consistently performed at a higher quality rate than '*Wireless day access,'* to the MS-Ro-BAC database network

F. *Summary of Previous Iridium Research*

This research tested a different approach to Secure RBAC in a post 911 environment. In order to defend the United States against enemies foreign and domestic; it is crucial that combat forces are equipped with the best communication equipment available. 21 years technology experience in a military environment supports this research and contributes an ideology that could revolutionize the government communication process. "Ref. [5]." The "Mobile Secure Role Base Access Control" (MS-Ro-BAC) Network is a system designed to access secure global databases via wireless communication platforms. "Ref. [2]." This research involved the development of a (MS-Ro-BAC) Database, for wireless vs. wired access connectivity. Proprietary software was developed for authenticating with the server system. 40 database access scenarios were tested during day and night hours, for performance. The Iridium Satellite gateway was ineffective as a communication service to efficiently access the database in a global strategic environment. "in press. [6]."

V. CURRENT INMARSAT & GLOBALSTAR RESEARCH

A. *Research Vision and Ideology*

The current research goals are to develop a standardized template and model to initiate data communications between a secure mobile-client and database server by satellite network. This investigation attempts to abstract the most efficient topology design and to maximize bandwidth and quality. By conducting comparison analysis research via the Inmarsat and Globalstar technology, measurable data and results can ensue; thus, supporting the international community with reusable models and templates for any satellite-ATM system to send and receive data in a global environment. "Ref. [13]." Because of the various implementation techniques to mobile connectivity using GEO & LEO satellite data communications, the quality and throughput of either technology is not utilized to its full potential.

B. *Research Inquiries*

Current experiments will satisfy a minimum of 5 questions in response to section II of this manuscript: 1. what is a credible standard for the client and server database configuration? 2. How will the client interface appear to the user while interfacing with the database? 3. What architectural model is used for hardware and software compatibility analysis? 4. What standard template(s) will measure the quality and throughput of each system? 5. What are the test scenarios for measuring the quality of GEO & LEO data communication system of the 21st Century? "in press. [13]."

C. *System Components*

The system case is lightweight and very durable where field use is applicable. A small keyboard and GUI is available to send and receive data. There are two USB ports to assist with uploading and downloading of files. A proprietary operating system (OS) that is similar to the *Microsoft* Pentium 4 processor New Generation Secure Computing Base (NGSCB) will control the mobile device. A wireless network radio will sustain MEO satellite connectivity. "Ref. [12]." Biometric thumbprint and retina scan requirements are part of the access authorization process when initiating the boot process. The device is capable of connecting directly to a static computing base that is not secure or as an independent system. Standard Wi-Fi communication electronics are standard in the hardware architecture; this authorizes the user to communicate with other MS-Ro-BAC users through satellite connectivity. This device should remain in a secure location with trusted personnel only. The immediate contribution to the computing community is the ability of this device to connect fellow users in an ad-hoc global network environment. To contact higher-level individuals, the lower level user can set flags for senior users to contact them in a hierarchical structure. "Ref. [13]." Encrypted software and hardware technology in the device require authentication with the operating systems at all times during data transmission. The proprietary software will support chat abilities, instant messaging and file transfers through secure VPN encrypted format. "Ref. [9]."

C. *Interface Device*

"Fig. 3," Illustrates the MS-Ro-BAC device physical attributes. The case measures approximately 132 sq. inches and 1.5 inches thick. Position 1 indicates the rear panel input areas for network and fiber optic connections. Position 2 indicates the USB ports. Position 3 indicates the areas for an external monitor and keyboard connection. Position 4 designates the thumbprint (T) and retina scan (R) location. Position 5 indicates various system indicators and control buttons. Position 6 depicts the satellite antenna for MEO Satellite device operations. Position 7 (C) portrays a digital camera. In the future, field commanders can securely network with higher headquarters and subordinates as soon as each individual's device has entered the MEO network and successfully authenticated their systems hardware and software. After a satisfactory handshake, the secure connection is made. "in press. [6]."

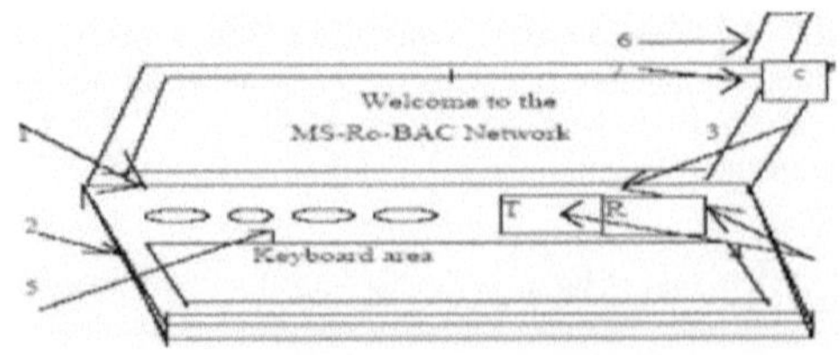

Fig. 3. MS-Ro-BAC network device

This device has three modes: (1) Deployed independently for MEO satellite connectivity from any location in the world. (2) Configured for normal unsecured use not connected to a wired or wireless network. (3) The least favorable use of the device is coupling with non-trusted static computer peripherals; keyboards, monitors and external storage devices. The preferred implementation of the device is a standalone Satellite Virtual Private Network (VPN) communication system. Fig. 4. Illustrates how each device activates and automatically authenticates through inheritance of junior roles where access is lower than the requestor's clearance levels. Once initiated, the user must submit a thumb and retina scan, and then login to the interface with user-name and password. The network software will initiate the "tracker program" that will survey the entire network for fellow MS-Ro-BAC devices and begin the handshake process. After completing the system authorization process, the user will receive a graphical user interface that depicts all activated MS-Ro-BAC devices. "in press. [6]." Standard graphics and data come standard on every machine to decrease the message size, redundancy and increase the bandwidth speed during transmission. "Ref. [9]."

D. *Communication and Authorization Process*

"Fig. 4," provides insight into the methodology of the communication process and access authorization. This design ensures participants in one classification cannot penetrate data of higher authorization levels. In distributed Compartments (DISCOM) 1, 2 and 3 the letters stand for the following: s = Subject (users, databases), o = Objects (files, etc.), p = Privileges, r = Resource Pool (CPU, computing power), h = Handles (names or code names used for users). Notice that D1 has direct connectivity to D2 and D3; this gives direct control and access to both DISCOMs. "Ref. [6]."

The access control relation is then simply the composition of these relations:

$$AC := PA \circ UA,$$

i.e.,

$$(u,p) \in AC \quad :\Longleftrightarrow \quad \exists role \in Roles: (u, role) \in UA \wedge (role, p) \in PA.$$

To further reduce the size of these relations and support additional abstraction, RBAC also has a notion of hierarchy on roles. Mathematically, this is a partial order $\geq$ on the set of roles, with the meaning that larger roles inherit permissions from all smaller roles. Formally, this means that the access control relation is now given by the equation

$$AC := PA \circ \geq \circ UA,$$

Fig. 4. Basic Inheritance Formula for RBAC Relationships. "Ref. [8]."

The proprietary software instantly reads the RBAC information of other devices and places each device in the hierarchy structure in which they have access. Therefore, if four users logged in and the head Governor (D1) was not there, each user becomes a peer-to-peer connection. In a MS-R-BAC infrastructure, the main DISCOM in Washington DC is "Big Brother" (BB). Management of lower DISCOMs is the job of lower ranking Controllers. BB has authority over every DISCOM and its individual users. BB can immediately suspend any user's rights without the permission of their local DISCOM controller or governor. BB creates an instance of itself to share information and chat with subordinate leaders. Washington, in figure 5, resides in a monitoring position. The duplicate image of BB ensures covert channels do not exist to senior DISCOMs. This code design is transparent to the users. "Ref. [3]."

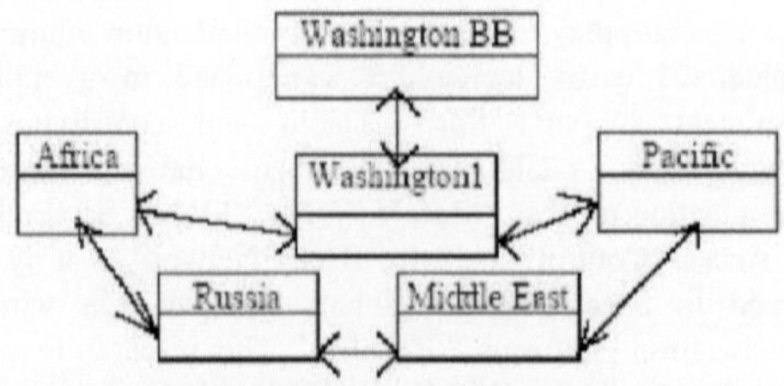

Fig. 5. B. Brother Conferencing with controllers

VI. PROPRIETARY TECHNOLOGY SYSTEMS

A. *Hardware and Software*

The MS-Ro-BAC Device will include various types of proprietary middleware, firmware and authentication programs to ensure file transfers are secure. Each device incorporates a retina and finger print scanner to identify the device user. A login name and password is required to access the systems application environment. The device includes encrypted conferencing software with integrated middleware to ensure authorized users are the only recipients and senders of secure information. The highly encrypted Object Oriented Data Module (OODM) ensures the "no write-up" restrictions of subordinates' users are enforced. Public Key Infrastructure software will digitally sign and encrypt files automatically before transmission. This dynamic approach to satellite communications allows several devices to correspond at anytime without the supervision of higher level DISCOMs. An aggressive anti-virus defense algorithm will ensure the device maintains system integrity before initiating connectivity with other devices. Once the user accesses the network, secure tracking and discovery software locates other devices available in the system. "Ref. [3]"

The software is designed for the senior most roles to acquire the permissions of their junior users. However, this ideology extends beyond inheritance of permissions. The hierarchical model is responsible for the theory of Role-sets of authorized users and permissions. Role-sets are objects grouped together under one class that authorizes multiple role positions to the selected users of that set. The permissions assigned to that role are basic and dynamic as needed by the RBAC Governor or BB. "Ref. [7]" The basic essentials in a Core RBAC interaction are: (1) Users (USERS), (2) Roles (ROLES), (3) objects (OBS), operations (OPS), and permissions (PRMS). Senior managers assign roles and permissions to each user. The user may be another device automatically working at the highest level in Washington. "Ref. [9]" and "Ref. [7]"

Fig.5. illustrates the Middle East DISCOM and four local stations within its command sector. *D1.0* is an instance of *D1*. In the diagram, a one directional arrow symbolizes the 'no write up' rule for preventing covert channels to unauthorized devices: *D1*. There is a two-way communication channel, depicted by a two-headed arrow, between *D1* and its instance *D10*. This illustrates the required procedure for *D1* to receive information from subordinate objects. Objects *D1.1*, through *D1.4* are examples of other countries in theatre: Iraq, Iran, Kuwait and Jordon. "Ref. [4]."

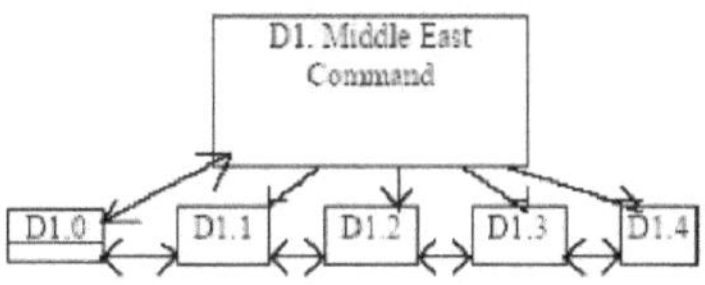

Fig. 6. A MEO network with 4 subordinates.

B. Advantages

At the time of this manuscript's publication, there isn't a communication system and network, which provides the operability of the Mobile Secure Role Based Access Control Device. The conventional and Special Operations community has not incorporated such a device as well. There are existing systems that support one or two aspects of the MS-Ro-BAC implementation; however, they are not capable of instantly linking individual users in a Virtual Private Network around the world. This device will not require a constellation of new celestial bodies; it will correlate with existing government MEO or civilian network satellites that are presently in orbit. "Ref. [11]." Information security through hardware and software authentication provides a sound way to ensure only authorized devices can receive and send data. The MS-Ro-BAC Device incorporates AI biometric systems to maintain the integrity of the authorized and unauthorized users. "Ref. [4]."

C. Disadvantages

The negative aspects of the MS-Ro-BAC system do not over-shadow the positive advantages of providing crucial information around the world in a timely manner. Lack of financial support will prohibit the adoption and full development of the global accessible system. The government's desire for a new system to perform the task that four separate machines can presently accommodate, may not be feasible. Funding has always inhibited the progress of new evolving technology. "Ref. [7]." Another disadvantage is poor reception during electrical storms or the absence of a 24 hour satellite 'foot print'. Such problems could prove disastrous for field agents or military maneuvering units. The required bandwidth and throughput to support the MIS-Ro-BAC Device would degrade the current satellite's ability to support their existing communication tasks. Most importantly, if unauthorized individuals acquire the system, it is possible to freeze the current state of all registers and reverse-engineer some aspects of the hardware and obtain classified information. However, AI security software will hopefully detect and defeat such attempts. "Ref. [11]."

VII. CONCLUSION

Continued research and development of the MS-Ro-BAC device is currently in progress, which includes proprietary software program design and testing. The significance of this research was to investigate different areas of RBAC, with the intent of producing a comprehensible proposal that will enhance the transmission and reception of information in a timely manner. The research has suggested a secure design and architectural framework for a Mobile Secure RBAC Device. The momentous principles of this manuscript are strategic security and information processing in a post 911 environment. The advantages of implementing a device with such operability would revolutionize the IT industry and secure combat forces and Department of Defense employees around the world. The MS-Ro-BAC Network will ensure portability and ease of data transfers within overt and covert government organizations or private corporations on a global scale. Critical areas of desired research and development are MEO satellite technology that will support Wi-Fi MS-Ro-BAC communications and a visual system for private viewing of classified information while in a non-classified environment. "Ref. [5], [7], [11]."

ACKNOWLEDGMENTS

This research was sponsored by Nova Southeastern University, Methodist College and the United States Army.

REFERENCES

[1] R.W. Baldwin, "Naming and Grouping Privileges to simply security" *Of the Symposium on Security and Privacy*, IEEE Pess, Los Alamitos, Calif, 2002, pp. 116-132.

[2] D. F. Ferraiolo, R. Sandhu, R, and Chandramouli. Proposed "NIST Standard for Role-based Access Control", *ACM Transactions of Information System Security, , August 2001, Vol. 4, No. 3, pp. 224-274.*

[3] S. J. Greenwald. "A New Security Policy for Distributed Resources Management & Access Control", *ACM New Security Paradigm Workshop Lake. Arrow Head CA*, 2001, pp. 4-6.

[4] T. C. House, Mobile instant secure role base access control (MIS-Ro-BAC) network, Presentation track C: *Proceedings of the IEEE Annual computer security application conference. Ma*rch 2004

[5] T. C. House, "Mobile secure role base access control device", *Proceedings of the IEEE SoutheastCon: Mobile devices and communications track. 2005,* pp. 542.

[6] T. C. House, "Client Server Access: Wired vs. wireless LEO satellite-ATM connectivity; a (MS-Ro-BAC) experiment" *CIS 2005, Part II, LNAI 3802, IEEE.* pp. 719-724, December 2005.

[6] M. Nynchama and S. Osborn. "Access Rights Administration in Role Based Security Systems". *Database Security, VIII: Status and Prospects,* 2000, pp. 37-56.

[7] G. Neuman. "Design and Implementation of a Flexible RBAC-Service in an Object Oriented Scripting Language" *ACM Workshop on Role Based Access Control,* 2001, pp. 12-18.

[8] Robert A. & Nelson, P.E. "President of satellite engineering research corporation". *Interview with a satellite engineering consulting firm,* Bethesda, MD. 23 April 2005.

[9] R. Sandhu. "Role-Based Access Control", *Proceedings of the 10th IEEE, Conference on Computer Security Applications December20,* pp. 3-6.

[10] R. Sandhu, "Role Activation Hierarchies", *Of ACM Workshop on Role-Based Access Fairfax VA,* 2001, pp. 11-12.

[11] R. Simon, And R. Zurko. "Separation of Duty in Role Based Access Control Environments", *New Security Paradigms Workshop.* 2001, pp. 11-17.

[12] D. J. Thomsen, "Role-Based application Design and Enforcement", *Database Security, IV: Status and Prospects*, 2002, pp. 151-168.

[13] S. J. Westfolds, E. Horvitz, S. Srinivase, C. Rouokangas. "A decision-theoretic approach to the display of information for time-critical decisions: The Vista project", *Proceedings of SOAR-92 Conference on Space Operations Automation and Research*, 1992, pp. 19-21.

Queueing Behavior of Hashing Methods Used to Build URL Routing Tables

Zornitza Genova Prodanoff
Department of Computer and Information Sciences
University of North Florida
4567 Saint Johns Bluff Road
Jacksonville, Florida 32224
zprodano@ unf.edu

Keywords: **CDN, HTTP, hashing, self-adjusting hashing.**

Abstract **- This study pioneers a new way to study the performance of hashing algorithms by evaluating their queueing behavior. We examine hashing as applied to URL routing used in Content Distribution Networks (CDNs). URL routing tables can be built as self-adjusting hash tables to reduce network bandwidth consumption in CDNs and at the same time take advantage of the high temporal locality of Web document requests. We show through a trace-driven simulation that the existing Aggressive hashing (based on move-to-front rule) improves upon H1 hashing (based on the transposition rule). Surprising results were found: even though the expected improvement in look-up time over H1 hashing is modest, the improvement from a queueing perspective appears to be very significant. The long range dependence in look-up time of H1 hashing contributes to its extreme queueing delay. This shows that using queueing delay as a means of evaluating hashing algorithms can be very insightful.**

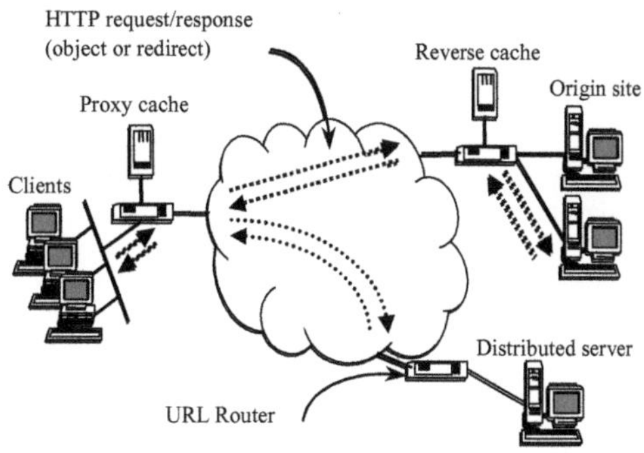

Fig. 1 **URL router forwarding or redirecting a request**

I. INTRODUCTION

Content Distribution Networks (CDNs) are overlay networks on the Web that allow for a single document or service to be contained in multiple servers and caches throughout the Internet. CDNs facilitate best server (*source*) selection for a given HTTP (Web browser) request. The criteria for best source selection include minimizing network load while satisfying a response time requirement for the client. Hence, most often the client accesses the "nearest copy" of a document or service. Client based methods (e.g., [3]) use a ping type approach to find the fastest responding server for a given request by generating excessive and redundant overhead traffic. Anycasting and centralized approaches have been studied ([5],[14]) as well, where routers or specialized services determine the best server by using high overhead or requiring infrastructure changes. A URL router based CDN [6] can be implemented overcoming these disadvantages.

The operation of a CDN is similar to Internet caching systems, where sharing of URL lists enables routing or forwarding of requests between the cache sites ([4],[13]).

Figure 1 depicts a CDN, where a URL router [6] at some distributed server intercepts a client request sent through a proxy and determines that the requested content is not available (locally). Hence, it redirects the HTTP request to the origin server, where all currently serviced content is found.

Each static or dynamic document in the Web is identified by a variable length URL string. Often a single URL maps to multiple content sources. The number of URLs at a givenserver (i.e. origin server or distributed server) is typically very large (1000's to 100,000's). Moreover, due to aging-out, documents can be discarded (i.e. the corresponding files are deleted). In contrast, IP addresses are typically few and fixed for a given Web site.

The URL routers are front-end nodes to edge-based servers and caches. A URL router does the following:

1. Establishes a TCP connection with a client (initiated by the client).
2. Receives and parses an HTTP request from a client.
3. Looks-up the requested URL in a routing table and determines the IP address of the best content source.
4. Spoofs or splices the connection with the content source if the content source is local to the URL router.

K. Elleithy et al. (eds.), Advances in Computer, Information, and Systems Sciences, and Engineering, 171–177.

5. Sends the client a redirection message containing the new URL of the content source if the content source is remote to the URL router.

The URL router must maintain a *URL routing table* from updates received from other URL routers or directly from content sources. The URL routing table is built from *URL lists* where a URL list contains the URL and the host IP addresses of the server which stores the requested by a browser Web document identified by this URL.

Sharing of URL lists, along with knowledge of network and server state, is necessary in order to build URL routing tables in the URL routers. Hence, methods of reducing the size of URL routing tables are investigated. Good methods will both reduce the size of the routing table and enable fast routing look-ups.

Cyclic Redundancy Check (CRC) codes have been investigated as a means of reducing a URL string to a 32-bit integer signature [6]. CRC32 can be computed efficiently in software or hardware. Properties of CRCs make it possible for a single CRC hardware circuit to compute CRCs over multiple fields in a packet. Thus, using CRC32 (with low computational complexity and "free" availability in existing network hardware) is motivated. The CRC32 URL signatures lead naturally to the use of hash tables for implementing a URL routing look-up. CRCs are uniformly distributed random values where any sub-sequence of bits from a CRC is also uniformly random and thus can be used as "pre-computed" hash function value. We investigate the queueing behavior of several hashing algorithms as applied to URL routing.

II. Hashing Used to Build Routing Tables

Providing for an efficient type of searching, hashing can be used to build URL routing tables. To guarantee access to data in a hash based URL routing table, collisions need to be resolved. A collision resolution mechanism is needed to make two or more records accessible from within the same hash table location.

table is $N/|hash\_indexes| = N/2^K$ (or the mean chain length), where N is the number of keys.

B. The Performance of Self-organizing Hashing

Colliding keys are placed in a linked list pointing out from the index location where the keys collided. When a (colliding) key is accessed the nodes in the list are traversed one at a time, starting with the head and advancing towards the tail of the list. It is easy to see that look up time is then proportional to the number of nodes traversed until the searched for key is found.

It has been shown that a static hashing scheme yields longer average key access times than a self-organizing scheme [37]. Let keys $k_1, k_2,..., k_N$ are located $1,2,..., N$ positions away from the list head and N is the cardinality on

Simple chain hashing is hashing method with collision resolution. An input key (URL identifier such as a CRC32 signature) is hashed into a hash index, where the hash index points to a chain of key-value pairs in the hash table. The input key is then compared to the stored keys in the hash chain and when a matching stored key is found, the value (e.g. server IP address) stored in association with that key is returned.

A. Self-organizing Hashing Methods

Self-organizing methods improve over simple chain hashing by using chained collision resolution aiming to reduce the average key access time by reordering the keys in each resolution chain on the fly. If a self-organizing method is used a key is repositioned after it is accessed so that prior knowledge of key accesses is used to reposition most frequently accessed keys closer to the list head. Requests for Web content will have very high temporal locality. This suggests that self-organizing hashes that can put popularly requested URLs at the head of a hash chain may be well suited for improving look-up time.

There are several known methods for reorganizing such colliding hash keys. In this study the queueing performance of hash methods is evaluated employing two of them: the *move-to-front* and *transposition* rules [10].

H1 hashing [15] is a method that rearranges the hash table chains based on the *transposition* rule. When a key is requested its address in the hash table is found and swapped with the address of the nearby key located next to it, one position closer to the head of the chain. As a result the chain gets rearranged, but the same hash function could be used to produce addresses on future accesses. If a single entry chain is accessed, no swap is done.

Aggressive hashing [16] is similar to H1 hashing, but is based on the *move-to-front* rule instead of transposition. In Aggressive hashing, the currently accessed record is moved to the head of its chain. The average worst case look-up time for the entire the set of keys. Then the optimal average time to do successful search for the static scheme will be

$$\overline{C}_N = 1p_1 + 2p_2 + ... + Np_N, \tag{2.2}$$

where key accesses are independent, with the probability of accesses for key k_i being p_i. If a self-organizing scheme based on the move-to-front rule is used, the average number of comparisons needed to find an item will be

$$\tilde{C}_N = 1 + 2\sum_{1 \le i < j \le N} \frac{p_i p_j}{p_i + p_j}. \tag{2.3}$$

It has been shown that $\tilde{C}_N$ is less than $\frac{\pi}{2}$ times the optimal value for $\overline{C}_N$ [12]. It has been shown that given any probability distribution (on the set of key access probabilities) and independent key accesses, an optimal reorganization scheme should employ the transposition rule [17]. In this same study chain reorganization instances are viewed as permutations on the link list nodes. Given an optimal set of such permutations exist, it is shown that the transposition rule provides the optimal number of comparisons. It is not shown however that an optimal set of request sequences (seen as permutations) exists.

Two other studies evaluated the performance of transposition and move-to-front based on amortization heuristics rather than a probabilistic proof [2],[20]. An amortization heuristic takes into consideration the worst case access times for any key access sequence rather than looking at the probabilities of accesses to keys. Rather than looking at worst case number of comparisons for a single access (which is order of list length) the worst case is considered over a sequence of accesses to several keys in turn, where the assumption that key accesses are independent is dropped. The obtained results show that when the independence assumption is dropped, move-to-front performs better than transposition even though probabilistic analyses imposing this assumption lead to concluding that move-to-front has better average search time performance. Some empirical studies have been performed to compare reorganization methods. To the best knowledge no studies have been found which examine *uniformly distributed* keys. Empirical evaluation has already been done on keys extracted from natural language text files, which have been noted to have Zipf's distribution [10]. In this study, the queueing performance of keys that are uniformly distributed – CRC32 codes – is evaluated by investigating their application to self-adjusting hash methods that use CRC32 codes as keys to build URL routing tables for CDNs.

III. EVALUATION

For a hashing algorithm, the response variable of interest is the look-up time to find a value for a given input key. This look-up time is the product of the number of memory accesses and memory access time (the "speed" of the memory). In this evaluation, memory access time was normalized to 1.0. In a hash list a single look-up requires one or more memory accesses to find the value associated with a key. For example, if the value associated with a 32-bit key is the second entry in a Simple hash chain then two memory accesses are required: one access (and comparison) to determine that the head of the chain is not the key; and a second access (and comparison) to determine that the second entry is a match to the key. The value associated with the second entry in the table is then read. The size of the key, its hash index, and associated value also affect the number of memory accesses. In this evaluation, a 32-bit wide memory was assumed. Thus, if a key was of size 32-bits only one memory access was needed to compare it to stored keys, and two memory accesses are needed to traverse and compare a single chain node. A larger key would require multiple memory accesses for comparison with the stored keys. A URL list that uses 32-bit signatures to represent the URL strings (the keys) reduces memory accesses when comparing keys for a match. Hash table size is measured in number of chains.

The three hash chaining algorithms were implemented in the C programming language with a single main program and three called functions – each function a different hashing algorithm. The hash table was always stored in main memory using dynamically allocated memory (that is, from a C code `malloc()` statement). The main program took as input:

- Hash table size (K),
- File name of a URL list to populate the hash, and
- File name of an URL access list of keys to look-up in the hash table.

The URL lists existed in both URL string and signature formats. The URL access lists contained either a URL string or URL signature for the key. The program used the URL access lists to look-up values in the created hash tables where the number of memory accesses was counted for each key looked-up. The simulation ensures that all look ups were successful. That is, all looked up keys are known in advance and they are keys already stored in the hash. For a given hash table size, URL list, and URL access list the minimum, maximum, mean, and standard deviation of the number of memory accesses were determined and reported.

A URL router can be modeled as a single server queue where the look-up is the service. Where hashing is used as the look-up method, the hashing algorithm defines the service center for the queue. Figure 2 shows a single-server queue where the service center is the hash table look-up (based on the number of accesses in the hash table) and the arriving customers are keys. The departing customers are the values resulting from the hash table look-up. The input to the queueing model are interarrival times (from the access list or exponentially distributed as described below) and number of memory accesses for a given URL associated with the interarrival time. The number of memory accesses was generated from the experiments in the previous section. This models a URL router where the hash table look-up is the service time for a request to be re-directed. The goal of the queueing study is to evaluate the performance of the hashing algorithms with respect to application in a queueing system – such as a URL router. This is a new way of studying the behavior of a hashing algorithm.

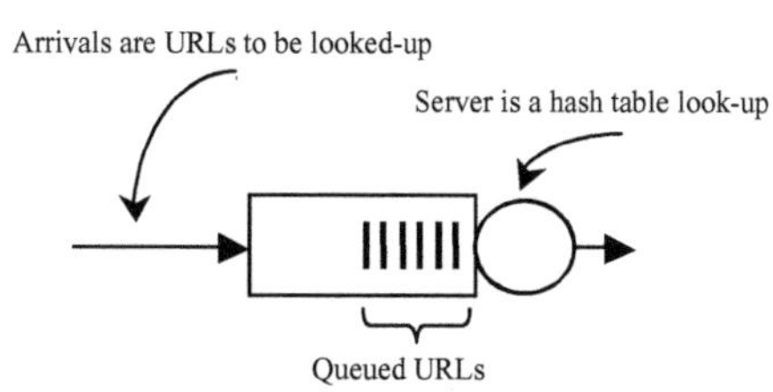

Fig. 2 Single-server queueing model for hashing algorithm evaluation

The source code for this evaluation is using the CSIM18 simulation model. CSIM18 is a process-oriented simulation engine that is a C function library [19]. Thus, CSIM18 models are written in C.

This source code models a trace-driven single-server queue. The `generate()` process reads the tuples from the trace file, holds for an interarrival time, and then queues the service demand (service time is proportional to the number of accesses) to the `queue1()` process. The `queue1()` process reserves, holds, and releases a server facility. CSIM18 maintains statistics counters internally and the final statistics results are output at the end of the main program (named `sim()` in CSIM18 programs).

For a single server queue, the response variable of interest is queue length. Given a queue length and a known arrival rate, the queueing delay can be directly obtained from Little's Law

$$W = \frac{L}{\lambda} \quad (1)$$

where L is mean queue length, λ is mean arrival rate, and W is mean wait or delay). The utilization of a queue, ρ (also U), is the arrival rate divided by service rate, μ,

$$\rho = U = \frac{\lambda}{\mu} \quad (2)$$

Queue length is affected by the mean, variance, and correlation properties of the customer interarrival times and service times. Buffer size was infinite in all simulation models. The control variables for the hashing algorithms evaluated for queueing behavior are the same as for the look-up time evaluation with the addition of the following:

- Utilization (U) of the server
- Request interarrival time distributed as such:
  - Exponentially distributed and
  - Truncated maximum interarrival time ($T_{\max}$) from the URL access list
- Ordering of requests (unshuffled or shuffled)

The utilization of a server is controlled by the modeled time for a memory access. For a desired target utilization, U, the modeled memory access time, T_m, is determined as follows. For a given access list the total time for all accesses to arrive, T_{all}, is known (T_{all} is the sum of the interarrival times in the URL access list). The total number of look-ups, N_{total}, required to look-up all the keys in the access list is known from the previous look-up time evaluation. Then,

$$T_m = \frac{U \cdot T_{all}}{N_{total}} \quad (3)$$

where T_m is set as a control variable in the simulation model to achieve a desired server utilization, U. A method of truncating interarrival times was used to control the degree of burstiness of the interarrival times (measured by the coefficient of variation) in the URL access lists. By controlling the degree of burstiness, the queueing evaluation could be run for greater utilization levels. This method entailed setting a maximum interarrival time, $T_{\max}$. All interarrival times greater than $T_{\max}$ were truncated to $T_{\max}$. This resulted in reducing the burstiness of the interarrival times as measured by the coefficient of variation. Shuffling is used as a means of breaking long range dependence (LRD) in the order of accesses. Shuffling is implemented by selecting each entry in an URL access list and swapping it with a randomly chosen entry in the list. This shuffling is repeated for several passes through the URL access list. LRD is well-known to affect queueing delay [12]. Autocorrelation is a measure of LRD. A slowly declining autocorrelation signifies the existence of LRD.

A. Access Lists Used in the Evaluation

For the evaluation of the simple, H1, and Aggressive hashing algorithms, two representative access logs that contained time-stamps (of access interarrival times) were used. The NASA [7] and Clark [8] access logs, were chosen due to their large size and use in other studies (that is, [1]). Tables 1 and 2 summarize the two access logs. In Table 1 the list size (1) is for the full URL string and list size (2) is for CRC32 signatures. The size of the CRC32 signature table is the number of URLs multiplied by four bytes. Tables 2 shows that the NASA access list represented one month of activity and the Clark list one week of activity. These access logs are representative of requests in a CDN where content can be distributed to both servers and caches. The large size of the logs (many millions of accesses) and time of collection are significant.

The three hashing algorithms were compared for queueing delay. The HTTP access and URL lists were used only in signature format. For the queueing evaluation, only the NASA access list was used since the results in the previous section for the two lists were very similar. For the experiments, the control variables were hash table size (K), utilization of the queue (U), distribution of requests interarrival times, and ordering of requests. For distribution of interarrival times a maximum interarrival time ($T_{\max}$) is used in some experiments. In other experiments, the interarrival times are synthetically generated from an exponential distribution. The synthetically generated interarrival times replace the interarrival times taken from the access list in the trace file. The order of requests is used in two ways: as taken from the access list (unshuffled) or shuffled to break long range dependence in request order.

TABLE 1 Characterization Of HTTP Access Lists For Size

Access list name	# of accesses	# of URLS	Mean URL length	List size (1)	List size (2)
Clark	1,673,750	37,266	35.06 bytes	1,306,546 bytes	149,064 bytes
NASA	1,569,898	15,700	35.58	558,606	62,800

TABLE 2 CHARACTERIZATION OF HTTP ACCESS LISTS FOR TIME

Access list name	Duration	Mean interarrival time	Stddev interarrival time
Clark	605,776 sec	0.36 sec	5.62 sec
NASA	2,592,007	1.65	39.79

Figure 3 shows the autocorrelation for the NASA access list for the number of accesses for the three algorithms. The autocorrelation was computed as $r(k)$ for a lag k.

$$r(k)\frac{E[(x_i-\mu)(x_{i-k}-\mu)]}{\sigma^2}, \quad (4)$$

where x_i is the value that the random variable x takes at time i, μ is the mean number of look-ups for the evaluated time series, and σ^2 is the variance of the number of look-ups. The results in Figure 3 show that Simple chain hashing and H1 hashing have a much higher autocorrelation than does Aggressive hashing.

Experiment #1: Evaluation of the effect of K on queue length for a fixed service rate such that Simple chain hashing is U = 80% and exponentially distributed interarrival times of requests. Mean queue length is measured.

Experiment #2: Evaluation of the effect of T_{max} on queue length for a fixed service rate such that Simple chain hashing is U = 80% and K = 8. Mean queue length and the utilization of H1 and Aggressive are measured.

Experiment #3: Evaluation of the effect of T_{max} on queue length for a fixed U = 80% and K = 8. Mean queue length is measured.

Experiment #4: Evaluation of the effect of autocorrelation (unshuffled and shuffled ordering of requests) on queue length L for a fixed U = 80% and fixed hash table size K = 8. Mean queue length is measured.

Experiment #5: Evaluation of the effect of autocorrelation (unshuffled and shuffled ordering of requests) on queue length for a fixed service rate such that simple is U=80% and fixed hash table size K=8. Mean queue length is measured.

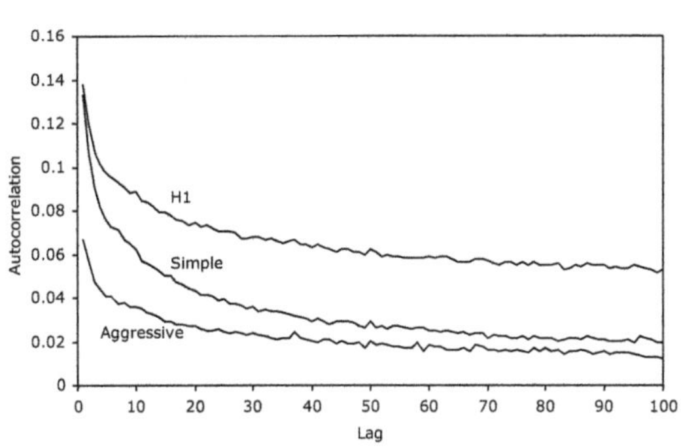

Fig. 3 Autocorrelation for 100 lags

IV. RESULTS

Results from Experiment #1 are shown in Figure 4. For smaller hash table sizes (K=8 and K=10), resulting in dense hash tables, Aggressive hashing outperforms the other methods: Simple chain (factor of 27) and H1 (factor of ~2). H1 performs 13 times better than Simple chain hashing. When the mean chain length for all hash index positions gets closer to 2 the performance of all methods is close. With increase in K all methods converge as expected.

Results for experiment #2 are shown in Figure 5. Increase in burstiness levels for all methods resulted in increase in mean queue length. The mean queue length for Simple was magnitudes greater that that of H1 and Aggressive, ranging from 5500 to 34,000 requests. The observed utilization levels for H1 were about 22%, while Aggressive resulted in a drop down to 18%.

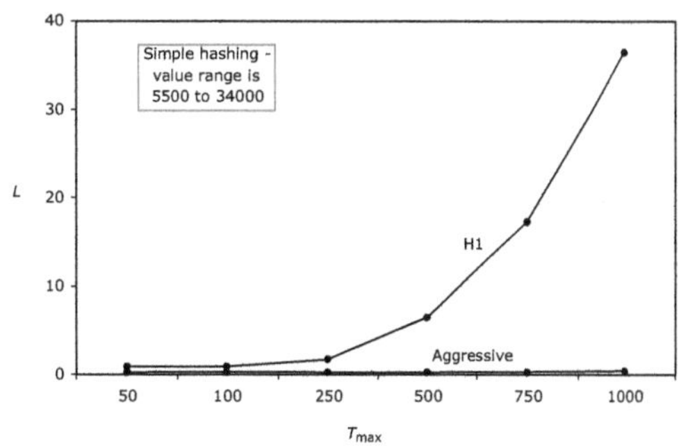

Fig. 4 The effect of hash table size (2^K) on mean queue length (L)

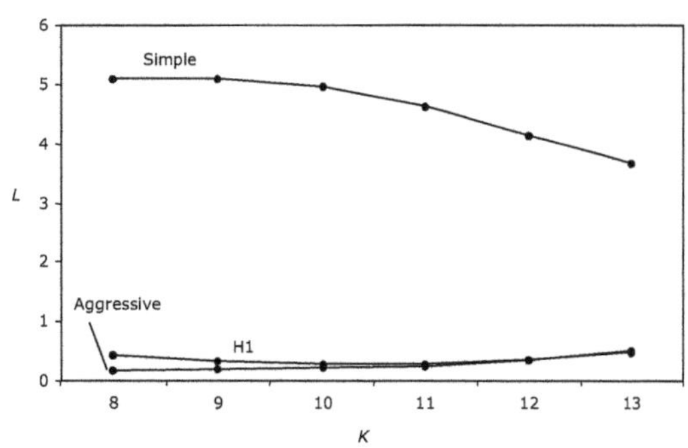

Fig. 5 The effect of burstiness (T_{max}) on L for K=8

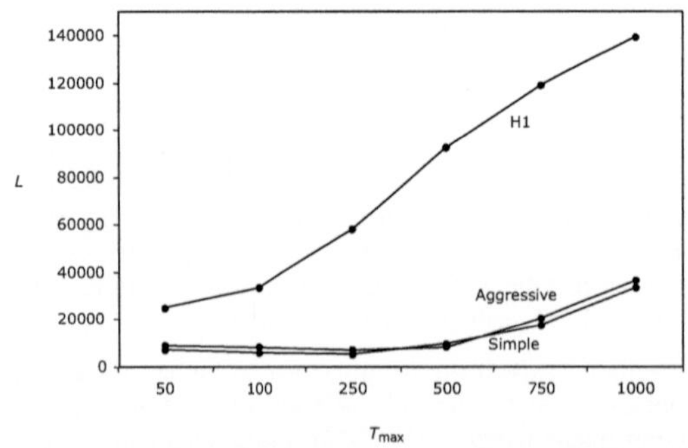

Fig. 6 The effect of $T_{\max}$ on *L* for *K*=8 and *U*=80%

The results from experiment #3 are shown in Figure 6. H1 resulted in larger (by a magnitude) and faster increasing mean queue length than Simple and Aggressive hashing. This outcome is surprising, because H1 and Aggressive produce comparably smaller mean look-up delay as compared with Simple chain hashing, but the queueing performance of Aggressive is much better than that of H1. This outcome indicates that there are other factors than mean number of look-ups that determine mean queueing delay and motivates the next two experiments.

Table 3 shows the results for experiment #4. The results include a theoretical M/G/1 queueing result for *L* using the Pollaczek-Khinchin formula [9]:

$$L = \frac{2\rho - \rho^2 + \rho^2 C_s^2}{2(1-\rho)} \quad (5)$$

where $\rho = U = \lambda/\mu$ and C_s is the coefficient of variation of the service time. As seen in Figure 5 the M/G/1 results agree very closely with the shuffled results showing that the shuffling of the ordering resulted in an independent series.

Table 4 shows the results for experiment #5. The greater efficiency of H1 and Aggressive results in a much lower utilization than Simple chain hashing. H1 is about four times more efficient than Simple chain hashing (that is, four times less look-ups are needed to find a value) and Aggressive is a factor of 1.7 as efficient as H1. At the lower utilizations, mean queue lengths are very small as would be expected.

TABLE 3 MEAN QUEUE LENGTH RESULTS FOR EXPERIMENT #4

Algorithm	unshuffled	shuffled	M/G/1
Simple	5.20	3.15	3.13
H1	29102.01	8.58	8.57
Aggressive	294.09	9.93	9.76

TABLE 4 MEAN QUEUE LENGTH RESULTS FOR EXPERIMENT #5

Algorithm	*U*	Unshuffled	Shuffled
Simple	80.0%	5.20	3.15
H1	21.7	0.43	0.36
Aggressive	12.9	0.19	0.18

Aggressive hashing improves upon H1 hashing. The improvement in look-up time was modest, a factor of about two for both the evaluated traces. As expected with an increase in the hash table size, the mean hash chain length for all methods decreases. This result is intuitive, since a larger index space will result in decrease in mean number of collisions. The queuing performance of H1 and Aggressive hashing for sparse hash tables is also similar. When the number of hash index locations is sufficiently large to prevent collisions of more than two keys at an index, both self-adjusting methods will have the same mean chain length and exactly the same look-up performance. A trend of merging the performance of these two methods with increase in table size was an expected observation since chain node rearrangement for chains shorter than two nodes in length is the same. Simple chain hashing will have the same performance as the other two methods if there are no collisions in the hash table. This will occur when the size of the index space matches the cardinality of the CRC32 space. Hence, a trend of merging the performance of all methods with increase in table size was an expected result, as well, since chain node rearrangement for chains shorter than two nodes in length is the same.

In the second set of queuieng evaluation experiments, where table size is fixed to match a densely populated table, and traffic burstiness varies, it was shown that using queueing as a means of evaluating hashing algorithms can be very insightful. This simulation–determined difference in mean look-up time between H1 and Aggressive hashing was 1.7 times for a fixed utilization of 80%. The difference in mean queueing delay was 6 times. When the requests were shuffled, all the hashing algorithms resulted in M/G/1 predicted queueing delay. This demonstrates that autocorrelation plays the role in queueing delay in excess of that predicted by the M/G/1 model and, very significantly, makes H1 hashing very unsuitable for use in routing tables. It is a remarkable result that only a very slight difference in mean look-up time can "hide" such a considerable difference in mean queueing delay. For any hashing implementation that is part of a queueing system, it is very important to evaluate queueing behavior. This paper has shown this for the first time.

V. CONCLUSION

Content Distribution Networks (CDNs) are a rapidly growing service on the Internet. CDNs distribute and co-locate content to be geographically close to the users (of the content). This content mirroring 1) reduces the load on the origin server, 2) reduces traffic on the Internet, and 3) improves response time to the users. For CDNs to be feasible, methods of routing of HTTP requests originating from users are needed. URL routers need routing tables similar to IP routing tables in IP routers. However, the methods used for IP routing cannot scale to URL routing.

A. Specific Contributions from This Research

This study has addressed how to build compact and efficient (in look-up) URL routing tables by using self-adjusting hash tables. The key contribution of this paper, however, extend beyond the scope of URL routing. This study pioneers a new way to study the performance of hashing algorithms - by evaluating their queueing behavior. Surprising results were found in how the existing H1 hashing algorithm has very high autocorrelation in look-up times resulting in mean queue lengths two to three magnitudes greater than that of Aggressive hashing. This magnitudes difference in performance would have not been recognized, if the queueing behavior of these two methods were not studied, since the mean look-up delay performance of both methods has been previously shown to be of the same magnitude.

B. Directions for Future Research

The use of signatures based URL routing tables can have applications beyond URL routing. For example, Napster-like peer-to-peer applications [18] that use a centralized directory for locating distributed content are in need of methods to reduce directory size and look-up time. Large directory size resulting in long look-up times caused a performance bottleneck at the centralized Napster server in 2000 [11]. Peer-to-peer networks without centralized directories that share directories are limited in scalability by the amount of traffic they generate. A future direction of research is thus investigating the application of signatures to open problems in improving the scalability of peer-to-peer networks.

ACKNOWLEDGMENT

The author would like to thank Kenneth J. Christensen from the University of South Florida for providing invaluable insights on the topics investigated in this study as well as improving the overall quality of this work.

REFERENCES

[1] M. Arlitt and C. Williamson, "Web Server Workload Characterization: The Search for Invariants", *Proceedings of the ACM SIGMETRICS*, pp. 126-137, April 1996

[2] J. Bentley, C. McGeoch, "Amortizing Analysis of Self-organizing Sequential Search Heuristics," *Communications of the ACM,*, Vol. 28, No. 4, pp. 404-411, 1985

[3] R. L. Carter, M.E., "Dynamic server selection in the Internet," *Proceeding of Third IEEE Workshop on the Architecture and Implementation of High Performance Communication Subsystems (HPCS95)*, 1995, URL: http://citeseer.nj.nec.com/crovella95dynamic.html

[4] L. C. Fan, P.; Almeida, J.; Broder, A.Z., "Summary Cache: a scalable wide-area Web cache sharing protocol," *Proceedings of SIGCOMM'98 Conference*, Vancouver, BC, Canada, 1998

[5] Z.-M. B. Fei, S.; Zegura, E.W.; Ammar, M.H., "A novel server selection technique for improving the response time of a replicated service," *Proceedings of IEEE INFOCOM '98*, San Francisco, CA, USA, 1998

[6] Z. Genova and K. Christensen, "Using Signatures to Improve URL Routing," *Proceedings of IEEE International Performance, Computing, and Communications Conference*, pp. 45-52, April 2002

[7] HTTP requests to the NASA Web server, URL: ftp://ita.ee.lbl.gov/traces/

[8] HTTP requests to the ClarkNet Web server, URL: ftp://ita.ee.lbl.gov/traces/

[9] L. Kleinrock, *Queueing Systems: Theory, Volume 1*, John Wiley & Sons; 1975

[10] D. Knuth, *The Art of Computer Programming: Sorting and Searching*, Vol. 3, Second Edition, Addison Wesley Longman, 1998

[11] J. Kurose and K. Ross, *Computer Networking: A Top-Down Approach Featuring the Internet*, Second Edition, Addison Wesley, 2003

[12] W. Leland, M. Taqqu, W. Willinger, and D. Wilson, "On the Self-Similar Nature of Ethernet traffic (Extended Version)," *IEEE/ACM Transactions on Networking*, Vol. 2, pp. 1-15, February 1994

[13] B. S. N. Michel, K.; Reiher, P.; Lixia Zhang, "URL forwarding and compression in adaptive Web caching," *Proceedings of IEEE INFOCOM 2000,. Nineteenth Annual Joint Conference of the IEEE Computer and Communications Societies*, Tel Aviv, Israel, 2000

[14] K. Moore, J. Cox, and S.Green, "SONAR — a network proximity service", IETF Internet-Draft, 1996, URL: http://icl.cs.utk.edu/projects/sonar/draft-moore-sonar-03.txt

[15] L. Pagli, "Self-adjusting hash tables," *Information Processing Letters*, Vol.21, Iss. 1, pp. 23-25, July 1985

[16] Z. G. Prodanoff and K. Christensen, "Managing Routing Tables for URL Routers in Content Distribution Networks," *International Journal of Network Management*, vol. 14, iss. 3, pp. 177-192, 2004

[17] R. Rivest, "On Self-Organizing Sequential Search Heuristics," *Communications of the ACM*, Vol. 19, pp. 63-67, 1976

[18] S. Saroiu, P. K. Gummadi, and S. D. Gribble, "A measurement study of peer-to-peer file sharing systems," *Proceedings of Multimedia Computing and Networking*, San Jose, CA, January 2002

[19] H. Schwetman, "CSIM18-The Simulation Engine," *Proceedings of the 1996 Winter Simulation Conference*, pp. 517-521, December 1996

[20] D. Sleator, R. Tarjan, "Amortized Efficiency of List Update and Paging Rules," *Communications of the ACM,*, Vol. 28, No. 2, pp. 202-208, 1985

FASMAC: A Low Latency and Energy Efficient MAC Protocol for Wireless Sensor Networks

Ali Abu-El Humos[1] and Bassem Alhalabi[2]
[1]Department of Computer Science
Jackson State University, Jackson, MS 39217
E-mail: ali.a.humos@jsums.edu
[2]Department of Computer Science & Engineering
Florida Atlantic University, Boca Raton, FL 33431
E-mail: alhalabi@fau.edu

Abstract

In this paper, we propose an enhancement on the SMAC protocol to improve its latency performance without compromising its energy consumption. The original SMAC provides an efficient solution for the energy consumption problem in Wireless Sensor Networks (WSNs) due to idle listening, but it has some latency problem especially in low duty cycle applications. TMAC was proposed to further reduce the energy consumption in SMAC and introduced the FRTS packet to solve the *early sleep problem* observed in TMAC. Later, Adaptive SMAC was proposed to reduce the latency problem in SMAC by at least 50% at light traffic load. Our enhanced protocol, FASMAC, combines the advantages of both adaptive listening and the usage of FRTS packet in TMAC to further reduce the latency of SMAC. In FASMAC, a packet can travel at least three hops away from its source node within one time cycle. This results in at least 67% reduction in latency at light traffic when compared with the original SMAC.

1. Introduction

The wide spectrum of applications that require remote control and monitoring made the research area of Wireless Sensor Networks (WSNs) attractive to many researchers. Military, health care, homeland security, marine science, habitat monitoring, road traffic monitoring and home automation are just some of many applications where WSNs have been the target platform. This would not have been possible without the latest advancement of enabling technologies, where microprocessors, sensing transducers, and transceivers are mass produced at an incredibly low cost and small size [4]. WSNs are, therefore, becoming the cornerstone to accomplish the ultimate goal of ubiquitous computing.

The main requirements of a traditional wireless network MAC protocol are per-node fairness, low delay and high throughput. Network lifetime is not considered critical, as battery might be replenished, although power-efficient schemes are recommended. In contrast, for a WSN, an energy-efficient design is considered as the top priority. The design should also be scalable in order to accommodate large networks. Latency and bandwidth utilization become secondary goals, while per-node fairness is not considered a critical requirement. The main requirements of a MAC protocol for WSNs can be summarized as follows [10, 11]:

Energy efficiency. The most important requirement of a MAC protocol is to be energy-efficient. Power consumption is a critical issue in WSNs since sensor nodes are battery operated and replacing or recharging the battery is usually infeasible.

Scalability. A MAC protocol should adapt and run properly in a dynamic and scalable topology. The WSN is expected to contain a large number of sensor nodes. Moreover, nodes may be added, die, or move to different locations.

Latency. Depending on the application, some latency delay can be tolerated in general. The latency and energy conservation are two conflicting attributes in the MAC layer design for WSN. Choosing the shortest path without considering the current energy level of intermediate nodes might prematurely deplete power resources of some nodes and trigger network partition.

Fairness. Usually fairness is not considered as a critical parameter since all the nodes in a WSN serve the same application. For example, when nodes in the same vicinity have similar data to transmit to the upper stream nodes, unfairness among nodes in the same vicinity is acceptable.

Bandwidth utilization. Designing a MAC protocol that efficiently uses the bandwidth is a desirable factor to enhance the WSN performance.

2. Related work

MAC protocols for wireless sensor networks can be divided into two broad categories: scheduled and contention based protocols.

The scheduled based protocols are basically TDMA based protocols such as LEACH [3] and HIT [1]. TDMA based protocols have the advantage of providing nodes with maximum possible sleeping time in which nodes consume

K. Elleithy et al. (eds.), Advances in Computer, Information, and Systems Sciences, and Engineering, 179–183.

negligible amount of power. However, if the number of nodes in the network is huge, the network must be divided into clusters, and each cluster manages its own nodes schedule. Intra-cluster communication is easily handled, but inter-cluster communication is very hard to handle. Also, if the number of nodes changes (nodes added or nodes removed) it is very hard to modify the network schedule to accommodate these newly added/removed nodes. Thus, the TDMA based protocols have the disadvantage of not being scalable and adaptive to the network changes, where contention based protocols provide much better scalability and can easily accommodate (adapt to) changes in the network.

The contention based MAC protocols are basically CSMA based protocols where nodes contend for the medium first before transmitting their data. The IEEE 802.11 [12] is optimized for Ad hoc wireless network, but cannot be adopted for WSNs because it has the idle listening problem, which is a major source of energy waste. SMAC [10, 11] was developed for WSN to overcome the problem of idle listening in the IEEE 802.11. However SMAC has the following drawbacks: The SYNC overhead packets are required to establish and maintain synchronization among nodes which may cause multiple synchronization schedules to develop in the network. The sleep delay is also a major drawback in SMAC. TMAC [8] was introduced after SMAC to further save energy by turning the nodes off during their listen time if no activity is noticed in the network. This approach enhanced the energy savings in SMAC but yet caused the early sleep problem for which the authors of TMAC proposed the FRTS packet as a remedy. Adaptive SMAC [11] was later introduced to alleviate the latency problem in SMAC. It reduced the latency of SMAC by 50% at light traffic conditions. WiseMAC [2] enhanced the energy savings in SMAC by bypassing the broadcast of SYNC packets and eliminating the need for the RTS and CTS control packets. Latency is still a problem in WiseMAC since nodes are not synchronized to wakeup together. DMAC [6] reduced the latency of SMAC significantly by staggering the wakeup/sleep schedule of the nodes in the data gathering tree according to their depth in the tree. DMAC works fine only under the specific data gathering tree scenario for unidirectional communication flow from multiple sources to a single sink [6]. The work in [5] addressed the multiple schedules problem in SMAC by introducing the global scheduling algorithm. It also makes the latency of SMAC close to the No sleep protocol by applying the fast path algorithm to the network. In [9] multiple CSMA scenarios were compared. Also a flow control technique was introduced to achieve fair bandwidth allocation between all nodes in the multi-hop network. In [7] a two-radio scheme MAC protocol was proposed to save energy consumption in WSNs. The two-radio schemes, however, are undesirable because of the hardware complexity added by requiring two radios on the sensor node board.

3. FASMAC Protocol

SMAC with adaptive listening reduced the latency of SMAC by at least 50% at light traffic [11]. This is explained in

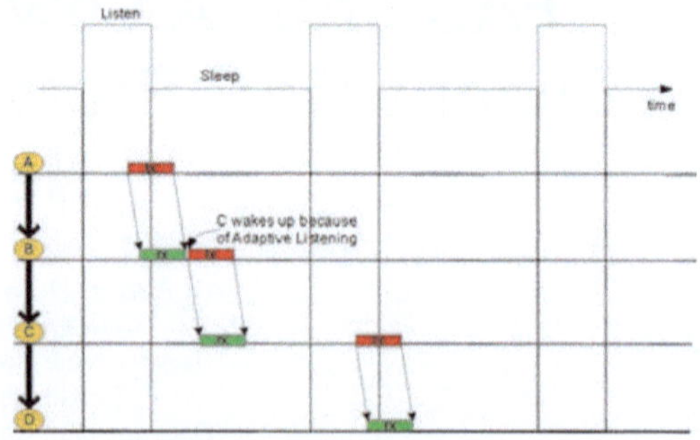

Figure 1: In adaptive SMAC, a packet can travel two hops away from its source node within one time cycle

Figure 1 where a packet sent from node A can travel two hops away from A within one time cycle. Notice that, in order to further reduce the latency, the FRTS packet (introduced in TMAC) is incorporated into SMAC with adaptive listening where a packet sent by node A can now travel a third hop farther away from its source (node A) within the same time frame as shown in Figure 2. Figure 3 explains how the FRTS packet can be incorporated into adaptive SMAC so that node D wakeup exactly when node B transmission to node C is over, and hence node C can forward the received packet to node D within the same time frame. Notice that when node C overhears the CTS packet sent from node B to node A, it will immediately send the FRTS packet to node D so that node D can schedule itself to wakeup later on to receive the packet from node C. Node A will pause (wait) after receiving the CTS packet from node B for a duration of time equal to the time required to transmit the FRTS packet to avoid data collision at node B with the FRTS packet transmitted by node C. After pausing (waiting) for FRTS packet duration, node A can send its data immediately to node B. Once this transmission is over, node C will wakeup (due to adaptive listening) and node B can start its transmission to node C.

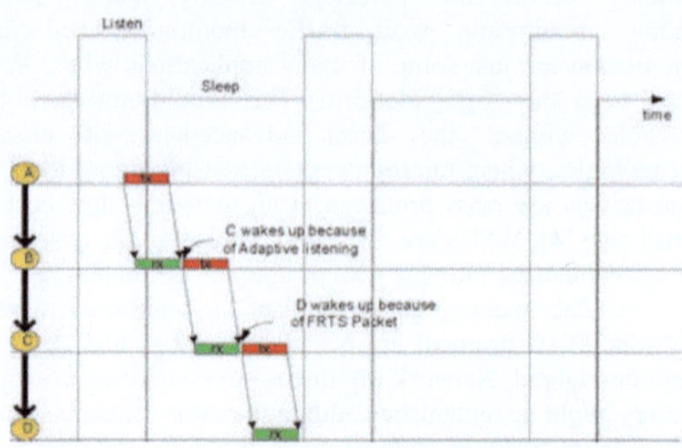

Figure 2: Latency can be reduced further if FRTS packet is added to adaptive SMAC

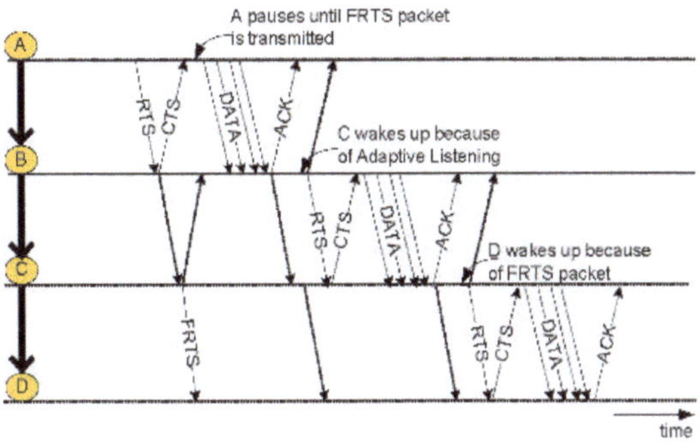

Figure 3: FRTS packet is incorporated into SMAC with adaptive listening

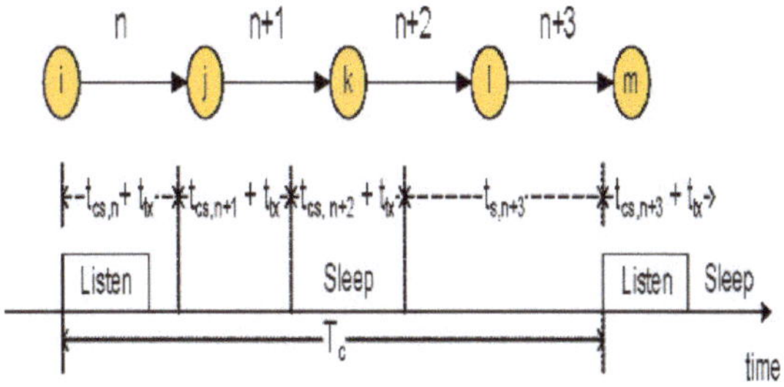

Figure 4: Adaptive Listening and FRTS packet can reduce SMAC latency by 67%

When B transmission to C is finished, node D will wakeup (This time, due to FRTS packet) and start receiving node C transmission. However, when node C transmission to node D is over, node D cannot forward its packets to its upper stream node because it is going to be sleeping. In this case, node D has to wait to the next frame cycle to forward its packets to its upper stream node.

3.1 Latency Analysis

In this analysis, we will follow the same methodology used in [11] to show the latency enhancement of FASMAC. Under light traffic, there is no queuing delay and backoff delay. Also, the propagation delay and processing delay are ignored. Only carrier sense delay, transmission delay and sleep delay are considered. If there are N hops from source to sink, $t_{cs,n}$ represents the random carrier sense at hop n with mean value equals to t_{cs} [11]. The transmission packet delay is constant and is equal to t_{tx}. It is shown in [11] that the average latency over N hops in MAC without sleep is:

$$E(\,D(N)\,) = N(t_{cs} + t_{tx}\,) \quad (1)$$

It is also shown in [11] that the average latency of SMAC without adaptive listening is given by:

$$E(\,D(N)\,) = N_{Tc} - Tc/2 + t_{cs} + t_{tx} \quad (2)$$

Where:

$$Tc = \text{listen time} + \text{sleep time} \quad (3)$$

We are assuming that $Tc >> t_{cs} + t_{tx}$ (which is the case in low duty cycle applications), thus periodic sleep introduces additional latency at each hop [11].

It is also shown in [11] that the average latency of SMAC with adaptive listening is given by:

$$E(\,D(N)\,) = NTc/2 + 2(t_{cs} + t_{tx}) - Tc/2 \quad (4)$$

Equation 4 shows that adaptive listening reduces the latency of SMAC by almost 50% at light traffic load.

In order to analyze FASMAC latency, we have first to make the following two assumptions:

The duration of FRTS packet is very small when compared to t_{tx}, and

$$Tc >> t_{cs,n} + t_{cs,n+1} + t_{cs,n+2} + 3\,t_{tx} \quad (5)$$

From figure 4, it can be shown that:

$$Tc = t_{cs,n} + t_{tx} + t_{cs,n+1} + t_{tx} + t_{cs,n+2} + t_{tx} + t_{s,n+3} \quad (6)$$

Where $t_{s,n+3}$ is the sleep delay at hop n+3.

From Figure 4 we can see that at node i the delay is $t_{cs,n} + t_{tx}$. Because of adaptive listening the delay at node j is $t_{cs,n+1} + t_{tx}$. When node k receives a packet from node j, it will immediately transmit it to node l, since node l will be awake because it had previously received the FRTS packet from node j. At node k the delay is $t_{cs,n+2} + t_{tx}$. Node l cannot forward the packet immediately to its upper stream node m because node m is sleeping and has to wait for the next frame cycle. Thus, the delay at node l is $t_{s,n+3}$.

Therefore, the latency over N hops can be calculated as

$$\begin{aligned} D(N) = {} & t_{s,1} + t_{cs,1} + t_{tx} + t_{cs,2} + t_{tx} + t_{cs,3} + t_{tx} + t_{s,4} \\ & + t_{cs,4} + t_{tx} + t_{cs,5} + t_{tx} + t_{cs,6} + t_{tx} + t_{s,7} \\ & + \ldots + t_{cs,N-2} + t_{tx} + t_{cs,N-1} + t_{tx} + t_{cs,N} + t_{tx} \end{aligned} \quad (7)$$

Substituting 6 into 7 and simplifying will result

$$D(N) = t_{s,1} + (N/3 - 1)\,Tc + t_{cs,N-2} + t_{cs,N-1} + t_{cs,N} + 3\,t_{tx} \quad (8)$$

Assuming the sleep delay at the first hop $t_{s,1}$ is a uniformly distributed random variable with expected value $Tc/2$, therefore, the average latency over N hops is

$$E(\,D(N)) = (N/3)\,Tc + 3\,t_{cs} + 3\,t_{tx} - Tc/2 \quad (9)$$

Equation 9 shows that adding FRTS packet to SMAC with adaptive listening reduces the latency of SMAC by almost 67% at light traffic load.

4. Performance Evaluation

We used the ns2 simulator [13] to simulate our proposed protocol. Our experiment setup was a multi-hop network with one source and one sink communicating over seven hops as shown in Figure 5.

Figure 5: Simulation Topology: 7-hop linear network with one source and one sink

Each node is within the transmission range of its neighboring nodes i.e. node 2 can send/receive packets to/from node 1 and send/receive packets from/to node 3. We assume that each node knows in advance the routing to the final destination by using the NO Ad-Hoc Routing Agent (NOAH) routing package [15]. We tested different traffic scenarios: light, medium and heavy traffic. For each scenario, twenty packets were generated at the source node (node 1) destined to the sink node (node 8). Four different protocol scenarios were compared: IEEE 802.11-like protocol (No sleep), SMAC without adaptive listening, SMAC with adaptive listening, and FASMAC, in terms of energy consumption and latency. Table 4.1 summarizes the radio parameters used in our simulation. Notice that, the txPower, rxPower and the idlePower are the same values of the TR3000 radio transceiver from RF Monolithics [14].

Table 1: Radio characteristics and parameter values

Radio bandwidth	20Kbps
Radio Transmission Range	22m
Radio Interference Range	139.6m
Packet Length	50Bytes
Transmit power (txPower)	24.75mW
Recieve power (rxPower)	13.5mW
Idle Power (IdlePower)	13.5mW
Sleep Power (sleepPower)	0W
Antenna gain factor(Gt,Gr)	1
Antenna height above the ground (ht,hr)	1m
Signal wavelength (λ)	0.69284m

4.1 Energy Consumption

Figure 6 shows the aggregate energy consumption on the nodes transceivers in the entire network to send a fixed amount of data from the source node to the sink node. FASMAC shows better energy savings than just SMAC with adaptive listen especially at light traffic load.

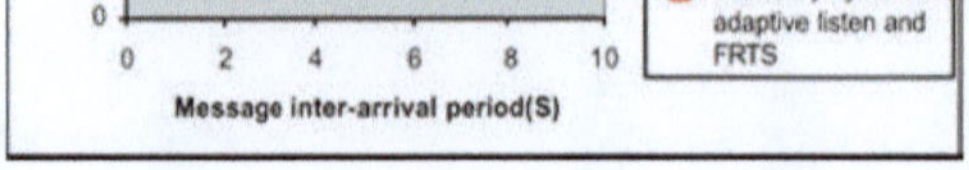

Figure 6: Aggregate energy consumption by the 8-nodes transceivers

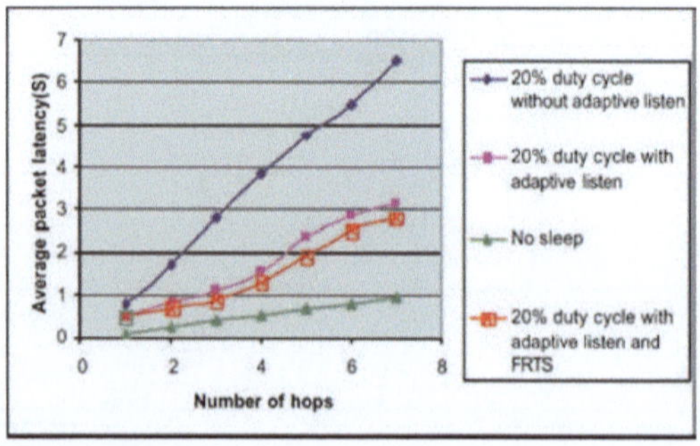

Figure 7: Mean message latency on each hop under low traffic load

4.2 End-to-End Latency

Figure 7 shows the mean average latency on each hop at the lowest traffic load. FASMAC has better latency performance than SMAC with adaptive listen. Figure 8 shows the mean message latency on each hop at a high traffic load. Again, FASMAC has a better latency performance than SMAC with adaptive listen.

4.3 End-to-End Throughput

Figure 9 shows the measured throughput from source to sink for different message inter-arrival times at the source node. As expected FASMAC shows better throughput performance than SMAC with adaptive listen since it has lower end to end latency.

4.4 Energy-Time Cost per Byte

Figure 10 shows the Energy-time cost per byte for different message inter-arrival periods at the source node. FASMAC has better performance than adaptive listen at all traffic loads. However, at very heavy traffic load the No sleep protocol has better performance than FASMAC because of the overhead (SYNC packet, latency, and FRTS packet) associated with FASMAC.

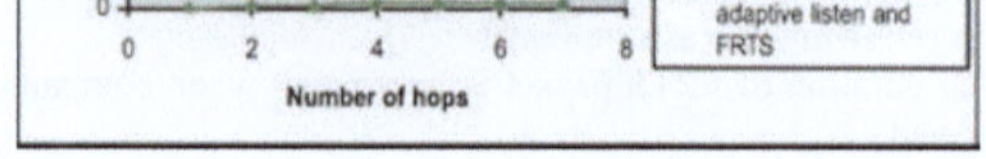

Figure 8: Mean message latency on each hop under a high traffic load

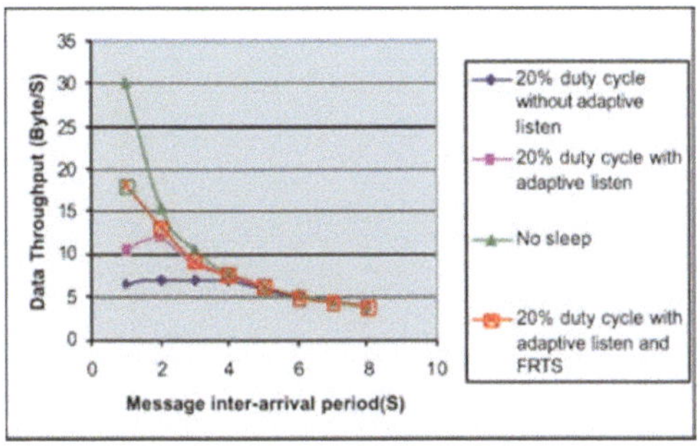

Figure 9: Throughput over 7 hops under different traffic load

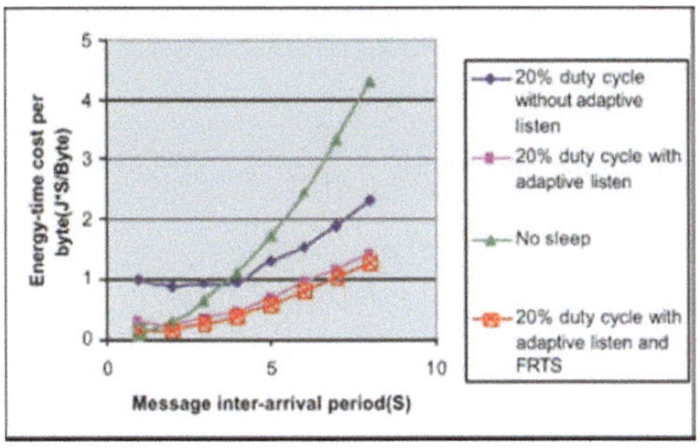

Figure 10: Energy-time cost per byte on passing data from source node to sink under different traffic load

5. Conclusion and Future Work

This paper has proposed FASMAC, a low latency and energy efficient MAC protocol for WSNs. FASMAC enhanced the latency performance of SMAC with adaptive listening by using the FRTS packet introduced in the TMAC protocol. In FASMAC a packet can travel three hops away from its source node within one time cycle. We have evaluated FASMAC performance through ns2 simulation and compared its energy, latency and throughput with three scenarios of SMAC, namely, SMAC with adaptive listening, SMAC without adaptive listening, and SMAC without sleep. FASMAC showed better energy savings and lower latency especially at light load. For future work, we would like to implement FASMAC on the actual testbed motes and evaluate its performance through real experiments.

Acknowledgment

We would like to thank the authors of SMAC who made their code available to the research community.

References

[1] J. Culpepper, L. Dung, and M. Moh, "Hybrid Indirect Transmissions (HIT) for Data Gathering in Wireless Micro Sensor Networks with Biomedical Applications," in IEEE CCW 2003 (Computer Communications Workshop) October 20-21, 2003.

[2] A. El-Hoiydi and J. Decotignie, "WiseMAC: An Ultra Low Power MAC Protocol for Multi-hop Wireless Sensor Networks," In Algorithmic Aspects of Wireless Sensor Networks: First International Workshop (ALGOSENSORS'04), Springer, (Jul. 2004).

[3] W. Heinzelman, A. Chandrakasan, and H. Balakrishnan, "Energy-Efficient Communication Protocols for Wireless Microsensor Networks," Proc. of the Hawaii International Conference on Systems Sciences (HICCS 2000), (Jan. 2000).

[4] J. Hill and D. Culler , "Mica: A Wireless Platform for Deeply Embedded Networks," IEEE Micro., vol 22(6), Nov/Dec 2002, pp 12-24.

[5] Y. Li, W. Ye, and J. Heidemann, "Energy and Latency Control in Low Duty Cycle MAC Protocols," IEEE Wireless Communications and Networking Conference (WCNC'05), (Mar. 2005).

[6] G. Lu, B. Krishnamachari, and C. Raghavendra, "An Adaptive Energy-Efficient and Low-Latency MAC for Data Gathering in Sensor Networks," In Workshop on Energy -Efficient Wireless Communications and Networks (EWCN '04), (Apr. 2004).

[7] M. Miller and N. Vaidya, "Minimizing Energy Consumption in Sensor Networks using a Wakeup Radio," IEEE WCNC'04, (Mar. 2004).

[8] V. Tijs and K. Langendoen, "An Adaptive Energy-Efficient MAC Protocol for Wireless Sensor Networks," In Proc. of the First ACM Conference on Embedded Networked Sensor Systems, (Nov. 2003), pp. 171-180.

[9] A. Woo and D. Culler, "A Transmission Control Scheme for Media Access in Sensor Networks," in Proc. ACM/IEEE Int. Conf. Mobile Computing and Networking, (Jul. 2001), pp. 221-235.

[10] W. Ye, J. Heidemann, and D. Estrin, "An Energy-efficient MAC Protocol for Wireless Sensor Networks," Proc. IEEE INFOCOM 2002, (Jun. 2002), pp. 1567-1576.

[11] W. Ye, J. Heidemann and D. Estrin, "Medium Access Control with Coordinated Adaptive Sleeping for Wireless Sensor Networks," IEEE, ACM Transactions on Networking, Vol 12, No 3, (Jun. 2004), pp. 493-506.

[12] LAN MAN Standards Committee of the IEEE Computer Society, Wireless LAN Medium Access Control (MAC) and Physical Layer (PHY) Specification, IEEE Std. 802.11-1997 Edition, 1997.

[13] The Network Simulator ns-2, http://www.isi.edu/nsnam/ns.

[14] RF Monolithics Inc., http://www.rfm.com/, ASH Transceiver TR3000 Data Sheet.

[15] http://www.icsi.berkeley.edu/~widmer/mnav/ns-extension/

Multilayer Traffic Engineering Based on Transmission Quality and Grooming in the Next Generation Optical Internet

W. Colitti*, L. Tran**, A. Nowé* and K. Steenhaut*
*Vrije Universiteit Brussel
ETRO Department and COMO lab.
Pleinlaan 2, B-1050 Brussel (Belgium)
{wcolitti, asnowe, kris}@info.vub.ac.be
Tel. +32 2 629 29 06
Fax +32 2 629 28 70

**Hanoi University of Technology
Department of Communication Engineeering
01 Dai Co Viet Road, Hanoi, Vietnam
ngoclan@netnam.org.vn
Tel. +84 4 8680974
Fax +84 4 8692241

***Abstract* – An integrated IP over Optical Network controlled by the Generalized Multi-Protocol Label Switching (GMPLS) common control plane is believed to be a promising solution for the Next Generation Optical Internet. In this scenario, Multi-layer Traffic Engineering (MTE) can be improved in terms of transmission quality and traffic grooming. This paper proposes an integrated MTE scheme able to route IP traffic considering the potential power penalty of all-optical connections (for transmission quality purpose) and the network resource usage (for grooming purpose). The routing algorithm is executed on an integrated graph formed by both the physical and the logical topologies. This graph representation enables integrated Resource Reservation Protocol (RSVP) sessions that decrease the latency time. Via extensive simulations, the performance of this scheme in terms of connection requests success probability, resource usage and transmission quality is evaluated.**

***Keywords* – GMPLS, MTE, power penalty, all-optical communication, resource usage, transmission quality, RSVP-TE.**

I. INTRODUCTION

Wavelength Division Multiplexing (WDM) and transparent all-optical on-demand connections (i.e. lightpaths) have drastically improved throughput and resource usage efficiency of IP over optical networks [3].

Most studies on optical networks are based on ideal assumptions where physical impairments affecting all-optical communications are not taken into consideration [1, 2, 7, and 8].

The optical fiber is a nonlinear medium and in a transparent all-optical pipe no signal transformation and reshaping at intermediate nodes occurs. Noise and signal distortions caused by non ideal transmission devices are accumulated along the physical path and the Optical Signal to Noise Ratio (OSNR) could be considerably decreased, resulting in a high Bit Error Rate (BER) [3, 4]. This suggests that the optical transmission quality is a critical issue that needs to be addressed when routing lightpaths to accommodate IP traffic.

Next generation Internet is believed to be an integrated IP over optical network controlled by the unique Generalized Multi-Protocol Label Switching (GMPLS) control plane [1, 2].

By integrating high speed IP routers with Optical Cross Connects (OXCs), Multi-layer Traffic Engineering (MTE) becomes more efficient. Packet–Label Switched Paths (P-LSP) can be routed over Lambda-Label Switched Paths (λ-LSPs) using a unique Traffic Engineering Databases (TED).

A unique TED facilitates simplified cross-layers designs which in turn facilitate the communication between IP and optical components without requiring a User to Network Interface (UNI).

Routing and rerouting can be executed on an ideal integrated graph formed by the physical topology and the logical one. The physical topology consists of the fiber set having available room (i.e. free wavelengths) to setup new ligthpaths. The logical topology is formed by the set of lightpaths having available room (sufficient capacity) to accommodate IP traffic.

As a result, two issues can be improved when implementing MTE in such an integrated scenario:

1. **Transmission quality:** physical impairments of all-optical connections can be considered when routing/rerouting IP traffic.

2. **Traffic grooming:** grooming policies depending on the real time network resource usage can be implemented.

K. Elleithy et al. (eds.), Advances in Computer, Information, and Systems Sciences, and Engineering, 185–191.

In this paper we present an MTE scheme for the new generation optical Internet. The scheme accommodates IP traffic considering the potential signal power penalty of all-optical communications (for transmission quality purpose) and the real time network state (for traffic grooming purpose).

The shortest path decision is made by means of a heuristic called Maximum Transmission Quality Algorithm (MTQA). The heuristic considers the lightpath's noise and propagation delay and the number of wavelength and transmitters currently used.

Mixed (lambda and packet) Label Switched Paths (LSPs) are established by integrated Resource Reservation Protocol (RSVP) sessions that decrease the latency time.

We will show, via simulations, the performance of the proposed MTE scheme in terms of connection requests success probability, resource usage and provided transmission quality.

The rest of the paper is organized as follows. In section II, we give an overview of the new generation optical Internet, focusing on the architectural model and on the GMPLS node model. In section III, we describe the integrated network graph representation used in this work. In section IV we introduce the proposed MTE scheme focusing on the MTQA cost metric. In section V, the integrated RSVP-TE and its advantages are discussed. Section VI shows the performance evaluation and Section VII draws some conclusions.

II. NEW GENERATION OPTICAL INTERNET

The next generation Optical Internet will consist of high speed IP routers integrated with Optical Cross Connects (OXCs) functionalities, as shown in Fig. 1.

This architecture is enabled by the multi-granularity GMPLS node described in [1] and represented in Fig. 2.

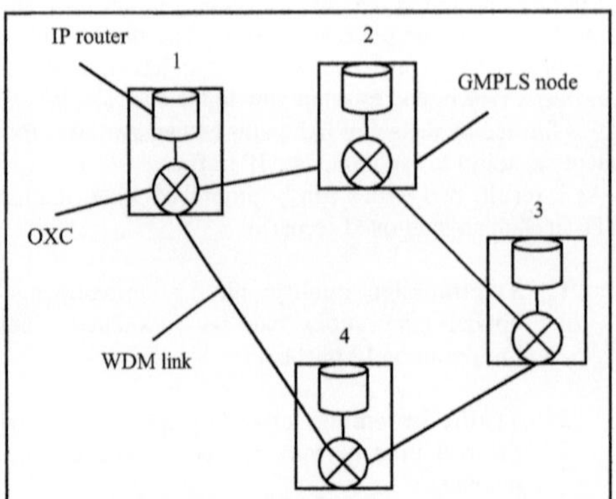

Fig. 1. Architectural model of the new generation optical Internet.

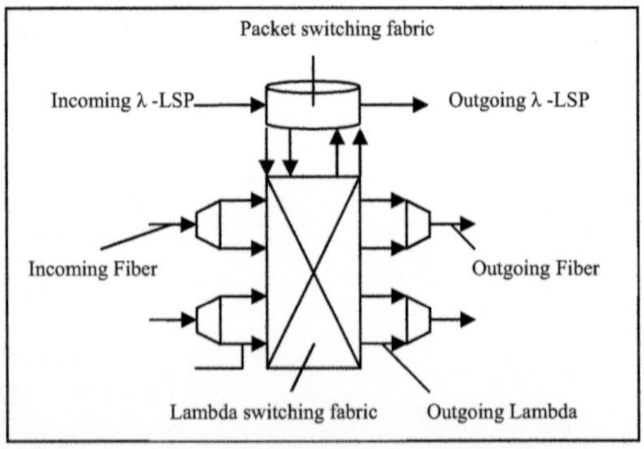

Fig. 2. Architecture of the GMPLS node used in the network model.

By integrating both IP/MPLS packet switching and wavelength switching fabrics, a GMPLS node can concurrently be end point or switching point for both (electric) P-LSPs and (optical) λ-LSPs. A node can act as an OXC or as an IP router. Lambda switching functionalities are needed when traffic has to be switched without leaving the optical domain (i.e. when traversing a λ-LSP); IP packet switching functionalities are needed when traffic needs to be switched from a λ-LSP to another (i.e. switching in the virtual topology). In the latter case, an optical-electronic-optical signal conversion is needed.

Dynamic traffic grooming/bypassing and online lightpath provisioning are guaranteed by a continuous information exchange among the GMPLS node controllers via Open Shortest Path First – Traffic Engineering (OSPF-TE) protocol. The network status information collected are passed to the Path Computation Element (PCE) which is capable of dynamically accommodating a connection request taking into account both electric and optical layers' state. This tight integration facilitates the use of a unique TED that reduces the information exchange between the two layers, enabling easier cross-layer designs and interworking strategies [2]. As a result, higher network performance and reduced network costs can be achieved.

Two parameters characterize each node: the number of transceivers in the OXC and the number of ports in the IP router. Before running the routing algorithm, the source node of a connection request checks whether there is at least one available port in the IP router (to setup a P-LSP) and one available transmitter in the OXC (if a λ-LSPs has to be setup). In this case, a path can be installed, otherwise the request is rejected.

III. NETWORK GRAPH MODEL

The GMPLS nodes are interconnected by optical fibers, each one carrying a limited number of wavelengths (W). The integrated routing algorithm is executed on a layered graph where two layers are considered: the wavelength layer (*WL*) and the lightpath (i.e. λ-LSP) layer (*LL*). Each edge in the graph representation is characterized by the parameters *c* and *w*, where *c* is the edge capacity and *w* is the edge weight calculated with the formula described in section IV.*A* and used for the shortest path computation. Each edge belongs either to the *WL* or to the *LL* according to the following definition:

- An edge *e* between nodes *i* and *j* is a *WL-edge* if there is at least one available wavelength on the physical fiber between nodes *i* and *j* and one available transmitter in the OXC at node *i*. The capacity *c* is the maximum capacity *m* provided for a λ-LSP. Note that nodes *i* and *j* are physically connected.
- An edge *e* between nodes *i* and *j* is an *LL-edge* if there is a λ-LSPs already installed between nodes *i* and *j* and if it has an amount of available capacity greater than the one required by the connection request. The capacity is defined by $c = m - \sum_{i=1}^{S} c_i$, where *S* is the number of connection requests already accommodated on *e*, each of them requiring a capacity amount equal to c_i, respectively. Note that nodes *i* and *j* can possibly not be physically connected.

The routing algorithm is executed on a graph represented by a superposition of both WL and LL. The resulting graph is called integrated layer (*IL*) and it is formed by the WL-edges and the LL-edges. Fig. 3 shows a simple example to clarify the graph model construction before the execution of the routing algorithm.

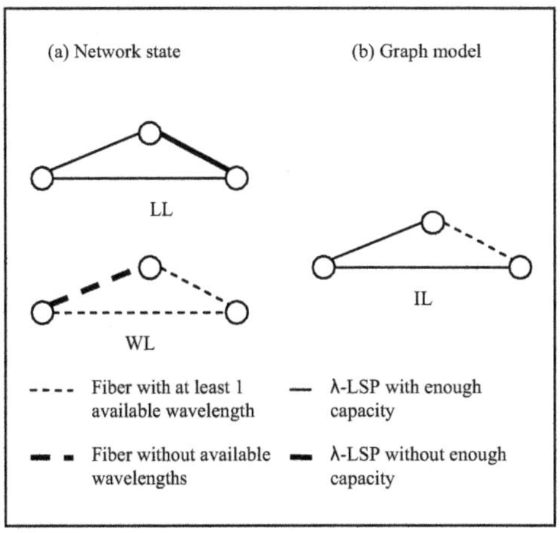

Fig. 3 Construction of the network graph model before the execution of the routing algorithm.

Fig. 3 (a) represents the network state when a new connection request arrives. In the LL, there are two λ-LSPs with enough capacity to accommodate the request and one λ-LSP with an amount of available capacity that cannot guarantee the request's requirement. In the WL, there are two fibers with at least one available wavelength and one fiber that cannot be used for a new λ-LSP setup because of lack of free lambdas. The resulting graph (i.e. IL) is formed only by WL-edges and LL-edges (Fig. 3 (b)).

Note that when between the same nodes pair there are both a WL-edge and an LL-edge as candidates for the graph construction, the system choose the LL-edge. In other words, the traffic grooming on already established λ-LSPs is preferred whenever possible. This feature will be better discussed in the description of the cost metric used in the routing algorithm (see section IV.*A*).

IV. AN INTEGRATED MTE SCHEME

A. Cost Metric

When calculating the shortest path to accommodate a connection request, the cost assigned to the generic edge *e* between nodes *i* and *j* in the graph model is evaluated according to the following formula:

$$C^{ij} = \frac{T^i W^{ij}}{T^i_{avail} W^{ij}_{avail}} \left[C^{ij}_{noise} + C^{ij}_{delay} \right]. \quad (1)$$

Basically, C^{ij} is the sum of two contributions: C^{ij}_{noise} is due to the power penalty introduced in the optical communication and C^{ij}_{delay} is the contribution due to the signal propagation delay along the optical fibers (both the values are associated to the edge between nodes *i* and *j*). The different values assigned to the parameters **α** and **β** can determine the higher or lower influence of one contribution with respect to the other one. T^i_{avail} represents the number of available transmitters at the OXC of node *i* and W^{ij}_{avail} is the number of available wavelengths between the physical fiber connecting nodes *i* and *j*. T^i and W^{ij} are the total number of transmitters at node *i* and the total number of wavelengths of the fiber *ij*, respectively. For an LL-edge, we force the relation $\left(T^i W^{ij} / T^i_{avail} W^{ij}_{avail}\right) = 1$. This means that when calculating the shortest path only the cost of a new wavelength is weighed by the amount of used resources, in order to avoid that new lightpaths are always preferred when routing new connection requests. This also implies that if between the nodes *i* and *j* there are both a WL-edge and an LL-edge with a sufficient amount of available bandwidth to route the new request, the WL-edge is surely dominant in the graph and is not considered by the system. Consequently, the traffic aggregation on existing lightpaths between two physically connected nodes is preferred whenever possible.

The two contributions to C^{ij} are defined as follows:

$$C^{ij}_{noise} = \frac{\sum_{k=1}^{N^{ij}} l_k}{S} + N^{ij} - 1 . \quad (2)$$

$$C^{ij}_{delay} = \sum_{k=1}^{N^{ij}} l^{ij}_k . \quad (3)$$

In (2), N^{ij} is the total number of optical fibers forming the edge *e* between nodes *i* and *j*; l_k is the length of each fiber and *S* is the fiber span (i.e. the distance the optical signal has to pass through before being amplified in an optical amplifier). In this paper we suppose that *S* has the same value on each physical link.

Note that if *e* is a WL-edge (i.e. it is not a lightpath), *N* is equal to 1. C^{ij}_{noise} embodies the noise introduced by the optical amplifiers $\sum_{k=1}^{N^{ij}} l_k \Big/ S$ and the one introduced by the OXCs, wavelength converters and filters when switching the optical signal, namely by the number of hops $N-1$. For simplicity, we assume that each optical fiber is equipped with the same kind of optical amplifiers.

In all-optical communications there are additional sources of noise not considered in our cost metric. We only take into account the contribution of noise coming from the optical amplifiers, which is considered the predominant source for OSNR penalty and degradation, and the contribution due to the number of hops, which better differentiates a WL-edge (i.e. no multiple optical hops) from a LL-edge (i.e. multiple optical hops).

In (3), *N* and *l* have the same meaning as explained above. C^{ij}_{delay} represents the optical signal propagation delay and only incorporates the length of the optical path.

When analyzing the transmission quality, the proposed scheme does not consider the impact of the electronic multi-hops paths (i.e. P-LSP formed by more than one lightpath and needing an electronic switching operation at the end point of each lightpath traversed). In doing so, we ignore queuing delays and optical-electronic-optical (o-e-o) signal conversions. This is not a stringent assumption considering that the high speed integrated GMPLS nodes have significantly lower queuing delays [6] and that the delay introduced in o-e-o conversions are roughly compensated by the OSNR increase during the signal reshaping occurring when switching the signal electronically. Actually, o-e-o conversions and signal reshaping processes should be accurately analyzed in order to find out their incidence on the overall signal quality, but this is outside the scope of this paper.

B. Basic Operations

The basic operations of the integrated MTE scheme proposed in this paper after a connection request arrival are described as follows:

- According to the network state information collected by the GMPLS controller via the OSPF-TE agent, the layered graph (see sec. III.*B*) is built. During this operation, first the existing lightpaths that cannot guarantee the requested bandwidth are removed from the graph description and then a weight, calculated according to the MTQA cost metric described in sec. IV.*A*, is assigned to each edge.
- The output of the previous calculation along with information about the destination nodes of the request is passed to the PCE which runs a Dijkstra-like algorithm to search for the shortest path.
- If the path is available, the output is passed to the RSVP-TE agent, which in turn launches a hard resource reservation process. After a resource reservation confirmation, the data communication phase can start and the OSPF-TE agents of each node along the path update the network TEDs by flooding Link State Advertisements (LSAs).
- If the path is not available, the request is rejected.

V. Integrated RSVP-TE

The unique GMPLS control plane is able to setup, maintain and tear down P-LSP over λ-LSPs using the same control protocols.

In this paper, for example, a unique Resource Reservation Protocol with Traffic Engineering extensions (RSVP-TE) has been implemented for setting up/maintaining/tearing down mixed label switched paths.

In Fig. 4, we represent how a simple integrated RSVP-TE session is activated when a suitable path has been found for accommodating a connection request.

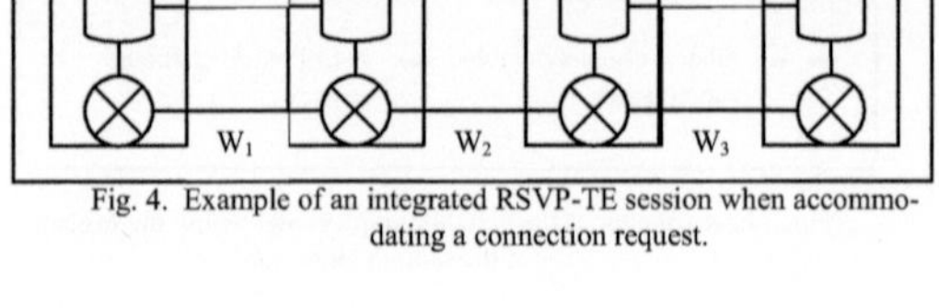

Fig. 4. Example of an integrated RSVP-TE session when accommodating a connection request.

We suppose that the logical topology is formed by two λ-LSPs already installed: L_1 between nodes 1 and 2, L_2 between nodes 3 and 4. The physical topology is formed by the links W_1, W_2 and W_3, which are supposed to be optical fibers carrying only one wavelength. Suppose that a connection request from node 1 to node 4 needs to be accommodated and the system has already found the available path $L_1 - W_2 - L_2$ (note that L_1 and L_2 are supposed to have a sufficient amount of capacity to route the connection request and W_2 is supposed to be a free wavelength). Node 1 starts a Packet-RSVP-TE (P-RSVP-TE) which involves the λ-LSPs L_1 and L_2 and responsible of installing a P-LSP between nodes 1 and 4. When the request arrives at node 2, it automatically calls a Lambda-RSVP-TE (λ- RSVP-TE) which is responsible of installing a λ-LSP between nodes 2 and 3 using the free wavelength W_2.

Once the integrated RSVP-TE session is terminated, the communication between nodes 1 and 4 can start and the network state can be updated via OSPF-TE protocol.

An integrated RSVP-TE significantly decreases the latency delay and improves the network performance.

VI. PERFORMANCE EVALUATION

We discuss the performance evaluation of the integrated MTE scheme proposed in this paper via simulation and comparison using as reference the scheme proposed in [4]. Unlike our scheme, the approach in [4] considers two different service classes with different QoS requirements. Nevertheless, the choice to take the scheme in [4] as reference work for our comparison was made because the authors significantly demonstrate the performance improvements when optical metrics are taken into account in the implementation of an MTE system. As far as our knowledge, there are no better approaches reported in literature.

The network under test is the Italian IP over Optical topology used in [7] and illustrated in Fig. 5. It consists of 21 nodes (each one supposed to have the architecture discussed in section II) and 36 bidirectional optical fibers. We suppose that each node can be source/destination of maximum 6 λ-LSPs (6 transceivers per OXC) and source/destination of an unlimited number of P-LSPs (infinite number of ports per IP router). This is due to the fact that the number of ports is not taken into account in the routing decision. The nodes are assumed to have full wavelength conversion capability and the wavelength conversion is made optically, meaning that no signal reshaping occurs during a switching operation in an OXC. A fiber carries 6 wavelength channels, each one having a capacity of 10 Gbps.

Both our approach and the reference scheme have been implemented and extensively simulated with the Queuing Network Analysis Package (QNAP) [9]. In [4] two different algorithms can be used to route a high priority request, depending on the grooming policy considered.

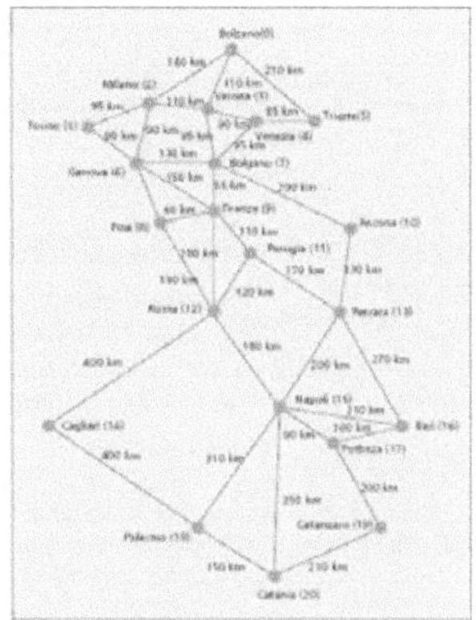

Fig. 5. Network topology

In the performance comparison, we only selected the one called Virtual Topology first (VT-first). Firstly, because the blocking probability is not significantly different from the other approach (i.e. Physical Topology first) and secondly because it is the one that better performs in terms of resource usage. It tries to exploit the virtual topology as much as possible before establishing new lightpaths. The compared results give us a more solid idea of the MTQA's quality.

The traffic demand is modeled with connection requests generated according to a Poissonian process with an average rate λ and connection holding time exponentially distributed with mean $1/\mu$. The network load is given by $\gamma = \lambda/\mu$ Erlangs. Connection requests have a bandwidth demand uniformly distributed between 1 and 3 lightpath capacity units. For each value of the network load, we consider the statistical values calculated over 10 runs. In each simulation we generate 10000 connection requests between randomly chosen node pairs.

The metrics considered for the performance evaluation are success probability, average λ-LSP length, average number of used wavelengths, average number of used transmitters and average number of lightpaths created. In the MTQA cost metric, we chose the fiber span S equal to 40 Km, the number of node's transmitters T equal to 6 and both the parameters α and β are set to 1. α and β represent the relative impact of noise and propagation delay. The impact of the setting of these parameters is outside the scope of this paper.

MTQA has been designed for a non DiffServ scenario. For this reason the comparison with the chosen reference is first made considering a version of VT-first without a preemption algorithm.

In Fig. 6, the success probability of the two compared schemes is shown. Our approach has a high success probability that remains significantly higher than the one measured for high and low priority (HP and LP, respectively) requests in the VT-first approach.

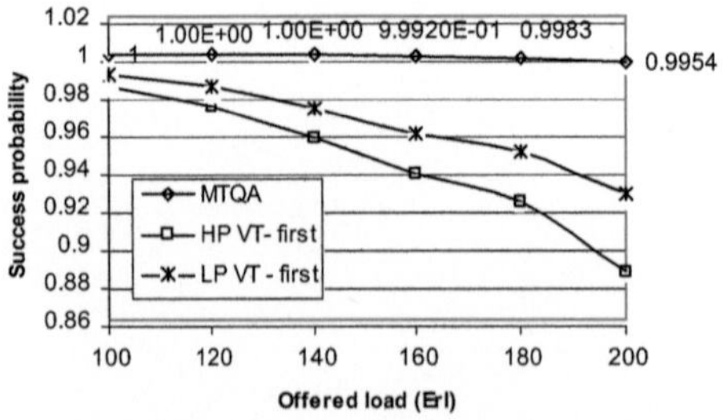

Fig. 6. Mean values of success probability of MTQA and VT-first. HP and LP refer to high priority and low priority of the reference scheme, respectively

The performance evaluation of MTQA with respect to the resource usage is shown in Fig. 7. The average number of both used wavelengths and transmitters is lower then the ones measured for the reference scheme, for any value of the offered traffic. This is a consequence of our choice to introduce in the routing decision, information concerning the resource usage instead of a priori including a grooming policy. This leads to a resource saving that decreases the blocking probability.

The results given in Fig. 7 are in line with the ones in Fig. 8. As a consequence of the low resource usage, the average number of λ-LSP created is lower when running MTQA.

In order to evaluate the performance of the proposed MTE scheme in terms of provided transmission quality, we measured the average λ-LSP length. The measure of this metric gives an idea about the quality of the all-optical connections installed. A long λ-LSP has to undergo higher propagation delay and higher signal degradation. The BER could be significantly degraded by the higher number of OXCs and optical amplified traversed.

The results illustrated in Fig. 9, show that the average length of the total number of λ-LSP setup during the entire simulation is considerably low when using MTQA.

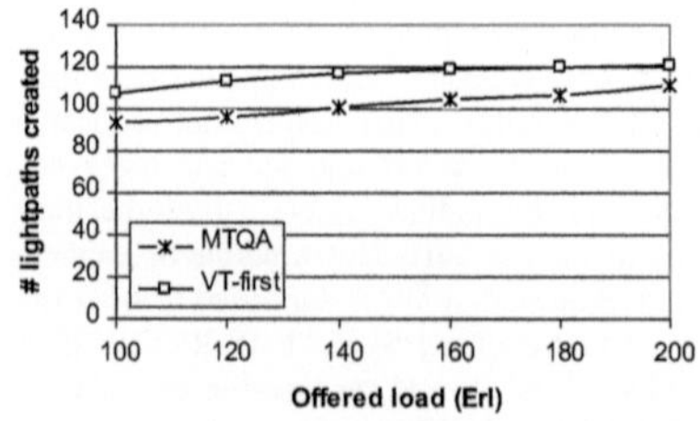

Fig. 8. Average number of created lightpaths.

Although we do not give an estimation of the signal transmission quality in terms of BER (because accurate measurements on the optical signal quality are required and this is outside the scope of this paper), we can conclude that the transmission quality provided by the proposed MTE scheme is significantly higher than the one provided by the reference approach.

We can also notice in Fig. 9, that the curve relating to our approach remains much more stable than the one relating to VT-first. This is due to the more optimized load balancing executed by using MTQA in the route selection.

To have an idea concerning the MTQA's extendibility to a DiffServ scenario and provide QoS guarantees, we ran some simulations of the reference scheme VT-first in [4] when it also includes a preemption algorithm. Obviously, the blocking probability of HP requests drastically decreases, as a consequence of the higher amount of available room when preempting LP requests.

Nevertheless, the average value of HP and LP blocking probability (shown in Fig. 10) remains higher or equal (only for a traffic load equal to 300 Erlangs) than the one calculated for MTQA. This result can be seen as an indication that our approach can give good performance also when implemented in a DiffServ scenario. The implementation of MTQA approach in a DiffServ network is one of the future works which will extend this paper.

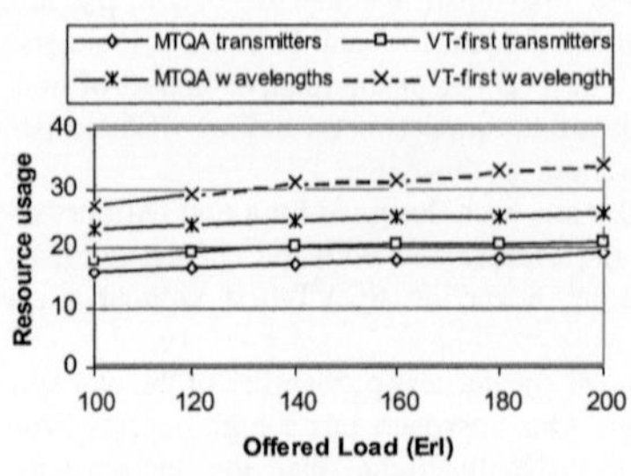

Fig. 7. Resource usage of MTQA and VT-first. The measurements refer to average number of used wavelengths and transmitters.

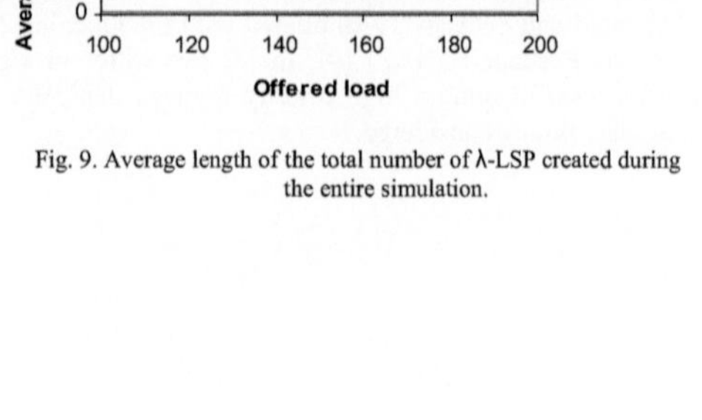

Fig. 9. Average length of the total number of λ-LSP created during the entire simulation.

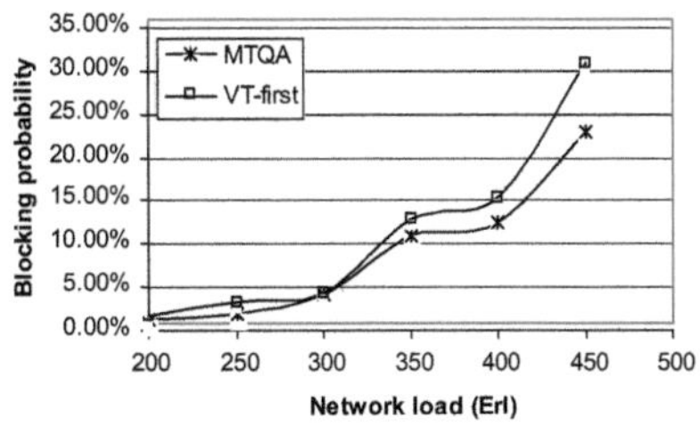

Fig. 10. Blocking probability of MTQA and VT-first with a preemption algorithm. The curve related to VT-first represents the mean values of HP and LP blocking probability.

VII. Conclusions

In this paper a new scheme for integrated MTE in the next generation Optical Internet has been proposed.

The next generation Optical Internet will consist of high speed IP routers integrated with Optical Cross Connects (OXCs) functionalities. A GMPLS integrated IP over optical network is able of dynamically routing Packet–Label Switched Paths (P-LSP) over Lambda-Label Switched Paths (λ-LSPs) using a unique protocols set.

In this scenario MTE can be improved in terms of transmission quality and traffic grooming.

Our scheme is based on a heuristic that calculates the shortest path considering the propagation delay and power penalty of all-optical communications and the real time network resource usage.

Aknowledgement

The authors would like to thank the FWO and the VLIR-IUS for their support.

IP traffic can be routed guaranteeing sufficient bandwidth, high transmission quality and low resource usage, while keeping the connection request success probability high. Mixed GMPLS LSPs are installed by integrated RSVP-TE sessions that decrease the latency time.

Simulation results demonstrated that the proposed scheme outperforms the reference scheme.

References

[1] E. Oki, et al."Dynamic multilayer routing schemes in GMPLS-based IP + optical networks," *IEEE Commun. Mag.*, vol 43, issue 1, pp. 108-114, January 2005.

[2] W. Wei, Q. Zeng, Y. Ouyang, and D. Lomone, "Differentiated integrated QoS control in the optical Internet," *IEEE Opt. Commun.*, vol 42, No 11, November. 2004.

[3] R. Cardillo, V. Curri, and M. Mellia, "Considering transmission impairments in wavelength routed networks," *Proc. Opt. Network Design and Modelling (ONDM)*, Milano, Italy, February 2005.

[4] E. Salvadori, and R. Battiti, "A traffic engineering scheme for QoS routing in G-MPLS networks based on transmission quality", *Proc. Opt. Network Design and Modelling (ONDM)*, Ghent, Belgium, March 2004.

[5] B. Puype, Q. Yan, S. De Maesschalck, D. Colle, M. Pickavet, and P. Demeester, "Optical cost metrics in multi-layer traffic engineering for IP-over-optical networks," *Proc. Int. Conf. on Transparent Opt. Networks (ICTON),* Wroclaw, Poland, 2004.

[6] R. Ramaswami, and K. N. Sivarajan, "Design of Logical Topologies for Wavelength-Routed Optical Networks," *IEEE J. Select. Areas Commun.*, vol. 14, No 1, June 1996.

[7] M. Ali, "Shareability in Optical Networks: Beyond Bandwidth Optimization," *IEEE Opt. Commun.*, vol. 42, No 2, February 2004.

[8] M. Kodialam, and T.V. Lakshman, "Integrated dynamic IP and wavelength routing in IP over WDM networks," *Proc. INFOCOM*, vol 1, pp. 358-366, Anchorage, Alaska, 2001.

[9] "The Queuing Network Analysis Package user guide"

A Remote Online Monitoring and Diagnosis System Incorporated With Wireless Sensor Network*

Liu Youyuan
Visiting scholar in EISLAB, CSEE
Luleå University of Technology, Sweden
Email: Daniel.Liu@ltu.se
Logistics Automation Department,
College of Logistics Engineering
Wuhan University of Technology
Wuhan 430063,China
Email: liuyysy@gmail.com

Deng Anqing
Logistics Automation Department,
College of Logistics Engineering
Wuhan University of Technology
Wuhan 430063,China
Email: lzrlover@163.com

***Abstract*-This paper introduces a web-based remote online equipment monitoring and diagnosis system incorporated with the wireless sensor network. The system architecture is discussed first, which is composed of wireless sensor nodes, wireless base stations, a host computer and some necessary intranet/Internet facilities that implements remote access to the sensory information, followed by the hardware and software design of the system. The wireless sensor network will implement the on-time data collection and wireless data-transmission from the spot, and the web-based software will perform the data display, malfunction diagnosis and some other tasks. Based on the results of the research, a tentative remote online monitoring system for an important blower fan in an iron and steel company is established and tested.**

***Keywords*-Wireless sensor network, Remote monitoring, Diagnosis, Web-based**

I. Introduction

In practical applications, data collection and transmission is frequently required for the monitoring and controlling process. In China, many big state-owned companies nowadays have been employing modern technologies to improve the production efficiency and reliability, and to realize production automation. On line and remote monitoring and diagnosis for key equipment is among these improvements. Most of such systems use cables as the data transmitting means for its stable and fast data rate. However, in some cases, the layout for cables is handicapped by the practical situations, which results in the difficulties for the construction of the remote online monitoring and diagnosis system.

Thanks to the emergence and the rapid development of the wireless sensor networks (WSN) in the past several years, many complicated practical problems can be solved using the wireless sensor networks now. Because of their low cost, easy deployment and convenient maintenances, WSNs have already found a wide range of applications in different fields, including environment monitoring, intelligent transportation, industrial and manufacturing automation, distributed robotics, building and structures monitoring, and so on [1,2]. Each different application has different constraints, or parameters, that drive its development and implementation, such as bandwidth, delay, connectivity, etc. Sensor networks may consist of many different types of sensors such as seismic, low sampling rate magnetic, thermal, visual, infrared, acoustic and radar, which are able to monitor a wide variety of ambient conditions. Many previous researches have been taken using the wireless sensor networks with different sensors. Toshio Hori, Yoshifumi Nishida, et al [3] presented a networked sensor system to monitor the human behavior. They use Ethernet, RS-232C, Rs-485 and Responsive Link for communication between sensors or between sensors and their controllers. They developed a software framework, named middleware to conquer the difficulties due to the change of the physical and/or logical topology of the network. Bill Nickerson and Rick Lally [4] introduced a smart wireless networking sensor for aircraft engine health management. The kernel component of the system is the Intelligent Component Health Monitors (ICHM™), which uses the Bluetooth as the wireless communication means. While the Engine Area Manager acts as a base station to perform communications with ICHMs (the sensor nodes) and the information network (the users). This is actually a centralized wireless network and is quite suitable for industrial automation. Based on the web technology, U. Kunze [5] developed a remote monitoring and diagnosis system for power plant. His efforts are mainly focused on the web-based software. Considering the existing situations of communication field in industries, D. Miorandi and S. Vitturi [6] studied the feasibility of hybrid wired/wireless implementations of Profibus DP based on Ethernet and Bluetooth. Their research results show that, in normal operating conditions, such an implementation is able to cope with the strict real time requirements of industrial communication systems. In many cases, the sensed data are preferred to being sent to the remote users through the Internet or intranet. Because Internet-based interfaces are another recent innovation, allowing for remote control of systems and access to sensory information. Karl Crowley, June Frisby et al [7] developed a wireless sensor network for

* This study was supported in part by the national natural scientific foundation project of China (No. 60475031) and by the Iron Making Plant, Wuhan Iron and Steel Corporation of China.

K. Elleithy et al. (eds.), Advances in Computer, Information, and Systems Sciences, and Engineering, 193–198.

real-time temperature monitoring of whelk catches, associated with the web-based technology. The wireless sensor network is composed of several wireless sensor nodes and a base station that communicates with a remote data server by GSM modem. The system is autonomous and can be controlled through web-based software, which allows for functional querying of real-time and archived temperature information.

To date, much effort is focused on the sensor network protocols and efficient routing algorithms. These approaches are essential for the wireless sensor networks. For the applications in industrial automation, however, special requirements and conditions must be considered as well. The construction of the wireless sensor nodes is one of the major subjects. Network configurations based on both the existing wired system and the newly introduced wireless system must be investigated carefully as well.

This paper introduces a web-based remote online equipment monitoring and diagnosis system incorporated with a wireless sensor network. The wireless sensor network is composed of a number of specialized sensor nodes and several base stations. It is autonomous and can be controlled through web-based software, which allows for remote online access to the sensory information and equipment condition analyzing. After the analysis for the field conditions and possible difficulties faced during the construction of the system, some requirements for the system are specified first. Then, the solution to the whole system is put forward, including hardware and software solutions. Based on the results of the research, a tentative application system for a large blower fan in an iron and steel company in China is established and tested.

II. Requirements Analysis for the System

When a wireless sensor network is utilized for an industrial automation system or an equipment monitoring system, some essential requirements should be satisfied first, including the communication protocols, communication rate, power consumptions, routing algorithms and cost. In addition, several particular constraints should be taken into account as well. They are the field conditions of the system, the special interface requirements for different sensors, and the adaptability to the existing monitoring devices or other low-level automation devices.

The working field of such a system in industries, especially in such places as iron and steel making factories, is usually in adverse conditions, high temperature, high noise and full of dust. Thus the wireless sensor nodes should be able to put up with such adverse conditions. Since most sensors used in industries nowadays can provide standard current (analog) or standard voltage (digital), but they usually have to be installed on the surface of the equipment or inside the equipment, therefore, we have to design special interface to the wireless transceiver for them to implement the sensor nodes. When the adaptability to the existing monitoring devices needs to be concerned, hybrid system is a good solution using Ethernet at both the physical and data link layers of the wired segments and, in the same way, the Bluetooth radio system for the wireless segments.

According to the analysis above and the characteristics of the WSN system for the remote online monitoring, the design tasks and requirements are specified as follows.

- The wireless transceiver module should be designed to meet the requirements of communication quality, distance and data transmission rate, low power dissipation, and anti-interference capability.
- It is convenient for different sensors to interface with the wireless transceiver module to form wireless sensor nodes or base stations.
- The wireless sensor network can be easily constructed.
- The base stations communicate with the host computer through the USB interface.
- The web-based software system can provide a series of functions for data collecting, displaying, monitoring and diagnosing.

III. System Architecture

The condition data of the key equipment are collected by the wireless sensor nodes and then transmitted to the base stations. The base stations forward the data form the wireless sensor nodes to the host computer through the USB interface, while the host computer performs a series of functions such as displaying the condition data in a visualized style, monitoring and diagnosing the working conditions of the equipment using specified algorithms. Meanwhile, through the intranet/Internet, the monitoring and diagnosis system can provides corresponding remote functions for the authorized users such as technicians, department managers and so on. The system architecture is shown in figure 1.

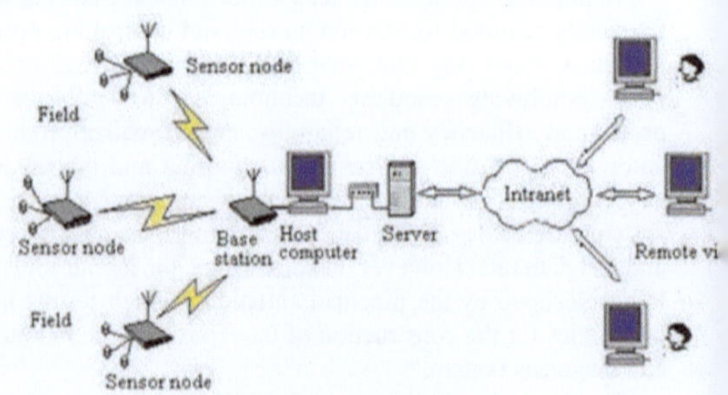

Fig. 1 System architecture

A. Construction of the Wireless Sensor Network

It can be seen from figure 1 that the wireless sensor network in our system is composed of several sensor nodes and one or two base stations, and they play a very important role in the whole system. Since the working principles are similar for both sensor nodes and base stations, and both employ a wireless transceiver module as the basic platform, thus the same framework is adopted for the hardware design.

Whether the basic platform is used as a base station or a sensor node is determined by the predetermined software settings. The hardware framework of the basic transceiver platform is shown in figure 2.

The microprocessor selected is MSP430F149. It is a 16-bit RISC architecture with 60KB+256B flash memory, and 2KB RAM on chip as well as a fast 12-bit A/D converter, ADC12, and is ideal for sensor systems and industrial control applications. The supply voltage is 3.3V, which is consistent with that of the core chip nRF905. MSP430 works with its peripheral modules cooperatively to implement the following functions.

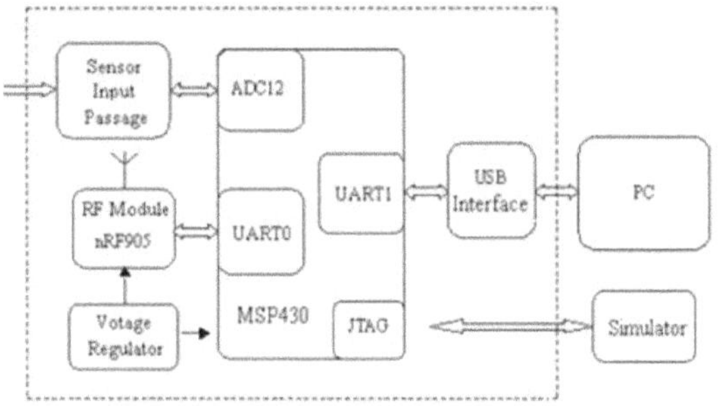

Fig. 2 Hardware framework of the basic platform

- Control the RF module to complete data transmitting and receiving and communicate with other nodes as well;
- Perform the data converting by the built-in ADC12;
- Receive the instructions from the host computer and exchange data with it;
- Perform on line programming through JTAG interface.

A RF chip, nRF905, manufactured by Nordic is used as the RF module. It is a single-chip radio transceiver for the 433/868/915 MHz ISM band. The transceiver consists of a fully integrated frequency synthesiser, receiver chain with demodulator, a power amplifier, a crystal oscillator and a modulator. The ShockBurst™ feature makes it possible to use the high data rate offered by the nRF905 without the need of a costly, high-speed micro controller (MCU) for data processing/clock recovery. By placing all high speed signal processing related to RF protocol on-chip, the nRF905 offers the application micro controller a simple SPI interface, the data rate is decided by the interface-speed the micro controller itself sets up. In ShockBurst™ RX, Address Match (AM) and Data Ready (DR) notifies the MCU when a valid address and payload is received respectively. In ShockBurst™ TX, the nRF905 automatically generates preamble and CRC. Data Ready (DR) notifies the MCU that the transmission is completed. All together, this means reduced memory demand in the MCU resulting in a low cost MCU, as well as reduced software development time [10].

In addition, the modulation of nRF905 is Gaussian Frequency Shift Keying (GFSK), which results in a more bandwidth effective transmission-link compared with ordinary FSK modulation. The data is internally Manchester encoded (TX) and Manchester decoded (RX). By using internally Manchester encoding, no scrambling in the micro controller is needed. This helps to shorten the development time and to reduce the design work for the additional external circuits.

There are totally 10 important digital input/output pins in nRF905 that need to interface with MCU. The interface is divided into three main functions, configuration, mode control and communication. Figure 3 illustrates the connections between nRF905 and the MSP430.

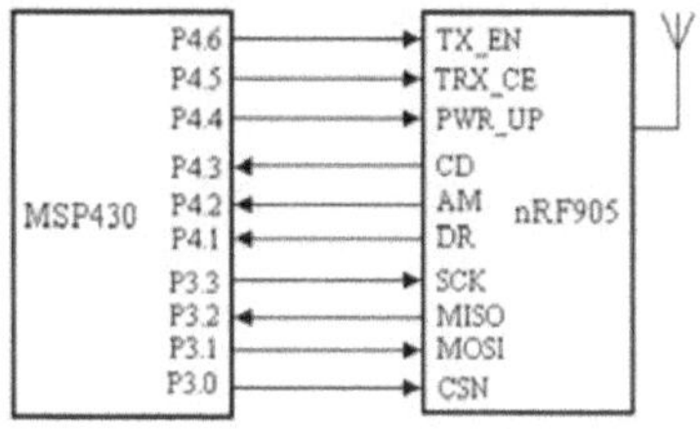

Fig. 3 Connections between MSP430 and nRF905

B. Construction of Sensor nodes

Sensor nodes are constructed by interfacing different sensors with the basic transceiver platform. There are many different kinds of sensors used in the industrial monitoring systems. The most common sensors are accelerometer sensors, temperature sensors, pressure sensors, camera sensors and so on.

Accelerometer sensors in our system are TM0782A-K accelerometer kit from PredicTech, Inc. They are mounted on the bearing house to measure the vibration of the shaft of the rotary machines, and connected with the Transmitter-Monitor TM101 with three-meters-long cables respectively. The condition data are output from TM101 in the way of 4~20mA current. Through the sensor input passage, the condition data of the rotary machines measured by the TM0782A is converted to the analog voltage and sent to the ADC12 inside the MSP430, and then is converted to the digital signal by the ADC12. The interface between the accelerometer sensor and the basic platform is shown in figure 4.

Thermocouple sensors we use are mounted inside the two bearing houses of the rotary machines with signal cables connecting to the sensor input passage. The interface between the temperature sensor and the basic platform is similar to that of the accelerometer sensor node.

In most cases, it is quite helpful to get image data on the spot so as to achieve direct vision effect of the spot. This can be achieved by constructing appropriate camera sensor nodes. Figure 5 shows the interface between the camera sensor and the MSP430 used in our system.

The camera sensor module used is the OV528 manufactured by OmniVision. It integrates camera sensor, serial interface and JPEG image compression & decompression engine into one single chip, which facilitates the development greatly.

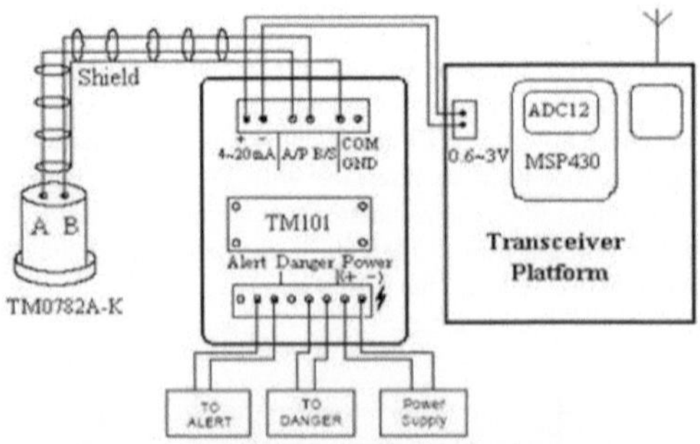

Fig. 4. Construction of the accelerometer sensor node

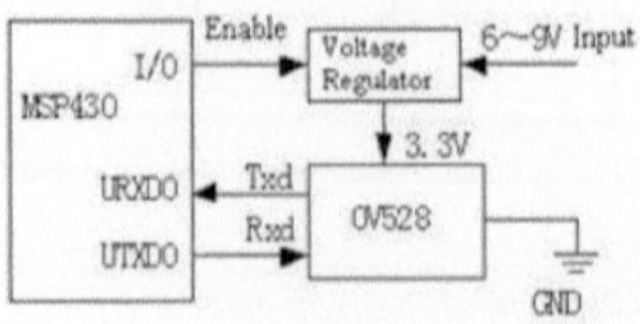

Fig. 5 Interface between the camera sensor and MSP430

IV. SOFTWARE DESIGN

Software design involves two major parts, the monitor and control program system for the basic transceiver platform and the web-based management, monitoring and diagnosis program system in the host computer.

A. Software solution for the basic transceiver platform

When the system is put into operation, the communication between the terminals (sensor nodes) and the base station is established as the first level of communication. The data connected by the sensors are transmitted to the base station in accordance with the instructions. General speaking, one base station will communicate with several terminals, which is a point to multi-point (PMP) communication mode. In this case, cyclical enquiry communication method is employed. That is, the base station detects every communication channel that is corresponding to a specific terminal during a specific interval. When the communication is established, the base station will send the transmitting instruction to the terminal and control the whole transmission process. In our system, the maximum terminals for a base station are 4.

In order to implement the vibration analysis for the bearings of the equipment so as to diagnose the possible faults, sufficient discrete data at one moment should be collected and transmitted to the host computer. In our system, the number of the discrete data needed is 512. That is to say, these 512 figures obtained by one of the accelerometer sensors make up of one complete packet. Only one complete packet for one sensor channel is transmitted completely, the other channels can be transmitted one after another. Hence, the communication protocol between the base station and the terminals takes the form shown in figure 6.

Pre-amble	Receiver ID	Transmitter ID	System Keyword	Op Cmd	Length	Data	CRC

Fig. 6. Protocol for the communication

Receiver ID and Transmitter ID are two 4 bits number and used to represent different base stations and terminals. System keyword is used to distinguish different communication units composed of one base station and several terminals, and the length is 1 byte. Op_Cmd is a control instruction used to control the operation mode of the nRF905, and the length is 1 byte. As for the Length, it is 1 byte and used to set up the data length.

Combining the protocol above with the characteristics of the nRF905, we can finish the control software for the communication process.

1) Transmitting process of the terminals

S1: Initiating. The MCU sets the nRF905 to the receiving status and the default channel is set to the channel 0.

S2: The MCU detects the carrier of the base station in the current channel until the communication between the base station and the terminal is established, then procedure goes to the next step, if no carrier is detected within 5 seconds, then the communication channel hops to the next channel and procedure goes to step 2.

S3: The MCU is waiting for the system keyword from the base station.

S4: If corresponding system keyword is received within 5 seconds and matches with both the receiver ID and the transmitter ID, then the procedure goes to the next step, or else, goes to step 2.

S5: MCU sets TRX_CE and TX_EN high, this activates a nRF905 ShockBurst™ transmission.

S6: If the whole packet is transmitted, the procedure goes to the step 1, or else goes to the step 5.

2) Receiving and controlling process of the base station

There are several basic functions for the base station.

- Receiving the data from the terminals;
- Sending corresponding instructions to the terminals;
- Communicating with the host computer.

Based on this, the software procedure for the base station is as follows.

S1: Let i=0,j=0, where i is the counter for the serial number of the terminals and j is the times that the base station has not received the signal from the terminals. The default terminals are 4.

S2: The MCU sets the base station to the receiving status and the default channel is set to the channel 0.

S3: The MCU detects the status of the current channel.

S4: If there are some disturbing noise, then the communication channel hops to a better channel, or else, goes to the step 5.

S5: The MCU controls the nRF905 to send 30 times of the

system keywords to the terminals so as to establish the connection.

S6: If any instruction from the host computer is received, the procedure goes to the step 8, or else goes to the step 7.

S7: the MCU controls the nRF905 to send 10 times of the system keywords to the terminals so as to keep the current connection, and then the procedure goes to the step 6.

S8: The base station calls the i-th terminal, then set the nRF905 to ShockBurst™ receiving, i=(i+1)%4.

S9: If nRF905 has not received data within 500 ms, then the procedure goes to the step 10, or else, the procedure goes to the step 11.

S10: If j++<20, then the MCU sets the nRF905 to ShockBurst™ TX, and the procedure goes to the step 7, or else, let j=0, and the procedure goes to the step 3.

S11: The nRF905 receives the data from the current terminal. When the receiving process is finished, the MCU sets the nRF905 to ShockBurst™ TX.

S12: If the received data are the corresponding sensor's data, then the procedure goes to the step 13, or else, the procedure goes to the step 14.

S13: Transmit the data to the host computer, and then the procedure goes to the step 6.

S14: Receive the control instruction from the corresponding terminal, and perform the instruction, then the procedure goes to the step 2.

3) Data exchange between the base station and the host computer

The data exchange between the base station and the host computer is implemented through the USB interface, and the process is controlled by the host computer. The basic process is as follows.

S1: The initiation process, including the setting of the communication parameters and the system buffers.

S2: Delay 20 seconds until the wireless sensor network operates stably.

S3: Sends the request instruction to the base station for the data transmission so as to obtain the data from the sensors.

S4: Waiting for the response of the base station, if no data received within 5 seconds, then control the base station and the terminals reset once and the procedure goes to the step 2, or else starting to receive the data and processing the data from the sensors, such as saving and displaying. When the processing is finished, then the procedure goes to the step 3.

B. *Monitoring and diagnosis software for the host computer*

Web-based system monitoring and diagnosis software is developed using C# on MS.net platform. The software package is composed of several modules shown in figure 7. Each module performs some specific functions.

- Communication management controls the data exchange process between the base station and the host computer.
- Operating status display module displays the data obtained from the sensors in curves dynamically and monitors the operating status of the equipment at same time.
- Condition analyzing module employs several analytical algorithms, including frequency analysis and wavelet analysis, to analyze the vibration conditions of the equipment so as to detect the potential faults.

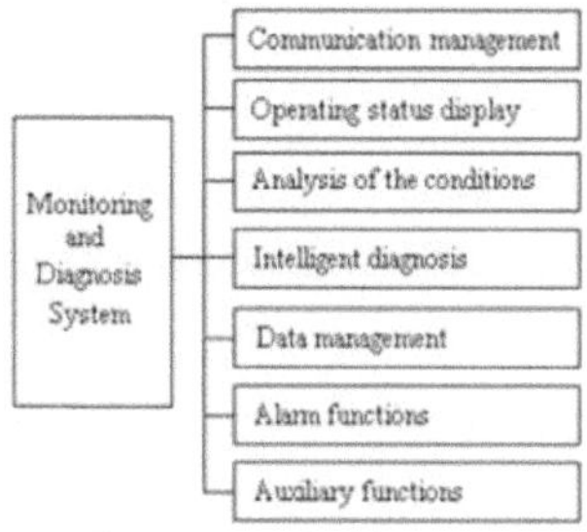

Fig. 7 Framework of the system software

- Intelligent diagnosis module utilizes the expert system to diagnose the possible faults of the equipment.
- Data management module manages and maintenances the database. It helps to establish technical parameters and documents, to test and track specific data, to enquiry history data etc.
- If operating status display module detects some abnormal phenomena, the system will start the alarm process so that the technicians will take some measures for further actions.
- The system also provides some auxiliary functions such as printing of the diagnosis reports, remote visiting authorizing for specific users and so on.

With the help of monitoring and diagnosis functions of the system, in particular, with the help of the expert experience and the theoretical analysis, potential malfunctions of the equipment can be detected as soon as possible. This will enable different operators, technicians, and managers know about the operating conditions of the equipment at any time. It will improve the equipment management level and equipment availability and utilization coefficient, and reduce failure rate and maintenance cost greatly as well.

V. APPLICATION AND CONCLUTION

Based on the discussion above, a tentative application system for remote online monitoring and diagnosis of the blower fan is built and tested. The blower fan is a set of important equipment for dust cleaning for the blast furnace during iron making process. Since it is very difficult to layout cables to build a remote monitoring and diagnosis system, the wireless sensor network is utilized for data collection and transmission instead. With the help of the WSN, the remote monitoring and diagnosis system can be easily constructed to fulfill data collection and transmission from the spot. This lowers the cost for the construction of the system and improves the operation efficiency of the equipment greatly due to the reduction of the unnecessary halting intervals for the maintenances.

Figure 7 is one of the operating interfaces of the system (in Chinese). Initial operating results show that the wireless sensor networks have a good application prospect in the complicated environment in industry, for example, the web-based remote online monitoring and diagnosis systems, because of their low cost, mobility and easy to construct and maintenance..

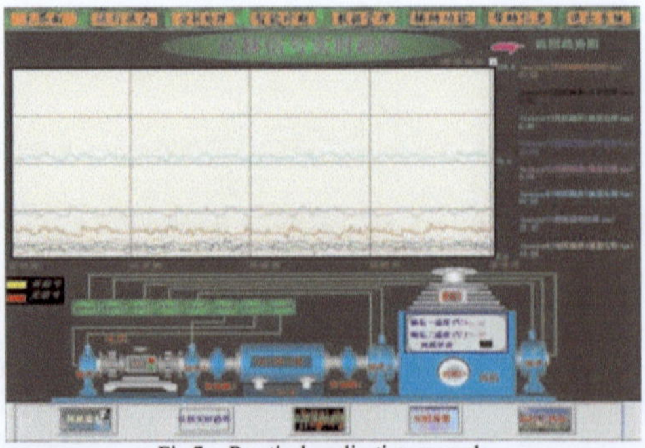

Fig.7. Practical application example

As a new application, of course, there are still some more challenges to be solved better. Anti-interference ability and reliability is among these challenges. In addition, real time response ability is also need to be improved further.

REFERENCES

[1] I.F. Akyildiz, W. Su , Y. Sankarasubramaniam, E. Cayirci, Wireless sensor networks: a survey, Computer Networks, 38 (2002) 393–422

[2] Chee-Yee Chong, Srikanta P. Kumar, Sensor Networks: Evolution, Opportunities, and Challenges, Proceedings of the IEEE, Vol. 91, No. 8, August 2003, pp1247~1256

[3] Toshio Hori, Yoshifumi Nishida, et al, Networked Sensors for Monitoring Human Behavior, Proceedings 2003 IEEE International Symposium on Computational Intelligence in Robotics and Automation, July 16-20,2003, Kobe, Japan, pp900~905

[4] Bill Nickerson, Rick Lally, Development of a Smart Wireless Networkable Sensor for Aircraft Engine Health Management, 0-7803-6599-2/01, 2001 IEEE, pp7-3255~7-3262

[5] U. Kunze, Condition Telemonitoring and Diagnosis of Power Plant Using Web Technology, *Progress in Nuclear Energy,* Vol. 43, No. 1-4, pp. 129-136, 2003，http://www.elsevier.com/locate/pnucene

[6] Daniele Miorandi, Stefano Vitturi, Hybrid wired/wireless implementations of Profibus DP:a feasibility study based on Ethernet and Bluetooth, Computer Communications, 27 (2004) 946–960

[7] Karl Crowley, June Frisby, S′eamus Murphy et al, Web-based real-time temperature monitoring of shellfish catches using a wireless sensor network, Sensors and Actuators A 122 (2005) 222–230

[8] Raul Morais, A. Valente, C. Couto, J.H. Correia, A wireless RF CMOS mixed-signal interface for soil moisture measurements, Sensors and Actuators A 115 (2004) 376–384

[9] Jochen Schiller, Achim Liers, Hartmut Ritter, ScatterWeb: A wireless sensornet platform for research and teaching, Computer Communications 28 (2005) 1545–1551

[10] Product specification, nRF905, 430MHz-950MHz Single Chip RF Transceiver, Nordic VLSL ASA, 2005

Online Privacy Principles

Kostas Zotos, Andreas Litke and George Stephanides
Dept. of Applied Informatics,
University of Macedonia,
54006 Thessaloniki, GREECE

***Abstract-* The Internet and e-commerce have helped transform economy into a prosperous New Economy. They are, and can continue to be, the nation's high-octane engines of economic growth, productivity improvement and high-wage job-creation. The issue of online commerce and privacy is incredibly complex and does not lend itself to simplistic and premature legislative solutions. More education among policy makers needs to be undertaken to fully understand the impact of legislation and to avoid any unintended consequences that may follow. This paper present some new privacy principles and rules.**

I. INTRODUCTION

Privacy is the ability of an individual or group to stop information about themselves from becoming known to people other than those they choose to give the information to. Privacy is sometimes related to anonymity although it is often most highly valued by people who are publicly known. Privacy can be seen as an aspect of security one in which trade-offs between the interests of one group and another can become particularly clear.

The right against unsanctioned intrusion of privacy by the government, corporations or individuals is part of many countries' laws, and in some cases, constitutions. Almost all countries have laws which in some way limit privacy, for example taxation normally requires passing on information about earnings. In some countries individual privacy may conflict with freedom of speech laws and some laws may require public disclosure of information which would be considered private in other countries and cultures.

Privacy is one of the areas of security where trade-offs become very clear and apparent. For the collection of taxes it is in the interests of government and, probably, the rest of your society if your earnings and income are well known. On the other hand, that same information may be used to select you or your family as a good target for kidnapping. In these narrow terms, our interest is to keep the information private whilst the government's interest is to obtain that information. In most countries this risk is reduced, but not eliminated, by limiting the number of people with access to taxation information. On the other hand, in some countries, mostly places with a low risk of kidnapping such as Finland, such information is directly accessible to anyone who wishes it.

II. PRIVACY PRINCIPLES

The basic privacy principles are the following:

- Provide Individuals with Notice

Web sites that collect personally identifiable information should provide individuals with clear and conspicuous notice of their information practices at the time of information collection. Individuals should be notified as to what type of information is collected about them, how the information will be used, and whether the information will be transferred to unrelated third parties.

- Ensure Consumer Choice

Consumers should have the opportunity to opt out of the use or disclosure of their personally identifiable information for purposes that are unrelated to the purpose for which it was originally collected. Consumers should be allowed to receive benefits and services from vendors in exchange for the use of information. It is important that the consumer understands this use and be able to make an informed choice to provide information in return for the benefit received[1].

- Leverage Market Solutions

Private sector privacy codes and seal programs are an effective means of protecting individuals' privacy. Lawmakers should recognize and build upon the self-regulatory mechanisms the private sector has put in place and continues to build. These mechanisms are backed by the enforcement authority of the Federal Trade Commission and state

K. Elleithy et al. (eds.), *Advances in Computer, Information, and Systems Sciences, and Engineering*, 199–203.

attorneys general. Public policies also should allow organizations to implement fair information practices flexibly across different mediums and encourage innovation and privacy enhancing technologies.

- Ensure National Standards

The Internet is a new and powerful tool of interstate commerce. Public policies related to Internet privacy should be national in scope, thus avoiding a patchwork of state and local mandates. This uniform framework will promote the growth of interstate e-commerce, minimize compliance burdens, sustain a national marketplace and make it easier for consumers to protect their privacy.

- Protect Consumers in the Public and Private Arena

Government and non-profit organizations collect a tremendous amount of personally identifiable information about citizens. The need to foster consumer confidence applies to private and public sector activities. Government agencies and non-profit organizations that collect personally identifiable information should be required to follow fair information practices imposed on the private sector by law or regulation[2].

- Don't Discriminate Against the Internet

Consumers should have confidence that their privacy will be respected regardless of the medium used. Similar privacy principles should apply online and offline. Public policy should not discriminate against electronic commerce by placing unique regulatory burdens on Internet-based activities[3].

- Utilize Existing Enforcement Authority

With the imposition of notice requirements, the Federal Trade Commission should use its existing authority to enforce the mandates of federal legislation. Legislation should not create any new private rights of action.

- Avoid Conflicting or Duplicative Standards

In cases where more than one government agency seeks to regulate the privacy practices of a particular organization or industry, those agencies should offer a single coordinated set of standards[4].

III. How to Achieve Compliance

In practical terms the first step toward compliance is to designate one or more individuals responsible for privacy compliance. This privacy officer must have the training, resources, authority and budget to develop and implement compliant policies and procedures. Organizations must then determine precisely what personal information they collect, use and disclose. This will involve an internal audit the breadth and extent of which will depend upon the complexity of the organization and the work it undertakes[5]. Larger organizations might consider establishing a working group representing input from various areas of the organization. A basic audit should identify the following:

- Personal information about customers and employees collected and retained. Points where personal information may be routinely collected include: point-of-purchase, contests, e-mail, surveys, video cameras, audio tapes, marketing lists, loyalty programs, delivery services, warranties, returns, application forms, websites, call centres, technology enablers, employment applications and benefits applications[6].
- What personal information is used in carrying out business, for example, in sales, marketing, fundraising and customer relations.
- What personal information the organization obtains from, or disclose to, affiliates or third parties, for example in payroll outsourcing or benefits provider.
- That appropriate protocols are in place to ensure continued protection where personal information is disclosed to a third party.
- For what purpose personal information is collected.
- To whom personal information is disclosed.
- What forms of consent are employed.
- How the business plan addresses the privacy of personal information.
- Adequate resources are allocated for developing, implementing and maintaining a privacy program.
- What privacy policies the organization has established with respect to the collection, use, disclosure, retention and destruction of personal information.
- How policies and procedures for managing personal information are communicated to employees.
- How management and employees with access to personal information are trained in privacy protection.

- How personal health information is collected, used, disclosed, stored and destroyed.
- That appropriate forms and documents are fully developed.
- Mechanisms are in place to ensure affected individuals are aware of the organization's privacy policies, including the rights to access personal information and if necessary to correct it.
- The organization efficiently and effectively identifies and locates personal information about an individual.
- To comply with established privacy policies, what specific objectives have been set for the organization[7].
- To what extent have appropriate privacy control measures been identified and implemented.
- How the effectiveness of the privacy control measures is monitored and reported.
- What mechanisms are in place to deal effectively with failures to properly apply the established privacy policies and procedures.
- How the organization will benefit from a comprehensive assessment of the risks, controls and business disclosures associated with personal information privacy.

IV. THE INTERNET LAW AND POLICY FORUM

The Internet Law & Policy Forum ("ILPF") convened a working group in Brussels, Belgium on 23 April 1999 to consider the high level principles presented in the ILPF's Legislative Principles For Electronic Authentication and Electronic Commerce (the Legislative Principles) within the current international context. The earlier Legislative Principles found wide use as the public dialogue over the recognition of electronic authentication began in earnest. The ILPF convened this international working group after publishing its survey of the trends in legislative proposals to address electronic authentication around the globe (the International Survey) taking into account extensive international dialogue on this complex topic thus far.

This International Working Group included experts from national governments and international intergovernmental bodies, European universities, the private practice of law, and the private sector, including providers of electronic authentication services. Discussion leaders and participants are listed at the end of the resulting International Consensus Principles.

The following principles reflect the views of business, as represented by those who participated in the discussion of 23 April and as informed by the other expert views expressed around the table. The term "electronic authentication" was meant in the broadest sense, to include a variety of authentication technologies and tools. The International Working Group continued to recognize the need to address electronic authentication consistent with existing legislative frameworks. In addition, perhaps in recognition of the challenges of establishing a detailed uniformity across a variety of national jurisdictions, the group reaffirmed a principle of party autonomy. They suggested that wide recognition of the traditional legal principle of freedom of contract, that is, respecting the rights of parties, both businesses and consumers, to agree to their own requirements for electronic authentication (so called "system rules"), subject only to paramount public interests, could reduce the need for detailed national legislation. The group itself did not have an opportunity to discuss in detail what particulars might be included in the carve out for matters of paramount public interest.

The group continued to underscore the value of harmonization of national legal frameworks and asked states to recognize that technological innovation and the growth of a vigorous global market in authentication products and services will best serve users.

These international Consensus Principles are intended to facilitate global electronic commerce based on the creation of a predictable legal environment for electronic authentication which protects users and reflects their needs. The working group process is intended to provide a neutral forum for an exchange of views on a complex set of issues.

The following principles were distilled from the discussion of the International Working Group convened in Brussels on 23 April 1999 and represent the consensus of the group and commentary by the Internet Law & Policy Forum.

- Governments should identify and remove legal barriers that hinder the recognition of electronic authentication. An electronic authentication should not be denied legal effect solely because of its electronic form. Governments should address formal writing, signature, and authentication requirements in law, regulation and policy to ensure that, where appropriate, electronic authentications (including electronic and digital signatures) are legally recognized. In many cases, whether or not existing laws or regulations impose legal barriers to the recognition of electronic authentication techniques may not be altogether clear. One important goal of legislation in this area should be to remove any such uncertainty. The approach for removal of such barriers differs between legal systems and national jurisdictions. Removing legal barriers to the recognition of electronic authentication techniques and the removal of any related uncertainty should be the first goal of legislation in this area - to the extent legislation is required at all. Governments should not introduce requirements for electronic authentication where none exist for traditional authentication.
- To the fullest extent possible, national laws and jurisdictions should recognize and give full legal effect to contractual agreements concerning the use and recognition of electronic authentication techniques. Governments should respect party autonomy (freedom of contract) as it relates to electronic authentication. For example, parties to an electronic transaction should be permitted to define their use of and reliance on specific technology and security requirements. Exceptions may apply if there is a demonstrable and compelling public policy interest for a government to provide a moderating or protective role, for example, to protect the public safety or prevent fraud.
- Legal rules relating to electronic authentication should be made to operate collaboratively and provide consistent results across jurisdictions to promote the growth of electronic transactions and establish a predictable legal environment for the use and recognition of electronic authentication methods. A consistent, harmonized legal framework will promote the growth of global electronic commerce. (Consistency or harmonization is not synonymous with uniformity.) However, the need for legislative or governmental action to harmonize legal rules may be reduced if users within a "closed" system are able to set their own rules for recognition (so-called "system rules"). Because closed systems are likely to grow significantly in number and size, particularly at the international level, and because these systems have the ability to create global interoperability so long as they are left unimpeded by national regulation, a harmonized legal framework can be fostered by providing a reasonable level of certainty that system rules will be recognized and enforced internationally. Neither authentication service providers nor users will be served by a premature designation of technologies or rules which are created for the sole purpose of erecting a uniform framework.
- Governments should recognize that their actions with respect to electronic authentication can create barriers to trade. Governments should not unreasonably discriminate against electronic authentication methods or providers from other jurisdictions or erect improper non-tariff barriers to trade. Electronic commerce occurs over an inherently global medium. Users of electronic authentication can profit from free and fair competition among service providers, including those located outside their national jurisdiction. In addition, governments should not unreasonably discriminate against electronic authentication technologies that originate in other jurisdictions or against communications that are authenticated by entities located in other jurisdictions. Governments should

furthermore not impede global competition (for example, in limiting the provision of electronic authentication services from foreign suppliers), and should avoid erecting improper non-tariff barriers (for example, in improperly impeding the recognition of foreign electronic signatures).

- Governments should not require or unduly promote the use of particular electronic authentication means or technologies. Governments should anticipate that authentication means (including both the use of business practices and technology solutions) will change over time in response to technological developments and market demands. They should avoid any action likely, directly or indirectly, to preclude or discourage innovation in authentication technologies or new applications for those technologies. In particular, when a government acts as participants in the marketplace, engaging in transactions with citizens and other parties, it should not "lock in" particular electronic authentication means through the force of its presence in the marketplace, but rather should allow for changing market standards and applications for existing and future technologies.
- Standards for use of electronic authentication methods or technologies should be market-driven to meet user needs. Governments should avoid laws that force the private sector to designate a particular technology for electronic authentication. Standards (for example, for technical interoperabililty) should evolve in response to needs in the commercial market, not by the requirements of government.

V. CONCLUSIONS

All of us (lawmakers, business and consumers) need to learn more about online privacy and how we all can work together to protect it. Companies will also schedule briefings for national and state policymakers to showcase new privacy-protecting technologies. The issue of online privacy is incredibly complex and does not lend itself to simplistic and premature legislative solutions. While we are encouraged to see that legislators from both parties are beginning to examine the impact of such regulation on the Internet economy, more education among policy makers needs to be undertaken to fully understand the impact of legislation and to avoid any unintended consequences that may follow[8]. Federal legislation should protect consumers without imposing burdensome, impractical new requirements. Poorly crafted legislation will translate into higher consumer costs, fewer online services, and less free content -- thus hurting the same consumers such legislation intends to benefit. It is important to keep the Internet open and allow consumers to enjoy the full benefits of the new economy's innovations.

REFERENCES

[1] Giorgos Samoladas, "Oikonomikes arxes", ekdoseis Zigos 1999, pages 50-60
[2] Manolis Laourtakis, "E-Society",Ekdoseis Zafeiriou 2002, pages 15-17
[3] Sakis Mpogdanos, "To internet", Anatolikes ekdoseis 2004, pages 112-114
[4] Ioannis Maldanes, "Basikes arxes Internet", Makedonikes ekdoseis 2004, pages 28-29
[5] Joan Birt, "Privacy Control", Somet 2002, pages 46-49
[6] http://www.privacy.gov.au/publications/npps01.html
[7] http://www.privacy.org/
[8] Martin Wilson, "Internet legislation", Feral 2000, pages 16-18

Efficient Support of Wireless Video Multicast Services in 3G and beyond

Kostas E. Psannis, Marios Hadjinicolaou and Yutaka Ishibashi

Abstract—**In this paper, we present an efficient method for supporting wireless video multicast services. The method is based on storing multiple differently encoded versions of the video stream at the server. The corresponding video streams are obtained by encoding the original uncompressed video file as a sequence of I-P(I) frames using different GOP pattern. Mechanisms for controlling the multicast service request are also presented and their effectiveness is assessed through simulations. Wireless multicast video services are supported with considerably reduced additional delay and acceptable visual quality at the wireless client –end.**

Index Terms—**Multicast Services, Multimedia Broadcasting, Wireless Video**

I. INTRODUCTION

The rapid growth of wireless communications and networking protocols will ultimately bring video to our lives anytime, anywhere, and on any device. Until this goal is achieved, wireless video delivery faces numerous challenges, among them highly dynamic network topology, high error rates, limited and unpredictably varying bit rates, and scarcity of battery power. Most emerging and future mobile client devices will significantly differ from those used for speech communications only; handheld devices will be equipped with color display and a camera, and have sufficient processing power to allows presentation, recording, and encoding/decoding of video sequences. In addition, emerging and future wireless systems will provide sufficient bit rates to support video communication applications. Nevertheless, bit rates will always be scarce in wireless transmission environments due to physical bandwidth and power limitations; thus, efficient video compression is required [1]-[2].

In the last decade video compression technologies have evolved in the series of MPEG-1, MPEG-2, MPEG-4 and H.264 [3]-[6]. Given a bandwidth of several hundred of kilobits per second, the recent codecs, such as MPEG-4, can efficiently transmit quality video.

An MPEG video stream comprises intra-frames (I), predicted frames (P), and interpolated frames (B) [3]-[5]. According to MPEG coding standards, I-frames are coded such that they are independent of any other frames in the sequence; P-frames are coded using motion estimation and each one has a dependency on the preceding I- or P- frame; finally the coding of B- frames depends on the two "anchor" frames - the preceding I/P frame and the following I/P frame. An MPEG coded video sequence is typically partitioned into small intervals called GOP (Group Of Pictures).

Streaming of live or stored video content to group of mobile devices comes under the scope of Multimedia Broadcast/Multicast Services (MBMS) standard [7]. MBMS standardization is still in process. It seems that its pure commercialization will need at least three more years. Some of the typical applications are subscription to live sporting, events, news, music, videos, traffic and weather reports, and live TV content. MBMS has two modes in practice: broadcast mode and multicast mode. The difference between broadcast and multicast modes is that the user does not need to subscribe in each broadcast service separately, whereas in multicast mode, the services can be ordered separately. The subscription and group joining for the multicast mode services could be done by the mobile network operator, the user him/herself or a separate service provider. The current understanding about the broadcast mode is that the services are not charged, whereas the multicast mode can provide services that are billed. Specifically MBMS standard specifies transmission of data packets from single entity to multiple recipients. The multimedia broadband-multicast service center should be able to accept and retrieve content from external sources and transmit it using error resilient schemes.

In recent years several error resilience techniques have been devised [8]-[15]. In [8], an error resilience entropy coding (EREC) have been proposed. In this method the incoming bitstream is re-ordered without adding redundancy such that longer VLC blocks fill up the spaces left by shorter blocks in a number of VLC blocks that form a fixed-length EREC frame. The drawback of this method is that the codes between two synchronization markers are dropped, results any VLC code in the EREC frame be corrupted due to transmission errors. A rate-distortion frame work with analytical models that characterize the error propagation of the corrupted video bitstream subjected to bit errors was proposed [9]. One drawback of this method is that it assumes the actual

Kostas E. Psannis is with the School of Engineering and Design, Department of Electronic & Computer Engineering , Uxbridge, Middlesex, UB8 3PH, UK (e-mail: mobility2net@gmail.com, Kostas.Psannis@brunel.ac.uk).

Marios Hadjinicolaou is with the School of Engineering and Design, Department of Electronic & Computer Engineering , Uxbridge, Middlesex, UB8 3PH, UK

Yutaka Ishibashi is with the Nagoya Institute of Technology, Graduate School of Engineering , Department of Computer Science and Engineering, Nagoya 466-8555, Japan

K. Elleithy et al. (eds.), Advances in Computer, Information, and Systems Sciences, and Engineering, 205–210.

rate distortion characteristics are known, which makes the optimization difficult to realize practical. In addition the error concealment is not considered. Error concealment has been available since H.261 and MPEG-2 [4]. The easiest and most practical approach is to hold the last frame, that was successful decoded. The best known approach is to use motion vectors that can adjust the image more naturally when holding the previous frame. More complicated error concealment techniques consist of a combination of spatial, spectral and temporal interpolations with motion vector estimation. In [10] an error resilience transcoder for General Packet Radio Service (GPRS) mobile accesses networks is presented. In this approach the bit allocation between insertion error resilience and the video coding is not optimized. In [11] optimal error resilience insertion is divided into two subproblems: optimal mode selection for macroblocks and optimal resynchronization marker insertion. Moreover in [12] an approach to recursively compute the expected decoder distortion with pixel –level precision to account for spatial and temporal error propagation in a packet loss environment is proposed. Both of these methods [11] and [12], inter frame dependencies are not considered. In MPEG-4 video standard [5], application layer error resilient tools were developed. At the source coder layer, these tools provide synchronization and error recovery functionalities. Efficient tools are Resynchronization Marker and Adaptive Intra frame Refresh (AIR). The marker localizes transmission error by inserting code to mitigate errors. AIR prevents error propagation by frequently performing intra frame coding to motion domains. However AIR is not effective in combating error propagation when I- frames are less frequent.

A survey of error resilient techniques for multicast applications for IP-based networks is reported in [13]. It presents algorithms that combine ARQ, FEC and local recovery techniques where the retransmissions are conducted by multicast group members or intermediate nodes in the multicast tree. Moreover video resilience techniques using hierarchical algorithms are proposed where transmission of I- P- and B frames are sent with varying levels of FEC protection. Some of the prior research work on error resilience for broadcast terminals focuses on increasing FEC based on the feedback statistics for the user [14]. A comparison of different error resilience algorithms for wireless video multicasting on wireless local area networks is reported in [15]. However in the literature survey none of the methods applied error resilience techniques at the video coding level to support multicasting services.

Error resilient (re)-encoding is a technique that enables robust streaming of stored video content over noisy channels. It is particularly useful when content has been produced independent of the transmission network conditions or under dynamically changing network conditions.

This paper focuses on signaling aspects of mobile clients, such as joining or leaving a multicast session of multimedia delivery. Developing error resilience technique which provides high quality of experience to the end mobile user is a challenge issue. In this paper we propose a very efficient error resilience technique for MBMS services. Similar to [16] by encoding separate copies of the video, the multicast video stream is supported with minimum additional resources. The corresponding interactive version is obtained by encoding every (i.e. uncompressed) frame of the original movie as a sequence of I- P(I)- frames using different GOP pattern.

The paper is organized as follows. In Section II the Multimedia Broadcast/Multicast Service standard is briefly discussed. In Section III the problem of supporting Multimedia Broadcast/Multicast Services over wireless networks is formulated. In Section IV the preprocessing steps required to support efficient multicast streaming services over wireless networks is detailed. Section V presents the extensive simulations results. Finally conclusions are discussed in Section VI.

II. MULTIMEDIA BROADCAST MULTICAST SERVICE

Third Generation Partnership Project (3GPP) has standardized four types of visual content delivery services and technologies

-- Circuit-switched multimedia telephony [17]
-- End-to-End packet-switched streaming (PSS) [18]
-- Multimedia messaging service (MMS) [19]
-- Multimedia broadcast/multicast service (MBMS) [7]

The first three mobile applications assume the point-to-point model, where two single end-points (e.g. client-server) are communicating one another. As its name indicates, MBMS has two modes in practice: broadcast mode and multicast mode.

A broadcast service can be generalized to mean a unidirectional point –to-multipoint service in which data is transmitted from a single source to multiple terminals in the associated broadcast service area. On the other hand, a Multicast service can be defined as a unidirectional point-to-multipoint service in which data is transmitted from a single source to a multicast group in the associated multicast service area. Only the users that are subscribed to the specific multicast service and have jointed the multicast group associated with the service can receive the Multicast Services. As a difference a Broadcast Service can be received without separate indication from the customers. In practice multicast users need a return channel for the interaction procedures in order to be able to subscribe to the desired services.

Similar to (PSS) and (MMS), two type applications of (MBMS) standard are anticipated:

- MBMS download: To push a multimedia message to clients
- MBMS streaming: Continuous media stream transmission and immediate playout

The protocol stack is designed to accommodate the above applications as illustrated in Fig.1

<table>
<tr><td>Streaming Applications</td><td colspan="2">Download Applications</td></tr>
<tr><td>RTP playload (Codec)</td><td>3GPP file download</td><td>Service Announcemnet</td></tr>
<tr><td rowspan="3">RTP</td><td colspan="2">FLUTE</td></tr>
<tr><td colspan="2">ALC/FEC</td></tr>
<tr><td colspan="2">LCT</td></tr>
<tr><td colspan="3">UDP</td></tr>
<tr><td colspan="3">IP-multicast</td></tr>
<tr><td colspan="3">Multimedia Brocast Multicast Service (MBMS) bearer(s)</td></tr>
</table>

Fig. 1. Protocol stack view of MBMS.

The streaming stack is very similar to PSS [18]. On the other hand the download stack is unique in terms of its adoption of IETF reliable multicast/broadcast delivery in error-prone environments. As protocol, FLUTE is fully specified and built on top of the Asynchronous Layered Coding (ALC) protocol of the layered coding transport (LCP) building block. File transfer is administrated by special-purpose objects, file description table (FDT) instances, which provide a running index of files and their essential reception parameters in-band of a FLUTE session. ALC is the adaption protocol to extend LCT for multicast. ALC combines the LCT and FEC building blocks. LCT is designed as a layered multicast transport protocol for massively scalable, reliable, and asynchronous content delivery. An LCT session comprises multiple channels originating at a single sender that are used for some period of time to carry packets pertaining to the transmission of one or more objects that can be of interest to receivers. The FEC building block is optionally used together with the LCT building block to provide reliability. The FEC building block allows the choice of an appropriate FEC(e.g., Reed-Solomon) code to be used with ALS, including using the no-code FEC scheme that simply sends the original data using no FEC coding [7].

III. PROBLEM FORMULATION

The MBMS system introduces a new paradigm from the traditional internet or satellite based multicasting system due to mobility. The system has to account for wide variety of receiver conditions such as handover, speed of the receiver, interference and fading. Moreover the required bandwidth and power should be kept low for mobile devices.

Since mobility is expected during session there is typically significant packet loss during handover. If the packet loss occurs on an I-frame, it would effect all the P- and B-frames that predict from the I- frame. In the case of P- frames, the error concealment techniques could mitigate the loss however the distortion would continue to propagate until an I-frame is found. These could also be managed using intra block refresh rate technique. On the other hand, loss of B- frames limits the loss to that particular frame and does not result in error propagation

When a mobile joints an existing multicast session, there is a delay before which it can get synchronized. This delay is proportional to the frequency of I- frames as determined by the streaming server. Since I- frames require more bits that the P- and B- frames, the compression efficiency is inversely to the frequency of I- frames. Assume that I_{number} is the frequency of I- frames, F_{rate} is the frame rate of the video compression. The worse case initial delay in seconds can be computed as follows

$$Delay = \frac{1}{I_{number}} - (\frac{1}{F_{rate}}) \tag{1}$$

where

$$I_{number} = \frac{F_{rate}}{N}$$

N is the distance between two successive I frames defining a "Group of Pictures" (GoPs). N can be defined as follows:

$$N = \begin{cases} \alpha \times M, & M>0,\ \alpha>0,\ I-P-B\,frames \\ N=\alpha & M=1,\ \alpha>0,\ I-P\,frames \\ N>0, & M=0,\quad I-frames \\ M, & N=M>0,\quad I-B.frames \end{cases} \tag{2}$$

where , M is the distance between two successive P frames. (usually set to 3) and a is nonnegative constant $(a \geq 0)$.

Fig.2 depicts the worse delay in seconds for different combination of frame rate and the number of I- frames in a Group Of Pictures.

The graph in Fig. 2 shows that the delay is proportional to the frequency of I- frames. The application would also require more frequent transmission of I-frames so as to allow the use to joint the ongoing session. However this would result in requiring more bandwidth.

Assuming that the ratio of frame sizes for I- P- and B-frames is 5:3:2.The MPEG bitstream used for simulation is the "Mobile" sequence with 180 frames. The average bandwidth

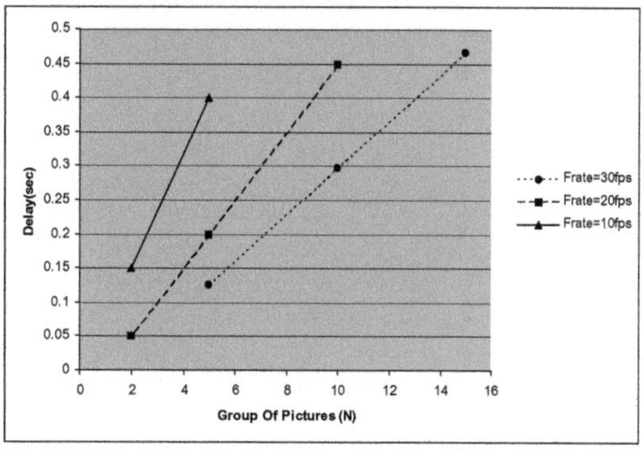

Fig. 2. Relative increase in the delay as a function of Group Of Pictures (N)

is given by [20]

$$Bandwidth = F_{rate} \times Average\,(IP)\ Size \times 8\,{}^{bits}\!/_{byte}$$

where

$$Average(IPB)Size = \frac{I_{average}}{N} + P_{average}x(\frac{1}{M} - \frac{1}{N}) + B_{average}x(1 - \frac{1}{M})$$

Fig.3 shows the increase in the network bandwidth as a function of Group Of Pictures. It can be seen from this graph that, more I- frames in a GOP results in increase in the network bandwidth.

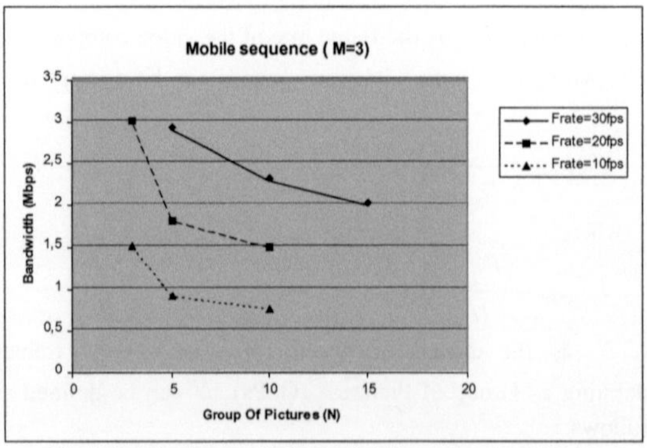

Fig. 3. Relative increase in the Bandwidth as a function GOP (N)

One other tool that is effective against error propagation is Intra block refresh technique [5]. In this technique, a percentage of P- and B-frames block are intra coded and criterion for determining such intra clock is dependent on the algorithm. However the intra blocks refresh technique is not effective in combating error propagation when I- frames are less frequent.

Apart from traditional broadcasting/ multicasting techniques, the MBMS system requires new technologies for error resilience. This is because MBMS does not allow retransmissions and the temporal fading conditions of wireless channels could result in corruption of certain frames. Due to the frame dependency within hybrid coding techniques the errors propagate until an I-frame is decoded.

IV. Proposed Technique

In a typical video distribution scenario, video content is captured, then immediately compressed and stored on a local network. At this stage, compression efficiency of the video signal is most important as the content is usually encoded with relatively high quality and independent of any actual channel characteristics. Note that heterogeneity of client networks makes it difficult for the encoder to adaptively encode the video contents to a wide degree of different channel conditions. This is especially true for wireless clients. It should also be noted that that the transcoding (decode-(re)encode) of stored video is necessary the same as that for live video streaming. For instance, pre analysis may be performed on stored video to gather useful information. If the server only has the original compressed bitstream (i.e., the original uncompressed sequence is unavailable), we can decode the bitstream.

The problem addressed is that of transmitted a sequence of frames of stored video using the using the minimum amount of energy subject to video quality and bandwidth constraints impose by the wireless network.

Assume that I- frame is always the start point of a joining multicast session. Since I- frames are decoded independently, switching from leaving to joining multicast session can be done very efficiently. The corresponding video streams are obtained by encoding the original uncompressed video file as a sequence of I-P(I) frames using different GOP pattern ($N = 5, M = 1$)

P(I) are coded using motion estimation and each one has a dependency only on the preceding I- frame. This result that corruption of P-frame do not affect the next P-frame to be decoded. On the other hand, it increases the P(I) frame sizes.

We consider a system where source coding decisions are made using the minimum amount of energy $min\,E^{q(i)}\{I,P(I)\}$ subject to minimum Distortion (D_{min}) at the mobile client and the available Channel Rate (C_Rate) required by wireless network. Hence

$$min\,E^{q(i)}\{I,P(I)\} \leq C\_Rate \tag{3}$$

$$min\,E^{q(i)}\{I,P(I)\} \leq D_{min}$$

It should be emphasized that a major limitation in wireless networks is that mobile users must rely on a battery with a limited supply of energy. Effectively utilizing this energy is a key consideration in the design of wireless networks. Our goal is to proper select a quatizer q(i) in order to minimize the energy required to transmit the sequence of I- P(I) frames subject to both distortion and channel constraints.

A common approach to control the size of an MPEG frame is to vary the quantization factor on a per-frame basis [21]. The amount of quantization may be varied. This is the mechanism that provides constant quality rate control. The quantized coefficients $QF[u,v]$ are computed from the DCT coefficients $F[u,v]$, the quantization_scale, $MQUANT$, and a quantization_matrix, $W[u,v]$, according to the following equation.

$$QF[u,v] = \frac{16 \times F[u,v]}{MQUANT \times W[u,v]} \tag{4}$$

The normalized quantization factor $w[u,v]$ is

$$w[u,v] = \frac{MQUANT \times W[u,v]}{16} \tag{5}$$

The quantization step makes many of the values in the coefficient matrix zero, and it makes the rest smaller. The result is a significant reduction in the number of coded bits with no visually apparent difference between the decoded output and the original source data [22]. The quantization factor may be varied in two ways.

- Varying the quantization scale (*MQUANT*)
- Varying the quantization matrix ($W[u,v]$)

To bound the size of predicted frames, an P(I)-frame is encoded such that its size fulfils the following constraints

$$BitBudget\{I, P(I)\} \leq C\_Rate \qquad (6)$$
$$BitBudget\{I, P(I)\} \leq D_{min}$$

The encoding algorithm in the first encoding attempt starts with the nominal quantization value that was used to encode the preceding I-frame. After the first encoding attempt, if the resulting frame size fulfils the constraints (6), the encoder proceeds to the next frame. Otherwise, the quantization factor (quantization_matrix, $W[u,v]$) varies and the same frame is re-encoded. The quantization matrix can be modified by maintaining the same value at the near-dc coefficients but with

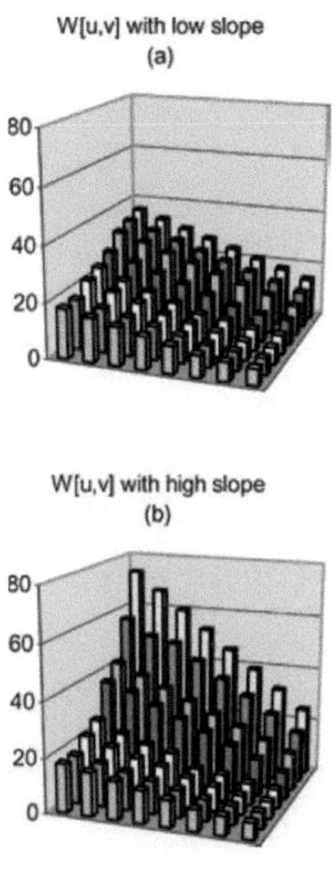

Fig. 4. Two normalized quantization matrices w[u,v] both with MQUANT=8 (a) W[u,v] with low slope; (b) W[u,v] with high slope

different slope towards the higher frequency coefficients. This procedure is repeated until the size of the compressed frame corresponds to (6). The advantage of this scheme is that it tries to minimizes the fluctuation in video quality while satisfying channel condition.

Fig. 4 shows two matrices both with the same value at the near-dc coefficients but with different slope towards the higher frequency coefficients. In other words, the quantization_scale is fixed *MQUANT* and the quantization_matrix $W[u,v]$ varies.

V. SIMULATIONS RESULTS

There are two types of criteria that can be used for the evaluation of video quality; subjective and objective. It is difficult to do subjective rating because it is not mathematically repeatable. For this reason we measure the visual quality of the interactive mode using the Peak Signal-to-Noise Ratio (PSNR). We use the PSNR of the Y-component of a decoded frame.

The MPEG-4 bitstream used for simulation is the "Mobile" sequence with 180 frames, with a frame rate of 30fps. The GOP format was N=5, M=1.We consider a set of allowable channel rate, C_Rate= (300Kbps, 200Kbps, 100Kbps). In order to illustrate the advantage of the proposed algorithm we consider a system where source coding decisions are mode without any constraints, using the same GOP format (N=5, M=1). Figure 5 shows the PSNR plot per frame obtained with the proposed algorithm and the reference scheme for the allowable channel rate. Clear the proposed algorithm yields an advantage of 1.42db, 1.39db, 1,35db, for the allowable channel rates 300kbps, 200kbps, 100kbps respectively.

Figure 6 depicts the performance of the proposed algorithm compared with the performance of MPEG-4 simple profile [5] codec during frame loss. The percentage of frames that are dropped is varied and it is clearly seen that the proposed approach maintains the quality. On the other hand the MPEG-4 simple profile codec, with different Intra (I)-VOP periods (N), during frame loss, the quality degrades with increase in frame loss percentage. The "Mobile" sequence was used for these experiments with bit rate 100kbps and frame rate 30fps.

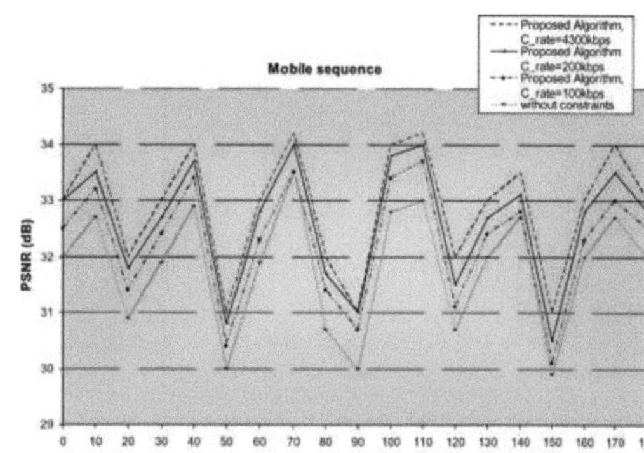

Fig. 5. PSNR plot per frame obtained with the proposed algorithm and the reference scheme

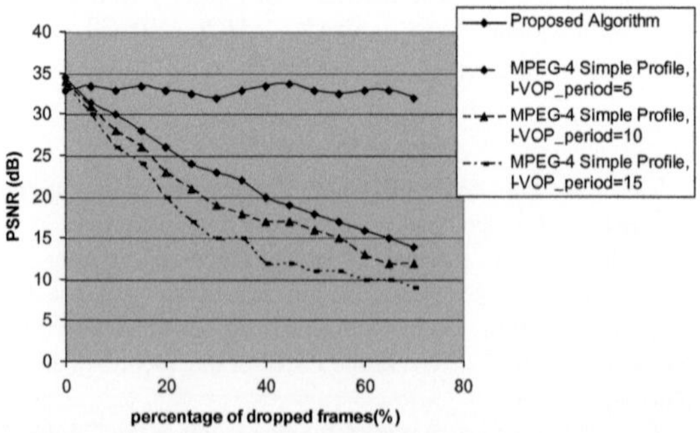

Fig. 6. PSNR as function of frames dropped

VI. CONCLUSIONS

Error resilient (re)-encoding is a key technique that enables robust streaming of stored video content over noisy channels. It is particularly useful when content has been produced independent of the transmission network conditions. In this paper, we investigated the constraints of supporting multimedia multicast services in wireless clients. In order to overcome these additional resources we proposed the use of differently encoded version of each video sequence. The differently coded sequences is obtained by encoding frames of the original (uncompressed) sequence as a I-P(I) frames using different GOP pattern. The server responds to a multicast request by switching from leaving to joining multicast session very efficiently. By proper encoding versions of the original video sequence, multicast video streaming services can be supported with considerably reduced additional delay and acceptable visual quality at the wireless client –end. Our future work includes developing a multilevel Quality of Services framework for wireless full interactive multicast video services.

REFERENCES

[1] M. Ethoh and T. Yoshimura, "Advances in Wireless Video Delivery", Proc. IEEE, vol.93, no.1, 2005, pp. 111-22
[2] Kostas E. Psannis, Marios Hadjinicolaou:, "MPEG based Interactive Video Streaming: A Review", WSEAS Transaction on Communication, vol.2, April 2004, pp 113-120
[3] Coding of Moving Pictures and Associated Audio for Digital Storage Media at up to About 1.5Mbits/s, (MPEG-1), Oct. 1993
[4] Generic Coding of Moving Pictures and Associated Audio. ISO/IEC 13818-2, (MPEG-2), Nov 1993.
[5] Coding of Moving Pictures and Associated Audio.MPEG98/W21994, (MPEG-4), Mar.1998
[6] T. Stockhammer and N. M Hannuksela, "H.264/AVC Video for wireless Transmission, IEEE Wireless Communications, August 2005, pp 6-72.
[7] 3GPP, "S-CCPCH Performance for MBMS, "Tech. spec, Group Radio Access Network , Mar 2004, TR 25.803 v.1.3.0.
[8] R. Swann and N.Kingsbury, "Transcoding of MPEG-2 for enhanced resilience to transmission errors", Proc IEEE Int. Conf. Image Proc, vol.2 Oct 1996, pp 813-16.
[9] G.de los Reyes et.al., "Error- Resilient transcoding for video over wireless channels", IEEE JSAC, vol 18, no.6, June 2000, pp 1063-74
[10] S.Dogan et al, "Error- Resilient Video Transcoding for Robust internetwork Communications using GPRS", IEEE transaction Circuits Sys. Video Tech, vol 12, no 6, June 2002, pp 453, 564
[11] G. Cote , S. Shirami, and F. Kossentini,, "Optimal Mode Selection and Synchronization for Robust Video Communications over error-phone networks", IEEE JSAC, vol 18, no 6, 2000, pp. 952-65
[12] R.Zhang, S.L Regunathan and K Rose , "Video Coding with Optimal Inter/Intra Mode Switching for Packet Loss Resilience", IEEE trans Multimedia, vol 3, no.1 , May 2001, pp 18-32.
[13] G. Carle and E. Biersack, "Survey of error recovery techniques for IP based audio visual multicast applications", IEEE Network, Dec 1997, pp 38-45
[14] Ge, Peng, McKinley P. K, "Experimental evaluation of error control for video multicast over wireless LANs", Proc IEEE Multimedia Networks Systems, 2001, pp 27-32.
[15] Ge, Peng, McKinley P. K, "Comparisons of Error Control Techniques for Wireless Video Multicasting ", Proc IEEE Computing and Communications, 2002, pp 88-95.
[16] Kostas E Psannis, Marios Hadjinicolaou and Argy Krikelis "MPEG-2 Video Streaming with Full Interactive Content", IEEE transaction on Circuits Sys. Video Tech, (23rd August 2005, accepted for publications)
[17] 3GPP, "Codec for Circuits Switched Multimedia Telephone Service; General Description ," Tech.spec., Group Services and Systems Aspects, June 2002, TS 26.110.
[18] 3GPP, "Transparent End-to-End Packet-switched streaming service stage 1, "Tech. spec. Group Services and System Aspects, Mar. 2002, TS 22.233.
[19] 3GPP, "Multimedia Messaging Service (MMS); Stage 1," 3rd generation partnership project; tech. spec., Group Services and System Aspects, Dec 2002, TS 22.140.
[20] Kostas E. Psannis, Marios Hadjinicolaou "Transmitting Additional Data of MPEG-2 Compressed Video to Support Interactive Operations". In: Proc. IEEE ISIMP01, Hong Kong, May 2001,pp 308-311.
[21] Kostas. E. Psannis, Yutaka Ishibashi and Marios Hadjinicolaou, "A Novel Method for supporting Full Interactive Media Stream over IP Network", International Journal on Graphics, Vision and Image Processing, Vol. 4, pp 25-31, (2005)
[22] Kostas E. Psannis, Marios Hadjinicolaou and Argy Krikelis, 'Full Interactive Functions in MPEG-based Video on Demand Systems", In Nikos E. Mastorakis and G Antoniou editors, Recent Advances in Circuits, Systems and Signal Processing Book Series, Published by WSEAS Press, pp 419-424 (2002).

Designing a pervasive architecture for car pooling services

Pier Luca Montessoro, Sandro Di Giusto, and Davide Pierattoni, *Member, IEEE*

Abstract— **The essential car pooling principle is to optimize the transport utilization by the ride sharing among people who usually cover the same route. Car pooling technique is really promising for rationalizing the private use of transports. Moreover, the economic and environmental advantages resulting from this kind of management are considered positively by several public and private international bodies and administrations. This article introduces a methodology for the definition of groups composed by people who usually cover the same route. It's a very innovative system, since it's based on pervasive computing resources. Therefore, here is introduced the distributed architecture of the pervasive car pooling system, with its constituent elements, its algorithm and the information flows shared by all the entities of the distributed application. This model appears to be easily put into practice, to be cost-effective, to be efficient, to be a safe way of communication and to protect the privacy of all the information treated by the system.**

Index Terms— **Car pooling, pervasive computing, public key encryption, wireless communication.**

I. Introduction

JUST think about how many times you can come into contact with people who potentially have a lot in common with you, and you are unaware of that! In particular, consider all the people you could share your way to workplace with, everyday. It's natural to guess all the advantages that such a knowledge can offer.

First of all, you can gain many economical advantages, thanks to the sharing of fixed costs which allows the individuals' retrenchment of expenses. Moreover, the natural environment would be protected through the reduction of circulating polluting vehicles and a lower fuel consumption. Besides, don't ignore all the social benefits, such as the acquaintance of new people, less psychophysical stress for traffic and the reduction of the average journey time [9]. Car Pooling concept is based on these very reasons: to optimize the transport utilization by the ride sharing among people who usually cover the same route.

In this article, we suggest the innovative idea of automating and simplifying the knowledge sharing among all the potential users of the car pooling service, through a widespread distribution of electronic infrastructures. The suggested architecture[1] uses some little portable devices – a key chain, for instance: such device will be called with the general term of Pervasive Object (PO) from now on – which can locate the average route covered by some interested and trustworthy users at the same time, in a safe and automatic way. This is the Pervasive Car Pooling approach.

Section 2 discusses previous related work on this subject, Section 3 describes the general characteristics of the Pervasive Car Pooling structure in short, while the Section 4 illustrates the overall working model in detail.

Sections 5 and 6 carry the working tasks of the system and the features of the different entities. Section 6 proposes some possible measures to be adopted for warranting the absolute security of communication and the users' privacy; Section 7 contextualizes some practical examples in some possible future backgrounds.

In the end, Section 8 provides some conclusions and considerations about the project proposed.

II. Related Works

Several countries have been stimulating and financing the mobility management policies for a long time, and they invest money into the planning to spread the car pooling [1][10].

In accordance with an Italian law, such principles should be spread, rationalized and standardized through the funding of development and research in this very field. Several projects of this kind have already been developed and studied from an academic point of view [2], and some web experimentations are already working [3][4][5]. An example of ride sharing experience in the United States is shown in [6], while in Canada such culture is spreading wide, as shown in [7].

Car Pooling could be seen as a particular instance of network-based introduction systems. References to Mobile Social Software can be found in [8], presenting the Social Serendipity MIT project, a project that is a good candidate to support a Car Pooling oriented introduction system. However, the Car Pooling application presents some critical issues that are not addressed by general purpose models, such as the

P. L. Montessoro is with the Department of Electrical, Management and Mechanical Engineering, University of Udine, 33100 ITALY. E-mail: montessoro@uniud.it.

S. Di Giusto is with the Department of Electrical, Management and Mechanical Engineering, University of Udine, 33100 ITALY. E-mail: sandro.digiusto@uniud.it.

D. Pierattoni is with the Department of Electrical, Management and Mechanical Engineering, University of Udine, 33100 ITALY. E-mail: pierattoni@uniud.it

This work is partially funded by Eurotech S.p.A., Amaro (UD), Italy.

K. Elleithy et al. (eds.), Advances in Computer, Information, and Systems Sciences, and Engineering, 211–218.

Serendipity approach. First of all, the privacy issue: in the MIT project, an ID related to the user is exchanged between the pervasive devices (cellular phones or PDAs) and a centralized server stores all the information about the user, including the private ones. ID broadcasting does not provide privacy and the server security can become highly critical; moreover, if applied to a large population (a region, for instance) this architecture would not scale adequately. Another obstacle faced by our approach is related to the background collection and progressive refinement of affinity messages, instead of Serendipity's immediate notification to the user. This is not a limitation: in fact, our approach enhances the effectiveness of the final application. Setting up a daily Car Pooling arrangement among ride sharers can and should be done at a later time in respect of the occasional proximity encounter, especially to benefit of the user authentication and trustiness validation by a Certification Authority. Last, our architecture requires ad hoc, cheap devices that can better manage the power supply and the transmission mode, requiring only a anonymous connectionless unacknowledged service. The information exchange in our model, which is later described as unidirectional among Pervasive Objects, can be already applied on existing wireless communication, like Bluetooth. Nevertheless, its advantages will get more appreciable when Pervasive Objects will embed energy-saving wireless technologies.

The security of the contents related to the shared information, together with the privacy protection of all the personal information stored by the system are a crucial point for this project. Some interesting investigations on the trust in uncertain environments are shown in [11]. This approach can overcome the initial suspicion among users to allow secure collaboration to take place; nevertheless, the absence of a server-based control reveals a difficult applicability in our particular field of interest.

III. System Description

A. Basic Idea

The basic idea for Pervasive Car pooling is that people who daily cover the same route at the same time, have a good chance to meet close together sometimes.

This kind of event is often ineffective because people do not exchange information about their daily trip, and most of the time they do not communicate at all (think about an elevator or a reception hall of a large enterprise).

A pervasive agent can automatically discover all the possible affinities. Moreover, the Pervasive Car Pooling here described always guarantees the privacy protection and security, until the final decision to share the car at least.

B. Main Functionalities

The proposed Pervasive Car Pooling architecture provides the following functionalities:

1. user registration at a trusted certification authority (may include user advisability evaluation);
2. daily work-trip (datatrip later on) definition; it can be user-defined or automatically discovered, depending on the available infrastructures;
3. encrypted datatrip information sharing among Pervasive Car Pooling users
4. decryption of collected datatrips and analysis at the trusted certification authority and searching for full or partial affinities
5. notification of the existence of any potential sharing car users.

As will be further investigated, the second functionality can be performed in both automated and manual fashion. In fact, the datatrip can be evicted and refined automatically by the PO entity, collecting events from GPS or WiFi access points or TAGs "on the road", and thus discovering the path. Nevertheless, a user is also allowed to specify his/her own preferred, "hand-made" datatrip to be shared among the service users, without requiring any additional infrastructure.

C. Technological Requirements

The proposed system needs also some technical and technological characteristics to realize a good system prototype and to allow Pervasive Car Pooling becoming a great social success:

1. cheap objects (on the user's side) and scalable network deployment (on the organization's side);
2. low power consumption;
3. logical autonomy (devices will not require a complex interaction with the user);
4. reliability and efficiency in matching affinities;
5. easiness of maintenance.

With current technologies such a system can already be produced, provided a sufficient scale production needed to keep costs lower .

The Pervasive Car Pooling need also some guarantees on behalf of system and users' security, such as:

1. privacy and secure treatment of personal information;
2. anonymous data sharing among users;
3. overall robustness of the whole architecture.

IV. Operation

The system proposed by this article consists of four main entities, which are clearly represented in Fig. 1. These are the Pervasive Objects (PO, which can be shaped as key cases or trinkets or can be integrated with any other devices of common use, such as mobile phones), a personal computer or docking station for the PO with user interface functionality (the User Interface), the certification authority (CA) and optionally a widespread TAG network to locate automatically all the routes sent by every PO.

On configuring his/her PO, every user must save (maybe after a self learning stage) an encrypted description of his/her own daily way to workplace with its respective time band

[1] Patent pending

(datatrip). This information is autonomously sent from every PO to all the other POs it comes into close neighbourhood with (and it can also get their encrypted datatrip, at the same time). Therefore, every PO will get a daily fair number of datatrip from other users. However, this encrypted information cannot be decoded immediately, so as to protect the users' privacy and security.

So, every user is required to send to the CA his/her own datatrip and the ones he/she got through the UI . The certification authority will decrypt all the data and seek any possible compatibility among the users who have come into contact with each other. In case of success, the CA will notify the interested users of the possible analogous routes to workplace, by respecting specific check and data control procedures.

Should a TAG locating network be situated in surroundings where the PO is working, the PO can analyze the user's movements and extract some typical recurring datatrip automatically. Likewise, if provided with a suitable software, the PO can get some further information about the movements by other current wireless systems, such as GPS and WiFi access points.

V. ENTITIES

The system we illustrate in this article refers to four entities (PO, UI, CA and eventually TAGs, see Fig. 1) and passes through a series of activities that are the user's certification, the definition and exchange of the datatrip, the incoming-outgoing affinities analysis made by the CA. Hereby, we describe the specifications for the entities in detail, while the next section is devoted to the activities.

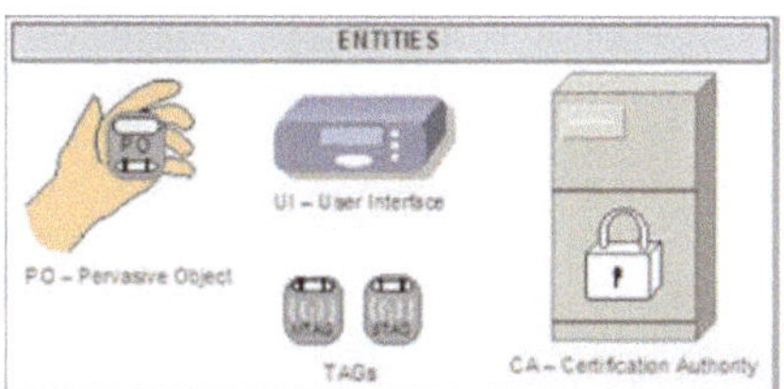

Fig. 1 Entities in the pervasive car pooling architecture

A. *Pervasive Object*

The POs are instruments planned for the spreading of the datatrip information.

Shape. The POs can be thought as small, pretty trinkets which don't need any interaction forms with their owners (without any digit keys, any keyboards, any displays nor leds, but provided with just an on/off switch or a tiny display for the battery charge control, instead). Every user should wear the pervasive object (he/she can also keep it inside his/her poket/handbag) without any particular care. Of course , its functionalities can be even integrated with other devices, such as mobile phones (so as to provide a service which is deeply integrated into pre-existent objects).

Power. The power consumption is a critical aspect when considering any mobile devices , particularly the pervasive ones (which necessarily require wireless intercommunication systems). In that very case, the PO should be provided with a rechargeable battery working for several days without any need to be soon recharged.

Moreover, the battery should be able to supply enough power to the circuitry so as to allow the necessary communication. The recharging system should require the minimum care and effort to the user (through an inductive system, for instance).

Technology. The PO must integrate a fair number of electronic components, since it's created to perform several different functions:

- Wireless communicator for the datatrip interchange among POs;
- Communication system through the UI (it could be the same system used for the trinkets communication, or a wired system associated to the recharging one, maybe through an USB interface)
- Good calculus and storing capacity, since it's required to store a large quantity of information during the learning stage , so as when harvesting and disseminating ones, as hereafter explained.

Should the environment be provided with TAGs to allow the PO to acquire the datatrip information, some further elements are required:

- An internal timing system (to supply the software with the right time and date when events occur)
- A wireless communication system to communicate with TAGs
- A geographical localization system, like GPS or such.

Functions. The functions to be performed by the PO while working are:

- Interaction with UI for the storage of the encrypted datatrip.
- The sharing of that encrypted datatrip with other POs (sending) and the storage of the datatrip received from other POs .
- Interaction with the UI for the transfer of the received datatrip.
- Possible dialogue with the TAGs and storage of the information extracted, also integrating it with other data about the geographical localization.

B. *User Interface*

The UI is the necessary means of interaction between your own PO and the certification/matching searching system , that is the CA. This entity is necessary to warrant the security, to optimize the CA working system and to make the work of the PO less hard. According to the current vision of the project, the UI is a docking station for the PO that allows the connection with a Personal Computer (through an USB interface, for instance) and provides for the charge of the internal batteries. The software can arrange for the automation of most functions, particularly for the interaction with the CA through Internet.

Functions. The UI has many important functions, including:

- defining the characteristic datatrip and transferring them to the PO;
- obtaining the collection of encrypted datatrip from the PO;
- sending the datatrip to the CA through a secure connection;
- management of the defining system for the datatrip information (manual or by autonomous self-learning from the TAGs) ;
- management of the power supply system for the batteries of the PO.

Seeking algorithm. As the PO collects some information about the user's movements through a TAGs network , the UI can process this data. This way, it can extract the user's typical datatrip.

The main concept of the system is to store the information got by the TAGs (even integrating it with others, if necessary) into a database. The PO stores just the most significant data for the location of every characteristic datatrip. Thus, it's possible to save both the necessary memory space and the computing complexity of the seeking algorithm (which doesn't overcrowd the PO anyway, since it's implemented on the UI). Substantially, the algorithm questions the database and processes the content, by managing the stored events as inputs of a finite state machine . Then, any possible location of the PO is associated to every different state (home, car, bus,…). Actually, the algorithm reconstructs the user's daily datatrip from the collection of data that it everyday gains access to. Afterwards, it strikes an average of the data so as to find out any possible recurrent routes covered by the user, and in that case it stores the output in the datatrip. Such datatrip will be confirmed or stored again by the user so as to be shared with other POs thanks to the pervasive technology. Evidently, the algorithm needs a plenty of data to assure an acceptable statistical value. Moreover, the reliability and the precision of the data are proportional to the length of the learning stage (a seven days' learning stage can probably assure very reliable outcomes , especially as for the users who cover the same roads recurrently).

C. Certification Authority

The CA is an element that assures the necessary level of security for the Pervasive Car Pooling service. It's a centralized server or a federation of servers (for organizational reasons, or for a matter of scale and redundancy) that the UI can page to through Internet.

Functions. The CA is required to perform the following functions:

- To manage the new users' entries for the pervasive car pooling service, by storing a minimum set of private information (which will be never spread but with the express consent of the owners) and by defining the parameters for the PO encryption.
- To keep the UI waiting for the connection and, when required, to check the right identification of the requesting UI. Then, to set a secure connection.
- To get from the duly connected UI the associated user's (encrypted) datatrip together with all the (encrypted) ones which have been gathered by him/her before.
- To decrypt the user's own datatrip and the ones belonging to the others, even by questioning other CAs to get the decryption keys.
- To seek for any total/partial match among the user's datatrip and the one(s) which have been collected before.
- In case of found matching, to signal such correspondence to all the interested partners, by keeping the mutual anonymity until their very express authorization to proceed .

The aspects concerning the users' certification and the datatrip encryption will be treated in section 6.

D. TAGs

TAGs can be represented as mini electronic devices with wireless communication facilities, little computational and storage capacity and large energetic autonomy. Two types of TAGs can be foreseen, depending on their collocation: MTAG (Mobile TAG) and STAG (Stationary TAG). The differences between these two types of TAG concern the transmitted data and the power supply.

Deployment. The TAGs should be installed in crucial spots of the geographical areas that is going to be covered with the Pervasive Car Pooling service. Particularly, these devices are used for informing the close PO about its absolute/relative location. That's why they should be installed:

- At the user's house and inside his/her car/motorcycle
- On the buses working across the covered geographic lines.
- At the railway station and the subway station.
- Inside some very crowded public premises (inside the post offices, for instance).

Such arrangement is an interesting compromise between a minute geographical cover (being partly dynamic) and a low social cost (a small number of devices is sufficient, all things considered) by warranting a wide availability of information and a considerable self-learning capability from the PO.

Power saving. The TAGs can be required to communicate wireless with the POs at any time . Therefore, they need a considerable amount of energy at their disposal. It's necessary to point out that our system provides for an exclusively one-way communication scheme from the TAGs to the PO (the TAGs get no information from the POs). This allows us to think that TAGs could be implemented through some passive technologies, which get their power from the light (by following closely the half-passive/passive RFID devices), instead of getting it from an autonomous source. Anyway, the places where TAGs should be installed (cars, motorcycles, buses, railway/subway stations, …) will have an energy source, that's why we can consider it as a secondary detail. On the other hand, an optimum implementation of the PO can be supported by self-supplying energy TAGs. Moreover, it is possible that the energetic characteristic of the PO proves to be incompatible with the activation of passive RFIDs.

Data. In this very case is also necessary to divide the fixed TAGs (stations, public offices, malls, ...) from the mobile ones (cars, motorcycles, buses, ...):

- Fixed TAGs – they can easily send the coordinates of their own location (geographical coordinates, for instance) which can be stored in the ROM at the time of installation
- Mobile TAGs – they send information about the means of transport/buildings they're installed in.

In both cases, the TAGs broadcast also a dataset of further information about the system (as the standard of the TAG, the kind of installation and so on).

One more comment is to be made about the non-specific TAGs: the system enables them to get some useful information even from devices that don't belong to their same network, like WiFi hotspots, for instance. This way it's possible to achieve a wide information by using preexistent devices.

VI. Tasks

Fig. 2 points out the sequential operation of the system and the tasks performed by the single entities during every stage.

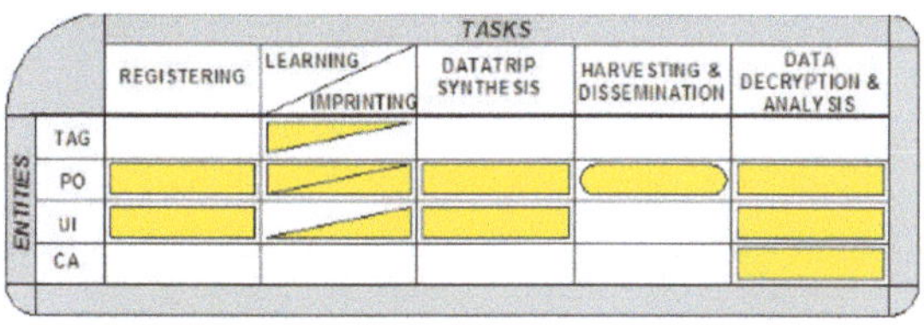

Fig. 2 Entities in the architecture and related tasks

A. Registering

This is the first stage every new user goes through. Every user is provided with a PO, an identifying key and a public (PO_PUK) or private (PO_PRK) couple of cryptographic keys, unless he/she doesn't already have one.

The public key of the CA (CA_PUK) will be stored in the UI program, which encrypts the sending datatrip and makes it accessible just from the CA.

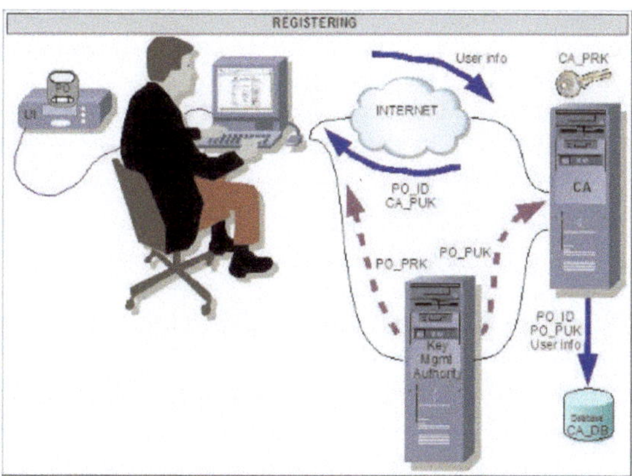

Fig. 3 Registering: a new user joins the system

B. Learning and imprinting

Learning. The learning stage takes place only if both the involved area and the means of transport are provided with TAGs. Such a stage includes the period of time that starts with the user's registration and ends with the datatrip definition (it usually takes some days' work). In this stage, the PO communicates only with the TAGs and with other preset devices (WiFi hot spots, for example), then it stores anything helpful for locating its owner's datatrip. As the learning stage has been settled, it falls to the Datatrip synthesis phase, during which the user is required to confirm or not the obtained datatrip.

Imprinting. For lack of TAGs, the Imprinting stage replaces the Learning one and allows the user to set his/her datatrip by hand.

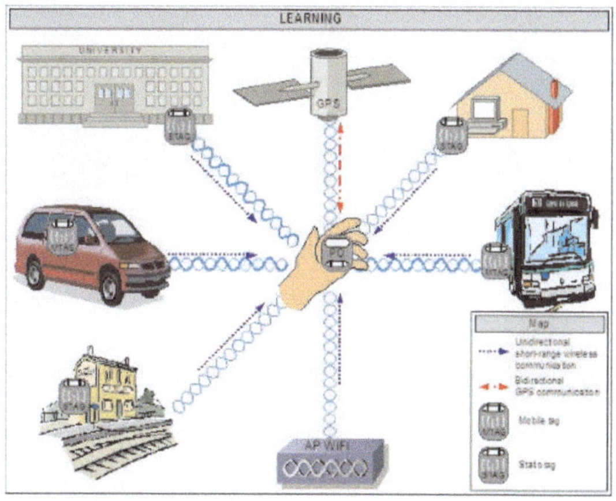

Fig. 4 Learning: PO discovers the datatrip

C. Datatrip synthesis

Any information about the user's movements that the PO can get from the TAG network is soon transferred to the UI and processed. As before specified, the datatrip can be otherwise set by hand. In any case, at this point the datatrip is encrypted and transferred to the PO so as to be broadcast to all the other POs during its working.

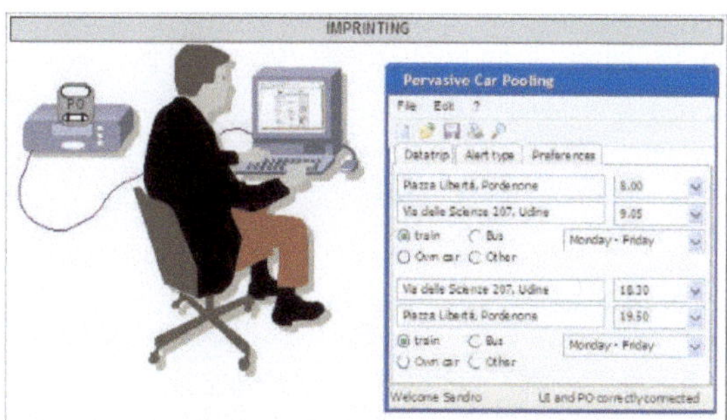

Fig. 5 Imprinting: user inserts the datatrip

D. Harvesting and Dissemination

After concluding the Datatrip synthesis stage, the PO can deliver its own encrypted datatrip (dissemination) and gather the encrypted datatrip sent by the POs it gets in contact with (harvesting). Thus, this stage is based on the wireless communicative power of the PO and on the users' mobility. Two different requirements clash in this phase: whereas on the one hand the datatrip is required to be really undecipherable for both the untrusted users and any possible eavesdropper (or whenever the CA doesn't intervene), on the other hand it's required to avoid the storage of the datatrip that don't match ours. For this purpose, we suggest to create a limited and incomplete version of the datatrip to be broadcast as a clear text (for example, a version including just some schedule, without either any geographical references or user ID) so as to carry out an early filtering stage directly to the PO.

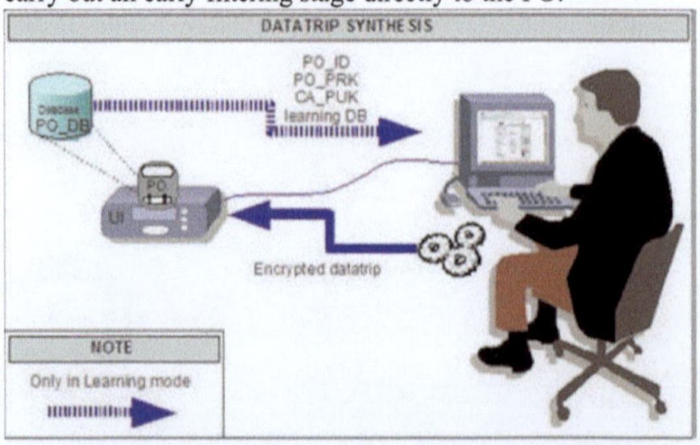

Fig. 6 Synthesis of the datatrip by collected informations

It' s to be remarked that during the Harvesting task, the POs don't set up a pure communication with each other but they simply send and receive a data package that is broadcast through a connectionless and unacknowledged one-way channel.

Fig. 7 Harvesting and dissemination of datatrips among users

E. Data Decryption and Analysis

The data which have been gathered from the PO during the Harvesting stage (encrypted datatrip) are sent to the CA through the UI (therefore, by a PC and through a secure connection) during this task. Together with these data, also the user's own encrypted datatrip is to be sent. As a matter of fact, such datatrip will be the main term of comparison for the matching . The CA can get some further information about other users even by questioning other federate CA servers. Therefore, it starts seeking any matches. The outcomes of the research are transmitted (still remaining anonymous) to the couples or groups of users who have similar datatrip. At this point, the users can authorize the CA to spread their own data inside the group. However, the CA will broadcast the users' names and datatrip only after express consent by all the users composing the couple or the groups.

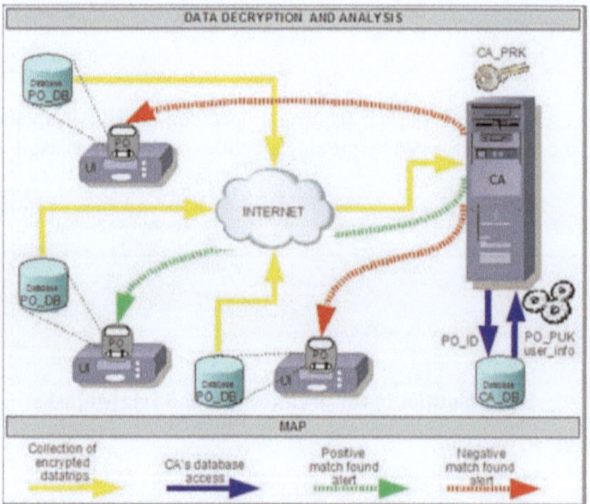

Fig. 8 Decryption and analysis of collected datatrips; searching for affinities

VII. Privacy and Security

The security of the contents related to the shared information, together with the privacy protection of all the personal information stored by the system are a crucial point for this project. Certainly, a device broadcasting indiscriminately the information about its own daily displacements would have poor chances to be successful. That's why our project is based on digital signature and coding of the datatrip.

A. Datatrip coding scheme

As previously explained, the POs can't decrypt autonomously the possible datatrip that have been gathered from other POs. They have to use the UI and transfer the datatrip to the CA which will manage to do that. The proposed system, that is illustrated in Fig. 8, is based on a double level coding scheme, in which every level is double key (asymmetrical coding).

Every CA is provided with its own couple of keys: a public key CA_PUK (which is made known to all the POs) and a private one CA_PRK (which is guarded only by the CA). Moreover, every PO has both its own public key PO_PUK, that the CA will store when registering the user or acquire if necessary from a trustworthy source of digital certificates and its private one PO_PRK that will be stored only inside the UI related to the very PO.

From now on, the symbol S is to mean the hash function, C stands for encrypting function ,D stands for decryption and S^{-1} stands for the hash verifying function. The levels to be referred to are illustrated in Fig. 9.

Creation of the encrypted datatrip

After being defined (Learning or Imprinting task), the datatrip is signed by the hashset (with some first-level added information, such as the date and time of the data processing) through the private key associated to the PO:

$$HASH = S_{PO\_PRK}\left(INF_1 + DATATRIP_{Clear}\right)$$

After calculating the signature, the first-level package is created. This package is composed by the datatrip in clear text, by the first-level information and by the signature which is calculated on the following parameters:

$$DATATRIP_1 = \left(HASH + INF_1 + DATATRIP_{Clear}\right)$$

Therefore, the package is encrypted by using a double key system, through the public key of the CA:

$$DATATRIP_2 = C_{CA\_PUK}\left(DATATRIP_1\right)$$

After getting the 2nd-level package, a third-level header is to be added . It must be in clear text and containing all the information related to the CA the package is addressed to , so as to be decoded and processed:

$$DATATRIP_3 = \left(INF_3 + DATATRIP_2\right)$$

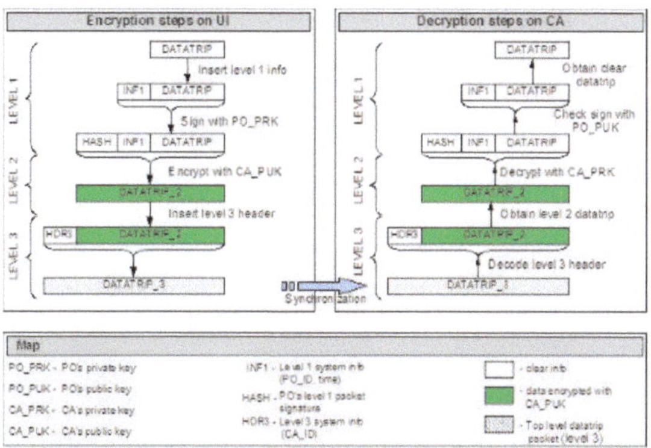

Fig. 9 Multi-level encryption and decryption process

Datatrip decoding

On the other side, after getting the encrypted datatrip, the CA will have to decode every third-level information so as to obtain the second-level encrypted datatrip.

$$DATATRIP_2 = \left(DATATRIP_3 - INF_3\right)$$

At this very point, the CA will raise its own private key to decrypt them and to obtain the clear text first-level package.

$$DATATRIP_1 = D_{CA\_PRK}\left(DATATRIP_2\right)$$

The CA gets the PO_ID from decoding the first-level datatrip . By such knowledge, it will be able to find out the associated user's public key in its database. Thus, it's necessary to check the authenticity and the integrity of the datatrip by comparing the calculated hashset to the one that belongs to the package:

$$S^{-1}_{PO\_PUK}(HASH) \overset{?}{=} \left(INF_1 + DATATRIP_{Clear}\right)$$

In affirmative case, you can get the original datatrip and the first-level information in clear text. On the contrary in an unsuccessful case (different hashsets) you should better to keep the outlines of the event and to watch out for any possible attempt to force the security of the system.

This structure warrants nobody can decipher the content of an encrypted datatrip, except for your CA. Evidently, an high security level is required when implementing such a system and when managing the CA, since its integrity leads to the security of the system. Particularly, this aspect is very important when considering the private key (which allows the datatrip information to be decoded) is unapproachable on the non-members' side.

VIII. Further Application Scenarios

The proposed system can be developed in the future thanks to the integration between the PO and the ordinary-used devices (such as mobile phones, iPODs™, watches and so on).

In particular, the following benefits could be obtained:

- Lower costs, especially for developing the hardware necessary to the POs
- less stress: users don't have to wear any further device
- a wider and faster spreading of the system.

The "pervasive sitter" is a further collateral application to the pervasive car pooling system. In a few words, the question is to make the most of the same network of devices used for the pervasive car pooling. It's sufficient to bring a little adjustment to the software so as to verify that the route followed by the PO is consisted with a prearranged one. In case of evident mismatches, the PO could provide for alarm signaling it . As a matter of fact, such system can prove to be really helpful for monitoring the displacements of children who wander about the urban centers. Also for this application is necessary to set a TAGs network.

IX. Conclusion

This article is about a project for supporting the car pooling services. The methodology for the identification of the users' groups according to their pooled route to workplace is really innovative, since it's based on the use of a pervasive architecture. The most important aspects related to this model are its simplicity of use, its secure communication way, its modest costs ,its efficiency and the protection of the privacy related to the information treated by the system. Such system can be carried out through the technologies which nowadays are at our disposal, even if it can also benefit by the future innovations both in the field of the power supplying for portable devices and in the wireless connections one. The constituent elements, the algorithm and the information flows which are exchanged among several entities are strongly connected to the application scenery; nevertheless, they represent a starting point for the next evolutions of the architecture and for the integration and the support of new services.

ACKNOWLEDGMENTS

The work described in this article is partly funded by Eurotech S.p.A., Amaro (UD), Italy.

Authors wish to thank dott. Giampietro Tecchiolli and dott. Roberto Siagri from Eurotech S.p.A. for their interest in the pervasive car pooling idea and their support in this work.

REFERENCES

[1] D. Hirasuna, "Congestion policy: a literature review," Minnesota House of Representatives Research Dept., 1999. Available: http://www.house.leg.state.mn.us/hrd/hrd.htm

[2] Keenan, P.B., and Brodie, S., "A Prototype Web-based Carpooling System," Proc. of Americas Conference on Information Systems (AMCIS), Association for Information Systems, Long Beach, California, USA, 2000, pp. 362-364.

[3] PadovaNet Car Pooling Project, 2004. Available: http://www.padovanet.it/carpooling/

[4] Mobility Management in Provincia di Bologna, 2004. Available: http://www.provincia.bologna.it/mobility/carpooling.html

[5] Sustainable Mobility Project in Regione Lombardia, 2003. Available: http://www.arpalombardia.it/mobilita/

[6] San Diego Carpool Information, 2003. Available: http://www.ridelink.org/car_pool.html

[7] Commuter Connections Society, 2001. Available: http://www.carpool.ca/

[8] Eagle, N., and Pentland, A., "Social Serendipity: Mobilizing Social Software," *IEEE Pervasive Computing*, vol. 4, no. 2, April-June 2005.

[9] Little, C., and Luscher, D., "ITS Societal Impacts: Current Knowledge and Research Needs," U.S. Dept. of Transportation – Research and Special Programs Administration, 1995. Available: http://www.its.dot.gov

[10] "Assessment of the Seattle Smart Traveler," U.S. Dept. of Transportation – Federal Transit Administration, 1996. Available: http://www.itsdocs.fhwa.dot.gov/jpodocs/repts_te/8R401!.pdf

[11] Cahill, V., et alii, "Using Trust for Secure Collaboration in Uncertain Environments", IEEE Pervasive Computing, vol. 2, no. 3, Jul-Sep 2003

Pier Luca Montessoro is professor of Computer Science at the Engi-neering Faculty of the University of Udine, Italy. Pier Luca Montessoro has been head of the Computer Center of the University of Udine for seven years, until October 2002. Currently, he is head of the Electrical, Management and Mechanical Engineering Department at University of Udine, rector's delegate for the computer networks and telematic sys-tems of the University of Udine. Together with Silvano Gai and Pietro Nicoletti, Pier Luca Montessoro wrote a popular book ("Reti locali: dal cablaggio all'internetworking", SSGRR, in Italian) on structured cabling systems, LAN and internet-working. Author of many research papers, his research interests are actually telecommunications and computer networks, networks and pervasive computing, e-learning and multimedia systems.

Sandro Di Giusto is Ph.D. student at the Electrical, Management and Mechanical Engineering Department of the University of Udine, Italy. He received the Dr. Eng. Degree (laurea) in Electronic Engineering cum laude in 2004. His research is focused on pervasive computing, sensor networks, and network operating systems.

Davide Pierattoni is researcher at the Electrical, Management and Mechanical Engineering Department of the University of Udine, Italy. He received the Dr. Eng. Degree (laurea) in Management Engineering in 1999 and the Ph.D. degree in Industrial and Information Technology Engineering in 2005 from the University of Udine. His research inter-ests are traffic analysis and network simulation environments, routing protocols, communication security, wireless networks and pervasive computing, distance learning and multimedia teaching systems. He is member of the IEEE and the IEEE Computer Society.

Design, Analysis and Implementation of a Cyber Vote System

Khaled Elleithy and Ihab Rimawi
Computer Science and Engineering Department
University of Bridgeport
Bridgeport, CT 06601
elleithy@bridgeport.edu

Abstract

In this paper, we present the design and implementation of the CyberVote system. The CyberVote system gives voters a reliable and highly secure environment to cast their votes using Internet terminals, such as PCs, handheld devices, and mobile phones. Furthermore, it relies upon an innovative voting protocol that uses cryptographic tools.

The cryptographic protocol ensures authentication of voters, notification, and confirming the process afterwards, in a random time, in order to eliminate any probability of identity theft. Also, it guarantees the privacy and integrity of each vote during the registration process, transformation through the Internet, and during the counting and auditing process.

Furthermore, the CyberVote system includes a tool that resets the database before the voting process starts. When the elections time ends, voters cannot cast their votes, and they are not allowed to access their accounts during counting process. The counting process is fully automated, giving no chance for manipulation of the results. When the results are ready, it will be posted on the official election website along with statistic studies and graphs that explain the results.

1. Introduction

1.1. Background

The 2000 presidential elections forced most of the organizations around the world to think of a better way of voting that enables as much people to cast their ballots in a convenient way [1]. It was suggested to introduce a system that enables voters to use the internet to cast their ballots, since the internet is a cheap, fast, and effective way to transfer data. Many researchers have studied the benefits and threats of using internet for voting [2]. Most studies raised concerns about voters' authentication, security of ballots during transfer through internet, and maintaining the secret ballot, which means to separate the identity of the voter from the ballot [3].

Online voting is not a new concept. In the last three decades, studies tried to come up with a way to create online voting systems that enable peoples to vote while they are at their homes. It was not until the late 90's when the idea became practically possible. Internet revolution helped the idea to grow. Studies expect that online voting will replace the present voting systems sooner or later [4].

1.2. Advantages and problems of online voting systems

1.2.1. Advantages of Online voting systems

Online voting provides convenience, cost-saving, and saves the voters from dealing with heavy traffic, bad weather conditions, and postal service issues. Furthermore, it helps millions of disabled and blind people to cast their votes without any assistance [5].

1.2.2. Problems face applying Online voting systems

1.2.2.1. California Task Force

California formed a committee called California Internet Voting Task Force to study the feasibility of replacing the current voting system with remote online voting system. The taskforce released its report on January 2000. The report suggested that using a remote internet voting system will increase the number of voters due to the convenience it provides. It added, "However, technological threats to the security, integrity and secrecy of Internet ballots are significant" [2]. The taskforce divided remote online voting into two different types:

a- Using computers and LANs controlled by the state.

b- Using computers and LANs controlled by voters.

The task force concluded that the first type can be easily used, while the second needs more analysis and study because of the threat of Trojan horses, and back doors [2].

K. Elleithy et al. (eds.), Advances in Computer, Information, and Systems Sciences, and Engineering, 219–225.

1.2.2.2. Technical Problems

There are many technical problems that appear when applying remote online voting systems. The first issue that can be a threat is a coordinated network flood [7]. Another problem that prevents the implementation of online voting is the security on the other side of the system, the voter terminal. Even if we manage to secure the server and we have a secure ballot transmission over the internet, we can not secure the voters terminal. This is probably the toughest challenge facing the developers of those systems. Giving voters the chance to vote using internet will affect the secrecy of ballots, since other people can watch the voting process from behind the voters back [7]. Having access to computers and internet is another important issue, since access to internet varies from one class to another in the society which might not give accurate results [7].

1.2.2.3 Solving technical problems

Flooding any network can be solved by having a backup server that picks up as soon as the other server stops working. Furthermore, replacing major servers with smaller servers placed according to population of voters in different areas will help stopping any kind of coordinate network flooding [8].

Securing all the voters terminals could be an impossible mission, simply because voters run multiple applications on their terminals that might be capable of changing or manipulating the voting process. The most practical solution is to have an online scanning tool that scans the voters' computer from potential threats before the voters get to cast their votes. Another solution that can be more time efficient is transmission a tool to the voter's computer after that stops all the processes running on the machine, except for the processes the system runs. This solution will reduce the risk of having security problems on the voters' terminals, but it does not eliminate it. Having access to the internet should not have a huge impact on the voting process if we keep the old fashion way by going to polls where terminals are available, located in the areas where computer access is not available as much as other areas.

1.2.2.4 Identity Theft

Identity theft is a problem that affects any voting process no matter what kind of technology is used. Remote online voting will make identity theft task easier than before. In order to overcome this problem, the voter should be asked challenging questions that help authentication. Also, sending a physical notice to voters confirming the voting process associated with the time of voting, can reduce the risk of identity theft [8].

1.3. CyberVote

CyberVote is a remote online voting system that tries to overcome the problems of the present remote voting systems. It enables voters to cast their votes from their own Internet terminals, in a highly secured environment.

CyberVote uses different methods of encryption, which ensures the security of the data given to the system, and it ensures voters authentication. Separating the database server from the CyberVote server gives the system another security advantage. Furthermore, CyberVote follows the secret ballots criteria, which separates voter's identity from the ballot, which maintains the privacy of the voting process.

2. Design

2.1. Requirement Analysis

The CyberVote system was designed to fulfill the following functional requirements:

- Administrator specifies registration times and date(s)
- Administrator specifies voting times and date(s)
- Administrator adds candidates to CyberVote
- Administrator has access to population table in database
- Voters can register during registration period
- CyberVote System checks voters' eligibility to vote
- CyberVote stops accepting new registrations after its period is over.
- Voters can log into their accounts
- CyberVote authenticates users
- Voters cast their votes using official CyberVote website
- CyberVote encrypts received data.
- CyberVote saves the vote in a highly secured database
- CyberVote arranges confirmation process by different techniques such as Phone messages, email, SMS messages, or mail.
- CyberVote can deletes the voters who voted from the database
- CyberVote stops Voting process according to pre-specified date and time
- CyberVote counts the votes
- CyberVote posts results on official website.

Furthermore, CyberVote fulfills three non-functional requirements which are:

1. Reliability,
2. Usability,
3. Security.

2.2. System Design

2.2.1Design Diagrams

Figure 1 shows the UML diagram which explains explain how CyberVote system works.

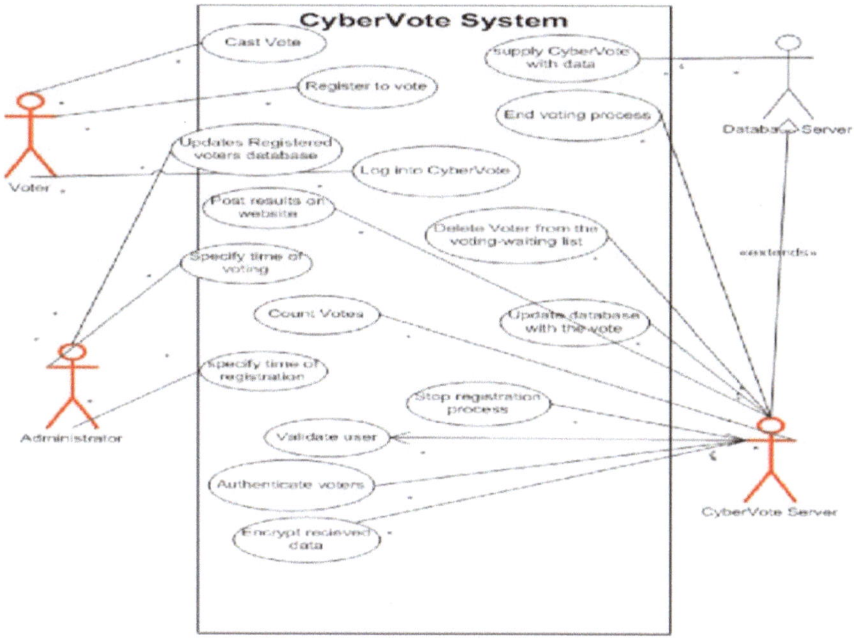

Figure 1: UML diagram of CyberVote system

2.2.2. Database Design

In order to separate the identity of voters from their ballots, a new technique is introduced in CyberVote system that deletes the data of voters from the registered Voters table as soon as they hit on vote button. Their vote will be stored in *casted* table, and their social security number will be stored in a separate table called confirm. That table will send the voters confirmations of the process, then the system will store in the population database that voter X voted in voting number Y, and so on.

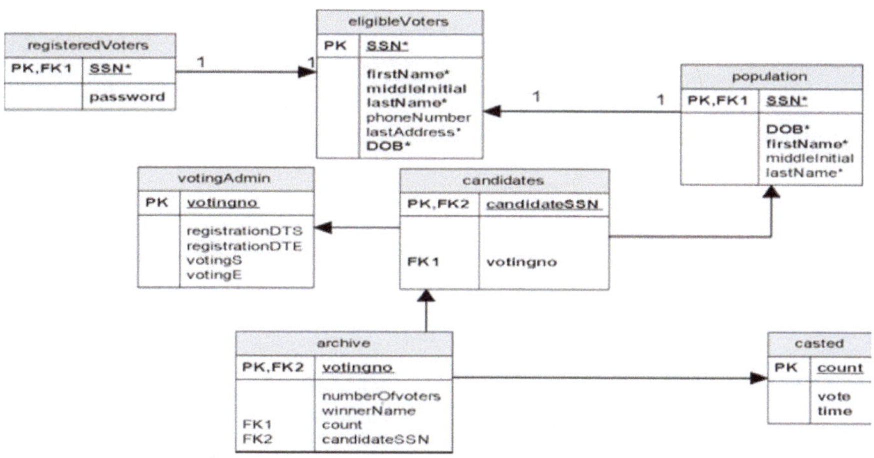

Figure 2: Relationships among the tables in CyberVote database.

2.2.3. Encryption Methods

CyberVote system uses different types of encrypting algorithms such as Secure Hashing Algorithm (SHA1), Message Digest Algorithm (MD5), SHA256, and SHA512. When a registered voter hits on register button, a random number will be generated. This number will determine the encryption algorithm. It will be encrypted afterwards, and will be added to the encrypted data. The technique provides high degree of security to the data of voters.

2.2.4. Classes

CyberVote uses different classes for security that Microsoft Visual Studio .Net provides, the main classes used are as follows:

1. SHA256, which provides SHA256 encryption method.
2. SHA512, which provides SHA512 encryption method.
3. FormsAuthentication, which provides SHA1 and MD5 encryption methods.

We used FbConnection class that provides the connectivity to the firebird database.

2.3. Implementation

The online voting systems use very complicated algorithms and methods trying to provide the security and privacy to voters. This complexity gives some doubts of having back doors in those systems. CyberVote system represents a very simple open source code system that ensures privacy and security of voters by providing the implementation phases step by step:

2.3.1. Component Diagram

Figure 3 shows the component diagram of the relationships among different components that comprise the system.

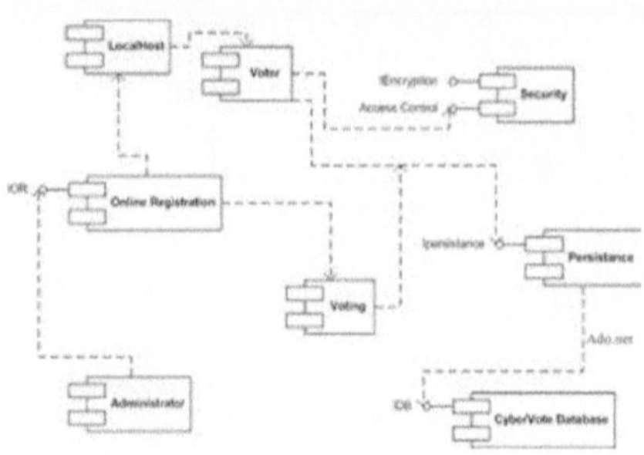

Figure3. Component diagram of CyberVote

We have implemented CyberVote using Firebird 1.5 database. Choosing this database was based on its open source code, which eliminates any probability of back doors. Also, it was implemented with ASP.Net due to the availability of encrypting algorithms that provide the required security for the system.

2.3.2. Administrator Authentication

CyberVote has different levels of authentication for different users, especially the employees controlling the population data. The emplyees they decide when and how to add new entries into the population, when to delete an entry, and when to alter an entry. CyberVote uses one level password given by two employees, i.e. the supervisor has to enter his/her password along with the employee in order to enable the employee work on adding new entries into the database. Altering the population database requires two level passwords from the employee, where altering can only change the information related to an entry, but not the social security number of an entry. Deleting requires three level passwords, which means that a supervisor and two employees should enter their passwords and usernames in order to enable the employee to work on deleting records from the database, any action will take place either adding entries, altering entries, or deleting entries will be documented in a separate table in the database, which consists of the action was done, Identification number of the employee working on this action, and a timestamp that shows the date and time of the update.

Another feature CyberVote is to ensure safety of its database is prompting the employees to change their passwords on regular bases. This feature increases the security of the system.

2.3.3. Layout of CyberVote

CyberVote has a simple interface with a minimal number of frames that is compatible with the handheld and mobile devices. This feature also helps the beginner users of the internet to browse CyberVote without any help.

2.3.4. Source code

CyberVote has an open source code that is released when starting voting in order to ensure the transparency of the voting process maintaining a high degree of security.

2.3.3 Screen shots of CyberVote

The PC version of CyberVote has some light flash animation as the banner shows on the login page. Furthermore, it has a help link that comes in two different ways, text or animated.

2.3.3.1 Client Side

Web pages designed for voters should be easy to understand and use by all levels of society. It should consider other devices like handheld devices and mobile phones. The main logon page is shown in Figure 4.

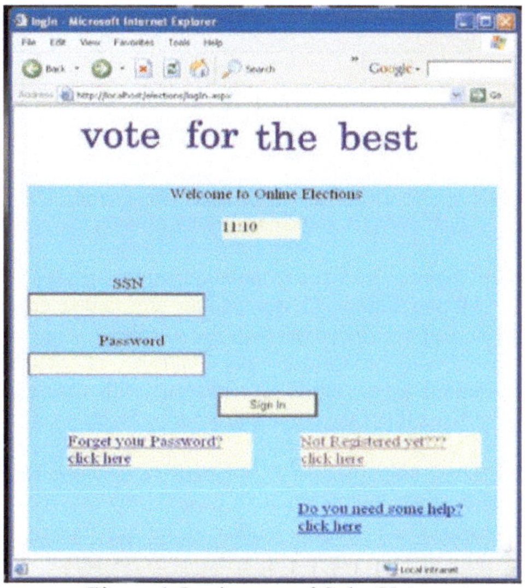

Figure 4. Login page of CyberVote system

If the voter is not registered yet, and registration period did not expire yet, voter should use the registration link that will invoke the registration module as shown in Figure 5.

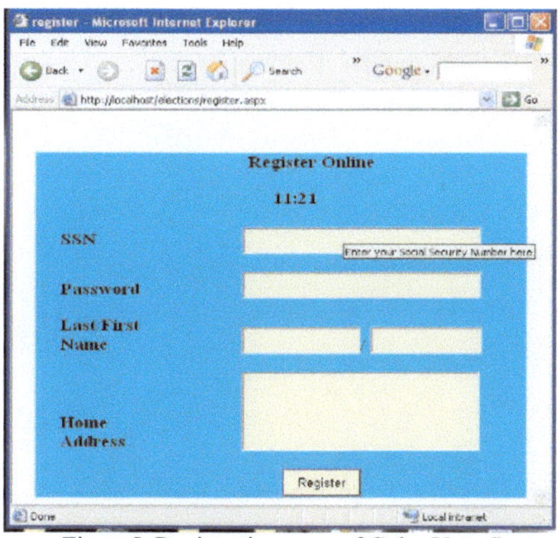

Figure5. Registration page of CyberVote System

After validating the data provided by the voter, the system will transform into the second step of registration asking for a secret question in case of the voter forgot the password as shown in Figure 6.

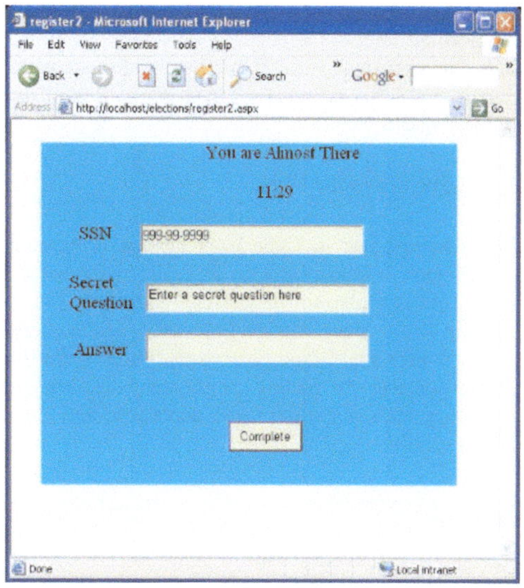

Figure 6. Registration page 2 of CyberVote System

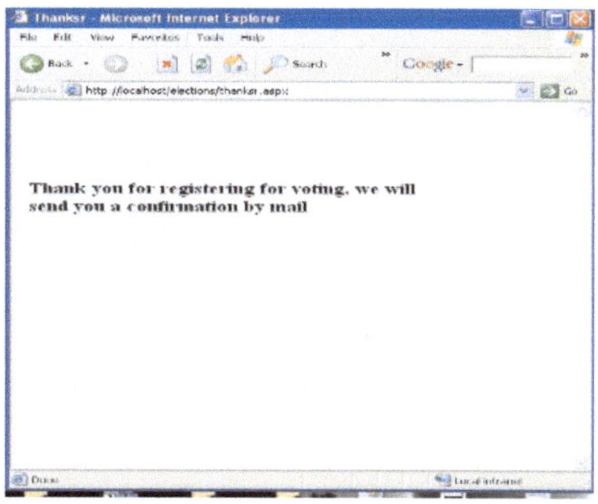

Figure7. CyberVote Ending Registration process

After the registration process is complete, as shown in Figure 7, CyberVote system will issue a letter to confirm the registration process and will also send an email confirming the process.

Furthermore, if a user tries to login before the voting process starts, as shown in Figure 8; the system will not allow the voters to access their pages, which enhances the security features of CyberVote security enhancement.

During the voting process, after signing into CyberVote, voters should choose the candidate out of the list as shown in Figure 9. Also, they get the chance not to choose any of the candidates.

When hitting on Vote, as shown in Figure 10, votes' data will be deleted from CyberVote database. The social security number will be added to the confirmation list, which is a list consists only of social security numbers that will be notified later on, depending on a random time and methods the system determine, such as automated phone call, email, SMS text message, or even by mail confirming the voting process.

2.3.3.2. Administrator Side

The administrator in CyberVote system is more than one department. Administrator is an entity responsible for deleting from population databases, adding to it, and altering on it. Since those processes are extremely important, it uses multistage passwords. Other department will be responsible for adding candidates, deleting them, setting voting and registration times and dates, and altering the data of candidates.

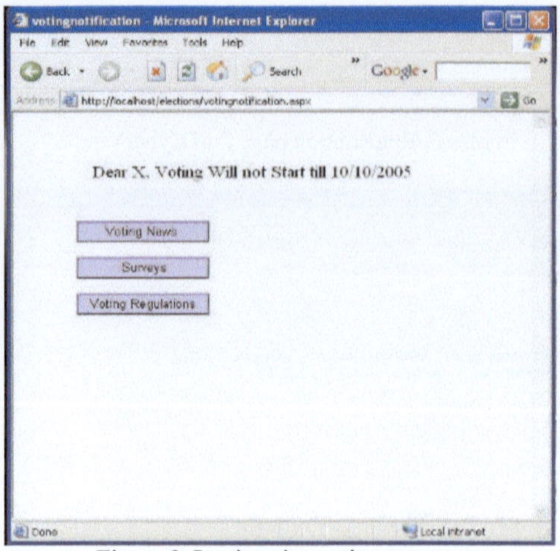

Figure 8. Login prior voting process

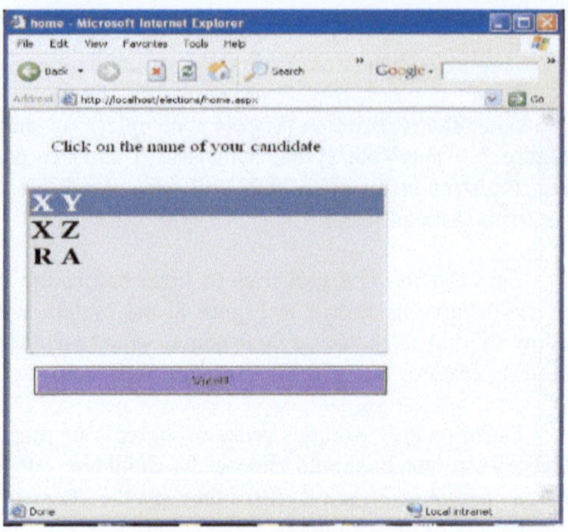

Figure 9. Voting home page

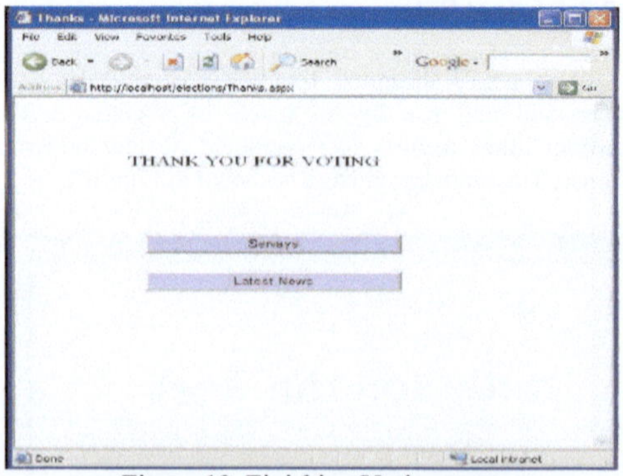

Figure 10. Finishing Voting process

When logging into the population table, as shown in Figure 11 and Figure 12, for adding any new entry, an employee and his / her supervisor are required to sign. This procedure will minimize any chance of database manipulation; every action is registered with timestamps along with the ID of the employee working on it.

After the login process is complete, as shown in Figure 13, the administrator home page will show. The Administrator will choose either to set the voting and registration periods, add entry, delete entry, or alter an entry as shown in Figure 14. The previous page, will enable the administrator to specify the times and dates for both registration and voting processes.

When adding a candidate, as shown in Figure 15, the candidate will be validated according to the voting regulations, and then will be added to the candidate database.

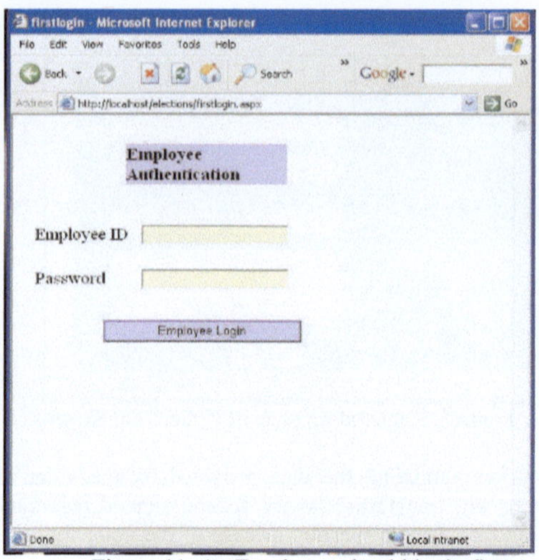

Figure 11. An Employee should login first

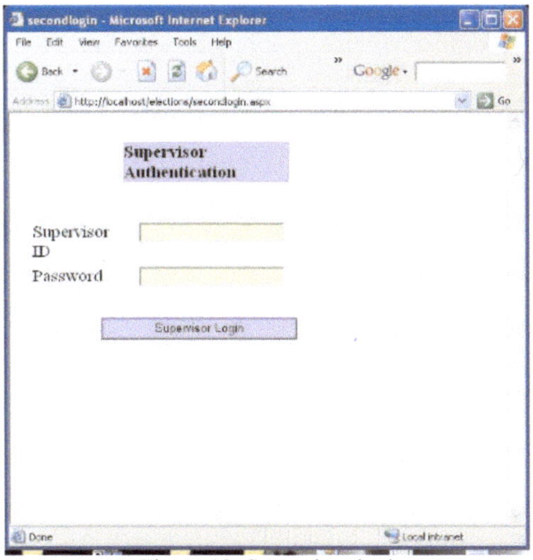

Figure 12. Supervisor login page

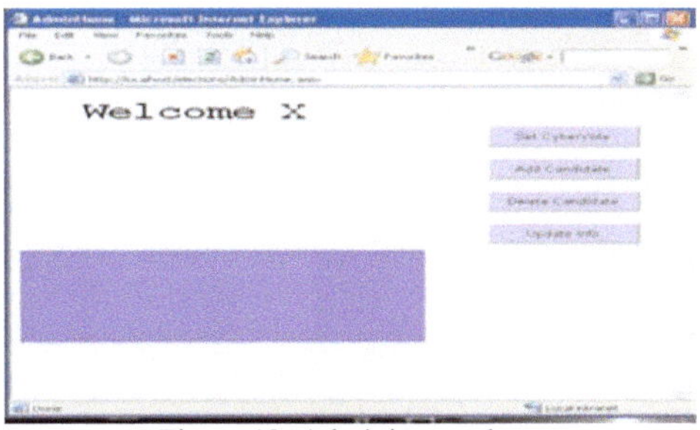

Figure 13. Administrator home page

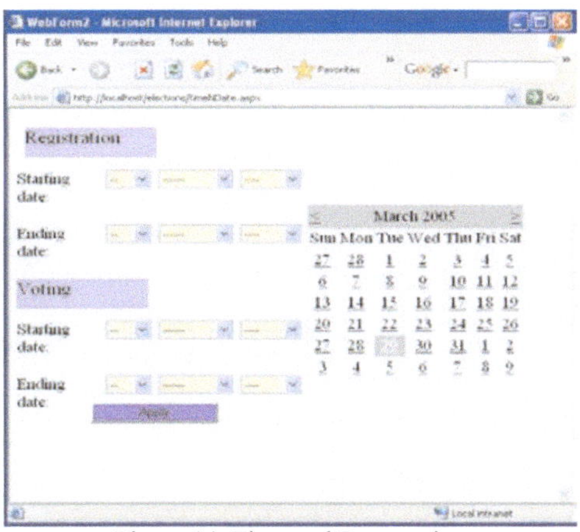

Figure 14. Time and Date setting

3. Conclusion and future work

In this paper we presented CyberVote system. CyberVote is a new step in the development of remote online voting. In this technique we separate the voters from their ballots. Similar systems will replace the traditional techniques of voting eventually. Furthermore, studies show that implementing such systems is feasible and practical with the modern technology resolving the security challenges.

Biometric authentication methods will be implemented in future work, such as adding the keyboard stroke fingerprint and use the iris scan via webcams.

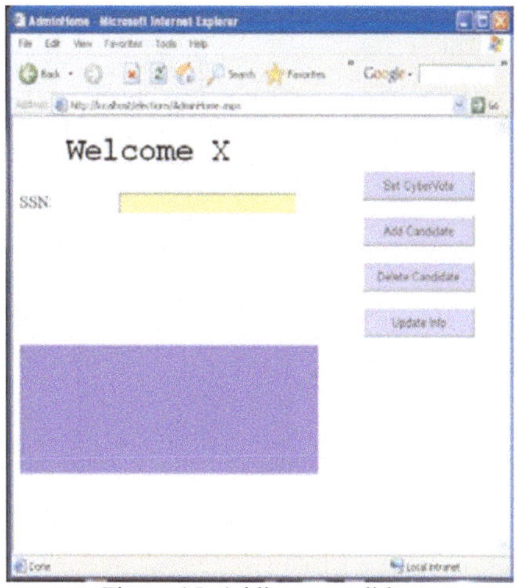

Figure 15. Adding a candidate

4. References

[1] Rebecca Mercuri, "Inside risks: voting automation (early and often?)", *Communications of the ACM*, 43, November 2000, page: 176.

[2] California Internet Voting Task Force, "A report on the feasibility of Internet voting", www.ss.ca.gov/executive/ivote/home.htm

[3] Workshop on Privacy in the Electronic Society, *Proceedings of the 2004 ACM workshop on Privacy in the electronic society,* Washington DC, USA, pp.: 33 – 34, 2004.

[4] Computers, Freedom and Privacy, *Proceedings of the tenth conference on Computers, freedom and privacy: challenging the assumptions,* Toronto, Ontario, Canada pp. 219 – 223, 2000.

[5] L. Hoffman and L. and Cranor, "Internet Voting for Public Officials: Introduction", *Communications of the ACM*, 44, January 2001, pp. 69-71.

[6] Tom Pender, UML Bible, John Wiley & Sons, 2003.

[7] Phillips and Spakovsky, "Gauging the risks of internet elections", *Communications of the ACM,* 44, January 2001, pp. 73-85.

[8] Joe Mohen and Julia Glidden, "The case for internet voting", *Communications of the ACM,* 44, p.44.

[9] Aviel D. Rubin," Security Considerations of remote electronic voting", *Communications of the ACM,*45, December 2002, pp. 39- 44.

DNPSec: Distributed Network Protocol Version 3 (DNP3) Security Framework

Munir Majdalawieh[1,2], Francesco Parisi-Presicce[2], Duminda Wijesekera[2]

{mmajdalalfparisildwijesek}@gmu.edu
[1]American University of Sharjah
[2] George Mason University

Abstract. Distributed Network Protocol Version 3 (DNP3) is an open and optimized protocol developed for the Supervisory Control and Data Acquisition (SCADA) Systems supporting the utilities industries. The DNP3 enables the Master Station to request data from Substations using pre-defined control function commands and Substations to respond by transmitting the requested data. DNP3 was never designed with security mechanisms in mind and therefore the protocol itself lacks any form of authentication or encryption. Discussion so far has been centered on two solutions to provide security for SCADA: cryptographic technologies placed at each end of the communication medium, or security enhancements placed directly in the protocol. This paper recommends a new Distributed Network Protocol Version 3 Security (DNPSec) framework to enable confidentiality, integrity, and authenticity placed directly in the DNP3. Such framework requires some modifications in the data structure of the DNP3 Data Link layer. Our main goal is to address the threats related to confidentiality, integrity, and authenticity in the DNP3 as part of SCADA architecture, with a minimum performance impact on the communication link; and without requiring modification to the much more expensive Master Station and Substation devices and the applications supporting them.

Keywords: SCADA, DNP3, DNPSec

1 Introduction

SCADA systems are used extensively throughout the utilities industries to monitor and control processes that are deployed in many sites. SCADA depends on many networks to support communications between its components, including microwave, PSTN, satellite, frame relay, wireless networks, and private fiber networks. The utilities have recently begun to leverage the Internet as a communication network to support SCADA systems. The Internet is a cost-effective communication tool to handle utilities application traffic as SCADA communication becomes more complex and requires more bandwidth. However, the Internet has opened SCADA, and the systems they support, to new vulnerabilities. The vulnerabilities of SCADA systems are widely recognized and the improving of their security has become an important priority for the utilities [2], [20], [16], [22], [24], [25], [26]. For example, approximately one-third of electric power utility control communications traffic is carried on public networks [7]. According to a Newton-Evans report, nearly 40% of the utilities responding either currently use or plan to use the Internet for SCADA [17].

Distributed Network Protocol Version 3 (DNP3) is used by SCADA systems to communicate between the Master host and the Slave units. This infrastructure is open and effective authentication or encryption mechanisms do not exist. Although the utilities have increased their attention on improving the security and reliability of the SCADA systems in recent years, many owners and operators do not yet have the technology, tools, capabilities, and/or resources needed to secure their systems. Discussion so far has been centered on two solutions to provide security for SCADA: Encryption/decryption technologies placed at each end of the communication media, or security enhancements placed directly in the protocol [3]. To solve the problem of providing a solution placed at each end of the communication media, American Gas Association (AGA 12) [1] developed a standard for "Cryptographic Protection of SCADA Communications." Our paper recommends a new security framework to enable confidentiality, integrity, and authenticity directly in the DNP3. Such a framework requires some modifications in the data structure of the DNP Data Link layer.

The main goal behind our approach is to address the threats related to confidentiality, integrity, and authenticity in the SCADA Systems using the DNP3 with a minimum performance impact on the communication link, and without requiring modification to the much more expensive Master Station and Substation units and the applications supporting them. Our framework is built along the lines of the IPSec standards [10], [11], [12], [19], but has some unique features to main-

K. Elleithy et al. (eds.), Advances in Computer, Information, and Systems Sciences, and Engineering, 227–234.

tain the specifications and the requirements of the DNP3 and SCADA architecture.

The remainder of the paper is arranged as follows. Section 2 summarizes related work. Section 3 provides an overview of SCADA systems. Section 4 describes the DNPSec framework by proposing security enhancements placed directly in DNP3. Section 4 concludes our paper. We added 4 appendices to the paper to help in providing background on SCADA and DNP. Appendix A provides a brief overview of SCADA systems. Appendix B provides a brief overview of DNP3. Appendix C provides threats analysis for SCADA systems using DNP3. And finally, Appendix D provides an analysis of our approach.

2 Related Works

Few publications are available on SCADA security, such as the American Gas Association Report No. 12 (AGA 12) [1]. AGA 12 recommends practices designed to protect SCADA's Master-Slave serial communication links from a variety of active/passive cyber attacks. One of these standards is AGA 12-1, Cryptographic Protection of SCADA Communications. The solution protects against hijacking or modifying the communication channel.

AGA 12 requires the installation of multi-channel SCADA Cryptographic Modules (SCM) on a communications channel between the SCADA unit (e.g., host, RTU, IED) and the modem. A SCM receives and transmits SCADA messages on two communication ports: plaintext port and ciphertext port. The plaintext port is used to receive and transmit plaintext messages from a SCADA unit to a SCM, and the ciphertext port is used to transmit and receive ciphertext messages from a SCM to its peer.

SCM immediately begins transmitting a ciphertext message header to its peer as soon as it receives the first SCADA message characters. When it receives enough characters to fill its cipher block, it encrypts and transmits a block of ciphertext. When it finished transmitting all message blocks, it transmits a trailer that includes a Message Authentication Code (MAC).

At the receiving SCM, an incoming ciphertext message header signals the start of a new message. Each time enough characters are received on the ciphertext port to fill a cipher block, the SCM decrypts the block and immediately begins forwarding the decrypted characters via its plaintext port to the receiving SCADA unit. When the trailer of the ciphertext message is received, the SCM computes and checks the MAC. By this time, the decrypted SCADA message may have already been forwarded in its entirety to the receiving SCADA unit. If the authentication check fails, it is too late to prevent forwarding the unauthentic message. Thus the authentication code only alerts the SCM to a possible failure of data integrity [27]. Such solution is limited and expensive. The standard does not protect an attack from a compromised field site or control center. In addition, SCADA owners need to install AGA 12 compliance multi-channel SCADA Cryptographic Module (SCM) and Key Management Appliance in the SCADA Control Center; and SCM and Maintenance Cryptographic Module attached to every Remote Terminal Unit (RTU). Moreover, AGA 12 is still in the early stages from a system implementation standpoint. Key management is a key component of the standards and is still in the development stage.

In another research, Graham and Patel [18] examined three security enhancements in SCADA communications to reduce the vulnerability of cyber attacks to include: (1) solutions that wrap the DNP3 protocols without making changes to the protocols, (2) solutions that alter the DNP3 protocols fundamentally, and (3) enhancements to the DNP3 application. One of the research directions they identified is to secure the DNP3 protocol which is the focus of this paper. They provided high level description of possible solutions to protect SCADA communications and analysis for existing solutions such as DNP3 over IPSec or DNP over SSL/TLS. The main purpose of the paper is to identify the possible solutions to secure SCADA messages for further research work to model and proof these solutions. The discussion about providing security for the DNP protocol is theoretical and describes the features of the proposed protocol at a very high level.

3 Supervisory Control and Data Acquisition (SCADA) Systems

SCADA systems are real-time process controls that enable a user from a single Master station to collect and analyze data received from one or more Slave units, send limited control instructions, monitor and control equipment status, and open and shut valves or motors. It is used to control power generation, electricity transmission & distribution, electric utilities, natural gas utilities, oil & gas production facilities, water utilities, and gas stations/gas pumps.

A SCADA system consists of four major components: (1) the Field Instrumentations that measures the environmental conditions (read temperatures, pressures, flows, voltages, currents, frequencies, or other physical quantities) and controls valves, circuit breakers, or other devices that influence the physical processes, (2) the Remote Stations (Remote Terminal Unit), (3) the Control Center, and (4) the Communication links.

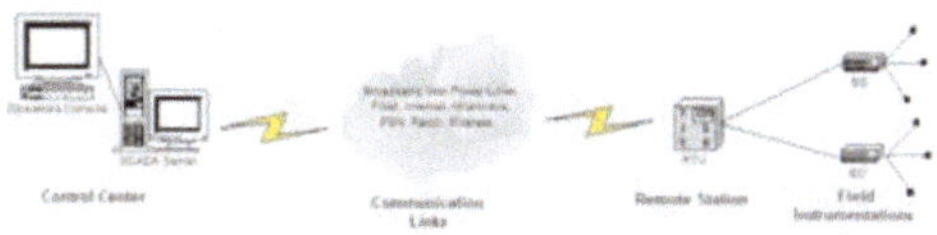

Figure 3-1: SCADA system architecture

A general architecture of SCADA components is shown in Figure 3-1. The RTUs located in the remote station poll many direct-wired field instruments. The RTUs are used for reporting data and controlling the remote devices. The RTUs collect and concentrate the data from the field instruments for transfer to the Master Station. The SCADA server using Human Machine Interface located in the control center polls the RTUs at a user defined polling rate over several communication links (Fiber, Radio, Modem, Microwave, Telephone, Wireless, Powerline Carrier, and Internet). The SCADA server is programmed to collect and reformat data from the remote sites and detect alarm messages reported by the RTUs. Alarm messages are reported when a status of an end point changed unexpectedly.

Several protocols are used to transmit messages between SCADA components. The following is a list of some of these protocols:

- Distributed Network Protocol Version 3, to be discussed in section 4
- Highway Addressable Remote Transducer (HART) Protocol [6],
  - A global standard for smart instrument communication
  - Designed to support the traditional 4-20mA analog signaling
  - Support three layers stack: Application, Data Link, and Physical
- Inter Control Center Protocol (ICCP) [8],
  - Also known as TASE.2
  - Developed to allow real-time data, schedule, and control command exchanges to occur between two or more utility control centers, regional control centers, and power pools.
- International Electrotechnical Commission (IEC) 61850 [9],
  - Intra-substation communications
- Modicon Bus (Modbus), Modbus-Plus [14]
  - An application layer messaging protocol, positioned at level 7 of the OSI model, which provides client/server communication between devices connected on different types of buses and network.

SCADA messages comprise commands, responses, acknowledgments, negative acknowledgments, keep-alive messages, etc. We will discuss these further in the next section.

4 Distributed Network Protocol V3

DNP3 supports two kinds of data: static data and event data. Static data is called class 0 data. Event data can have three different classes or priorities: 1 (high priority), 2 (medium priority) and 3 (low priority). Also, it supports several data types, e.g. "binary input" and "analog input", and the corresponding events, e.g. "binary input change" and "analog change".

DNP3 protocol uses two message sets. The Master set, contains the valid commands for Master initiation a request (polling) or issuing command confirmation, and the Slave set, contains the valid commands to provide a response or initiation unsolicited messages. Messages can be sent between the DNP3 Master (Control Center) and the Slave (RTU) by one of the following communication operating modes [4]:

- Quiescent Operation. In this mode the Master does not poll the Slave. The Slave can send unsolicited report-by exception messages and the Master can send application layer confirmations to the Slave. During quiescent periods the device can be placed in an idle state.
- Unsolicited Report-by-Exception Operation. The communication is basically, unsolicited but the Master occasionally sends integrity polls for Class 0 data to verify that its database is up to date.
- Polled Report-by-Exception Operation. The Master regularly polls for event data and occasionally for Class 0 data.
- Static Report-by-Exception Operation. The Master polls only for Class 0 data or the specific data it requires.

The Master can address individual Slaves, or can initiate a broadcast message to all Slaves. Slaves return a message (response) to requests that are addressed to them individually. The DNP3 protocol establishes the format for the Master's request message by placing it into the Slave (or broadcast) address, a function code defining the requested action, any data to be sent, and an error-checking field. The Slave's response message is also constructed using DNP3 protocol. It contains fields confirming the action taken (if requested), any data to be returned, and an error-checking field. If an error occurred in receipt of the message, or if the Slave is unable to perform the requested action, the Slave will construct an error message and send it as its response.

Figure 4-1 shows the communication process between a Master and a Slave. The figure illustrates a Master initiates a request of data from a Slave; this could be a poll for current data. Also, the figure illustrates the communication sequence between a Master and a Slave with message direction shown between them.

The request message is contained in the application layer information within the message. A confirmation (acknowledge) response is required to this message. The Slave station sends an ACK message to the Master.

Since the last transaction contained an application level request for the transmission of data, the Slave station then performs the action requested and initiates a communication with the requested data.

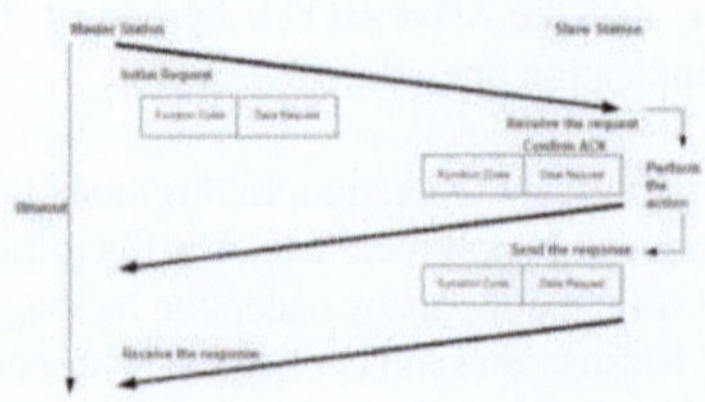

Figure 4-1: Master/Slave Communication Process

The DNP3 stack has three layers: the application layer, the data link layer, and the physical or adaptation layer. To support advanced Remote Terminal Unit (RTU) functions and messages larger than the maximum frame length, DNP3 added a transport pseudo-layer to be used with the Data Link layer (See figure 4.2).

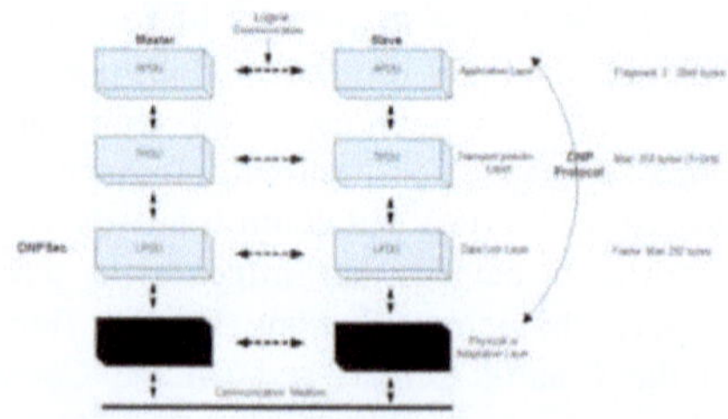

Figure 4-2: DNP3 Protocol Stack [23]

6 Distributed Network Protocol Version 3 Security (DNPSec) Framework

DNPSec authentication and integrity framework capabilities will verify the frame origin, assure that the frame sent is the frame received, assure that the network headers have not changed since the frame was sent, and give anti-replay protection. DNPSec confidentiality framework capability will encrypt frames to protect against eavesdropping and hide frame source by applying encryption methods. Some modifications to the DNP3 LPDU or frame structure are required to provide these capabilities.

CRC is a common technique used in DNP3 for detecting data transmission errors. CRCs occupy 34 bytes out of 292 bytes of the DNP3 LPDU for integrity. These bytes will be utilized in a different way in the DNPSec framework.

There are two main components of the DNPSec. The first is the DNPSec structure to construct the frame and transfer data in secure mode between the Master and the Slave. The second is the key exchange established during the installation and connection setup between the Master and the Salve.

DNPSec structure

DNPSec consists of five fields (as shown in Figure 6-1) the new header, the key sequence number, the original LH header, the payload data, and the authentication data.

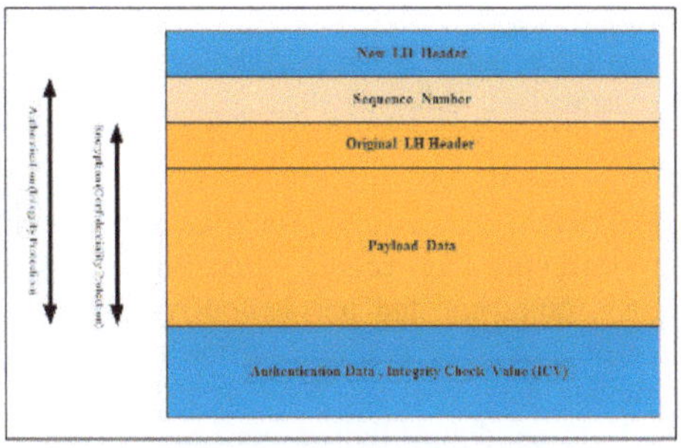

Figure 6-1: Protocol Structure - DNPSec

The **new header** is an unsigned 4 bytes field containing the destination address (DA) which occupies 2 bytes; the MH flag bit to recognize if the message is coming from the Primary Master host (0) or from the Secondary Master Host (1); the SK flag bit to indicate to the Slave if the message contains the new session key (1) or to decrypt the message using the session key in the S-keydb (0); and 14 bits reserved.

The **key sequence number** is an unsigned 4 bytes field containing a counter value that increases by one for each message sent by the Master. Each time a Master sends a message it increments the counter by one and places the value in the key sequence number field. Thus, the first value to be used is 1. The Master must not allow the key sequence number to cycle past $2^{32} - 1$ back to zero. If the limit of $2^{32} - 1$ is reached, the Master should terminate the session key and send a new key to the Slave using the value zero in the frame sequence number. This kind of functionality will guarantee that the Master and the Slave will continue establishing a new session key even if the connection is always open. Moreover, the security policy should indicate that a new session key must be established between the Master and the Slave in case the Slave used the same session key for a certain time period. DNPSec uses the variable key-session-life-time to keep track of the life span of the session key.

The **original LH header** (DNP3 data link header without the 2 CRCs) and the payload data is protected by encryption and composed of 264 bytes field containing , 8 link protocol data unit header bytes, 250 Transport Protocol Data Unit bytes, and 6 padding dummy bytes. 264 bytes is a multiple of 4 bytes, which provides alignment of 4 bytes boundary and provides boundaries of 64 bits to support the encryption algorithms. For example, Data Encryption Standard (DES) specifies that the plaintext is 64 bits in length and the key is 56 bits in length. Longer plaintext are processed in 64-bit blocks.

The **authentication data** field containing an Integrity Check Value (ICV) computed over the key sequence number, the original LH header, and the payload data fields. ICV provides integrity services and is provided by a specific message authentication algorithm (MAC) such as, HMAC-MD5-96 or HMAC-SHA-1-96. The integrity algorithm specification must specify the length of the ICV and the comparison rules and processing steps for validation. DNPSec required 20 bytes for the authentication data field. To simplify the key management process, we are recommending to use the Master/Slave encryption/decryption session key to calculate the authentication data.

The DNPSec fields are as follows:

0 – 3 New Header (4 bytes)
DA: 0-1 Destination Address (2 bytes)
MH: 2(bit 0) 0: Primary Master Host,
1: Secondary Master Host
SK: 2(bit 1) 0: Fitch the database for the session key,
1: The frame contains a KSN value from the Master.
2(bits 2-7)-3 Reserved (2 bytes)
4 – 7 Key Sequence Number (4 bytes)
8 – 15 Original LH Header (8 bytes)
8 – 9 Sync (2 bytes)
10 - 10 Length (1 byte)
11 – 11 Link Control (1 byte)
12 – 13 Destination Address (2 bytes)
14 – 15 Source Address
16 – 271 Payload data (256 bytes)
16 – 265 TPDU data
266 – 271 Padding dummy data
272 – 291 Authentication Data (20 bytes)

Key management

The key management operations in DNPSec are very simple to accommodate the static nature of the SCADA environment. They occur during the configuration of the Primary Master host, the Secondary Master host, and the Slaves to establish the initial connection between them; after the re-initialization of the Key Sequence Number (KSN) to generate and distribute a new key to the hosts; and after the timeout of the usage of the session key.

The Master host generates and manages a secure database "M_Keydb" for the shared session keys with the Slaves (see Figure 6-2). The database consists of four fields: the Slave address used as an index key to the database, the shared session key, the time stamp used to limit the usage of the shared key for a certain pre-defined time period, and the Key Sequence Number. The Master calls "M_GenKey" to generate a unique session key when the old session key expired. "M_PutKey" is the function used to insert the new session key into the database. And the M_PutKSN is the function used to insert the new KSN into the database.

The Slave needs to maintain two session keys, one for communicating with the Primary Master host and the other for the Secondary Master host (see Figure 6-2). It manages a secure database "S_Keydb" for the shared session keys with the Master hosts. The database consists of three fields and two records: (0, Primary Master Session Key, Key Sequence Number and 1, Secondary Master Session Key, Key Sequence Number). The simple "S_PutKey" is the function used to update the database with a new session key and the "S_PutKSN is the function used to update the database with a new KSN.

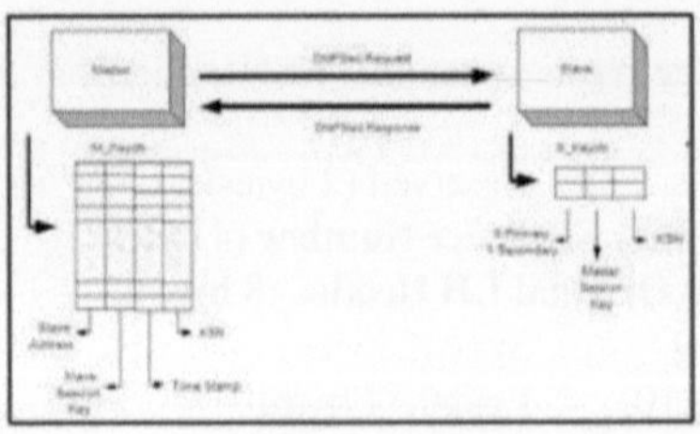

Figure 6-2: DNPSec Request / Response link communications

See appendix E for our approach to implement the DNPSec protocol.

7 Discussion

Analysis of our approach

Reliability and time to delivery of DNP3 frames are very important requirements for SCADA/DNP3 Systems. These requirements are vital to market acceptance of a particular DNP3 security implementation. Reliability, as per DNP Group, is provided by Cyclic Redundancy Code (CRC) function in the Data Link Layer. CRC is non-cryptographic mechanism for detecting transmission errors. DNPSec added more efficient reliability and security capabilities by introducing cryptographic and authentication capabilities in the DNPSec framework. Such capabilities introduced new challenges related to time to delivery of the frames. But we will show soon that these challenges are not significant and our framework maintains a good balance between security, reliability and time to deliver since most of the frames in the SCADA systems are not real-time sensitive.

As we described in section 4, DNP3 provides several different means of retrieving data. These methods for retrieving data require different means of efficiency, quiescent and unsolicited report-by-exception operation requires real-time efficiency. The time of retrieving data from the Slave or the time the Slave needs to send unsolicited messages to the Master should not be significantly delayed by the implementation of DNPSec. More detailed performance analysis related to the implementation of DNPSec needs to be conducted.

Several performance studies on the effect of cryptography on the set-up time and the delivery of the messages from one end to the other indicate that the delay is not significant based on the advanced technologies in the communication networks, processing power at the end systems, and the cryptographic algorithms [5], [15], [21].

In a study by Kim and Montgomery [13] they examined the dynamic behavior and relative performance characteristics of large scale VPN environments based upon IPSec and IKE. The results of their study are summarized in the following table:

Operation	DES	3-DES
Encryption Speed (Kbit/s)	10508 kbit/sec	4178 kbit/sec
Decryption Speed (Kbit/s)	10519 kbit/sec	4173 kbit/sec

Based on the performance information above, we will calculate the worst case scenario to measure the time of delivery for the unsolicited message from the Slave to the Master, which required real-time delivery. Although, the numbers are far from exact, they should be usable as a first approximation. The total time to deliver such message is the sum of the encryption speed (ES), the decryption speed (DS), encryption key set up (EK), decryption key set up (DK), and the transmission time (TT).

Unsolicited delivery time = ES + DS + EK + DK + TT

We assume that the size of the DNPSec message is 292 bytes, Triple DES is the algorithm of choice with 112 bit key, the network bandwidth is 1.5 Mbps, and the performance speed is measured in kbit/s. The EK and DK are not applicable in our case since we are assuming that we are using manual distribution of the session keys during the installation of SCADA components. The table below shows the performance of each operation:

Operation	Performance	Time
Encryption Speed	4178 Kbit/sec	.00007 sec
Decryption Speed	4173 Kbits/sec	.00007 sec
Transmission Time	1.5 Mbit/sec	.0002 sec

As a result the unsolicited delivery time is equal to .00034 sec. Even if we double this number to accommodate for the authentication calculation time, we believe that this is a very minimum time to have an effect on the delivery time of the unsolicited messages in the SCADA systems. Accordingly, adding the operations above to include cryptographic and authentication operations will not affect the efficiency and the speed of delivery of DNP3 messages.

8 Conclusion

DNP3 was not designed with security capabilities in mind. The SCADA vendors can build such capabilities by utilizing the DNPSec framework with a minimum time and cost without a major impact on the systems components and the application supporting them.

This paper has discussed what DNPSec framework is, the components that constitute DNPSec, how DNPSec works, and provided analysis to our approach. The framework enables confidentiality, integrity, and authenticity in the DNP3. Such a framework requires some enhancements in the data structure of the DNP3 Data Link layer, without requiring modification to the Master Station and Substation devices and the applications supporting them. Confidentiality and integrity are achieved by encrypting frames between the Master and the Slaves using a common session key assigned at the setup time of the SCADA components. A new session key is established when the frame sequence number reaches the value $2^{32} - 1$ or when the time period for the use of the session key is expired. Authentication is achieved by applying authentication techniques to assure that the sender of the frame is what it claims to be. Proof of concept by testing or simulation could be a future topic worth investigation.

Our ongoing work address performance issues are related to the implementation of DNPSec.

References

1. American Gas Association (AGA), Draft 4,AGA Report 12, November 2004, Cryptographic Protection of SCADA Communications Part 1: Background, Policies and Test Plan, http://www.gtiservices.org/security/AGA12Draft4r1.pdf]
2. Brian Broyles and Frank Kling, "Is there anything new under the SCADA sun?", December 2003 http://www.rigzone.com/news/insight/insight.asp?i_id=32
3. Jim Coats, President Triangle MicroWorks, Inc. DNP3 Protocol AGA/GTI SCADA Security Meeting August 19, 2002 / Washington DC
4. DNP3 Users Group, http://www.dnp.org
5. Gordon Clarke, Deon Reynders, "Practical Modern SCADA Protocols, May 2003 http://cs.gmu.edu/~menasce/papers/IEEE-IC-SecurityPerformance-May-2003.pdf
6. Highway Addressable Remote Transducer (HART) Protocol, http://www.hartcomm.org
7. Information Assurance Task Force of the National Telecommunications Advisory Committee. "Electric Power Risk Assessment." Alpha Communications Integration. December 2004 http://www.aci.net/kalliste/electric.htm
8. Inter Control Center Protocol, Electric Power Research Institute, http://www.epri.com
9. International Electrotechnical Commission (IEC) 61850 http://www.ucausersgroup.org
10. S.Kent, R.Atkinson, "Security Architecture for the Internet Protocol," IETF RFC 2401, November 1998
11. S.Kent, R.Atkinson, "IP Authentication Header." IETF RFC 2402, November 1998
12. S.Kent, R.Atkinson, "IP Encapsulation Security Payload (ESP)," IETF RFC 2403, November 1998
13. Kim and Montgomery, Behavioral and Performance Characteristics of IPSec/IKE in Large-Scale VPNs, http://w3.antd.nist.gov/pubs/cnis-perf-vpns-ikev1.pdf. Proceedings of the IASTED International Conference on Communication, Network, and Information Security, pp. 231-236, December 2003
14. Modicon Bus (Modbus), Modbus-Plus, http://www.modbus.org
15. Erich Nahum, Sean O'Malley, Hilarie Orman, and Richard Schroeppe, "Towards High Performance Cryptographic Software" ftp://ftp.cs.arizona.edu/reports/1995/TR95-03.ps
16. National Science Foundation (NSF) workshop, SCADA/IT for Critical Infrastructure Protection Workshop, Minneapolis, MN. 20-21 October 2003 http://www.adventiumlabs.org/NSF-SCADA-IT-Workshop/
17. Newton-Evans Research Company, "The World Market for Substation Automation and Integration Programs in Electric Utilities: 2002-2005"
18. Sandip C. Patel and James H. Graham "Security Considerations in DNP3 SCADA Systems" . 17th International Conference on Computer Applications in Industry and Engineering. November 17-19, 2004. http://www.louisville.edu/speed/cecs/facilities/ISLab/tech%20papers/ISRL-04-01.pdf
19. J.Postel, "Internet Protocol" IETF RFC 791, September 1981
20. Riptech, Inc. January 2001 "Understanding SCADA System Security Vulnerabilities", http://www.iwar.org.uk/rerources/utilities/SCADAWhitepaperfinal1.pdf
21. Schneier, Bruce, Performance Comparison of the AES Submissions, http://www.schneier.com/paper-aes-performance.pdf
22. Sauyer, Joe St, Ph.D. University of Oregon. SCADA Security, http://darkwing.uoregon.edu/~joe/scada/
23. Transporting DNP V3.00 over Local and Wide Area Networks, http://www.dnp.org
24. U.S. Department of Energy, "21 Steps to Improve the Cyber Security of SCADA Networks," http://www.ea.doe.gov/pdfs/21stepsbooklet.pdf
25. U.S. General Accounting Office, "Critical Infrastructure Protection: Challenges and Efforts to Secure Control Systems" http://www.gao.gov/new.items/d04354.pdf
26. William F. Young, Jason E. Stamp, John D. Dillinger, and Mark Rumsey. "Communication Vulnerabilities and Mitigation in Wind Power SCADA Systems" http://www.sandia.gov/wind/other/031649c.pdf
27. Andrew Wright, John Kinast, and Joe McCarty, "Low-Latency Cryptographic Protection for SCADA Communications", http://scadasafe.sourceforge.net/security.pdf

Approximate Algorithms in Mobile Telephone Network Projects

Adriana Aparecida Rigolon
University of Fortaleza
MA in Applied Computer Science
arigolon@ubbi.com.br

Luiz Otavio Ribeiro Afonso Ferreira
Faculdade Integrada do Ceará (College)
lotavior@yahoo.com.br

Plácido Rogério Pinheiro
University of Fortaleza
MA in Applied Computer Science
placido@unifor.br

Elder Magalhães Macambira
Federal University of Paraíba
Departament of Statistics
elder@de.ufpb.br

***Abstract-* The present work offers a comparative study of Lagrangean relaxation in simple subgradient and space dilatation methods as applied to generalized assignment problems in diversity and capacity constraints (GAP-DC). This problem, which occurs when setting up a mobile telephone network, consists in attributing radio base stations to hubs at a minimal cost, so that the demands on each station and on the capacity of each hub can be met. The GAP-DC are known to be NP-hard. Computational results indicate that although simple subgradient algorithms are capable of producing good lower bounds, computational tests based on space dilatation methods resolve instances in less computational time as well as in the first iteration.**

Key words- **generalized assignment problem, mobile telephone network, Lagrangean relaxation, space dilatation.**

I. INTRODUCTION

A mobile telephone network in its basic form is made up of one or more Mobile Switching Centers (MSC), as well as various Radio Base Stations (RBS). These RBS's are distributed in such a manner so as to reach all areas covered and at the same time provide a minimal level of radio frequency signals in order to guarantee acceptable conversation quality levels. Each station has a defined capacity in terms of numbers of calls that can be made simultaneously. This capacity is determined by the number of subscribers and channels designated to carry voice and radio control loads to the MSC.

As the number of mobile telephone network users increases, the need to add more base stations to the network to meet the demands of the new users and the consequent cost of transmission from the base stations to the MSC also increases. In the face of these problems, telephone operators often attempt to select radio base stations to act as hubs. However, this solution creates other difficulties in guaranteeing data reliability when sending or receiving information, and thus problems ensuring the survival of the network in the event of a fail in the hub. One possible solution to this problem is to define a tree topology. This allows the physical diversity of path to the RBS's and, consequently, guarantees that data traffic is equally routed to all hubs to which the RBS is connected. It is within this context that generalized assignment problems in diversity and capacity constraints (GAP-DC) occur.

In effect, GAP-DC use a topology in which base stations belong to different trees and the MSC represents the root node of each tree. The trees are thus made up of MSC's and hubs. Figure 1 illustrates a mobile telephone network with a tree topology for GAP-DC with 3 hubs and 4 RBS's.

Consequently, generalized assignment problems in diversity and capacity constraints consist in attributing radio base stations to hubs at a minimal cost so that the demands of each base station can be met and the capacity of each hub respected. In Kubat *et al* [5] it is proved that GAP-DC are NP-hard problems.

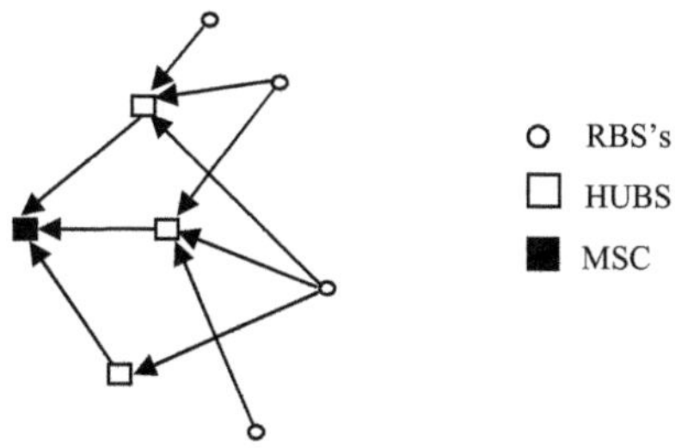

Fig 1 – Mobile Telephone Network with GAP-DC tree topology

Thus, the objective of this work is to propose a model for mixed integer linear programming as an extension of the research presented by Kubat et al. [5]. This extension includes an additional communication channel for each hub. The resulting problem, hereafter referred to as GAP-DC*, is particular to GAP-DC, and remain NP-hard problems. To solve this problem, Lagrangean relaxation techniques in simple subgradient and space dilatation methods are applied.

Diverse approaches can be proposed for GAP-DC and their variants. In Kubat et al. [5], the authors present various greedy heuristics and suggest the application of Lagrangean

K. Elleithy et al. (eds.), Advances in Computer, Information, and Systems Sciences, and Engineering, 235–242.

relaxation techniques. In Bose et al. [2] and Kubat et al. [4] GAP-DC are considered as multi-period problems. Consequently, the authors sought a model which might deal with certain alterations that occur in networks with time. In [2] new integer formulas and greedy heuristics are proposed to solve this problem while in [4] the authors suggest a Lagrangean heuristic and a branch-and-cut algorithm.

Due to the success of the Lagrangean relaxation technique in resolving GAP-DC, principally in setting up mobile telephone networks (c.f. [4, 5]), this work proposes the use of algorithms based on Lagrangean relaxation techniques in simple subgradient and space dilatation methods in solving GAP-DC*.

The present paper is organized in the following manner: Section 2 proposes a model for GAP-DC*. In Section 3 a Lagrangean relaxation with a simple subgradient method and another with a space dilatation method are presented as a solution to GAP-DC*. In Section 4 several computational experiments are made with Lagrangean relaxation, using LINGO 8.0, for small to medium-sized instances. Finally, in Section 5, a comparative study is made of applied methodologies and future research is discussed.

II. A MATHEMATICAL MODEL FOR GAP-DC*

To present a mathematical model for GAP-DC*, it is first necessary to describe the proposed model for GAP-DC:

- $M = \{1, ..., m\}$, hub node indices;
- $N = \{1, ..., n\}$, radio base station node indices;
- MSC, a mobile switching center;
- $V = \{v_1, v_2, ..., v_n\}$, connected radio base station nodes;
- $H = \{h_1, h_2, ..., h_m\}$, hub nodes;
- $R = H \cup \{MSC\}$, the hub nodes plus the MSC;
- $Q = \{T_1, ..., Tq\}$, tree nodes, such as $T_i \cap T_j = \emptyset$. Each tree has its root in the MSC;
- S_i, the diversity of each RBS $i \in V$;
- D_i, the fixed demand in number of circuits DS0 (64 Kbps) of each base station $i \in V$;
- K, the hub link capacity to the MSC in number of circuits DS0 (64 Kbps);
- a_{jl}, a constant, such as $a_{jl} = 1$ if link $l \in H$ is on the path from the root to the hub $j \in H$ and $a_{jl} = 0$, otherwise;
- c_{ij}, a fixed cost to connect a base station $i \in V$ to the hub $j \in R$;
- b_l, the fixed cost to acquire more capacity.

As the network uses a tree topology, and each base station generates a demand D_i, that will be sent to a node of that tree (hubs), a decision variable x_{ij}, such as $x_{ij} = 1$ if station $i \in V$ is connected to the hub $j \in R$, and $x_{ij} = 0$ otherwise, can be defined for each RBS. In addition to this, another variable can be defined as y_l, which indicates the quantity of hubs to be acquired so that the demand sent to the network can be met. In this manner, the GAP-DC can be formulated according to the following mixed integer programming problem, as presented in [5]:

(M)

$$Min \sum_{i \in V} \sum_{j \in R} c_{ij} x_{ij} + \sum_{l \in H} b_l y_l \tag{1}$$

$$\sum_{j \in R} x_{ij} = S_i, \qquad \forall i \in V, \tag{2}$$

$$\sum_{i \in V} \sum_{j \in R} \frac{D_i}{S_i} a_{jl} x_{ij} \le K y_l, \qquad \forall l \in H, \tag{3}$$

$$\sum_{j \in Tq} x_{ij} \le 1, \qquad \forall i \in V, q = 1, ..., Q, \tag{4}$$

$$x_{ij} \in \{0,1\}, \qquad \forall i \in V, \forall j \in R, \tag{5}$$

$$y_l \ge 0, \qquad \forall l \in H. \tag{6}$$

The objective function (1) is to minimize the cost of connecting the RBS's to the hubs and to acquire greater capacity, as needed. Constraints (2) guarantee a defined diversity or rather that if a call connected to a determined hub fails, all traffic will not be lost. The possible values used for diversity S_i, of each station i, are 1, 2 and 3. Constraints (3) establish the amount of capacity to be acquired for each link. Constraints (4) stop the cell from being connected to the hubs from the same tree. Constraints (5) correspond to constraints on the variables' integrality x_{ij}, and finally, to constraints (6) which correspond to non-negative variable constraints y_l.

Thus, a modification in the model (M) must be realized in order to render the modeling of the GAP-DC* possible. This modification is necessary due to the increase in capacity resulting from another link being added to each hub, a phenomenon which will be detailed in the following model.

To facilitate comprehension of the proposed mathematical model for GAP-DC*, the following auxiliary notation is to be considered:

- $D1_i$: data demand on each cell $i \in V$;
- $D2_i$: voice demand on each cell $i \in V$;
- $y1_l$: amount of capacity for the data link to be acquired, that being $l \in H$ and $y1_l$ is an integer;
- $y2_l$: amount of capacity for the voice link to be acquired, that being $l \in H$ and $y2_l$ is an integer;
- $B1_l$: cost of acquiring each additional data capacity unit;

- $B2_l$: cost of acquiring each additional voice capacity unit.

The GAP with constraints of diversity and capacity with two communication channels for each hub can be expressed by the following mixed integer linear programming problem:

(MS)

$$Min \sum_{i \in V} \sum_{j \in R} c_{ij} x_{ij} + \sum_{l \in H} (B1_l y1_l + B2_l y2_l) \quad (7)$$

$$\sum_{j \in R} x_{ij} = s_i, \quad \forall i \in V,$$

$$\sum_{i \in V} \sum_{j \in R} \left(\frac{D1_i + D2_i}{s_i} \right) a_{jl} x_{ij} \le K1 y1_l + K2 y2_l, \quad \forall l \in H, \quad (8)$$

$$\sum_{j \in Tq} x_{ij} \le 1, \quad \forall i \in V, \forall q \in Q,$$

$$\sum_{l \in H} K1 y1_l \ge \sum_{i \in V} D1_i, \quad \forall l \in H, \forall i \in V, \quad (9)$$

$$\sum_{l \in H} K2 y2_l \ge \sum_{i \in V} D2_i, \quad \forall l \in H, \forall i \in V, \quad (10)$$

$$x_{ij} \in \{0,1\}, \quad \forall i \in V, \forall j \in R, \quad (11)$$

$$y1_l \ge 0, \quad \forall l \in H, \quad (12)$$

$$y2_l \ge 0, \quad \forall l \in H. \quad (13)$$

Now, instead of one unique transmission link for data and voice, there are two separate links, one transmitting data and another only voice. Consequently, the model is altered in its objective function (7), in that the goal is to minimize the cost of connecting the cells to the hubs, however there is an additional cost of acquiring greater capacity for voice and data links. Constraints (8) establish that the total demand in data and voice must be met. Constraints (9) and (10) ensure that data as well as voice demands will be met. Finally, constraints (11) correspond to integrality constraints and to constraints (12) and (13) guaranteeing that the amount of the capacity to be acquired will be is an integer.

III. PROPOSED ALGORITHM

This section presents an algorithm to solve GAP-DC*. As previously discussed, although GAP-DC* are NP-hard problems, finding an optimum solution can require substantial computational time. Thus, the necessity to obtain good solutions, in a reasonable amount of computational time, led us to carry out a comparative study of simple subgradient and space dilatation methods.

The solution quality obtained for combinatorial optimization requires the generation of lower and upper bounds. In minimization problems, the upper bound offers a feasible solution which can be obtained using a heuristic. The lower bound can be obtained immediately by relaxing some of the more difficult constraints of the problem [10].

One of the techniques for generating lower bounds is Lagrangean relaxation (see [1, 6], for more details). This technique consists in relaxing the more difficult problem constraints and launching them as additional terms in the objective function. The violation of these constraints is penalized by the use of the Lagrangean multipliers associated with each of the constraints relaxed. The problem of calculating Lagrangean multipliers, which in turn maximizes the relaxation problem, is referred to as Lagrangean duality. To solve this relaxation problem, the subgradient method is commonly used.

Consequently, it is necessary to discuss the generation of lower and upper bounds in relation to GAP-DC*. Lower bounds are obtained by Lagrangean relaxation and upper bounds using a heuristic. Here, it is important to point out that updating of Lagrangean multipliers is accomplished using subgradient methods.

A. Lower Bounds

In the mathematical model (MS) there are two sets of constraints, (2) and (4), which can be considered as difficult in resolving the problem. Therefore, relaxation can be adopted in each of these cases.

Let λ_i, for all $i \in V$, be the set of multipliers associated with constraint (2). The Lagrangean relaxation obtained for the model with a definition of λ_i, after reorganization of the terms of the objective function, is thus stated as:

$$Z_{RL}(\lambda) = Min \sum_{i \in V} \sum_{j \in R} c_{ij} x_{ij} + \sum_{l \in H} (B1_l y1_l + B2_l y2_l) + + \sum_{i \in V} \lambda_i \left(s_i - \sum_{j \in R} x_{ij} \right) \quad (14)$$

$$\sum_{i \in V} \sum_{j \in R} \left(\frac{D1_i + D2_i}{s_i} \right) a_{jl} x_{ij} \le K1 y1_l + K2 y2_l, \quad \forall l \in H,$$

$$\sum_{j \in Tq} x_{ij} \le 1, \quad \forall i \in V, \forall q \in Q,$$

$$\sum_{l \in H} K1 y1_l \ge \sum_{i \in V} D1_i, \quad \forall l \in H, \forall i \in V,$$

$$\sum_{l \in H} K2 y2_l \ge \sum_{i \in V} D2_i, \quad \forall l \in H, \forall i \in V,$$

$$x_{ij} \in \{0,1\}, \quad \forall i \in V, \forall j \in R,$$

$$y1_l \ge 0, \quad \forall l \in H,$$

$$y2_l \ge 0, \quad \forall l \in H.$$

Thus $\alpha_l \ge 0$, for all $l \in H$, denotes the set of multipliers associated with constraints (4). Therefore, the Lagrangean relaxation obtained with this set of multipliers α_l is stated as:

$$Z_{RL}(\alpha) = Min\sum_{i\in V}\sum_{j\in R} c_{ij}x_{ij} + \sum_{l\in H}(B1_l y1_l + B2_l y2_l) +$$

$$+\sum_{i\in V}\sum_{q\in Q}\alpha_{i,q}\left(1-\sum_{j\in Tq}x_{ij}\right) \quad (15)$$

$$\sum_{j\in R} x_{ij} = s_i, \qquad \forall i \in V,$$

$$\sum_{i\in V}\sum_{j\in R}\left(\frac{D1_i + D2_i}{s_i}\right)a_{jl}x_{ij} \le K1y1_l + K2y2_l, \qquad \forall l \in H,$$

$$\sum_{l\in H} K1y1_l \ge \sum_{i\in V} D1_i, \qquad \forall l \in H, \forall i \in V,$$

$$\sum_{l\in H} K2y2_l \ge \sum_{i\in V} D2_i, \qquad \forall l \in H, \forall i \in V,$$

$$x_{ij} \in \{0,1\}, \qquad \forall i \in V, \forall j \in R,$$

$$y1_l \ge 0, \qquad \forall l \in H,$$

$$y2_l \ge 0, \qquad \forall l \in H.$$

For any set of Lagrangean multipliers λ_i and α_l, the Lagrangean relaxation yields a solution, $Z_{RL}(\lambda)$ or $Z_{RL}(\alpha)$, which is always less or equal to the solution of the original problem [3]. Thus, the greatest possible solution value yields the maximum lower bound for GAP-DC*. As previously stated, the problem in finding the set of multipliers to maximize the solution to the Lagrangean problem is referred to as Lagrangean duality. In the case of GAP-DC*, since two Lagrangean problems were defined, there are two dual Lagrangean problems to be solved:

$$Z_{LD}(\lambda) = Max\{Z_{RL}(\lambda) \mid \lambda \text{ is unrestricted in sign}\}, \quad (16)$$

$$Z_{LD}(\alpha) = Max\{Z_{RL}(\alpha) \mid \alpha \ge 0\}, \quad (17)$$

B. Simple Subgradient Methods

The subgradient method is the most commonly used in solving Lagrangean duality. It is thus necessary to give a succinct description of how this method functions after which, the subgradient method will be applied to solve problems (16) and (17).

Given a set of Lagrangean multipliers, the relaxation problem is solved by generating a lower bound, then calculating the subgradients which correspond to the relaxed solution. Thus, subgradients are used to update multiplier sets, in view of obtaining a new lower bound that is higher than the previous one. The process is repeated until the lower and upper bounds are equal or until a duality gap occurs.

The Lagrangean multiplier λ, in the (k+1)-th iteration, is stated as:

$$\lambda_i^{k+1} = \lambda_i^k + t^k g\lambda_i^k(x^k), \quad \forall i \in V, \quad (18)$$

where t^k is a positive scale, or step size, $g\lambda_i^k(x^k)$ is the subgradient vector of $Z_{RL}(\lambda)$, and x^k represents the optimal value of variables x_{ij} in the solution of problem $Z_{RL}(\lambda)$.

Thus the second Lagrangean multiplier α, in the (k+1)-th iteration, is stated as:

$$\alpha_i^{k+1} = Max\{0, \alpha_i^k + s^k g\alpha_i^k(x^k)\}, \quad \forall i \in V, q \in Q \quad (19)$$

where s^k is a positive scale, or step size, $g\alpha_i^k(x^k)$ is the subgradient vector of $Z_{RL}(\alpha)$, and (x^k) represents the optimal value of variables x_{ij} and y_l in the solution to problem $Z_{RL}(\alpha)$.

The vectors of the subgradients mentioned in equations (18) and (19) are defined, respectively, as:

$$g\lambda_i^k(x^k) = \sum_{i\in V}\left(1-\sum_{j\in R}x_{ij}\right), \quad \forall i \in V, \quad (20)$$

$$g\alpha_i^k(x^k) = \sum_{i\in V}\sum_{q\in Q}\left(1-\sum_{j\in Tq}x_{ij}\right), \quad \forall i \in V, q \in Q, \quad (21)$$

The step size, t^k, to be taken in the direction of the subgradient when relaxing constraint (2) is:

$$t^k = \pi[Z_{ub} - Z_{RL}(\lambda_k)] / \| g\lambda_i^k \|^2, \quad (22)$$

where π is a defined scale in interval $0 < \pi \le 2$ and Z_{ub} represents the known upper bound. The value π is used to regulate the step size, and consequently, the velocity of the convergence of the subgradient method.

Thus in constraint (4), the step size, s^k, is defined as:

$$s^k = \pi[Z_{ub} - Z_{RL}(\alpha_k)] / \| g\alpha_i^k \|^2, \quad (23)$$

After describing all the basic points of the subgradient method used, the algorithm adopted in solving the GAP-DC* will now be presented. First, a version of the algorithm using Lagrangean multiplier λ will be given, and then necessary alterations with respect to Lagrangean multiplier α will be mentioned. Both algorithms are based on the description in [1].

Algorithm 1

1. Start LI = $-\infty$, LS = ∞, k = 0 e $\pi = 2$.
2. Set Lagrangean multiplier λ_i^k, for all $i \in V$.
3. Increment k.
4. Calculate lower bound $Z_{RL}(\lambda)$.
5. If ($Z_{RL}(\lambda)$ > LI)
6. Then LI = $Z_{RL}(\lambda)$.
7. Else go to step 13.
8. Calculate upper bound Z_{ub}.
9. If (Z_{ub} < LS)
10. Then LS = Z_{ub}.
11. If (LI = LS) or (LS – LI < 0.00001)
12. Then return (LI).
13. Calculate the subgradient $\| g\lambda_i^{k+1} \|$.
14. If $\| g\lambda_i^{k+1} \| = 0$

15. Then STOP.
16. If LI is not incremented to 10 iterations
17. Then $\pi = \pi / 2$.
18. Calculate step size t^k.
19. Update the Lagrangean multipliers λ_i^{k+1}.
20. If ($\pi < 0.001$)
21. Then STOP.
22. Else go to step 3.

Algorithm 2 is obtained by carrying out the necessary changes for Lagrangean multiplier α in steps 2, 4-7, 13-15, 18 and 19 of algorithm 1. Both algorithms use value $1.05Z_{ub}$ in equations (22) and (23). This constant is an empirical value exposed by Beasley [1] in the implementation of Lagrangean relaxation. The same empirical observation can be made for value 10 in step 16.

Next, the space dilatation subgradient method proposed for GAP-DC* will be detailed.

C. Space Dilatation Methods

The convergence of algorithms 1 and 2, is very slow, in general, and depends on the chosen step size and set of relaxation constraints. In [8], one of the numerous variations for the previous algorithm is developed to improve the relative convergence. In [7, 8], the authors suggest a modification in the direction of the movement with the aim of decreasing the angle between the subgradient and the direction of the optimal IP solution. This modification is called the space dilatation method, which will be described hereafter.

Consider the problem (16), i.e.,

$$Z_{LD}(\lambda) = Max\{Z_{RL}(\lambda) \mid \lambda \text{ is unrestricted in sign}\},$$

Suppose that iteration number K of the simple subgradient method produced a change in variables $\lambda = B_k\sigma$, where B_k is a nonsingular matrix (identity matrix) with the appropriate dimension. The following subgradient method of the function $\psi(\sigma) = \varphi(B_k\sigma)$ is applied. This implies carrying out a spatial movement of variables λ in the corresponding direction:

$$d = B_k B_k^t g\lambda_i^k \tag{24}$$

where $g\lambda_i^k$ is the subgradient of φ in the actual point λ^k.

Assume that t^k correspond to the chosen in step size of the applied algorithm $\psi(\sigma)$. So, we should have:

$$\sigma^{k+1} = \sigma^k + t^k B_k^t g\lambda_i^k \tag{25}$$

and, consequently, it is possible to find $\lambda_i^{k+1} = B_k\sigma^{k+1}$ by using the formula:

$$\lambda_i^{k+1} = \lambda_i^k + t^k B_k B_k^t g\lambda_i^k(x^k), \quad \forall i \in V, \tag{26}$$

Thus, the second Lagrangean multiplier α, in the (k+1)-th iteration, is calculated by:

$$\alpha_i^{k+1} = \alpha_i^k + s^k B_k B_k^t g\alpha_i^k(x^k), \quad \forall i \in V, \forall q \in Q, \tag{27}$$

The following definition, aims at updating matrix B^k:

$d \in \Re^n$ being such that $\| d \| = 1$. We shall call the linear operator $R_\alpha(d)$ to be defined by:

$$R_\alpha(d)\lambda = \lambda + (\alpha - 1)\lambda_d, \quad \forall \lambda \in R^n \tag{28}$$

A dilatation operator in the direction of d with a dilatation coefficient of $\alpha \geq 0$, where $\lambda_d = (\lambda^t d)d$ is the projection of λ in the direction of d.

Operator $R_\alpha(d)$ has, however, the property of leaving all orthogonal vectors of d intact, but multiplies by $\alpha \geq 0$ all collinear vectors with d. $R_\alpha(d)$ can also be put in the formulation:

$$R_\alpha(d)\lambda = \lambda + (\alpha - 1)(dd^t)\lambda, \quad \forall \lambda \in R^n \tag{29}$$

And yet, the matrix associated with this linear operator is:

$$I + (\alpha - 1)(dd^t), \tag{30}$$

where I is the identity matrix of order *n*.

In spatial dilatation methods, the matrixes B_k are updated according to formula:

$$B_{k+1} = B_k R_{\alpha k}(d_k) \tag{31}$$

where direction d_k is calculated from the difference between the two successive subgradients [9]:

$$d_k = r_k / \| r_k \|, \tag{32}$$

where

$$r_k = B_k^t \left[g\lambda_i^{k+1} - g\lambda_i^k \right] \tag{33}$$

As all the basic points of the space dilatation method have been described, the algorithm proposed to solve GAP-DC* can now be presented. First, a version of this algorithm shall be outlined using Lagrangean multiplier λ, then necessary alterations in cases where Lagrangean multiplier α is used will be given. Both algorithms are based on the description in [7, 9].

Algorithm 3

1. Start LI = $-\infty$, LS = ∞, k = 0 e $\pi = 2$.
2. Set the Lagrangean multipliers λ_i^k, for all $i \in V$.
3. Increment k.
4. Calculate the lower bound $Z_{RL}(\lambda)$.
5. If ($Z_{RL}(\lambda)$ > LI)
6. Then LI = $Z_{RL}(\lambda)$.
7. Else go to step 13.
8. Calculate the upper bound Z_{ub}.
9. If (Z_{ub} < LS)
10. Then LS = Z_{ub}.
11. If (LI = LS) or (LS – LI < 0.00001)
12. Then return (LI).
13. Calculate the subgradient $\| g\lambda_i^{k+1} \|$.
14. If $\| g\lambda_i^{k+1} \| = 0$
15. Then STOP.
16. If LI is not incremented to 10 iterations
17. Then $\pi = \pi / 2$.
18. Calculate the step size t^k.
19. If $t^k = 0$,
20. Then $\lambda^{k+1} \leftarrow \lambda^k$
21. $g\lambda_i^{k+1} \leftarrow g\lambda_i^k$
22. $g\lambda_i^k \leftarrow g\lambda_i^{k-1}$
23. and go to step 26
24. Else
25. $\lambda_i^{k+1} = \lambda_i^k + t^k B_k B^t_k g\lambda_i^k(x^k)$
26. $r_k = B^t_k[g\lambda_i^{k+1} - g\lambda_i^k]$
27. $d_k = r_k/\|r_k\|$
28. $B_{k+1} = B_k[I + (\alpha - 1)d_k d^t_k]$
29. $k \leftarrow k + 1$
30. If ($\pi < 0.001$)
31. Then STOP.
32. Else go to step 3.

Algorithm 4 is obtained by consequent changes to Lagrangean multiplier α in steps 2, 4-7, 13-15, 18-22 and 24-26 of Algorithm 3. Both algorithms use the value of 0.9 for α in the equation (28). This constant is an empirical value which was exposed by Shor [8, 9] for the implementation of space dilatation methods.

D. Upper Bound

In this section the heuristic used to obtain the upper bound will be described (Z_{ub}), which supports the calculation of the step size described in (22) and (23).

As we are working with small to medium-ranged instances, where the optimal IP solution is known, the calculation of the upper bound increases by a mere one percent, and varies from 0.01 to 0.1, above the optimal IP solution. This confirms that the value obtained is feasible, otherwise decreased (or increased), until a feasible upper bound is found.

IV. COMPUTATIONAL RESULTS

In this section, the computational results found with the Lagrangean relaxation model using simple subgradient and space dilatation methods, as proposed for GAP-DC*, will be presented. The experiments were conducted on an IBM-PC computer, with a 1.2 MHZ Pentium processor and 256Mb of RAM memory. A Microsoft Visual Basic 6.0 environment and a LINGO 8.0. commercial optimizing package were used to solve the problem.

The instances used in the experiments were generated randomly. The characterization of each instance will be carried out using three parameters: the number of RBS's, the number of hubs and the number of MSC's. The set of test instances is presented in Table I.

TABLE I
CHARACTERIZATION OF THE INSTANCES USED IN THE EXPERIMENTS

Instances	#RBS's	#Hubs	#MSC's
1	10	3	1
2	15	3	1
3	40	7	1
4	33	8	1
5	120	14	1

The results obtained by the test instances using LINGO 8.0 are presented in Table II. The first column of the table indicates the instance number, the second the linear relaxation value, the third the optimal IP solution, the fourth the gap between value z* of the optimal IP solution and value z of the linear relaxation. The fifth column represents the computational time and the sixth the amount of iterations.

$$\Delta = \left[\left(z^* - z\right)/ z^*\right] * 100 \qquad (34)$$

TABLE II
EXECUTION RESULTS OF LINGO 8.0

Instance	Z	Z*	Δ (%)	Time	#Iter
1	1158.0291	1159.0	0.08377049	0.00000	6
2	377.0000	377.0	0.00000000	0.00000	2
3	1741.3002	1746.0	0.26917535	0.00000	2253
4	1658.6000	1659.0	0.02411091	0.00000	107
5	3533.6000	3534.0	0.01131862	0.00000	987

Tables III and IV, respectively, show the results obtained by Lagrangean relaxation for the instances described in Table I using algorithms 1, 2, 3 and 4. The significance of the columns in these tables is the following: the first column indicates the number of instances, the second the value obtained through Lagrangean relaxation using simple subgradients – LI(S), the third the value obtained through Lagrangean relaxation using spatial dilatation methods – LI(D), the fourth the optimal IP solution , the fifth the gap between value z* of the optimal IP solution and the LI(S) value of the Lagrangean relaxation, or,

$$\Theta = \left[\left(z^* - LI(S)\right)/ z^*\right]*100 \qquad (35)$$

The sixth column indicates the gap between value z* of the optimal IP solution and value LI(D) of the Lagrangean relaxation , or,

$$\mu = \left[\left(z^* - LI(D)\right)/ z^*\right]*100 \qquad (36)$$

Column numbers 7 and 8 represent the computational time spent (in seconds) to execute the simple subgradient and space dilatation methods, respectively, and 9 and 10 the amount of iterations.

TABLE III
RESULTS OBTAINED THROUGH LAGRANGEAN RELAXATION AND EXECUTED BY ALGORITHMS 1 AND 3

Instance	LI(S)	LI(D)	Z*	Θ (%)	μ (%)	Time(s)	Time(D)	#Iter LI(S)	#Iter LI(D)
1	1147.6189	1151.7026	1159.0	0.9819000000	0.6296000000	0.0000231481	0.0001504630	50	100
2	377.0000	377.0000	377.0	0.0000000000	0.0000000000	0.0000000000	0.0000000000	1	1
3	1614.2747	1613.0472	1746.0	7.5444055999	7.6147098871	0.0003587963	0.0014699074	182	194
4	1576.6492	1576.5913	1659.0	4.9638811804	4.9673737586	0.0002199074	0.0005208333	173	123
5	3387.3848	3453.1878	3534.0	4.1487036109	2.2867072513	0.0083449074	0.0059953704	414	68

TABLE IV
RESULTS OBTAINED THROUGH LAGRANGEAN RELAXATION AND EXECUTED BY 2 AND 4

Instance	LI(S)	LI(D)	Z*	Θ (%)	μ (%)	Time(S)	Time(D)	#Iter LI(S)	#Iter LI(D)
1	1148.0000	11480000	1159.0	0.9490940466	0.9490940466	0.0000000000	0.0000347222	10	1
2	354.0000	354.0000	377.0	6.1007957560	6.1007957560	0.0000810185	0.0000115741	170	1
3	1586.7370	1586.7370	1746.0	9.1215938775	9.1215938775	0.0020601852	0.0007638889	250	1
4	1585.6000	1585.6000	1659.0	4.4243520193	4.4243520193	0.0003240741	0.0000694444	161	1
5	3406.6000	3406.6000	3534.0	3.6049801924	3.6049801924	0.0021643519	0.0007520000	120	1

Analysis of the results of Tables II, III and IV, confirms the fact that in addition to linear relaxation (Table II) obtaining better lower bounds in less time, a high amount of iterations in small-range instances can also be observed. In Table III, it is confirmed that relaxation using simple subgradients solved the majority of gap instances, in less time and with fewer iterations in respect to spatial dilatation methods. However, when the set of constraints (4) are relaxed simple subgradients as well as space dilatation methods present exactly the same gap, whereas spatial dilatation was solved in less time and in the very first iteration.

V. CONCLUSIONS

In this work a mixed integer programming model and algorithms for problems which occur in mobile telephone operators is proposed. The problem in question, GAP-DC*, is an extension of generalized assignment problems in diversity and capacity constraints (GAP-DC). GAP-DC* are NP-hard problems. Thus, due to the complexity of this problem, it was necessary to develop two Lagrangean relaxation algorithms using simple subgradient and space dilatation methods to solve the GAP-DC*. These algorithms were obtained in compliance to the constraints of the relaxation model.

The results obtained with the use of the proposed space dilatation methods demonstrate the efficiency of the method for our objective function, since the amount of iterations obtained with these algorithms was better than those obtained using linear relaxation. Furthermore, in relaxation using space dilatation methods, gap and acceptable computational time is encountered in the very first iteration, which leaves us very optimistic with respect to solving high-range problems using this technique.

This opens up new prospects for future research in this area such as implementing and testing new heuristics to determine upper bounds and testing the algorithms presented in higher-range instances.

REFERENCES

[1] Beasley, J. E. "Lagrangean Relaxation", Modern Heuristic Techniques for Combinatorial Problems (editor C. Reeves), Blackwell Scientific Publications, 1993.
[2] Bose, I., Eryarsoy, E. e He, L. "Multi-period design of survivable wirelles access networks under capacity constraints". Decision Suport Systems, 2005.
[3] Geoffrion, A. M. "Lagrangean relaxation ends its uses in integer programming". Mathematical Programming Study, 2:82–114, 1974.
[4] Kubat, P e Smith, J. MacGregor. "A multi-period network design problem for cellular telecommunication systems".

European Journal of Operational Research, 134:439-456, 2001.

[5] Kubat, P., Smith, J. MacGregor e Yum. C. "*Design of celullar networks with diversity and capacity constraints*". IEEE Transactions on Reability, 49:165–175, 2000.

[6] Minoux, M. Mathematical Programming, Theory and Algorithms. John Wiley and Sons, 1986.

[7] Rodrigues, S. I. M. Relaxação Lagrangeana e subgradientes com dilatação de espaço aplicados a um problema de grande porte. RJ, 1993.

[8] Shor, N. Z.Utilization of the operation of space dilatation in the minimization of convex functions. Cybernetics, 1:7-15, 1970.

[9] Shor, N. Z. Zhurbenko, N. G. A minimization method using the operation of extension of the space in the direction of the difference of two successive gradients. Cybernetics, 7(3):450-459, 1970.

[10] Wolsey, L. A. Integer programming. John Wiley & Sons, 1998.

Sensor Network Applications: a Module for Monitoring and Remote Control of Physical Variables Using Mobile Devices

A. F. Muñetón, J. M. Saldarriaga, A. de J. Montoya, D. L. Aristizábal
Scientific and Industrial Instrumentation Group
Universidad Nacional de Colombia sede Medellín
Medellín-Colombia
Autopista Norte, Carrera 65
A.A 3840
amontoya@unalmed.edu.co

***Abstract*- A module was developed to monitor and to control physical variables by means of mobile devices like a cellular telephone or PDA (Personal Digital Assistant). This one is made up of two parts: an application client accessible from a mobile device and an application Web in charge to manage requests of the application client. This last one, in interaction with the application Web via Internet, allows to visualize the state of the physical variables and to activate or to deactivate control systems with the purpose of altering one or more of these variables. The module was developed in the language OO Java combining platforms J2ME and J2EE for the application client and Web respectively. In addition the methodology of software engineering Un-method of the School of Systems of the National University of Colombia - Host Medellín was used.**

***Key words*- monitoring, control, physical variables, mobile device, language OO, Java, J2ME, J2EE**

I. INTRODUCTION

The mobile technologies have surpassed widely their initial functionalities, as they are it the transmission of voice and its use like digital agendas. They have entered the world of the data, and in addition, adding an insurmountable added value: to equip the people with the information whom they require at any time and place by means of a small device with the capacity of processing of a desktop computer of a pair ago of years.

In this work a system appears that allows the monitoring and remote control of physical variables, like temperature and humidity, using mobile devices working in networks GSM (Global System of Mobile Communication) that operate in the frequencies of 900, 1800 and 1900 MHz [1].

The solution is oriented to the differentiated monitoring from ecosystems (zones of culture, conservatories, forests), that considers the land like a heterogenous system; this unlike techniques and technologies normally applied, like the satelite images, that do not allow excellent distinctions between zones, losing therefore the opportunity to optimize the use of resources like fertilizers, to predict with more exactitude the production of cultures, to know conditions more located of the ground, to take statistical data for a future decision making, between many other advantages of the differentiation. The developed system was reconciled to a network 1-wire of sensors and controller in a small conservatory allowing to make demanding tests of its functionality and stability. The methodology of software engineering Un-method was used [2] of the National University of Colombia - Host Medellín, and the platforms Java J2EE (Java Enterprise Edition) [5] and J2ME (Java 2 Micro Edition) [6], [7].

II. DESCRIPTION OF THE SYSTEM

The module centrally was reconciled to a denominated unit "Station" composed by a small network 1-wire of sensors and controller administered by a contracted device called TINI (Tiny InterNet Interfaces) [3]; this one, every certain time, sends using protocol TCP/IP, the measurements taken by its sensors. These received measurements are stored in a database through one applications server (Tomcat 5.5.7 and MySQL 4.1). Additionally the TINI receives and responds to indications sent from the servant for the change of state of some of their controllers. The state of the physical variables and the function of control are acceded remotely using movable devices with access to Internet. The development of the network of sensors and the handling of the TINI comprise of a complementary module to the mentioned one in this project [4].

III. REQUIREMENTS

A. *Objectives*

The system was designed and implemented to manage:

- Different stations with its actuators and sensors, in addition to the measurements taken at these last ones.
- The operators, specially its activity within the system.

B. *Data storage*

The system stores the following data:

- Stations: name, state and direction IP of the station.

K. Elleithy et al. (eds.), Advances in Computer, Information, and Systems Sciences, and Engineering, 243–245.

- Sensors and controller. Identifier, state and variable on which act.
- Variable Physical: name of the variables, measurements done on these and the dates in which the measurements become.
- Operators: identification, name and movements in the system.

C. *Actors*

-Cliente J2ME (Operador): only user of the system. He is the one in charge of the administration of the Stations, which includes the monitoring and control. He has access to the system by means of a mobile device.

D. *functional requirements*

The application in the movable device must be formed, indicating the URL to him of the application Web. The validation is obligatory to have access to the monitoring functions and control. When the user enters his authentication code, this Web is sent to the application, which will return the result, true or false, of the validation (see fig 1)

In the monitoring the user has available a listing of the areas (stations) in which the ecosystem was divided, and when selecting one of them it will be able to visualize the measured physical variables with the most recent value in the data base.

After reviewing the state of the physical variables, the user can request the visualization of the controllers. When selecting a controller, if she wishes, asks for his change of state, in which case, a request is sent to the servant, who gives the instruction of control to the TINI.Esta will respond indicating the change of state, and sending a confirmation of control to the movable device.

IV. ANALYSIS AND DESIGN

A. *Technical Architecture*

The system was developed following architecture J2EE multilayer [5].

Client layer, is represented by MIDlets or applications developed with the Java platform for movable devices (J2ME), which are executed in PDAs or cellular telephones with access to Internet and superior profile MIDP 1,0 or higher.

In layer Web they are servlets responsible for the processing of requests of the clients, for which they lean in

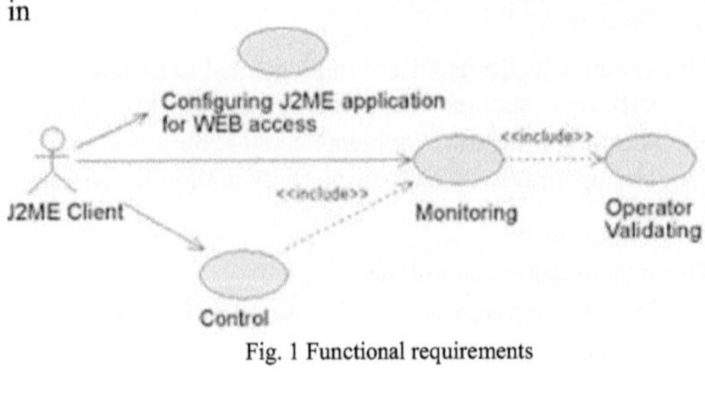

Fig. 1 Functional requirements

the elements of the inferior layers. The container of servlets is Tomcat 5.5.7.

The organizations that contain the logic of the business find in Intermediate Capa, distributed in the servant of applications and card TINI and communicating through the network using special units called "packages".

The layer EIS (Enterprise Information Service) presents/displays the data base, administered with MySQL 4.1.

B. *J2ME Application*

From application J2ME, the operator can review the value of the physical variables of each station and, according to these values, request the activation or turn on of some controller who causes a change on the system. He allows in addition the storage to certain data that do not vary frequently, such as the list of stations, thus avoiding unnecessary delays by the reception of data and connections to the network. The communication with the application Web is done using protocol HTTP [4].

C. *Web Application*

Meeting point between the clients and the network of sensors; reason why two types of interaction can be defined: Interaction Web - TINI and Interaction Web - Clients J2ME.

Interaction Application Web - TINI: bidirectional interaction through the network by means of serialize objects called packages. Two types of packages exist: Monitoring and Control. Every certain time, the TINI sends a sequence of packages of monitoring (so many as sensorial have the station) that indicate the sensor that made the measurement, the measured physical variable, the date of the measurement and direction IP of the station (TINI). The application Web stores each measurement in the data base and obtains the average of each physical variable; this is, if station X sends 8 packages, of which 4 contain measurements of temperature and the rest of humidity, then they would obtain two values average for the station, one of temperature and another one of humidity. The control packages are used to request to the TINI the change to him of state of some controller, of such form that, for example, ignites the lights of a station to control the temperature. Once fact the change, the TINI returns a package of control that indicates the present state of the controller, which allows to the application Web to maintain updated the data base with the real states of each element.

Interaction Application Web - Clients J2ME: Clients J2ME are registered in the data base so that the application Web identifies and maintains them a registry of its actions in the system. The requests done by the clients include a series of parameters that indicate to the application Web to him who is the client, which is the goal (monitoring or control) and if it has stored the listing of stations. This last one is useful to avoid the unnecessary shipment of information that does not undergo frequent changes, as it happens with the stations.

In order to conserve the structure of the application and to

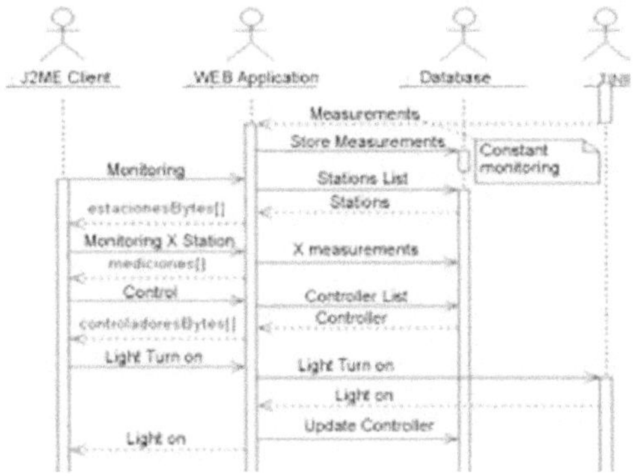

Fig. 2 Normal sequence of Monitoreo and Control. Even though the application Web is receiving measurements of the TINI, can take care of requests of the clients. and controller

encapsulate of the data, the application Web sends to the clients objects "Station", "Measurement" and "Controller", in representations of adjustments of bytes. The objects are not properly serializados then is not counted on this option in platform J2ME [6],[7]. In figure 2 these interactions are synthesized.

V. CONCLUSIONS

The mobile devices and the movable and wireless technologies equip the systems that implement of competitive advantages forehead to the traditional systems.

The systems of this type still present/display serious difficulties due to the technology on which they lean, as it is the case of the network of mobile operators. An improvement in the communications net of data of these operators in the country is expected, because the reception and shipment of information, as much for the monitoring as the control, are quite slow in comparison with the equipment of traditional access to Internet (desktop computers). This would allow, for example, the possibility of counting on information and statistical graphs in "real time".

The system was proven in a single station, nevertheless the form as it were designed and implemented it makes possible his connects to multiple sensorial stations each one with manifold

VI. ACKNOWLEDGMENTS

To NERIO MONTOYA and LAURA GIRALDO, Physical Engineers of the National University - Host Medellín who are the designers and developers of the network 1-wire of sensors and the applications of embedded system (TINI).

VII. REFERENCES

[1] S. Buckingham, GSM World [web] http://www.gsmworld.com/technology/gsm.shtml [Last access, 16 September 2005].

[2] C. Ramírez, "Guía Metodológica de Desarrollo o Adaptación de Software" [web] http://www.unalmed.edu.co/~csjimene/pub2/software.zip [Last access, 16 September 2005].

[3] MAXIM - Dallas Semiconductor, *Application Note 708: Exploring Tiny InterNet Interfaces (TINI)*, [web] http://pdfserv.maximic.com/en/an/app708.pdf, [Last access, 16 September 2005].

[4] N. Montoya y L. Giraldo, "Monitoreo y control remoto de variables de un invernadero", Unpublished.

[5] J. Keogh, *J2EE: Manual de Referencia*, McGraw-Hill , 2003, 803 p.

[6] J. Keogh, *J2ME: The Complete Reference*, McGraw-Hill / Osborne, 2003, 745 p.

[7] A. F. Quintas y P. J. Cárdenes, *Java 2 Micro Edition (J2ME). Manual de Usuario y tutorial*, RA-MA Editorial, , 2004, 572 p.

A NEW METHOD FOR STEGANOGRAPHY IN HTML FILES

Mohammad Shirali Shahreza

Computer Science Department
Sharif University of Technology
Tehran, IRAN
mohammad@shirali.ir http://mohammad.shirali.ir

ABSTRACT

In line with the expansion of the Internet, the issue of information security has gained special importance. Many methods have been provided in order to establish hidden communications. In the meanwhile, steganography is one of a relatively new method. A method is introduced in this paper for exchanging information by steganography on HTML pages. The main idea in this method is to hide coded data in the ID attribute of the HTML document tags. Since numerous HTML pages are on the web, this method can be widely used. The suggested method is implemented by using the Java language.

***Keywords*: Steganography, HTML, Internet Security, Information Hiding, Watermarking.**

1. INTRODUCTION

With the expansion of the Internet and extensive use of the Internet in various aspects of life and work, the issue of information security has gained special importance. One of the grounds discussed in the realm of information security is hidden information exchange. To this end, a variety of methods such as cryptography, steganography, coding, etc have been used.

Steganography is a method that has received attention in recent years. While implementing this method, the main goal is to hide information in cover media so that other people will not notice the existence of information. This is a major distinction of this and the other methods of hidden information exchange because, for example, in the coding method, individuals will notice the existence of information by seeing coded information but they will be unable to comprehend the information. In steganography, however, nobody can notice the presence of information in the sources [1].

Most steganography jobs have been carried out on images [2], video clips [3], music and sounds [4]. However, very little work has been done on steganography in text [5]. In addition, steganography has been implemented in a variety of systems such as the computer and the mobile phone [6].

Recently, however, information security has been considerably improved by combining the steganography with the other methods. In addition to application in hidden exchange of information, steganography is used in other fields such as copyright protection, prevention of e-document forging, etc.

One of the most frequently used types of information exchanged on the Internet is HTML (Hyper Text Markup Language) documents used for creating web pages. Therefore, hiding information on these pages can be a good option for hidden exchange of information.

Therefore, this paper deals with a method for steganography on web pages.

2. PREVIOUS WORKS

HTML is a simplified version of the SGML that offers information in the network using the HTTP (Hyper Text Transfer Protocol) [7].

HTML documents were first studied in 1989 by "Tim Burners Lee". This language is actually used for forming and communicating text pages. Using special symbols, this language tells the browser how to show and format texts [8].

HTML pages consist of tags that determine how page components are to be presented and combined. With the help of the tags, you can format text, create tables, place images, play sound, show film, title the page, etc. In fact, the tags cerate the page objects.

Every tag is known with a private name that has special members of its own. The attributes constitute one of the elements of any object. Each attribute is applied for activating an attribute of the object to which it is attributed [9].

A tag is placed between the two 'less than' (<) and 'greater than' (>) signs. First, the tag name and then the members of the tag are defined. Figure 1 shows a sample tag that introduces an image object on the page. The name of this tag is IMG. It has an image address attribute (SRC) as well as the height and width attributes of the image.

```
<IMG SRC = "image.gif" HEIGHT = "120" WIDTH = "160">
```

Fig1. A HTML tag

K. Elleithy et al. (eds.), Advances in Computer, Information, and Systems Sciences, and Engineering, 247–251.

In this part, I will study some works that have been done on HTML document steganography.

2.1. Hiding in Comments [9]

We can add comments on HTML pages as with other languages. On HTML pages, the comments begin with this symbol "<!--" and end with this symbol "-->". An easy way of steganography on HTML pages is to place data as comments.

As the browser does not show such information, the regular user cannot view the comments. However, by seeing the source of HTML document, he will be able to see the comments.

2.2. Hiding at the End of HTML Files [10]

HTML pages begin with an <html> tag and end with an </html> tag. One of the other simplest methods for steganography on HTML pages, similarly to the method used in steganography at the end of images and documents [11, 12], is to place them after the </html> tag at the end of the page. As the browser does not show such information, the user cannot see the information. However, if the user refers to the contents of the HTML document, he can easily see the information at the end of the page.

2.3. Steganography by Creating White Spaces [13, 14]

On HTML pages, the existence of space in tags is not important and the browser does not pay attention to such spaces. This feature is used for hiding information by placing certain white spaces among tag members. Then, while extracting information from the page, information is extracted by calculating the amounts of spaces and by using the appropriate function. In this method, the document size increases considerably because of the many spaces in the document. Also, the HTML document loses its normal form and may attract attention.

2.4. Hiding by Changing the Case of the Letters

A method that I used for steganography in HTML documents was steganography by change of case.

From among the other features of HTML documents is their case-insensitivity of tags and their members. For example, the three tags <BR>, <Br> and <br> are equally valid and are the same. As a result, one can do information steganography in HTML documents by changing the case of the letters in document tags. To extract information, one can extract information by comparing words in normal case and by using the appropriate function.

Since it is very difficult (and even impossible!) to type letters in lower cases and upper cases, the presence of information on the page will be noticed and the page will attract attention, esp. because typing words in lower cases and upper cases has become common among hackers.

2.5. Hiding by Permuting Attributes of Tags [15, 16]

One of the features of HTML documents is the insensitivity of such documents to the order of the tag members. Therefore, one can hide information in HTML documents using the same feature.

To extract information from the document, the order of the tag members is compared with the sorted state of the members and the hidden information is thus extracted.

Despite the advantages of this method, such as no change of the document size, very little information can be hidden in this method.

3. THE SUGGESTED ALGORITHM

One of the attributes that exists in almost all tags on HTML pages is the ID attribute. Based on this attribute, each tag is given a unique ID. Our purpose in this algorithm is to hide information in this tag attribute.

To this end, first the file containing the information is compressed so that the size of the information is reduced. Then, the bytes of the compressed file are converted by the following function to a combination of letters and digits so that they can be used in the ID attribute. This function converts two bytes into four characters.

Function converting two bytes into four characters:

A byte has a code from 0 to 255. To convert this byte into two characters, first we save the last digit of the byte code exactly as it is. The remaining two digits are between 0 and 25, which exactly corresponds to the number of English letters. Therefore, we save the letter corresponding to these two digits. We carry out the above action exactly for the second byte. Then we put the two produced letters and the two produced digits, respectively, besides each other so as to obtain a proper code (Fig. 2).

After converting the bytes to characters, the tag ID is created by a combination of the object name, title of the HTML page and, finally, the four coded characters. The method to create this ID is as follows:

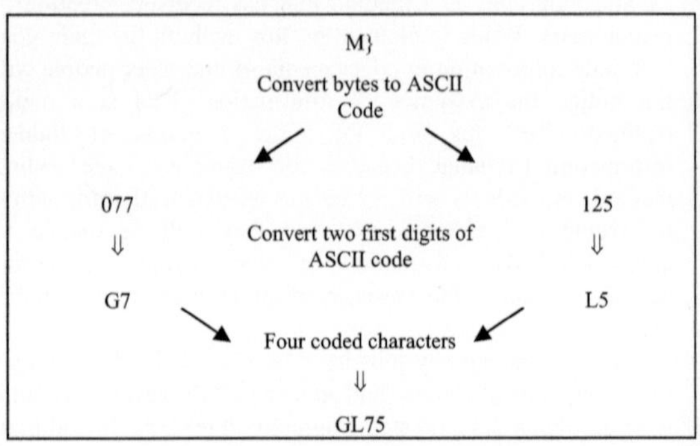

Fig.2 Example of converting two bytes into four characters

First, the first two letters of the object name are selected. Then the first three letters of the HTML page title are selected. Now, the ID is created by placing the two selected letters of the object name, placing a zero (the reason for which will be provider later on), placing the three selected letters of the page title and, finally, placing the four coded characters, respectively. The same procedure applies for hiding the entire information. Please note that placing the ID is done only for tags that do not contain an ID attribute. Therefore, the amount of information can be hidden varies and does not depend just on the number of the tags. It rather depends on the number of tags without an ID attribute which are capable of adding an ID attribute.

The reason for placing a zero after the two letters of the object name is the possibility of similarity of the two ID. If the two ID are exactly similar, in the second ID, instead of zero, a 1 is put. Similarly, if a larger number of ID is similar, the numbers will increase. If, while selecting the three letters of the page name or the two letters of the object name, the number of the letters is less than three or two, the last letter is repeated. For example, if the page title is AB, the letters ABB will be considered while selecting the first three letters of the page title.

Figure 3 provides an example of creating the ID attribute for a tag.

For extracting information from the HTML document, the following phases are undergone:

First, all the tags having an ID with the following features are identified:

A. The size of the tag ID must be 10 characters.
B. The first two letters of the tag ID must be the same as the first two letters of the tag name. (In the case of tags with a name consisting of fewer than two letters, the last letter must be repeated.)
C. The first three letters of the tag ID must be the same as the first three letters of the page title. (In the case of pages with a title consisting of fewer than three letters, the last letter must be repeated.)
D. The seventh and eighth characters must be only letters.
E. The ninth and tenth characters must be only digits.

Now, the first 6 characters of the ID of the identified tags are deleted and only the last four characters remain. In this phase, these four characters are given to the following function to be decoded and converted into two bytes.

HTML page title = Contact

Tag name =

Two bytes of data = M}

ID created with function = BR0ConGL75

Fig.3 Example of ID attribute creation for a tag

Function to convert four characters into two bytes:

First, the first letter and the first digit are placed beside each other. Then the digit corresponding to the appropriate English letter is identified. (Letters A to Z correspondent to 0 to 25, respectively). This number is put behind the previous number. The obtained number (which may consist of 1 to 3 digits) is the ASCII code of the byte in question. The same action is carried out for the other remaining letter and number. (Fig. 4)

The entire compressed file is obtained by putting the decoded bytes beside each other. Finally, information is obtained by extracting the compressed file.

4. ADVANTAGES AND DISADVANTAGES

4.1. Advantages

A. With the rapid growing of the Internet, there are presently billions of documents on the Internet. In the meanwhile, many of these documents are saved in the HTML language because of its simplicity and ease of use. As a result, it is hardly possible to be able to identify HTML pages that contain hidden information from among the large number of HTML pages on the Internet.
B. The ID attribute is one of the commonly used attributes in tags and most website designers use this attribute for classifying and marking objects in the HTML document. Therefore, use of this attribute for hiding information is a good choice.
C. Almost all HTML tags have an ID attribute. However, the other attributes that exist in HTML documents are specific to a certain group of tags and, therefore, have a more limited use than the ID attribute. Consequently, by hiding information in the ID attribute, one can hide more information in HTML pages.

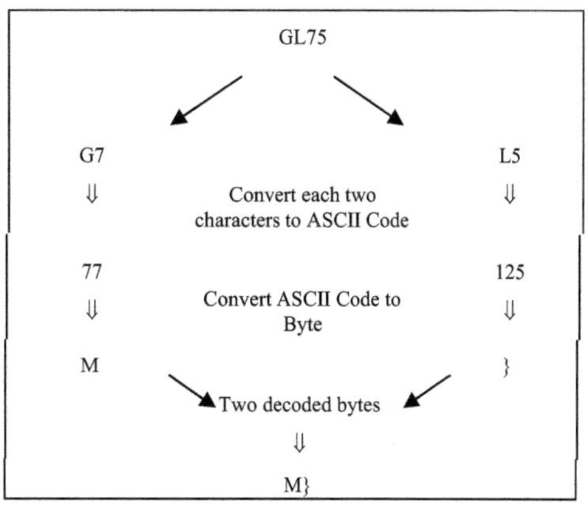

Fig.4 Example of converting four characters into two bytes

D. In the other HTML steganography methods, the hidden information is changed by adding a tag to the page. However, in the suggested method, adding a tag to the page most probably will not result in any change in the hidden information because the appropriate program decodes only the valid ID that follow the above mentioned algorithm.

E. Some webpage editor programs and web browsers change the case of letters while saving the web pages. This deletes hidden information in HTML pages, esp. in the 2.4 method. However, in the suggested method, this will not result in loss of hidden information because information is saved in the ID attribute of tag and such programs will not change the quantity of the attributes, rather they change the case of the letters of the attribute (e.g. SRC to src).

In addition, such programs sometimes delete or add white spaces in the HTML document in order to reduce the size of the HTML document or formatting the HTML document. In methods such as the 2.3, this eliminates information hidden in the HTML document. However, as it has been already mentioned, since such programs do not change the amount of the attributes, information hidden in the suggested method is not lost due to deleting or adding the white spaces.

4.2. Disadvantages

A. According to the algorithm used in the plan, in order to hide two bytes of attribute in an HTML document tag, by considering the very attribute of the ID and the equality sign (=) and the quotation mark ("), it is necessary to add 15 bytes to the tag. As a result, by hiding information in the HTML document, the document size is increased and, therefore, the rate of transfer and display of the page is reduced.

B. In designed HTML pages, page designer usually assign ID attribute to some tags. The number of these tags with an ID attribute varies according to the type of page and the taste of the page designer. Consequently, if the page designer has already given an ID attribute to many of the tags, a small number of tags will remain that do not have the ID attribute. As a result, a small amount of information can be hidden in these pages.

C. On the other hand, there are few tags in some HTML pages. Occasionally, despite the large size of the page, only a few objects such as Java Applets or flash files are used for designing the page. Consequently, the number of tags in the page is few in these cases as well. Therefore, not much information can be hidden in HTML pages.

5. EXPERIMENTAL RESULTS

This project deals with the steganography of files and information in web pages by using the above mentioned algorithm. To this end, some text, image and executive files were chosen. The files were then compressed so as to reduce their size. Then, first the HTML pages were selected and, by using the program, the amount of information that could be hidden in these pages was calculated and the pages that had the capacity to hide the compressed information were selected. Then, with the help of the steganography program, the files were hidden in the HTML pages, using the decoding program, the hidden information was then extracted from the HTML pages and then the files were uncompressed. After comparing the output files with the input files, it was seen that both files were similar. In addition, the ID's used in the HTML file was proportionate to the algorithm and had the same format.

After carrying out this action, we studied the capacity of some well-known web pages for steganography purposes. The website addresses and the amount of information can be hidden in each page are provided in Table 1. The main homepages of each site were used for this purpose.

All of the above pages were visited by using the Mozilla Fierfox 1.0.2 browser and saved in HTML format. The amount of information witch could be hidden in the pages was then calculated by the appropriate program. All these pages were seen and saved on 03 August 2005.

6. CONCLUSION

In this paper dealt with a new method of steganography. With this method, one can exchange hidden information through the Internet.

Some intelligence can be added to the program to reduce hacker attention, for example, by searching the pages for finding ID's previously used in the page and creating similar ID's, one can help the pages look more logical.

Since the Internet has reached such other means of communication as the mobile phone, the home TV, etc, this method can be used in such systems for exchange of information.

In addition to HTML, there are also other formats for web pages such as XML, ASP, etc. One can extend and develop the above mentioned algorithm on such pages.

TABLE I
CAPACITY OF HTML FILES

Web Address	Capacity (Byte)
http://www.altavista.com/	278
http://www.bbc.co.uk/	1194
http://www.cisse2005.org	590
http://www.ebay.com/	1342
http://www.google.com/	78
http://java.sun.com/	1422
http://www.microsoft.com/	516
http://www.yahoo.com/	908
http://www.intel.com/	2298
http://www.nvidia.com/	1470
http://www.amd.com/	1142
http://www.mozilla.org/	624
http://www.adobe.com/	490

5. ACKNOWLEDGMENT

I would like to thanks Mr. Noorollah Hadi, my English teacher in Sharif University of Technology, for editing the final version of this paper.
Finally, I would like to thank my mother for her consistent care, support and encouragement during the process of performing this research.

7. REFERENCES

[1] J.C. Judge, "Steganography: Past, Present, Future", SANS white paper, November 30, 2001, "http://www.sans.org/rr/papers/index.php?id=552".

[2] R. Chandramouli, and N. Memon, "Analysis of LSB based image steganography techniques", Proceedings of the International Conference on Image Processing, vol. 3, 7-10 Oct. 2001, pp. 1019 - 1022.

[3] G. Doërr, and J.L. Dugelay, "A Guide Tour of Video Watermarking," In Signal Processing: Image Communication, vol. 18, no. 4, April 2003, pp. 263-282.

[4] K. Gopalan, "Audio steganography using bit modification", Proceedings of the IEEE International Conference on Acoustics, Speech, and Signal Processing, (ICASSP '03), vol. 2, 6-10 April 2003, pp. 421-424.

[5] A.M. Alattar, and O.M. Alattar, "Watermarking electronic text documents containing justified paragraphs and irregular line spacing", Proc. of SPIE, Volume 5306, Security, Steganography, and Watermarking of Multimedia Contents VI, June 2004, pp. 685-695.

[6] M. Shirali Shahreza, "An Improved Method for steganography on Mobile Phone", WSEAS Transactions on Systems, Issue 7, Vol. 4, July 2005, pp. 955-957.

[7] D. Goodman, Dynamic HTML: The Definitive Reference, Second Edition, O'Reilly & Associates, Inc., 2002.

[8] Elizabeth Castro, HTML 4 for the World Wide Web, Fourth Edition: Visual QuickStart Guide, Peachpit Press, 4th Edition, 1999.

[9] HIPS Systems, "ShadowText", "http://home.apu.edu/~jcox/projects/HtmlStego1", last visited: 29 July 2005.

[10] Adrien Pinet, "SecurEngine Professional 1.0", "http://securengine.isecurelabs.com", last visited: 20 September 2005

[11] CamouflageSoftware, "Camouflage", "http://camouflage.unfiction.com/", last visited September 20, 2005

[12] Fazle Arefin, "Stego Machine", "http://www.fazle.info/", last visited September 20, 2005

[13] NeoByte Solutions, "Invisible Secrets 4", "http://www.invisiblesecrets.com/index.html", last visited: 20 September 2005

[14] -, "wbStego4open", "http://wbstego.wbailer.com/", last visited: 20 September 2005

[15] S. Forrest, Introduction to Deogol, "http://wandership.ca/projects/deogol/", last visited: 20 September 2005.

[16] Corinna John, "Steganography 13 - Hiding binary data in HTML documents", "http://www.codeproject.com/csharp/steganodotnet.asp", last visited: 20 September 2005

Dynamic Admission Control for Quality of Service in IP Networks

J. Kasigwa, V. Baryamureeba, and D. Williams

***Abstract*—**The goal of admission control is to support the Quality of Service demands of real-time applications via resource reservation in IP networks. In this paper we introduce a novel Dynamic Admission Control (DAC) mechanism for IP networks. The DAC dynamically allocates network resources using the previous network pattern for each path and uses the dynamic admission algorithm to improve bandwidth utilization using bandwidth brokers. We evaluate the performance of the proposed mechanism through trace-driven simulation experiments in view point of blocking probability, throughput and normalized utilization.

***Keywords*—**Bandwidth broker, dynamic admission control (DAC), IP networks, quality of service, real-time flows.

I. Introduction

ADMISSION control in IP networks has become an important area of interest for most network researchers and network managers in the provisioning of end-to-end Quality of Service (QoS) [1], [2]. Admission control is a key component for QoS delivery in IP networks because it determines the extent to which network resources are utilized and whether the contracted QoS requirements are actually delivered.

Although significant progress has been made in providing scalable QoS guarantees to a wide range of real-time applications, little attention has been paid to the dynamic aspect of admission control in IP networks [3]. Current research has ignored an important issue of admission control in IP networks, particularly when traffic is traversing multiple IP network domains [4]. Real-time applications need more stringent QoS requirements in terms of bandwidth, latency and loss. QoS does not create bandwidth, but manages it more effectively to meet the wide application requirements [5].

To satisfy these requirements the IETF proposed Differentiated Service (DiffServ) architecture in the end of 1990. DiffServ provides a simple and coarse method of classifying services of various applications. There are currently three Per Hop Behaviors (PHB) defined that effectively represent three service classes such as Expedited Forwarding (EF), Assured Forwarding (AF) and Best Effort (BE). DiffServ simplifies the function of core nodes and distributes the complex control function to edge nodes to provide scalability [6]. The bandwidth broker is introduced for the control and management of end-to-end QoS to reduce the complexity of QoS control plane [7].

The static admission control methods such as PBAC and MBAC operate well in the management of simple IP networks. However, the static admission control methods cannot cope with dynamic changes of input traffic and network topology since it reflects the IP network status. In this respect, the bandwidth broker model provides the dynamic provisioning. The dynamic admission control allocates the network resources to the requested user flow according to the network situation. However, this mechanism also introduces its own scalability issue to handle large volumes of flows as the network system scales [8][9].

In this paper, we propose the Dynamic Admission Control (DAC) mechanism. The DAC allocates the network resource using the previous traffic pattern to each path. Edge nodes do not need communication with the bandwidth broker if the allocated bandwidth for the path is sufficient to satisfy the flow's request. In the proposed DAC mechanism, "dynamic" means that the bandwidth broker allocates the resource to the path dynamically and the amount of bandwidth allocated to the path is not fixed but is variable upon the traffic flow's QoS requirements.

The remainder of the paper is organized as follows. In section 2, we first summarize the previous work of the admission control. In section 3 and 4, we describe detail procedure of the DAC mechanism and present the simulation results and the performance evaluation analysis. We conclude the paper in Section 5.

II. Dynamic Admission Control Mechanism

The proposed Dynamic Admission Control (DAC) mechanism in IP networks is divided into two phases: *Path-level admission control* and *Link-level admission control.* The bandwidth broker manages only the path-level bandwidth resources within the domain and path-level admission control is performed at the ingress edge node. For the link-level admission control, MBAC is performed for management of the link-level bandwidth along the path. For the DAC mechanism, a two-phase measurement-based admission control method is used [10].

The bandwidth broker calculates the path bandwidth within the domain and sends it to each ingress edge node. The ingress edge node then performs the path-level admission control using the path bandwidth. If the path bandwidth is fully occupied, the ingress edge node performs the movable boundary bandwidth management and extends the path bandwidth of the Expedited Forwarding (EF) or Assured Forwarding (AF) service. It also performs the link-level admission control using MBAC. On the other hand, if congestion occurs in the core node along the path, a congestion notification is sent to each ingress edge node through the bandwidth broker [11]. The ingress edge node then performs only link-level admission control using DAC if there is enough path bandwidth.

The proposed DAC framework simplifies the bandwidth calculation in the bandwidth broker by performing only path-level resource management. It also has an advantage in that the ingress edge node can perform the admission control for

K. Elleithy et al. (eds.), Advances in Computer, Information, and Systems Sciences, and Engineering, 253–257.

itself within the path-level bandwidth regardless of any communication with the bandwidth broker. Furthermore, when the path-level bandwidth allocated initially is fully occupied, the proposed mechanism performs dynamic bandwidth management with dynamic admission control to extend the allocated bandwidth at the edge node for the link-level admission control. Another advantage of the proposed framework is that the bandwidth broker need not maintain any link-level bandwidth information for the link level admission control.

MBAC has been studied by several researchers; some noteworthy contributions are those by Gibbens et al. [13] and Jamin et al. [14]. MBAC alleviates the burden on the users to provide accurate traffic models by shifting the task of traffic specification from the application to the network. The end result is that admission decisions are based on traffic measurements instead of an explicit traffic specification, (i.e. instead of assuming a statistical or worst case model for the traffic).

Grossglauser et al. [15] designed an MBAC scheme in which the basic model used was to consider a buffer-less single link with capacity C, and that traffic flows arrive over time requesting service. Once admitted, the bandwidth requirement of a traffic flow fluctuates over time while in the system. The assumptions made in this design are that the traffic flow holding time in the system is exponentially distributed with a certain mean, and that departures of the traffic flows are independent of the bandwidth processes. The MBAC scheme makes decisions of whether to accept or reject a new traffic flow requesting service based on observation of the past traffic flow. Resource overload occurs when the instantaneous aggregate bandwidth demand St exceeds the link capacity, C; and the QoS was measured by the steady-state overflow probability p_f. The goal of this admission control scheme is to meet a desired end-to-end QoS objective while maintaining a high average utilization of the link.

Many different researchers working on this subject area have proposed and worked on a variety of different Admission Control algorithms. Each of these algorithms has two processes: a *measurement* process that produces an estimate of network load; and a *decision algorithm* that uses this load estimate to make admission control decisions. Every MBAC has two components, namely: a *measurement procedure for estimating current network load*, and *an algorithm that uses the load estimate to make admission decisions*.

III. Dynamic Admission Control Algorithm

In this section, we describe the basic operation model and the algorithm of the proposed dynamic admission control mechanism. Dynamic admission control allocates the network resources to the requested user flow properly according to the network situation. It means that this algorithm has to react fast to the unexpected traffic pattern changes. The efficient network resource management of IP networks depend on how ingress edge node can perform the distinct functions according to service classes instead of simple operation in core node. The traffic violation information of network should be gathered immediately by the bandwidth broker and feedback given to the ingress edge node.

In the proposed DAC mechanism, the bandwidth broker initially allocates provisioned bandwidth to each path. The state of a path maintains the amount of bandwidth that has been allocated to the path. When a flow reservation set-up request along a path arrives, the ingress edge node only needs to check the corresponding path whether the amount of bandwidth allocated to the path is sufficient to satisfy the flow's request. If the answer is positive, the flow request is accepted. Only when the allocated bandwidth of the path is less than the requested bandwidth, the ingress edge node requests the additional bandwidth to the bandwidth broker. The amount of the additional bandwidth is calculated in the ingress edge node based on the previous traffic pattern information.

The bandwidth broker will check each link along the path to see whether there is available bandwidth at all the links except already allocated bandwidth for each path. If available bandwidth is sufficient, the bandwidth broker allocates the additional bandwidth to the path and informs each node along the path to change allocated bandwidth information for that path. Therefore, this mechanism simplifies the bandwidth calculation in the bandwidth broker by limiting the resource management to the only path level. That is, the bandwidth broker can minimize the communication between the bandwidth broker and each node in the domain. Also, it has some advantages that the ingress edge node performs the admission control for itself in path level, the estimation of additional bandwidth, and the dynamic bandwidth allocation.

A more formal and detailed description of the scheme is presented by the pseudo-code in Fig. 1 and Fig. 2.

$p(s,d)$: A path from ingress edge node s to egress edge node d
$BW^{p(s,d)}$: Reserved bandwidth along the path
$UBW^{p(s,d)}$: Used bandwidth along the path
c : Requested bandwidth for a new call
a : Allocated bandwidth from BB
M : Initial bandwidth
M' : Additional required bandwidth to be used in next period
MAX_BW : Maximum requested bandwidth from edge node to BB
MIN_BW : Minimum requested bandwidth from edge node to BB
$Total\_BW$: Link capacity
$\sum UBW_i^{p(s,d)}$: Total used bandwidth along the path
min : The minimum available bandwidth among the links along the path

Fig. 1 Notations

```
BW^p(a,d) = M
if  BW^p(a,d) >= UBW^p(a,d) + c   then
    UBW^p(a,d) += c
    return ACCEPT
else
    recompute M' = BW_i+1 - BW_i
    MAX_BW = M'
    MIN_BW = UBW^p(a,d) + c - BW^p(a,d)
    if BB_Request(MAX_BW, MIN_BW) == OK then
          BW^p(a,d) = a              /* return value from BB */
          UBW^p(a,d) += c
          return ACCEPT
    else
          return REJECT
```

Fig. 2 Dynamic admission control at the ingress edge node

Fig. 2 describes the admission control in the ingress edge node. If the reserved bandwidth is sufficient for new flow set-up request, the ingress edge node accepts the flow request. If not, the ingress edge node computes the additional bandwidth considered to be needed in next period and requests bandwidth allocation to the bandwidth broker. That additional bandwidth, BW_{i+1} denotes the amount of bandwidth that the ingress edge node will request to bandwidth broker in time T_{i+1}, is estimated based the previous traffic pattern such as follows:

$$BW_{i+1} = \frac{UBW_i - UBW_{i-1}}{T_i - T_{i-1}} . \Delta t \qquad (1)$$

BW_i is the allocated bandwidth in time T_i, and UBW_i is the used bandwidth in time T_i.

Also, Δt means that the average value of time interval from T_{i-1} to T_i. Through this Δt, we can reflect the previous traffic patterns to the next bandwidth allocation.

$$\Delta t = \frac{\sum_{k=0}^{i} T_k - T_{k-1}}{i-1} \qquad (2)$$

If the minimum available bandwidth among the links along the path is greater or equal to the maximum requested bandwidth from the edge node to the bandwidth broker, then additional bandwidth is allocated. In the case that the available bandwidth in each link along the path is sufficient to satisfy the requested additional bandwidth from the ingress edge node, the bandwidth broker will allocate *MAX_BW* or *min* to the ingress edge node as well as to each node in domain. Otherwise, the flow set-up request is rejected.

IV. PERFORMANCE EVALUATION

To evaluate the proposed DAC mechanism, we perform extensive simulation experiments to ascertain the efficacy of the proposed solution. To measure the accuracy and effectiveness of the DAC algorithms, we simulate different admission control mechanisms and the obtained results are compared with the DAC mechanism.

A. *Simulation Model and Results*

In this section, we conduct simulation experiments to evaluate the accuracy of the aforementioned admission control algorithms by performing a set of admission control experiments. We used the NS-2 (network simulator-2) simulator and the peer-to-peer simulation model.

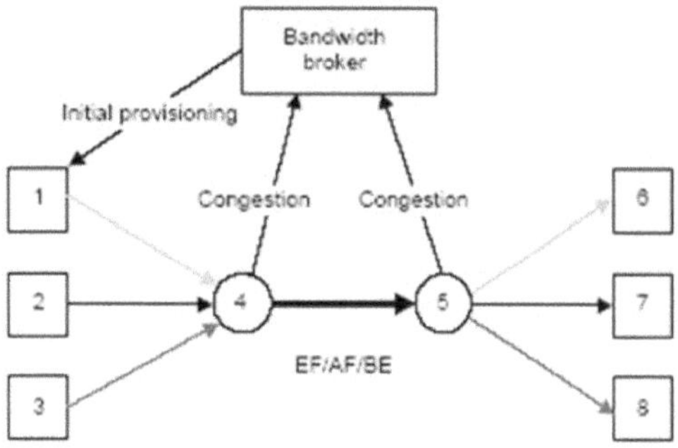

Fig. 3 Simulation model

The workload consists of a set of twenty 30-minute traces of MPEG and JPEG compressed video of Expedited Forwarding (EF) and Assured Forwarding (AF) traffic flows. We used the actual traces of MPEG-compressed video traffic for the experiments.

In the implementation of the proposed dynamic admission control algorithms, we perform a set of trace-driven simulation and admission control experiments for a wide variety of traffic mixes and network capacities. To achieve this, we first implement a number of admission control algorithms from the aforementioned traffic classes and determine their respective admissible regions for various traffic mixes and QoS parameters.

Each path has an initial bandwidth provisioning to EF/AF/BE traffic services. We evaluate the blocking probability of each ingress edge node and packet loss probability of the core node of DAC and compare them with the measurement-based admission control (MBAC) method.

Simulation time is always 1200 seconds of which the first 600 seconds are discarded as "warming period". The particular area of interest is the blocking probability, throughput and packet loss probability of different admission control methods. The node is loaded with exponentially distributed inter-arrival and holding times. The mean holding time is 30 s, the mean inter-arrival time is set to achieve load factor ρ. We define the input traffic load factor (ρ) = mean holding time/call attempt time interval. The target loss probability of the AF service is 10^{-3}, the end-to-end delay is 50 ms.

Our simulation scenario consists of a first-come-first-serve multiplex with buffer size *B*, MPEG and JPEG real-time traffic traces as the workload, and the link capacity of *C* = 45 Mbps, unless otherwise noted. In all experiments, we set the buffer size *B* equal to *C* times the required delay bound so that

delay violation probability is the same as *packet loss probability*. i.e.
Buffer size = Link capacity x delay bound i.e. B = Cd

B. Result Analysis

An admission control test's effectiveness is ultimately determined by its ability to correctly decide whether or not a new traffic flow can be admitted while still satisfying the QoS constraints of all established flows plus the new flows. We calculate 95% confidence intervals for each probability estimated via simulation using the method of batch mean [12].

Fig. 4 shows the blocking probability of the proposed DAC mechanism compared to the static MBAC for EF and AF traffic services.

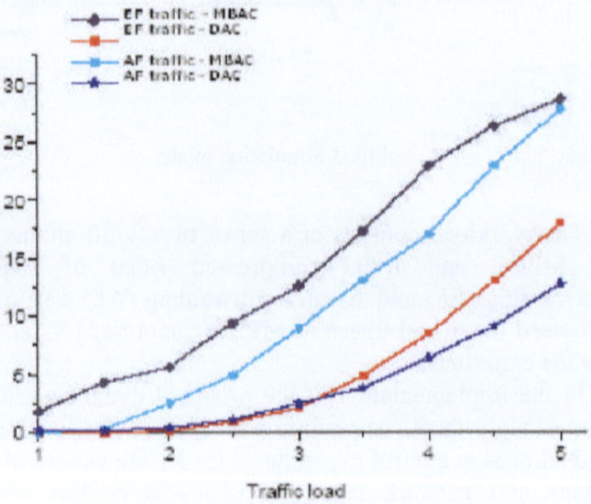

Fig. 4 Blocking probability of proposed DAC

This simulation result shows that the proposed DAC mechanism has less blocking probability than static measurement based admission control (MBAC) mechanisms even with a high traffic load. As the proposed mechanism can use the unused bandwidth of BE service through the dynamic admission control method and can accept more requested flows. On the other hand, the static admission control allocates a fixed bandwidth of the link to each service class and this allocated bandwidth is not changed.

From the simulation experiments, we can see from Fig. 5 that AF3 connections experience higher end-to-end packet delay with MBAC as compared to the proposed DAC.

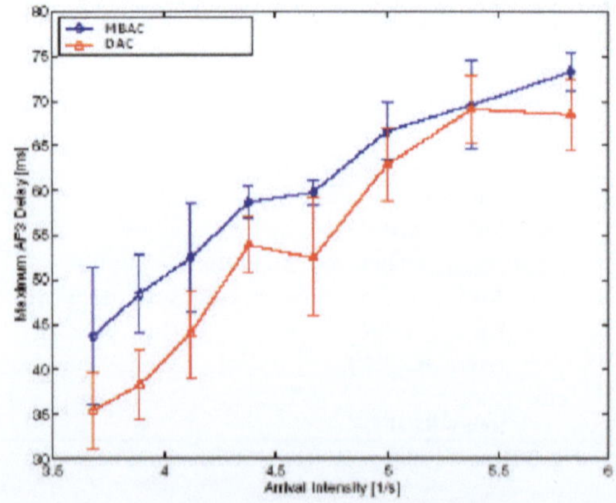

Fig. 5 AF3 traffic bottleneck link delays

DAC mechanism gives better end-to-end throughput performance for real-time applications at different packet arrival intensities. This admission control test is able to capture non-exponential relationship between packet losses and delays.

Fig. 6 shows the packet loss probability of the proposed mechanism for AF service when the traffic load is 5 and the buffer size is 500. We can also see the effect of the admission control scheme of the proposed mechanism. If a mechanism has no proper admission control scheme, the target loss probability cannot be satisfied when traffic load is 5.

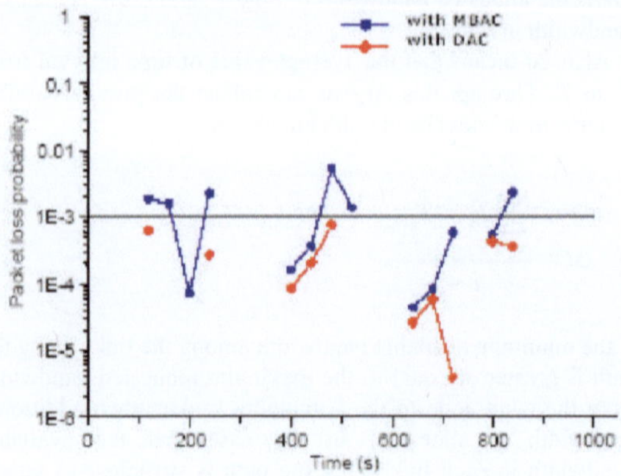

Fig. 6 Effect of dynamic admission control

From Fig. 6, it can also be seen that even though the proposed DAC mechanism accepted more flows than static MBAC, the target loss probability (10^{-3}) of the AF service is guaranteed.

V. Conclusion

In this paper, a novel Dynamic Admission Control mechanism for IP networks has been designed and developed for scalable support of end-to-end QoS guarantees in IP networks. The proposed DAC mechanism simplifies the calculation of the bandwidth data table in the bandwidth broker by performing only path-level resource management.

Both the theoretical and simulated results of the proposed DAC mechanism provide better end-to-end QoS guarantees than the existing admission controls methods. With DAC, the demands of real time traffic do not override all other traffic requirements, but rather traffic link resources are shared in a more dynamic and flexible manner.

For the performance evaluation of the proposed mechanism, we presented simulation results in terms of blocking probability, loss probability, and throughput, comparing the proposed DAC mechanism with static MBAC, comparing our mechanism with the existing dynamic provisioning mechanisms using an ns-2 simulator. Through simulation results, we proved that the proposed DAC mechanism guarantee users QoS requirements and provide bandwidth efficiency.

References

[1] Elek, V., Karlsson, G. and Ronngren, R. (2000). *Admission Control Based on End-to-End Measurements.* In: Proceedings of the INFOCOM 2000, 19th Annual Joint Conference of the IEEE Computer and Communications Societies, Stockholm Sweden, 2:622-630.

[2] Samir, C., and Jongbok, B. (2002). *Quest for the end-to-end network QoS.* Eighth Americas Conference on Information Systems, 1919-1925.

[3] Eleni, M., Charalampos, C., Panos, G., Danny, G. and Griffin, D. (2003). *Admission Control for Providing QoS in DiffServ IP Networks. The TEQUILA Approach.* IEEE Communications Magazine, 38-44.

[4] Mei, Y., Yan, H., Jaime, K., Meejeong, L. Tatsuya, S. and Matsubara, D. (2004). *An End-to-End QoS Framework with On-Demand Bandwidth Reconfiguration.* In: Proceedings of the IEEE INFOCOM, 0-7803-8356-7/04.

[5] L. Mathy, C. Edwards and D. Hutchison, *The Internet: A Global Telecommunications Solution?* IEEE Network Magazine, pp.46-57, July/August 2000.

[6] M. Albrecht, M. Kster, (2003). End-to-end QoS Management for Delay-sensitive Scalable., The Proceedings of the 25 Annual Conf. on Local Computer Networks (LCN'03), Tampa, FL.

[7] A.Terzis, *.The role of BB in Traffic Engineering.* Proc. of ICNP'99, IEEE Computer Society. Sec 3.4.1.

[8] Z.-L.Zhang et al., *On Scalable Design of Bandwidth Brokers*, IEICE Trans. Commun., Vol.E84-B, No.8, pp.2011-2025, August 2001.

[9] Rhee, W.S, Lee, J., Yang, M., Kim, S. (2003). *"Admission control mechanism using measurement based dynamic provisioning in differentiated service network,"* IEEE ICCS2003, Singapore.

[10] Rhee, W., Kim, H., Park, K., Kim, K., and Kim, S. (2001). *Two Phase Edge-to-Edge Distributed Measurement Based Admission Control Mechanism in Large IP Networks.* In: Proceedings of the IEEE GLOBECOM 2001, 2571-2575.

[11] Zhiqun, Z., Xu, S., and Wei, D. (2004). MPLS ATCC. *An Active Traffic and Congestion Control Mechanism in MPLS.* In: Proceedings of the International Conferences on Infotech and Infonet (ICII2004, 5(3):234-241.

[12] P. Bratley, B. Fox, and L. Schrage, *A Guide to Simulation.* New York: Springer-Verlag, second ed., 1987.

[13] Gibbens, R., Kelly, P., and Key, P. (2002): "A decision-theoretic approach to call admission control in IP networks," IEEE Journal on Selected Areas in Communications, 13:1102-1114.

[14] Jamin, S., Danzig, P., Shenker, S., and Zhang, L. (1997): "A measurement based admission control algorithm for integrated services packet networks," IEEE/ACM Trans. Networking, 5(1):56-70.

[15] Grossglauser, M. and Tse, D. (1999): "A Framework for Robust Measurement-Based Admission Control", IEEE/ACM Transactions on Networking, 7(3):.293-309.

Author Bibliography

Justine Kasigwa holds a Masters Degree in Computer Science and Information Technology (2002) and Bachelor of Sciences Degree with a major in Mathematics (1999). She is currently pursuing her PhD. in Computer Science from Makerere University. Her research interests are Quality of Service in Networks and Data Security in IP networks.

Dr Venansius Baryamureeba received his Ph.D. in Computer from the University of Bergen (UiB) in Norway 2000. He was awarded the prize of the best Ph.D. student for the year 2000 in recognition of his overall academic Excellency at UiB. His also the Director of Makerere University Institute of Computer Science and also the Chairman and Managing Director of ICT Consults Ltd., the leading ICT consulting firm in the Uganda, and several other assignments and tasks both in Government and non-government organizations.

Dr. Ddembe Williams received his PhD in 2003 from London South Bank University. He is the director, Centre for Advanced Systems Modelling and Simulation, Faculty of Business, Computing and Information Management, and Senior Lecturer in Decision Sciences at the same University in the UK.

Design of a priority active queue management

Hattab **Guesmi**, Belgacem **Bouallegue**, Ridha **Djemal**, Habib **Youssef** & Rached **Tourki**
Laboratoire d'électronique et de micro-électronique, faculté des sciences de Monastir, Tunisia
e-mail : Hattab.guesmi@fsm.rnu.tn

Abstract

In this paper we present a fast, a reconfigurable and scalable priority active queue managemnt (AQM) to be used in high-performance switches supporting fine-grained quality of service (QoS) guarantees. Priority queues are used to implement highest-priority-first scheduling policies. Our hardware architecture is based on a new data structure called priority circular linked list (P-CLL). This data structure enables the reconfiguration of the AQM. In addition, the architecture have to scale very well to a large number of priority levels and to large queue sizes. In this respect, we will give a detailed description of our proposed new data structure, the associated algorithms and the adequate hardware implementation.

I. Introduction

Quality of service (QoS) guarantees in term of delay, throughput, and loss rate can be provided using a service discipline at switching nodes in packet switching networks. In this paper, we limit our discussion to the class service disciplines specified for output-queued switch such as Class Based Weighted Fair Queueing, Weighted Fair Priority Queueing, and First Comet First Served Service. In these service disciplines, packets are assigned transmission deadlines (or priority indices), and the packet with the earliest deadline (or highest priority) is served at first. This approche requires a priority queue manager for sorting priority levels and managing queues. By using service disciplines and priority queues, packets from various connections are interleavingly served so that each connection's QoS requirements can be guaranteed. However, in order to achieve QoS guarantees in practice, the priority queue manager must be able to support a large buffer size in high-speed networks. Therefore, it is crucial that the design of the AQM is reconfigurable according to the network domain to increase the performance: switches must be scalable.

The design of an AQM that implements the deadline-ordered service disciplines with a large number of priority levels, a large buffer size, and a large number of input links is not easy to acheive for high-speed packet switching networks. In fact, stored-priority disciplines must always choose the packet with smallest deadline or highest priority among all prioritized packets in the queue to serve first. If the number of packets stored is M, then the complexity is *O*(logM). So the bottleneck comes from the high cost of maintaining a high-speed sorting mechanism associated with implementing these service disciplines. Moreover, to improve the network performance, a large number of input links to an AQM is necessary. However, many proposed priority queue designs have limitations in term of priority levels number, buffer size, and input links number or bandwidth.

Extensive research has been done on the next generation high-speed routers. Nick Mckeown's group at Stanford University did intensive research on high speed switching [1, 2, 3], R. Bhagwan proposed a Design of a high-speed packet switch for fine-grained quality of service guarantees and a fast and scalable priority queue architecture for high-speed networks switches[4]. H. Jonathan et al. have proposed a design for packet-fair queuing schedulers using a RAM-based searching engine [5]. Many other papers and proposals are dealing with some key issues in high-speed routing, such as high-speed routing table lookup, real-time packet scheduling, fine grain QoS control, high-speed switches for the data path and so on [9, 10, 11]

In our approach, we propose a PQM architecture for deadline-ordered service disciplines that can be used in output-buffered switches to provide QoS guarantees in high-speed packet switching networks. In this respect, we have to use a RAM-based search mechanism to find the highest priority cell instead of commonly used shift and insert mechanisms.

II. Data Structure of the priority circular linked list (P-CLL)

1. Description of the active queue manger

Our queue management requires a specific organized data structure related to our scheduling algorithm which represents the subject of this section. So, this data structure depends primarily on the implemented scheduling algorithm, since it presents its data base. It must be structured in a way that all stored packets are visible and accessible by the scheduler so that the decision have to be fast and effective according to the QoS parametres required for each flow. As we have to show

K. Elleithy et al. (eds.), Advances in Computer, Information, and Systems Sciences, and Engineering, 259–265.

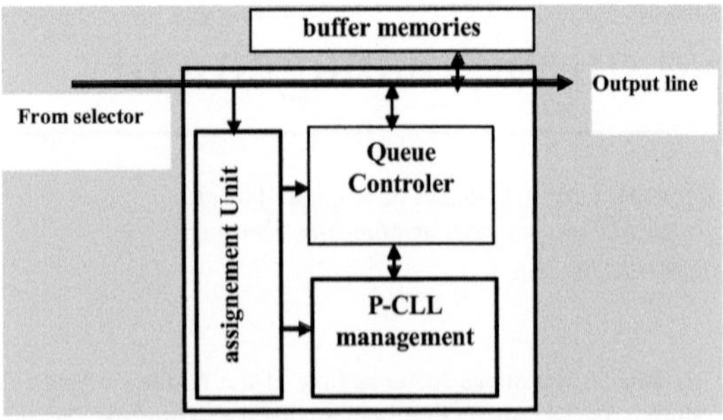

Figure 1: the output queue manager architecture

the QoS parameters are optimized by the scheduling mechanism and the queue management in [19]. Thus, basic operations like creation, access and release of the flow are achieved in a dynamic way according to the number of declared flows and the free memory spaces. Consequently, the memory spaces have to be adjustable according to the dynamic parameters of the loss management algorithm. In the beginning the event of congestion of a service class we have to borrow the memory space if the other classes are not congested, especially if the congested class does not tolerate any more data losses. The number of flows per class is not fixed statically, but can be managed dynamically according to the allocated memory capacity. To meet these needs, we have used the circular linked list which ensures the dynamic queues management. The active queue manager consists primarily on a priority assignement unit, a P-CLL manager and a queue controller as depicted in the figure 1. In this respect, the priority asignement unit stamps the incoming packets with a certain priority, which is decided depending upon the scheduling algorithm in use where the P-CLL manager contains the P-CLL priority queue. Eah of the overall elements in the P-CLL holds a priority value, responsible for the queue managmenet. The queue controller maintains a lookup table with entries corresponding to each priority value. Each entry consists on a pointer to a list of packets of the same priority. We refer to this list as a priority list. Thus, our proposed structure is one of per-priority queueing rather than per-flow queueing, and is more general in the sense that it can handle many priorities within the same flow. It should be noted that at any point of time, the P-CLL contains only the priority values for witch the priority list is non-empty and which contains the active priority values.

2. Data Structure of the active queue management

This implementation is one among the implementations per-priority queuing, rather than an implementation of priority per class of flow and it is more general to handle various priorities for same flow. At every moment of time the P-CLL contains only the values of active priorities (filled out lists of priority).

In order to increase the design performance of the per-priority queueing, we have retained the circular linked list combined with a dedicated data strtucture and which implements then priority assignment. In this respect, data related to each class of service is stored in a per-connection data queues, so that each data type can be served seperately.

When a new packet needs to be inserted into the queue, the priority assignment unit stamps the packet with a suitable priority value. The queue controller determines whether a priority list already exists for the stamped priority value. If it does, it simply adds the new packets to the corresponding priority list. However, if the list does not exist, the queue controller creates a new priority list. It also signals the P-CLL manager to perform an enqueue operation, which inserts the new priority value into the P-CLL in a sorted manner. This operation is done to make sure that the highest priority stays at the top of the P-CLL so that when a dequeueing of a packet is required, the priority list with the highest priority can be rightly accessed. In the other hand, when a packet needs to be removed from the queue, the P-CLL manager determines the non-empty priority list with the highest priority by looking at the topmost element of the P-CLL and sends this priority value to the queue controller. The queue controller accesses the corresponding priority list and removes a single packet from it. When the priority list become empty, the P-CLL manager initiates a dequeue operation which removes the topmost element from the P-CLL while making sure that the P-CLL remains sorted.

The P-CLL data structure maintains various priority lists sorted in the circular chained list. Each enqueue operation requires N times to be serviced, where N is the size of the P-CLL. However, the emulation of the circular linked list which implements the priority queueing mechanism is easy to make and it provides a big save in term of time constraints. Furthermore, this structure allows a pipelined implementation compared with the conventionnal circular linked list and provides a constant time operations. In order to improuve the performance of our P-CLL, we implement the method of research based on a content addressable memory where each priority queueing operation is performed in only one time.

P-CLL based Priopritу queueing operation

The motivation behind defining a new data structure is to be able to pipeline the enqueue and the dequeue operations on it. The conventional circular linked list operations, though very simple to implement, execute in O(n) steps, where n is the number of elements in the circular linked list, and they cannot be easily pipelined. On the other hand, architectures like the systolic array and binary heap can be pipelined, but have extremely high hardware requirements. Our purpose is to design a modified circular linked list data structure and associates algorithms with it, which while being simple and easy to implement, can be easily pipelined to provide constant time priority queueing operation at a low hardware costs [7, 8].

1. The enqueueing operation

To enqueue a new value of priority in the circular linked list of the P-CLL, we need to find a free cell. This operation is achieved by exploring the valid data path from the table T values which allows us to determine the inserted cell address or the address pointer. If the pointer of the cell was determined, we modify the flag in table T with one (priority list non-empty), which indicates the state of the priority list. Then, we can accede to the free cells list to write the priority value in this free cell and we carry out the updated list. Finally, we add this cell to the circular linked list P-CLL in a sorted manner to determine the position of insertion of the cell and then to update the list. The insertion position is identified by determining the first least weak priority level of non-empty priority list and by reading the table T from the inserted value address. If we determine the pointer of the cell which at the least weak priority we will determine the insertion position of the priority level. Thus, the operation of addition consists on reading and writing the bond pointers of the two cells and updating the class descriptor service list. The operation of research requires N time where N is the priority levels number. However, if we use a content addressable memory, the operation of research needs only the conversion time of the priority level into address. We can summarize the required setps for inserting a new priority list as the following:

Step1:
Request an available free cell
Read the pointer at the table address
Step2:
Write the value in the free cell
Update the flag in the table
Update the free cell list
Step3:
Determine the insertion position
Insert the priority list

2. The dequeue operation

The dequeueing operation consists on extracting the cell pointed by the header pointer from the P-CLL since it has the highest priority value making the cell inactive and hence decreasing the file capacity by one. The dequeue operation are summarized below:

Step1:
Read the head pointer of the circular linked list
Add the free cell to the free cell list
Step2:
Update the circular linked list
Update the status flag
Decrease the file capacity

IV. Implementation of the active queue management

The active priority queue management is one of the main components in our IP switch. Our implementation of the proposed active priority queue management includes three modules: the priority assignment Unit, the P-CLL manager Unit and the queue controller Unit [12, 14].

1. Priority assignment Module (PA)

This module stamps the packets by a label calculated according to the implemented scheduling algorithm [15, 16]. This label determines the transmission order of the data packets to the output line and needs to know the three following factors:

- The weight allotted to the concerned class of service: the concept of weight determines the band-width percentage that the class is seen allotting.
- The length of the packet of this service class;
- The interaction with the other concurrent active classes.

In order to stamps packets by labels, this unit calculates the virtual finish time and the virtual start time according to the scheduling algorithm (WFQ). The calculated values related to each assigned paquet level represent the transmission order (packets priority). According to these values, packets are stored in suitable queue or class of service. Since our architecture implements N_{cs} classes of services, the calculation of priority values needs a fluid system which is implemented in a data base in order to describe the state of each service class and its interaction with the others. This data base is constructed in the P-CLL manager module to be consulted and updated by this unit after any emission/reception operation in order to provide the correct priority values.

2. P-CLL manager module (PM):

This module implements the selected data structure which is based on the priority circular linked list P-CLL. This list consists on a circular linked list based on the service classes' descriptors where each one of these descriptors inter-connects a priority circular linked list of priority levels (class list) referred to the same class of service. Each cell of these descriptors of the priority levels has an active priority level which gathers all packets having the same priority level in a circular linked list which is called priority list (figure 2). Every cell related to the priority list contains two fields: a pointer pointing to its priority list and an address pointer pointing the memory space which stores the packet in the buffers. The cell of the class list contains several fields which are: a head and a tail pointers associated to the priority lists (packets of data having the same value of priority), a pointer pointing to its class list and an additionnal field which describes the priority level value. In the other hand, the service class descriptor cell contains the associated control parameters related to each class such as the queue average size, the probability of reject.etc...

In this respect, our design of the P-CLL manager unit uses two separated memories: the first one contains the class of service descriptors where the second it contains the free priority values list. Thus, the first memory implements the service class linked list which is organized as Ncs blocs (Ncs: number of service classes), where each bloc has Npc fields (Npc: number of control parameters of each service class). The second memory which implements the free priority values linked list is organized as Npv * 4 fields (Npv: number of priority values; one field for the priority value, one for the pointer and two fields for the head and tail pointer). The variation of the memory size depends only on the service number class for the first memory and the priority values numer belonging to the second memory. The size of these memories varies in a linear way according to the increase of the number of the priority level managed by this architecture and the service class number.

We are used two separated memories in our design for three reasons: First, our architecture will be scalable by increasing the memory size with a simple adding of memories in ordrer to increase simple adding space memory to increase the number of priority values related the number of service class. Second, the number of priority values belonging to each service class is reserved in a dynamic way according to the need for each class of service (number of flows treated in each service class): allowance on request or dynamic adjustment of the band-width between the N_{Cs} classes of service. And finally, we wish to facilitate the architecture reconfiguration according to the domain in which the switch will be installed (generic architecture) what enables us to reconfigure our architecture according to three operating modes: DiffServ mode, IntServ mode and high-speed networks [17, 18].

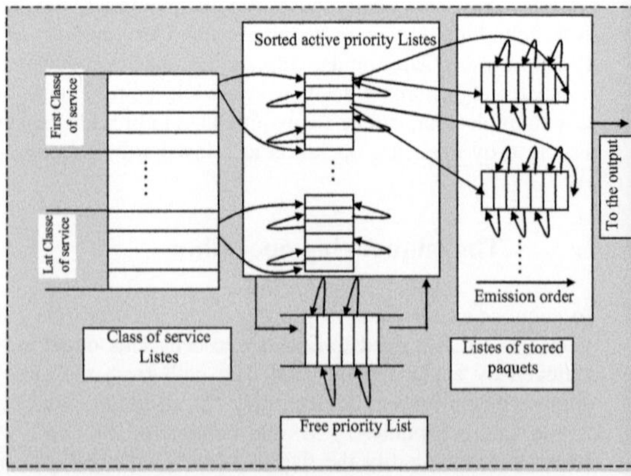

Figure 2: structure of the priority circular linked list

3. Queue controller module (QC):

This module is used to control all priority levels (queues, or priority lists) related to the active queue management: which can be switched between the active and the inactive state. It is represented by a table which stores all the priority values with their associated parameters. When an assigned packet is received, this module seeks in the table if the priority value is active. If the priority is active so this module launches an addition operation requested by this packet with its corresponding priority list, like the determination of the available address of the priority list by reading the field address pointer in the table. In the other case, this module inserts this priority value in its suitable site in the sorted circular linked list. During the insertion operation, QC issues an available address to the priority value. This available address is appended to a linked list for the class list. During the emission, this module outputs pointers of the packet to be transmitted. If this package is the last in the list, therefore it launches an extraction operation to make the level of priority inactive in the table. The extraction operation is achieved by the QC which outputs the pointer to the cell buffer at the head of the queue which has the highest priority level found (figure 3).

The design of the queue controller module uses a common memory organized as Ncs blocs (Ncs: number of service classes) where each bloc contains Npv * 3 fields (Npv: number of priority values of each class). In this respect, each field represents the three following parameters related to the priority value:

- Address pointer to the priority list,
- The priority value,
- The indicator of the occupancy of the queue.

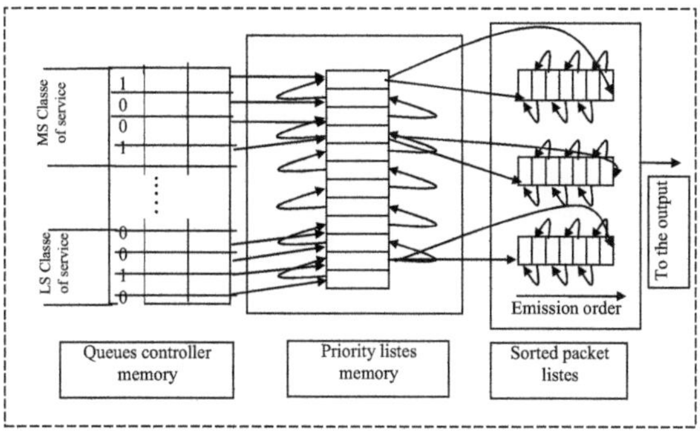

Figure 3: structure of the queue controller module

The memory size of this component is equal to Ncs * (3 * Npv) depending only on the service classes number and the priority values number; this capacity increases if the number of priority values increase because they represent the number of flows treated by this architecture [6, 13].

Timing Issues

The AQM performance is acheived by three computation threads: (1) the first thread for records priority levels of incoming packets and finds the highest priority and performs the P-CLL Manager (PM) updating; (2) the second thread is reserved for managing the priority queue; (3) the last thread is devoted for writing and reading packets to and from tampon memories. The on-chip clock runs at a clock period of Toc s, which is equal to the cycle time of all on-chip SRAM's where the cell module (AQM) runs at a clock period of Tc s, which is the cycle time of standard off the shelf SRAM's plus a 2ns margin. We remarks that the thread (1) includes times for recording the priority levels of all incoming packets for searching the highest priority, and for encoding the least significant half of the highest priority found. If the number of incoming packets is N, then Tp = (N + m + 2) * Toc, because the search and encoding need (m + 2) cycles respectively. Unlike thread (1), QC needs 3 cycles for a write and 4 cycles for a read operation. Thus, thread (2) takes: Tq = (N * 3 + 4) * Toc. Where the cell module requires N write cycles for input packets, 1 read cycle for the highest priority packet, and 1 dead cycle between read and write operation (assuming late-write SRAM). Thus Tm = (N + 1 + 1) * Tc. In addition, an update operation (U) is needed when the queue becomes empty as a result of an extract operation. This operation can starts after Tq and only takes one cycle (Toc).

VI. Description of the implementation of the IP switch architecture

For the implementation of our IP switch architecture, a description in VHDL is carried out at the RTL level and the simulated using ModelSim simulator. This approach has permitted to evaluate the behavior of each component alone as well as the intercation of the overall architectuecture with all interconnected components. The implemented components can integrate basic standard elements with a known behavior such as adders, comparators, MUX, etc, or other elements where their behaviors have to be defined like sequencer, specific interface and telecommunication operators. Our proposed solutions have to give answers for all questions related to the design specification and the time constraints required by the application context. In order to emulate the environement of our architecture, we have used some file of test containing the entries stimuli with the overal VHDL description to constitute the validation vitual prototyping IP switching system [19, 20].

We remarks that our architected uses an hybrid model of control which combines the control dominated and the memory dominated model. This target architecture has to respond to both the time constraints and the memory saving for the averall data computing operations. Furthermore, in order to make better the organization of our architecture in order to be easily extensible, we have control part, the data path part and the interface part. This separation is also justified to automate the process of design thereafter. Consequently, our proposed architecture of the integrated protocols related to IP switch consists on a set of interface and control components. This architecture is capable to transmit data and control information in two directions between distant communication entities. This architecture includes three different parts as depicted in Figure 4:

- The network interface unit,
- The memory part unit,
- The main part of the integrated protocol architecture.

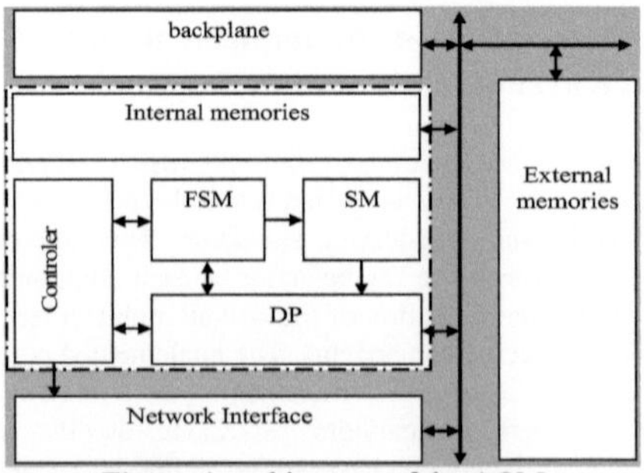

Figure 4: architecture of the AQM

The memory part

As mentionned in the above section, our architecture is memory dominated which means that we have to take into consideration the critical aspect of the melmory management. In this respect, we have proposed a dynamic QoS management based architecture which allows us to define the suitable data structure and to choose the appropriate technique. This dynamic bandwidth allocation on demand is based on the circular linked list technique. This technique presents many advantages. Besides the optimization of resources in memory, it reduces the time to access to data. Furthermore, the use of the linked list technique avoids the multiple copies of packets and consequently optimizes the transfer delays as well as the space used for buffering. Also this technique makes the architecture reconfigurable to optimize the QoS differentiation. Memories are used to implement selected techniques implemented in our architecture.

FSM The finite state machine: it receives the primitive of the system (type of data, procedure of emission of a cell, procedure of reception of a cell, etc.) and it defines the primitive execution order to be applied to the controller.

The sequencer (SM): it allows the generation of all necessary commands and control signals for the data path operation.

The controller: To reconfigure the architecture according to the need for the QoS management, and to administer these memories a general controller is designed to assure the dynamic QoS management. The controller performs the following operations:

At the reception side, the controller determines the nature of the packet and activates the adequate procedure such as:

- Linked list initialization;
- Data cell buffering according to the conformities with the negotiated parameters,
- Connection parameters update to which belongs this packets.

At the emission side, the controller activates the procedure of emission of the current packet

In addition the controller permits to generate the appropriate signals for the data path components.

The data path (DP): This module contains essentially a register interface and dedicated and many standard operators like counters, adders, CRC operators and so on, to perform all operation required by the entire architecture. It's also responsible for the generation of the signals of readings and of writings for memories.

The network interface: This unit is responsible for the reception of data from the network interface to be stored into receive buffer. It's also responsible for data transmission from the interface to the network

V. Conclusion

QoS guarantees in high-speed packet switched networks still remain a huge challenge for network designers. Since most of service disciplines for providing QoS are deadline-ordered and specified for output queues, AQM becomes an essential component for high performance packet networks. The AQM must be designed not only to support many priority levels, large buffer size, many input links, and high bandwidth but also to support scalability and reconfigurability.

In this paper, we have proposed a scalable and a reconfigurable AQM architecture for IP switches that can handle a large number of priorities in different network domains. In addition, we have proposed three architectures in which scalable AQM's can be reconfigured to support the QoS guarantees. The IP switch can therefore be used efficiently in a high-speed packet switch providing fine-grained quality of service guarantees. In addition, to be very fast and reconfigurable, the architecture also scales very well to a large number of priority levels and to large queue sizes. Our current implementation is described in VHDL language at RTL levels to be synthesized in a low level.

References

[1] N. Meckeown "a fast switched backplane for gigabit switched router" *journal*, Business communication review, December 1997.

[2] S-T. Chang, A. Goel, N. Mckeown, and B. Prabhakar "matching output queueing with a combined input/output switch", *journal*, IEEE journal on selected areas in communications, ISACEM, ISSN 0733-8716, volume 17, number 6, p-p 1030-1039, juin 1999.

[3] P. Gupta and N. Mckeown "algorithms for packet classification", *journal*, IEEE Network, mars 2001.

[4] R. Bhagwan and B. Lin "Design of a high-speed packet switch for fine-grained quality of service guarantees", *journal*, IEEE International conference on communication (ICC'00) 2000, New Orleans, vol3, pages 1430-1434

[5] H. Jonanthan Chao, Yau-Ren Jenq, Xiaolei Gue and Cheuk H. Lam "design of packet-faie queuing schedulers using a RAM-based searching engine" IEEE *journal* on selected areas in communications. Vol. 17, No. 6, June 1999, pp. 1105-1125.

[6] P. Giaccone "Queueing and scheduling algorithms for performance routers" *thesis* de l'université de polytechnique de Torino, Itali, février 2002.

[7] S. Floyd, R. Gummadi, and S. Shenker "adaptive RED: an algorithm for increasing the robustness of RED's active queue management" *rapport*, longer technical report, 1 août 2001.

[8] S. FLOYD, V. Jacobson, "random early detection gateways for congestion avoidance", *journal*, IEEE/ACM transaction on networking, vol1, No.4, pp397-413, aout 1993.

[9] N. Ni, L-N Bhuyan "fair scheduling and buffer management in internet router" article, conference on computer computing, INFOCOM2002, vol3, New York, juin 2002.

[10] A. C-K. Kam "efficient scheduling algorithms for quality of services guarantees in the internet" these de l'institut de technologie de Massachust, avril 2000

[11] Michael V. Lau, S. Shieh, P-F. Wang, B. Smith, D. Lee, J. Chao, B. Shung and C-C Shih, "Gigabit Ethernet switches using a shared buffer architecture", article, IEEE Communication magazine, p-p 76-84, December 2003.

[12] Neal Seitz, NTIA/ITS, "ITU-T QoS standards for IP-based networks" article, IEEE communication magazine, pp 82-89, june 2003.

[13] James Aweya, "On the design of IP routers part 1: router architecture", article, journal of systems architecture, 46 (2000) 483-511.

[14] Seshasayi Pillalamarri, S. Ghosh, "high-speed networks: definition and fundamental attributes", article, computer communications 28 (2005) 956-966.

[15] Zoubir Mammeri, "framework for parameter mapping to provide end-to-end QoS guarantees in IntServ/DiffServ architectures", article, computer communications 28 (2005) 1074-1092

[16] Fengyuan Ren, C. Lin, B. Wei, "a robust active queue management algorithm in large delay networks", article, computer communications 28 (2005) 485-493.

[17] Sebastian Wallner, "A configurable system-on-chip architecture for embedded and real-time applications: concepts, design and realization" article, Journal of Systems Architecture, Volume 51, Issues 6-7 , June-July 2005, Pages 350-367.

[18] Hattab. Guesmi, R. Djemal, B. Bouallegue, J-P. Diguet, R. Tourki, "high performance architecture of integrated protocols for encoded video application", article, computer standards and interface 26 (2004) 301-315.

[19] H. Guesmi, R. Djemal, B. Bouallegue, H. Youssef et R. Tourki, "Architecture d'un routeur IP de haute performance optimisée pour la différentiation de qualité de service", Premier Congres International IEEE de Signaux, Circuits Et Systèmes (SCS'04), Du 18 - 21 Mars 2004, Monastir - Tunisia.

[20] A. Baghdadi, « Exploration et conception systématique d'architectures multiprocesseurs monopuces dédiées à des applications spécifiques », thèse de l'institut national polytechnique de Grenoble, 14 mai 2002.

Enhancing QoS Support in Mobile Ad Hoc Networks

Nityananda Sarma, Sukumar Nandi
Indian Institute of Technology, Guwahati
Guwahati – 781039
{nitya,sukumar}@iitg.ernet.in

***Abstract* • Due to resource constraints and dynamic topology of Mobile Ad Hoc Networks (MANETs), supporting Quality of Service (QoS) in MANETs is a challenging task. Although lots of research have been done on supporting QoS in the Internet and other networks, they are not suitable for the MANET environment. This paper provides a brief review of the current researches on QoS support in MANETs and proposes a new model for enhancing QoS support in MANETs. Through extensive simulation, we observe and analyze the performance of the network in terms of throughput and average delay. A thorough investigation of the performance of higher and lower priority traffic is made under different traffic loads and mobility scenarios. The simulation results show an improvement of the performance of our QoS model as compared to SWAN[7].**

I. INTRODUCTION

Mobile Ad Hoc Networks consist of wireless mobile hosts that communicate with each other without the presence of a fixed infrastructure, forming dynamic autonomous networks. Due to its non-dependency on a centralized infrastructure, it can be rapidly deployed and can dynamically be reconfigured over time [14].

With the increase in Quality of Service (QoS) needs in evolving applications and the widespread use of wireless and mobile devices, it is also desirable to support QoS in MANETs. The resource limitations and variability further add to the need for QoS support in such networks. However, the characteristics of these networks make the QoS support a very complex process [1, 2, 3].

QoS is usually defined as the set of service requirements that need to be met by the network while transporting a packet stream from a source to a destination [15]. The network is expected to guarantee a set of measurable pre-defined service attributes to the users in terms of end-to-end performance, such as delay, bandwidth, probability of packet loss, delay variance (jitter), etc. Power consumption and service coverage area are two other QoS attributes that are more specific to MANETs. Furthermore, the QoS metrics could be defined in terms of one of the parameters or a set of parameters in varied proportions. The multi-constraints QoS aim at optimizing multiple QoS metrics while provisioning network resources, and is an admittedly complex problem.

A lot of work has been done in supporting QoS in the Internet, but unfortunately none of them can directly be used in MANETs [2]. To support QoS, the link state information such as delay, bandwidth, cost, loss rate, and error rate in the network should be available and measurable. However, getting and managing the link state information in MANETs are very difficult because the quality of a wireless link is apt to change with the surrounding circumstances. Furthermore, the resource limitations and the mobility of hosts make things more complicated. Some of the issues and difficulties for supporting QoS in MANETs [1] are: *Unpredictable Link Properties*, *Node Mobility*, *Limited Battery Life*, *Hidden and Exposed Terminal Problem*, *Route Maintenance* and *Security*. These entire issues make provisioning QoS functionality in MANETs a challenging task.

In this paper, we briefly review the current approaches for QoS and their effectiveness to deploy in MANETs. We present a simple QoS model for MANETs as an extension over SWAN [7,18]. Our model considers priority within real-time and best-effort packets. We have analyzed the performance of the new model in terms of throughput and average delay through simulation and it is found to be efficient in comparison with SWAN.

The paper is organized as follows: Section II describes the related works. Existing QoS model SWAN as a basic model is discussed in section III. Section IV discusses our proposed model. Simulation setup and simulation results are discussed in section V. Finally, section VI gives conclusions.

II. RELATED WORKS

Many research works are going on for QoS support in MANETs[10]. The major approaches are QoS Models [2], QoS Resource Reservation Signaling [1], QoS Routings [3] and QoS Medium Access Control (MAC) [2]. These approaches are closely related. A QoS model specifies the architecture in which some kinds of services can be provided in MANETs. All other QoS components (such as QoS signaling, QoS routing and QoS MAC) must cooperate together to achieve this goal. Next we discuss the above-mentioned QoS approaches in brief.

A. *QoS Models*

The QoS model [2] specifies the architecture for providing some kind of services in the network. A QoS model for MANETs should consider the challenges of MANETs, e.g. dynamic topology and time-varying link capacity etc. In addition, the potential commercial applications of MANETs require connection to the Internet. Thus QoS model for MANETs should also consider the existing QoS characteristics in the Internet.

K. Elleithy et al. (eds.), Advances in Computer, Information, and Systems Sciences, and Engineering, 267–273.

A.1 IntServ/RSVP

IntServ [4] architecture allows sources to communicate their QoS requirements to routers and destinations on the data path by means of a signaling protocol such as RSVP [13]. In this model the flow-specific states are kept in every IntServ-enabled router. IntServ proposes two service classes in addition to Best-Effort Service. One is Guaranteed Service; the other is Controlled Load Service. The Guaranteed Service is provided for applications requiring strict delay bound. The Controlled Load Service is for applications requiring reliable and enhanced best effort service. IntServ/RSVP is not suitable for MANETs due to the following factors:

- Huge amount of storage and processing overhead, so the problem of scalability.
- Signaling packets will consume a substantial amount of bandwidth in MANETs.
- Performing of admission control, classification and scheduling by every node will put a heavy burden for the resource-limited mobile hosts.

A.2 DiffServ

Differential Service (DiffServ) [5, 25] architecture avoids the problem of scalability by defining a small number of per-hop behaviors (PHBs) at the boundary routers and associating a different DiffServ Code Point (DSCP) in the IP header of packets belonging to each class of PHBs. The interior routers along the forwarding path use DSCP to differentiate between different QoS classes on per-hop basis. Thus, DiffServ is scalable but it does not guarantee services on end-to-end basis. This hinders DiffServ deployment in the Internet as well as in MANETs. In DiffServ, we can identify three different service classes: *expedited forwarding*, *assured forwarding*, and *best effort*. DiffServ may be a possible solution to the MANET QoS model, because it is light-weighted in interior routers. But it also has some weaknesses as follows:

- It is difficult to predefine boundary and interior routers.
- The concept of *Service Level Agreement* (SLA) in the Internet does not exist in MANETs which is indispensable for the functioning of DiffServ.

A.3 FQMM

Flexible QoS Model for MANETs (FQMM) [6] tries to take the advantage of both the per-flow service granularity in IntServ and the service differentiation in DiffServ. This model defines three types of nodes: an ingress node which sends data, an interior node which forwards data to other nodes, and an egress node (destination). Obviously, each node may have multiple roles. FQMM selectively uses the per-flow QoS guarantees of IntServ for applications with high priority, and the service differentiation of DiffServ for applications with lower priorities. FQMM is the first attempt at proposing a QoS model for MANETs with the following problems:

- Without an explicit control on the number of services with per-flow granularity, the scalability problem still exists.
- To make a dynamically negotiated traffic profile is a very difficult problem.
- It is difficult to code the PHB in the DS field of IP, if the PHB includes per-flow granularity considering the DS field is at most 8 bits without extension [2].

A.4 SWAN

Service Differentiation in Stateless Wireless Ad hoc Networks (SWAN) [7] is a stateless network model which uses distributed control algorithms to deliver service differentiation in mobile wireless ad hoc networks in a simple, scalable and robust manner We will discuss the details of SWAN in section III as our work is related to this model.

B. QoS Signaling

QoS signaling [2] is used to reserve and release resources (e.g., buffer space, bandwidth etc.), set up, tear down, and negotiate flows in the networks. Next, we briefly discuss two QoS signaling protocols.

B.1 RSVP and dRSVP

Resource reSerVation Protocol (RSVP) [4,13] is an out-of-band signaling system and is adopted as the signaling system in the Internet. Here, with the exchange of PATH and RESV messages resources can be reserved (if possible) from receiver to sender. However, it is not suitable for MANETs because of its heavy signaling overhead. Also, it is not adaptive to time-varying topology, because it has no mechanism to rapidly respond to the topology change in MANETs. To solve these problems, a Dynamic QoS concept is proposed in [16] and a dynamic ReSource reSerVation Protocol (dRSVP) is defined to provide the flexibility needed in dynamic environments such as MANETs. The dRSVP extends the basic RSVP to support dynamic QoS in MANETs. The reservation requests are specified as a range of values (with minimum and maximum levels). It uses class-based queuing and the controlled-load service. The use of out-of-band signaling is a problem in bandwidth limited MANETs.

B.2 INSIGNIA

INSIGNIA [9] is a simple in-band signaling mechanism which can be combined with a variety of routing protocols such as DSR, AODV, OLSR etc. The primary goal of INSIGNIA QoS framework is to support adaptive services, depending on the availability of the resources. It is designed to adapt user sessions to the available level of service without explicit signaling between source-destination pairs. The INSIGNIA protocol provides per-flow QoS by piggybacking soft-state reservations onto data packets in an IP option field. The idea behind this approach is to avoid explicit signaling and hard state reservations in order to deal with the dynamics of mobile ad hoc networks in a flexible way.

C. QoS Routing Protocols

QoS routing protocols try to find a path that has a good chance of meeting the QoS requirements [3,11,12,22]. QoS routing is difficult in MANETs. Firstly, it has too high overhead due to maintenance and updation of link state information. Secondly, maintaining the precise link state information is very difficult. Thirdly, a feasible path may not guarantee a QoS path. Here, we discuss two of the QoS routing protocols.

C.1 CEDAR

CEDAR [11], a Core Extraction Distributed Ad hoc Routing algorithm is proposed for small to medium size ad hoc networks. It dynamically establishes the core of the network, and then incrementally propagates the link states of stable high-bandwidth links to the core nodes. The route computation is on demand basis, and is performed by the core nodes using only local state information. CEDAR has three key components: *Core Extraction - to* select a set of nodes to form the core that maintains the local topology of the nodes in its domain, and also performs route computations. *Link State Propagation-* QoS routing in CEDAR is achieved by propagating the bandwidth availability information of stable links to all core nodes. *Route Computation* - Route computation first establishes a core path from the domain of the source to the domain of the destination. Using the directional information provided by the core path, CEDER iteratively tries to find a partial route from the source to the domain of the furthest possible node in the core path satisfying the requested bandwidth.

C.2 Ticket-based Probing

A distributed multipath QoS routing scheme for MANETs, called ticket-based probing is proposed in [12] to tolerate imprecise state information. This method uses tickets to limit the number of candidate paths searched during route discovery. When a source wishes to find QoS paths to a destination, it issues probe message with some tickets. One ticket corresponds to one path searching; and one probe message should carry at least one ticket. At an intermediate node, a probe with more than one ticket is allowed to split into multiple ones, each searching a different downstream sub-path. Hence, when an intermediate node receives a probe, it decides, on the basis of its available state information, whether the received probe should be split and to which neighbors the probe(s) should be forwarded. Some questions must be answered in the route search approach: *First*, how many tickets should be issued by the source? *Second*, when an intermediate host receives a probe message of *n* tickets, it will have to make the decisions regarding - ticket splitting and probe forwarding rules. *Third*, how are multiple-paths dynamically maintained?

C.3 PANDA

Positional Attribute-based Next-hop Determination Approach (PANDA)[24] discriminates the next hop nodes based on their location or capabilities. When a route-request is broadcasted, instead of using a random rebroadcast delay, the receivers opt for a delay proportional to their abilities in meeting the QoS requirements of the path. The decisions at the receiver side are made on the basis of a predefined set of rules. Thus the end-to-end path will be able to satisfy the QoS constraints as long as it is intact. A broken path will initiate a QoS route discovery process.

Besides above mentioned QoS routing algorithms, Multipath Routing, DSDV,OLSR, DSR, AODV etc. routing algorithms can be modified with facility for specifying resource requirements to work as QoS routing algorithms.

III. SWAN MODEL

The SWAN model [7] considers TCP traffic as best-effort traffic and UDP traffic as real-time traffic requiring QoS assurances. It tries to maintain delay and bandwidth requirements of real-time traffic by admission control of UDP traffic and rate control of TCP and UDP traffic.

A. Rate Control of Best-effort Traffic

Each node in the mobile ad hoc network independently regulates best effort traffic. The rate controller determines the departure rate of the shaper using the Additive Increase Multiplicative Decrease (AIMD) rate control algorithm based on feedback from MAC. This feedback measure, used by the rate controller, represents the packet delay measured by the MAC layer. SWAN uses existing best effort IEEE 802.11 DCF mode MAC specifications in MAC layer. These packet delays, based on a certain threshold, trigger the operation of AIMD. It aims to maximize the transmission rates of best effort traffic under the constraint of packet delay, while providing sufficient bandwidth for real time traffic. Since the requirements of real time traffic change are dynamic, AIMD is constantly drawing feedback from the system in the form of packet delay to provide the QoS requirements of the system

B. Source-based Admission control of Real-time traffic

Admission control applies radio resource monitoring and uses packet delay as feedback result to decide if a flow can be admitted into the network. Signaling, through the use of probe request and probe response packets, is used to determine if the flow can be supported by the network, specially the nodes through which the flow will be directed, without sabotaging the support provided to satisfy present QoS requirements of already existing flows. Excessive delays will result to all flows, if a flow is taken on by the network for which it has insufficient bandwidth. The algorithm works expressly to prevent such instances from arising, and while it works perfectly well in most scenarios, it can fail in certain scenarios.

C. Dynamic Regulation of Real-Time Traffic

Mobility and *false admission* are two conditions that violate this simple approach to source-based admission control thereby complicating the delivery of soft-QoS assurances in stateless wireless networks. To resolve these problems, SWAN dynamically regulates the real-time traffic when a node experience congestion /overloading (due to the re-routing of admitted flows and/or false admission). SWAN uses explicit congestion notification (ECN) to regulate UDP traffic. ECN bits in IP header of packets are set indicating the source to re-initiate the flow starting from admission control. A failed admission control causes the source to drop its flow. Thus, SWAN can provide only soft QoS assurances.

D. Limitation of SWAN

SWAN differentiates traffic into two classes, best-effort and real-time traffic [7,8]. It treats all UDP (real-time) traffic with equal priority so they will be treated similarly during network congestion. In practice, some connections may be more important than others. For example, a voice conversation between a director and a manager may need higher priority treatment than a voice conversation between two salesmen. Also traffic belonging to different classes may have different resource requirements. For example, voice traffic need stringent delay requirements and a lower steady bandwidth. A live video on the other hand, has high bandwidth requirements, though small amount of jitter can be tolerated. Therefore, SWAN can not be a complete QoS solution for a network. Our model, which will be discussed in the next section tries to remove the drawback of SWAN and at the same time achieves performance gain in most of the traffic load conditions.

IV. THE PROPOSED QOS MODEL

From the limitation of SWAN discussed in the previous section, it is found that SWAN cannot be a complete QoS solution. We propose a hybrid model over SWAN by incorporating service differentiation concepts from DiffServ model. The proposed model uses the feedback from the network to dynamically regulate its parameters as done in SWAN and support multiple classes of traffic as supported by DiffServ. We consider the priority and resource requirement aspects of applications to provide multiple classes of traffic. Accordingly, we consider some service profiles for prioritizing the TCP and UDP traffics. We observe the performance gain in the new model in comparison to SWAN model through simulations.

A. The Model

Fig. 1 shows the proposed QoS model. There is a classifier and a shaper that operate between the IP and MAC layers. The classifier classifies the packets as real-time or best-effort. The shaper controls the flow of best-effort packets using the AIMD rate control mechanism used in SWAN[18]. The rate controller calculates the rate based on the observed MAC delay. In this algorithm different parameters (*d*, *c* and *g*) are adjusted according to [19].

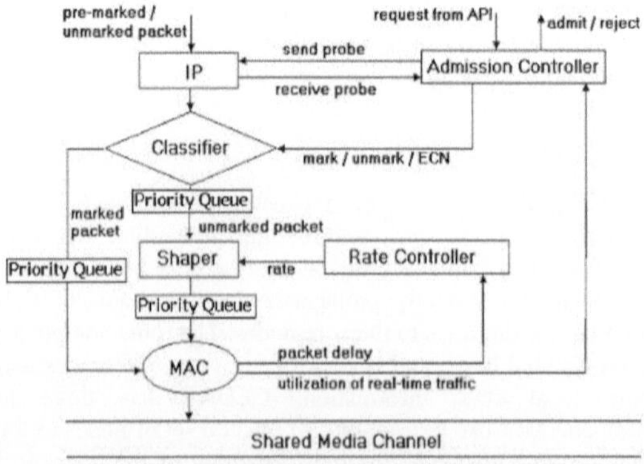

Fig. 1. The Proposed QoS Model

TABLE I: The Rate Controller Algorithm

```
Procedure update_shaping_rate()
        /*called after every T seconds*/
Begin
if(n>0)             /*one or more packets have delays greater
                    than the threshold delay d sec*/
 s ← s * (1- r/100) /*multiplicative decrease by r •*/
else
 s ← s + c          /*additive increase by c Kbps*/
if ((s-a)>a * •g/100)/*diff. between actual rate and shaping rate is
                          greater than g% of actual rate*/
s← a• * (1+g/100)/*adjust shaping rate to match actual rate*/
end
```

Where, *n* is the number of packets that have delays greater than threshold delay *d*, *s* is shaping data rate, *a* is actual data rate, *d* is threshold delay, *c* is additive increase bandwidth and *g* is difference bandwidth.

Every ***T*** seconds, each mobile device increases its transmission rate gradually (with rate ***c*** kbps) until the packet delays become excessive (exceed threshold ***d***). As soon as the rate controller detects excessive delays, it back off the rate by ***r****%*. The shaping rate is adjusted every ***T*** second which should be small enough to be responsive to the dynamics of MANETs. When the difference between the shaping rate and the actual transmission rate is greater than ***g****%* of the actual rate, then the rate controller adjusts the shaping rate to be ***g****%* above the actual rate.

The packets are queued based on priority before entering the shaper. So the best-effort packets of higher priority have less probability of getting dropped. When a source is to send a real-time packet, its admission controller sends a probe message to destination that contains the bottleneck bandwidth along the path. Based on the response from the destination, the node's admission controller admits or rejects the new session for the source. Two RED[17] queues with multiple physical queues for different classes of traffic are maintained. One between the shaper and MAC for best-effort packets and the other is between the classifier and MAC for real-time packets. To incorporate priority in the packets, we are using 3 bits of the TOS field of IP. As in SWAN model, 2 bits of TOS field in IP are used to mark whether a new session is admitted/rejected, or a session is re-probed. As we are using 3 bits, so the maximum priority of a packet can be 7. The packets are already classified either as real-time or best-effort packets, so priority range for both the types of packets can be from 0 to 7. In overloaded/congestion situations, due to *mobility* or *false admission*, the nodes mark the packet's ECN field for real-time traffic regulation. The traffic belonging to low priority classes are regulated more aggressively as compared to the traffic belonging to higher priority. Priorities are assigned to the packets at the time when they are first generated at the source. The packets retain this priority till they reach the destination. The routing packets and probe/reply packets are always given top priority in our model.

V. SIMULATION

We have simulated both proposed model and SWAN, which run on the same set of traffic models and mobility scenarios.

A. Simulation Environment

We use network simulator version-2 (2.26) [23] as the simulation tool. The simulation environment consists of 8 nodes moving in a 670 × 670 square meters area. The nodes move in a random fashion with the maximum speed of 20 m/s. The density of the nodes is sufficient to maintain connectivity at all times as we are not handling QoS in disconnected network. UDP packets (CBR) of 512 bytes and 80 bytes size are taken as real-time video and real-time voice packets respectively. FTP packets of 32 bytes size are taken as best-effort packets. The simulation is done for a total time of 100 seconds. We use AODV in our simulations. The results are taken for *Shaper Bandwidth, Actual rate of best effort packets, MAC delay* and *Channel utilization by real-time packets*. We use an arbitrary test scenario file as given below. Traffic load is high and the nodes send packets in the following manner (800 connections):

Node 2 and Node 7: Low priority best-effort (LPBE) packets and packet transfer starts randomly in between 0-40 sec. and continues up to 100 sec. **Node 3:** LPBE packets as in node 2 and node 7 and High priority best-effort (HPBE) packets: Packet transfer starts randomly in between 50-90 sec. and continues up to 100 sec. **Node 5:** HPBE packets as in node 3. **Node 0 and Node 1:** Low priority real-time (LPRT) packets where packet transfer starts randomly in between 0-40 sec. and continues up to 100 sec. **Node 4:** LPRT packets as in node 0 and node 1, High priority real-time (HPRT) packets where packet transfer starts randomly in between 50-90 sec. and continues up to 100 sec. **Node 6:** HPRT packets as in node 4.

B. Comparisons and Analysis

We perform a comparative study of the proposed model and SWAN[21]. We analyze these two models for a different type of traffic scenarios. But, here we give the analysis for the scenario file described in the previous section.

We observe from the results that there are no significant variations in throughput of best-effort packets in the two models. In both the models, the throughputs are less. This is because the best-effort packets are using the unused leftover bandwidth. As such, increase in real-time packets causes the throughput for best-effort packets to be lower in both the models. Another reason for this is the node mobility.

Significant improvements are seen in the MAC-delay for the nodes transferring high-priority packets in our model. It is because the probabilities of dropping high priority packets are very less compared to low priority packets in our model compared to SWAN. Therefore, the overall MAC delay has reduced for higher priority traffics. Fig. 8 for MAC delay of Node 5 shows the result. Similarly, fig. 10 for MAC delay of node 6 shows comparatively lower delay for proposed model. But for the nodes transferring low priority packets, the MAC delay does not seem to show much improvement in our model compared to SWAN as can be seen in fig. 3, fig. 4 and fig. 6.

Significant improvement is seen in the throughput of different classes of real-time packets with in our model as can be seen in fig. 2, fig. 5, fig. 7, and fig. 9. This difference is mainly because of the improved scheduling and dropping policy of the high priority packets.

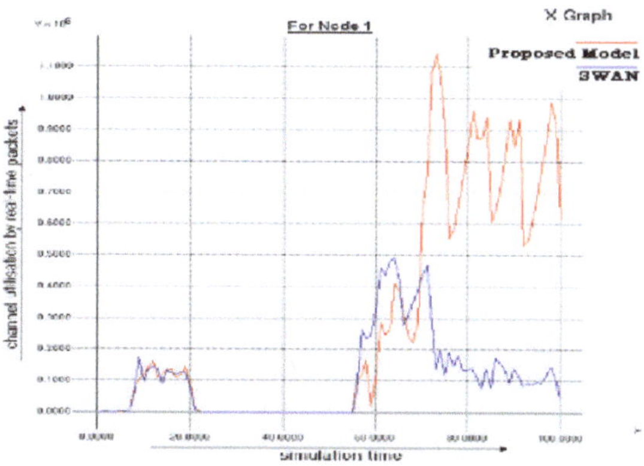

Fig 2. Channel utilization by real-time packets vs. simulation time (node 1)

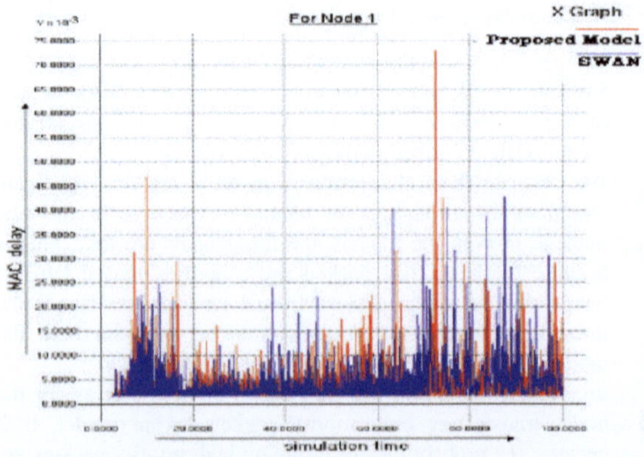

Fig. 3. MAC delay vs. simulation time (node 1)

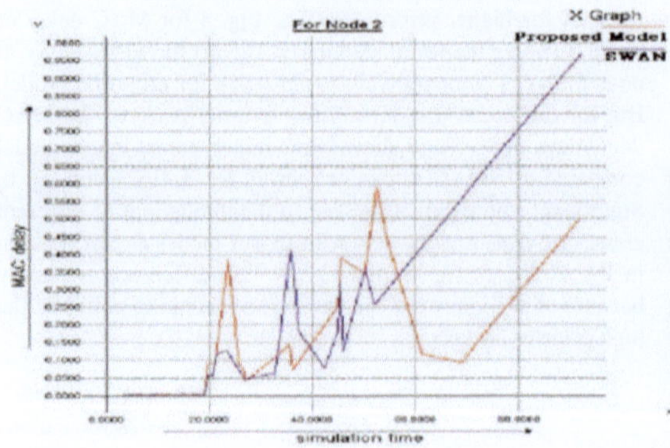

Fig 4. MAC delay vs. simulation time (node 2)

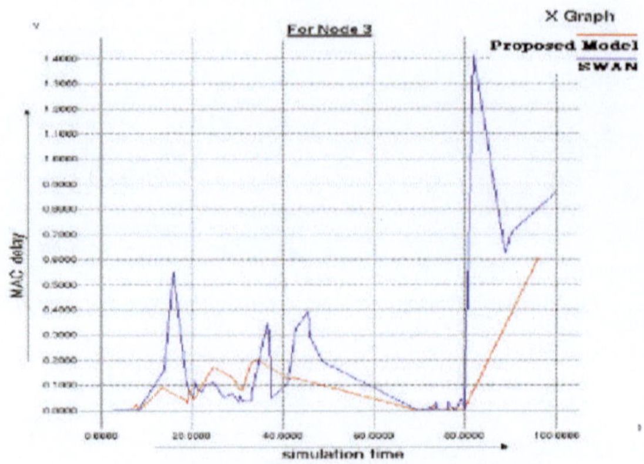

Fig. 5. MAC delay vs. simulation time (node 3)

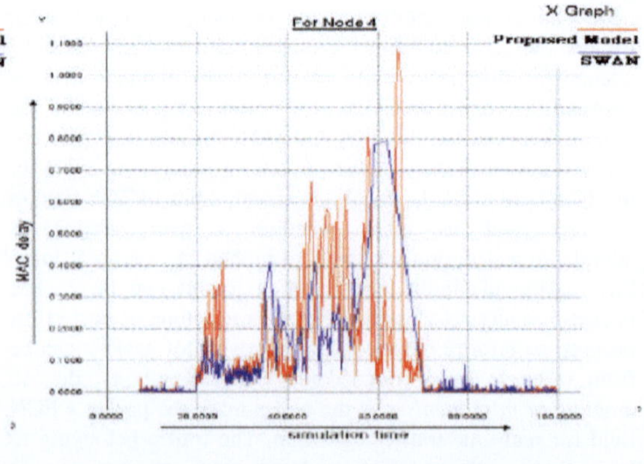

Fig. 6. MAC delay vs. simulation time (node 4)

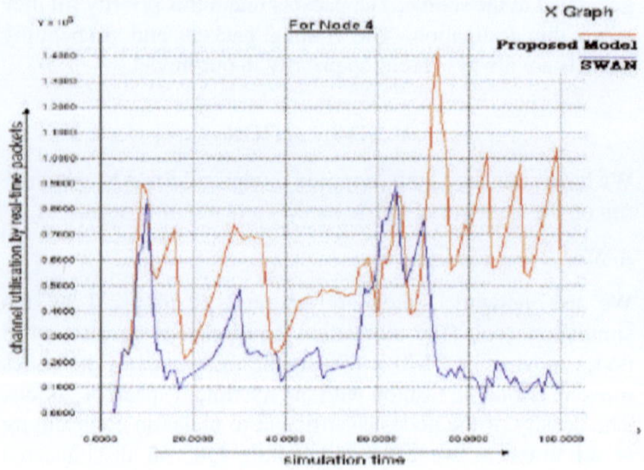

Fig. 7. Channel utilization by real-time packets vs. simulation time (node 4)

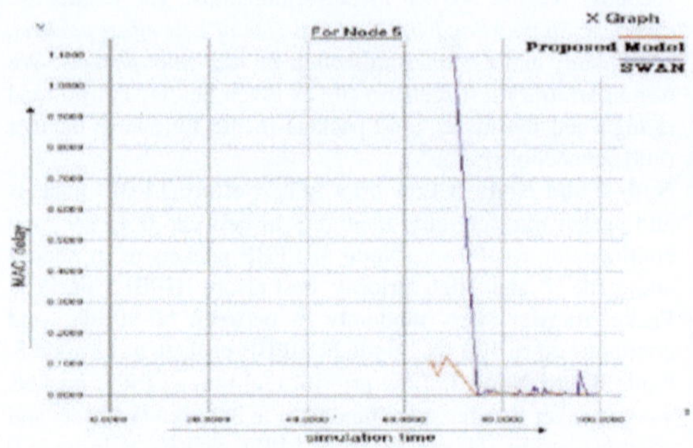

Fig. 8. MAC delay vs. simulation time (node 5)

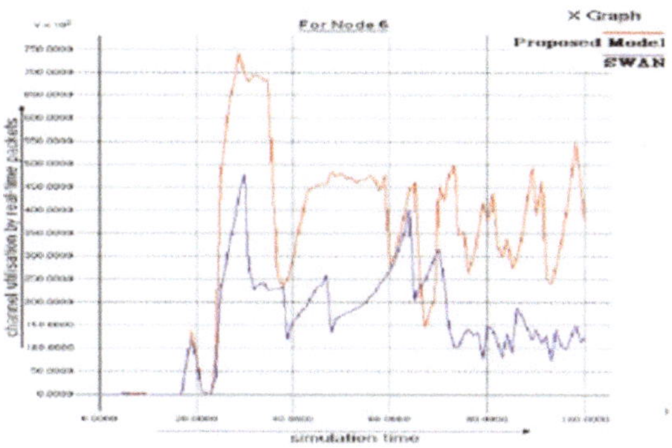

Fig. 9. Channel utilization by real-time packets vs. simulation time (node 6)

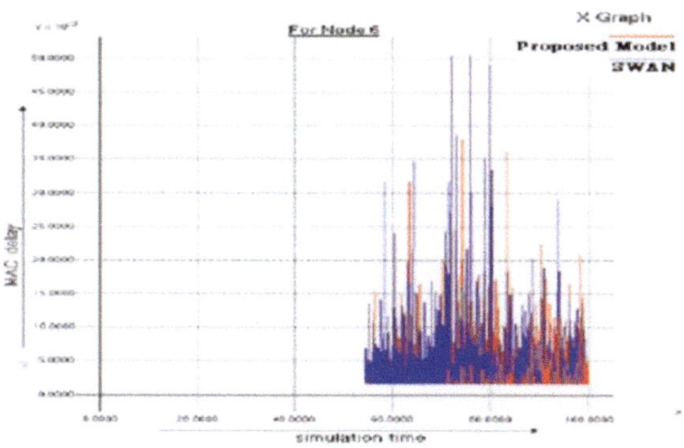

Fig 10. MAC delay vs. simulation time (node 6)

From the above analysis, we can observe that the proposed model is more suitable for enhancing QoS support in MANETs and is efficient compared to SWAN

VI. CONCLUSIONS

In this paper, we present the various QoS issues related to MANETs. Our main focus is on analyzing the feasibility of different QoS models in MANETs. We found that SWAN is a good candidate for a QoS model in MANETs but it still has some problems. Based on our analysis, we propose a new QoS model as an improvement over SWAN to provide more suitable QoS supports for all kinds of traffics. We evaluate through simulations the feasibility of our model as a QoS model for MANETs. Our main concern in the proposed QoS model is to provide more QoS assurances to higher priority real-time packets and at the same time utilizing the leftover bandwidth efficiently by the best-effort traffic. For that, we incorporate service prioritization to give resource priority to certain traffic types. Our model gives 70-75% efficiency (delay) in Real-Time traffics at the cost of 15-20% degrades in performance (throughput) of Best-Effort traffic. The effect of incorporation of QoS routing and QoS MAC layer and the impact of incorporation of different scheduling and buffering mechanisms in this model needs further verifications.

REFERENCES

[1] P Mohapatra, J Li, and C Gui, QoS in Mobile Ad Hoc Networks, IEEE Wireless Comms., June 2003, pp. 44-53.

[2] K Wu and J Harms, QoS Support in Mobile Ad Hoc Networks, Crossing Boundaries- an interdisciplinary journal, Vol. 1, No. 1, Fall 2001.

[3] S Chakrabarti and A Mishra, QoS Issues in Ad Hoc Wireless Networks, IEEE Communication Magazine, February 2001.

[4] R. Braden, D. Clark, and S. Shenker, Integrated Services in the Internet Architecture- an Overview, IETF RFC 1633, June 1994.

[5] S. Blake, An architecture for Differential Services, IETF RFC 2475, December 1998.

[6] H. Xiao, W. K. G. Seah, A. Lo, and KC Chua, A Flexible Quality of Service Model for Mobile Ad-Hoc Networks, IEEE VTC2000-spring, Tokyo, Japan, May 2000.

[7] G S Ahn, A T. Campbell, A Veres and L H Sun, SWAN: Service Differentiation in Stateless Wireless Ad Hoc Networks, Proc. IEEE INFOCOM '02, 2002.

[8] H Arora, L Greenwald, U Rao and J Novatnack, Performance Comparison and Analysis of Two QoS schemes: SWAN and DiffServ, Drexel University Res Day 2003, April 2003

[9] S-B. Lee and A.T. Campbell, INSIGNIA: In-band signaling Support for QoS in Mobile Ad Hoc Networks", proc. MoMuC, 98, Berlin, Germany, October 1998.

[10] Z Y Demetrios, A Glance at Quality of Services in Mobile Ad-Hoc Networks, Technical Report, University of California-Riverside, 2001.

[11] P. Sinha, R. Sivakumar, and V. Bharghavan, CEDAR: a Core-Extraction Distributed Ad Hoc Routing algorithm, IEEE JSAC, Aug. 1999, vol. 17, no. 8, pp. 1454-65.

[12] S. Chen and K. Nahrstedt, Distributed Quality-of-Service Routing in Ad-Hoc Networks, IEEE JSAC, Vol. 17, No. 8, August 1999.

[13] Zhang L, Deering S, Estrin D, RSVP:A new resource reservation protocol, IEEE nw,1993.

[14] Goldsmith A. J. and Wicker S. B.,Design challenges for energy-constrained ad hoc wireless networks, IEEE Wireless Communications, 2002.

[15] Crawley E, Nair R, Rajagopalan B, Sandrick H, A Framework for QoS Based Routing in the Internet, RFC 2386, August 1998.

[16] Mirhakkak M, Schultz N, Thompson D, Dynamic QoS for Mobile Ad Hoc Networks, The MITRE corporation, April 2000.

[17] M. May, T. Bonald, and J. Bolot, Analytic Evaluation of RED Performance, Proc. INFOCOM 2000, March 2000.

[18] G S Ahn, A T Campbell, A Veres and L H Sun, Supporting Service Differentiation for Real-Time and Best-Effort Traffic in Stateless Wireless Ad Hoc Networks (SWAN), IEEE Communication Magazine, September 2002.

[19] A Veres, A. T. Campbell, M. Barry and L-H. Sun, Supporting Service Differentiation in Wireless Packet Networks Using Distributed Control, IEEE JSAC, Vol. 19, No. 10, pp. 2094-2104, October 2001.

[20] S. Floyd and V. Jacobson, Random early detection gateways for congestion avoidance, IEEE/ACM Transmissions on Networking, Volume:1 Issue:4, pp. 397-413, August 1993.

[21] SWAN: http://comet.columbia.edu/swan

[22] S. Corson and J. Macker, Mobile Ad hoc Networking (MANET):Routing Protocol Performance Issues and Evaluation Considerations, RFC-2501, January 1999.

[23] Network Simulator: http://www.isi.edu/nsnam/ns

[24] J. Li and P. Mohapatra, PANDA: A Positional Attribute-based Next-hop Determination Approach for MANETs, Tech. Report, Dept of Computer Science, University of California, Davis, 2002.

[25] K. Nichols, V. Jacobson and L. Zhang, A Two-bit Differentiated Services architecture for the Internet, RFC-2638, July-1999.

The Design, Modeling and Simulation of Switching Fabrics for an ATM Network Switch

Dmitriy Molokov, Muhammad Shaaban,
James Heliotis, Andreas Savakis.
Rochester Institute of Technology.
Department of Computer Engineering.

Keywords: ATM, communication, switch, fabric, traffic.

ABSTRACT

The requirements of today's telecommunication systems to support high bandwidth and added flexibility brought about the expansion of Asynchronous Transfer Mode (ATM) as a new method of high-speed data transmission. Various analytical and simulation methods may be used to estimate the performance of ATM switches. Analytical methods considerably limit the range of parameters to be evaluated due to extensive formulae used and time consuming iterations. They are not as effective for large networks because of excessive computations that do not scale linearly with network size. One the other hand, simulation-based methods allow determining a bigger range of performance parameters in a shorter amount of time even for large networks. A simulation model, however, is more elaborate in terms of implementation. Instead of using formulae to obtain results, it has to operate software or hardware modules requiring a certain amount of effort to create. In this work simulation is accomplished by utilizing the ATM library -- an object oriented software tool, which uses "software chips" for building ATM switches. The distinguishing feature of this approach is cut-through routing realized on the bit level abstraction treating ATM protocol data units, called cells, as groups of 424 bits. The arrival events of cells to the system are not instantaneous contrary to commonly used methods of simulation that consider cells as instant messages. The simulation was run for basic multistage interconnection network types with varying source arrival rate and buffer sizes producing a set of graphs of cell delays, throughput, cell loss probability, and queue sizes. The techniques of rearranging and sorting were considered in the simulation. The results indicate that better performance is always achieved by bringing additional stages of elements to the switching system.

1 INTRODUCTION

The purpose of this work is to evaluate the performance of switching fabrics for ATM network switches. Simulation seems to be the only effective way to observe the cooperation of the sequence of elements in an ATM switch and to predict its behavior under selected conditions. Since there are several factors affecting the switching fabric behavior and many parameters need to be evaluated, such an evaluation cannot be made analytically. The most appropriate way of simulation is software simulation.

Different approaches were involved in the simulation of switch components and particularly switching fabrics. The common way of simulating networks, as many sources suggest, is to make use of a synchronous network, in which at each clock cycle messages move from stage i to stage i+1. This method was proposed in [1]. Networks are usually constructed of "software chips" and the organization of the simulation model is object oriented. The examples from sources [1] and [4] served as the starting point in the creation of the simulation tool implemented in C++ called "ATM library".

This discrete time simulation software package makes it possible to construct models of all the ATM switch components and to organize communication among them, thus reproducing the ATM switch operation. The ATM library was carefully elaborated, adapting many useful features of "Arch library" [5] used for simulation of computer architectures. The ATM library simulates switching elements, queues, connectors, and two modules for interfacing with input sources and output locations, as components of an ATM switch. The interfacing modules called sockets substitute the input and output modules by reading the data from the input files and writing it to the output files. An input socket is also able to accept traffic from dynamic traffic generators based on [4].

This simulation tool models a synchronous network, where time is measured by clock cycles equal to the intervals, during which one bit propagates between two stages of a network. Worm-hole or cut-through routing of ATM cells represented by streams of bits makes a cell arrival event not instantaneous, thus extending one arrival to a time slot equal to 425 clock cycles. Hence, the traffic arrival rate has an upper bound. Common simulation models treat message arrivals as instantaneous events, which mimics store-and-forward routing. The simulation results with non-instantaneous arrivals are only comparable with the results obtained analytically or by usual way of simulation when very large number of simulation traces is

K. Elleithy et al. (eds.), Advances in Computer, Information, and Systems Sciences, and Engineering, 275–281.

generated to make one time slot infinitely small relatively to the overall simulation time.

The sections of the paper are grouped as follows. The second section gives the overview of the basic switching methods used in the ATM traffic management, followed by the description of the switching part of the simulation model in section 3 and the traffic generation part in section 4. The fifth section offers the simulation results of the various network architectures, and the last sixth section summarizes the results of this work.

2 SWITCHING FABRICS AND TECHNIQUES OVERVIEW

Modern communication systems are built from transmission links interconnected with routers. Routers rearrange the connections among the links according to a packet or a circuit switching regulation. Particularly, ATM standard accepts switching of cells preserving their order within routers. A router consists of a set of input and output port controllers connected with an interconnection network. In order to establish a path within a router, called switch in the ATM standard, a sequence of commutations is performed in the switching elements forming the interconnection network.

Various types of network architectures exist. A network may be blocking or non-blocking. Blocking networks may deny requests to set up new connections due to conflicts with already established connections. Non-blocking architectures are further divided into strict sense non-blocking -- always allowing establishing a new connection, wide sense non-blocking -- preventing conflicts through proper policy of allocating the connections, and rearrangeable non-blocking -- giving the way for a new connection by rearranging already existing paths.

A banyan network presented in Figure 1 is a blocking architecture. This type of architectures is the simplest and the most scalable comparatively to all other types. All the architectures of banyan type are subject to internal conflicts at the intermediate stages that substantially reduce their performance. To avoid this, lots of alternatives to banyan networks reducing contention in the fabric, i.e. set of switching elements, are used. One group of alternatives is completely interconnected networks, such as knockout or crossbar tree with each input connected to each output. Another group is fully connected networks where every element is connected to every element in the contiguous stage. The network with the largest number of switch-boxes is the crossbar switch. It eliminates all internal conflicts. Other techniques of conflict reduction are network replication, link dilation, and extended general shuffle. Conflicts at the output stage between cells directed to the same output port cannot be eliminated.

Of the variety of conflict reducing techniques, in this work the methods of rearranging and sorting are investigated. Usual multistage interconnection networks

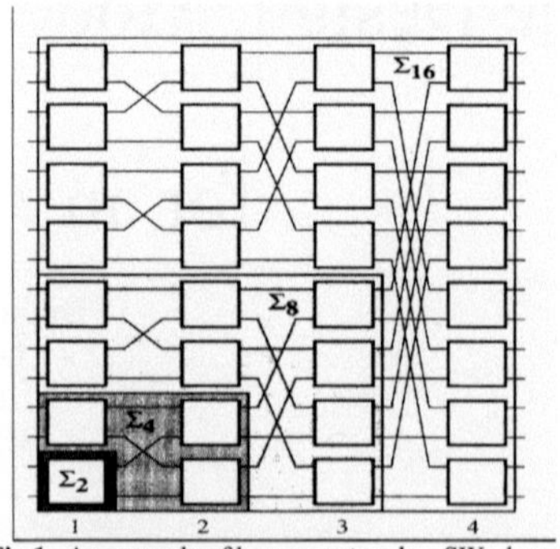

Fig 1. An example of banyan network -- SW - banyan [3].

provide only a single path between each input port and each output port, which leads to internal conflicts. Rearrangeable networks have alternative ways to route cells. For example, Benes rearrangeable network simulated in this work is derived from omega network by the horizontal extension technique discussed in [3]. According to this technique, m additional stages are incorporated into the available banyan network architecture of n stages to increase the number of paths from any input to any output from 1 to 2^m, where m is no greater than n-1. But even with rearranging, the number of paths may not be sufficient enough to eliminate all the internal conflicts as opposed to sorting. The sorting techniques are even more effective and sometimes allow complete eliminating of internal conflicts, although at even higher cost. The simulation model discussed in the next section evaluates performance of these techniques.

3 SIMULATION APPROACH

The simulation methodology in this work is based on the ATM library. This tool allows evaluating the time in clock cycles required to pull a certain amount of traffic through an ATM switch. It is possible to obtain the average and the maximum queue sizes in all the objects. Additionally, the absolute maximum queue size and the overall average queue size are determined for all the objects of the same type: Queue's and SwElement's separately. These two classes are further defined in this section. The other evaluated parameters are the number of conflicts in all the switching elements, the total number of lost cells in all objects, and the cell loss probability. For each output port, the minimum, average, and maximum cell delay is determined, and for all output ports, the absolute maximum and minimum delays and the overall average delay are calculated. The ATM library uses as input data either artificial traces or traffic produced by traffic generators discussed in section 4. UML diagramme of the ATM library without the traffic generation part is in Figure 2. The ATM library has a class called BaseClass from which

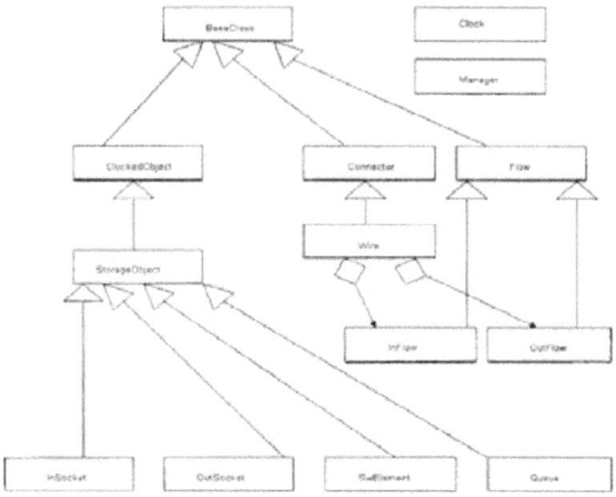

Fig 2. Simplified ATM library UML diagramme.

many other classes inherit. BaseClass keeps the count of all the objects and contains some common functions of the polyfunctional classes. When all the objects inheriting from BaseClass are destroyed, it invokes the *death()* functions of Clock and Manager to display the final results of the simulation. Clock and Manager are the two static classes carrying out some controlling functions; no objects of these classes need to be instantiated. Class Clock drives the system and records the simulation time. This class has the public function *tick()* to invoke the three driving functions *phase1(), phase2()* and *phase3()* in the objects inheriting from the pure virtual class ClockedObject. One of the Manager's functions is to attach routing tags and route the cells towards the output. Class Manager creates the routing table for the ATM switch at the first clock cycle by exhaustive traversing of the switch network graph. It looks for the optimal path for a cell or assigns the first available path according to the specified option. The optimal path is computed by comparing the number of possible internal conflicts in all paths. When the simulation is stopped because all the traffic on the input side has been serviced, destructors in all the objects are invoked. The destructors call the Manager to record the results of the simulation in a special file. Classes InSocket, OutSocket, SwElement, and Queue represent the building blocks of an ATM switch. These four classes have many common functions; therefore, class StorageObject is used to contain all the functions shared by them. StorageObject itself is derived from class ClockedObject; consequently, all of its subclasses are so-called active classes.

Class SwElement describes the switch-boxes and the sorting boxes as the building blocks of an ATM switch. A switching element examines the destination address of an incoming cell and establishes a temporary connection between the input and the output for the time, during which the cell is being transferred. A sorter forwards two incoming cells to the corresponding outputs based on comparison of their destination addresses.

An object of class SwElement can have six functional options. The basic function of this class is switching without buffering (option 0). In this case a cell head when encountering a busy path offers backpressure by sending a signal to the elements back in the chain for the bits to stop propagating through the fabric and another signal to resume motion when the path is free. An SwElement puts the cells in a queue in the case of a conflict with option 1. If the queue overflows, the incoming cells are dropped. Size of the queue is set by the parameter and must be an integer greater than 0. The smaller is the queue size, the greater is the cell discard probability. All the cells have the same priority in this library and are discarded equally. With options 3 and5 an SwElement becomes a sorter with usual queues, whose sizes are set by the input parameter. While only a few bits that belong to the tag of a cell are used in the process of sorting to determine the output address. In option 3 the pin for the minimum output address has the connection ID of 0, and for the maximum address -- connection ID of 1. In option 5 the pins for the minimum and maximum output addresses are reversed. Options 2 and 4 make an unbuffered sorter from SwElement with queues holding only the bits of the tag. Option 2 corresponds to option 3, and option 4 corresponds to option 5 of the buffered sorter. With these two options the queues hold only the bits of the routing tag, and if the queue overflows, the sorter offers backpressure by sending the stall signal to the backward elements. This technique is an attempt to imitate the delayed push-out method [9]. . Options 3 and 5 are reserved for future use because it is possible to reorder cells in a buffered sorter.

Class Queue is a special modification of class SwElement. Objects of this class also accumulate bits of ATM cells, but they do not perform switching. If there is more than one pair of connections, the Queue becomes shared. Currently, the queue is implemented as a doubly linked list. Classes SwElement and Queue perform statistical calculations in their destructors. Class InSocket accepts the traffic on the input side. Class OutSocket writes the output traces to the files and evaluates the delay for a cell to get from the input to the output side. Class Wire establishes permanent connections among the objects of StorageObject's subclasses.

There are two ways to build and simulate a switch using the ATM library. One of them is to declare all the objects and describe all the interconnections manually, and the other one is to use special constructors. For the automatic construction and execution a setup file needs to be created. Figure 3 gives an example of how the objects of a baseline switch with 4 rows and 3 stages are connected. This switch is built with a constructor using setup file called "q_baseline3x4.des". Here the switching elements are marked with letters "SE", strings "IS" symbolize input sockets and letters "OS" denote output sockets. Letter "Q"

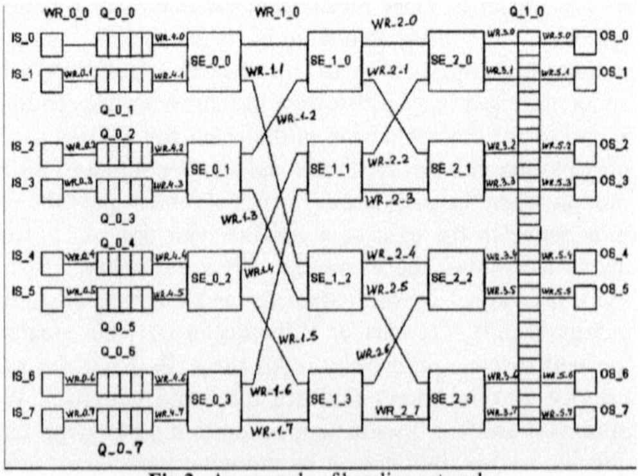

Fig 3. An example of baseline network.

represents queues, and wires are indicated by "WR". In order for the ATM-library to be suitable for the simulation of multibus architectures, some modifications need to be made to it.

4 Dynamic Traffic Generation

The ATM library allows simulating switch fabrics under the traffic, automatically generated by the traffic generation part based on the source code of [4]. Recalculating interarrival times after each arrival event produces Poisson, Pareto, Bursty, and Zeta distributions. InSocket's now take input data from the plugged in traffic generators of class TGen controlled by their managing objects of class TManager. Different groups of input ports take inputs from the traces or are regulated by separate objects of class TManager. The traffic generation part has its own clock synchronized with the switching system's clock. To imitate speedup of the switching system, the clock of the traffic generation part may be slowed down.

In this work, the switch fabric is only simulated under Poisson distribution traffic. The original formula determining the probability that the next arrival time is lower or equal to some threshold T is:

$$p(t_a \leq T) = 1 - e^{-\lambda T}, \qquad (1)$$

which is called cumulative distribution function with λ representing the mean arrival rate [12]. Starting from formula 4.1, the formula for the Poisson distribution making the mean interarrival time equal to 1/λ was calculated according to the inverse transform method suggested by [11] as:

$$\Delta t = -(\ln U)/\lambda, \qquad (2)$$

where U is a random variable generated on the interval (0, 1]. In a real simulation system the mean interarrival time is not always equal to 1/λ. It approaches 1/λ when a large number of simulation traces is generated. The implementation of other types of distribution is also available. They were realized by restructuring of the source code of [4] to be used with the ATM library. There is also a possibility to read the interarrival times from a group of input files, thus simulating any arbitrary distribution.

This section concludes the description of the ATM library functions allowing effective evaluation of various network parameters. In the next section the results of simulations are provided for the 16-output and 256-output ATM switches.

5 Simulation Results

For the purpose of simulation, models of seven types of ATM switches have been constructed for sixteen outputs, which form the first group. They are marked as q_4_cube4x8, q_omega4x8, q_baseline4x8, q_banyan4x8, q_sorter_omega4x8, q_benes7x8, and q_crossbar16x16. Prefix "q_" in the setup file names states that the networks are complemented with two stages of queues -- one stage of input queues and one shared output queue. The layout of q_omega4x8 switch without queues is shown in Figure 1. Switch architecture q_sorter_omega4x8 was constructed by linking a special sorting stage and the omega subtype of the banyan switching stage. The Benes switch was tested in the usual mode, when an incoming cell engages the first path leading to the output, and with the enabled rearrangeability. All of them forward the traffic from 16 input files to 16 output files. The other group of switches for 256 outputs consists of q_omega8x128, q_baseline8x128, q_4_cube8x128, and q_banyan8x128. The performance evaluation of the first group was made with the artificial traffic patterns in the unbuffered version. The second group of networks was tested with dynamic traffic generators under varying arrival rate.

Groups of integers representing ATM cells with arbitrary destination addresses placed into the input files contained in one directory form the traffic patterns. The name of the directory corresponds to the name of the traffic pattern. The arrival rate λ in a simulation with traffic patterns is always equal to 1, although it is possible to decrease the arrival rate by placing additional characters different from end of file between any two ATM cells. The first group of traffic patterns up to PAT5 was created only to observe the performance characteristics of all the types of fabrics without any intention to test the special features. Traffic pattern PAT5A was specially created to analyze the performance of Benes rearrangeable network. It produces a lot of internal conflicts in some networks without rearrangeability, but no conflicts at the final stage since the cells are forwarded to different output ports. To examine the performance of the sorting network, traffic patterns PAT7, PAT8, and PAT9 were created one after another. Each new pattern was an attempt to find a better correlation of output addresses producing as much as possible internal conflicts, but allowing significant reduction of the number of conflict occurrences and their duration by means of sorting. Such a pattern could give a substantial variation of results of sorting versus usual networks. The parameters evaluated are basically the same in each type of simulation and are related. Of the two results, measured on two different networks, one is better than the other, if it has a lower simulation time, which means that it takes less time to pull the traffic through the fabric. It is likely that the network with lower clock cycle time will produce fewer conflicts, cell delay will be lower, and the queues will be smaller. Though only throughput graphs are presented in this paper. Throughput is inversely proportional to the simulation time unless the switch has discarded some cells. For all fabric types, throughput is defined as the ratio of the number of cells returned by the system to the simulation time in time slots. One slot is equal to 425 clock cycles, i.e. the time to admit one cell to the system. Under saturated load and with no conflicts, the throughput approaches the number of output ports in the system, but does not reach it due to additional clock cycles spent in the routing process.

5.1 SIMULATION OF UNBUFFERED NETWORKS

In this simulation step seven switch architectures forming the first group were simulated without internal buffering turned on. All the switch-boxes are built with option 0 as described in section 3, and the sorters are also unbuffered. Figures 4 shows the throughput of the first group of switches. The point of these simulation results is to demonstrate that with varying traffic patterns different switch architectures behave differently. Taking to account various parameters obtained from the simulation of each switch architecture, it can be noted that banyan4x8, omega4x8, baseline4x8, and 4_cube4x8 occasionally produce identical results because they are topologically equivalent, and omega4x8and 4_cube4x8 always produce identical results because they are functionally equivalent. Benes network without the enabled rearrangeability option always performs worse because it has the largest number of switch-box stages. This architecture was tested in order to show the difference in the performance when rearrangeability is enabled and disabled. Benes switch 7 x 8 with turned on rearrangeability option performs better with some traffic patterns than the mentioned above four switches because it has alternate paths allowing to reduce the number of internal conflicts.

Rearrangeable networks are useful when the traffic supplied to the inputs is highly correlated and the particular correlation model produces a large number of conflicts in the networks without rearrangeability. At the same time, in some other networks without rearrangeability the number of conflicts may be no greater than in the rearrangeable ones. For example, the number of conflicts and the throughput, are usually equal in the pairs of 4_cube -- omega and banyan -- baseline. This can be explained by the fact that in those pairs, cells from the same input ports enter the corresponding switch-boxes. If the destination

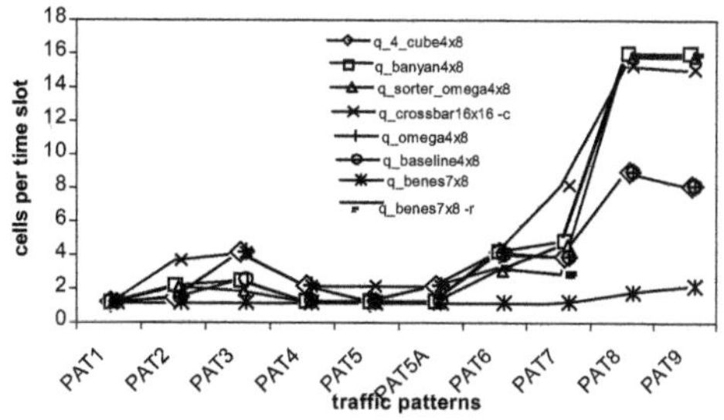

Fig 4. Throughput in unbuffered networks.

tags of any two cells differ only in one pair of bits (0000 and 0001), then these cells are in conflict in three stages in one pair. In the other pair, these cells may not be in conflict at all, because they enter different switch-boxes. The effect of rearrangeability is demonstrated in Figure 4 Here with traffic pattern 5A pair banyan -- baseline gives the throughput of 1.06 cells per slot, while pair 4_cube -- omega produces the throughput of 1.99 cells per slot, which is twice as much. The rearrangeabe network Benes generates 1.99 cells per slot that is the same as the above mentioned pair. This traffic pattern was specially selected to produce maximum number of conflicts in the pair banyan -- baseline. Another correlated traffic pattern could cause a greater number of conflicts, and therefore lower throughput, in the other pair 4_cube -- omega. The rearrangeable network helps to reduce the duration of a conflict using the special algorithm performed by class Manager. Although the number of conflicts itself is not reduced significantly, the conflict duration is lower and the overall throughput is increased. Therefore, if the traffic pattern correlation is not known in advance, rearrangeable network is practical.

The crossbar switch performs better than the rest of the switches, unless a particular banyan switch is well suited for a given traffic distribution or special features such as sorting are involved. For example, with traffic pattern PAT5 the throughput of the crossbar switch is twice greater than that of the other switches (see Fig. 4). Its performance is much less affected under varying traffic patterns. Since there are always enough available paths in a crossbar switch, only those cells forwarded to the same destination simultaneously are in conflict. One drawback of a crossbar switch is that it has significantly larger amount of switch-boxes than that of any other network.

Another performance evaluation is done with the sorting network. The sorting network performs better than the majority of the other networks without special features when one of the specially selected traffic patterns 7, 8 or 9 are loaded. Sorting networks reduce the number of internal conflicts by properly arranging the traffic supplied to the input of the switching stage. Rearrangeable networks, on the other hand, route the cells that are in conflict by

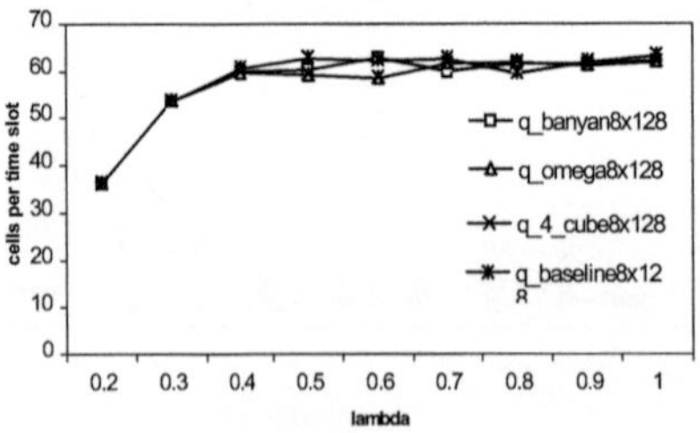

Fig 5. Throughput of large networks with 256 outputs.

alternative paths. There is one instance, namely pattern PAT7, where the sorting network performs better than the rearrangeable one, because its throughput is higher. This can be explained by the fact that the rearrangeable network may not have enough additional paths to route all the cells that are in conflict, while the sorting network eliminates all the conflicts, however at a price of a greater number of switch-boxes.

5.2 SIMULATION UNDER VARYING ARRIVAL RATES

This set of tests was run to measure the performance of networks under varying arrival rate of Poisson distribution. Four. basic types of switches of the second group were tested. This was the longest simulation step. The arrival rate of the traffic applied to each input port is spread on the interval from 0 to 1. The output addressing of the cells is random. Figures 5 shows the throughput and the cell delay obtained by this simulation. The simulation results show that all parameters increase monotonically with the arrival rate changing from 0 to 1. The reason for this is the increased saturation of the system with the increasing arrival rate. The results for the networks vary from one point of simulation to another, except for the pair q_omega8x128 and q_4_cube8x128, and there is no rule according to which the results vary because the output addressing is not systematic. The throughput stays low at low arrival rates because the system is idling. This simulation is conducted under non-instantaneous arrival events; therefore, the graph is saturated at the arrival rate equal to 1. Otherwise the simulation graph would asymptotically approach the vertical line corresponding to the arrival rate of 1. It is shown in the various graphs of analytical and simulation evaluation in [10].

5.3 SUMMARY OF RESULTS

The simulation results in this section suggest that the performance of each particular network is unique with each traffic distribution. However, the evaluation of several networks occasionally gives the same result because of the isomorphism among them that leads to the functional equivalency. If the order of input connections were changed, some other isomorphic networks would also behave similarly. But the problem is that input connections are permanent. The performance of one type of network may be better than that of another type with some particular traffic distribution. The main difficulty in the routing of cells is the case when two cells with the same output address enter the switching fabric. In this situation it is not possible to eliminate the conflict and the waiting time is involved. Various techniques are employed in reducing of the internal conflicts before the final stage, among them are rearrangeability and sorting. Any attempt to create a network that could handle various traffics equally is costly as far as hardware implementation is concerned. Usually extra stages are added to a network to make it capable of handling changing traffic distributions without conflicts. The better performance is desired, more overhead must be added to the network hardware.

The crossbar switch is considered to have the best and the most stable performance, however it is not very scalable. The number of switch-boxes is in the order of square of output destinations. On the other hand, banyan-type multistage interconnection networks are scalable, but very inefficient in handling traffic with unknown in advance distribution of interarrival time and destination addresses. Simple banyan networks may be designed for predetermined traffic distribution only; otherwise additional hardware overhead is necessary to improve the process of routing.

This section described the results of the simulation of various switching fabric types using the ATM library. These results are comparable with the other ones obtained analytically and by alternative simulation methods. The next section summarizes this work and gives suggestions regarding the utilization of different types of switching fabric.

6 CONCLUSION

The outcome of this work is the creation of the ATM library -- a software package for the simulation of ATM switch architectures. The library is a set of control driven "software chips" describing the basic elements of an ATM switch. It is implemented as a discrete time simulation tool simulating synchronous networks. Utilizing this library, it is possible to evaluate a set of basic parameters describing the behavior of a generic interconnection network employed in an ATM switch. These parameters include cell delay, queue length, cell loss probability, throughput, and the number of conflicts. This tool is extremely flexible.

The set of tests was run on a series of banyan networks and a crossbar network. The features of rearrangeability and sorting improving the ATM switching fabric performance were explored as well. Rearrangeability reduces the number of internal conflicts in a network by selecting the least busy path for a cell. Sorting eliminates the internal conflicts by arranging the cells entering the network into an ordered sequence. While the two features improve network performance, they also introduce additional overhead into the system. As the simulation results showed, simple banyan networks are extremely scalable and performed well with traffic loads known in advance, but they are not adaptable to random traffic distributions. On the other hand, enhanced networks, such as rearrangeable, sorting, crossbar, and many other types, are adjustable to varying traffic loads but their implementation is complex and costly. The main goal of the simulation is to find a compromise between performance and complexity for a particular topology and a traffic distribution; therefore, the use of complex components must be justified for each specific network design. There is no network that would be cheap and effective under all conditions, otherwise the theory of switching would not exist.

REFERENCES

[1] Josepmaria Malgosa-Sanahuja, Jordi Castells-Cuscullola, Joan Garcia-Haro, "An ATM Switch Simulation Tool Based on the C++ Object Oriented Programming Language." 1997 IEEE Pacific Rim Conference on Communications, Computers and Signal Processing. Vol. 2.

[2] Andrew Tanenbaum, "Computer Networks." Prentice Hall, Inc., New Jersey, 1996.

[3] Achille Pattavina, "Switching theory: architecture and performance in broadband ATM networks." Chichester, West Sussex, England; New York: J. Wiley, c1998

[4] Adam Lange-Pearson, "Self-Similarity in a Multi-Stage Queuing ATM Switch Fabric." Graduate thesis. Rochester Institute of Technology, May 1999.

[5] James Heliotis, "ARCH library." Software tool for the simulation of computer architectures. Rochester Institute of Technology, January 1994.

[6] Raj Jain, "A Survey of ATM Switching Techniques". http://www.cis.ohio-state.edu/~jain/index..html

[7] Patrick G. Sobalvarro. "Analytical Modeling of Multistage, Multipath Networks". IEEE Transactions on Parallel and Distributed Systems. Vol. 7, №. 10, October 1996.

[8] Arif Merchant. "A Markov chain Approximation for the Analysis of Banyan Networks". ACM Sigmetrics and performance Int'l. Conference on Measurement and Modeling of Computer Systems. Volume 20, number 1, June, 1992. Newport, Rhode Island, USA.

[9] Abhijit K. Choudhury, Ellen L. Hahne. "A new Buffer Management Scheme for Hierarchical Shared Memory Switches". IEEE/ACM Transactions on Networking. Vol. 5, number 5, October 1997.

[10] Thomas G. Robertazzi. "Performance Evaluation of High Speed Switching Fabrics and Networks". Department of Electrical Engineering. State University of New York at Stony Brook. 1992.

[11] Averill M. Law, W. David Kelton. "Simulation Modeling and Analysis". McGraw-Hill, Inc.

[12] Kenneth Reek. "Course materials for the Data Communications and Networks class". Rochester Institute of Technology. May, 1999.

Reliability of Telecommunications Laws and Regulation

Jack Freund
Lucent Technologies
jfreund@lucent.com

***Abstract*-This paper describes the laws of the telecommunications industry in terms of the Weibull and Gompertz functions. The laws and regulations of the industry are presented and analyzed from this perspective. The results show that the regulations fail according to the Gompertz functions.**

I. INTRODUCTION

Rules dictate behavior. In business, the rules are often laws, and these laws describe which behaviors are allowable in certain contexts. This is extremely useful for understanding how to compete and how to position a company's product offerings for best competition in the marketplace. Certain businesses such as telecommunications, aerospace, and nuclear power providers find themselves constantly under extreme regulation.

Telecommunications has an extensive history with regulation in the United States. Here, legislative and judicial leaders have developed laws for decades that try to offer the best solution between the needs and rights of companies and that of the individual consumer. There have been so many regulations applied to telecommunications businesses that some may wonder whether a desired effect has been achieved at all. The ability for a law to achieve its intended goals is subject to a myriad of different factors. However, choosing one of these that is relatively unbiased is somewhat of a challenge. Law is fundamentally a political ally. It use can further a politician's agendas.

For businesses, choosing the right course of action can be difficult if the law changes too quickly. It can be even more difficult to choose a long-term strategy if the law will not permit long-term thinking. For these reasons, it can be useful to have a quantitative measure to evaluate telecommunications regulation. Knowing when a next major change in law may occur could be valuable in deciding a new strategy, or redeveloping a whole approach to conducting business in a regulated environment.

Central to this problem is the question of how reliable is the law. Without a solid bedrock of rules upon which to conduct business, the whole discipline becomes ill-conceived. The reliability of the telecommunications business environment, specific to legality, is the subject of this work. The next sections discuss an experiment designed to reveal the reliability of telecommunications regulation. The theoretical grounding for the design is reviewed in the study of reliability of mechanical engineering and later applied to humans.

II. BACKGROUND

On reliable systems

In developing mechanical entities, there is a need to ensure that components and systems perform as designed. Several methods are used to help guarantee this design. Factory testing helps to make sure the components and systems work the way they should. Components are designed with redundancy that helps to ensure continued and smooth operation. Fail-safe systems ensure safe operation. However, despite all of this, mechanical and electrical things still end up aging and cease to work as they should.

The theory of reliability explains how mechanical and electrical elements fail and expire [1-3]. It helps to optimize these devices for longevity. Maximizing the lifespan of the devices is the primary concern. If one can understand how things fail, there is a better basis for determining how to build them for longer life spans. Reliable systems are defined by the two variables of age and failures. The older the system is, the more likely it is to die. This is a direct function of the system's failures. Failures, paradoxically, are the systems way of maintaining life without dying. Consider a computer with a redundant hard disk drive function. If one disk fails, the system can remain online because there is a failover installed. However, if there is only one disk, and it fails, the system would be unavailable. It would be dead. Its death in this case is a direct result of not having redundant systems to prevent death. In this example, it takes one failure to cause death. In the dual-hard drive example above, it can withstand two failures before succumbing to death. However, after a single failure, the system is not performing the way that it should. The system has aged. If it were not for redundancy, the system would be dead. It can therefore be said that aging is a direct result of a system's redundancy. This is not necessarily a bad thing, as one often associates with aging—the alternative to aging in this scenario is death.

System failures are random. More directly, there is a random probability that a particular component in a system may fail. Failure, and its correlation to death, is probabilistic in nature [4]. The Gompertz and Weibull functions articulate this probability. In 1825, Benjamin Gompertz proposed an equation to help life insurance actuaries in predicting death better for their insured [5-7]. Gompertz's equation indicated that the logarithm of mortality rates increased linearly with age. Simply put, the older one is the more likely they are to die. This equation is most often attributed to human and other biological mortality rates. Waloddi Weibull's equation is similar but more germane for mechanical failures [8-11].

K. Elleithy et al. (eds.), Advances in Computer, Information, and Systems Sciences, and Engineering, 283–287.

Weibull showed that the logarithm of failure rates increases linearly with the logarithm of age. This power law function shows that failures accelerate with age. Weibull's work applies directly to the failure and death rates of machines, and mechanical or electrical entities. These laws are closely related and are complementary in nature. One's own experiences with life validate this relationship between failure and death. It is only logical then that failures increase with age, as does the probability of one of those failures causing death [12-14].

Variance occurs in the mortality rates for different subsets of humans. Women and men die at different rates, as do men and women from different countries or ethnic backgrounds. Any number of factors can be attributed to these discrepancies. Nevertheless, despite these initial variances, over time, this evens out. Consider men and women. As aging occurs, men are more likely to encounter a failure that causes death. However, as the two sexes approach the 100-year marker, there mortality rates converge. In fact, this same phenomenon is experienced for almost all subsets of humans. Despite country of birth, sex, ethnicity or any other factor, mortality rates all converge once one reaches 100 years in age. This behavior is referred to as the compensation law of mortality. Additionally, humans experience another mortality trend at age 100. Prior to age 100, one is more likely with each passing year to die. For every age marker after 100, there is no longer a linear relationship between death and age. That is, there is an equal chance of experiencing a failure that causes death at age 101, 102 and so on, as there is at age 100. This is known as the late-life deceleration law [2].

Another characteristic of human mortality is infant mortality. There is an extreme likelihood of death immediately after birth. In fact, one is almost nearly as likely to die as an infant as they are at age 40 or so. This likelihood drops off sharply until about age 10 which marks a small period of optimum operation. It is at this age that humans are operating with the least probability of defect. After this, it is a slow rise in mortality until the 100-year marker. A Gompertz graph of this reveals a bathtub-shaped human mortality rate: high at birth, a drop-off during childhood, and then a slow rise towards 100 years. This curve is not found in the Gompertz graphs of mechanical deaths. Factory testing eviscerates the infant mortality rate in electronics and mechanics. Even so, there is still the occasional out-of-box failure. However, this argument is improperly worded. The infant mortality never disappears—instead testing mitigates it. The testing reveals the failures before coming into the hands of the customer. Therefore, it can also be said that this newborn mortality exists, but not for the consumer of these products.

The functions outlined here provide a unique opportunity for application to telecommunications law. A need for reliability data for these regulations has been argued in the first section. The Gompertz and Weibull functions show exactly the ways in which mortality occurs. For that, they are perfectly applicable to this study. Further, the need for an unbiased approach to evaluating these laws reveals itself. The regulations can find themselves fairly appraised by measuring their life span, as opposed to their effectiveness.

On telecommunications regulations

The landmark pieces of telecommunications regulation are successive in their breadth. In other words, they subsume all the previous laws and incorporate their facets, or replace them with newer thinking. This structure of legislation and judicial decree leant itself very well to evaluation by the reliability standards outlined above. In this section, the pieces of regulation are covered and the cause for their demise is outlined. The phrase demise here is intended to mean failure or death in reliability terms. The author reviewed the history of each item, and compiled them for inclusion in this study [15-18].

The main drivers of telecommunications history in the United States begins with the Mann-Elkin Act. This act placed the regulation of interstate telecommunications under the purview of the Interstate Commerce Commission (ICC). It is replaced later with the Kingsbury Commitment. Although still under ICC purview, AT&T was excused from certain behaviors. The then President of AT&T Nathan Kingsbury agrees that they will no longer purchase competitive exchange carriers. Up until now, this was a major way that AT&T solidified itself as a lead phone service provider. Since nobody was allowed to connect to the AT&T network besides AT&T, it only makes sense that people would be more included to do business with a carrier that allows them to call a greater amount of people. This is a "rich-get-richer" scenario, although it was ultimately for the greater good of the developing network. Therefore, the AT&T network grew larger, and encompassed a greater amount of territory. AT&T was becoming the monopoly some feared, and some desired. However, the country was getter more comfortable with a telephone monopoly. The Willis-Graham Act of 1921 excused AT&T from the provisions of the Sherman Antitrust Act, thus making AT&T a sanctioned monopoly. In June of 1934, Congress passed the Communications Act of 1934. This act acknowledged the ineffective job the ICC was doing in regulating broadcast radio and interstate telecommunications by creating the Federal Communications Commission (FCC), and allowing for action under the Sherman Antitrust laws. The effects of the 1934 Act is modified in 1956 with the legislation heard around the world—the breakup of AT&T into its Baby Bells with the 1956 Consent Decree, also called the Final Judgment. The final judgment remained in effect until 1982, when Judge Harold Greene amends this to become the Modified Final Judgment. Lastly, the Telecommunications Act of 1996 supersedes the MFJ. It is important to note that each law represented here does completely subsume its predecessors. Many aspects of the Communications Act of 1934 are still in effect by virtue of the 1996 Act. This is simply the nature of law—a new law will reinforce, or not addresses elements in previous law, allowing them to stand. This behavior is inconsequential to the notion of reliability. For this work, if a law subsumes a previous one, it is because it has failed in some way. This failure is represented in the data as a death date.

During this period there where several other legislative strings formed. The Rural Electrification Act was passed in

1936. Its primary goal was to provide power to rural parts of the United States where running a power infrastructure was going to be costly. The act provided low cost loans to build out the power facilities. Part of this was an arrangement with AT&T to use many of the telephone poles already in place. This is referred to as right-of-way and is solidified in later telecommunications acts. The REA is amended in 1949 and 1971. The tenets of its telecommunications background are solidified in the Telecommunications Act of 1996.

Attaching equipment to the network was for a long time the domain of only AT&T. The phone company came to install a phone in one's home and that was what was used. Several industrious inventors created devices that are meant to complement the phone in several ways. This fact was contested in court beginning with the 1956 decision on the Hush-A-Phone case. Here a device is connected to the phone that allows one to whisper. The user's voice is amplified and sent over the network. The court ruled that these types of equipment were indeed lawful. Later, a device known as the Carterphone was contested for being attached to the network. This was a device that allowed ham radio operators to make long distance communications connections in conjunction with their ham radios. It too was deemed lawful for use. The death date for the Carterphone decision is difficult to ascertain. Because the entire basis of these cases depended upon AT&T owning Western-Electric, the 1996 act encouraged the divestiture of manufacturing from AT&T, and is therefore set as its death date.

The last string of regulation is the Computer Inquiries I, II, and III. These FCC investigations were begun to determine how computer equipment should be regulated in the telecommunications environments. While originally not ubiquitous, these two devices became synonymous. In the days of human, mechanical and then electro-mechanical switching, computers where thought to be separate. Later as they became part of the network, they became regulated as well. These three investigations were done in order between the consent decree and the 96 act. Computer Inquiry I was conducted from 1971 to 1981, II from 1981 to 1986, and III from 1986 to 1996.

One last interesting piece of legislation was passed during the First World War. The government assumed control of AT&T for roughly one year (1918-1919) during this time. Then President Woodrow Wilson felt it was important to maintain control over the phone network during the time of war. He had AT&T report to the postmaster general, who returned ownership to private interests in 1919.

III. DESIGN AND IMPLEMENTATION

These experiments were designed to model telecommunications regulation as a human life, and then evaluate it using the Gompertz and Weibull functions. Each regulation was researched for its historical significance. If the law was referenced specifically in telecommunications history texts, then it was included in the study. After compiling a list of regulations, data regarding its "birth" and "death" date needed to be obtained. For this, the date the regulation became effective was used as the birth date. In evaluating the death dates, several factors had to be considered. Many laws ran in straights. This indicates that many belonged to a serious of laws or judicial rulings meant to regulate a particular thing. Consider the chain referenced above beginning with the Mann-Elkins Act and concluding with the 1996 Telecom Act. The laws and rulings between them are successive in nature. That is, there is a root law that is subsumed by later ones. This determination is largely a matter of history. However, there is some subjectivity in choosing the death date. The death date for each of the items corresponds to the birth date of a subsuming regulation. The rationale here is that when the law goes out of effect is the same time that a new law takes its place or vacates its intents.

After the raw data about the regulations was compiled, the next step was to code them based on their origin. For this work, the two code categories were legislation and judicial. Regulations that were created because of court hearings were clearly judicial. However, items that came about because of negotiation with the Justice Department were also included in this category. Laws originating in Congress were of course labeled legislative. Items that were the result of FCC inquiry are also considered legislative as the FCC reports to the Congress of the United States, thus being a member of the legislative branch of government.

The next step was to build the charts that represented this data. In order to simulate death rates of men versus women as experienced in the compensation law of mortality and the late-life deceleration law, the data was graphed as sets according to its coding. This chart showed the death rate of the laws versus time. The last step was to graph this data in the format of the Gompertz and Weibull functions. In order to support this, the calculated values of life span and population were created. Life span is the difference between the birth and the death date. The population entry is the sum total of laws in effect at the time a law dies. Population is a key factor in computing the mortality rate, or average number of deaths normalized by the population figure. This date was graphed according to the scales set up by Gompertz and Weibull. For the Gompertz graph, the scale was set at the log value of the mortality rate versus year of life. Weibull used the log value of failures versus the log value of year of life. It is interesting to note that for the Weibull function, failures were needed. Since none of the laws considered had redundant features, when they were replaced, they were considered dead. Therefore, they only failure they were able to incur was the one that would cause their death. Therefore, the mortality rate was substituted for the failure rate data.

IV. RESULTS AND OUTCOMES

The overarching feature of a Gompertz graph is that the human death rate is linear—it increases with age. Weibull shows us that failures in machines are linear over time. It was reasoned that regulation being a product of humans, it would fail as their organic makers do. This hypothesis was supported by the conclusions.

Fig. 1 describes the death rate of each type of regulation versus time. In this, the judicial and legislative laws are graphed separately to mimic the varying deaths rates between men and women. What this chart shows is particularly interesting. For a given regulation, it has a higher death rate within the first eight years of its life than legislation. After eight years, legislation has a higher death rate and continues this trend. It does not appear to experience late-life mortality deceleration or the compensation law of mortality.

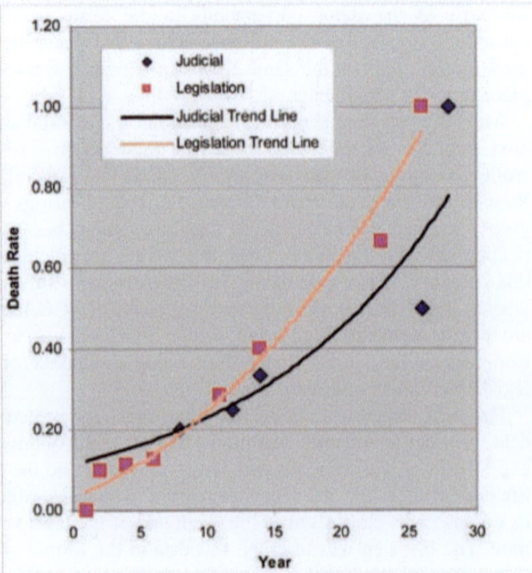

Fig. 1. Telecom Regulation Death Rates -- Judicial vs. Legislative.

Fig. 2 shows us the Gompertz graphs. This graph shows us the log value of mortality rate versus the year. When the trend line is added, one can see there is a linear relationship between the two values. This illustrates telecommunications regulations adhering to the Gompertz law of mortality. That is, telecommunication regulation dies as an organic being.

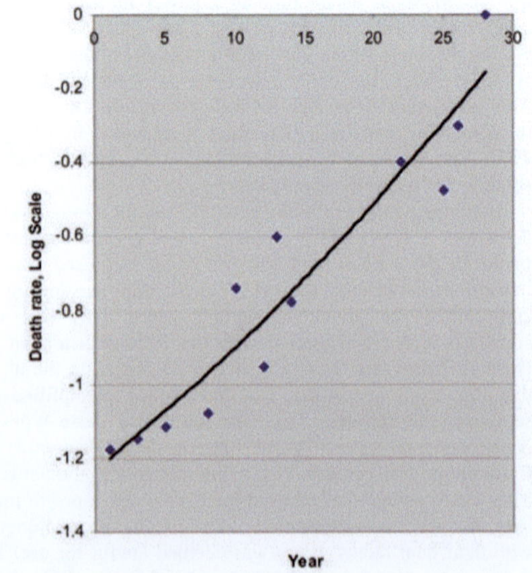

Fig. 2. Telecom regulation Death Rates Gompertz Graph.

Lastly, Fig. 3 illustrates the log value of failures (mortality in this case) versus the log value of year. This trend line is not linear, thus it can be concluded that the data does not adhere to the Weibull observations on mechanical failures.

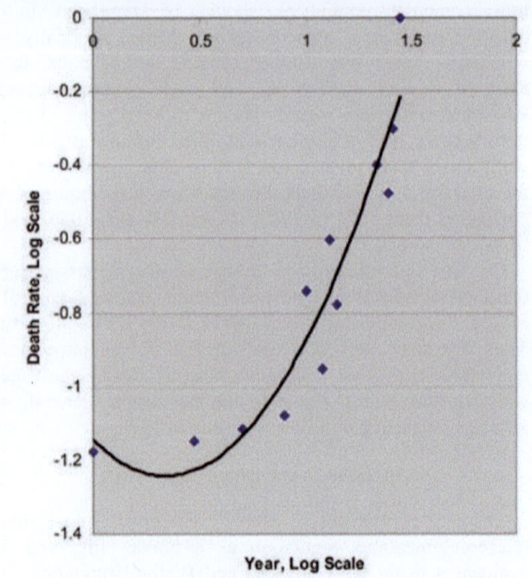

Fig. 3. Telecom Regulation Death Rates Weibull Graph.

V. CONCLUSIONS AND RECOMMENDATIONS

The data sets here support telecommunications regulations, both legislative and judicial, failing and dying according to organic scales. One can reason that this is rational, given that regulation is a human creation, thus subject to human mortality and failures. Alternatively, it may be because of humankind's need to control their environments. Such conclusions are interesting conjecture, but not scientific. The reason for the similarity between humans and telecommunications legislation cannot be concluded from this study. Instead, one can take away several interesting observations. First, there is a relatively limited life span of judicial decrees. Given the makeup of the relationship between the branches of government in the United States, it makes sense that this would be the case. However, short-term judicial decisions enjoy less death than legislation for approximately the first eight years they are in effect.

Caution should be used when applying the results of this study to businesses operating in the telecommunications sector. There is a very limited number of data points used in this experiment. The data used does show that there is a trend that supports the hypothesis. Further research needs to be conducted with an expanded set of data points to include many minor revocations of laws and judicial decisions. Further research to expand this experiment can only help to improve the results. Other researchers in behavioral science disciplines may want to explore why the regulations failed as humans do. It would be interesting to see if the law fails, or if humans cause it to fail. Despite the caution, the results can be extrapolated to a current example. The Telecommunications Act of 1996 is currently the main telecommunications law in effect in the United States. Almost all of the chains in the data sets ended with this law. There is presently speculation that the congress and/or the FCC will begin to retool the Act or completely replace it. The current outgoing head of the FCC, Michael Powell has expressed his support of this idea. If the life span trend holds true, the Telecom Act has passed the critical mark of eight years. It is currently in its ninth year of effect. Therefore, it has a much better chance of survival than any judicial motions in place. In other words, it is likely to outlive any court cases currently in effect. Despite this, each passing year causes the law to become more prone to death. Because of this, any telecommunications corporation basing major strategy on the tenets of this law may be wise to hold off until a new law changes the legislative landscape.

The world of telecommunications regulation is tumultuous. There has been a myriad of complex rulings and decrees over the years that have allowed the industry to become the highly regulated being it exists as today. Those seeking to find stability here will not find it. In spite of this, the mortality laws for humans described by Samuel Gompertz over a century ago appear to also describe this industry. This fact can offer solace to some. It is especially comforting to business managers who are working to place their telecom company at the forefront. If there is anything to be learned from the history of telecommunications, it is that being at the forefront makes you a target for regulation.

REFERENCES

[1] L. Gavrilov and N. Gavrilova, *The Biology of Life Span: A Quantitative Approach*. New York: Harwood Academic Publisher, 1991.

[2] L. Gavrilov and N. Gavrilova, "The reliability theory of aging and longevity," *Journal of Theoretical Biology*, vol. 213, pp. 527-545, 2001.

[3] L. Gavrilov and N. Gavrilova, "Why we fall apart," *IEEE Spectrum*, vol. 41, pp. 31-35, 2004.

[4] W. Feller, *An Introduction to Probability Theory and its Applications*, vol. 1, New York: Wiley and Sons, 1968.

[5] B. Gompertz, "On the nature of the function expressive of the law of human mortality and on a new mode of determining life contingencies," *Phil. Trans. Roy. Soc. London A*, vol. 115, pp. 513-585, 1825.

[6] C. E. Finch, *Longevity, Senescence, and the Genome*, Chicago: University of Chicago Press, 1990.

[7] W. M. Makeham, "On the law of mortality and the construction of annuity tables," *Journal of the Institute of Actuaries*, vol. 8, pp. 301-310, 1860.

[8] R. E. Barlow and F. Proschan, *Statistical Theory of Reliability and Life Testing: Probability Models*, New York: Holt, Rinehart, and Winston, 1975.

[9] G. Hirsch, R. J. Williams, and P. Mehl, "Kinetics of medfly mortality," *Experimental Gerontology*, vol. 29, pp. 197-204, 1994.

[10] H. R. Hirsch and B. Peretz, "Survival and aging of a small laboratory population of a marine mollusc, Aplysia californica," *Mechanisms of Ageing and Development*, vol. 27, pp. 197-204, 1984.

[11] C. Janse, W. Slob, C. M. Popelier, and J. W. Vogelaar, "Survival characteristics of the mollusc Lymnaea stagnalis under constant culture conditions: effects of aging and disease," *Mechanisms of Ageing and Development*, vol. 42, pp. 263-174, 1988.

[12] M. J. Crowder, A. C. Kimber, R. L. Smith, and T. J. Sweeting, *Statistical Analysis of Reliability Data*, London: Chapman and Hall, 1991.

[13] S. E. Rigdon and A. P. Basu, *Statistical Methods for the Reliability of Repairable Systems,* New York: John Wiley and Sons, Inc., 2000.

[14] R. E. Barlow, F. Proschan, and L. C. Hunter, *Mathematical Theory of Reliability*, New York: John Wiley and Sons, Inc., 1965.

[15] R. J. Bates and D. Gregory, *Voice and Data Communications Handbook*, Berkeley, CA: McGraw-Hill Companies, 2001.

[16] L. T. Knauer, R. K. Machtley, and T. M. Lynch, *Telecommunications Act Handbook: A Complete Reference for Business*, Rockville, MD: Government Institutes, Inc., 1996.

[17] D. Massey, *Bell System Memorial*, retrieved 14 October 2004 from http://www.bellsystemmemorial.com.

[18] M. Cole, *Telecommunications*, Upper Saddle River, NJ: Prentice Hall, 1999.

Improving authentication in voice over IP infrastructures

Francesco Palmieri

Università degli Studi di Napoli Federico II – Centro Servizi Didattico Scientifico
Via Cinthia 45, Complesso Universitario di Monte S. Angelo, 80126 Napoli, Italy

***Abstract* - Now the Internet is not only an information platform but also a convergent services platform where voice over IP (VoIP) technologies are promising as a silver bullet for future voice communications and transport services. The well-known benefits of VoIP are cost savings, open standards, multi-vendor interoperability and integrated voice and data services. On the other side, VoIP comes with completely different operational and security settings than the old telephone network: the physical location of clients is not fixed and great flexibility is required to provide enhanced services to end users, thus strong authentication is required as a basis for authorising correct service use. The above security concerns are actually the major barrier that may prevent the wide deployment of VoIP technologies, and coping with them is a truly challenging task. Consequently we developed a novel but completely standard-based architectural framework for strong authentication in the VoIP/IP Telephony environment based on the introduction of digital certificates and a hierarchical organization of public key infrastructures to enforce subscribers' identity evidence and no-repudiation throughout the whole voice communication system.**

***Keywords* – VoIP, IP Telephony, SIP, H.323, X.509**

I. INTRODUCTION

With the increased deployment of high-speed ("broadband") Internet connectivity, a growing number of businesses and individuals are starting to use the Internet to transport voice telephony, according to the so-called Voice Over IP (VoIP) and IP telephony paradigms. Voice transport and call management are now being treated as usual IP network applications and long distance and local service telephone providers already run or are arranging to run VoIP for an increasing portion of total traffic to save on circuit and toll costs. Under the IP telephony paradigm phone numbers can become location independent, phone calls totally distance independent and several innovative IP-based applications like enhanced call routing and integrated messaging are being developed for the IP network to add further value to the existing telephony services and foster the deployment of the new IP based technologies. In a few words, VoIP generally means a win-win for both the consumer and the service provider in terms of costs and available features. Unfortunately, as VoIP deployments become more widespread, the technology becomes a more attractive target for hackers, increasing the potential for harm from hostile activities, such as session hijacking, man-in-the-middle and denial of service attacks. Without proper safeguards, an attacker could intercept a VoIP call and modify the call parameters/addresses. This would open the call up for spoofing, impersonation/identity theft, call redirection and other attacks. Clearly, the emergence of these attacks will likely occur as attackers become more familiar with the technology through exposure and easy access. In this scenario, when delivering a voice service over IP, the first and most fundamental security issue is authentication as a prerequisite for authorization: each further call processing or billing activity will be dependent from the initial identification/authentication of the involved user. Strong and irrefutable way of identifying the users of the IP telephony system is needed to avoid they may later find themselves with huge bills for calls they did not make. Unfortunately, unlike in traditional circuit-based telephony service, the physical network end points cannot be used any more, unless the added value of location and network-independence of VoIP clients is sacrificed. Options to consider are to authenticate users, or their terminals. Since physical location as in Mobile Telephony is irrelevant, the only realistic choice seems to use suitable credentials, that can be associated directly to users or most likely to their terminals. In this paper we present a novel framework for strong authentication in the VoIP/IP Telephony environment based on the introduction of digital certificates and a hierarchical organization of public key infrastructures of to enforce subscribers' identity evidence and no-repudiation throughout the whole telephony system. The enhanced security features of the most common VoIP signaling protocols such as H.323 and SIP have been properly exploited to provide the basic security features to the proposed framework. Moreover, our objective is providing insights towards building a federated and worldwide deployable authentication/authorization model for secured VoIP communications.

II. SECURITY ISSUES IN VOIP INFRASTRUCTURES

Privacy and Security are mandatory requirements in any telephony system and service network. Although not perfect, along the years a certain level of security has been achieved with traditional *plain old telephony*-based (POTS) networks. On the other hand, IP telephony-based networks, which will be a core part of our worldwide Telephony infrastructure in the near future, introduces caveats and security concerns which traditional POTS-based networks do not have to deal with, long forgot about, or learned to cope with. Delivering voice services over packet-switched IP has radically different security characteristics than circuit-switched operation since the overall security of the system was based on the (physical) security of the core network components and the last mile. The "*principals*" involved in a connection were practically the sockets in the walls on the customer's premises, and access to a socket implies authorisation to use the network. This model works (pragmatically seen) sufficiently well, largely due to the

K. Elleithy et al. (eds.), Advances in Computer, Information, and Systems Sciences, and Engineering, 289–296.

fact that attacks hardly scale in a distributed environment that requires physical access to exploit vulnerabilities. The security characteristics of VoIP networks are also significantly different from the POTS: they typically use the public Internet and the TCP/IP protocol stack and do not provide the same "physical wire" security as telephone lines. In more detail, a typical VoIP-based infrastructure is built on the following key elements:

– the participants to a call (the caller and the callee in case of a two-person call) with their terminal devices (e.g. IP telephones, PCs) which are used to initiate and receive calls;
– call signalling and control devices such as gateways and servers (e.g. gatekeeper in H.323 and proxies and location/registration services for SIP, as will be better explained in the following sections), which refer to all kinds of intermediate devices that are needed during a phone call;
– communications media, i.e. data links connecting gateways and terminal devices together, hence forming an end-to-end communications path for VoIP packets to travel. These data links can be either based on wires (copper, fibre), or they can be wireless when radio waves are used.

By gaining access and monitoring network traffic at certain parts on a network infrastructure (such as to/from a VoIP gateway), an attacker could capture, modify, reassemble and re-inject VoIP signalling/control packets to subvert the security behaviour of the telephony system. Furthermore, the lack of a central network/routing control makes it hard to embed security within the network, and, due to the open nature of a shared medium, traffic is accessible (and attackable) for all parties unless dedicated protection measures are applied. Unfortunately, also the risk factors associated with IP telephony-based networks are far greater compared to traditional telephony-based networks. The security concerns associated with IP telephony-based networks are overshadowed by the technological hype and the way IP telephony equipment manufacturers push the technology to the masses. In some cases IP telephony-based equipment is being shipped although the manufacturer is well aware of the clear and present danger to the privacy and security of the IP telephony-based network its equipment is part of.

A. Authentication in IP telephony systems

Authentication is the core of VoIP security, and is absolutely indispensable both for users and providers (e.g. for billing). The technical challenge is to securely identify users/customers in a scenario that provides (network) mobility to them. From a security perspective, this is very similar to the GSM/UMTS settings, with the additional threat of permanent, high-bandwidth Internet-connections and (most likely) highly vulnerable terminal devices. Without the introduction of proper security mechanisms in the IP telephony environment, preferably based on strong cryptographic techniques, neither the telephone service provider nor the other participants in a call can really trust the identity of a specific subscriber/participant. The user's identity, the "caller-ID" information (e.g., a phone number, an URL or other means to identify a subscriber in IP telephony-based networks), is easily spoofable in a wide variety of scenarios. An identity-related attack might occur anywhere along the path signalling information is taking between call participants. A malicious party might use properly crafted software, available into the arsenal of attacker's tools, to perform digital impersonation and spoof the identity of another subscriber and charge him with a lot of unwanted international phone calls or perform some malicious/hostile activity through the telephone network and the unaware victim will be responsible of it. Unlike IP telephony-based networks, spoofing identities with the PSTN is a much harder task to perform, and however only at the endpoints, where, for example, someone other than the intended subscriber answers the subscriber's phone or a calling party claims to be someone he/she is not, since each user access to the network is implicitly authenticated through the subscriber identification and addressing number, or caller-id, that is statically associated at the exchange level to the subscriber endpoint. Unfortunately, in the IP telephony environment things are much more difficult, due to the totally dynamic association between the physical location on the network and the user identification and addressing scheme, which, as already asserted before, is crucial to achieve user mobility. That's worse, the different signalling protocols actually available for implementing VoIP networks provide very different authentication facilities and identification/addressing schemes.

B. Identification and addressing in VoIP

In the old telephone network there was not a clear distinction between customer identifiers, mainly used for authentication and billing, and addresses (or routing numbers), used by the network to route calls, since they usually coincided and were both based on the E.164 specification. Real customer names were usually only reported in telephone directories and provider service databases realizing their mapping with the corresponding E.164 identifiers. On the other hand, in IP networks there is a clear distinction between naming and addressing schemes. Names take the well known hierarchical form of domain names separated by dots, e.g. netlab.unina.it. Addresses are 32 bits long (or 128 bits in the case of the new version of the IP protocol, IPv6) and identify an interface of a host connected to an IP network (multi-homed hosts or routers have as many addresses as they have interfaces). The association between names and addresses is at IP level and is resolved using the Domain Name System (DNS) application, which is a world wide distributed directory. VoIP infrastructures based on IP networks need their own naming and addressing schemes defining all the rules necessary to make resolutions and call routing possible, that is, to define the functions that allow the conversion between subscriber logical names and their actual IP addresses at the network communication layer. Resolution can be carried out in very different ways and by different network entities depending on the signalling protocol used for call setup and management. SIP [1], for example, that is the main IETF VoIP signalling proposal, uses SIP-URLs to identify "users at hosts". They take a form similar to mailto URLs (e.g., `mailto:user@domain`), for example:

- sip:fp@unina.it, which is the simplest form of SIP-URL and identifies a user (fp) in a domain (unina.it),
- sip:fp:secret@unina.it, which is the same as the previous example but includes a password,
- sip:+39081676518:1234@gw.unina.it;user=phone, which identifies a telephone number which is accessible through a gateway located at the domain gateway.com (1234 specifies a password).

There are also other naming schemes based on URL formats being used in VoIP. The most common one is the tel-URL scheme [2], that allows the specification of E.164 numbers inside a URL (e.g., tel:+39081676518). There are also formats to specify fax and modem numbers. Finally, in ITU-T H.323 standard signalling framework [3], calls are directed to hosts where H.323 applications reside. That is why H.323 bases its naming and addressing scheme on IP names and addresses. However, in order to add flexibility, H.323 includes a complex system of aliases that allows users to make calls using, for example, e-mail addresses, E.164 numbers or even the new H.323 URLs defined in the latest version 4 of H.323. The H.323 URL will allow entities to access users and services in a consistent manner, much like other defined URLs allow for other IP-based services. The form of the H.323 URL is very similar to the SIP URL previously explained: h323:user@host, where "user" is a user or service and "host" might be the Gatekeeper that can translate the URL into a call signalling address. The Gatekeeper, which is the entity that controls H.323 systems in a zone, is in charge of translating the aliases into the network address for the destination terminal, using its internal translation tables or external directories like DNS. Clearly, a common scheme, or a flexible scheme that allows full interoperability between the existing identification and addressing paradigms will be a fundamental prerequisite for basic subscriber identification and further strong and irrefutable authentication in complex and highly interoperable VoIP networks.

III. STANDARD AUTHENTICATION FEATURES IN VOIP SIGNALLING PROTOCOLS

The IP telephony-related protocols (see Fig. 1) were not designed with security as their first priority or as a prime design goal. Some of those protocols added more sophisticated security features, such as strong authentication and no-repudiation only when newer protocol versions were introduced. Other IP telephony protocols introduced some security mechanisms only after the IETF threatened not to accept a newer version of the protocol if security was not part of it. Despite such demands and an effort to introduce "decent" security mechanisms within some IP telephony protocols during their design phase, in some cases inappropriate security concepts were adopted only to satisfy the IETF. Some of those security mechanisms were simply not enough, regarded as useless or impractical and giving a false sense of security to the users of these IP telephony protocols. Consequently all the above protocols contain several design flaws – not easily identified, but costly – that would allow an attacker to cripple an IP telephony network. One flaw, for example, is a signalling protocol that does not maintain knowledge about changes made to the media path during a call. If one is able to abuse the media path, the signalling path will remain unnotified and clueless about the changes performed to the media path. Another example is a signalling protocol that does not have easily applicable, or does not have at all, strong authentication or integrity checking mechanisms. In the following we focus, after a brief explanation of the basic protocol features, on the authentication features provided by each of the most common signalling protocols that may be useful as the basic building blocks of our enhanced VoIP security framework.

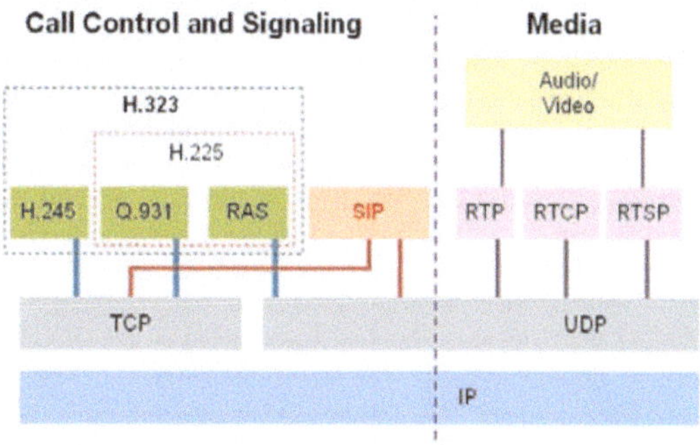

Fig. 1. IP telephony protocol framework

A. H.323

H.323 is an ITU (International Telecommunications Union) standard signalling framework for real-time voice and videoconferencing over packet networks, including LANs, WANs and the internet. It is a very comprehensive standard that supports voice, video, data, application sharing and whiteboarding, the parts relating to audio protocols are widely used for IP telephony applications and has been the first VoIP standard to transport audio and video over IP networks. H.323 consists of a family of sub-protocols used for call set-up, call termination, registration, authentication and other functions and defines the various functional elements of a VoIP system that are endpoints, media gateways, gatekeepers and multipoint control units (MCUs). Endpoints (telephones, softphones, IVRs, voicemail, video cameras. etc.) are the subscriber devices needing access to the communication system. Gateways (gateways and controllers) handle signalling and media transport, and typically serve as the interface to other types of networks such as ISDN, PSTN and or other H.323 systems. Gateways which focus primarily on converting between IP and other forms of media (such as PSTN) are termed *Media Gateways*. Gatekeepers are the logical entity with which endpoints register and are administered. They also manage the most important call-control functions such as call setup, teardown, and status and can assist in address resolution. MCUs are designed to support multi-party conferencing.

Earlier version of the H.323 specification defined a sub

protocol, H.235 version 1 [4], for supporting security services. This protocol provided very basic security guidelines and thus lacked well-defined security procedures, which lead to interoperability headaches between vendors. H.323v4 attacked the security problems encountered in previous architectures by reworking the H.235 sub-protocol. From that work, version 2.0 of the H.235 specification emerged, bringing with it two new security procedures between the endpoints and the gatekeeper, named annex D and annex E. Basically three different approaches can be classified - passwords with symmetric encryption, passwords with hashing and public-key mechanisms:

- Passwords with symmetric encryption: This approach is based on the idea that both communicating entities share a secret - e.g. a password. - and that each endpoint has a unique *generalID* that has been configured before outside from the protocol and that is known to both endpoints. The endpoint that wishes to be authenticated generates a CryptoToken which consists of the sender's generalID, the receivers generalID, a timestamp and a random number (that is monotonically increasing, making messages with the same timestamp unique) encoded with the secret key (derived from the shared password). Encryption can be done by a selection of mechanisms such as DES, 3-DES or any other algorithm that is registered in ISO/IEC 9979. It is also possible that a manufacturer can use other algorithms although this will not be interoperable.
- Passwords with hashing: this is a similar approach, where CryptoHashedToken is computed using the generalID and a timestamp to be passed through a hashing function like HMAC-MD5 or algorithms defined in ISO/IEC 9797.
- Public-key mechanisms: this approach also generates a CryptoToken but uses asymmetric encryption. This enables the use of signature cards and digital certificates that allow users to prove their identity in electronic transactions and can be grated to in a secured way from external system.

Annex E is an enhancement on the annex D security protocol that enables certificate procedures for usable key management to enable key modifications to be handled by the end user with RAS and H.225.0. The basic difference between this security procedure and the H.235/annex D is that digital signatures (certificates) are used instead of the shared secret that annex D defines allowing much advanced scalable and enhanced security solution. The algorithms used to create digital signatures can be either RSA-SHA1 or RSA-MD5. The secret key exchanged can be used for advanced scalable user authentication by a gatekeeper and for media stream confidentiality for the users participating in the session.

B. SIP

SIP (Session Initiation Protocol) is an IP telephony signalling protocol developed by the IETF (Internet Engineering Task Force) for creating, modifying and terminating sessions between two or more participants. Primarily used for VoIP calls, it can also be used for video or any media type (e.g. SIP has been used to set up multi-player Quake games, and also for instant messaging). Due to its simple and fast session setup mechanism, the SIP protocol has quickly made large inroads into the VoIP market previously dominated by implementations adhering to the rather complex H.323 standard. Recently, SIP seems to be going to dominate the market since it has also been selected as the standard for signalling in UMTS Release 5, and most providers that offer VoIP-connections for consumers base their solutions on it. Whereas H.323 is closely modelling a traditional ISDN Layer 3 call set-up and uses ASN.1-coded binary messages for signalling, SIP is based on an HTTP-like request/response transaction model using human-readable ASCII messages, with a syntax nearly identical to HTTP/1.1, and MIME containers which make it suitable and very flexible for integrated voice-data applications. Designed for real-time transmission, it uses fewer resources and is considerably less complex than H.323, and, as explained in the previous section, is based on a very simple and readable addressing scheme that uses URLs. SIP relies on the session description protocol (SDP) for session description and the real-time transport protocol (RTP) for actual transport. Typical SIP architecture includes a registration/location server, which handles signalling associated with multiple calls, receives registration messages from endpoints regarding current user location and maps the SIP addresses with the physical location(s) in the domain where the endpoint is located; a proxy server, which forwards the SIP messages to multiple proxy servers, creating a search tree, in order for the SIP messages to reach their destination; and a redirect server, which helps endpoints to find the desired address by redirecting them to try another server. The former authentication mechanisms provided by SIP are challenge-response based. There exist two simple (and rather weak) alternatives, which are basic authentication and Digest authentication, but since the SIP message structure is a straight derivation from the HTTP request/response model, all security mechanisms available for HTTP [5] can also be applied to SIP sessions to achieve stronger authentication mechanisms. More precisely, the use of MIME containers within SIP messages suggests the potential use of email security mechanisms like PGP [6] or S/MIME [7], and of course similar to a https: URI, a corresponding sips: URI will try to build up a secure transport layer tunnel using TLS [8]. Last but not least IP security (IPsec) [9] can be used as a general purpose mechanism to encrypt all IP based communication right on the network layer. Two of the previously presented methods, namely HTTP basic authentication and PGP have been deprecated by version 2 of the protocol so the S/MIME coupled together with X.509-based digital certificates is considered the approach of choice to secure MIME bodies either on a hop-by-hop basis or end-to-end from user agent to user agent.

IV. SECURE AUTHENTICATION FOR VOIP NETWORKS

According to the above considerations, a satisfactory authentication system for the upcoming VoIP infrastructures, offering the same (or increased) security guarantees than the legacy POTS system without sacrificing the user mobility and enhanced service flexibility requirements, has to be likely based on subscriber credentials, preferably stronger ones, such as digital certificates, rather than on user-chosen passwords. A smart card-based solution similar to today's mobile networks would be very suitable: anything relying on platform security of clients will probably be broken quickly under "Internet circumstances".

A. *The authentication paradigm*

To ensure the authenticity and integrity of the information exchanged between the different entities participating in the VoIP network, each terminal endpoint has to be identified in an irrefutable way on its initial registration to the network and however must digitally sign each significant signalling message it generates. In addition, to allow network intermediate devices to validate the signature, a strong verification against a known key associated to the originating subscriber must be performed by H.323 gatekeepers or SIP registration/location servers on each received message. These signatures can be used to validate each vital signalling information flowing between the subscribers and the signalling control elements in the VoIP system. Clearly, modern standardized digital signature schemas relying on Public Key Cryptography are highly appropriate in this environment. The public key cryptography is the popular synonym for asymmetric cryptography, i.e. cryptography using two matched keys in which a pair of users does not share a single private key. Instead, users have their own key pairs and each key pair consists of a matched private and public key. Moreover, the process of computing one key from another is not possible. Asymmetric cryptography can perform digital signature, thus offering data origin authentication and data integrity, secure transmission or exchange of secret keys and data privacy (encryption and decryption). To accomplish the above functionalities on our protocol messages, each message has to be signed using a private key d, to produce an authenticated message $(m; t)$, where t is the tag on m, and this authenticated message can be verified by the sender's public key e. In relation to digital signature schemes, the tag generated with this method can be seen as a message signature and is known as a *Message Authentication Code (MAC)*. A MAC is used by the receiver to verify that the message originated by the presumed sender and has not been tampered with. Hash functions can be used for providing data integrity in digital signature schemes, where the message is typically hashed first, then the hash-value is signed in place of the original message. Each signing entity only knows its private and public key whereas each entity performing verification need to be able to identify the server and know its public key. Examples of public key cryptography standards are DSS (Digital Signature Standard) and RSA (Rivest, Shamir, and Adleman).

B. *Key trust management and distribution*

The only disadvantage of asymmetric cryptography is the fact that the public keys must somehow be introduced and trusted by all the participating parties. In other words, a trustable binding it is needed between the user's identity and his public key. The best solution is given by the third trusted party (TTP) model. In this model a third party - commonly agreed to be trusted by all interested parties – authenticates users to each other. The X.509 ITU-T Recommendation defines a feasible (and widely adopted) TTP model/framework to provide and support data origin authentication and peer entity authentication services, including formats for public-key certificates, and certificate revocation lists (CRL). The X.509 certificates [10] are cryptographic structures used for binding a user's identity to its public key. This binding is sealed by means of a digital signature performed by the Third party Trust Authority (usually a certification authority or CA) that issues, trusts and digitally signs a digital certificate. Hence, an X509 CA has the roles of issuing, distributing and revoking, whenever necessary, its public-key certificates. Regarding issuance of certificates, the CA can decide to delegate the task of user identification to an optional separate entity, a registration authority (RA), that has the responsibility for recording or verifying some or all of the information needed by a CA to issue certificates and CRLs and to perform other certificate management functions. The distribution of certificates can be achieved in at least three ways: the users can directly share their certificates with each other, or the certificates can be distributed via HTTP protocol or, better, via LDAP [14] directory service. In what concerns revocation of certificates, the CAs have the duty to publish at certain time intervals the CRLs, the black lists on which the revoked certificates are enlisted, together with the date and reason of revocation. In our specific context some form of scalable certificate management structure will be needed to handle mutual authentication on a global scope. For the connection of phone terminals to the network, certificate exchange among different certification authorities must be possible to allow different registration servers to mutually trust themselves in a worldwide scenario. Also, a unified mechanism for certificate revocation has to be in place. Accordingly, to ensure sufficient scalability to the above model through clever engineering of its complex components the concept of public key infrastructure (PKI) needs to be introduced. A PKI is usually defined as a set of cooperating CAs that issue, distribute and revoke certificates according through a hierarchical trusting policy. Typically, a per PKI-authority is established to approve or coordinate certification practice statements, which are security policies under which the components of a PKI operate. This organization reduces the problem of verifying everyone's public key to verifying just one (or a few) public keys. In a typical PKI, an authorized third party issues certificates according to a well defined hierarchal structure. A certificate binds an entity with its public key and is signed with the authorized third party's own private key. The recipient of a certificate use the authorized third party's public key to authenticate it. Consequently, using digital signature for trusting call control information implicitly assumes that there should exists a hierarchy of cooperating Trust Authorities or a PKI to which the entire VoIP infrastructure domain refers. A Trust Authority participating into the above PKI has its own private/public key pair and all the bindings between the referring user terminals and their public keys. Each terminal will be configured with its own pair of private/public key and each registration server needs to know the public key of the Trust Authority, necessary to verify the trustiness of all the information about the public keys of the registering entities obtained from the authority, that will be used for signature verification on each received message. On the other side digital certificates are very useful to bind subscriber identities to key pairs for strong authentication sake, but they do not necessarily assure trustworthiness of the authenticated principals: For instance, a certificate will usually not provide any assurance about the security policy or the protection against malicious intrusion the authenticated principal has in place. In other words, mutual authentication on the entire route between user agents does not guarantee a standard of security functionality at the gatekeeper, proxies or redirect servers level. Furthermore, experience with huge PKIs has shown that large scale deployment is a major undertaking under "real life conditions", and when feasible it has to be done very carefully.

V. IMPLEMENTATION DETAILS

The first and most important security barrier in our model is the user registration phase. User registration involves mapping of E.164 phone numbers or some other human understandable identifiers such as user names, email addresses or generic SIP/H.323 URLs, to network addresses. By allowing users to be reached by location-independent identifiers, User registration provides personal mobility. In this scenario user registration servers are the entities in the network which store user registration information such as SIP registrars and H.323 gatekeepers. For instance, a call destined at sip:bob@mydomain.com reaches user Bob no matter what IP address he might currently be using. In SIP, proxy and redirect servers access a location server, often also a registrar that receives user registration information. A server at mydomain.com will map all the addresses of the form sip:xyz@mydomain.com to the appropriate IP addresses depending on where xyz is currently logged in. In H.323, the same functionality is performed by the H.323 gatekeeper hierarchy. Since initial user registration is propedeutical to al their future operations on the network we propose the introduction of the first and most important authentication check at this level. More precisely, the accepting registration server will strongly verify, against a trusted CA the identity of the terminal endpoint requiring access to the network and accomplish the registration only if the verification is successful. The verification will be based on a digital signature scheme relying on Public Key Cryptography whose fundamental idea is to add digital signature, that will be transported through the standard security features introduced in the previous session III, to each protocol message, and thus use an hash function such as message digest (like keyed MD5) to protect all exchanged message data. The originator will sign each message before sending it and the signature will stay within the message body during all the message life. This wil protect the message integrity and provide authentication for all the transported data, ensuring that the data really does come from their legitimate source and has not been modified in transit. Obviously, the modification of any of the fields belonging to the signed protocol messages, including the signature information itself will result in a failure in signature verification, that implies, on intermediate devices supporting this enhanced security feature, a message discard. Furthermore, in the case where incorrect data is distributed by a faulty device, the signature facility provides a way to trace the problem to its source. No proprietary extensions are needed to the standard signalling protocols and devices supporting this digital signature facility behave, except for the signature verification activity, when needed, as normal VoIP devices for all the protocol functionalities and so they can be straightforwardly mixed, ensuring backward compatibility, with other devices not implementing this feature. In these cases, all signature information received will be silently ignored without affecting the overall protocol behaviour. Let us examine the detailed operating scenario.

A. The certificate schema

For key distribution and management sake, there should be a consistent and unique way of mapping an SIP or H.323 subscriber identifiers to their X.509 certificates in the contest of a complex PKI infrastructure. By standardizing the schema used to represent the user certificate, different protocol can be deployed together to create an overall application environment. SIP URLs in the form *user@host* will make our life almost easy: we generate the distinguished name (DN) by simply copying the SIP *user* verbatim into the X.509 common name (CN) and the *host* domain components into multiple DC identifiers. In this case we will use SIPidentity as the organization unit (OU) value. Representing an H.323 address into a X.509 certificate may be more difficult since multiple representations (e.g., E.164) need to be handled. In the easiest case, the alias contains an H.323 URL in the form user@host than can be handled as above, with the only difference of using an OU with the value "H323identity". In the other cases the E.164 number can be used directly as the DN value.

B. H.323 authentication scenario

The proposed authentication paradigm in H.323 call set-up scenario works as follows. Suppose the H.323 terminal A wants to talk to terminal B. A will be registered with its own gatekeeper in the H.323 network and the gatekeeper who knows its IP address, conveyed via the RRQ registration request message will respond with a success confirmation message (RCF) only if it successfully verifies the signature contained in the H.323 RRQ, otherwise it sends a reject (RRJ) message. Neither the confirmation or reject messages nor any backward notification from the network intermediate device to the terminals will be signed, and hence verified, to avoid the necessity of storing, at expense of memory, many digital certificate on the terminal, aiming to keep them as simple and light, in terms of resource needs, as possible. Moreover, we followed a network-centric approach so our choice is to protect the network from its users and not vice-versa, consequently all the network intermediate devices participating to the communication infrastructure are assumed to be trusted. The problem of a malicious device, masquerading as an intermediate signalling server and spoofing its responses has the only effect of breaking the setup of some regular calls or generating some fictitious or partial/incomplete call attempts. This will be essentially a Denial of Service problem that is not an issue to be addressed in our work. The initial verification implies an LDAP request from the gatekeeper to the referred CA to obtain the other party public key certificate needed for all the following signature checks. After successful registration each further signalling message flowing from the terminals to the gatekeeper will be verified against the subscriber real identity, this ensures that no signalling messages an be spoofed from malicious users to subvert the network operating or billing behaviour. Clearly A and B can be registered on different gatekeepers, referring to different cooperating PKIs, and participating in a complex H.323 hierarchy (the scenario in fig. 2 has been simplified). In this case the gatekeeper does not have the address mapping for B and multicasts the location request (LRQ) for B to the other gatekeepers participating in the VoIP network. Anyway, the authentication paradigm will be essentially the same, since, from our initial assumption about the trustiness of the internal network, the whole authentication process is however restricted from the users to their nearest gatekeepers and its global consistence is guaranteed from the cooperating PKIs, so no other interaction is needed. From now on, all the following call set-up activity will proceed according to the usual flow depicted in fig. 2, with the A party signing its admission request (ARQ) message and the B party signing the last confirmation (ACF) message, both verified from the associated gatekeepers. The signatures in H.323 messages are stored according to the H.235 annex E facility. In detail each signature transported in a request message is computed over the specified fields identified by a *ClearToken* sub-field with a value of *toBeSigned* in the *cryptoSignedToken* field. The *tokenOID* within the *cryptoSignedToken* is set to "B" indicating that only the specified subset fields in the *ClearToken* are signed. The token in *cryptoSignedToken* has *algorithmOID* set to "V" indicating use of MD5-RSA or "W" indicating use of the SHA1-RSA signature algorithm and the *params* field set to NULL. The signature will be stored in the token sub-field of the *cryptoSignedToken* field.

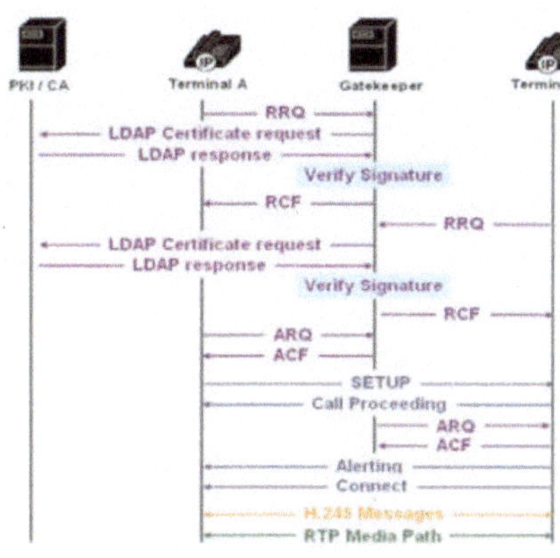

Fig. 2. H.323 authentication scenario

C. SIP authentication scenario

Authentication in the SIP scenario works in essentially the same way. The initial signed REGISTER message, originated from each terminal endpoint and passing through a SIP proxy server reaches a location/registrar server, where it has to be verified against the subscriber public key obtained via LDAP from a trusted CA. Only if the verification succeeds the registration is completed and a success notification message (200 OK) is sent, without signature, back to the user, passing through the corresponding SIP proxy server. All the remaining signalling activities in the call set-up scenario will flow in the usual way, as depicted in fig. 3, with all the messages originated from the terminals through the network that will be digitally signed, passed through the proxy server and verified in the location/registrar server. In more detail, the initial INVITE, the backward ringing (180) notification, the setup confirmation (200 OK) and its acknowledgement (ACK) will be signed from their senders, to prove their identities, and verified at the location server level. Also the call teardown scenario will behave in the same way, with the release (BYE) and acknowledgement (200 OK) messages signed by their originators and checked on the location/registrar servers. Note that to allow the complex and flexible relocation features handled with the interactions between the locators and the proxy servers all the authentication verification activity will be done at the location server level.

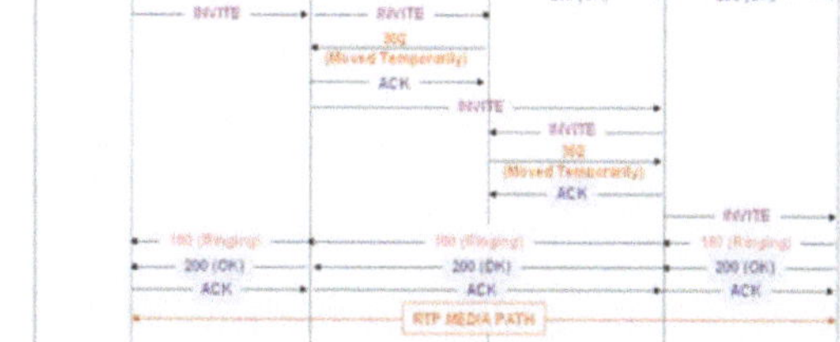

Fig. 3. SIP authentication scenario

Due to the inherent simplicity of the SIP protocol, the transport of signatures in signalling messages, handled through the S/MIME support feature is straightforward. Any message can be signed using the multipart/signed method [10]. The signature plus optionally the X.509 certificate of the signer is contained in the binary *application/pkcs7-signature* structure which is attached after the MIME object to be signed. As an alternative a binary PKCS#7 *signedData* structure [9] could be used which transports both the data to be signed and the signature within a single attachment. Furthermore, the SDP MIME attachment embedded in a SIP

INVITE request can be also encrypted (and of course signed) using S/MIME. The *application/pkcs7-mime* binary *envelopedData* structure can encapsulate the symmetrically encrypted SDP payload and also contain the symmetric key which is encrypted with the public key of the recipient, as it can be seen in the following figure (fig. 4).

```
INVITE sip:bob@unina.it SIP/2.0
Via:SIP/2.0/UDP 143.225.26.139:5060;branch=z9hG4bK4129d28b8904
To: Bob <sip:bob@unina.it>
From: Alice <sip:alice@unina.it>;tag=dwa21162
Call-ID: 392c3f2b568e92a8eb37d448886edd1a@143.225.26.139
CSeq: 1 INVITE
Max-Forwards: 70
Contact: <sip:alice@kalimero.unina.it:5060>
Content-Type: multipart/signed;boundary=992d915fef419824;
micalg=sha1;protocol=application/pkcs7-signature
Content-Length: 3088
--992d915fef419824
Content-Type: application/pkcs7-mime;
smime-type=envelopeddata; name=smime.p7m
Content-Disposition: attachment;handling=required;filename=smime.p7m
Content-Transfer-Encoding: binary
<! envelopedData object encapsulating encrypted SDP attachment not shown !>
--992d915fef419824
Content-Type: application/pkcs7-signature;name=smime.p7s
Content-Disposition: attachment;handling=required;filename=smime.p7s
Content-Transfer-Encoding: binary
<! signedData object containing signature not shown !>
--992d915fef419824--
```

Fig. 4. SIP encryption and signature example

VI. CONCLUSIONS

As VoIP is based on normal IP networks, VoIP applications implicitly inherit the known and unknown security weaknesses that are associated with the IP world. The resulting security problem domain is considerably larger compared to that of a traditional Public Switched Telephone Network. Therefore, many essential security requirements for providing voice services in public networks cannot be easily met in today's VoIP network architectures. Particularly, the very basic problem of strong user authentication is insufficiently addressed. The reason for this is, simplified, that VoIP provides user/terminal mobility, a "feature" which lacks "compensation" in terms of authentication technology. Evolving VoIP based voice communication infrastructures to a solution which comes with security characteristics similar to e.g. today's mobile telephone networks is a significant undertaking, which did not even start yet; the technical differences are significant: signalling, just to mention one example, is done inbound in an open network for VoIP, whereas a separate signalling channel is used in POTS/GSM for decades due to good security reasons. Technically seen, only an enhanced security architecture based on some proven security technologies (like asymmetric cryptography and digital signatures) seems a realistic option for achieving a security level that would bring VoIP close to that of the traditional telephone system. Accordingly we focused our work on the introduction of X.509 digital certificates, within an organization of co-operating public key infrastructures, to enforce subscribers' identity evidence and no-repudiation throughout a complex large-scale IP telephony system. However, other aspects, like availability, would still be missing and seem to be even harder to provide. Overall, we believe that since the various aspects of security in VoIP are little understood today and the implications to businesses, end users, and governments can be (and probably will be) significant, further work on the issues we touched in our paper is clearly needed.

REFERENCES

[1] Rosenberg, Schulzrinne et al., "SIP: session initiation protocol", IETF RFC 3261, 2002.

[2] A. Vaha-Sipila. "URLs for Telephone Calls". IETF RFC 2806, 2000.

[3] ITU-T, "ITU-T Recommendation H.323, Packet based multimedia communication systems, 2003

[4] ITU-T, "ITU-T Recommendation H.235, Security and encryption for H-Series (H.323 and other H.245-based) multimedia terminals", 1998

[5] J. Franks et. al., "HTTP authentication: Basic and Digest Access Authentication", IETF RFC 2617, 1999.

[6] M. Elkins, "MIME Security with Pretty Good Privacy (PGP)", IETF RFC 2015, 1996

[7] B. Ramsdell, "S/MIME Version 3 Message Specification", IETF RFC 2633, 1999.

[8] T. Dierks, C. Allen, "The TLS Protocol Version 1.0", IETF RFC 2246, 1999.

[9] B. Kaliski, "PKCS #7: Cryptographic Message Syntax Version 1.5", IETF RFC 2315, 1998.

[10] R. Housley, W. Ford, W. Polk, D. Solo, "Internet X.509 Public Key Infrastructure Certificate and CRL Profile", IETF RFC 2459, 1999.

[11] M. Wahl, T. Howes, S. Kille, "Lightweight Directory Access Protocol (v3)", IETF RFC 2251, 1997

[12] J. Galvin et al., "Security Multiparts for MIME: Multipart/Signed and Multipart/Encrypted", IETF RFC 1847, 1995.

Adoption of Hot Spot Game Playing

Anita Garling
Division of Engineering Psychology
Lulea University of Technology, Sweden

***Abstract*-The aim of this study was to assess the acceptability of a new mobile service; namely hot spot game playing. The more specific objectives were to reveal current mobile phone and new mobile phone service usage, and to relate this to domain-specific innovativeness as well as to willingness to pay. Moreover, to study how a service like this will affect current general consumer behavior. Subjects were female and male volunteers living in the city of Lulea in the northern part of Sweden recruited from the Website www.testplats.com. The results show that the subjects had good knowledge about both mobile telephony and game playing which imply that the knowledge barrier should not be too high to overcome and, hence, the actual adoption of the new service should be promising. However, due to the somewhat immature technologies used, this was not realized in this study. Though, the study in itself seemed to have had an impact on shopping behavior and gave valuable insights to further elaborate upon regarding acceptability of new mobile phone services.**

I. INTRODUCTION

Identifying adopter categories and studying innovators, the first adopters of an innovation, is considered important for marketing purposes [1] and [2], respectively. Understanding and identifying innovators is also important in marketing new products, or services. Moreover, future commercial success of new products, or services, heavily depends on whether these are accepted, and thereby used, by intended users [3]. For a product, or service, to be accepted, achieving critical mass with a community of users becomes critical. If critical mass is achieved the product, or service, is likely to be, more or less, universally adopted and if not, the technology will most probably be abandoned [4]. Besides the distribution of adoption thresholds and the actions of early adopters, adopter expectations that a technology will eventually be widely accepted can become a self-fulfilling prophecy [5]. Another barrier to adoption is the knowledge barrier [6]. Some technologies can just not be bought in a poke, but rather, have to be tried out in their intended context. In line with this, the present study was conducted to assess the acceptability of a new mobile service; namely hot spot game playing. The more specific objectives were to reveal current mobile phone and new mobile phone service usage, and to relate this to domain-specific innovativeness as well as to willingness to pay. Moreover, to study how a service like this will affect current general consumer behavior.

II. METHOD

A. Subjects

The subjects were female and male volunteers living in the city of Luleå in the northern part of Sweden recruited from the Website www.testplats.com. In a first round 44 volunteers were targeted and in a second 168. In the former, 5 females and 15 males accepted to participate in the study and in the latter 12 females and 41 males. Hence, altogether 17 females aged 13 to 41 years (M=26.2 years) and 56 males aged 11 to 44 years (M=24.7 years) completed the first two parts of the questionnaire. However, during the subsequent game trial set-up 16 (2 females and 14 males) of the subjects decided to leave the study. Of the 57 subjects remaining, only 29 were identified by the game server and, hence, only the results from these questionnaires were analyzed under the heading "Questionnaire 3: Experiences from the Trial Set-Up". More of the subjects in the Lifestylers/Ericsson condition than in the BotFighters/Nokia condition left the study/were not identified by the game server, $F(1, 72)=7.9$, $p<.01$, and more due to the non-working technology than to other reasons, $F(1, 71)=4.4$, $p<.05$. Moreover, those who left the study were less interested than those who stayed in new functions in the area of mobile telephony, $F(1, 72)=5.1$, $p<.05$, and they were also less interested to pay more than SEK 10 per minute for game playing, $F(1, 70)=5.8$, $p<.05$.

B. Questionnaires

Three questionnaires were prepared and administered. In the first, the subjects were asked about sociodemographics such as age, gender, mobile phone ownership and use, mobile phone subscription type, and game playing habits as well as shopping mall visiting habits (e.g. how often and why they visited a specific shopping mall located in downtown of Lulea (i.e. the same as where the trial set-up was conducted)). The subjects were also asked to state their mobile phone number and their e-mail address. In the second, administered right before the game trial set-up, a further developed version of an innovativeness scale [7] was used. Questions about willingness to pay a game subscription fee as well as for each play session were also posed. In the third questionnaire, administered right after the game trial set-up, questions about usage of the specific mobile phone and experiences of the game trial set-up were posed.

K. Elleithy et al. (eds.), Advances in Computer, Information, and Systems Sciences, and Engineering, 297–302.

C. Procedure

The Website www.testplats.com is open for private persons and there are no specific requirements to join this test bed community. Questionnaires with different research aims are administered weekly on the website. All who have joined the test bed community are given the possibility to participate in the research studies presented at the website. Participators are compensated through an incentive system. A call for subjects to the present study was displayed on the above mentioned Website for 10 weeks (10/11/03 to 12/20/03[1]) and, then, withdrawn. The only requirement to join this particular study was to fill out the first part of the questionnaire, which was displayed together with the call on the mentioned website. From the filled out questionnaires a random sample of subjects was chosen for the first round[2]. The subjects were then targeted through a SMS asking them to meet with the research staff at a local shopping mall at a given day and time. At this point, the subjects who actually turned up signed up for the leasing of a mobile phone (Ericsson P800) to be used for playing a hot spot game named Lifestylers[3]. The leasing contract stated the subjects' responsibility for the mobile phone. Moreover, they also filled out the second part of the questionnaire and were instructed to actively participate in the coming hot spot game trial set-up. After the game trial set-up the subjects were asked to answer the third and last part, of the questionnaire. The first round lasted for 8 weeks (11/11/03-6/1/04). A second round was also run in which the same procedure as in the first was utilized. However, in this second round another brand of mobile phone (Nokia 3650) was used as well as another hot spot game (BotFighters[4]). The second round lasted for 10 weeks (4/12/03-11/2/04).

III. RESULTS

A. Questionnaire 1: Mobile Phone and Computer Usage

Altogether 17 females and 56 males completed the first questionnaire. Overall, the subjects' ages ranged from 11 to 44 years (Table 1). The subjects in the Lifestylers/Ericsson condition were on average somewhat older than subjects in the BotFighters/Nokia condition (25.9 and 24.3 years, respectively). Overall, 23.5 % of the subjects were females and the percentage of females was somewhat higher in the Lifestylers/Ericsson condition than in the BotFigthers/Nokia (26.5 and 20.5 %, respectively).

All subjects owned a mobile phone. The Nokia brand was the most common brand among the subjects and Ericsson the next most (54.2 and 37.5 %, respectively). A somewhat higher percentage of females and those in the BotFighters/Nokia condition than of males and those in the Lifestylers/Ericsson condition owned a mobile phone made by Nokia (62.5 and 56.4 % to be compared with 51.5 and 51.5 %). Overall, there was almost no difference between subscription types, but the males and subjects in the BotFighters/Nokia condition subscribed to a somewhat higher percentage through rental cards than females and subjects in the Lifestylers/Ericsson (51.8 and 53.8 % to be compared with 47.1 and 47.1 %).

All of the subjects used their mobile phones to make and receive voice calls. About 94 % of the subjects used their mobile phone to send and receive SMS, about 55 % to play games, about 25 % to WAP, about 5 % to chat, and about 10 % to send/receive e-mail. No females stated that they used the mobile phone to chat or to send/receive e-mail while the corresponding percentages for males were 7.1 % and 12.5 %, respectively. About 12 % of the females stated that they used the mobile phone to WAP while the corresponding percentage for males was about 30 %. Moreover, females used the mobile phone less to play games with than males (35.3 % and 62.5 %, respectively). There was no difference between conditions regarding use of mobile phones to CHAT. Subjects in the Lifestylers/Ericsson condition used their mobile phones somewhat more than subjects in the BotFighters/Nokia to send/receive SMS while the opposite was true for wapping

Table 1. Mobile phone and computer usage.

	Lifestylers/ Ericsson P800 (n=34)	*BotFighters/ Nokia 3650 (n=39)*
Age (M, range)	25.9 (11-44)	24.3 (18-42)
Females (%)	26.5	20.5
Mobile phone ownership (%)	100.0	100.0
Fixed mobile phone rental (%)	52.9	46.2
Mobile phone usage (%)		
Make/Receive voice calls	100.0	100.0
Send/Receive SMS	94.1	94.9
Chat	5.1	5.1
Send/Receive e-mail	11.8	7.7
WAP	23.5	28.2
Game playing	52.9	59.0
Mobile phone game playing, at least once a week (%)	55.9	56.4
Computer game playing, at least once a week (%)	58.8	66.7
Game playing together with others (%)	35.3	48.7
Shopping mall visits, at least once a week (%)	67.6	51.3
Shopping mall visits to meet with others (%)	14.7	7.7

[1] During this period of time the subjects were able to withdraw their filled out questionnaires any time they themselves chose.
[2] Questionnaires filled out by subjects not living in the area of Lulea and subjects associated with the research institute responsible for the study were excluded from the list of subjects.
[3] Information about the game can be found at www.aspiro.com
[4] Information about the game can be found at www.BotFighters.com

and playing games 11.8, 23.5 and 52.9 % to be compared with 7.7, 28.2 and 59.0 %). Overall, somewhat more than half of the subjects stated that they played mobile phone games at least once a week while the corresponding percentage for playing computer games was somewhat higher. A higher percentage of males than of females stated that they played mobile phone games as well as computer games at least once a week (62.7 and 69.9 % to be compared with 41.2 and 53.6 %) and subjects in the BotFighters/Nokia condition stated to a higher percentage than subjects in the Lifestylers/Ericsson condition that they played mobile phone games as well as computer games at least once a week (56.4 % and 66.7 % to be compared with 55.9 % and 58,8 %).

When the subjects played games about 40 % played together with others. Females did so more often than males (50.0 % and 35.3 %, respectively) and this difference between the genders was also significant, $F(1, 72)=5.9$, $p<.05$. The subjects in the Lifestylers/Ericsson condition played less often than the subjects in the BotFighters/Nokia condition together with others (35.3 % and 48.7 %, respectively). If females played with others they most often played with only 1 other person while males, if they played with others, they most often played with 2 to 4 other persons. Regardless of condition the subjects, if they played with others, they most often played with 2 to 4 other persons.

Finally, Table 1 shows that about 60 % of the subjects visited the targeted shopping mall at least once a week. A somewhat higher percentage of males and subjects in the Lifestylers/Ericsson condition than of the females and subjects in the Botfighters/Nokia condition visited the targeted shopping mall at least once a week (60.7 % and 67.6 % to be compared with 52.9 % and 51.3 %). Only about 12 % of the subjects stated that they visited the targeted shopping mall to meet with others and males and those in the Lifestylers/Ericsson condition so to a higher percentage than females and those in the other condition (14.5 % and 14.7 % to be compared with 7.9 % and 7.7 %). Somewhat more than 40 % of the subjects stated that they thought that they were suitable as subjects in this study because they were genuinely interested in new innovations/technologies, about 15 % because they liked to play all kinds of game, and about 10 % because they thought it seemed exciting and funny or because they considered themselves as competent players. There was no difference between genders with respect to genuinely interested in new innovations/technologies but males considered themselves to a somewhat higher percentage than females as competent players (9.4 and 6.3 %, respectively) while the opposite was true for liking to play all kinds of games (18.8 % and 11.3 %, respectively). Subjects in the BotFighters/Nokia condition stated to a higher percentage than subjects in the Lifestylers/Ericsson condition that they would be a most suitable subject because of their genuine interest in new innovations/technologies (51.4 % and 34.4 %, respectively) while the opposite was true for liking to play all kinds of games, exciting and funny as well as competent players. Other stated reasons for being a suitable subject were being a good evaluator, having plenty of leisure time, and being smart, handsome, and nice.

B. Questionnaire 2: Innovativeness and Expected Use of Different Mobile Phone Functions

The same subjects who answered the first questionnaire also answered the second. At this point they also met with a person from the research team to fill out the leasing contract for the mobile phone. The second questionnaire contained several statements about innovativeness and of mobile phone usage as well as about willingness to pay for accessibility to new innovations/technologies and expectations of the coming game trial set-up.

Table 2 shows that, overall, the subjects were possible to characterize as belonging to the so-called first adopters of new innovations/technologies. Except from the statements *"Among my friends I'm one of the first to try out mobile phone with new functions"* and *"I do always look for new functions in the area of mobile telephony"* the subjects, more or less, hit the upper end of the scale. A significant difference was found between genders for the statement *"In the future I'll, in my work, use more of the functions available in the mobile phone"*, $F(1, 72)=4.2$, $p<.05$. Males stated to a higher degree than females that they would, in their work situation, use more of the functions available. No significant differences were found between conditions.

It was quite important, overall, for the subjects to keep up to date with the latest general news and to have

Table 2. Innovativeness (Scale from 1(low) to 7 (high)).

	Lifestylers/ Ericsson P800 (N=34)	*BotFighters/ Nokia 3650 (n=39)*
Among my friends I'm one of the first to try out mobile phone with new functions (M)	5.5	5.3
I do like to try out new technologies (M)	6.7	6.7
I do always look for new functions in the area of mobile telephony (M)	5.5	5.6
It is easy for me to get the mobile phone to work as I want it to work (M)	5.9	6.1
There are more usable functions in a mobile phone than making/receiving voice calls and sending/receiving SMS (M)	6.1	6.1
In the future I'll, as a private person, use more of the functions available in the mobile phone (M)	6.1	6.0
In the future I'll, in my work, use more of the functions available in the mobile phone (M)	6.0	5.8

access to the Internet as well as to their favorite music regardless of own geographical location (Table 3). On the other hand, it was not perceived as that important to keep up to date with the latest general news within the area of medicine, to see the person they spoke with on the phone, or to have the same make and brand on various products as their friends. No significant effects were found between genders and conditions.

Table 3. Expected usage of different mobile phone functions (Scale from 1(low) to 7(high)).

	Lifestylers/ Ericsson P800 (n=34)	*BotFighters/ Nokia 3650 (n=39)*
It is important for me to keep up with date to the latest general news (M)	6.0	5.7
It is important for me to keep up to date with the latest news within the area of medicine (M)	4.6	4.3
It is important for me to have access to the Internet regardless of my location (M)	5.8	5.8
It is important for me to have access to my favorite music regardless of my location (M)	5.4	5.4
It is important for me to see the person I speak to on the phone (M)	3.5	3.5
It is important for me to have the same make and brand as my friends on various products (M)	4.8	4.4

More than half of the subjects stated that 100 SEK (about 14 US$, September 2005), or less, was a suitable cost for mobile phone game playing monthly subscription and that 10 SEK (about 1.4 US$, September 2005), or less, was a suitable cost per minute for mobile phone game playing. A higher percentage of males and subjects in the BotFighters/Nokia condition than of females and subjects in the Lifestylers/Ericsson condition stated that 100 SEK, or less, was a suitable cost for a fixed monthly subscription (64.3 % and 60.6 % to be compared with 41.2 % and 44.1 %). Regarding cost per minute for mobile phone game playing females and subjects in the BotFighters/Nokia condition to a higher percentage than males and subjects in the Lifestylers/Ericsson condition stated that SEK 10, or less, was a suitable cost (76.5 % and 65.8 % to be compared with 59.3 % and 60.6 %). Finally, overall, most of the subjects expected to, through their participation in the study, learn something new, get a glimpse of future mobile telephony, to enhance usability of their own mobile phone, or just to have an extra-ordinary experience. A somewhat higher percentage of males than females expected to learn something new (41.7 % and 38.5 %, respectively) while the opposite was true for the other stated expectations.

C. Questionnaire 3: Experiences from the Trial Set-Up

At the same as the subjects handed in the mobile phones the third and last questionnaire was administered. In this questions about actual usage of the mobile phone and shopping behavior as well as about their opinions of the game they had played were posed. According to loggings from the game server only about 40 % of the subjects that answered the first and the second questionnaires actually played the games. The subjects that participated in the Lifestylers/Ericsson condition were identified on average 92 times (23-267) while the corresponding average for the subjects in the BotFighters/Nokia condition was 165 (1-2191). Of these, 1 female aged 24 years and 6 males in the ages 20 to 39 years (M=27.3) participated in the Lifestylers/Ericsson condition and 5 females in the ages 20 to 34 years (M=26.2) and 17 males in the ages 20 to 42 years (M=23.7) in the BotFighters/Nokia condition.

As Table 4 shows the leased mobile phone was mainly used by the subject her/himself. If the mobile phones had been lent to somebody else it was most probably to a daughter or a son and so more often by males and subjects in the BotFighters/Nokia condition than by females and subjects in the Lifestylers/Ericsson condition. Moreover, the leased mobile phone was most often used for making/receiving voice calls and only in about 12 % of the time for playing games. Almost no difference was found with respect to genders or conditions.

All subjects stated that they had visited the targeted shopping mall to play during the trial set-up period but only about 20 % at least once a week. Females and subjects in the BotFighters/Nokia condition stated to a higher percentage than males and subjects in the Lifestylers/Ericsson condition that they had visited the targeted shopping mall at least once a week (33.3 % and 28.6 % to be compared with 21.7 % and 12.5 %). The most often stated reason for not visiting the targeted shopping mall more often was that they at their first

Table 4. Actual usage of mobile phone and changed shopping behavior.

	Lifestylers/ Ericsson P800 (n=7)	*BotFighters/ Nokia 3650 (n=22)*
Mobile phone usage by subject (%)	99.7	94.1
Game playing (% of usage)	11.6	13.1
Shopping mall visits for game playing at least once a week (%)	12.5	28.6
At least 1 h at game playing at a time (%)	100.0	90.5
Played with unknown (%)	85.7	44.4
Mobile phone purchase willingness (Yes and Yes, maybe) (%)	28.6	22.7
Changed shopping behaviour at targeted shopping mall (More often) (%)	25.0	23.8

visit(s) were met with a technical breakdown. Other reasons were lack of time and not in the area during the game trial set-up. About 95 % of the subjects spent, at least, 1 h playing the game at a time and females as well as subjects in the Lifestylers/Ericsson condition to a somewhat higher percentage so than males and subjects in the other condition. More than half of the subjects stated that they played together with earlier unknown individuals and males and subjects in the Lifestylers/Ericsson condition more so than females and subjects in the BotFighters/Nokia condition (60.0 % and 85.7 % to be compared with 40.0 % and 44.4 %). Only one male subject who participated in the BotFighters/Nokia condition believed that he would keep up the relationship with the earlier unknown individual after the study was completed.

The subjects were, furthermore, asked about how willing they were to purchase the leased mobile phones, namely the Ericsson P800 and the Nokia 3650. The alternative answers to this question were "Yes", "Yes, maybe", and "No". The subjects were also asked to motivate their answers. Two subjects who had borrowed an Ericsson P800 and 5 who had borrowed a Nokia 3650 stated "Yes or Yes, maybe" (about 25 % of the subjects). The main reason for this was, regardless of brand, mainly the camera application. Consequently, most of the subjects stated that they were not at all interested in purchasing a mobile phone of the same brand and make as the mobile phones they had leased. The main reasons stated for this lack of interest were, overall, the immature technology of the leased mobile phones (lost voice calls and SMS, hang-ups, bad graphic, short battery life, etc.) and their low level of usability (exterior design, size, button placement and design, interface design, etc.). Finally, it was obvious that the study in itself had an impact on the subjects' shopping behavior. Overall, their shopping at the targeted shopping mall increased with about 25 %. Responsible for this result was mainly males participating in the Lifestylers/Ericsson condition.

D. Experiences of Lifestylers and BotFighters

According to the makers Lifestylers is a relation-based community game where the player takes on the role of a fictive game character shaped in personality by the player while BotFighters is a location-based mobile game where the player shapes a robot and battle against other players' robots. The games provide opportunities to communicate with the real person behind the characters or the robots. The target group for Lifestylers is 14 to 25 year and for BotFighters 20 to 25 (in both cases for both females and males).

From the stated experiences of the games an overall conclusion was that a majority of the subjects did not believe that it was easy to understand how to play the game and, hence, only less than 3 % of the subjects were actually able to exactly shape desired characters/robots (Appendix 1, Table 1). Actually, none of the females succeeded in shaping a desired character and none of them perceived shaping a character/robot as an easy task. About 40 % of the subjects, mainly males, who participated in the Lifestylers/Ericsson condition (the questions about characteristics were not applicable to the BotFighters/Nokia condition) wanted to have had more characteristics to handle and most often more interest characteristics. About 15 % of the subjects, mainly females, perceived the graphics as fantastic. Two males and 1 female stated an interest in continued playing after the completion of the study and males were more interested in frequent (e.g. more than once a week) mobile phone game playing. A competition involved in the game playing attracted only males. About 50 % of the females and about 45 % of the males had the belief that others usually are involved in this kind of game playing, even though, they themselves did not exactly feel as a part of a larger game community when playing. One male subject in the Lifestylers/Ericsson condition stated that he sometime flirted when playing and one female subject stated that she did CHAT. About 30 % of the subjects were more amused by playing with others and about 50 % wanted to get their friends involved in the game playing. Finally, overall the subjects in first-hand preferred shopping malls and in second-hand beaches as hot spot game zone locations.

III. DISCUSSION

The present study was conducted to assess the acceptability of a new mobile service; namely hot spot game playing. In doing this, the subjects' current usage of mobile phones was analyzed. The results show that all subjects owned a mobile phone which most often was used for making/receiving voice calls and SMS, and next most for playing built-in games. Furthermore, that the subjects were frequent players of both mobile phone and computer games. Hence, they could be characterized as having quite good knowledge in the area of both mobile telephony and game playing which might imply that the knowledge barrier should not be too high to overcome and, accordingly, the actual adoption of the new service should be promising.

Unfortunately, the trial set-up did not work out properly which resulted in many drop-outs, about 60 % of the subjects left the study and did so mainly due to the immature technologies used for the set-up. The lesson learned from this is that it is of vital importance to in beforehand make sure that the technologies that will be used really works, otherwise this will most certainly affect the results of the set-up negatively. The subjects who completed the study were, maybe because of this, not too flattering in their opinions about the set-up. They did not use the lent mobile phone that much for hot spot game playing and very few were interested in buying the mobile phone in question. Moreover, the experiences of the used hot spot games were not too positive, though, it must be kept in mind that these games were, according to the makers, intended for somewhat younger players than those who participated in this study. Moreover, very few of the subjects were able to create the characters/robots they wanted, they were not at all satisfied with the graphic of the used game, and very few were interested in continuing playing after the study was completed. However, the belief was, though, that many others do play hot spot games which in the longer run might produce an illusion of an existing critical mass. This might, in turn, according to the power of the self-fulfilling prophecy facilitate the adoption of the service of hot spot gaming.

Although the subjects initially were quite frequent visitors at the targeted shopping mall the study in itself seemed to have had, from the storekeepers' point of view, a positive effect in that the visiting frequency, overall, increased. Furthermore, according to the subjects' choices of preferred location for a hot spot game zone the choice of a shopping mall seemed just right.

Given that so many subjects left the study in beforehand and that there was almost no variation regarding domain-specific innovativeness, or willingness to pay for hot spot game playing, relating hot spot gaming experiences to these variables did not turn out that fruitful. One reason for this seems quite obvious taken into account that the subjects participating in this study were recruited from a Website designed as a meeting place for those interested in trying out new mobile and stationary services and products such as phones, computers, and computer games. Of course, this will, mainly, attract individuals already biased positively towards the topics in question. Another reason was, of course, that those who left the study in beforehand were those less interested in this area than those who stayed. Hence, future studies should avoid both using immature technologies and biased samples in order to really get to know what intended users' desire and demand. The presented study gave, however, some valuable insights to further elaborate upon within this area of research.

ACKNOWLEDGEMENT

Without the help and support from Vinnova, Centre for Distance-Spanning Technology, SonyEricsson, Nokia Sweden, Tadlys, Aspiro, Botfighters, and Konsum Smedjan, Lulea this study would never have been possible to accomplish.

REFERENCES

[1] Rogers, E. M. *Diffusion of innovations*, 3rd Ed. NY: Free Press, 1995.

[2] Flynn, L. R., & Goldsmith, R. E. Identifying innovators in consumer service markets. *The Service Industries Journal, 13*, 97-106, 1993.

[3] Fichman, R. G. *Information technology diffusion: A review of empirical research*. Technical Report, MIT Sloan School of Management, Cambridge, MA, 1992.

[4] Markus, M. L. Toward a "critical mass" theory of interactive media: Universal access, interdependence and diffusion. *Communication Research, 14*, 491-511, 1987.

[5] Farell, J. & Saloner, G. Competition, compatibility and standards (pp. 940-955). In H. L. Gabel (Ed.), *Product standardization and competitive strategy*. Netherlands: Elsivier, 1987.

[6] Cohen, W. M. & Levinthal, D. A. Absorptive capacity: A new perspective on learning and innovation. *Administrative Science Quarterly, 35*, 128-152, 1990.

[7] Goldsmith, R. E. & Hofacker, C. F. Measuring consumer innovativeness. *American Journal of Sociology, 83*, 1420-1443, 1991.

APPENDIX 1

Table 1. Experiences of Lifestylers and BotFighters.

	Lifestylers/ Ericsson P800 (n=7)	*BotFighters/ Nokia 3650 (n=22)*
I was able to exactly shape the character/robot (%)	0.0	4.8
It was easy to shape the character/robot (%)	37.5	23.8
It should have been better with more characteristics to choose between (%)	62.5	15.0
First choice of characteristics:		
Age (%)	16.7	
Optional (%)	16.7	
Nationality (%)	16.7	
Social status (%)	0.0	
Interests (%)	50.0	
It was easy to understand the game (%)	37.5	47.4
I think the graphic was fantastic (%)	25.0	5.3
I would like to continue to play the game (%)	12.5	10.0
I would like to play at least once a week (%)	37.5	35.0
I would play more often if there was a competition involved (%)	25.0	30.0
I do believe there are many who play this game by using a mobile phone (%)	62.5	60.0
When I play I do feel I am a part of a game community (%)	50.0	31.6
I did flirt with others when I played the game (%)	14.3	0.0
I did chat with others when I played the game (%)	12.5	0.0
It was funny to play together with other players (%)	37.5	26.3
It would be nice if my friends also could participate in this game (%)	42.9	68.4
First choice of preferred game location:		
Movie Theatres (%)	0.0	5.3
Beaches	42.9	15.8
Night Clubs	28.6	5.3
Sport Arenas	0.0	21.1
Art Academies	0.0	0.0
Shopping Malls	12.5	72.2

Research and Implementation of Telecommunication System Based on Inmarsat

SHEN Yi-ting, CAO Yan, JIANG Bu-yuan

(College of economic and management, Dalian Maritime University, Liaoning Dalian, 116026, China)

***Abstract*: From the requirement of shipping enterprises' communication, this paper presents a solution of safety telecommunication between ocean ships and land station and introduces the structure of system and the method of realization. It also designs a protocol based on satellite data communication and discusses some key technique in realization process.**

Key words: Inmarsat; Serial Communication; E-Mail

Ⅰ. Foreword

With the development of informationization, Internet has already broken through the restriction of the geography and penetrated to each field in the global range. Traditional telex, fax, telephone are being replaced by E-mail, SMS, e-fax or other network communication mode progressively because they are costly and limited.

More competition in shipping market makes the communication between shipping enterprises and ocean ships more and more frequent. People working on the sea hope to have convenient communication with family, friends urgently. So far the communication between ocean ships and land station can only rely on Inmarsat satellite system to realize, but satellite communication has some disadvantages such as low speed, long time taking, easy broken off and high cost. So shipping enterprises hopes eagerly that there can be a kind of safe, steady, reliable data transmission which can save the communication expenses effectively at the same time.

The work of this paper is based on this background, to build a telecommunication system between ocean ship and shipping enterprises with Inmarsat Mini-M based on C/S mode, and implement the compressed transfers and resume interrupted transfer support, in order to satisfy the requirement of shipping enterprise.

Ⅱ. The Design of Telecommunication System

2.1 System summary

The telecommunication system is based on Inmarsat Mini-M system which is a new conceptional satellite telephone terminal station and was introduced to the market at the end of 1996 by Inmarsat, it is small, light, convenient to carry and flexible in usage, feature in digital technology, clear conversation quality, shortest putting through time and high privacy[3], also offer the pronunciation, fax, datum business, can meet non-scientific investigation ships' communication requirement completely, so is adopted by most shipping enterprises at present.

Inmarsat Mini-M has modem-pool itself, so can dial directly and enter Internet to receive and dispatch E-mail, but current standard mail protocol does not support resume interrupted transfer firstly, secondly it will increase the actually transmitting data after encapsulation, It means that users will increase extra communication expenses in use for satellite communication with speed of 2.4kbit/s. At the same time standard fax transmitting with graphic format also increase actually transmission data, so we design the system to C/S mode, transmit mails, fax and messages with primitive data form from client to the server, and then send to Internet after server's encapsulation.

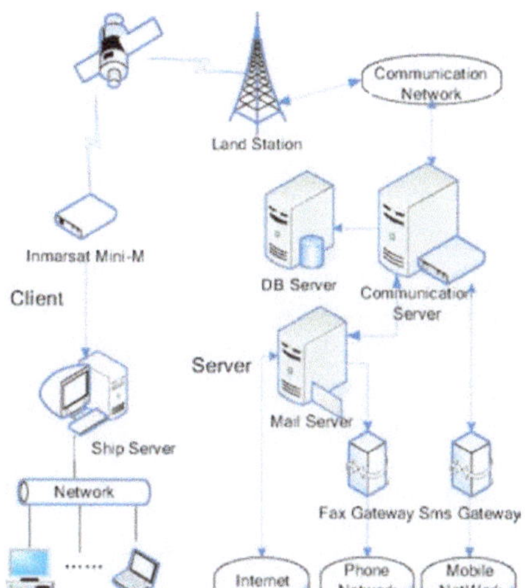

Fig.1 The Frame of System

K. Elleithy et al. (eds.), Advances in Computer, Information, and Systems Sciences, and Engineering, 303–306.

The frame of the whole system is as Fig. 1 shows. The end user's primitive mail, message and fax are compressed and kept in the form of binary file on the ship server, once connect to the server then transmit the data on the two-way with the communication server. After transmission the server encapsulation the received data, put the short message into SMS gateway directly, fax files regarded as mail accessories are sent to the mail server together with other standard mails. The fax gateway will automatically intercept fax from mail server by the third fax gateway and convert it into graphic then send to the telecom network. At the same time the communication server will query from mail server regularly, download new mail, decode it and write into the file, wait for the connection request from the client to transmit.

2.2 General Design of System

2.2.2 2.2.1 Server Design

The server is the core of the whole system, implemented communication hardware drive in order to communicate with ocean ships through satellite, and exchanged the mails with Internet at the same time. Its structure chart is as Fig. 2 shows:

（1）Communication Module. This module drive the communication device through serial port, once detect the connection request from client, automatically answer it and set up the connection, after validate the client's identity transmit the data to client. According to the characteristic of serial port communication, we have designed a set of point-to-point communication protocols based on frame transmission way, realized the two-way transfer and resume interrupted transfer function, furthermore, respond to two or more clients at the same time by standup of multi-thread technique to operate multi-serial port. Besides, the speed of serial port transmission is relatively slow, so we need to set up overtime system to prevent waiting too long to waste the communication expenses.

（2）Data Processing Module. This module includes receiving and sending of mail, decoding of mail, filtering of mail and database record. The server encapsulate the received data from client and send them to the SMS gateway or mail server, at the same time receive new mails from mail server, filter the mails by black and white list, decode the legal mail and save it in fixed format. Besides, this module is responsible for recording each communication course (including time, state, transmitting amount etc.) and backup the mail, message and fax file after successful transmission, add the data to the database for inquiring about.

（3）Management Module. This module is to operate

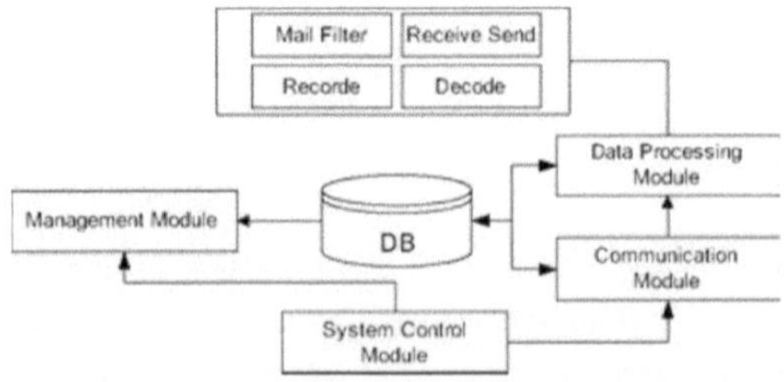

Fig.2 Structure of Server

database by administrator, including management of client accounts and user's mail account such as adding, deleting or prohibiting, and include the function of querying the communication record and statistics, at the same time maintain and backup the database.

（4）System Control Module. This module controls turning on and turning off of the server, and minimize to tray, initialize the communication apparatus through setting some hardware parameter, At the same time show the current state of communication apparatus. Also set up the parameter of connecting to the database server and mail server.

2.2.3 Client Design

The client is the interface of user's final operation, it needs to be simple and clear expression of the system function, so we can divide client into 3 pieces of module:

（1）Mail Processing. View, create, reply, delete and compress save the mail.

（2）Communication Module. Transmit the data in two-way with land station through Mini-M.

（3）Management Module. Manage the sub user on the ship, and provide data view function.

2.3 Communication Protocols Design

2.3.1 The communication flow

The serial port communication's speed is rather slow, need frame transmission, the data are sent through serial port one by one frame, the receive part reassemble the received data. According to communication requirement, the system designed user frame, breakpoint control frame, control frame, file info frame, data frame, finish frame and reply frame. First the client send user frame to server which contain user account information, after validated, send breakpoint control frame to server and the server also send breakpoint control frame to client at the same time, from which enter their own transmission procedure, realize the two-way transmission.

The breakpoint control frame shows whether there is incomplete file in the last receiving time, if there is, send the interrupted file's time mark and the breakpoint number to the other side, then switch to the interrupted file's transmission. Else if there isn't then send control frame directly to tell the other side the file number and data size want to transmit in this time. And then send the first file's info frame which contains the file index information such as time, size, and address, etc. Then send the data frame repeatedly until receive responsion of the file finish frame which prove this file has been sent successfully.

2.3.2 The format of frame

The both sides communication with each other complete the functions of circuit connecting, data transmission, etc. according to the frame format agreed on. Basic frame format as table 1 shows, (1 bit info, 1bit Affirm, 5 or more bits data and 2 bits verify). Data 1 and data 2 place different content according to the function of the frame in order to save user's account information, time marks of the interrupted file or data 3's info length, etc.. The frame is divided into send frame and reply frame according to the type, every frame is distinguished through different info bit (Table 2 shows).

Table 1

Basic Frame Format

Field	Len	Remark
Info	1	Frame kind
Affirm	1	Discern frame with Info bit
Data 1	5	Store different data according to the
Data 2		different frame
Data 3	N/A	File data or user's information
verify	2	CRC verify

Table 2

Frame Type

Info bit	Define	Answer bit
A1	User frame	5E
A2	Control frame	5D
A3	File Info frame	5C
A4	Data frame	5B
A5	Breakpoint frame	5A
A6	Finish frame	59

2.3.3 Security guarantee

In order to guarantee the reliability of transmission, every frame adopt the way send / reply / send again and fail, that is, one party send successfully only after receive right reply frame, then can continue sending the next frame. If receive error reply frame or not receive any reply frame, then send preview frame again, if resend for 3 times, means currently the communication circumstance rather bad and give up this transmission. The frame check adopts the circulation redundancy check (CRC), the sender add the parity bit at the last of every frame. The receiver recalculate every frame and compare the result to parity bit in order to guarantee every frame' exactness.

III The realization of key technology

3.1 Serial port read and write technology

Because Mini-M station is connected with computer through serial port, the system need operate the serial port. Under the environment of Windows, develop the serial port communication program generally use the Windows API communication function, Win32 API can support two kinds of operation: synchronous operation and asynchronous operation[4]. In the way of synchronous operation, I/O operation function can not return before the end of the operation, this can append the thread, until the end of I/O operation. In the way of asynchronous operation, take the advantage of multi-thread structure of Windows, the I/O operation of serial port can be run in back, the other part of the application program execute in the front.

Because of the large amount of communication data and the slow speed of satellite, in order to avoid the situation of no responding, we use the method of asynchronous read and write operation. Firstly initialize the DCB structure of serial port and start the surveillance thread function CommThread to scout serial port event, when the EV_EXCHAR event is found, read function is to be executed to complete the reception, and put the received data into the cache to be dealt with further, When the data is needed to be sent, just execute WriteFile function directly.

3.2 Use AT commands to set up connection

In order to set up the connection between the Mini-M station and the ground, we should send AT command to operate Mini-M station through serial port. Initialized AT commands include AT&F/r/n(load default setup)[5] and AT&C1&D2S7=65/r/n. C1 presented the load check status is low bit, D2 presented when DTR status changed from ON to OFF, the circuit will be cut off automatically. S7=65 presented that after completing the dial, the waiting signal overtime is 65 seconds. When received the OK result code, it means the initialization is correct, and you can send the dial command: ATDT+ telephone number /r/n. If you received the result code of CONNECT 9600[5], it means the connection is successful. At the same time, the server must set automatic answer status is ON, use ATS0=1/r/n.

When the transmission is over, we can send cut off command +++ ATH0/r/n, '+++' is to prevent mixing up the communication data and the command.

3.3 Multi-Thread synchronous technology

In this system, each serial port has its own scout thread, and when the serial port start communicating, it will start the additional receipt and send thread. Besides, it will also have a mail check thread to receive mail from mail server at a interval. So the problem of multi-thread synchronous is a problem which valued more attention.

Windows API and MFC class provide several kinds of synchronous methods[4]. We control the synchronous thread through the method of Event and Critical Section. We create a Event in the mail checking thread, when need to control synchronously, first execute WaitForSingleObject function, if find the user of this account is transmitting the data currently, then set the Event Object status not allowed thread passed. At the same time, create criticalsection in the data send and receive thread, use InitializeCriticalSection function initialize it. The criticalsection are the code blocks which need monopolization way to access share data and can only be monopolized by one thread. When objects in the criticalsection are monopolized by a thread, if another thread wants to use it successfully, it must wait until the front thread release the criticalsection.

By the two kinds of rule above, we can resolve the problem of multi-thread synchronization basically.

3.4 Mail Parse

Recently, generally used mail code includes 7bit, 8bit, binary, quoted-printable and base64 [2]. Basically, the code method above have included 90 percents of the ordinary mails, mail's title, text or attachment will be encoded probably, so decoding that will obtain every part's Encoding attribute must obtained at first while decoding, and then according to decoding the procedure to decode correspondingly. We had better pay attention to the "charset" attribute in mails; non-English mails need characters conversion to be show normally in domestic operation system.

So foreign mails need characters conversion operation after decoded.

Ⅳ. Conclusion

This text has set up the telecommunication system based on Inmarsat system according to the communication requirement between ocean ships and shipping enterprises, introduced the relevant knowledge of satellite communication and mail transfer at the same time, discussed some technology and method in realization process. Additional, realized the safety and reliably transmission of the data in real use, saved communication expenses for enterprises effectively.

Reference

1. Wang Bingjun etc. Modern Satellite Communication System. Beijing: Publishing House of Electronics Industry.2004.1
2. Wood, D.Programming Internet Email. Beijing: China Electronics Press.2001.10
3. Tang Jiang，Inmarsat — Wizard of Emergency Communication System，《Modern Communication》2004.10，23-24
4. Qiushi Technology，Tan Siliang, Zou Chaoqun etc. Navigation of Develop Serial port Communication Engineering by Visual C++. Beijing: Publishing House of People's Post and Telecom, 2003.4
5. Huaishi Studio, MODEM Manual. Beijing : China Electronics Press,2000

A Scheduler Based Architecture for QoS Provisioning in IEEE 802.11 MAC Protocol

Kiran Anna and Mostafa Bassiouni
School of Electrical Engineering & Computer Science
University of Central Florida
Orlando, FL 32816
{kanna, bassi}@cs.ucf.edu

***Abstract*-In this paper, we propose a scheduler based architecture to provide QoS support in IEEE 802.11. Our architecture extends Cali's dynamic p-persistent 802.11 by providing multiple traffic classes while using separate parameters for QoS and channel access functionalities. We used a weight-based fair scheduler to obtain throughput service differentiation at the node level. Compared to 802.11 EDCF, the scheduler-based architecture gives accurate throughput differentiation and easy tuning. We present the performance results of our scheme using nodes with symmetric traffic flows. The simulation results show that the scheduler based scheme consistently achieves the desired throughput service differentiation among traffic classes, better channel access latencies, incurs very little overhead, and gives higher system throughput than EDCF.**

I. INTRODUCTION

In the recent past, the deployment of WLANs has grown many folds for its flexibility and non-wired capability especially, in hotspot locations like bookstores, airports, shopping malls, etc. With the increasing interest of the integrated multimedia services over high-speed networks including wireless LANs, it is natural for us to expect that WLANs will support real-time applications with Quality of Service (QoS) guarantees at the same level as its wired counter part.

IEEE 802.11 employs a carrier sense multiple accesses with collision avoidance (CSMA/CA) mechanism called Distributed Coordinated Function (DCF) as a basic access method and is designed for best effort services only. Due to collisions and subsequent retransmissions, the IEEE 802.11 cannot achieve theoretical capacity bound [7] even though it is compatible with the current best effort service model in wired networks. IEEE 802.11 also suffers from variant transmission delay [18].

The lack of built-in mechanism to support real-time services makes it very difficult to provide quality of service (QoS) guarantees for the throughput-sensitive and delay sensitive multimedia applications. Therefore, modification of the existing 802.11 standards is necessary. Although, IEEE 802.11e is being proposed as the upcoming standard for the enhancement of service differentiation, QoS guarantee in 802.11 is still a challenging problem and needs further study [19].

Several QoS-enabling mechanisms have been proposed as extensions to DCF over the last few years. These proposals provide service differentiation among different traffic classes, but they also suffer from high throughput variability, inability to achieve maximum system capacity and lack of fairness, especially to traffic from a low priority class.

As wireless LANs may be considered as just another technology in the communications path, it is desirable that the architecture for QoS follow the same principles in the wireless network as in the wire-line Internet, assuring compatibility among the wireless and the wire-line parts [15]. In this paper, we present a scheduler-based architecture to support QoS in 802.11 by extending Cali's dynamic p-persistent MAC protocol [5].

A. *Related Work*

We find several research efforts in the literature to model the performance of legacy 802.11 DCF [5-11]. Bianchi [7] models the behavior of a station with a Markov chain model and obtains the stationary probability (τ) and analyzes the saturation throughput based on constant and independent collision probability (p). However, the work does not give any specific algorithm to determine the value of p. Cali *et al.* [5, 6] analyze the performance of the legacy IEEE 802.11 DCF protocol through the *p-persistent* version of IEEE 802.11 protocol. Based on the analytical model, they observe that the system throughput relies on the value of p and the number of active stations, M. Also, they suggest that the IEEE 802.11 protocol with its current settings can hardly achieve the theoretical capacity bound. They extended their work by dynamically tuning the value of p so as to achieve the capacity bound under varying network conditions. Ge [11] extended Cali's work to multiple queues to provide throughput service differentiation while achieving the maximum system capacity. But, the scheme works only in infrastructure mode.

Various Distributed Fair Scheduling algorithms are proposed in the literature to provide fairness of the bandwidth and service differentiation by assigning different weights to flows based on their priority [14-16]. Yang [12] proposed QPART (QoS Protocol for ad Hoc Real-time Traffic) which provides QoS guarantees by adapting the contention window sizes at the MAC layer. The

This work has been supported by NSF under Grant No. 0086251 and by ARO under grant DAAD19-01-1-0502

K. Elleithy et al. (eds.), Advances in Computer, Information, and Systems Sciences, and Engineering, 307–314.

architecture consists of a QoS Aware Scheduler and QoS Manager above a core MAC layer. The QoS aware scheduler realizes the functionality of QoS aware scheduling while the QoS manager implements admission control and conflict resolution. The Core MAC layer uses 802.11 EDCF and as a result, it doesn't provide maximum system capacity. Liu [13] proposed a MAC architecture where different adaptors can be added above the core MAC layer to provide service differentiation. Each adaptor works on adjusting only one of the parameters viz., CW, CWmin, CWmax, etc. But, the adaptors are very basic and do not tune based on the network conditions.

The rest of the paper is organized as follows. In Section II, we briefly review the basics of the IEEE 802.11 standard. The analytical model of IEEE 802.11 using p-persistent IEEE MAC protocol is presented in section III. In section IV, we present the scheduler-based architecture which provides the service differentiation for QoS traffic while guaranteeing high system throughput. The simulation results are presented in section V. Finally, we conclude our paper in section VI.

II. THE IEEE 802.11 MAC

In WLANs, IEEE 802.11 has become the standard medium access control (MAC) and physical layer (PHY) specification [11]. It provides two access methods: (i) the Distributed Coordination Function (DCF), and (ii) the Point Coordination Function (PCF). The DCF, also known as the basic access method, is a carrier sense multiple access with collision avoidance (CSMA/CA) protocol, whereas PCF is similar to a polling system and uses a centralized point coordinator to determine which station has the right to transmit. Since we are more interested in QoS in a distributed scenario, we only present the main features of DCF [1].

A node with a frame to transmit senses the channel to determine whether another node is transmitting or not. If the channel is sensed idle for a specified time period equal to Distributed Inter-Frame Space (DIFS), the node is allowed to transmit. If the channel is sensed busy either immediately or during the DIFS, the transmission is deferred until the node finds the channel to be idle for a DIFS. At this point, the node generates a random back-off interval to minimize the probability of collision with frames from other nodes.

DCF adopts an exponential back-off scheme. The back-off time is uniformly chosen in the range [0, CW], where CW is called contention window. The CW depends on the number of failed transmission attempts for the frame. The CW in initialized to CW_{min} (Contention Window Minimum) at the first transmission attempt of a frame and is doubled up to a maximum value of CW_{max} (Contention Window Maximum) for each subsequent unsuccessful transmission attempt.

The back-off timer is decreased as long as the channel is sensed idle, stopped when a transmission is in progress, and reactivated when the channel is sensed idle again for more than DIFS. As soon as the back-off timer expires, the node starts to transmit at the beginning of the next slot time.

Since the CSMA/CA protocol does not rely on the capability of the nodes to detect a collision, a positive acknowledgement (ACK) is transmitted by the receiver node to signal the successful frame reception. The receiver sends an ACK frame after a time period equal to Short Inter-Frame Space (SIFS), which is less than DIFS. If the transmitting station does not receive an ACK within a specified ACK Timeout, the data frame is presumed to be lost, and after *extended inter-frame space* (EIFS), a retransmission is scheduled. According to the standard, a maximum number of retransmissions (default=7) are allowed before the frame is dropped.

The two-way handshaking technique outlined above for the frame transmission is called Basic Access mechanism. In addition, DCF defines an optional four-way handshaking technique called request to send/clear to send mechanism (RTS/CTS), to solve the hidden terminal problem. In this mechanism, a node that wants to transmit a frame, waits until the channel is sensed idle for DIFS, follows the back-off rules explained above, and then transmits a special short frame called Request To Send (RTS) instead of the frame. When the receiving node detects an RTS frame, it responds with a Clear To Send (CTS) frame after a SIFS.

The transmitting station is allowed to transmit its frame only if the CTS frame is correctly received. The frames RTS and CTS carry the information of the length of the frame to be transmitted. This information can be read by any listening station, which is then able to perform virtual sensing through its Network Allocation vector (NAV) that contains information about the amount of time the channel will remain busy. The RTS/CTS mechanism is very effective in terms of system performance, especially when large frames are considered.

III. P-PERSISTENT MAC PROTOCOL

The p-persistent IEEE 802.11 protocol differs from the standard IEEE 802.11 protocol only in the selection of the back-off interval. Its back-off interval is sampled from a geometric distribution with parameter p (as opposed to the binary exponential backoff used in the standard DCF version). In the p-persistent 802.11 protocol, a station, at the beginning of each empty slot, transmits with a probability p or defers with probability 1-p. Hereafter, we refer to the p-persistent IEEE 802.11 protocol as PDCF. Due to its memoryless backoff algorithm, the PDCF is well suited for analytical studies [6]. It closely approximates the standard (at least from the capacity standpoint) when the average backoff interval is the same [5, 6].

Cali estimated the 802.11 MAC protocol's system capacity by developing an analytical model with finite number, M, of stations operating in asymptotic conditions. The model was based on PDCF's memoryless backoff algorithm. As a result, all the processes that define the occupancy pattern of the channel (i.e., empty slots, collisions, successful transmissions) are regenerative with respect to the sequence of time instants corresponding to the completion of a successful transmission [5]. The time interval between two successful transmissions is called as *virtual transmission time (VTT)*. Therefore, a virtual transmission time includes a successful transmission and may include several collision intervals (see Fig. 1). The protocol capacity is thus:

$$\rho_{max} = m_{avg} / E[t_v] \quad (1)$$

Where, $E[t_v]$ is the average virtual transmission time, and m_{avg} is the average transmission time for a frame. As shown in [5], $E[t_v]$ can be written as

$$E[t_v] = E[N_c]\ \{ E[Coll]_{|Collision} + \tau + DIFS\} + E[Idle\_p]\ .\ (E[N_c] + 1) + E[S] \quad (2)$$

Where,

$E[Coll]_{\|Collision}$	Average collision length provided that a collision occurs;
$E[N_c]$	Average number of collisions in a virtual transmission time;
$E[Idle\_p]$	Average number of consecutive idle slots;
τ	Propagation Delay;
$E[S]$	Time required to complete a successful transmission (including all the protocol headers)

It can be verified from the protocol behavior that $E[S] \approx m_{avg} + 2\tau + SIFS + ACK + DIFS$ [5]. The system capacity can be maximized by minimizing the virtual transmission time. Under static conditions, when M remains constant, VTT depends on p. We obtain the optimum p (p_{opt}) by minimizing VTT. Therefore, the standard 802.11 protocol can be tuned to reach its theoretical capacity by setting CW equivalent to the backoff interval corresponding to the p_{opt} of PDCF. However, in real situation, the network and load conditions are not constant. Therefore, the p_{opt} calculated under static conditions won't result in optimal results with changing network conditions.

Cali proposed a dynamic version of p_{opt} to support transient network conditions [5]. Each node monitors the number of idle slots and collision length (if there is one) in each of its transmission attempt. It then estimates the number of active nodes (M) in the network using the above two parameters and p_{opt} used in the previous transmission attempt. The current p_{opt} is calculated using the updated values of M, number of idle slots and collision length. A drawback of this scheme is that it works only for single class traffic. We extend Cali's scheme to multiple classes of traffic by means of a scheduler-based architecture.

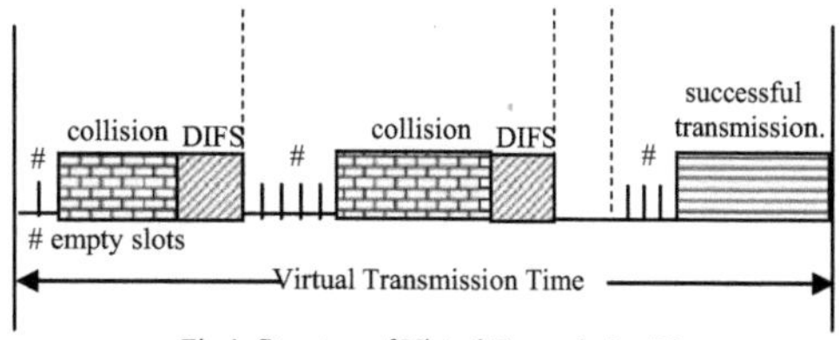

Fig.1. Structure of Virtual Transmission Time

IV. SCHEDULER ARCHITECTURE

The QoS support in the EDCF is provided by prioritizing the channel access among the various classes of traffic within a node. The channel access is differentiated by using different AIFS, CW_{min}, CW_{max}, and PF values at each traffic queue in the MAC layer. As a result, the two separate functional components, the QoS support and the channel access mechanism, are combined into one functional component increasing the complexity of the 802.11 MAC. In this paper, we present a scheduler-based architecture for QoS provisioning in IEEE 802.11 MAC.

Our architecture consists of a Scheduler and a core MAC layer. The scheduler is built above the core MAC layer and provides QoS support to the various traffic classes. The core MAC layer provides the channel access mechanism. We use Cali's dynamic p-persistent 802.11 MAC in our core MAC layer. It is well suited for our scheme and we can tune it to provide maximum possible system capacity. One additional feature with this architecture is that each component can be changed or improved independently of the other. Fig. 2 shows the block diagram of the architecture. We used a packet-based weighted fair queuing in our scheduler and as a result it provides throughput service differentiation at the node level. The scheduler, based on weights assigned to the queues, sends the next packet from a queue. When one of the queues is empty, it sends the packets available in the remaining queues while keeping track of the deficit in the empty queue. It tries to compensate the deficit once the empty queue is filled with packets.

Our scheduler-based architecture provides several advantages. In EDCF, each queue is provided with separate channel access parameters and as a result, each queue makes an independent attempt to achieve its own QoS requirements. Consequently, there is no easy way in EDCF to enforce certain differentiation ratio among different traffic classes in a node.

The scheduler-based scheme provides easier and more accurate control in providing QoS differentiation for all queues within a node. Also, with our architecture, there are only two parameters, P_{opt} and CW, at the channel access level. Moreover, our architecture is similar to the DiffServ architecture of the wired network thus, increasing the compatibility between the wired and wireless networks.

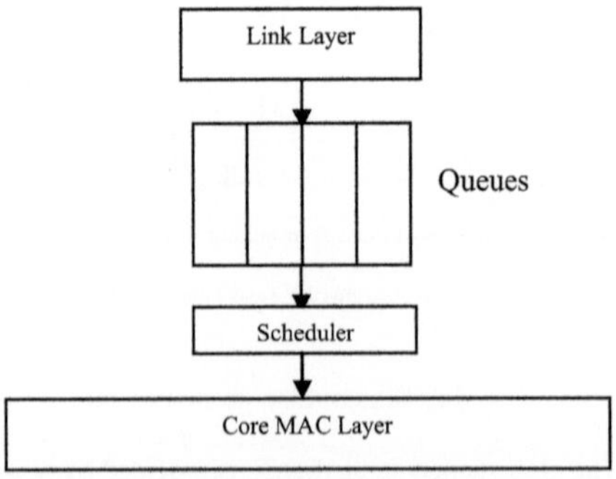

Fig.2. Scheduler-based MAC Architecture

V. PERFORMANCE EVALUATION

In this section, we carry out the performance evaluation of our proposed architecture. We show that the scheduler-based architecture provides desired throughput service differentiation according to the value set in the QoS parameter while providing high system throughput. We compared the performance of our scheduler-based architecture with adaptive EDCF (AEDCF) scheme [20]. AEDCF is a dynamic scheme that adapts to the varying network and load conditions to provide a better performance. It uses the same set of parameters for both QoS differentiation and channel access.

Our tests have shown that it is very difficult to set the AEDCF parameters to accurately achieve a desired throughput differentiation among the traffic classes of a node. For example, there is no clear way of how to set the parameters CW_{min}, CW_{max}, and PF in order to achieve throughput ratio 2:1 between two classes of traffic. In contrast, achieving exact throughput differentiation in the scheduler-based method is straightforward. Aside from throughput differentiation, the scheduler-based method is capable of achieving higher total system throughput than AEDCF. AEDCF provides a prioritized channel access among its traffic classes. Our performance tests showed that the scheduler based scheme provides better channel access latency than the AEDCF scheme for a given throughput ratio.

A. *Simulation Tool*

We used the NS2 simulator [21] in all the tests reported in this paper. We incorporated the scheduler-based architecture by extending the NS2 simulator by modifying its IEEE 802.11 DCF implementation and the protocol stack of the wireless node. We made changes to the code such that the scheduler can still provide the original 802.11 functionality. Fig. 3 shows the placement of the scheduler in the protocol stack. The scheduler is part of the PriQueue class and interacts with the MAC layer directly.

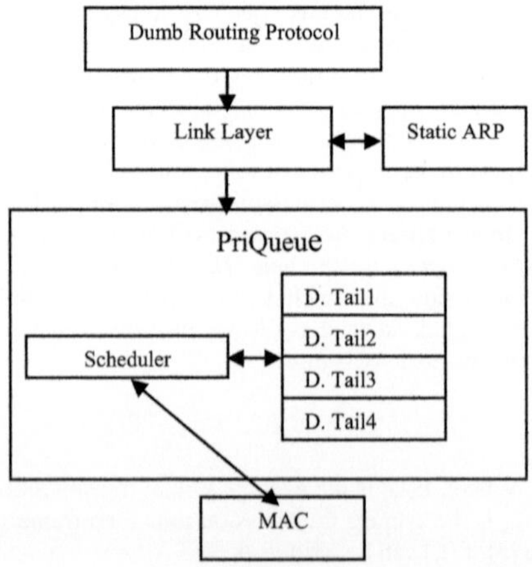

Fig.3. Modifications in NS2.

It decides the next frame to be sent from the drop tail queues and informs the MAC layer about the frame. Upon transmitting the frame, the MAC layer calls the scheduler (through the PriQueue class) to get the next frame. We also added a static ARP and dumb routing protocol to the NS2 code. The static ARP doesn't send any ARP packets and the dumb routing protocol does not use any routing mechanism. It just sends the data packets down the stack to the link layer. As a result, there is no overhead due to the routing protocol and ARP.

B. *Simulation Scenario*

In our study, we simulate a WLAN consisting of ten wireless stations, where each station can hear all other stations. The WLAN operates in IEEE 802.11a mode with each station operating at 54Mbps. Our simulation scenario consists of a circular topology, where node 'i' sends traffic from all its classes to node 'i+1', i<10, and node '10' sends its traffic to node '1'. Each station has two classes of CBR traffic with an average data packet size of 50 transmission slots (3037.50 bytes). The MAC layer system parameters are set as shown in Table I.

We ran each simulation for 50 seconds and we activated all the traffic sources at t = 0.000000001 seconds and deactivated them at t = 50 seconds. In each station, both the CBR traffic sources generate traffic at the same rate. In our study, we used CBR traffic rates from 0.25 Mbps to 3.5 Mbps to generate a system load of 5 Mbps to 70 Mbps.

C. *Evaluation of the Overhead due to the Scheduler on P-Persistent 802.11 MAC*

We start the presentation of the performance results by showing that the scheduler-based architecture has very low overhead. We conducted experiment to evaluate the impact of the overhead of the scheduler on the performance of the p-persistent 802.11 protocol. We compared the performance of Cali's single class dynamic p-persistent 802.11 with the multi-class scheduler-based scheme. In the scheduler-based scheme, we used 2, 3 and 4 classes of traffic in each node with asymptotic traffic flows. In all simulation tests, we used 12 flows in each of the ten nodes and the flows were equally divided among the available classes.

Since the purpose of the above experiment was to evaluate the overhead of the scheduler-based architecture and compare it with the single-class dynamic p-persistent 802.11, we used equal weights for all traffic classes managed by the Scheduler. Fig. 4 clearly shows that the addition of a scheduler has no extra overhead on the performance of p-persistent 802.11.

The x-axis in Fig. 4 is the average data size in slot times and the y-axis is the channel utilization. The performance of the scheduler-based scheme is similar to that of Cali's single class p-persistent 802.11 irrespective of the number of traffic classes used.

D. *Throughput Service Differentiation*

In the following experiment, we evaluated the performance of the scheduler-based scheme and AEDCF in terms of throughput service differentiation among traffic classes as well as total system throughput. In the scheduler-based scheme, the desired throughput service differentiation is easily and accurately achieved by simply setting the weights of the different traffic classes in the desired proportions. The AEDCF scheme, however, uses a set of parameters namely CWmin, CWmax, and PF for each class to provide QoS as well as channel access. It becomes a very complex operation to correctly set these parameters in order to achieve a desired throughput differentiation among classes. To further illustrate this important issue, let us assume we want to achieve a throughput service differentiation ratio of 2:1 between class1 and class2 traffic. To do so in the scheduler-based scheme, we simply set the weights of class1 and class2 to 2.0 and 1.0, respectively. In the AEDCF scheme, however, a tedious search process is needed to find the parameter values to achieve the required differentiation ratio. Setting each of the parameters CWmin, CWmax, PF for class1 to double the values of class2 does not result in a throughput differentiation ratio of 2:1. Fig. 5 and Fig. 6 plot the total system throughput, throughput of class1, and throughput of class2 versus time for the scheduler-based scheme and AEDCF scheme, receptively, at a system load of 50Mbps. Table II shows the MAC parameters used for class1 and class2 traffics in the AEDCF scheme. We designate AEDCF that uses the parameters given in Table II as AEDCF1.

As shown in Fig. 5, the throughput of class1 traffic in the scheduler-based scheme is reasonably close to two times that of the low priority class2 traffic. In Fig 6, the throughput of class1 traffic in AEDCF1 is about 3.0 times that of class2 traffic. Fig. 5 and 6 show that the total system throughput of the scheduler-based scheme is about 11.4% higher than that of AEDCF1.

In order to get a throughput ratio of 2:1 for the AEDCF scheme at 50 Mbps system load, we had to run several simulation tests with different parameter combinations. Table 3 gives the values of the parameters that reasonably approximated the desired throughput ratio. We refer to the AEDCF scheme that uses the parameters values given in Table III as AEDCF2. Fig 7 shows an approximate throughput differentiation ratio of 2:1 for class1 and class2 traffic flows of AEDCF2. The total system throughput of the scheduler-based scheme is about 5.5% higher than that of AEDCF2.

We next examine the performance of the two schemes under different CBR traffic loads. All the simulation conditions in this experiment are the same as in the previous one except for the traffic source load. The load rate for each flow is varied from 0.5 Mbps to 3.5 Mbps (giving a total system load of 5Mbps to 70Mbps). Fig. 8 shows that for a system load of up to 30 Mbps, there is no throughput differentiation in all the three schemes PDCF, AEDCF1 and AEDCF2. This is because there is no contention and the traffic load is not sufficient enough to differentiate between the two classes. As the system load is increased up to 70Mpbs, the scheduler-based scheme maintained the desired throughput differentiation ratio of 2:1 whereas, the AEDCF1 scheme gave less preference to class2 traffic. The AEDCF2 scheme provided a differentiation of slightly above 2.

Fig.9 shows that the scheduler-based scheme achieved better total system throughput than both AEDCF1 and AEDCF2 schemes. Even though the AEDCF scheme dynamically adapts its channel access parameters, it is still bound by the input access parameters that may not be the optimal values.

Notice that the AEDCF2 at a system load of 50.0Mbps gave the required throughput differentiation; but as the system load increased AEDCF2 gave slightly increasing preference to class1 traffic over class2 (exceeding the required 2:1 ratio). Thus, we infer that the same set of AEDCF parameter values cannot provide the same level of throughput differentiation at different system loads.

TABLE I
802.11 MAC PARAMETERS

t_{slot}	9 μs
SIFS	16 μs
DIFS	34 μs
Data Rate	54 Mbps
PLCP Data Rate	6 Mbps
Preamble Length Bits	144 bits
PLCP Header Length	48 bits
ACK Length Bytes	14 bytes
MAC Header Length Bytes	28 bytes

TABLE II
AEDCF1 MAC PARAMETERS FOR THE TWO TRAFFIC CLASSES

Parameters	Class1	Class2
CWmin	5	10
CWmax	200	400
AIFS (μs)	34	34
PF	2	4

TABLE III
AEDCF2 MAC PARAMETERS FOR THE TWO TRAFFIC CLASSES

Parameters	Class1	Class2
CWmin	6	5
CWmax	200	225
AIFS (μs)	43	34
PF	2.0	2.25

E. *Channel Access Latency Differentiation*

Our simulation results in the previous section showed that it's very difficult to set the parameters in AEDCF to obtain the desired throughput differentiation. The same set of parameters which are applied to both the QoS and channel access level mechanism resulted in different throughput ratios at different system load conditions.

In AEDCF scheme, the parameters for different classes of traffic provide differentiation in terms of channel access. In other words, provide differentiation on channel access latency. In this simulation test, we run the AEDCF scheme with the parameter values given in Table IV for different system loads. The parameters are chosen such that the AEDCF gives a throughput differentiation of 2:1 at each system load. The scheduler based scheme is set to provide a throughput differentiation of 2:1. We refer the AEDCF that uses Table IV as AEDCF3. Since, we now have the same throughput ratios in both the schemes; we compare their channel access latency for the two traffic classes.

TAB V
A DC MAC A AM T S TH TW T A C C ASS S

	C	CW	CW x	A S	
0 Mb	C 1	6	00		0
	C				
Mb	C 1	6	00		0
	C	6	00		
60 Mb	C 1	6	00		0
	C		00		
6 Mb	C 1	6	00		
	C		00		
70 Mb	C 1	6	00		
	C	6			

The simulation conditions for this test are same as in previous experiment. Fig. 10 and Fig. 11 show the channel access latencies for class1 and class2 traffic. The scheduler based scheme performed better than AEDCF3. In our scheme, the core MAC layer tries to maximize the system capacity by adapting its P_{opt} to the conditions. A greater throughput implies lesser channel access latency. However, in AEDCF scheme, the parameters used, provide prioritized channel access without emphasizing on optimal throughput. When we set the parameters to achieve a throughput differentiation of 2:1, the parameters are not optimal enough to provide better channel access latency as well as throughput as the same set of parameters serve the channel access as well as service differentiation.

In summary, our scheduler based scheme outperformed AEDCF in terms of system throughput, class throughputs and channel access latency.

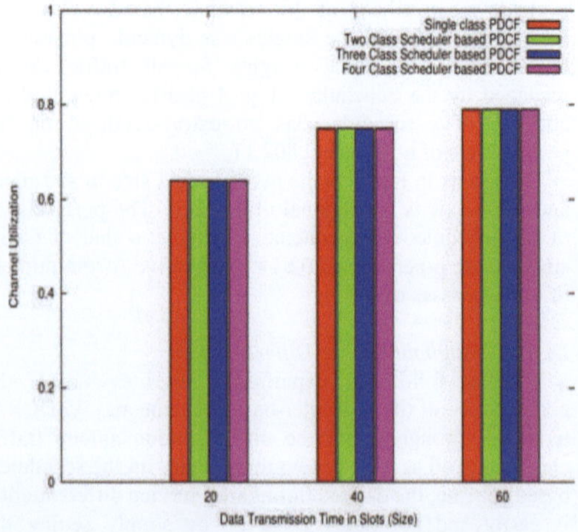

Fig. 4 Effect of scheduler on the p-persistent 802.11

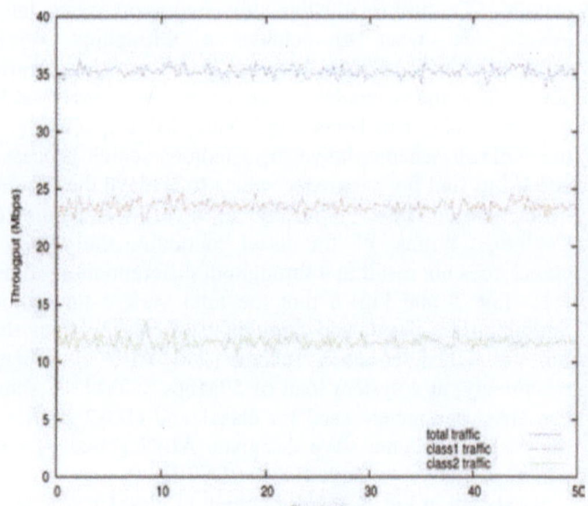

Fig. 5 Scheduler-based p-persistent 802.11 (CBR Traffic)

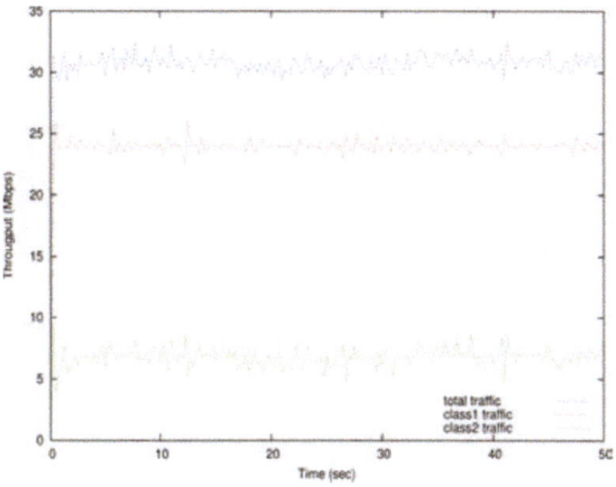

Fig.6. AEDCF1 802.11 (CBR Traffic)

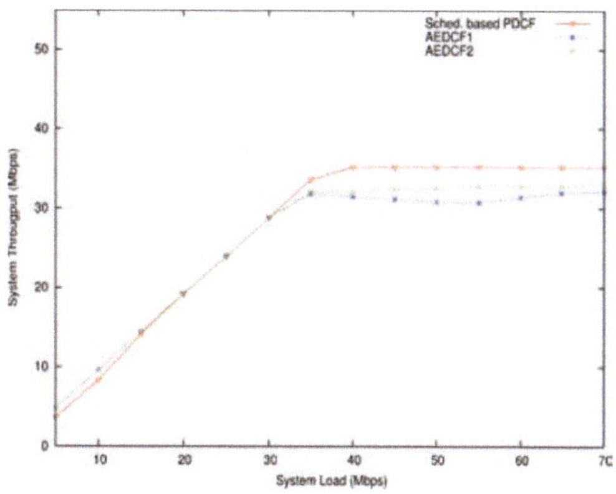

Fig. 9 System Throughput Vs Load (CBR Traffic)

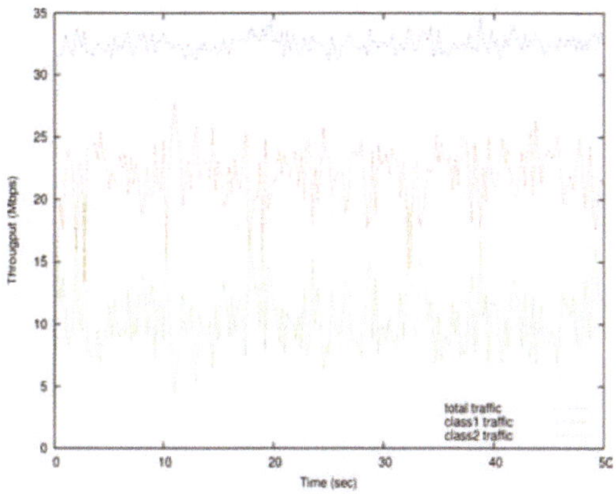

Fig. 7 AEDCF2 802.11 (CBR Traffic)

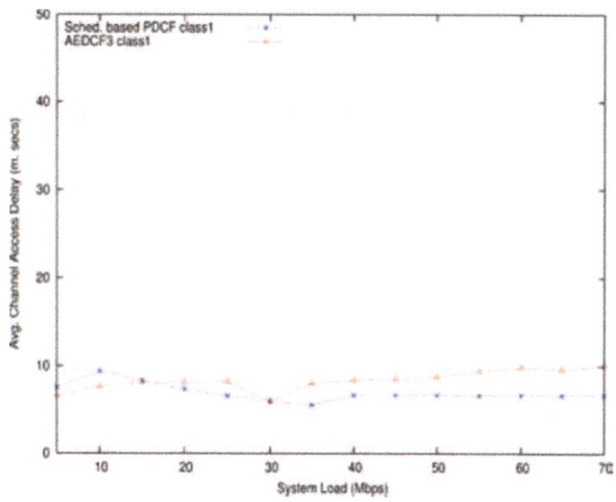

Fig. 10 Class1 Channel Access Latency Vs Load (CBR Traffic)

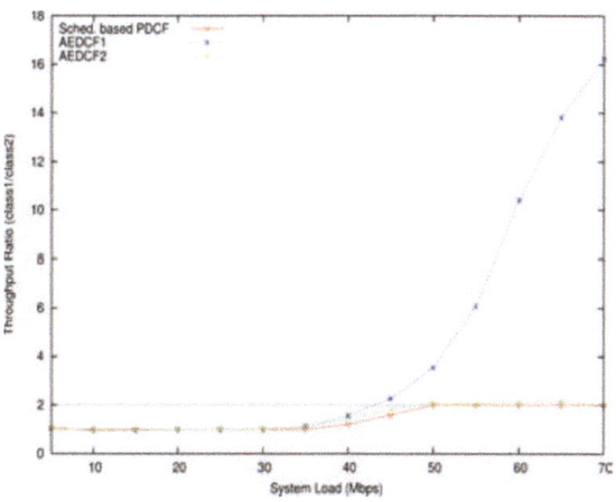

Fig. 8 Traffic Class Throughput Ratio Vs Load (CBR Traffic)

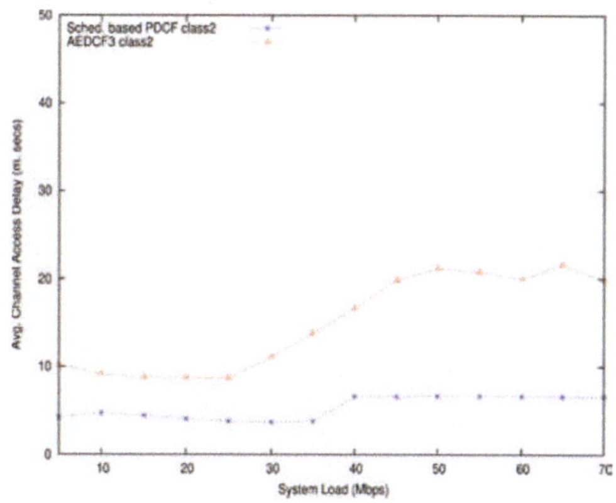

Fig. 11 Class2 Channel Access Latency Vs Load (CBR Traffic)

VI. Conclusions and Future Work

In this paper, we extended Cali's p-persistent 802.11 from single class to multiple classes by means of a scheduler-based architecture. The scheduler-based scheme separates QoS and channel access components in the MAC layer paving for better control on QoS. The scheme not only provides throughput service differentiation among the multiple traffic classes within a node, but also has very little overhead and achieves high system capacity.

Our future work includes extending the scheduler-based architecture to support system level differentiation for heterogeneous nodes with asymmetric traffic classes. Another extension is to design an admission control mechanism to help maintain the required QoS differentiation for throughput and end-to-end delay.

References

[1] IEEE Std. 802.11-1999, Part11: Wireless LAN Medium Access Control (MAC) and Physical Layer (PHY) specifications, Reference number ISO/IEC 8802-11:1999(E), IEEE Std. 802.11, 1999 edition, 1999.

[2] S. Blake, D. Black, M. Carlson, E. Davies, Z. Wang, and W. Weiss, "An Architecture for Differentiated Services," RFC 2475, December 1998.

[3] IEEE 802.11 WLANS, http://grouper.ieee.org/groups/802/11.

[4] IEEE 802.11e/D11.0, Draft Supplement to Part 11: Wireless LAN Medium Access Control (MAC) and Physical Layer (PHY) specifications: medium Access Control (MAC) Enhancements for Quality of Service (QoS), November 2004.

[5] F. Cali, M.Conti, and E. Gregori, "IEEE 802.11 Protocol: Design and Performance Evaluation of an Adaptive Back-off Mechanism", IEEE Journal on Selected Areas in Communications, vol.18, no.9, pp. 1774-1786, September 2000.

[6] F. Cali, M.Conti, and E. Gregori, "Dynamic Tuning of the IEEE 802.11 Protocol to Achieve a Theoretical Throughput Limit", IEEE/ACM Transactions on Networking, vol. 8, no. 6, pp. 785-799, December 2000.

[7] G. Bianchi, "Performance Analysis of the IEEE 802.11 Distributed Coordination Function", IEEE Journal on Selected Areas in Communication, vol.18, no. 3, pp. 535-547, March 2000.

[8] G. Bianchi and I. Tinnirello, "Kalman filter Estimation of the Number of Competing Terminals in an IEEE 802.11 Network", In Proceeding of IEEE INFOCOMM 2003, vol. 22, no. 1, pp. 844-852, March 2003.

[9] C.H. Foh and M. Zukerman, "Performance Analysis of the IEEE 802.11 MAC Protocol", In Proceedings of European Wireless 2002, pp. 184-190, February 2002.

[10] H. Wu, Y. Peng, K. Long, S. Cheng, and J. Ma, "Performance of Reliable Transport Protocol over IEEE 802.11 Wireless LAN: Analysis and Enhancement", In Proceedings of IEEE INFOCOM 2002, vol. 21, no. 1, pp. 599-607, June 2002.

[11] Y. Ge, "QoS Provisioning for IEEE 802.11 MAC Protocols", PhD Dissertation, Ohio State University, 2004

[12] Y. Yang and R. Kravets, "Distributed QoS Guarantees for Reatime Traffic in Ad Hoc Networks", In Proceedings of SECON 2004, pp. 118-127, October 2004.

[13] W. Liu, W. Lou, X.Chen, and Y. Fang, "A QoS-enabled MAC Architecture for prioritized Service in IEEE 802.11 WLANS", In Proceedings GLOBECOM 2003, vol. 7, pp. 3802-3807, December 2003.

[14] N.Vaidya, P. Bahl, S. Gupta, "Distributed Fair Scheduling in a Wireless LAN", In Proceeding of MOBICOM 2000, pp. 167-178, 2000.

[15] A. Banchs and X. Perez, "Distributed Weighted Fair Queuing in 802.11 Wireless LAN", In proceedings of ICC 2002, vol. 5, pp. 3211-3127, April 2002.

[16] W. Atikom, S.Banerjee and P.Krishnamurthy, "Starvation Prevention and Quality of Service in Wireless LANs", In Proceedings of the IEEE 5th Intl. Symposium on Wireless Personal Multimedia Communications (WPMC), vol. 3, pp. 1078-1082, October 2002.

[17] Z. Jiang, L. Chang, and N.K. Shankarayanan, "Providing multiple service classes for bursty data traffic in cellular networks", In Proceedings of IEEE INFOCOM 2000, vol. 3, pp. 1087-1096, March 2000.

[18] H. Kim and J. Hou, "Improving Protocol Capacity with Model-based Frame Scheduling in IEEE 802.11-operated WLANs", IEEE Journal on Selected Areas in Communication, vol. 22, no. 10, December 2004.

[19] H. Zhu. M. LI and I. Chlamtac, "A Survey of Quality of Service in IEEE 802.11 Networks", IEEE Wireless Communications, vol. 11 no. 4, pp. 6-14, August 2004.

[20] L.Romdhani, Q. Ni, T.Turletti, "Adaptive EDCF: Enhanced Service Differentiation for IEEE 802.11 Wireless Ad-Hoc Networks", In Proceedings of IEEE WCNC 2003, vol. 4, no.1, pp 1373-1378, March 2003.

[21] www.isi.edu/ns

An Integrated Environment for Network Design and Simulation

Muhammad Azizur Rahman*, Algirdas Pakštas*, Frank Zhigang Wang**
*London Metropolitan University, London, England
**Cranfield University, Cranfield, England
mam406@londonmet.ac.uk, a.pakstas@londonmet.ac.uk, f.wang@Cranfield.ac.uk

Abstract **—It is often useful to move network models between different network design, simulation and topology discovery tools in order to obtain different view of the network. Unfortunately, these tools are not compatible to each other and cannot be used together directly. This paper presents a tool, Network Design and Simulation Environment (NeDaSE) of version v2.0, developed for the purpose of integrating such different types of tools. Some experiments have been illustrated using topology generation tool `BRITE`, analytical tool `Delite`, simulation tool `ns-2` and wireless simulator `Glomosim`. A network description language LNMet-X has been used for the implementation.**

I. Introduction

There are currently available various tools created by the separate companies and researchers in academia which are intended for use as practical and/or educational tools for network design. Very often such tools are addressing different issues in the network design process [1, 2, 3]. One tool can be complementary to other tools for their distinctive features. Analytical tools may design and analyze the network models efficiently and provide solutions more quickly than simulation tools but do not always achieve the required accuracy. Most often, dynamic behaviour of the protocol cannot be visualized using the analytical tools but can be visualized using simulation. Network topology generators generate different topology based on different model, which can be analyzed by analytical tool or can be simulated by simulation tool. In this respect, there is a need for integrating analytical and simulation, topology generators and discovery tools all together.

A good tool should have the some features that are mentioned in [4]. Combination of these features is rarely found within one tool. Unfortunately, tools are usually not compatible to each other and cannot be used together directly. These tools store internal information in different ways as well as visualize outputs (if at all) in different ways. Therefore, there is a need to create integrated environment, which will be enabling shared use of tools and exchange of models between the tools and develop a common language for the tools. In the integrated environment, user/planner will have the facility to test network model in different views and can consider the output in different angles. This paper is looking at the methodology for implementation of a tool for integrated environment. Previous papers [5] and [6] show the integrated tool NeDeSE of Versions v0.0 (integrated tools Delite and Ns-2) and v1.0 (additional tool BRITE) respectively. This paper shows the latest version v2.0 of NeDaSE with extended capability.

The rest of the paper is organized as follows. Section II describes approach to integration of tools. Section III describes tools BRITE, `Delite`, `ns-2` and `Glomosim` which are chosen for case study with their integration approach. A comparison and incompatibilities among them is also shown here. In section IV some experiments are shown with small size models using NeDaSE tool. Finally, section V provides conclusions and future work.

II. Approach to Integration of Tools

An integrated tool, *Network Design and Simulation Environment (NeDaSE)*, has been implemented as a "Proof of concept" prototype that is based on the approach of using a meta-language. After analyzing features of a few tools, the most common features are selected as a "skeletion" for meta-language. Such language LNMet-X (*Language for Network Meta-description based on XML)* [7] was designed as a intermediate format for model transformation by NeDaSE (see Fig. 1 and Fig. 2). NeDaSE is implemented in programming language C.

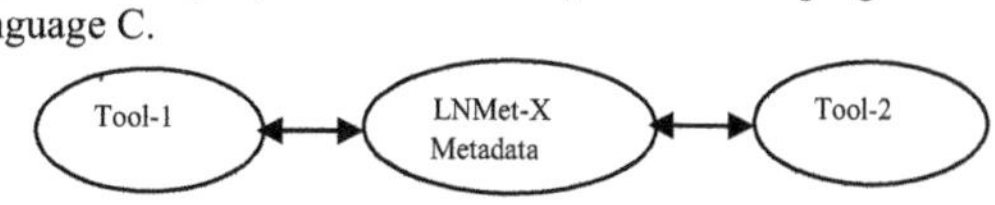

Fig. 1. Block diagram of transformation technique of the tool NeDaSE

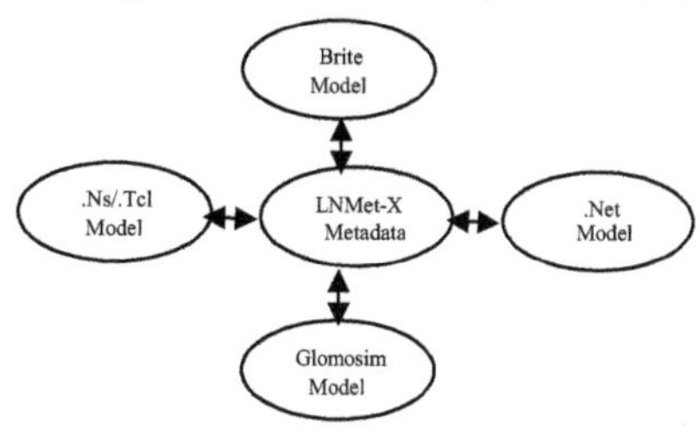

Fig. 2. Transformation of models among LNMet-X, Ns-2, Delite, Brite and Glomosim tools.

K. Elleithy et al. (eds.), Advances in Computer, Information, and Systems Sciences, and Engineering, 315–322.

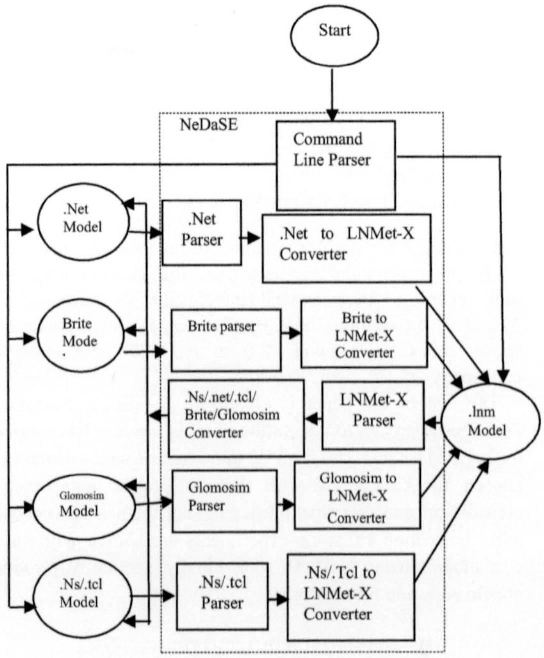

Fig. 3. Links between the modules of NeDaSE

To integrate selected tools, style, format, semantics, scaling, protocols supported, representation etc. incompatibilities had been considered. Fig. 3 shows the modular structure of NeDaSE and related external files as a whole.

The program starts from reading and parsing command line. The implementation of example of command line is as follows:

C:\>nedase prim20.ns prim20.net

This command instructs NeDaSE to read prim20.ns file used by ns-2 and convert it to prim20.net file (for use by tool Delite).

III. Overview of the Selected Tools

Four tools are selected for the NeDaSE prototype implementation. The comparative features of BRITE, Delite, ns-2 and Glomosim are shown in Table I from the meta-model perspective.

A. Overview of Topology Generator BRITE

Boston University Representative Internet Topology gEnerator, BRITE [8, 9] is a topology model generation tool. BRITE is designed to be a flexible synthetic topology generator which is producing data for network simulators SSF, JSim and Ns. BRITE reads the generation parameters from a configuration file that can be either hand written by the user or automatically generated using BRITE's GUI.

BRITE contains eight different generation models: flat routers models (Router Waxman, Router Barabasi), Flat Autonomous System (AS) models (AS Waxman, AS Barabasi), Imported file models (BRITE 1.0, GT-ITM, NLANR, Inet, SKITTER), Hierarchical Top-Down model, Hierarchical Bottom up model. There is a BRIANA Analysis Engine having a set of analysis routines to analyse the topology found in the `BRITE` generator.

At present, `BRITE` has only a single output interface: a `BRITE` format topology file that is stored in a specially formatted text file. This tool has no visualization output of its own.

B. Overview of WAN Design Tool Delite

DEsign tool LITE, Delite [10] is an educational and practical wide area network (WAN) design tool, which can produce network designs of limited size using a set of the embedded network design algorithms. Delite can produce graphical displays representing network nodes and links as well as some additional analysis data (delay analysis, reliability analysis, average delay analysis of the link, average number of hops, link utilization for every separate link between nodes, utilization of each node, overall network model utilization, etc.)

There are seven files that are handled by the Delite tool for each network design. However, the most important for the users are .gen (original node information, coordinates and available link types) and .net (additionally has table of links between nodes i.e. actual design) files. Links between nodes can be generated using a few design algorithms. Thus, various designs may have different costs, delays, reliability, average number of hops etc. Other files are used for different purposes: noeqip.tbl and param.tbl are predefined (used to supply some network parameters), .cst file describes the costs associated with link types (e.g. T1, T3, D96 etc.), .req file describes capacity of each link type, .inp file contains the names of files related to a particular model. The users can edit all of the files mentioned above as they are in ASCII text format. Fig. 4 shows the files and their relationship. Link direction of the arrow between file types means direction of a "call" for information. E. g. .net file is only using .inp files and .inp file is using all others.

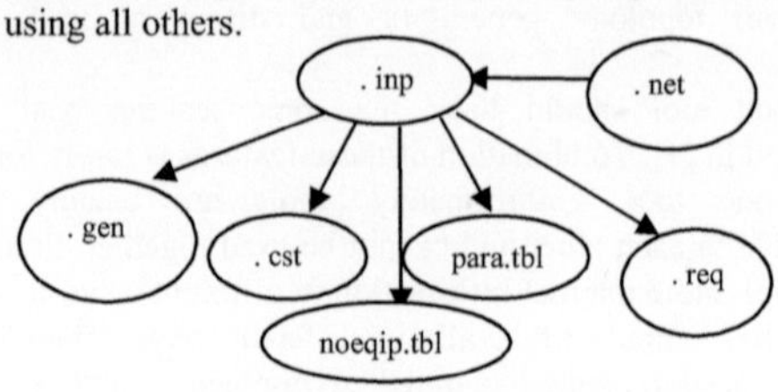

Fig. 4. Relationship of files used in Delite tool

C. Overview of Network Simulator ns-2

Network simulator, `ns-2` [11, 12], a VINT (Virtual Inter-Network Testbed) project from U.C. Berkeley/LBL/ Xerox PARC, is a discrete event simulator targeted at network research, which provides substantial support for simulation of TCP, routing and multicast protocols. The simulator is using a

Table I: Comparison among tools BRITE, Delite, ns-2 and Glomosim with respect to different features.

Criteria/Tools	BRITE	Delite	Ns-2	Glomosim
General features				
Scalability	Very large scale topology generator	Small scale network design	Medium Scale simulator	Very large scale simulator
Usages	Educational, Research	Educational	Educational, Research	Educational, Research
Target Network Model	LAN, MAN, WAN	WAN, LAN	WAN	WAN
Topology Design Algorithm(s)	Some predefined models	Some predefined algorithms	Manually designed by users	Currently only wireless network is implemented
Platform	Unix	Windows	Unix, Windows	Unix, Windows
Model Data Format	Specially formatted text file	Specially formatted text file	Text file (ns/tcl format)	Text formatted file
Outputs	Generated topology file	Graphical output, Link analysis, node utilization, overall utilization in percent, delay analysis	Packet level trace file, animation	Graphical output (using VT tool), Generate a file for statistics
Parameters supported	Size of the one side of the plane, size of the one side of a high-level square, Number of nodes, Model id, Alpha (Waxman-specific exponent), beta (Waxman-specific exponent), node placement, number of links per new nodes, how nodes joined the topology, maximum and minimum bandwidth assignment to links, Edge connection, Intra-domain bandwidth assignment distribution, min, max bandwidth for intra-domain links, inter-domain bandwidth assignment distribution, min, max bandwidth values for inter-domain links, assignment of routers to ASs, number of ASs "desired", how routers are grouped into ASs, inter-domain bandwidth assignment distribution.	Node number, Population, link type, delay, Alpha, seed, centre node, topology design algorithm	Scheduling, queuing, mobility of nodes, network topology, TCP parameters (maximum congestion window, max segment size), simulation time, bandwidth of link, propagation model, node placement, traffic source, errors.	TCP parameters, MAC parameters (max contention window size, payload size), simulation time, power range, packet reception model, bandwidth of links, propagation model, node placement, traffic source, Simulation area.
Traffic model	No	No	Yes	Yes
Network topology generation	Yes	Yes	No	No
Easy configuration of network parameters	No	Yes	Yes	No
Metrics to evaluate the performance or routing protocols	No	No	Yes	No
Extensibility	Yes	Yes	Yes	Yes
Implemented Language	C	C	C++ and OTCL	JAVA
Open Source Code	Not freely available	Not freely available	Freely available	Not freely available
Good documentation of the simulator (users' manual)	Yes	No	No	Yes
Simulation/Dynamic Features				
Easily modifiable simulation event tracing	No	No	Yes	Yes
Network and transport Protocols	No	No	IPv4, TCP, OSPF, UDP, HTTP, FTP, CBR, DSR, TELNET, Pareto, Dynamic Routing and other TCP family	TCP, CSMA/CD, UDP, HTTP, FTP, CBR, fisheye, LARScheme 1, ODMRP, WRP, DSR, MACA, TELNET, AODV
Graphical representation of simulation results	No	Yes	Yes	Yes
DiffServ	No	No	Yes	No
Support of distributed simulation	No	No	Yes	No
Dynamic reconfiguration of network topology	No	Yes	Yes	No
Simulation is easy to extend with new protocols or additions/changes to existing protocols	No	No	Yes	Yes
Run real application on the top of the simulated network (emulation)	No	No	Yes	No
Use of timeliness	N/A	N/A	Yes	Yes
Used simulator features should be known 'to work correctly'	No	No	Yes	No
Wireless features				
Mobility of nodes in wireless	No	No	Yes	Yes
Ad hoc connectivity in wireless	No	No	Yes	Yes
Support for modeling of power-consumption due to processing in nodes	No	No	No	Yes

Tcl/Otcl (Tool Command Language/Object Oriented Tcl) [13] as a command and configuration interface.

There are four types of files related to ns-2 simulator. Model is described in files .tcl or .ns which have common subset of commands but not exactly compatible between each other's. While simulator runs a model defined in .tcl/.ns file, simulation trace file (.tr) and animation file (.nam) are created during the session. Network Animator (.nam) files are used to visualise the behaviour of the network protocols and traffic of the model. Once created, users can play with the .nam file just like a media player and check the behaviour repeatedly. ns-2 facilitates the three broad themes of network research simulations: 1) selecting a mechanism from several options; 2) exploring complex behaviour; and 3) investigating unforeseen multiple-protocol interaction. The available file types and their relationship are shown is Fig. 5.

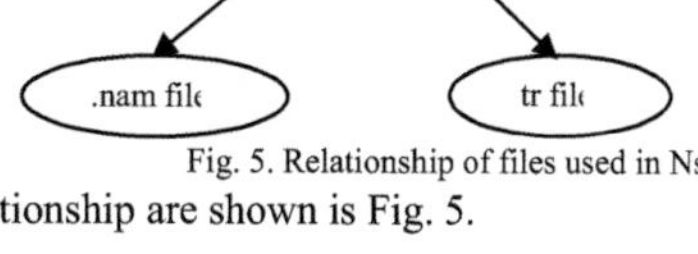

Fig. 5. Relationship of files used in Ns-2 tool

D. *Overview of Network Simulator Glomosim*

Glomosim [14, 15] is a scalable simulation environment for wired and wireless mobile network systems. It is designed using the parallel discrete-event simulation capability provided by PARSEC. The protocol stack includes models for the channel, radio, MAC, network, transport, and higher layers. It supports TCP, IEEE 802.11 CSMA/CA, MAC, UDP, HTTP, FTP, CBR, Fishleye, LARScheme-1, ODMRP, WRP, DSR, MACA, Telnet, AODV, etc protocols. There is a visualization tool VT designed to view and help debugging the protocols. Glomosim currently supports protocols for only wireless network. In the future, it is planned to add functionality to simulate a wired as well as a hybrid network.

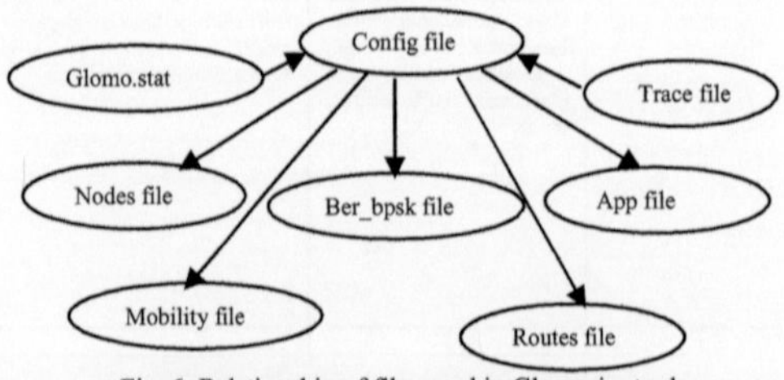

Fig. 6. Relationship of files used in Glomosim tool

There are eight files related to a network model designed in Glomosim network simulator as shown in Fig. 6. Six input files are used to execute a network model: Configuration file refers all other input files. It contains the basic configuration of a particular network model. Network protocols are defined here. Nodes file contains the node id and node coordinates. Node mobility (movement of the nodes and traffic) of the network model is defined in the Mobility file. In Routers file the packet routing is defined. Application protocols are defined in the App file. Bit error rate (bit per second in K) is defined in the Ber_bpsk file.

Two files are related to the output of the network model. All of the statistics (selected simulation events) is compiled together into a file called glomo.stat that is produced at the end of the simulation. Trace file is produced (if 'write trace' option is chosen) which can be played as many times as user wants and this is faster than real time. Basically, Glomo.stat and trace files store same set of data. But trace file can be simulated any time which is not possible with Glomo.stat.

IV. Experiments with Some Models

Some experiments have been performed using `NeDaSE` with these four tools `BRITE`, `ns-2`, `Delite` and `Glomosim`. As these tools are developed by different users/planners/developers, naturally, they have some incompatibilities [5]. After analyzing these incompatibilities suggested solutions have been implemented in tool *NeDaSE*.

A. *Transformation from .brite to .net/.gen .ns-2/.tcl and glomosim*

For the first experiment 10 nodes are chosen. After applying Waxman topology design algorithm in `BRITE` topology generator, the topology was created as shown in Fig. 7 which contains nodes id, nodes coordinates, the source and sink of links. There are generated 9 duplex links. Now, the model is converted into `Delite, ns-2` and `Glomosim` compatible format (see visualization examples in Fig. 8. Fig. 9 and Fig. 10 and see model description examples in [5]). In order to do this, incompatibilities have been removed and additional data are included which are needed for `Delite, ns-2` and `Glomosim.`

To have complete `ns-2` model, additionally to the relatively simple mappings of the node and link data, it is necessary to create certain amount of traffic (transport connections and traffic sources, simulation session). Current prototype is applying simple algorithm to generate traffic. Some parameters have been applied from template to `Delite` to make the model workable. Five files are generated (e.g. .gen, .net, inp, .cst, .req) by `NeDaSE` for `Delite` network model. To have complete model in Glomosim, five files have been generated (configuration, nodes, mobility, routes and app files).

B. *Transformation from .net to .ns-2/.tcl, .brite and Glomosim*

Here 8 nodes are chosen in a "square world" in `Delite`. Using Sharma algorithm a network model Sharma8 is designed (see Fig. 11). With `NeDaSE` tool transformed the Sharma8 model (`Delite`) into `ns-2, BRITE` and `Glomosim` formats (shown in Fig 12, Fig. 13 and Fig. 14). In Glomosim and Brite, nodes 1, 2 ....n are labeled as 0, 1,....(n-1) respectively.

```
Topology: ( 10 Nodes, 9 Edges )
Model (3 - ASWaxman):  10 1000 100 1  2  0.15000000596046448 0.20000000298023224 2 2 10.0 1024.0

Nodes: ( 10 )
0    192  565  1    1    0    AS_NODE
1    43   454  1    1    1    AS_NODE
2    418  158  1    1    2    AS_NODE
3    827  780  1    1    3    AS_NODE
4    249  847  1    1    4    AS_NODE
5    946  120  1    1    5    AS_NODE
6    470  419  1    1    6    AS_NODE
7    491  140  1    1    7    AS_NODE
8    721  947  9    9    8    AS_NODE
9    969  848  1    1    9    AS_NODE


Edges: ( 9 )
0   8   3   197.80040444852483   0.6597911293970071   846.9652508902105    8   3   E_AS   U
1   8   1   838.2917153354196    2.7962401753796606   603.9047449044509    8   1   E_AS   U
2   8   9   267.02996086581743   0.8907160728700434   357.8318759066749    8   9   E_AS   U
3   8   0   652.5067049463936    2.1765280864617136   836.2474527834161    8   0   E_AS   U
4   8   6   584.6238106680227    1.9500951243677471   543.5865956499515    8   6   E_AS   U
5   8   4   482.4769424542483    1.609369847637222    626.9905723413225    8   4   E_AS   U
6   8   2   845.1804541043291    2.819218534524738    657.7450511109523    8   2   E_AS   U
7   8   7   839.1358650421278    2.7990559557109598   372.34376211386393   8   7   E_AS   U
8   8   5   857.0612580206855    2.858848630610599    184.301105133698 8   5   E_AS   U
```

Fig. 7:Topology of Waxman model generated by BRITE tool (file Brite10.brite)

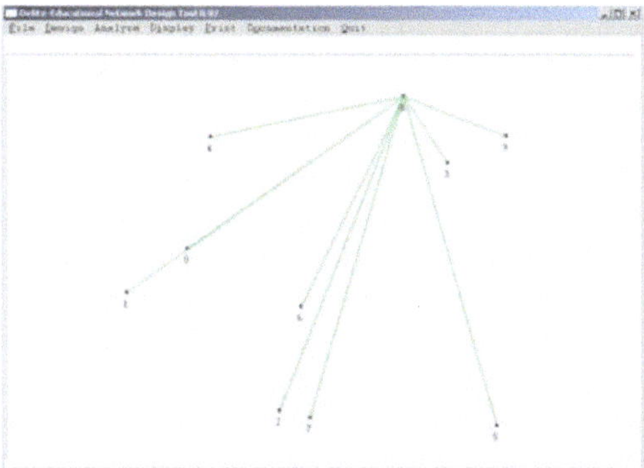

Fig. 8: Visualization of Brite10 model generated by NeDaSE in Delite

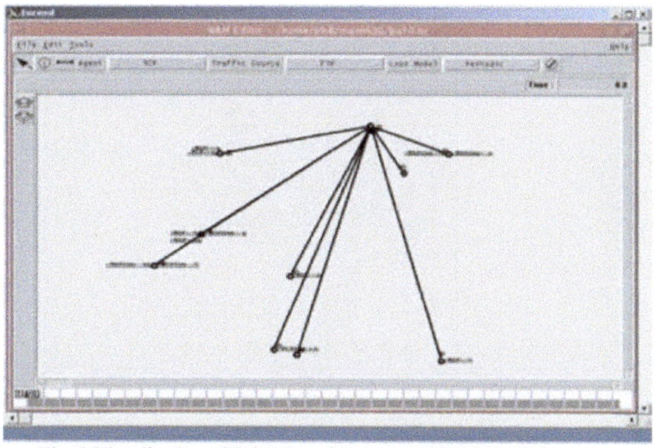

Fig. 9:Visualization of Brite10 model generated by NeDaSE in ns-2

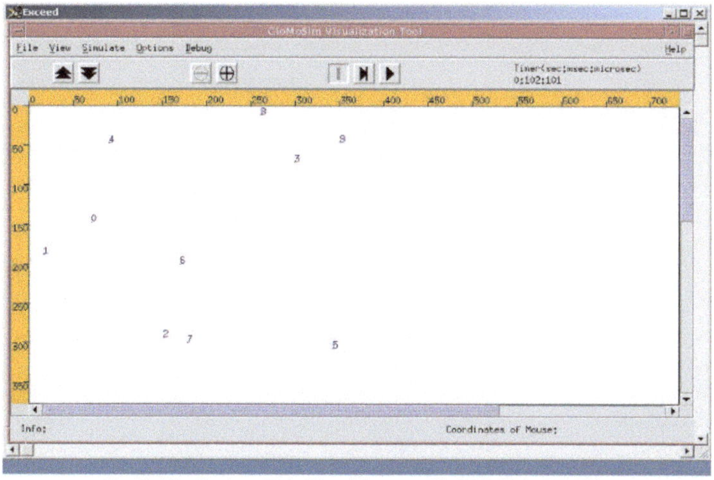

Fig. 10: Visualization of Brite10 model generated by NeDaSE in Glomosim

C. Transformation from .ns-2/.tcl to .net/.gen, .brite and glomosim

For this experiment, a N6 model with 6 nodes is designed manually for `ns-2` (see Fig. 15). The topology has five duplex links. Some transport protocols and traffic sources are included in this model. The models generated by `NeDaSE` in `Delite`, `Brite` and `Glomosim` format are shown in Fig. 16, Fig. 17 and Fig. 18. As there is no way to use application and transport protocols in `Delite` and `BRITE`, this information present in `ns-2` model is jus t ignored.

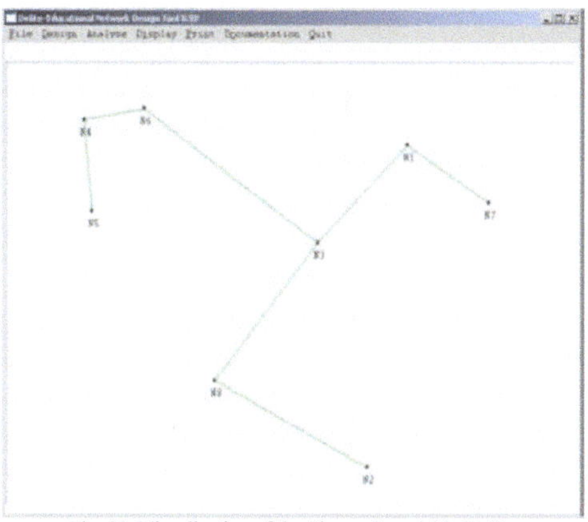

Fig. 11: Visualization of the Sharma8 model in Delite

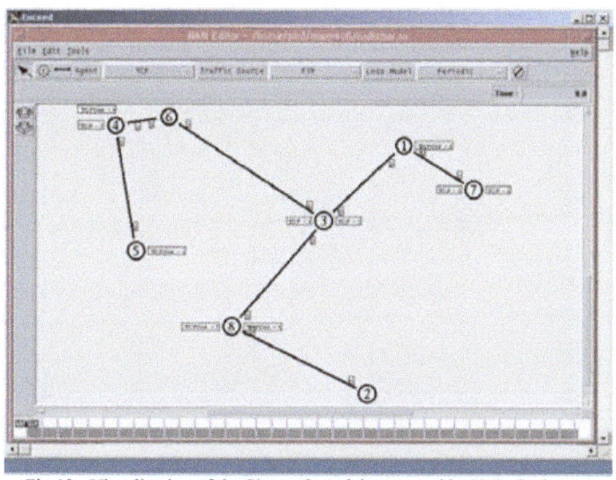

Fig.12: Visualization of the Sharma8 model generated by NeDaSE in ns-2

```
Topology: ( 8 Nodes, 7 Edges )

Model (3 - ASWaxman):  8 1000 100 2  2  0.15000000596046448  0.20000000298023224 1 1 10.0 1024.0

Nodes: ( 8 )
0   83  53  2   2   0   AS_NODE
1   80  30  2   2   1   AS_NODE
2   75  46  2   2   2   AS_NODE
3   56  55  2   2   3   AS_NODE
4   58  43  2   2   4   AS_NODE
5   61  56  2   2   5   AS_NODE
6   90  49  2   2   6   AS_NODE
7   67  37  2   2   7   AS_NODE

Edges: ( 7 )
0   7   2   10  2   10  8   3   E_AS U
1   6   0   10  2   10  7   1   E_AS U
2   5   2   10  2   10  6   3   E_AS U
3   4   3   10  2   10  5   4   E_AS U
4   3   5   10  2   10  4   6   E_AS U
5   2   0   10  2   10  3   1   E_AS U
6   1   7   10  2   10  2   8   E_AS U
```

Fig. 13: Sharma8 topology model in BRITE format generated by NeDaSE

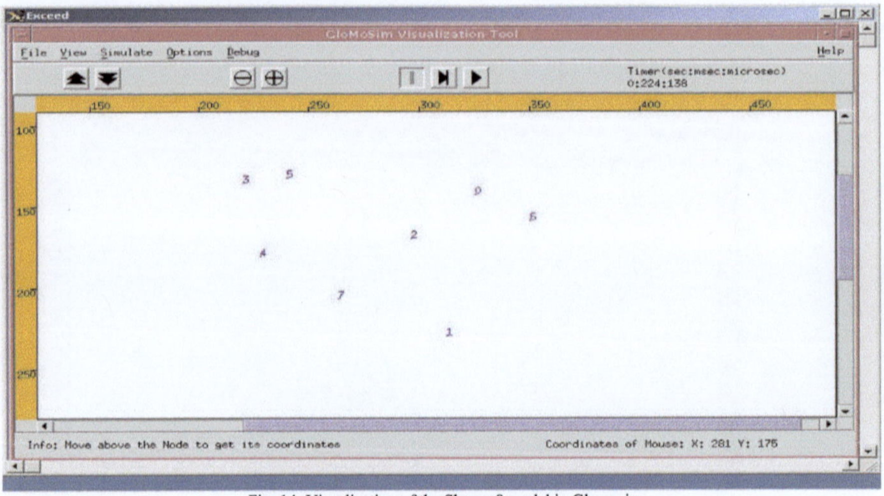

Fig. 14: Visualization of the Sharma8 model in Glomosim

D. *Transformation from.Glomosim to .ns/.tcl, .net/.gen and .brite*

Experiments were carried out with model from Glomosim (.in) to Ns-2, Delite and Brite. Fig. 19 shows a model with 5 nodes. It is transformed to Ns-2, .net as shown in Fig. 20 and Fig. 21 respectively.

As there is no link information in Glomosim (wireless network simulator), it cannot be transformed into brite and .net (other files of `Delite` can be created).

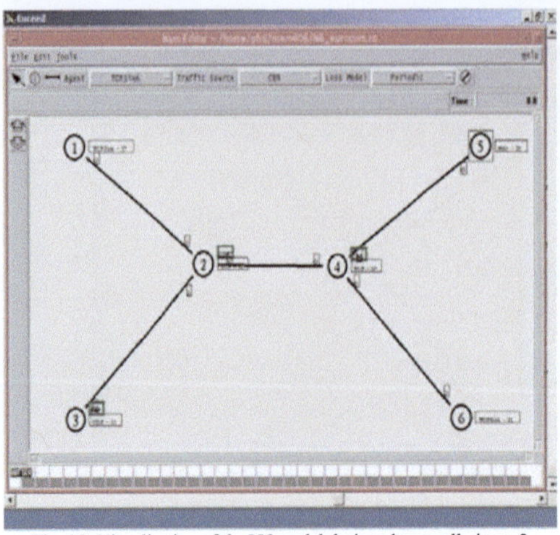

Fig. 15: Visualization of the N6 model designed manually in ns-2

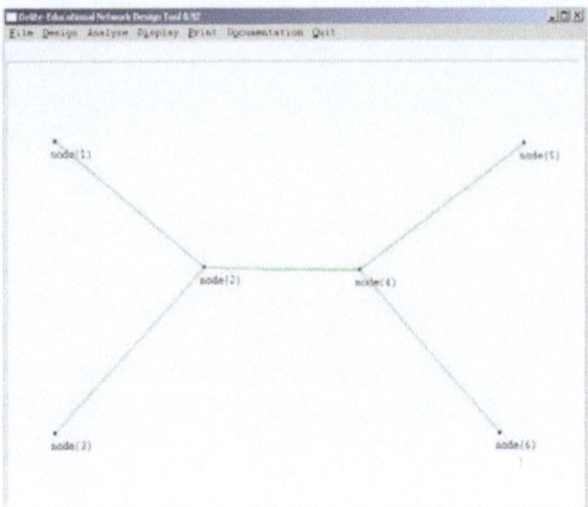

Fig. 16: Visualization of the N6 model generated by NeDaSE in Delite

```
Topology: ( 6 Nodes, 5 Edges )
Model (3 - ASWaxman):  6 1000 100 2  2  0.15000000596046448  0.20000000298023224 1 1 10.0 1024.0

Nodes: ( 6 )
0   90  73  2   2   0   AS_NODE
1   90  88  2   2   1   AS_NODE
2   84  80  2   2   2   AS_NODE
3   73  80  2   2   3   AS_NODE
4   67  73  2   2   4   AS_NODE
5   68  89  2   2   5   AS_NODE

Edges: ( 5 )
0   2   0   10  2   10  3   1   E_AS U
1   2   1   10  2   10  3   2   E_AS U
2   3   2   10  2   10  4   3   E_AS U
3   4   3   10  2   10  5   4   E_AS U
4   5   3   10  2   10  6   4   E_AS U
```

Fig. 17: N6 model in BRITE generated by NeDaSE

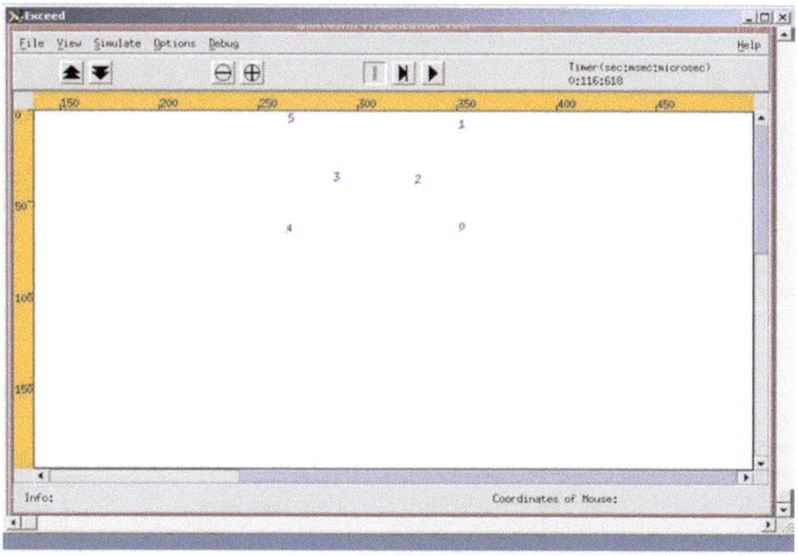

Fig. 18: Visualization of the N6 model designed manually in Glomosim

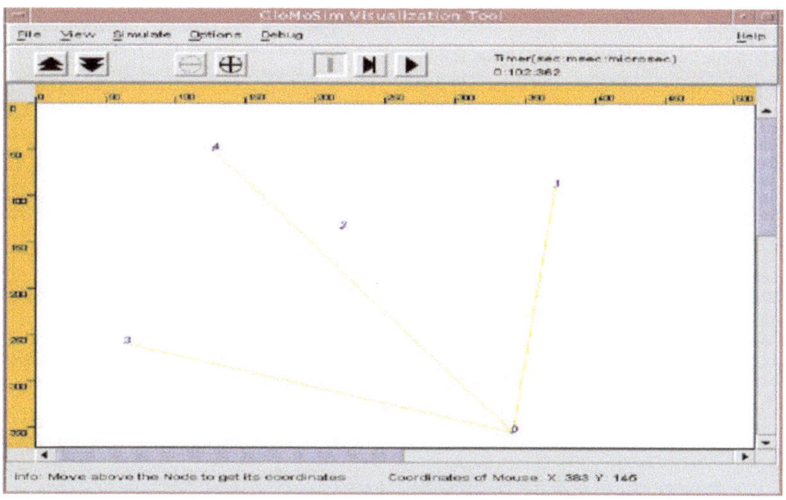

Fig. 19: Visualization of the G6 model designed in Glomosim

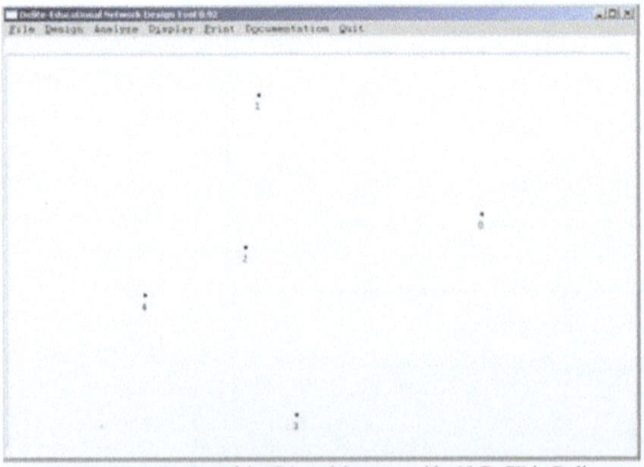

Fig. 20: Visualization of the G6 model generated by NeDaSE in Delite

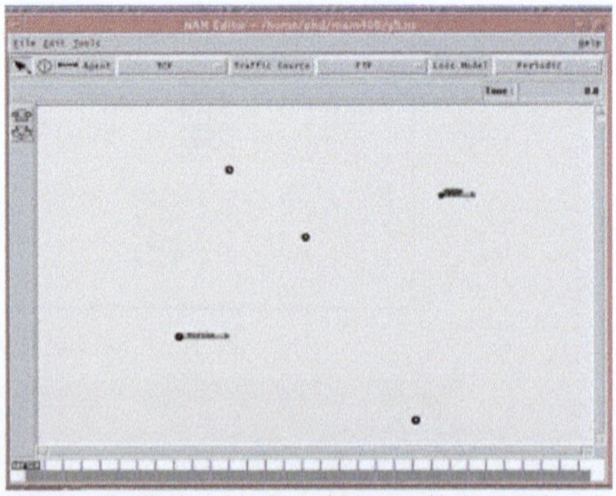

Fig. 21: Visualization of the G6 model generated by NeDaSE in ns-2

V. CONCLUSIONS AND FUTURE WORK

This paper presents a prototype of an approach to integration of network design, simulation and topology discovery tools. Because of the distinctive nature of the network design, simulation, topology generation and topology discovery tools and usefulness of them, it is beneficial to use them. But models used by them often differ in the expression, format and style. For an instance, this paper shows details about `BRITE`, `Delite`, `ns-2` and `Glomosim` tools. A lot of incompatibilities exist among them are solved for integration. Paper also demonstrates some experiments that are performed for the transformation of small sized models between `BRITE`, `ns-2`, `Glomosim` and `Delite` in all directions using an integrated tool, `NeDaSE`. Experiments with large models are performed and these work fine. Possibility to exchange models between tools shows the feasibility of such integration approach. The tool `NeDaSE` is currently used in the London Metropolitan University for teaching of the course on Network Planning and Management.

Future work will focus on studying most useful network design, simulation and topological discovery/analysis tools, their characteristics, their inconsistencies and integrate these tools to make a full phased integrated design and simulation environment.

REFERENCES

[1] T. G. Robertazzi, "Planning Telecommunication Networks", *IEEE Press*, 1999.

[2] R. V. Slyke, "Network Planning and Design", Polytechnic University, New York, Revised 21 January 2001. (http://cis.poly.edu/rvslyke/ndr1.pdf).

[3] A. W. Bragg, "Which network Design Tool Is Right for You?", *IT Professional*, September/October 2000, pp. 23-31.

[4] H. Fathi, H. Yomo, G. Kuijpers, H. Christiansen, "CNTK Deliverable: Simulator requirements - a comparative evaluation of tools", *CNTK Project*, Department of Communication Technology, Institute of Electronic Systems, Aalborg University, Denmark, September, 2003.

[5] A. Pakstas, M. A. Rahman, "Incompatibility of Network Design and Simulation Tools and An Approach to Integration", *Proc. of the 6th EPSRC Annual Postgraduate Symposium on the Convergence of Telecommunications, Networking and Broadcasting (EPSRC PGNet 2005)*, Liverpool John Moores University, June 27-28, 2005, Liverpool, England, pp. 529-535.

[6] M. A. Rahman, A. Pakstas, F. Z. Wang, "NeDaSE: A Bridge Between Network Topology Generator, Network Design and Simulation Tools", *Proc. of the EUROCON 2005 - The International IEEE Conference on Computer as a tool,* November 21-24, 2005, Sava Center, Belgrade, Serbia & Montenegro.

[7] M. A. Rahman, A. Pakstas, F. Z. Wang, "LNMet-X: A language for Network Meta-description Used for Integration of Tools", *Proc. of the 2005 International IEEE Conference on Software, Telecommunications and Computer Networks (SoftCOM 2005)*, September 15-17, 2005, Split, Marina Frapa, Croatia.

[8] A. Medina, A. Lakhina, I. Matta, J. Byers, "Brite: Universal topology Generator from a User's Perspective", *Technical Report, BUCS-TR-2001-003,* April 12, 2001, Boston University. (http://www.cs.bu.edu/brite/publications/usermanual.pdf)

[9] A. Medina, A. Lakhina, I. Matta, J. Byers, "Brite: An Approach to Universal Topology Generation", *Proc of the 9th IEEE International Symposium on Modeling, Analysis and simulation of computer an Telecommunication Systems (MASCOT)*, 2001.

[10] R. S. Cahn, *"Wide Area Network Design: Concepts and Tools for Optimization"*, Morgan Kaufman Publishers, San Francisco, 1998.

[11] L. Bresla, D. Estrin, K. Fall, S. Floyd, J. Heidemann, A. Helmy, P. Huang, S. McCanne, K. Varadhan, Y. Xu, H. Yu, "Advances in Network Simulation, The VINT Project", *IEEE Computer,* 2000, N5, pp. 59-67.

[12] K. Fall, K. Varadhan, The NS Manual, The VINT Project, 13 December 2003.(http://www.isi.edu/nsnam/ns/doc/ns_doc.pdf)

[13] B. B. Welch, *"Practical Programming in Tcl and Tk"*, Prentice Hall, Inc, 1995.

[14] Xiang Zeng, Rajive Bagrodia, Mario Gerla, "GloMoSim: A Library for Parallel Simulation of Large-scale Wireless Network", *Proc. of the 12th workshop on Parallel and Distributed Simulation,* Banff, Alberta, Canada, 1998, pp. 154-161.

[15] *GloMoSim Manual (ver. 1.2)*. 07 February 2001. University of California, Los Angeles. (http://pcl.cs.ucla.edu/projects/ glomosim/ GloMoSimManual.html)

Bringing DRM interoperability to digital content rendering applications

Carlos Serrão, Miguel Dias and Jaime Delgado

***Abstract*— Modern high-bandwidth and open networks are a good medium for digital content distribution. However, digital content distribution over open networks can raise an all set of different issues related to piracy and intellectual propriety infringements and circumventions. Therefore, in order to avoid and overcome such issues, a good content security mechanism must be placed online to control and manage how the content access is processed.**

There are some actual discussions and initiatives going on to provide such controlling mechanisms, but perhaps one of the most highlighted is Digital Rights Management (DRM). Nevertheless, current DRM solutions, although solving part of the problem, most of the times end up adding an additional problem – interoperability.

This paper will provide a description of a method and a system that is currently developing a new approach to the DRM interoperability issue. The method, which is described on the paper is part of a DRM system implementation whose description will also be provided here, and is based on the introduction of a middleware layer between the content rendering applications and the rights and content suppliers.

***Index Terms*— Interoperability, Digital Rights Management, Security, Digital Content, Network**

I. Introduction

Digital open networks, specially the Internet have changed many aspects of our live – technologically, economically and socially. In particular it has mostly changed the way we exchange (digital) information. This digital network has also influenced the way multimedia content is distributed and obtained [1]. However, the distribution of digital multimedia content over digital networks has some serious problems. One of the most common and important problems refers to the content piracy or copyright circumvention.

This is an important barrier to those who whish to use this contribution channel – especially to authors and content owners. To overcome some of the problems created by unauthorized digital content distribution and usage, some technologic measures were put in place. Common and simple technologic procedures include content encryption, scrambling and watermarking. More complex measures include the definition of business rules and content usage rules, managed by an infrastructure called Digital Rights Management (DRM).

DRM is a set of technological components which is capable of performing a variety of functions, such as content protection, content management, authentication, authorization and accounting, payment and rights management [2]. DRM can uphold specific conditions to specific groups of users or domains in terms of digital content usage. This is something that goes inline with the content owners and authors expectations, on what concerns piracy prevention, but sometimes it creates some obtrusiveness on the user side.

One of the main reasons for this user side obtrusiveness is mostly due to the fact that most of the user requirements on what concerns the digital content experience are disregarded and only content owners and authors are considered in the design of the actual DRM-solutions. This results, in most of the cases, that the user experience with DRM copyrighted and protected digital content is not the best and the most obvious alternative for users is to find alternative ways to enhance such experience – obtaining illegal pirated content on P2P networks, for instance.

Other important source of obtrusiveness resides in the fact that most of the existing DRM solutions are proprietary and vertical. This means that they have their own protected-content formats, specific protection tools and specific rights expression mechanisms. In practice, this means that a user that buys a music track from an online digital content store that uses a specific DRM protection scheme is unable (most of the times) to render that content on its music player application, and its limited to whatever music player application which implements the DRM protection scheme that is the result of a business partnership between the digital content provider and the DRM technology provider. Sometimes, this situation causes more problems than it solves.

The following section of the paper will start by presenting a short description of an open DRM platform - called Digital object Rights Management (DoRM). This presentation will focus mainly on how the platform generates and manages digital content usage rights [3]. Then the description of a middleware layer that manages the rights at the end user side is also presented.

Carlos Serrão and Miguel Dias are with Adetti/ISCTE - Ed. ISCTE – Av. das Forças Armadas, 1600-082, Lisboa, Portugal; (e-mail: Carlos.Serrao@iscte.pt , Miguel.Dias@iscte.pt.)

Jaime Delgado is with Universitat Pompeu Fabra, Departament de Tecnologia, Pg. Circumval·lació 8, E-08003 Barcelona, Spain (e-mail: jaime.delgado@upf.edu).

K. Elleithy et al. (eds.), Advances in Computer, Information, and Systems Sciences, and Engineering, 323–329.

II. THE DIGITAL OBJECT RIGHTS MANAGEMENT SOLUTION

Interoperability can be addressed from many directions. In what concerns digital multimedia content interoperability it can be seen in terms of content format interoperability, content protection methods interoperability, rights expression interoperability and others. For the sake of this paper, the interoperability refers to rights management interoperability, although some of the concepts may also be applicable to other types of interoperability.

The approach that will be followed to achieve such rights management interoperability will consist in two different components:

1. At the server-side, in the usage of a set of pre-defined rights templates that will enable digital multimedia content stores to align their own business model with the DRM enforcing layer;
2. At the client-side, the establishment of a DRM middleware layer, between the content rendering applications and the rights definition, allowing any content rendering application to ask to this middleware layer, the permission for conducting an operation over the content.

These two approaches are being developed as part of a DRM solution called Digital Rights object Management. DoRM is a service-oriented DRM platform [3][4], independent from the type of content, the content protection and the business model under which the content is used. It can be used with multiple communication protocols and is based on the emerging service-oriented paradigm (SOAP [13], WSDL and UDDI) approach [3], called Service Oriented Architecture (SoA). DoRM (Figure 2) can cover most of the content lifecycle phases: from content authoring, distribution and management of the related rights up to the final user.

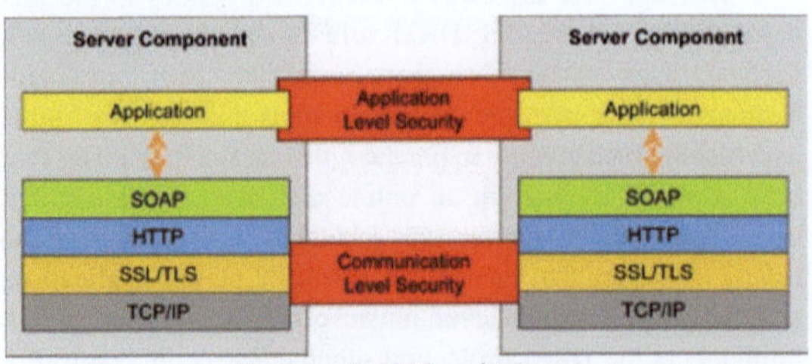

Figure 1 – DoRM Security Layers

The communication between the single components is typically performed over an insecure network. This introduces special needs regarding the security of this communication. DoRM uses two different security layers (Figure 1). The first security layer is established at the communication level, which will provide the necessary secure and authenticated communication medium that allows components to communicate with each other. This first layer uses SSL and X.509 certificates that are installed on the servers allowing this first authentication and ciphering layer. The second security layer is established at the application level, ensuring the integrity, authentication and non-repudiation mechanisms needed by the different components which provide the DoRM services.

This second layer, which corresponds to the exchange of SOAP messages internally between the different DoRM services, and between external actors and those same services, uses the possibilities offered by the W3C XML security-related recommendations: SOAP extensions to use XML Signature. This is part of the new XML security-related architecture (WS-Security).

The authentication between components is provided using a specific developed mapping of X.509 certificates in XML. Each of the DoRM platform services uses this two layered security model. One might think that this is an overlap, or even an overuse of security techniques, however it is not. While the underlying layer provides the authenticity of the web-servers (in which web-services are deployed) and the confidentiality of the communication channel between these servers, the upper layer provides only authentication, integrity and non-repudiation for the different services and actors that interact over the platform, which may be independent of the web-server itself. The other main reason for using SSL is because the DoRM communications are mostly point-to-point and without big scalability requirements.

The DoRM platform was designed having in mind concepts such as content adaptation and business models wide range applicability (download, super-distribution, streaming or even broadcasting). In a more technical approach, DoRM is composed by a set of external actors or elements and a set of internal components [3]. The internal components are oriented towards the service they supply, and are described in more detail in the next section. These internal components are self-descriptive, in the sense that they expose an open WSDL description of the services they provide, and any authenticated component can connect to it and the services it provides – DRM services. These internal components communicate with each other using SOAP messages [13]. The discovery and identification of services is currently being provided by a central configuration component, but in the future this service will be provided by an UDDI server which will provide information about the services provided by the platform and how to use them.

This DoRM platform has already been implemented and was already used in some specific R&D projects. However, there are still some on-going developments that will take place mainly to extend some of the functionalities and improve it in terms of robustness.

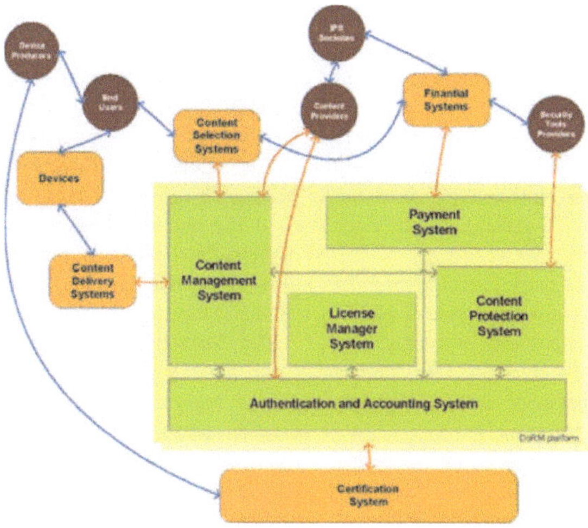

Figure 2 – OpenSDRM service-oriented architecture

III. DoRM ARCHITECTURE

Around the DoRM platform there are a set of actors and external systems (Figure 2).

A. External Actors

The external actors are: the End-User, the Device Provider, the Content Provider, the Security Tools Providers and the IPR societies.

The **End-User** represents an entity who wishes to use some multimedia digital content. This content may or may not be DRM-protected. However, the way to access and display such content may require the use of protected devices, software and licenses. The User will make requests to DoRM in order to: provide identification information, perform authentication, download licenses and use the content.

The **Device Provider** is an entity which provides either the necessary software and/or hardware (the Device) that can be used by the End-user to render digital multimedia content (DRM-protected or not).

The **Content Provider** is any digital multimedia content provider that feeds a Commerce Platform or a Content Management System, connected to the DoRM with digital content and optional associated metadata.

The **Security Tools Provider** is any organization that produces tools or technologies for encryption, scrambling, watermarking and others that can be applied to content protection. These tools are registered and made available to DoRM for use in content rights protection. These tools will need to comply with some guidelines, defined by the platform manager. These guidelines bound together with a subscription, are translated into a business relation that must exist between a given Content Provider and the Protection Tools Provider. A given producer and/or distributor of content, may want to choose which type of protection the content will have and, respectively, which tools can be applied to the content and from which supplier.

The **IPR Societies** are the entities which represent the legal owners of the multimedia content and that are responsible for managing royalty's distribution among all the digital multimedia content value chain.

B. External systems

The external systems which may interact with the DoRM platform are: the Devices, the Content Delivery Systems, the Content Selection system, the Financial System and the Certification System.

The **Devices** are client-side components that represent the software or hardware that will be used to render the content. This is a generic component with the particularity of being able to display/playback the appropriate content for which the necessary audio/video codec should be available (if this codec is not available it must be downloaded from a remote secure server).

The **Content Delivery Systems** are systems that are used to deliver the final content to the End-User through a Device –

these systems may be particular for each of the methods used to deliver the content.

The **Content Selection System** is an integration component responsible for establishing the liaison between the platform that actually supplies the content and the DRM platform. Normally, content is chosen via web browser, some very generic metadata might be consulted, information about the price is also available, and especially the content usage conditions might be established.

The **Financial System** facilitates the commercialization of digital content. DoRM plays an important role since it provides the services for handling electronic payments. The interface between DoRM and the Payment Infrastructure is generic and independent from the payment method, allowing therefore a multiplicity of payment systems.

The **Certification System** is responsible for receiving requests for and issuing credentials to entities. These credentials will be used by entities to authenticate themselves to each other, allowing the establishment of secure and authenticated communication channels between them (this is part of the establishment of one of the two DoRM security layers). All the components in the DoRM architecture communicate using the channel security provided by the SSL/TLS protocol [3]. This Certification Authority may be internal to DoRM, and therefore entirely managed by some entity, or it may be an external commercial entity, such as Verisign or Thawte [3][4].

C. *Internal Components & Interfaces*

The internal components of the DoRM platform include: Content Management System, License Manager System, Payment System, Content Protection System and Authentication and Accounting System.

The **Content Management System** is a server component whose role is to assign unique identifiers to content and to register metadata information for that specific content. The service assigns unique identifiers to content using the MPEG-21 [6] directives about Digital Item Identification (DII) [7], using a reduced version of the MPEG-21 DII Digital Object Identifiers [6][7][16]. This server component is also responsible for notifying the appropriate content servers that a given content has been requested and that needs to be feed to the final user. This Content Management System handles also the content preparation. It receives raw content from a specified source or sources and encodes it on a specified format, adds metadata and protects it. It is not implemented using the WS approach, although it uses some components that provide such approach.

The **License Manager System** is a server component responsible for house-keeping the rules associating a user, the content and his/her corresponding access rights. This component will accept connections from authenticated content rendering application clients for downloading licenses, which will be applied to the protected content through an appropriate protection tool. By default the licenses are XML formatted using Open Digital Rights Language (ODRL) [14], but other licensesformats, such as MPEG-21 Rights Expressions Language (REL) [6] can also be supported.

The **Payment System** is a server component responsible for verifying and validating the payment methods provided by the User to a Commerce Server.

The **Content Protection System** is a component responsible for registering new protection tools and for receiving authenticated client content rendering application requests for the downloading of a specific protection tool. It is also responsible for making protection tools available to the Content Preparation service to allow the protection of content.

The **Authentication and Accounting System** is a key-component of the system. It is responsible for authenticating all the internal services and components as well as some external actors to the DRM system. It validates the access rights of all of them working as a single sign-on point, registering and managing components and users on the system. It uses cryptographic XML credentials to authenticate both components and users in order to authenticate the transactions exchanged between them (XML Encryption and XML Signature) [10][11].

IV. DIGITAL CONTENT RENDERING APPLICATIONS DRM INTEROPERABILITY

As it was referred before on this paper, interoperability is a key issue in DRM and it is extremely hard to achieve. This section of the paper discusses only one of the many DRM-interoperability issues. The approach that is described and discussed on this paper is based on the establishment of a rights management interoperability layer at the client-side. The objective of this interoperability layer is to free the content rendering applications from the burden of having to support multiple rights expression languages processing and different authorization modules implementation. This interoperability layer provides such functionalities to all registered content rendering applications.

In order to achieve such interoperability level, a DRM middleware layer is defined and built at the client-side. The major goal of such interoperability layer is to allow that different content rendering applications can be abstracted from the inner DRM mechanisms that will uphold the content provider user rights at the end-user side. This doesn't reduce however the need for content rendering applications to support the necessary cryptographic mechanisms that will be needed to access to the content.

This DRM middleware layer is capable of managing the access to protected content by different Content Rendering Applications (CRA). Every time a CRA wishes to perform an operation over the content, it contacts the wallet that authorizes or not such operation according to what is specified on the rights expression. This layer allows the coexistence of many DRM-protected content files and DRM-enabled applications on a single client system, presenting a total horizontal approach to DRM. Most of the DRM approaches existing nowadays are vertical: examples include Microsoft

Windows Media Rights Management [5][8] or Apple iTunes [9]. While a solution like Microsoft Windows Rights Management is end-to-end Microsoft system-dependent (even at the client-side) relying on Windows Media Player to obtain the licenses and enforce them over the content [5][8][12], DoRM follows a more horizontal approach in which several content applications can share the access to content, mediated by the DoRM middleware layer. This represents in fact an important interoperability layer at the client-side (Figure 3).

A. Establishment of a DRM middleware layer

To establish such DRM middleware layer at the end-user side several steps need to be concluded. One of the most important steps that will need to be achieved is the end-user registration at the DoRM platform. This DoRM registration process occurs the first time the DRM middleware is boot up and uses the SSL/TLS protocol to establish a secure and authenticated channel with the DoRM platform servers. The process is composed by the following steps:

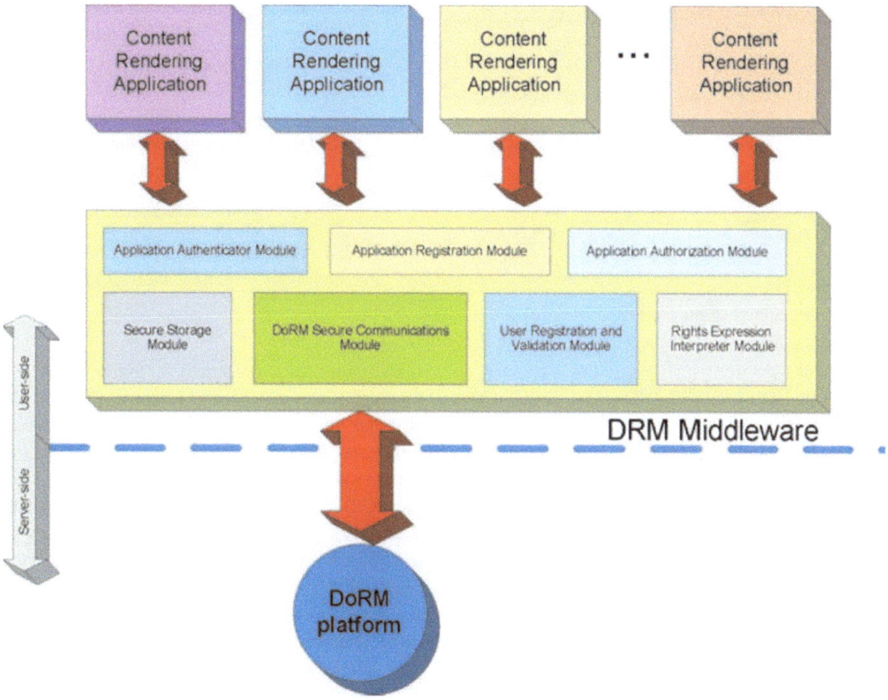

Figure 3 - DoRM client-side middleware

1. The DRM middleware layer software computes a key-pair ($Kpub_{mid}$, $Kpriv_{mid}$);
2. The $Kpriv_{mid}$ is internally stored on a secure repository (an encrypted file) which is ciphered with a key ($SSkey_{AES}$) which is computed hashing information of the pair username and password choose by the end-user and generic information collected from the device – $SSkey_{AES}\{Kpriv_{mid}\}$;
3. The user will introduce some more information to the DRM middleware interface and then this information is sent to the DoRM platform;
4. The DoRM platform registers the user and returns back a certificate that will validate this DRM middleware installation ($Cert_{DoRM}^{mid}$). This certificate contains the $Kpub_{mid}$ and the $Kpub_{DoRM}$, signed by the DoRM platform;
5. This certificate is received by the DRM middleware and is also stored in the secure storage – $SSkey_{AES}\{Cert_{DoRM}^{mid}\}$. This concludes the user registration process. Every time the DRM middleware boots up, the user has to authenticate to it – this DRM middleware supports more than one user, meaning that each of the users has to individually register to DoRM, repeating this five steps.

After a successful boot up the DRM middleware can receive multiple requests from different CRA. But before this occurs, the CRA needs to be registered at the DRM middleware – this is necessary to ensure that only registered applications are allowed to use the platform. Only registered and authenticated CRA can request content operations to the DRM middleware (this may include receiving content deciphering keys provided in the rights expressions). This means that any of the CRA that wishes to use this system will need to know how to execute the following two processes:

- Enrol to and request authentication to the DRM middleware, exchanging a set of credentials with the DRM middleware, to enable application authentication and the establishment of a secure channel between the application and the wallet. The process is performed in the following way:
    - The first operation that the CRA needs to perform is to compute a key pair ($Kpub_{CRA}$, $Kpriv_{CRA}$);
    - The KprivCRA should be stored securely by the CRA. The CRA sends $Kpub_{CRA}$ to the DRM middleware, to register it;
    - The DRM middleware, registers the $Kpub_{CRA}$, and generates a certificate to be returned for the CRA ($Cert_{mid}^{CRA}$) – this certificate contains also $Kpub_{mid}$ and its signed by the DRM middleware;
    - The certificate is received by the CRA and stored. The registration process is concluded with success.
- After the CRA is registered on the DRM middleware it can use it to request access clearance to perform DRM-protected content operations. This process starts with an authentication between the CRA and the DRM middleware to establish a common secret key to create channel between them. This process is performed in the following way:
    - The CRA sends its credentials to the DRM middleware: $Cert_{mid}^{CRA}$;
    - The DRM middleware validates the certificate and computes a secret session key ($SessKey_{AES}$). This session key is ciphered with the CRA public key ($Kpub_{CRA}$) and returns it to CRA;
    - CRA receives the session key and can use it ($SessKey_{AES}$).
- Request authorization to the DRM middleware to perform operations over the content. This process includes the extraction of content unique identifier (C_{ID}) and requesting the DRM middleware the permission to use the content. This middleware is responsible for getting the license from the server, parsing it, analyzing the rights associated to it before the approval or rejection of the operation over the content (this may include passing the decryption key to the application or the appropriate protection tool). This authorization request is expressed in a XML notation. This is processed in following way:
    - An authenticated CRA sends an authorization request to the DRM middleware: $SessKey_{AES}\{Authorization_{REQ}, C_{ID}\}$;
    - This request is received by the DRM middleware that verifies if there are on the system licenses for the C_{ID} and the User. If not, the DRM middleware connects to the DoRM platform and checks if there is a license for the User and CID specified. If this license exists, it is downloaded by the DRM middleware and securely stored. This license is interpreted locally at the DRM middleware, and the authorization that is requested by the CRA is checked against the rights expression present in the downloaded license. If the request authorization is a valid action over the content, the Content Encryption Key (CEK) is read from the license and returned to the CRA - $SessKey_{AES}\{CEK\}$;
    - The CRA receives the $SessKey_{AES}\{CEK\}$ and deciphers the CEK. This CEK is then use to perform the operation over the content. This authorization request is performed each time the CRA when an operation is conducted over the content.

This DRM middleware acts an intermediary between the CRA and the rights management imposed by the DoRM platform.

V. Conclusions

This paper has presented a comprehensive system for digital content rights management. This system uses a distributed service-oriented architecture that exchanges digital rights management related information over an open-network. To create secure and authenticated channels over this open-network, DoRM uses a two-layered system. A first one that is placed at the communication level supplied by the SSL/TLS protocol offering authentication and confidentiality of the services, and a second one, at the application level that offers authentication and non-repudiation of the messages exchanged between the services.

The DRM interoperability issue is a hot topic in the DRM World. Most of the existing DRM solutions do not interoperate – they are totally vertical. This results that each of the Content Rendering Applications has their own specific proprietary DRM system. This paper presented a possible solution, based on a middleware layer established between the CRA and the DRM platform that is used. These results in a system that is capable of supporting multiple CRA, multiple content types, multiple rights expression technologies and can even support multiple DRM platforms – this represents a total horizontal approach to the DRM interoperability problem.

A crucial issue on the system is the security of the Content Encryption Key. The details of the secure storage and the security of the communication channel between the CRA and the DRM middleware are also described.

This DRM interoperability layer is still at the prototype stage. Future plans include the support for several rights expression languages, the development a generic authorization model and the porting to several devices (PC, PocketPC and Mobile Phone).

REFERENCES

[1] Chiariglione, L., “Intellectual Property in the Multimedia Framework”, Management of Digital Rights, Berlin (2000)
[2] Duhl J., Keroskian S., "Understanding DRM Systems", IDC White Paper, 2003
[3] Serrão C., “Open Secure Infrastructure to control User Access to multimedia content”, WEDELMUSIC2004, September 2004, Barcelona
[4] Serrão C., Neves D., Kudumakis P., Barker T., Balestri M., "OpenSDRM - An Open and Secure Digital Rights Management Solution", IADIS 2003, Lisboa,
[5] Prunela A., “Windows Media Technologies: Using Windows Media Rights Manager to Protect and Distribute Digital Media”, MSDN Magazine, December (2001), http://msdn.microsoft.com/msdnmag/issues/01/12/DRM/default.aspx
[6] http://www.chiariglione.org/mpeg/mpeg-tech.htm#Digital_Item_Technologies
[7] ISO/IEC 21000-3 Information technology -- Multimedia framework (MPEG-21) -- Part 3: Digital Item Identification
[8] Microsoft, “Architecture of Windows Media Rights Manager”, Microsoft Corporation, http://www.microsoft.com/windows/windowsmedia/howto/articles/drmarchitecture.aspx, May 2004
[9] Lenzi, R. et al, “Apple iTunes Music Store”, technical report, The Interactive-Music Network, June 2003
[10] “XML Signature Syntax and Processing”, W3C Recommendation, February 2002, http://www.w3.org/TR/xmldsig-core/
[11] “XML Encryption Syntax and Processing”, W3C Recommendation, December 2002, http://www.w3.org/TR/xmlenc-core/
[12] Microsoft, “Scenarios for Windows Media DRM”, Microsoft Corporation, 2004, http://www.microsoft.com/windows/windowsmedia/drm/scenarios.aspx
[13] “SOAP Security Extensions: Digital Signature”, W3C Note, February 2001, http://www.w3.org/TR/2001/NOTE-SOAP-dsig-20010206/
[14] “Open Digital Rights Language (ODRL) Version 1.1”, W3C Note, September 2002, http://www.w3.org/TR/odrl/
[15] Iannella R., 2002, D-Lib Magazine, Volume 7 Number 6 ISSN 1082-9873, June (2001), http://www.dlib.org/dlib/june01/iannella/06iannella.html
[16] Dalziel, J., “DOI in a DRM environment”, White Paper, Copyright Agency Limited, 2004
[17] Rosenblatt, B. "Enterprise Content Integration with the Digital Object Identifier: A Business Case for Information Publishers", June 2002

DESIGN OF AN EDUCATIONAL SOFTWARE FOR SERVOMECHANISM EXPERIMENTS USING C-BASED GRAPHICAL PROGRAMMING

Author:Mirza Tariq Hamayun
Comsats Institute of Information Technology Lahore Campus
Email:Mirzatariq_hamayun@hotmail.com

Abstruct

This project deals with the computer control of a servomechanism using a C-based graphical programming. The speed and position of a servomechanism is controlled by a digital PID controller. The difference equation of the PI controller is obtained from the transfer function and the digital controller is implemented by the graphical Programming method. In the position control application of servomotor the proportional controller is used but in speed control application of servomotor proportional controller can reduce the steady state error if the load is not applied. In the loaded condition, the steady state error increases which is eliminated by the use of the integral controller. So by using the PI controller we can attain the actual speed equal to the desired speed in spite of the load variations. During the development of the graphical control software, the PCI-1200 DAQ card is used for the real time control. Square and triangular reference wave signals are produced by the proposed software, which greatly simplified the hardware configuration. The control system developed in this project is an alternative to very expensive commercial experimental sets, and this technique can also be used to develop computer based experimental sets for the other fields of engineering.

I Introduction

The main aim of this thesis is to develop a PC based experimental set using C-based graphical programming. The software developed in this project has similar features to that of commercial software such as feedback software. Apart from being a cheaper alternative the software developed in this project renders the use of some hardware components. Using this teaching ware software an implementation of a speed control of servomotor is made to investigate the properties of the software. The educational system of developed using C-based programming monitors any variations in the motor speed so that it can quickly return the speed to its correct value. In this project a fully computer controlled system is designed in which we can produce the reference input signal by using the proposed software, and control the frequency and amplitude of the control signal.

The use of a digital computer as a compensator (controller) device has grown during the past two decades as the reliability of digital computers has improved [1].
The digital computer in a system configuration receives the error in digital form and performs the calculations in order to provide an output in digital form. The computer may be programmed to provide an output, so that the performance of the process is near or equal to the desired performance. A digital computer receives and operates on signals in digital (numerical) form, as contrasted to continuous signals [2].

When we compare this teaching ware software with that of the commercial software such as feedback software, the proposed software produces similar performance with less complexity in hardware with a cheaper cost, because it produces the reference control signals within the computer and we don't need any boards for signal conditioning, unlike the commercial system. We know that cost is the most important factor in producing any product, so this software is more advantageous as compare to feedback software.

We can also compare our implementation work with a similar implementation made by [6]. That implementation requires an expensive package called LabVIEW, where as this software can operate standalone. Again cost is the factor, which is making our teaching ware software more preferable.

In one of the implementation using this proposed software digital PI controller is used to control the speed of the servomotor using a C-based graphical programming.

In the position control application of servomotor the proportional controller is used, but in speed control application of servomotor proportional controller can reduce the steady state error if the load is not applied. In the loaded condition, integral controller is used together with the proportional controller, because in loaded case, the steady state error increases which is eliminated by the use of the integral controller. So by using the PI controller we can attain the actual speed equal to the desired speed in spite of the load variations.

K. Elleithy et al. (eds.), Advances in Computer, Information, and Systems Sciences, and Engineering, 331–336.

The whole implementations work is done using the educational software and the PCI 1200 DAQ card.

II System Familiarization and Modeling

A. System Characteristics

The servo system under consideration consists of the Power Supply Unit and Mechanical Unit. Tachogenerator, which provides a voltage proportional to the speed, and a magnetic brake, which loads the system and potentiometer, which measure the position of the shaft, are available in the Mechanical unit. For more detail about the different functions available with these units see [3].

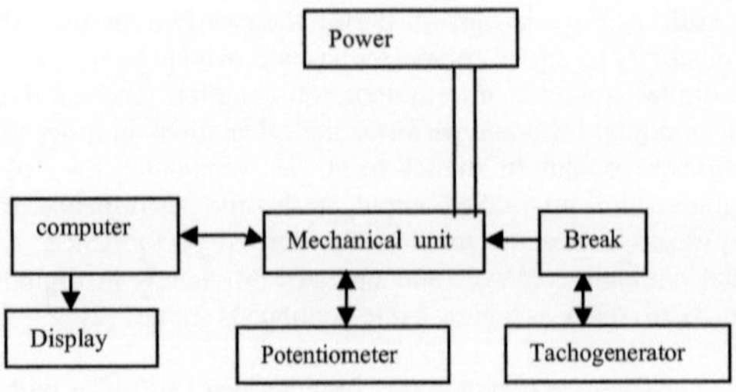

Fig. 1, Servo System

B. System Modeling:

The first step in analyzing a system is to establish a mathematical model of the system. In control engineering rather than dealing with hardware device whose components may be electromechanical, hydraulic, pneumatic or electronic we replace such devices or components by their mathematical models.

In obtaining a mathematical model however, we must make a compromise between the simplicity of the model and the accuracy of the results of the analysis.

Fig.2 shows the circuit diagram of the Dc servomotor, which is under consideration.

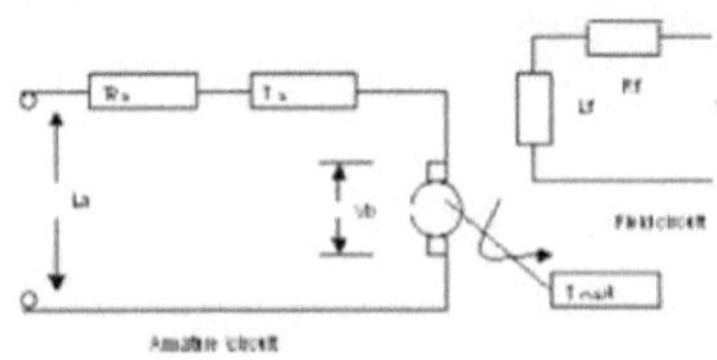

Fig. 2 Internal circuit diagram of the motor

The torque Γ generated by the motor [4] is

$$\Gamma(t) = K\, I_a(t)\, I_f(t) \tag{1}$$

The torque is used to drive a load through a shaft. The shaft is assumed to be rigid. Let J be the total amount of inertia of the load, shaft and the rotor of the motor, and θ be the angular displacement, and f the viscous friction coefficient of the bearing. Then the torque is

$$\Gamma(t) = J\,\frac{d^2\,\theta(t)}{dt} + f\,\frac{d\theta(t)}{dt} \tag{2}$$

The above equation is the relationship between the motor torque and the load angular displacement. For armature controlled Dc motor, field current I_f is kept constant or the field circuit is replaced by the permanent magnetic field, and if the input voltage u(t) is applied to the motor then (2) can be rewritten as

$$\Gamma(t) = K_t\, I_a\,(t) \tag{3}$$

Where $K_t = KI_f\,(t)$ is a constant. When the motor is driving a load a back Electro-motive force (back emf) voltage V_b will be developed in the armature circuit to resist the applied voltage. The voltage $V_b(t)$ is linearly proportional to the angular velocity of the motor shaft.

$$V_b\,(t) = K_b\,\frac{d\theta(t)}{dt} \tag{4}$$

Thus the armature circuit in Fig.2 is

$$R_a I_a(t) + L_a\frac{dI_a(t)}{dt} + V_b(t) = V_a(t) = u(t) \tag{5}$$

$$or\ R_a I_a(t) + L_a\frac{dI_a(t)}{dt} + K_a\frac{d\theta(t)}{dt} = u(t) \tag{6}$$

By applying the Laplace transform to the equation (2) and equation (6) together with (3) yields

$$K_t\, I_a\,(s) = Js^2\,\theta\,(s) + f\,\theta\,(s) \tag{7}$$

$$R_a\, I_a\,(s) + sL_aI_a\,(s) + sK_a\,\theta\,(s) = u(s) \tag{8}$$

Eliminating I_a (s) from these equations yields

$$G(s) = \frac{\theta(s)}{u(s)} = \frac{Kt}{[s(Js+f)(Ra+L_a s)+K_t K_b]} \qquad (9)$$

Where L_a is very small and can be negligible

$$G(s) = \frac{\theta(s)}{u(s)} = \frac{K_t}{[s(JR_a s + fR_a + K_t K_b)]} = \frac{K_m}{[s(\tau s+1)]} \qquad (10)$$

$$\text{Where } K_m = \frac{K_t}{[K_t K_b + fR_a]}$$

$$\text{and } \tau = \frac{JR_a}{K_t K_b + fR_a}$$

The transfer function from v to θ

$$\frac{v(s)}{\theta(s)} = k_l s \qquad (11)$$

and the combine transfer from v to u is

$$\frac{v(s)}{u(s)} = \frac{v(s)}{\theta(s)} \cdot \frac{\theta(s)}{u(s)} \qquad (12)$$

$$\frac{v(s)}{u(s)} = \frac{K_m}{[s(\tau s+1)]} \cdot k_l s \qquad (13)$$

$$\frac{v(s)}{u(s)} = \frac{k}{\tau s+1}. \qquad (14)$$

where v is the tacho output, u is the input voltage, k is the motor dc gain constant, and τ is motor time constant. No matter how carefully the experiments have been performed in order to compute a system transfer function, there are always uncertainties in the system [4]. So from [5] we have set the transfer function as follow,

$$P(s) = \frac{k}{\tau s+1}. \qquad (15)$$

Where $k \in [0.30, 0.733]$, $\tau \in [0.1, 0.325]$

III IMPLEMENTATION

The complete system which is used for implementation is shown in Fig.3

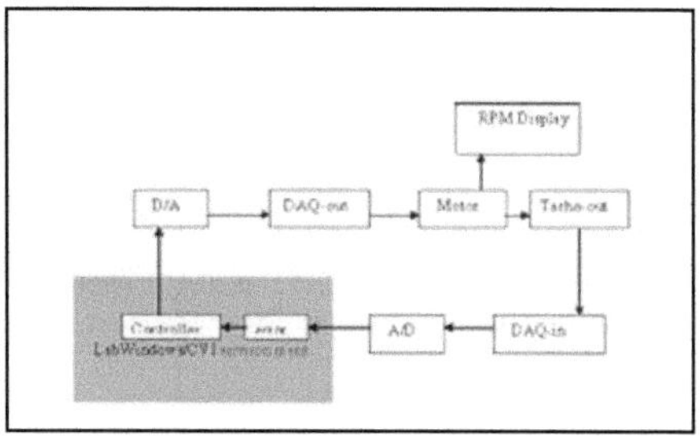

Fig. 3, complete servo- system block diagram

A. Digital PI Controller:

The transfer function of a PI controller is:

$$G_c(s) = \frac{U(s)}{E(s)} = K_p + \frac{K_i}{s} \qquad (16)$$

$$= K_p\left(1 + \frac{1}{Tis}\right) \qquad (17)$$

$$\text{where } T_i = \frac{K_p}{K_i}$$

$$= K_p \frac{T_i s+1}{T_i s} \qquad (18)$$

In the implementation of our application, the proportional controller can be used to control the speed of the servomotor with zero external load, but when the external load is applied (on the disc break), the speed of the motor will decrease, and the error will increase. So by using the proportional-Integral controller we can eliminate the steady state error.

The transfer function of the PI controller in z-domain is [14],

$$D(z) = K_p + K_i \frac{T}{2}\left[\frac{z+1}{z-1}\right] \qquad (19)$$

For the implementation purposes we need the difference equation of the PI controller, so the difference equation of the PI controller is,

$$u(kT)= \qquad (20)$$

$$\frac{1}{2}[2u[(k-1)T]+(2K_p+K_iT)e(kT)+(K_iT-2K_p)e[(k-1)T]]$$

where u(kT) is the controller output, K_p is the proportional gain, K_i is the integral gain, e(kT) is the error signal, and T is the sampling time.

B. Physical limitations

Consider the complete servo-system block diagram as shown in Fig. 3. During implementation a critical eye is kept on the following points:

1. PCI-1200 DAQ-output must be in the range of ± 4.99 volts. Since in our implementation we are dealing with the square wave as a reference input, so the DAQ should be configured as bi-directional. In this case the DAQ-output is limited to ± 4.99 volts.

2. Similarly DAQ-input must be in the range of ± 4.99 volts. Otherwise the inputs exceed the limit may damage the DAQ the PCI-1200 card. These are very important steps, which should be kept in mind while implementing our system; otherwise we can damage the PCI-1200 DAQ card.

3. The input to the motor must be in the range of ± 10 volts

IV Practical Results

Finally after designing the virtual instrument for speed control, we can control the servo system from the front panel of the user interface file. On this panel different controls are given, so by changing the control parameters we can see the effects of these changes on the system response. Changing the values of the controls we can see effects graphically on this panel, where the reference speed input, the controller output, the motor speed output response, and the speed error signal can be observed.

To observe the square wave response, the PCI-1200 is configured in the bipolar range [-3.5,3.5]. Motor speed response, when the reference input is switching from positive to negative is checked. When the positive input is applied the motor rotates in the clockwise direction, but for the negative input it moves in the anti-clock wise direction.

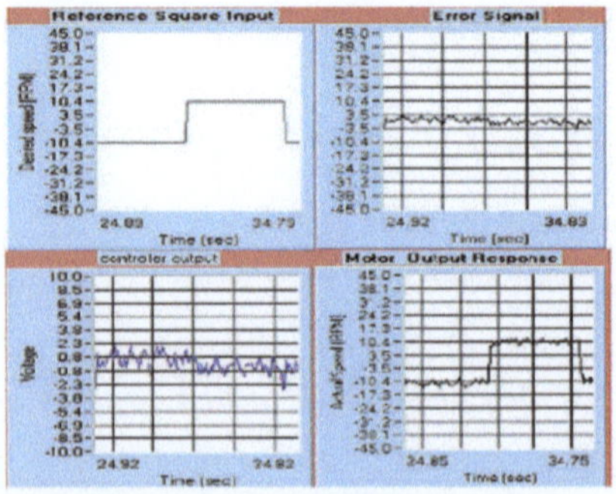

Fig. 4, Square wave response of the motor when the reference input signal is 10.4 rpm (no load)

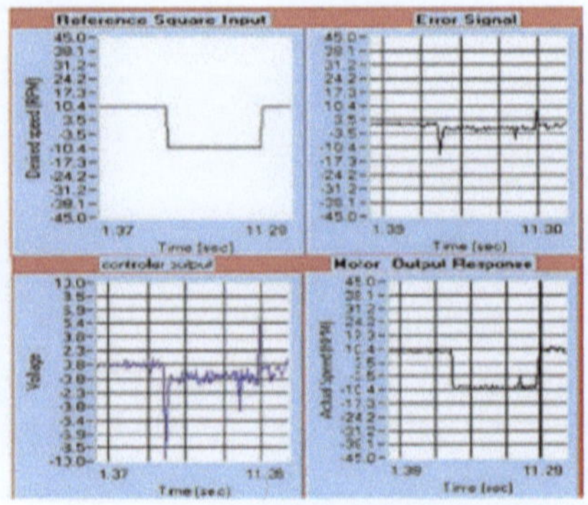

Fig. 5, Square wave response of the motor when the reference input signal is 10.4 rpm (with load)

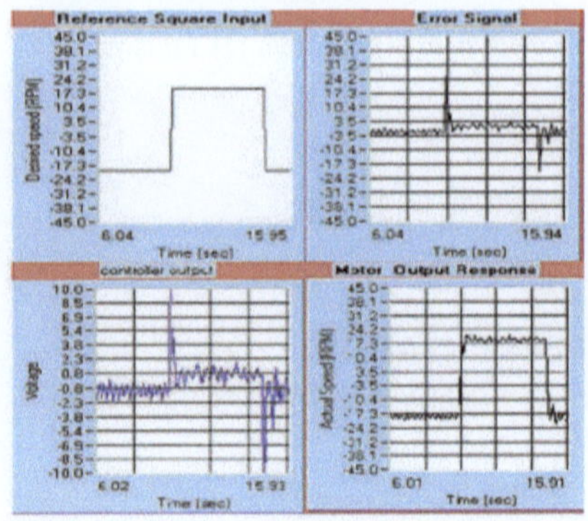

Fig.6 Square wave response of the motor when the reference input signal is 17.5 rpm (no load)

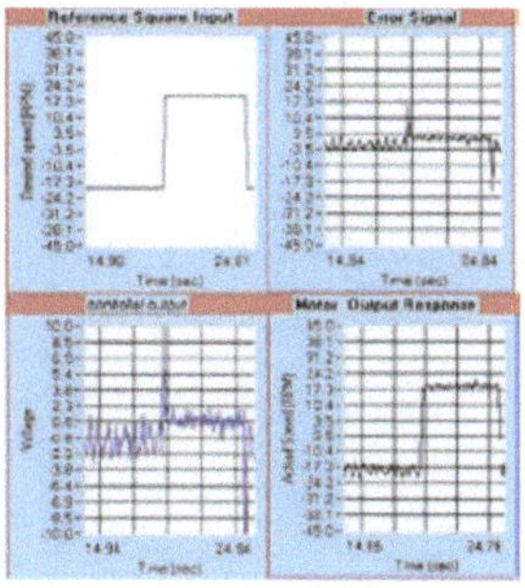

Fig.6, Square wave response of the motor when the reference input signal is 17.5 rpm (with load)

In one of the implementation we use a potentiometer as a position transducer. Potentiometer is actually a variable resistor, which produces a voltage depending on its angular position. In this implementation we will take the input from the external potentiometer, which is on the mechanical unit. By rotating the potentiometer in any direction, the motor will track that direction. In the following figure we can see that when the input potentiometer on the mechanical unit rotates in the clockwise direction, the waveform goes towards 180 degrees and the motor will also rotate towards that direction. Similarly motor moves towards –180 degrees direction when the potentiometer rotates anticlockwise.

Potentiometer produces a voltage proportional to the position of the shaft. This voltage is passed to the analog to digital converter and into the computer. The computer controls the position of the motor via the digital to analog converter. So we can control the position of the motor by this way.

Fig.7waveform showing the position of potentiometer

V. Conclusion

In this project, a complete servomechanism system is developed using the graphical C programming technique, which can be used for educational purposes in engineering schools. This project focuses on the importance of computer-controlled systems. By using the proposed software some implementations are made such as speed and positions control of servomotor. By using the proposed control technique, we can monitor how the system works in a sophisticated manner. On the computer screen we can see all operating values related to the control system.

This project concerns with the servo system, which includes a Mechanical unit (MU154C), so in order to get familiarized with the servo system, different experiments have been performed. Dead-zone, which is the nonlinear characteristic of the motor, is observed. Due to this characteristic, the system is insensitive to small input voltages in the range of [-0.7,0.7].

In one of the implementation of speed control of servomotor the proportional control reduces the steady state error, if the external load is not applied to the motor shaft. If the load is applied, the actual motor speed decreases, and the steady state error increases. To overcome this problem, we use integral controller together with the proportional controller, which eliminates the steady state error.

Using menu bars that are placed on the front panel of the graphical user interface, the whole system is controlled by pressing the relative command icon. The C-based measurement and control software developed in this project performs almost all functions of an expensive software produced by a commercial firm. Moreover it renders the use of some electronic boards by incorporating the functions of these boards in the software. This software can run standalone.

Using the proposed software step, triangular and square wave inputs are produced within the computer [7], which are used as reference input signals in speed and position control applications.

This software together with the mechanical unit MU154C can be used as a computer based experimental sets in engineering schools. Computer based experimental sets for the other disciplines such as electronics, telecommunications, microwaves, etc, can be designed and implemented with similar techniques using the C-based graphical programming.

VI References

[1] Karl J. ASTROM Bjorn WITTENMARK, Computer Controlled Systems Theory and Design 3rd edition Prentice Hall, 1997.

[2] R .G. Jacquot, Modern digital Control Systems Marcel Decker, New York 1995.

[3] Servo Fundamentals Trainer SFT154, Feedback Control & Instrumentation,
Feedback Instruments Ltd, 1996.
[4]Doyle. J. C., Francis. B., and Tannenbuam. A., Feedback Control Theory, Macmillan Publishing Company, 1992.
[5]Sarfaraz Ali Hussani, "Design and Implementation of a Robust H^{∞} Controller for the Speed Control of Servomotor" MS Thesis, EMU, 1999.
[6] Easy I/O for DAQ Library, in the LabWindows/CVI Standard Libraries
Reference manual, Part Number 320682D-01 1998.
[7] LabWindows/CVI Advanced Analysis Library Reference manual
Part Number 320686D-01 1998.

Mirza Tariq Hamayun received the MS degree in Electrical engineering from EMU TRNC Turkey in 2001, and received the MSIT degree from Stuttgart university Germany in 2003. Since 2004, he is working as Assistant Professor in Comsats Institute of Information Technology Lahore Campus.

Acknowledgement

I want to thank to Comsats Institute of Information Technology for their support.

The MIS Course and the Curriculum of IMIS Specialty in China

Jindong Li
Business School, Zhejiang Wanli University
8 Southern Qianhu Road, Ningbo, PRC 315100

Tel : 0086-574-88222299

Fax: 0086-574-88222823

E-mail : jindongl@mail.nbptt.zj.cn

***Abstract*-This paper describes some background about the MIS as a course, MIS as a specialty and the relationship between MIS and IMIS in China. Based on the survey of Chinese universities selected by random on Internet, it reveals the domestic arguments about MIS course in the curriculum in IMIS specialty. Having analyzed the development of this discipline in China, the paper proposes different ways to provide MIS course in IMIS.**

I. INTRODUCTION

In China higher education has developed rapidly and many new colleges and universities have been built, so the methodology of courses, majors and discipline development in Chinese university, especially for "Information Management & Information System"(abbreviated as IMIS hereafter), is paid much more attention because IMIS is considered as a new and prevailing specialty. Problems are then raised about the nature of IMIS education. Some will think it as one very technological specialty but others may argue it is a managemental major or even a kind of social science. The divergence of viewpoints brings about the differences in determining of required courses and the curriculum of IMIS. The IMIS in China is obviously not the same thing as Management Information System (abbreviated as MIS in the paper) abroad in the case of major, especially in USA, and in the case of course how to provide MIS in IMIS has been also doubted among the Chinese field. MIS, as one important and unique course of IMIS, also remains varied in its content, semesters chosen and types of experiments or even its name. Professor Gordon Davis once said that the MIS pays much attention on how to utilize the technological ability provided by the computing science and, in its basic process, has closer relationship to organization effectiveness than that to algorithm in computer. Scholars from many universities in China have been studying how to provide unique MIS course to the students from different specialties. There are imperative demands for IMIS graduates and other majors' students to establish a proper methodology teaching MIS.

II. SOME BACKGROUNDS ABOUT IMIS

According to the statistical data of 2004, in China there are three hundred and forty colleges or universities where undergraduate education of IMIS are introduced to students. In Zhejiang province, there are Zhejiang University, Zhejiang University of Technology, Hangzhou Dianzi University (the former Hangzhou Institute of Electronic Engineering), Zhejiang Gongshang University (the former Hangzhou Institute of Commerce), Zhejiang Finance and Economic University, Ningbo University and Zhejiang Wanli University etc. which provide different levels of program, including post-doctoral station, Philosophy Doctor, Master of Science, undergraduate program, and two-or-three-year college education, etc. To enhance the development of IMIS, Zhejiang Provincial government has established important major project to support the construction of IMIS in some universities, such as Zhejiang Gongshang University and Zhejiang Wanli University. These two universities have been ratified as the important major of Zhejiang Province.

Under the catalog of Chinese colleges and universities

K. Elleithy et al. (eds.), Advances in Computer, Information, and Systems Sciences, and Engineering, 337–342.

education, IMIS is undoubtedly characterized by its diversity of natures. With regard to classification of majors by Ministry of Education of China, being subordinate to IMIS, there should be at least three divisions: the first is called information management, which recruits students according to management science and economics, the second is information system, which sometimes overlaps with some technical and engineering specialties, such as computer, network and communication. The third is related to document, librarian, archives, and some specialized information, which recruit both social science and human arts students. Hence IMIS can be treated as engineering or science or even social science, mostly it is called cross- discipline. The curriculum of IMIS includes technological courses like Data Structure, Database Management and C++. Meanwhile it also teaches social science and human art to the students such as concepts about information, cataloging and organization, behavior etc. Some courses are between them, for example, MIS and other related classes.

The comparison and analysis about higher education organizations in both China and foreign countries show that the development of one major education usually keeps consistent with that of the research on such discipline. The education of IMIS has been only provided in China and develops in varied dimensions since it has different backgrounds. As one former division of IMIS, management information system, for example, remained uncertainty in its definition. Management information system was considered to be same with information system and might be included in some respects by information system. But the term "information system" was already used in other discipline such as electronic communication, and should not to be instead of "management information system"(Xue, 1999, p.4). Now in Chinese undergraduate degree education, MIS is not a name for a major formally but a name one unique special course. Such situation is just opposite to that in USA where information system is same with management information system but information system only differs from management information system since they are applied in the different managerial levels. According to Professor Stair information system is consisted of a set of interrelated elements that collect, manipulate, and disseminate data and information and provide a feedback mechanism to meet the object of organization, while management information system serves managers and decision makers by organizing a collection of people, procedures, software, database, and device providing operational information. (Stair, 2001, p.4 & p.22)

In China, the term "MIS" was used for the first time at the end of 1970's and it was in 1980 that Tsinghua University ran MIS as a major for trial. After 1990, MIS began to develop very quickly and received wide attention from every walk of life, especially from various scales of enterprises. After the overall adjustment of undergraduate specialty in 1998, MIS is only treated as one unique course and provided to students who major in management, computer and many other related disciplines. How to provide such course to the students with different majors and what the key points are that is required for IMIS students are not unified among the Chinese universities that provide the program. The computer science, as the most related discipline to IMIS, has already formulated a rather more mature and compact curriculum and methodology. In 2002, the Education Commission of China Computer Federation and The Computer Education Research Association of Chinese Universities set up a task force to carry on a special study on the curriculum of computer science and technology in one-year-term based on the forefront of ideas, concepts and methodology in such discipline world widely such as the CC 2001 by IEEE and ACM. The achievement of this work is CCC 2002 that provides a frame instructing to the curriculum and core course of each university. Such achievement and other approaches could help and guide us to cope with what problems we met in the IMIS program.

III. Survey on the MIS Course in Chinese Universities

Though there are three hundreds and forty universities in China where IMIS is provided to students, it is also very common to consider the education with MIS theory and application as required knowledge. Such knowledge is not taught only in a single course like MIS, but in some other relevant course as well. To investigate the MIS or relevant courses provided to IMIS students in Chinese universities, we carried out a survey through Internet in 2004. The method used in this survey is to search and browse the websites of forty Chinese universities in random, make it sure that they have IMIS major and then find out whether the MIS course or relevant courses are listed in their curriculums. Six of forty universities don't give their curriculum on websites, data are calculated statistically among the left thirty-four. The following conclusions would be deducted.

1) The name "Management Information System" is so common that is simply a standard term for such one unique course, but for system analysis and design course it has six related but different terms to name it which are listed from No. 2 to No. 7 in Table 1.

2) There are nine universities in total who don't provide courses like either MIS or system analysis and design, but in left twenty-five universities the number of universities providing MIS course is 16 (See Table 1 & 2), percentage is 64%, that is about two thirds.

3) There are twelve in eighteen universities to provide either MIS or Analysis and Design of MIS courses, about two thirds, and there are four universities offering both MIS and Analysis and Design of MIS courses (See Table 2). If we simply employ one term "system analysis and design" in Table 2 to mark all five related but different courses from No. 3 to No. 7 in Table 1, an interesting thing is coming up presence that nearly equal number of universities providing system analysis and design course to that providing MIS, e.g., seventeen in total.

4) Six universities only provide MIS but seven universities only offer system analysis and design for the students. There are ten universities which provide both MIS and system analysis and design and nine neither MIS nor system analysis and design.

TABLE 1
MIS COURSE AND RELEVANT COURSES PROVIDED BY IMIS IN CHINESE UNIVERSITIES

MIS and Relevant courses		Univer-sities	Percen-tage
MIS	Only MIS	6	17.5%
	With system analysis and design	6	17.5%
	With 《Analysis And Design Of MIS》	4	12%
Only 《Analysis And Design Of MIS》		2	6%
None of MIS	Only system analysis and design	7	20.6%
	Neither MIS nor relevant courses	9	26.5%
In Total		34	100%

TABLE 2
DISTRIBUTION OF MIS COURSE AND RELEVANT COURSES IN CHINESE UNIVERSITIES

No.	*Course Name*	*Number Of Universities*
1	《MIS》	16
2	《Analysis And Design Of MIS》	6
3	《Analysis And Design Of Information System》	5
4	《Analysis And Design Of System》	3
5	《Design Of Information System》	1
6	《Method to Develop And Design Information System》	2
7	《Method to Develop Information System》	1

The results have illustrated that MIS plays very important role in IMIS major in China. Meanwhile the curriculum of IMIS is not as unified as that in other similar specialties. The knowledge and content are also overlapped among MIS and other related courses, especially between MIS and system analysis & design, sometimes repeated several times in these courses of different semesters. Any mature specialized knowledge in a certain discipline has accumulated its basic elements, such as the concept, category, principle and theorem to a certain amount, and arrange the inner relationship, then become systematized, during a long difficult process before it comes into being, so it should be sole scientific system in that theory. In line with this perspective, the course we have introduced to students must retain stable in names, relationships with other courses and educational functions. Hence the questions we argued most often are among those: Is it necessary to provide MIS in IMIS curriculum or other majors? How should we introduce MIS to the students with different backgrounds and desires for their future careers? Could a newly built university push one specialty developing faster and how to do?

Ⅳ. The Programs of MIS in IMIS

We have also checked the IS 2002's Model Curriculum and Guidelines and other literature focused on information technology education, found little discussing the methodology to provide unique MIS course but some on the strategy and integration of the curriculum for MIS as a major. This is because the MIS is mostly a major or discipline in United States not a course. Adapted from IS'97, IS 2002's model provides an updated and standard curriculum for undergraduate students. All courses are consisted of eleven and divided into six curriculum presentation areas. There is no single course named MIS but equivalent one or two such as IS 2002.1 "Fundamentals of Information Systems" and IS 2002.3 "Information Systems Theory and Practice", but IS 2002 includes system analysis & design courses like IS 2002.7 "Analysis and Logical Design", IS 2002.8 "Physical Design and Implementation with DBMS" and IS 2002.9 "Physical Design and Implementation in Emerging Environments".

The broader, more general theories in IS field has also been searched out on information technology education. System thinking, Resources-Events-Agents model and so on doesn't only add the new content for IS courses but also becomes an alternative strategy (Kevin R. Parker *et a.*, 2005) and objectives for the IS discipline (Steven Alter, 2004). Anyway literature focusing on course teaching fell into our sights. The REA method, as a replacement of ER model and one new part of content in System Analysis and Design, can meet the undergraduates desires and help construct the students' day-to-day experience into a more structured framework (Kevin r. Parker *et al, 2005*). Since REA provides a means to semantically address other forms of modeling, it can be used to link value chain concepts, data and process modeling. In fact, System Analysis and Design has been treated as combined cross-listed course for both MIS and ACS undergraduates and been stretched on both business focus and solution focus to the students (Ken Surendran, *et al*). And such class can resemble the real business world where cross-functional teams are quite frequently used.

The MIS in Chinese universities also serves this kind of learning outcomes and actually the most part of MIS are overlapped to that of System Analysis and Design. This is the reason why we would notice from the survey that there are some universities saying no to MIS although it is evidently necessary to provide MIS course in IMIS.

Regardless of what name for such course to use, we first trace back to the research of MIS in USA, which is reflected in the fact to the degree education. We can see that it has three research mainstreams, Sloan School of Business of MIT and Harvard University pay more attention to information management system theory, and UMN and Indiana University are inclined to management science background and management application and information management, New York University (Stern) and the University of Arizona mainly focus on the research of management information system developing and implementing technology. The theoretical research determines the three features of content of MIS course, namely, management theory, mathematics means and computer tools. Five majors' areas are needed to organize the framework of MIS knowledge by business students (O'Brien, 2002, p. ix).

In Chinese universities the teaching techniques has generally been paid more attention than content of the course especially for the newly built and fast-developing universities. Zhejiang Wanli University has been keeping fast speed in developing for three years (as shown in Table 3) and its objective is to cultivate senior application-type professionals. But when one university developed to a certain scale, it needs a proper method to maintain the quality of the course teaching. To construct one course should be restricted by three elements like knowledge, society and students. Knowledge itself is in the process of constant expansion and evolution, so there are always a dynamic restriction to the level, range, profundity and scope of course content.

To ensure students being well prepared for the information technology management careers or advanced degrees, Daniel Brandon *et al.* in Christian Brothers University designed and implemented a capstone course. Such course should focus on a combined business and information technology management environment and be the concluding course to explore and put to practical use of the entire body of the knowledge. They finally developed ITM 455 as the capstone

TABLE 3
STUDENTS RECRUITED IN IMIS IN RECENT THREE YEARS IN ZHEJIANG WANLI UNIVERSITY

Academic Year	Students Recruited	Description
2002-2003	142	Recruiting plan separately
2003-2004	141	Recruiting Plan together with other two Majors
2004-2005	245#	Recruiting Plan together with other two Majors

#: This number includes all students who have chosen IMIS from three majors: IMIS, Accounting and Marketing.

course and formulated its educational outcomes (Daniel Brandon *et a., 2002*). Some other approaches to handle the capstone course, for example, Gupta's targeted assignment, is also addressed in Reference [19]. In our understandings MIS is so important that it has some natures of capstone course but it would rather be a fundamental course than summative one. The idea to setup a capstone course is good point and we need to design MIS from this perspective. We have made efforts to teach MIS in this way and the practice is illustrated below.

A. Arrange the Study Term of MIS Reasonably

Due to the fact that MIS has formed its own system after combining the thoughts and methods of different disciplines, consideration should also be taken to the requirement that this course should be kept properly in order when MIS course is provided to students. For the four-year undergraduates majoring in IMIS, MIS courses should be fundamental and arranged in the second semester of their secondary year at university. But for those majoring in Computer Science, the proper time to provide MIS will be one year later than the formal.

B. Avoid the overlapping among MIS and other relevant courses

For these relevant courses we divide them into the following three types, each of which has unique goal of their own.

1) Technique and Operational Course: Education of such course should be oriented to application, and be kept specific.

2) Discipline Course: The content of this type is always comprehensive, with a low differentiation extent, and specialization degree.

3) Academic Course: The academic courses are more abstract, more literal (laying more emphasis on writing compared with oral communication), and more individual (avoiding group work and cooperation), with less relationship with school knowledge. For different majors MIS would be treated as different types to introduce for students and the experiment must be arranged in different levels. For IMIS students we do not provide any experiment but for those majoring marketing or HRM will carry out some observing experiment and verifying experiment.

C. Emphasize Application in Methods and Tools

With regard to teaching methods, there are many means, such as questioning method, multimedia teaching, network teaching, cases teaching, and system illustrating, all of which can be alternated in the class. And we illustrate one example throughout the whole process of lecture. For example, while instructing theory and practical teaching, teachers may keep using the example of salary management system to motivate the students to think actively by themselves: how will analysis and design project alter after the demands change. While learning MIS, our students will always, under the instruction of teachers, go to enterprises and software companies, where they can learn and inspect the actual management information system and development situation. By doing that, the students can be led to touch society and make practice as early as possible. Teachers who teach MIS or relevant courses should also lead students to make development of actual system, which can not only train students, but also bring remarkable economic benefits.

We are trying to revise and perfect the syllabus allowing teachers to make adjustments within small scope for different students while making no change of integral content of the course and find out a balance point of the three course types—academic courses, discipline, and application. In the process of theoretical teaching, the students are divided into three types: management-type, information-type and computer-type. For the second and third type, more attention should be focused on the new features and behavior modes of management science theory in information age. But for the first type, more attention should be paid to the analysis means and technology development of information technology in management system. The most important thing is to choose a group of combined textbooks separately for all those relevant courses. The next step to develop the MIS and IMIS specialty will depend on these factors both in a domestic and international situation.

ACKNOWLEDGMENT

Many thanks to colleagues such as Mr. Zhou Zhidan, Mr. Hou Ancai, Mr. Chen Guohe and MS. Zhou Zhiying, Ms. Xu Ying etc. in Zhejiang Wanli University for their kind help in collecting literatures and data. Thank Mr. Tong Yiping and Wu Wei, who did the investigation on Internet instructed by myself and thank to my daughter and son who did typewriting of the paper.

REFERENCES

[1] ACM, AIS, and AITP. "IS 2002 Curriculum Guidelines for Undergraduate Degree Programs in Information Systems", http://www.acm.org/education/curricula.html#IS2002

[2] Steven Alter, "Desperately Seeking Systems Thinking in the Information Systems Discipline," *Twenty-Fifth International Conference on Information Systems*, 2004, pp. 757-770.

[3] Y-j. Bian, "The Development Trend of Management Information System Overseas," *Journal of Hehai University (Philosophy and Sociology Edition, 2:4)*, December 2000.

[4] H. Han and Q. Dong, "Increasing the Choice of Study: one of the Elements for the Curriculum Reform," *Journal of Higher Education (25:4)*, July 2004, pp.79-85.

[5] Fleming, S., Shire, J., Jones, D., McNamee, M., Pill, A., and Mcnamee, M. "Continuing professional development: suggestions for effective practice,"*Journal of Further and Higher Education (28:2)*, May 2004, pp. 165-177.

[6] M-s. Lai and L-y. Zhang, "Discipline Development and Education Problem of Information Science," *Journal of Information Science (22:2)*, February 2003.

[7] L. Liu and Z-z. Zhang, "Standardization and Unique Feature of Information Management Course Design," *Library and Information Service*, 2001,vol.8, pp.86-89.

[8] James O'Brien, A. *Management Information Systems (Fifth edition)*, McGraw-Hill, NY, 2002, p. ix.

[9] G-x. Qian, "Master Principles and Means of Major Construction," *Higher Scientific Education*, 1997, vol.1, pp.29-31.

[10] Ralph M. Stair and George W. Reynolds, *Principles of Information Systems: a Managerial Approach*, Course Technology-ITP, Boston, 2001.

[11] Study Team on China Computing Curriculum 2002 of Computer Science and Technology of China, *CCC 2002 of Computer Science and Technology of China*, Tsinghua University Press, August 2002.

[12] Y. Wang and K-h. Wu, "Construct New Course System of Management Information System," *Journal of North China Electronic Power University (Philosophy and Sociology Edition)*, 2000, vol.3.

[13] H-c. Xue, *Management Information System (the third edition)*, Tsinghua University Press, Beijing, 1999.

[14] M. Young, *The Curriculum of the Future*, translated by W-h. Xie and X-y. Wang, East China Normal University Press, Shanghai, June 2003.

[15] J. Zhang and Z. Sun, "Improve Competitive Power of Information Management Students' Employment," *Research of Scientific Higher Education (23:2)*, April 2004, pp.77-78.

[16] Kevin R. Parker, Cynthia LeRouge and Ken Trimmer, "Alternative Instructional Strategies in an IS Curriculum", *Journal of Information Technology Education*, Volume 4, 2005

[17] Ken Surendran, Ike C. Ethie and Chellappan Somarajan, "Enhancing Student Learning across Disciplines: A Case Example using a System Analysis and Design Course for MIS and ACS Majors", *Journal of Information Technology Education*, Volume 4, 2005

[18] Fred L. Kitchens, Sushil K. Sharma and Thomas Harris, "Integrating IS curriculum Knowledge through a Cluster-Computing Project---A Successful Experiment", *Journal of Information Technology Education*, Volume 3, 2004

[19] Daniel Brandon, James Pruett and Jim Wade, "Experience in Developing and Implementing A Capstone Course in Information Technology Management", *Journal of Information Technology Education*, Volume 1, 2002

Can a game put engineering students in an active learning mode? A first experiment in sustainable agriculture teaching

M. Michelin
ISIMA
Complexe des Cézeaux, F63173 AUBIERE CEDEX
S. Depigny, Y. Michelin
UMR METAFORT (CEMAGREF, ENGREF, ENITA, INRA), ENITA
Marmilhat, F63370 LEMPDES (France)

Abstract- **An experiment using educational games has been conducted in France with first year engineering students to develop their understanding of what sustainable farming is. We have devised a game that models the impact of grazing practices on landscape dynamics and compared a board version and a virtual one. The game appears to be more efficient in developing the desire to learn more and stimulating players' imagination than in teaching precise scientific knowledge. The game does not take the place of the classical courses. It introduces them. Finally, there is not any competition between the board game and the computerized one ; the board game is more relevant to start the educational process, the second allows more possibilities to experiment different ways of management.**

I. Introduction

After the Second World War, agronomists had to propose technical solutions to develop food production at the lowest price in order to feed an increasing population arriving in large cities. In Europe, they got very good results but with negative consequences on either the environment or the landscape quality. Nowadays, the society asks them to imagine new farming systems, more sustainable, at the same time profitable for farmers and able to take care of either the environment or the landscape.

A good example : the landscape management in French uplands areas

The evolution of the landscape's demand in the French Massif central provides a good illustration of the gap increasing between what people expect and what farmers do to maintain their activity, a gap that new agronomists will have to bridge.

A large part of the traditional summer grazing pastures of these mountains covered with grass and heath have disappeared from the end of the nineteenth century till the end of the twentieth century [13]. Most of them have been converted into woodlands (spruce and pine plantations, beech colonization, etc).

As a response of this evolution, the direct landscaping policies in the nineties aimed at stopping closing of the landscape. They generated few results. In fact, this major change, which was badly felt by the inhabitants, was not a voluntary landscape change but the result of five main factors, a demographic decrease in the population (landowners moved away and prefered to plant their land rather than rent it), a consequence of the national reforestation subsidy policy initiated in 1946, the disappearance of the common grazing system (Fig. 1), which was the best adapted way to give better value to these relatively poor pastures, a technical and economic improvement of agricultural systems, with a decrease in flocks that led to an abandonment of bad pastures which were not useful in these new farming systems and an ecological phenomenon of broom expansion furthered by a low stocking rate.

Fig. 1. Old shepherd and the common flock, Chaîne des Puys, 1980.

That is why it was so difficult to explain to local representatives or visitors that the best method for keeping these landscapes open was to have an economic policy that was favorable to livestock farmers. When this explanation was understood by the locals, the regional support of common flocks through the regional park action maintained enough animals to stem the expansion of broom [14].

Regarding this example, our hypothesis is that the misunderstandings between landscape producers (farmers, foresters, etc) and landscape users (inhabitants, tourists, etc) is mainly due to a lack of intelligibility.

Our students meet with the same difficulties. During the first three semesters, the courses are shared into different topics with very few links. Another difficulty comes from the academic way of teaching that starts with theoretical courses. Even if all the students in agronomy have a practical period on a farm after a one month course, they have difficulties to apply this theoretical knowledge in describing and analyzing actual situations. In front of the complexity, a simple explanation is not sufficient. Students have to "touch" the process, to see it in their mind. Later, when a systemic approach combining different kinds of information occurs, the students understand at least why we asked them to learn so many things but they have lost a part of their time and

K. Elleithy et al. (eds.), Advances in Computer, Information, and Systems Sciences, and Engineering, 343–349.

sometimes they have forgotten what they have learned before. That is why we devised starting the first semester by an explanation based on a demonstrative situation that put them in an active learning mode. For Bonwell and Eison [3], in active leraning, "*students are doing things and thinking about what they are doing*". In 1987, Stice [20] demonstrates that when learners remember only 50 % of what they see and hear, they remember 90 % of what they say as they do something. Regarding to this results, Stalheim-Smith [19] proposes the concept of meaningful learning, by putting students in small groups top create through cooperation with peers an active learning environment. But the concept of learning by doing is difficult to apply to agronomic studies because of the time needed to assess the consequences of what have been done. As the use of models allows the possibility of testing scenarios, Bareteau et al. [2] consider that it is possible to learn by simulating as well as by doing. With the same point of view, we thought that an educational game could involve the students in the educational process in an active way while giving them a clear meaning and a lively dimension of what they will have to learn before being able to apply it on actual situations.

II. Interests of a Gaming Process

The educational potential of the game has been known for many centuries [12]. However, this point of view was only studied and formalized in the early twentieth century. In 1958, Caillois proposed a theoretical classification based on four aims (competition, chance, mimicry and vertigo) and two attitudes (eccentric and ruled) which remain relevant to this day [4]. At the same period, Piaget [17] considered that the game theory could facilitate the teaching of social sciences, in a multidisciplinary way. More pragmatically, geography and history teachers [9] have devised many games using emulation and simulation. They have demonstrated that these attitudes offer strong possibilities on motivating pupils and students to become more active in the learning process. Similarly, these teachers consider games as useful teaching aids for helping students visualize the consequences of events decided or suffered by them during play. Simulation exercises and educational games are widely used as teaching resources in the USA and Canada [7]. However, very few techniques and methods have been developed in France especially for students of public schools [9]. One possible explanation for French teachers' mistrust is the confusion between educational and instructive games. Instructive games which are based on competition between players, are often unattractive and leave very few possibilities for involving players. On the other hand, educational games combine emulation and mimicry attitudes and create an illusion that puts players in an active mode and develops their will to learn [10].

Moreover, simulation games have been used by social workers to help poor people to express their problems and imagine solutions [8]. Today, different studies in easier social conditions are using simulation gaming. In the context of agricultural systems, Daily et al. [5] uses a software (GRASSGRO) to help students developing an appreciation of the interrelationships between soil, pasture growth and animal production. However, it is more a decision support program than a game. Except the Mejan Jeu, used in landscape planning to generate a discussion between farmers and stakeholders about the best way of managing the landscape of the Causse Méjan [6], very few techniques of gaming are used with students. Etienne *et al* consider the computerized version of the game as a very efficient tool for involving local actors. However, as it removes a material part of the representations, the same result may not be reproduced elsewhere. Thus, we preferred to start our experimentation using a concrete game with a board and 3D checkers, the SHRUB BATTLE before computerizing it (the GENIX software presented below).

III. Description of the Shrub Battle Game

A. Principles

The major educational goal of this game is to better understand the making of the landscape. Its gaming objective is the competition for land cover. We built the SHRUB BATTLE like a model by adapting the results of studies carried out by our colleagues describing the vegetation development as the consequence of livestock farming practices [15, 16]. We also took advantage of our own experience in farming system management [14] to identify socioeconomic events that could influence farmers' decisions. This represents a very complex system, combining different space-related factors (plant station, agricultural field, landscape point of view, etc) with different time-related factors (immediate decision and action, the farming year, ecological processes, etc). We decided to make a semi-realistic model and we defined simple, understandable and clear rules, less complex than in the actual world. On the other hand, we included recreational aspects into the model to make the game funny. Thus, we imagined an original situation: the plants can directly interfere in the decision process!

B. Concepts

The game is based on three main concepts that are linked in a global system (Fig. 2) :

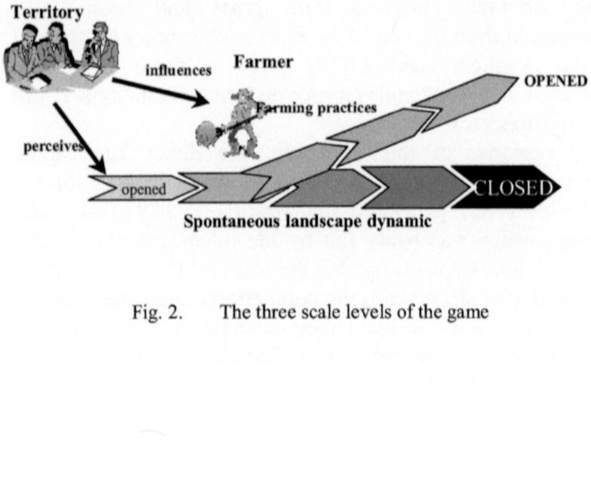

Fig. 2. The three scale levels of the game

- ***Natural vegetation expansion*** is determined by ecological rules.

Each species has a specific power of invasion that depends on their reproduction system (sexual or vegetative), their way of colonizing new areas (by spreading moving or heavy seeds, by producing rhizomes…), the time they need before being mature. We chose three representative plants of the Massif central uplands areas. In the game, each species has its own type of spreading and a specific color:

- *Rubus caesius* or Blackberry bramble expands by vegetative reproduction: it needs at least one full-grown plant to create a progressively larger patch. We chose this plant because it is highly resistant to farmers' practices.
- *Cytisus scoparius* or Broom often appears on abandoned plots due to significant seed supply in the soil. It expands by random and abundant germination: it can appear anywhere without a full-grown plant, and can quickly create landscape patterns. We chose this plant in relation to several questions asked to both farmers and researchers. Broom has got a short lifespan (about 15 years before natural degeneration) but its very high seed supply gives it significant germination capacities as soon as agricultural fields are left unused [16].
- *Pinus sylvestris* or Pine tree represents the woody stratum, the last stage of vegetation development. It produces seeds around itself, with a distribution proportional to distance [18]. Like blackberry it needs full-grown plant to colonize land.

- ***the farmers' practices*** modify the vegetation growth by limiting the shrub and trees expansion. Farmers have three different types of management: grazing practices that eliminate young plants and permit animals to increase their weight. Their landscape impact is seen only several years later, depending on the agricultural management system (under-stocking grazing, field intensification, etc…). Clearing practices (burning or crushing undesirable plants) and agronomic investments (fences and fertilizers) have an immediate impact on the landscape.

- ***natural and socio-economic events*** interfere with the farmers' decisions and the vegetation growth with favorable or unfavorable consequences for each player. The game offers two kinds of events. The socio-economical events can be compulsory (a law), inciting (subsidies, price levels) or negotiated (agro-environmental agreement). Natural events are unpredictable: climatic events can modify the vegetation dynamics, pest or disease attacks on herds can lead farmers to change their farming management system, etc.

C. Game structure and components

The SHRUB BATTLE is a board game shown by Fig. 3.

The playing board represents a small part of an agricultural field patterns. It is split into 144 small plots, representing decision units for farmers or plants. Each plot's status is characterized by a color (grass, fertilized grass, grazed grass, protected area, moorland area or woodland area). Small wooden checkers on the playing board mark the players' position. Each player has particular tools, which always combine three important components:

- Determinist behavior model, extracted from research studies.
- Unpredictable events, generated by a dice or draw cards.
- Player strategy: Each plant player can colonize the board with the combination of a toss of the dice (when the dice gives from 1 to 4) and a rules table, which represent spread types and germination capacities. For example, blackberry cannot spread itself without one adult plant, which can produce new plantlets while broom can appear anywhere. An originality of the SHRUB BATTLE game is that the plant

Fig. 3. A view of the play board, the pawns and the players

players have human decision-making potential. This property provided a way to represent ecological trusts, which try to protect the interests of Nature against human pressure. The fourth player represents a livestock farmer. The player is alone against nature, its main objective being to protect its grass field against plants invasion particularly from broom. The player does it by combining two activities : i) earning money from production practices: where their small herd (four or five animals) can graze and produce beef, and ii) spending money via clearing practices such as the slash-and-burn technique, fertilization, enclosure and clearing.

The last components of the game are cards. These cards represent two types of possible events. The first type lists several events that are dependent on the socioeconomic context (compulsory policies, the wishes of trusts, markets laws, etc.). The second type lists several unpredictable events (frost, drought, pests and diseases, etc.). When the dice gives a 5, the player takes an EVENT card and has to apply it, even if the result is bad for him. If the dice gives a 6, the player can choose to apply a socioeconomic event. If he prefers, he plays as if the dice gave a 4.

A game covers a six-year period. One year is finished when each player has played six times. Players have to record many parameters through a game: numbers of plants, events, farmers practices, etc. These data are used to build graphs, which can be used to compare the results from one game to another to show the different methods of livestock farming management or land management.

D. First results

After about ten experiments with various people, agronomy students[1] (250), countryside planning students[2] (300), agricultural researchers and technical experts (20) and farmers (2), we noticed a very good sensitive involvement. Each player was strongly involved in the game. Players quickly picked up an identification with a plant or an animal, and lived through the making of the landscape. This may have been partly due to the fact that the 'virtual' situation presents no risk for the player who can experience many possibilities and evaluate the results in a short time. A large majority of players were enthusiastic about the game, especially younger people untrained in agronomic principles. During one presentation with 94 students playing in 20 groups, we conducted a short survey to evaluate the impact of the game on their level of knowledge and comprehension of agri-environmental landscape management. We asked the same questions before and after the game. We did observe a better understanding of the impact of grazing management on protected plants. Before gaming, 70% of students thought that forbidding grazing on dry pasture could eliminate the orchids, compared to 83% after gaming. The best progression results were recorded for people who came from towns and urban areas. However, farmers' children did not change their opinion. The last question concerned the impact of fencing on vegetation dynamics. 25% changed their opinion during the game, but only 30% proposed a correct answer. Two factors can explain this moderate result. First, the question was too complicated to be well understood. Second, several groups did not use fences during the game, and so could not appreciate the result.

This short overview shows that SHRUB BATTLE can be considered as an attractive game for players. Despite the many simplifications, it produces a result that is not too far from reality. We think that this good result is due to an homogenous level of simplification and to the care taken to define the events and the relationships between factors. Only a few engineering students considered that the game was not serious enough for a course. Another slightly negative point came from certain agronomists or agricultural technicians who were well-trained in grazing management. They considered the game as too simplistic and took time to discuss the rules despite taking part in the game. However, if the game is funny to play, it has some educational limits. One is due to the fuzzy rules, specially those concerning the plants, because the randomization is difficult to obtain by hand. Another limit is a consequence of the time needed to register the different data during the play. In a three-hours period, it is only possible to play one time and many events don't occur and a part of the understanding of the processes is missed.

That is why we decided to computerize it. Both games have advantages and inconvenient that will be discussed in the last part.

IV. DESCRIPTION OF THE GENIX GAME

The principle of the GENIX game is a bit different from the board game's. Even if the rules are globally the same, it is not presented the same way.

Indeed, there is only one human player, the farmer, fighting against the plants, simulated by the computer. One can consider the GENIX game as a landscape maker simulator.

A. Technical aspects

From a technical aspect, there are two points to consider : the global structure and the execution progress.

To determine the best structure of the game, we had to identify all its components, their behaviours and actions, and the link they have with the others. Once the components identified, with their own role and actions, it was possible to fix the structure, that is an object one. We used the UML language to show as precisely as possible the objet structure of the game. In the state diagram (Fig. 4), classes have been

[1] Students of ENITA of Clermont-Ferrand. It is an Engineering National College of Higher Education Specialized in agricultural techniques. We use SHRUB BATTLE to introduce all the concepts, which can be tackled during the three years of the agronomic learning process.

[2] Students of ENGEES of Strasbourg. It is an Engineering National College of Higher Education Specialized in water and environment management. We use SHRUB BATTLE to show negotiation process and to explain agricultural points of view in rural area management.

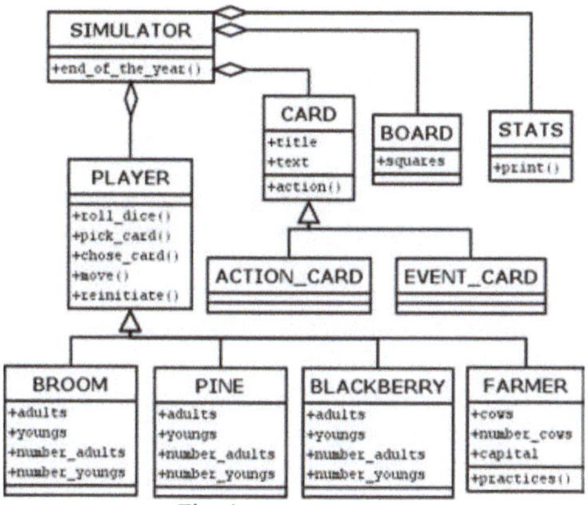

Fig. 4. State diagram of GENIX

created to represent the game, the players and the elements that make up the game, like the board or the cards. Each player can do the same kind of action like tossing the dice, picking a card, moving animals ... that is why there is a mother class that gathers all these actions. As each player has his own way to do these actions, each class reimplements them.

The simulator manages the game progress on the whole. It allows each player to play. It proposes choices to the farmer (practices or pasture). It reinitiates the board at the end of the year. It manages the statistics.
This game is a sequential, not an evenemential one. Each player plays one after the other. The simulator only stops when it waits for the farmer's action. From a dynamic point of view, the game progress is simple. It consists in two loops, one including the other. The first one, managing the years, embraces the months loop and the processing at the end of the year. The second loop, the months one, manages each player turn.

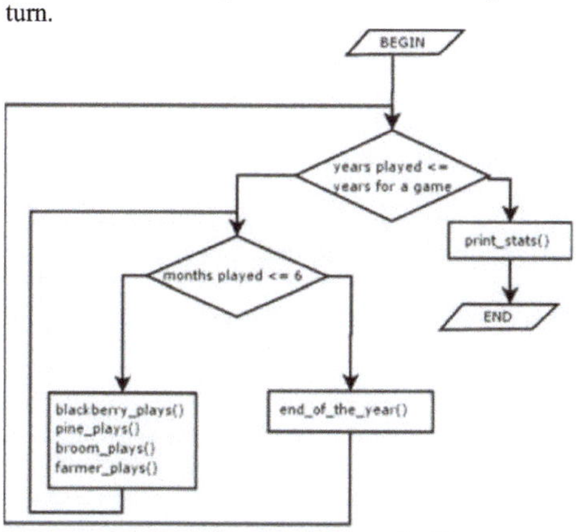

Fig. 5. Game process algorithm

The main difficulty lies in the fact that there are a lot of events that can happen and cumulate. For instance, a small plot can have different states and occupants. A square can be grassy, empty, fertilized, burnt or crushed. But at the same time, a plant, young or adult, a cow or a cow and a plant can be present on it. Moreover, the square can have a caps or a flower on it. So when an event appears, like fire, pasture ... all these aspects have to be taken into account to decide how the state and the occupants of the square will be modified.

B. Physical aspects

To obtain a user-friendly game, we decided to limit the resort to the keyboard. Indeed, the mouse is easier to use than the keyboard. Moreover, we wanted the graphical interface to be attractive, simple to play and understandable, and efficient too. Thus, the game window has been divided in two parts. The first one is the board, with all the 144 squares which compose it. It looks like the board game with more beautiful pictures. The second part of the window recapitulates the main information about the game : which year and month it is, the level of the farmer's capital, etc. It also shows the information about the present month and the available choices.

We developed this game with Linux and the development tools provided with this system, like GCC. As very few persons are using Linux, we decided to integrate this game to a live-CD saving players from installing anything. The environment is loaded in memory, so the game can be used everywhere, if the computer has enough memory.
This live-CD was elaborated from a Knoppix distribution, lightened in order to reduce the loading time. We added the game, with a script to launch it after the boot, and also 'openoffice' and a macro to generate the statistics at the end of the game. All these elements allow the game to work independently from the computer. One play simulating a six-years grazing period takes around half an hour.

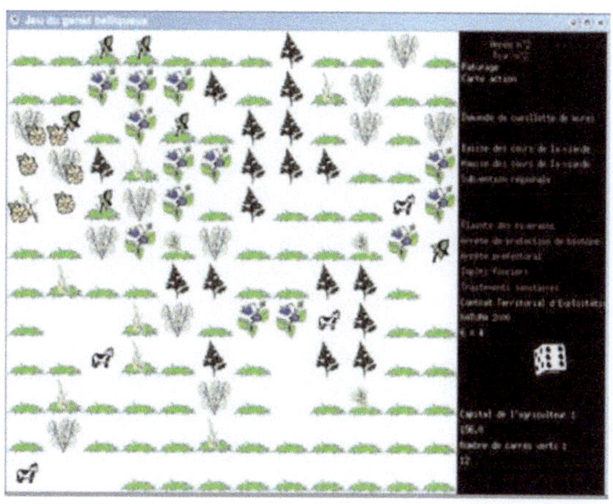

Fig. 6. GENIX game

The first version of the GENIX software is not yet customized. We have preferred to work on the simulator before spending time to make it more pleasant. However, we recently started a test with several students who have played SHRUB BATTLE, to better know which game they prefer and why. The first players consider that GENIX is as efficient as the SHRUB BATTLE to enlighten the relationship between farmers' practices and landscape evolution. They find the software version easy to use and quicker. They proposed several improvements (on line documentation, sounds and music, short presentation when the game starts, another window indicating which events are active…).

V. Discussion

Without mentioning about the lack of user-friendliness, we noticed some problems with the GENIX game. As it is a one-player game, it is difficult to use it with large groups. A part of the originality of the game (100 students playing at the same time in the same place) is lost. Another limit is due to the disappearance of concrete elements like pawns or small animals. The last problem comes from the speed of the game. A great deal of information arrive at the same time without any explanation and discussion, and the player could be lost at the beginning. In fact, this game is perceived by students a bit less funny and less original than SHRUB BATTLE. We think it is possible to solve these problems by developing a nice environment with more realistic pictures and by adding sounds, commentaries, or short videos.

However, the computerized game offers new opportunities. First, while the SHRUB BATTLE play-board needed a large table for each board and was difficult to transport, the GENIX software is an 'autorun CD', easy to transport and to duplicate. It is well adapted to small groups and to students who want to take time to play at their own speed. It takes less time to play and, as the data are automatically registered and converted into graphs, it allows new possibilities of analysis.

Secondly, during the computing process, we discovered some inaccuracies in the rules and some situations we had not considered before, so we had to modify the organization of the play. The passage from the board game to the computerized one helped us to think back of the processes that we wanted to explain by gaming. As a result, it has strengthened our understanding of the system modeled by the game.

Thirdly, the attitude of the students is rather different in front of the board than in front of the screen. When they play together with large groups, with pawns and small cows, they change their attitude. They become more active and then start to think differently. However, they do not gain a precise agronomic knowledge. They only discover that the different topics they will have to learn could be linked and crossed to answer practical questions. Alone or in small groups in front of a screen, they quickly start to experiment different grazing management systems. They still remain in an active learning mode but they go deeper in the process, they ask questions and try to imagine solutions. Of course, this attitude is only possible if they have first played the SHRUB BATTLE.

It is a great paradox to consider that while the game has become less concrete in its appearance, it is closer to the reality and explain it better. What the game has lost in gaming value was won in educational power. That is why we consider the board game as a tool for stimulating the curiosity of students early in their studies. The computerized game does not replace the board game, it extends it by allowing the player to experiment different systems of management that conciliate environmental dimensions and landscape projects with a satisfactory level of cattle production.

CONCLUSION

This short overview shows that a game can be used with engineering students who starts their studies. However, the game is sometimes felt as too childish so we had to give a scientific course before starting the play. This introduction, carried out in an academic way, presented a survey of the theoretical background and established a link with our research activity. This fact demonstrates that the game does not produce an educational result if it is not combined with a presentation of the context and the rules before gaming and a debriefing session after. The educational ability of the game manager is one of the most important keys to the success of the game, since explanations given while recording analyses may help to overcome misunderstandings. The game appears to be more efficient in developing the desire to learn more and stimulating players' imagination than in teaching precise scientific knowledge. We think that the game does not take the place of the classical courses. It introduces them in a constructivist way of teaching [1]. It could also be used to help students develop a more holistic, systemic vision than the basic education taught in secondary schools. Finally, there is not any competition between the board game and the computerized one. The concrete game is the first step of the educational process. It "opens the eyes". The virtual one allows students to access a second level : the experimentation of different landscape and grazing management systems. The third step is to go deeper into the subject during the courses. At this moment, a more realistic tool is needed, in order to maintain users interest, as it has been noticed by Hyltander with surgeon students, learning laparoscopy technique with a simulator [11]. At least, the practical period in a farm gives the students the opportunity to use the concrete situation of the farm where they stay to combine what they learn to better understand the farming system and its impact on the environment.

We propose two perspectives for future development. Firstly, assessment tools inquiries should be developed using short surveys to provide insight into what these games actually teach or not. A second way is to customize the computer version in order to propose a more attractive framework for this game.

ACKNOWLEDGMENT

The authors thank Jacques Verdier and Jacques Virmont for their help and the ENITA students Julien Coquillou, David Liautard, Mathieu Orth, Antoine Reulier for the time they spent to test the GENIX ver1.1.

REFERENCES

[1] Barker, S. (2002). Virtual learning environments for constructivist teaching in Biosciences to promote sustainable development in higher education, CAL-laborate, volume 8, june 2002, <http://science.uniserve.edu.au/pubs/callab/vol8/barker.html>

[2] Barreteau, O., LePage, C., D'Aquino, P. (2003). Role-playing games, Models and Negociation Processes, *Journal of Artificial Societies and Social Simulation*, vol. 6, No. 2, <HTTP://jasss.soc.surrey.uk/6/2/10.html>

[3] Bonwell, C., Eison, J.A., (1991). *Active learning : Creating excitement in the classroom*, ASHE-ERIC Higher Educaton Report No. 1., Washington, D.C.: The George Washington University, center for Faculty Evaluation and Developement.

[4] Callois, R., (1967). Les jeux et les hommes. Paris, Gallimard.

[5] Daily, H.G., Hinch, G.N., Scott, J.M., Nolan, J.V. (2000). The use of decision support program to facilitate the teaching of biological principles in the context of agricultural systems, TEDI conferences: Effective teaching and learning at university, Brisbane, the University of Queensland.

[6] Etienne, M., Le Page, C., Cohen, M., (2003). A step-by-step Approach to Building Land Management Scenarios Based on Multiple Viewpoints on Multi-agent System Simulations, . *Journal of Artificial Societies and Social Simulation*, vol. 6, No. 2, <HTTP://jasss.soc.surrey.uk/6/2/2.html>.

[7] François, P., (2001). La nature dans les jeux pédagogiques. *Géographes associés*, n°25.

[8] Freire, P., (1972). Pedagogy of the Oppressed. London, Penguin.

[9] Glemarec, R., Hochet, Y., Joffrion, Y., Sestier, D., (2004). Petite classification des jeux. Ludus network. [*online*] URL :http://www.discip.crdp.ac-cen.fr/histgeo/ludus/classification.html

[10] Grataloup, C., (1994). L'espace de la transition. Essai de géohistoire chorématique. Le laboratoire des jeux. Livre III, Thèse inédite.

[11] Hyltander, A., (2003). Simulation as a teaching alternative : Utopia or reality ?, CAL-laborate, volume 10, june 2003, <http://science.uniserve.edu.au/pubs/callab/vol10/hyltand.html>

[12] Mauriras-Bousquet, M., (1984). Théorie et pratique ludiques. Economica.

[13] Michelin, Y., (1995). Les jardins de Vulcain. Paysages d'hier, d'aujourd'hui et de demain dans la chaîne des Puys. Maison des Sciences de l'Homme.

[14] Michelin, Y., (1997). Gestion concertée du domaine pastoral: l'exemple des parcs régionaux auvergnats. Les parcs naturels, un concept de développement territorialisé et environnemental à l'épreuve du temps. Bulletin de la société languedocienne de géographie, 123-139.

[15] Orth, D., Chevillot, B., Teuma, M., Dulphy, J. P., Carrere, P., Michelin, Y., (2002). Combining multiple land use with shrub invasion management. *European Grassland Federation*, vol. 7, 1062-1063.

[16] Orth, D., Picon-Cochard, C., Prévosto, B., (2003). Dynamique d'invasion par des petits ligneux de prairies sous-exploitées de moyenne montagne du Massif Central : le cas du genêt à balai. Rapport final, Enitac Inra Cemagref.

[17] Piaget J., (1970). Psychologie et Epistémologie, Paris, Denoël

[18] Prevosto, B., Hill, D. R. C., Coquillard P., (2003). Individual-based modelling of Pinus sylvestris invasion after grazing abandonment in French Massif Central. *Plant Ecology*, vol. 168, 121-137.

[19] Stalheim-Smith, A., (1998). Focusing on active, meaningful learning, IDEA paper, No. 34, IDEA center, Kansas State University

[20] Stice, J.E. (1987). Using Kolb's learning cycle to improve student learning, *Engineering Education*, 77(5), 291-296.

Brain Wave Interactive Learning Where Multimedia and Neuroscience Converge

Paras Kaul
George Mason University
4400 University Drive, MSN 2F7
Fairfax, Virginia 22030
www.brainwavechick.com

***Abstract*-This paper presents research related to a new methodology for teaching, which combines a brain wave interface to the computer with multimedia to create neurological learning tools.**

I. INTRODUCTION

The objective of this paper is to introduce an educational process that combines multimedia and neuroscience to enhance current methodologies for teaching in order to more fully develop human perception. An increased ability to perceive expands cognition and utilizes a greater percentage of the brain's natural potential. Greater brainpower actualizes an increased frame of reference, which further develops the capability for nonverbal communication, remote viewing, and self-healing. Enhanced ability in these areas is necessary for our evolution as a species and will be needed to solve planetary problems that arise as challenges for survival in the future.

Michio Kaku explains that some physicists have speculated about the existence of a fifth force, which may be a paranormal or psychic force. He continues by explaining that each time a force has been mastered, human history has undergone a significant change, and he summarizes that the mastery of each force helped to create a new revolution in history. [1]

II. THE BRAIN WAVE INTERFACE

The brain wave interface system discussed in this paper is the Interactive Brain Wave Visual Analyzer System (IBVA4). IBVA has been on the market as a brain wave hardware and software system developed and programmed by Masahiro Kahata since 1991. The system has been demonstrated by this author at Special Interest Group for Computer Graphics and Interactive Techniques (SIGGRAPH) exhibitions, as well as presented in performance mode as brain wave controlled multimedia. Currently, the IBVA4 neural interface system has been significantly enhanced by Kahata, which has produced new software that enables development of applications for brain wave multimedia, thus increasing its value for educational purposes.

III. TECHNOLOGY ADVANCEMENTS

Contributing to the sophistication of the current version of IBVA4, is the fact that today's high-speed dual processing computer systems, produced by Apple computers, have increased computing performance to a degree that more closely matches neurological processing performance. Increased CPU speed enables the interface to collect, transmit, analyze, and convert brain wave data fast enough to facilitate programming needed for neural signaling to operate on multimedia elements in real time.

These hardware advancements, coupled with new software developments inherent to the OS X /Tiger operating system, such as the virtual midi translator and the Quartz Composer application, have facilitated more efficient audio and visual programming solutions, which enable designers, as well as programmers, to develop brain wave interactive multimedia applications. Using Tiger's Quartz Composer, text, graphics, movies, interactive objects, and IBVA4 wave functions are easily dragged into an editing window as patches. As seen below in "Fig. 1," these patches are connected and composed to create user interfaces with multimedia elements that are manipulated by brain wave data. Animated textures can also be mapped to graphics, which are animated by brain wave activity.

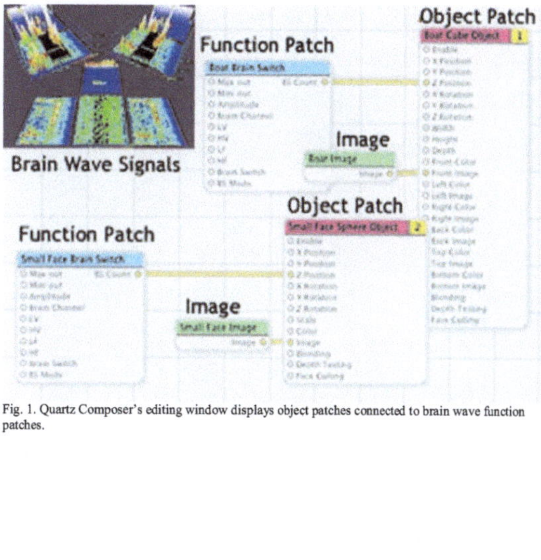

Fig. 1. Quartz Composer's editing window displays object patches connected to brain wave function patches.

K. Elleithy et al. (eds.), Advances in Computer, Information, and Systems Sciences, and Engineering, 351–355.

Since multimedia interfaces can be designed with the visual programming software built into Tiger's operating system, educational institutions no longer need to buy multiple software applications for multimedia development. What in the past would have taken several weeks to design, can now be developed by connecting object patches to function patches available for selection in Quartz Composer's patch library.

IV. NEUROLOGICAL MULTIMEDIA

IBVA4 outputs data from the left and right hemispheres of the brain, but unlike other brain wave interfaces developed for desktop computers, the interface also outputs coherence data between the two hemispheres. See "Fig. 2" below. The physiological connector between the left and right hemisphere is the corpus callosum, which is suggested to be responsible for the intuitive function of the brain. If this statement is true, it is possible that coherence data from the brain is a measurement of the intuitive function, which plays a critical role in human perception. Using IBVA4, one can manipulate multimedia elements with coherence data, for what may be, the visual feedback of human perception.

Brain wave states, are defined by wave amplitudes and gamma, beta, alpha, theta, and delta frequencies, as delineated by electroencephalographic (EEG) analysis of the brain wave signals. Unlike traditional key frame animation, quantum programming directs brain wave data into functions that create motion paths for objects. The function patches store calculations between the wave function and variable ranges of transformational values on the x, y, and z planes.

Using Apple's Quartz Composer application with an IBVA4 plug-in and function patches, artists and educators can learn to visually program educational interfaces with brain wave interactive multimedia elements. The new technologies facilitate the use of IBVA4 for the development of neurological learning tools (NLT's) for learning games, which will be discussed in the context of innovative instructional techniques.

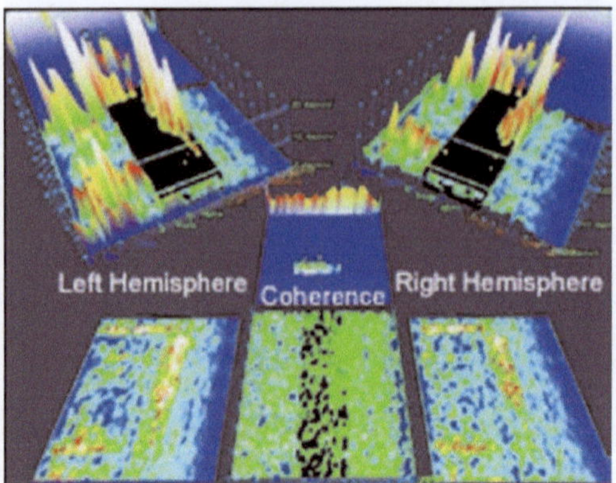

Fig. 2. Interactive view of the left and right hemisphere brain wave data and the coherence data.

V. NEUROLOGICAL LEARNING TOOLS

There is another level of complexity added to the learning process, which is that students also learn from the audio and visuals experienced in a neurological learning environment. According to Lynette Burmark, using multimedia in the context of education accelerates the development of visual literacy, that is, the ability "to interpret and create visual information—to understand images of all kinds and use them to communicate more effectively. "1

Dr. Renee Sandell, director of the MAT in Art Education Program at George Mason University, expresses that "students require capabilities that enable them to encode visual concepts through creating art as well as decode meaning by responding to the images, ideas, and media that permeate our increasingly complex visual world." 2

Students using a neurological learning tool in a virtual gaming environment also learn from the exercise of switching from one predominant frequency and amplitude to another, referred to as brain wave switching. A brain wave switch feature built into IBVA4 can be visually programmed to trigger functions that operate on multimedia elements. Brain wave switching becomes a mental exercise controlled by one's will. As with any part of the body, exercise increases the strength of the part being exercised.

Regular practice with an NLT strengthens one's ability to consciously switch between predominant frequencies, while experiencing changes in multimedia. Increases in the brain's strength, the brainpower, enable one to have greater control over the brain wave activity, to become more perceptive, and to more fully utilize the brain's inherent capabilities. Allen Newell says that the theory that intelligence is a scientific construct defined by tests that are created is in opposition to a notion of defining intelligence by cognitive theories and experiment. [2]

Jeff Hawkins argues that the brain does not function like a computer because the brain doesn't compute answers to problems. He explains that the brain retrieves answers from memory. [3] His argument suggests that there is latent information stored in our memories. Continued practice of brain wave switching may strengthen the brain's ability to access information from those memories, thus actualizing greater perceptual awareness.

1 Lynette Burmark, "Visual Literacy: Learn to See, See to Learn," *Association for Supervision and Curriculum Development,* Virginia: 2002, vii.

2 Renee Sandell, "Inspiring Pedagogy: The Art of Teaching Art, " *CAA Newsletter*, College Arts Assoc. September 2005.

Fig. 3. Brain wave controlled objects viewed from low frequency and low voltage signaling.

Music created from neural data is closely related to what is referred to as the harmony of the spheres, harmonics that are produced by the natural movement of planetary bodies. Similarly, brain wave harmonics are vibrations that resonate within our psyches. With IBVA4, the brain's harmonics, as spectral data can be converted and output as midi notes that can be input to a real or virtual keyboard to produce brain wave music.

Kaku's string theory supports the concept of the music of the spheres, as he explains that the heart of an electron is a vibrating string and not a point particle. He suggests that if these strings could be played, all the notes of the musical scale would be played. [4]

Audio and visuals, activated by low frequency, coherent brain wave signals, move in harmony with calm mental states and become reflections of how we think and feel without stress; whereas, audio and visuals activated by high frequency signals move in more random, haphazard paths, as reflections of agitated mental states. See "Fig. 3" and "Fig. 4" above from a neurological game. As human perception develops, there follows an increased awareness of how the body and mind function together. Learning to switch brain wave frequencies by will facilitates learning to focus the mind's attention.

Focused attention decreases mental noise and high stress levels. With less mental distraction, there is greater opportunity to focus inwardly on the condition of the body. When the body is out of balance, the mind learns to be aware of a problem and naturally focuses to identify and self-heal, or solve the problem.

As high frequency beta neural activity decreases, lower alpha and theta frequencies generally become predominant, which results in clear mental perception and focused mental activity. Learning begins to occur in an effortless manner, as learning from experienced reactions to audio and visual stimuli. While experiencing lower brain wave frequencies, one realizes an increased ability to problem solve; thus, one learns the advantages of existing in a clear, healthy perceptual state of mind.

Fig. 5. Brain wave controlled objects viewed from high frequency and high voltage signaling.

VI. IBVA4 IN THE CLASSROOM

The ideal classroom for neurological learning would have IBVA4 installed on Apple G5 computers running OS X/Tiger. At each computer workstation, there would be a headband with three electrodes to collect the brain wave data. Two small receivers placed at each workstation receive the neural data, input data to the computer, and direct the data through IBVA4 function patches in the Quartz Composer's editing window. In the editor, the IBVA4 patches are connected to object patches, which are viewed on a monitor as objects in a three dimensional gaming environment for learning.

Brain wave interaction with the objects provides real time feedback to students, which influences them to work to effect more significant changes in the multimedia events they are controlling. With practice, students begin to identify the kind of mental state that creates the most significant change in the multimedia events. They learn to control their brain wave switching. Using the interactive biofeedback measurement of their neural activity, they learn through experience at their own pace from their interaction with the multimedia elements

Greater mental attention and clarity of thought help to strengthen natural precognitive abilities, which enable students to more fully understand subtle aspects of perception: to hear what in the past was not heard and to see into the future with remote viewing. "Remote viewing has now taken a long step into the public domain with the formation of a professional association to propose standards, test performance, and promote peaceful use and development of this unique human mental capacity." 3

In his book, *The Conscious Universe*, Dr. Dean Radin explains that understanding the experience of remote viewing requires an expanded view of human consciousness. He suggests that the reality of psychic phenomena is currently understood as a result of new ways of evaluating massive amounts of scientific evidence collected over a century by researchers.

3 International Remote Viewing Association, http://www.irva.org, 2000-2004.

He also indicates that acceptance of the human ability for remote viewing as a form of non-conventional communication generally falls into two categories: one, perception of objects or events beyond the range of the ordinary senses and two, mentally causing action at a distance. [5]

VII. NEW TEACHING METHODOLOGIES

In order to adopt neurological gaming into learning environments, educators need to switch from past learning modalities, which program students to learn to think in prescribed ways. Neurological methodologies for learning encourage students to expand their own potentials by increasing the brain's ability to perform. Students are introduced to techniques for monitoring their own performance on an individual basis through self-analysis and self-regulation, which allows them to learn from experiencing their own brain wave patterns.

Educators monitor their students and determine a student's progress by comparing their early ability to brain wave control multimedia with their later ability to control the same multimedia events. As variables affecting the learning environment are monitored on a daily basis, students begin to see how their brainpower is influenced by these variables, and they learn to understand how to focus their attention to lower frequencies, which enables them to think and perceive more clearly.

Programming user interfaces for neurological learning environments can become part of the learning process, as a user's ability to effectively animate the audio and visual objects relative to their brain wave activity is relative to the programmer's ability to calculate the interaction of wave values with animation transformations associated with those objects. Students can work together with educators, or they can work independently to create the interactive user interfaces that are used in conjunction with IBVA4, that is to create neurological gaming environments. Producing the learning environment can become part of a student's educational program.

Development and deployment of neurological learning tools provides opportunities for cross-disciplinary interaction between visual artists, musicians, programmers, and performers working with images, animation, music, words, and design. Meaningful content from a variety of professional disciplines needs to be produced for intelligent neural interfaces that feature audio and visuals that are animated by the brain wave activity. The interfaces that are designed can be programmed with the idea of developing intelligent gaming environments that accelerate learning.

In addition to learning to control neural processes, students will learn from the multimedia events they are viewing and controlling with their neural activity. Careful planning needs to go into the production of multimedia interfaces that will be used in the classroom. Educators need to work together to develop visuals that are accurate reflections of the world. Virtual representations of the world as it really is can be developed as part of the expansion of human awareness. Increasing awareness of the world in its present condition enables mental focus to be directed toward planetary issues that are forthcoming, in order to influence intelligent problem solving.

Neurological learning tools allow students to tap into capabilities previously unknown, which opens up new avenues for acquiring information needed to expand the boundaries of what is known about how the brain functions and how the future is perceived.

Michio Kaku describes eleven steps or stages of the evolution of human intelligence in conjunction with the need for future planetary changes that may challenge the survival of intelligent life in the universe. [6]

VIII. CONCLUSION

New possibilities for human communication and interaction will result from expanding the intuitive function. With the acknowledgement of a fifth force, that is brainpower, coupled with neurological learning tools, future generations will see the evolution of more intelligent humans.

In an interview published on KurzweilAI.net June 26, 2003, Kaku generalizes that human intelligence will eventually transition from a Type 0, savage civilization to a Type I more intelligent civilization and that humans have centuries to create prosperity and knowledge on Earth. [7]

In his book, "Parallel Worlds," he describes eleven stages of the evolution of intelligence and concludes that consciousness is the dominant force that determines the nature of existence. [8]

Ray Kurzweil suggests that our purpose is the same as the purpose of the universe, which is to move toward greater intelligence. He states that we will infuse the solar system with our intelligence through self-replicating non-biological intelligence, which will saturate matter and energy in the universe with intelligence. [9]

Brain wave interactive learning provides a continuous research and development environment for educators and students across multidisciplinary platforms. Adopting new methodologies for teaching is a slow process that involves acceptance, by teachers and administrators, of new ways of thinking and using technology to enhance the learning process. The neurological learning tool provides a means of learning to perceive the future. Lifting the veil that covers the truth of our present existence is vital for our survival into the future. Visualizing the future we want to live in is the next step needed to have the future we see.

ACKNOWLEDGMENT

Masahiro Kahata (Psychic Labs, Inc., www.psychiclab.net), E-mail: pppf6@mindspring.com) is the programmer and developer of IBVA4 and provides technical support and direction for the development of Paras Kaul's neurological learning tools and gaming environments.

REFERENCES

[1] M. Kaku, *"Parallel universes, the Matrix, and superintelligence,"* KurzweilAI.net, Interview by Amara D. Angelica, June 26, 2003.

[2] A. Newell, *Unified Theories of Cognition*, Harvard University Press, Cambridge, by the President and Fellows of Harvard College, 1990, p. 89.

[3] J. Hawkins with Sandra Blakeslee, *On Intelligence*, An Owl Book, Henry Holt and Company, New York, 2004, p. 89.

[4] M. Kaku, *Parallel Worlds*, New York: DoubleDay, 2005, pp. 196-197.

[5] D. Radin, "The Conscious Universe," Harper Edge, 1997, "Chapter I: Introduction," online at http://www.irva.org/papers/RadinCU.html.

[6] M. Kaku, *Parallel Worlds*, New York: DoubleDay, 2005, pp. 321-342.

[7] M. Kaku, "Parallel universes, the Matrix, and superintelligence," Interview on KurzweilAI.net June 26, 2003.

[8] M. Kaku, *Parallel Worlds*, New York: DoubleDay, 2005, pp. 321-342.

[9] R. Kurzweil, *The Singularity is Near, the Penguin Group, New York, 2005, pp. 371-375.*

THE MODERN SCIENCE LAB: INTEGRATING TECHNOLOGY INTO THE CLASSROOM IS THE SOLUTION

Meetu Walia, Edwin Yu, Magued Iskander, Vikram Kapila, and Noel Kriftcher
Polytechnic University, Six Metrotech Center, Brooklyn, NY 11201

Abstract— Technology continues to profoundly impact daily lives. It is therefore imperative that all students receive comprehensive, high quality education in STEM subjects (Science, Technology, Engineering and Math) from adequately trained teachers. In order for students to pursue STEM career, achieving high scores on standardized science and math exams is critical. Unfortunately, science labs often make use of antiquated technology that fails to tap the potential of modern technology to create and deliver exciting lab content. As a result, students are turned off by science, fail to excel on standardized science exams, and do not consider STEM as a career option

"Revitalizing Achievement by using Instrumentation in Science Education (RAISE)" project is a partnership between Polytechnic University and four New York City (NYC) high schools. Project RAISE seeks to enhance students' academic achievement by using computerized data acquisition and sensor-based equipment in science labs. RAISE seeks to excite students about STEM and help them comprehend challenging scientific concepts.

The authors argue that the modern science lab must integrate technology into the curriculum. This paper presents a description of the RAISE project, together with some of the sensor-based experiments which are currently in use in the "living environment" science lab. Lessons learned from year-one of the program as well as improvements made for the second year are also presented.

I. INTRODUCTION

Today's students are attracted to new gadgets, such as iPods, video games and cell phones. To them, gadgets are *cool*, and this can be leveraged as a way to interest students in technology and to motivate them to excel in STEM disciplines. Unfortunately, for reasons of limited resources and a well-documented shortage of adequately trained high school science teachers, schools often present required science and math courses in an unimaginative manner [1]. They introduce basic scientific concepts as before, but fail to relate these concepts to science, as encountered by students in their daily lives. That is a major reason for students' losing interest in their science studies. Additionally, uninspiring laboratory experiments contribute to a general lack of interest in studying STEM disciplines. This results in poor achievement on standardized science and math exams and apathy about pursuing careers in engineering and technology.

Laboratory experiments commonly support learning in math and science in high school. By integrating modern sensor technology, into traditional labs, students can be introduced to lab experiments, which are not only informative, but also very exciting [2]. For example, in a living environment laboratory, a demonstration may involve measuring how plants generate energy by photosynthesis. To perform this experiment in a traditional lab, students use a lamp, a plant and a bicarbonate indicator, which serves as a traditional sensor. It would take at least one day to detect the change in concentration of carbon dioxide qualitatively as its color changes from purple to yellow. Students would have to wait for the color to change, but by then the average student's attention would be lost. In a sensor-based lab, as will be explained in greater detail later in this article, a few spinach leaves, a light source, an oxygen sensor and a carbon dioxide sensor are used. Real-time data acquisition hardware and software allows groups of students to monitor and compare the process of photosynthesis through changes in the concentration of carbon dioxide and oxygen. The measurements are easily accomplished in a 55-minute class period. Furthermore, students are able to control the experiment by changing the parameters that affect photosynthesis and monitor the consequences. For instance, they could alter the results of the experiment merely by changing the location of the light source or the numbers of plant leaves,.

Integrating modern sensing technology into science labs makes labs more appealing to students by having them use gadgets that are *cool* and by allowing students to visualize results graphically and in real time. This is beneficial for learners who rely on visualization. Modern sensors allow inductive and reflective learners to develop inquiry-based learning skills, by developing measurable recorded data. These skills are essential in an increasingly technological society [3,4]. In addition, those students who are most proficient in science and math, will be introduced to contemporary innovative ideas which, it is hoped, will make them more likely to be enticed by career paths that are related to science and engineering.

K. Elleithy et al. (eds.), Advances in Computer, Information, and Systems Sciences, and Engineering, 357–362.

II. PROJECT OVERVIEW

The "Revitalizing Achievement by using Instrumentation in Science Education (RAISE)" project is a partnership between Polytechnic University and four New York City high schools [5]. Its mission is to enhance student's academic achievement in STEM disciplines. This program is funded by a National Science Foundation (NSF) GK-12 grant. The high schools participating in Year-1 of this program were George Westinghouse, Marta Valle, Seward Park, and Paul Robeson. For Year-2, Seward Park High School was replaced by The High School of Telecommunications, Arts, and Technology.

The main purpose of RAISE is to increase students' achievement in science, as measured in part by their achievement on standardized exams. The project seeks to stimulate students by integrating sensing, instrumentation, and modern data acquisition and computing technologies into the curriculum used in Active Physics, Regents Physics, and Living Environment courses. Marine Science was added to the program in Year-2. Additionally, the RAISE Fellows serve as teaching assistants in the classroom, and as science resources to teachers and also make themselves available for extra tutoring.

The RAISE program is supervised by two engineering faculty, and one liberal arts faculty. Each Fellow is paired with a RAISE teacher who receives a stipend and attends a weeklong workshop during the summer to learn about modern sensing technology. Each Fellow spends a minimum of ten hours a week at his/her assigned high school as a science resource and at least five hours a week on campus preparing experiments and materials to be used in the high school classroom. The graduate RAISE Fellows are full time students and are required to make satisfactory progress in thesis research. They receive tuition remission as well as a stipend. The undergraduate RAISE Fellows receive a stipend only. An external evaluator evaluates the progress of the program.

RAISE fellows attend a weeklong professional development workshop conducted by an education specialist who formerly served the New York City school system as a trainer of teachers. The workshop was designed to enhance the Fellows' pedagogical, communication and presentation skills, and to help them prepare effective lessons. Many Fellows felt that this training program gave them an advantage in the classroom, especially with matters of classroom management, discipline and effective questioning techniques. Fellows learned practical approaches to anticipated challenges that may arise in the classroom and they also developed techniques of conveying academic material in a more effective manner.

The RAISE teachers underwent a weeklong training session on how to implement sensors in classrooms, effectively. The teachers also received a crash course on mechatronics and engineering, where they made their own robots [6]. The rationale behind this training was to encourage and excite the teachers to go beyond basic implementation of sensors in the classroom, and to present an opportunity for the teachers and Fellows to bond with one another. A secondary goal was for some teachers to be motivated to start a robotics club at their schools, where they could use what they learned during the training session and to motivate the students into joining the club.

III. DIFFICULTIES ENCOUNTERED IN YEAR-1

The program got off to a slow start in Year-I because of scheduling conflicts and a lack of appropriate equipment. Initially, it was difficult to attract graduate students to join RAISE in a hot job market despite the generous stipend ($30,000+ tuition). Out of the twelve fellows selected for Year-I, ten were undergraduate students. Since undergraduate classes are scheduled mainly during the daytime, this limited undergraduate Fellows' availability to serve in high schools. In the second year, the benefits of RAISE to the Fellows were evident: five of the ten undergraduate Fellows applied to continue as graduate RAISE Fellows and pursue their studies towards the master's degree. In Year-II of the program, there are eight graduate and five undergraduate Fellows, which makes scheduling considerably easier.

A second difficulty encountered in Year-1 was that the NSF GK-12 program provides generous stipends but does not allow for adequate equipment funding. Polytechnic University subsidized the program by providing each school with four biology and four physics sensor kits [7]. However, when Marta Valle was added to the program at the last minute, it had to share kits with Seward Park High School.

Schools were required to provide laptops. It was difficult for many schools to find funding to meet this requirement, at least in a timely manner. In Year-1, schools that were unable to provide laptops or computers used the sensors with graphing calculators. By Year-2, all schools were equipped with laptops or computers.

Despite receiving intensive training, problems were bound to arise during the pilot year of the RAISE program. These problems ranged from poor classroom discipline to students' inability to cope with new technology. For many students, it was the first time that they were being exposed to such equipment. However, eventually over the course of the year, students formed an appreciation for the sensors and were able to do many computations that were required of them.

By the start of Year-II, all of these problems had been successfully addressed and the RAISE program was ready to make a long-term impact on student learning. One dramatic contribution in Year-I, however, was the role of the Fellows at one of the schools when a teacher suffered a debilitating heart attack. The principal had enough confidence in the Fellows to have them provide instruction for the class in an effort to salvage the class although a replacement physics teacher could not be found. According to the NYC Board of Education rules, a certified teacher must be present in the classroom at all times. A substitute teacher was hired to meet that requirement, even though they did not play an active role in teaching. This was a noble effort, and although the class eventually had to be disbanded, the Fellows had proven their merit in this rescue effort.

New York City is in the middle of a school reform, which involves among other things breaking up large schools into several small ones occupying the same building. Seward Park is one of the schools being phased out. The principal and the assistant principal for science, as well as some science teachers, left to pursue a variety of professional and personal interests. It was difficult, therefore, to continue to maintain Seward Park in the program under the circumstances.

In the second year of the RAISE program, the mistakes that were made during the pilot year were corrected. Over the course of the second summer, the Fellows and teachers started planning for the upcoming school year in terms of scheduling, lab experiments, tutoring, and various other activities. The Fellows underwent a second summer workshop similar to the previous year in order to train them effectively for the classroom. During this 5-day workshop the fellows created a database of presentations and labs to be delivered in the classroom for instruction. Each Fellow has access to this database. Since most of the Fellows are experienced returning fellows, they enjoy a comfort level with their respective teachers. This familiarity promotes communication that can lead to more improvements in the program's infrastructure.

IV. BROADER IMPACTS OF PROJECT RAISE

A change in the students' attitude was observed over the course of the year. Many students looked forward to doing lab experiments. The tasks that they may have found daunting in the beginning of the year later became second nature. Furthermore, students were devising extensions to the lab experiments for a fuller learning experience. Some students started expressing interest in careers in science and engineering to Fellows, whom they looked up to as a source of inspiration, and career advice.

The RAISE program affected the Fellows as well. Fellows displayed an improvement in their communication and presentation skills. Additionally, many expressed that their comprehension of scientific principles has improved as a result of having to teach.

The impact of RAISE extended beyond the classroom. There were many events that took place during the first year of the program, which supported its goals and objectives:

1- Career Day

Polytechnic University hosted RAISE Career Day on April 20, 2005. Students from the four participating high schools were exposed to various disciplines of engineering via presentations and tours made by faculty and recent Polytechnic graduates. The purpose of this event was to arouse students' interest in science, engineering, and technology that may lead them to consider engineering as a possible career option.

2- NYC NSF GK-12 Grant Holders Meeting

The NYC NSF GK-12 Grant Holders Meeting took place on May 20, 2005. This meeting was hosted at Polytechnic University. Participants included representatives from the four GK-12 projects in NYC (Columbia University--two programs), CUNY Graduate Center, and Polytechnic University), as well as the NYC Department of Education. This meeting provided an opportunity for participants to discuss major challenges that are faced within collaborative programs, which seek to enhance K-12 learning in science and technology. In addition, it gave a chance for the participants of the different GK-12 projects to network and explore new methods of tackling similar problems. The GK-12 program team from NSF also attended the event.

3- Annual GK-12 Meeting

The Association for the Advancement of Science (AAAS) and the NSF hosted an annual meeting for the GK-12 project teams in Arlington, Virginia on March 4-6, 2005. Various GK-12 projects throughout the nation displayed their work, using poster boards and PowerPoint presentations, and gave participants a chance to network and exchange ideas.

4- Third Annual Convergence on Inquiry

On June 11, 2003, RAISE was invited to the Third Annual Convergence on Inquiry at the American Museum of Natural History to give a presentation to a group of teachers and other educators on how sensors and instrumentation can help students "ask more questions" related to science and technology. The goal of this conference was to discuss different means of getting the students to "inquire."

5- SMART Program

RAISE fellows also participated as science resources for Polytechnic's SMART program at which ten teachers from New York City schools were chosen to undergo intensive training in mechatronics for four weeks during the summer. The Fellows themselves sharpened their skills in mechatronics and were made available to assist the teachers in building their respective projects.

V. SENSOR-BASED LIVING ENVIRONMENT EXPERIMENTS

Living Environment is a class typically taken by freshmen in high school. Students take an exam organized by the state called the Regents exam. Each student must pass one Regents examination in science before graduating high school [8]. To earn a Regents-endorsed diploma, which is recommended as a minimum achievement for a student entering a college such as Polytechnic University, a student must pass three Regents exams in science. In NYC Living Environment tends to be the student's first exposure to "Regents science" in a high school.

The sensor-based experiments that were developed by the RAISE fellows modeled the New York States Regents Living Environment curriculum. The developed experiments incorporated sensors, which demonstrated concepts that originally would seem difficult to visualize and comprehend. The sensors helped students get a more interactive illustration of important concepts of living environment. Table 1 shows the Regents Living Environment experiments developed by the RAISE Fellows.

TABLE 1 - Living Environment Experiments

Experiment	Description
1. Acid Rain	A pH sensor is used to understand the relationship between pH and CO_2 concentration of distilled water and how different pH level affects living organisms.
2. Acid & Base	A pH sensor is used to measure the pH of different liquid that are acidic and basic and help understand the pH scale.
3. Aerobic Respiration	A CO_2 gas sensor is used to measure to measure the rate of respiration. Students compare the different rates of respiration for different energy drinks which help them conclude which drink is the most effective energy drink.
4. Anaerobic Respiration	A gas pressure sensor is used to measure the rate of respiration. Students compare the different rate of respiration for various energy drinks which help them conclude which energy drink is the most effective
5. Calorie Content in Food	A temperature probe is used to determine the energy content of a small sample of food. Students learn how energy is given off by food as it burns. They determine and compare the energy contents of different foods.
6. Conducting Solutions	A conductivity probe is used to measure the conductivity level of different solutions. Students have to understand how the number of ions of solutions relates to conductivity level.
7. Diffusion	A conductivity probe is used to measure the change in ionic concentrations in a solution over a period of time. Different factors affecting the rate of diffusion were studied.
8. Enzymes	A colorimeter sensor was used to study the functions of an enzyme present in various detergents.
9. Greenhouse effect	A temperature probe was used to measure the temperature changes within an environment to understand the greenhouse effect as a physical phenomenon
10. Monitoring Heart Rate	An EKG sensor was used to graph one's heart's electrical activity. Based on the EKG recording the heart rate can be calculated.
11. Photosynthesis	A carbon dioxide and an oxygen sensor was used to calculate the rate of photosynthesis and compare the affects of light on the rate of photosynthesis
12. Population Dynamics	A colorimeter was used to monitor a closed population growth of yeast by measuring the turbidity or cloudiness. Photons of light strike the yeast cell and reflect away from the photocell. The colorimeter monitors the light reflected by the photocells as absorbance which is proportional to the yeast present in the medium.

VI. EXAMPLES OF LIVING ENVIRONMENT EXPERIMENTS

1. Aerobic & Anaerobic Respiration

Some of the experiments designed went beyond just demonstrating key concepts of living environment. Many of the experiments related everyday life to the content of lab. In one living environment experiment, students were asked to identify which drink would give the most surge of energy. Some of the choices given were: Coca Cola, orange juice, Gatorade, milk, and water. The students tested each drink using a carbon dioxide sensor to measure the rate of aerobic respiration. To make the experiment analogous to the human body, yeast was used to consume and break down the sugar molecules of each drink and release energy. Energy is directly proportional to the amount of carbon dioxide released. Hence, the greater the amount of carbon dioxide released, the more energy is released, thus the greater the rate of respiration.

Based on their findings, students were required to compare the rates of respiration and conclude which drink would be the best option right before an athletic event. A similar process was done for anaerobic respiration except for two main differences.

- The mixture of the yeast and drink was sealed air tight, not allowing any contact with outside oxygen.
- A gas pressure sensor was used to measure the rate of respiration based on the concept that gas pressure increases as more gas is released.

Normally in a lab similar to this, the drinks would have been substituted with different sugars for example: Sucrose, Fructose, Lactose, and Glucose. These are the same sugars that are found in the drinks that were used in the experiment. Students don't fully understand why respiration is so important and how does it affect our bodies. However, by substituting the sugars with what students drink everyday allows them to fully appreciate the importance of respiration.

2. Monitoring EKG

One of the most entertaining and interactive experiments conducted during the year involved an Electrocardiogram (EKG) sensor. Many of the students were already exposed to what an EKG diagram looks like from television shows such as ER. Taking advantage of this the EKG lab was designed to have students determine their heart rate based on the EKG diagram produced using the sensor. The EKG sensor is hooked up to three electrode patches that are placed on a volunteer's arms. The students collected data while at rest and after performing fifteen jumping jacks. They were also required to determine their heart rate by measuring their pulse and then compared their results with what was calculated using the EKG sensor.

The students found this lab to be the most interesting experiment partially because the sensor was directly hooked up to their bodies. They were able to visualize what their EKG looks like in comparison to one of their classmates. Branching off from this experiment, students came up with many questions, such as:

- How would age, gender, and weight affect the EKG diagram?
- Why are the peaks of my EKG smaller, larger than my classmates?
- How can we use an EKG to diagnose various ailment of the heart?
- Why don't we get electrocuted if the EKG sensor measures the electrical impulses?
- How does the EKG sensor work?

This lab opened up a riveting discussion about different heart ailments and how useful an EKG is in diagnosing such illnesses, which brought lots of excitement to the classroom [9]. Not only did they form an appreciation for an EKG, but this experiment helped the students better understand the hearts functions.

3. Photosynthesis

Life requires energy to go on. Plants generate energy from photosynthesis with carbon dioxide. During the process of photosynthesis, carbon dioxide is consumed, while the plant generates oxygen and energy. In this experiment, students illustrate the process of photosynthesis by detecting the concentration of oxygen and carbon dioxide in a plant. To perform this experiment, students use a lamp, few spinach leaves, an oxygen gas sensor, a carbon dioxide sensor, and real-time data acquisition hardware and software. Use of sensors allows students to visualize the concentration of oxygen and carbon dioxide when observing the phenomenon. Thought-provoking questions are asked to reinforce the concept of photosynthesis; such as:

- What evidences do we have to conclude that the plants experience photosynthesis?
- What is the ultimate concentration of gases if the experiment would be continuously running without touching the setup?

Students investigate their hypothesis and discuss the results with their classmates. Such experiment excites students about the subject, motivates students, and fulfills their curiosity.

VII. CLASSROOM IMPLEMENTATION

Many of the teachers have limited resources available to them to support their classrooms. In addition, overcrowded classrooms limit the personal attention that is given to students. Thanks to funding from the U.S. Congress, NSF recognizes these shortcomings and has created the GK-12 fellowship program. What distinguishes raise from other GK-12 fellowships is the integration of technology into classroom activities. The designed experiments had to be implemented within the science curriculum in order to achieve the program's objectives and gain the support of high school teachers and administrators. Therefore, the developed experiments tended to be more sophisticated than the ones already in existence.

Usually before performing an experiment, the fellow would give a demonstration to allow the class a chance to gain familiarity with the lab. Then the students are split into groups of three to five to perform the lab. Since labs by their nature are interactive, each student is assigned a specific task, such as setting up the experiment, controlling the pace of the experiment, recording data, or performing calculations.

Starting in year-2, Fellows set up tutoring sessions to further assist students who require extra help. Students' receptivity to the Fellows is based on the positive interaction they have experienced, which gives the student an opportunity for a one-on-one learning experience with the Fellow. The presence of another science resource in the class helps alleviate the pressure on the teacher and encourages the students to ask more questions.

VIII. ASSESSMENT

1. Impact On High School Students

It is difficult to assess the impact of raise on student achievement, since many of the objectives need time to pan out. Nevertheless, most students were excited at the opportunity to use the new instrumentation available in the lab. More students felt that the laboratory component was their favorite aspect of science course. Teachers believe that concepts learned in sensors based lab were more memorable to students than in a traditional lab. However, it is impossible to verify this observation because a control group was lacking.

One of the goals project RAISE is to encourage high school students to continue their education at a college level in areas related to STEM. There are two problems with assessing this objective. First, none of the students is college bound yet. Second, it is difficult to determine if those students who are presently expressing interest in a STEM career would have chosen to study STEM anyway without RAISE. Finally, with the many changes associated with NYC school reform, it is difficult to verify the relative contribution of many initiatives.

All things considered the project team believes that sensor have made a positive impact on the students academic experience. The team also believes that the opportunity for high school students to interact closely with goal-oriented, engineering college students will help them, to develop academic goals for themselves.

2. Impact On Fellows

The effect of the program on RAISE fellows was found to be positive, as evidence by the fact that seven out of twelve fellows chose to continue the program. Fellows clearly improved their communication and technical skills as a result of frequent presentation of their work. Through classroom management techniques, fellows polished their leadership and management skills. These elements will serve the Fellows well in the future as they seek to become leaders in their fields, once they graduate and are employed.

IX. Pedagogical Lessons Learned

After working with students in Year-1, several lessons were learned and adopted to improve classroom management and teaching strategies:

Classroom Management

- Check and test the availability and condition of equipments (outlets, computers and projectors) before beginning the lab.
- Divide students into small groups.
- Students behave better when visitors are in the classroom, in addition to the teacher. Invite guests, including the principal/assistant principal to visit the class.
- Don't stop the entire class if a student comes late. Ask another student to explain what the class is doing.

Teaching Strategy

- Keep students engaged.
- Find the common ground between what the students are interested in and the curriculum.
- Don't assume students recall material from past courses. Keep reminding them what of what they have learned.
- Create rules to make students work effectively and systemically. For example, prepare lectures with built in exercises.
- Always encourage students, give positive reinforcement and build up their confidence.
- Hand out extra responsibilities for energetic students.

X. Conclusion

This paper argues that integrating modern technology into science lab is the answer to the fading interest in STEM disciplines among American high school students. Students found the labs interesting. It is however too early to determine if (1) students will do better on standardized exams, and (2) if they will be motivated to pursue STEM careers, as a result of RAISE. The Fellows have improved their technical and pedagogical skills. The RAISE project team strongly believes that, given time, the goals and objectives of the program will materialize.

XI. Acknowledgements

Revitalizing Achievement by using Instrumentation in Science Education (RAISE) is supported by the GK-12 Fellows Program of National Science Foundation under grant DGE-0337668. Significant financial support for the acquisitions of 13 sets each of Vernier LabPro Biology Deluxe Package and Vernier LabPro Physics Deluxe Package was provided by Polytechnic University. Significant discounts were provided by Vernier Software and Technology.

XII. References

1. *Losing the Competitive Advantage? The Challenge for Science and Technology in the United States*. American Electronics Association, February 2005. http://www.aeanet.org/publications/IDJJ_AeA_Competitiveness.asp.
2. G. C. Orsak *et al.*, "High-Tech Engineering for High School: It's Time!" *IEEE Signal Processing Magazine*, pp. 103—108, January 2004.
3. *Innovate America: Thriving in a World of Challenge and Change*. The National Innovation Initiative Final Report, December 2004. http://www.publicforuminstitute.org/nde/sources/NII_Final_Report.pdf.
4. T. L. Friedman, The World Is Flat: A Brief History of the Twenty-first Century. Farrar, Straus and Giroux, New York, NY, 2005.
5. Online: http://raise.poly.edu, website of the RAISE project.
6. 21. Online: HTTP://MECHATRONICS.POLY.EDU, website of Mechatronics @ Poly.
7. Online: http://www.vernier.com/soft/lp.html, website of Vernier's Logger Pro Software
8. Performance Standards: Science, Board of Education of the City of New York. 1999. ISBN: 1-55839-505-9.
9. W. S. Bacon (Ed.), Bringing the Excitement of Science to the Classroom: Using Summer Research Programs to Invigorate High School Science. Research Corporation, Tucson, AZ, 2000.

A Novel Computer Aided Learning Technique in Engineering Education

Habibullah Jamal, M. Zafrullah and M. M. I. Hammouda[#]
University of Engineering and Technology Taxila, Pakistan
[#]on leave from MED, Al Azhar University, Cairo, Egypt

Abstract - **In this paper we present a newly developed computer aided learning technique for the Engineering students. The proposed methodology assumes that each student has easy access to well equipped computer lab of the university or his own personal computer, connected to the internet. Document based teaching material is developed by a team of instructor and the software developer, which enables knowledge self-assessment. Typical examples demonstrate the applicability of this system. The purpose of this paper is to explore, with illustrations, the extent to which such benefits can be achieved in practice.**

I. Introduction

Co-operative work in a team is an advantage of computer labs. In the computer lab environment knowledge can be easily transferred, particularly, if the players are of similar age. Computer labs are always very animated as students are encouraged to challenge each other. Computers can be efficiently utilized to motivate and promote self study skills and to provide individuals with thinking.

The technology of today offers the opportunity for significant educational benefits. Educators can save their time and effort by the use of suitable computer aided learning softwares (CALS) [1, 2]. More efficient engineering education can be offered, should carefully CALS be prepared. In the context of well planned project work, involvement, enthusiasm and motivation of both faculty and students can increase. Educators/students contact time can be reduced. The number of students who can be taught simultaneously can increase. Assessment can also be automated [3]. Self-assessment of student's knowledge before final testing is also possible.

The production of CALS involves considerable knowledge, effort and skill. To be appreciated and widely used, such software must show superiority over the currently available media. They must improve achieving something that the main competitor textbook cannot. This pre-requisite can be satisfied in most engineering topics, particulary; those by nature necessitate the presentation of events varying with time. Interactive displays can be provided to show those dynamic events and conditions which are not possible in a few static images. Further, those software's can avail easy visualization of the behaviour of a system as a result of a change in any of its controlling parameters.

One may expect approaching an era of learning at a distance with well equipped labs, new knowledge assessment system, full course documentation and accreditation. The concept of virtual universities is, now, in mind. The world may be approaching the situation to have one universal department for each discipline. Enrollment, education and assessment will be electronic from everywhere in the world. Well equipped labs will be centralized and distributed all over the world for use whenever necessary. The level of the graduates will be standardized with a single description. Jobs will be internationally available. With those thoughts in mind, universities should be prepared [4]. To provide the best engineering education possible to their brightest students, engineering institutes should plan to have all its faculty working for professionally producing their courses to be taught by making use of nowadays available technology. This paper explores, with examples, the extent to which benefits can be achieved in practice by making use of this effort.

II. Proposed Methodology

A model is developed to promote the use of staff time to the best advantage and to challenge the traditional forms of taught effort. Flexible learning materials are produced to work with either in the faculty or at home. Interactive CALS are constructed and developed for some engineering topics. The development of CALS assumes the existence of a carefully prepared document based teaching aids in the form of study and instructor guides and lab manuals, if any. The developed CALS enable the process of self-assessment by the students. VISUAL BASIC environment was utilized as a tool to produce the present CALS. Examples of the above activity are given in the following sections. Those materials are available on the web so that Engineering students everywhere can benefit. More examples may be demonstrated during the presentation of that paper.

K. Elleithy et al. (eds.), Advances in Computer, Information, and Systems Sciences, and Engineering, 363–366.

III. INTERACTIVE CALS

All engineering topics may be addressed and interactive graphical PC-based CALS can be, correspondingly developed. To show their capability and demonstrate their benefits in improving the quality of teaching and learning, some examples are available for presentation

The assumption behind these CALS is to enable the students, at their own pace and time, studying an engineering system, exploring the effect of many variables controlling its behaviour and, then, self-assessing related knowledge. The students are assumed to have met the concepts of the considered topic in lectures and textbooks. Such software can be introduced in a lecture or tutorial. In this case, the lecturer is in control in order to stimulate interest. The students can, then, be directed to explore them more fully in their own time. The student can progress at whatever rate is appropriate. The student is also able to raise new relevant questions which can be self-answered.

Each document based module of a course can be carefully and professionally designed in the form of software consisting of a single page interface. That page has menus to demonstrate the corresponding topic, objectives and brief notes. Well prepared multiple choice questions (MCQs) are also contained with their checking. Knowledge self-assessment is included with messages conveyed to the students to show their points of weakness and to accordingly give appropriate recommendations. Further, the terminologies related to the module are explained just by moving the cursor over the required one. Examples relevant to the curriculum of Mechanical and Electrical engineering undergraduate studies can be demonstrated by visiting the website of UET Taxila, Pakistan. The link may be established by moving the cursor over the following address, pressing the control key on the keyboard and, then left-clicking the mouse

http://www.uettaxila.edu.pk/CAL.htm

A combo box will be visible to enable the choice of one of the university departments, one of the eight academic terms, one of the courses lectured to the corresponding students during the term and finally one of the prepared lessons within that course. Should the lesson be appropriately chosen, its page interface will be displayed for interaction. Few examples are curtailed below.

IV. EXAMPLES

A. Charging and Discharging of a Capacitor

This module is tailored for a lesson within the syllabus of the course EE-107, Circuit Analysis-I lectured to the undergraduate Electrical Engineering students of second term. The topic of that lesson is about charging and discharging of an electrical capacitor in an electrical resistance as will be given on the top of the displayed page. It is abbreviated in the combobox lessons as Lesson-1 RC. The page is designed as previously described. Text boxes exist on the top right corner of the page to type the values of the variables relevant to the problem, i.e. battery voltage, capacitance, resistance of light emission diode and discharging resistance. On activating one of the items of the menu located at the top of the page, the corresponding data is displayed in the text box located at the right bottom corner. That text box is also utilized to demonstrate relevant information whenever the cursor moves over any of the control tools within the page. Six command buttons are located at the left bottom corner of the page to sequentially maneuver the scenario tailored for the lesson. At the start, the buttons are disabled. When the controlling variable values are correctly entered, the upper command button is enabled, activated on clicking to have the charging-discharging circuit shown in the opened mode just above the buttons and to sequentially enable the other five buttons for (1) closing the capacitor charging circuit, (2) showing the derivation of the equations governing the charging process, (3) opening the charging circuit, (4) closing the circuit for discharging the capacitor in the resistance and (5) displaying the derivation of the equations governing the discharging process. Throughout the display of the charging and the discharging processes, and whenever required, a grapgh may be displayed in the middle of the page or at its top to demonstrate the corresponding variation of electrical charge, current and voltage with time.

B. Gear Meshing

This lesson belongs to the syllabus of the course ME-301, Mechanics of Machines-II given to the Mechanical Engineering students in their 5th term. The topic of that lesson is about the meshing of two spur gears. It is abbreviated in the combo box lessons as Lesson-1 Gear Meshing. The page is designed as previously described in the first example. Here, the data necessary to run the application are the number of teeth of each gear and the teeth module. The target may be either the recognition of the terminology commonly used for the present problem or visualizing the gears rotating in both available directions. The options of having either one gear or the two gears in meshing are available. The interacting meshed teeth can be focused. This last action shows superiority in explaining and easily understanding the locus of the point of contact between two teeth while they are meshing. In relevant textbooks, more than two pages are consumed with the aid of some static figures in a trial to convince the reader that this point is always along the pressure line of the two gears. Lecturers do the same in classrooms spending considerable time and effort and at the end they may fail to convey the message.

C. Examples on Vibration

Three examples are tailored similar to the above two examples with the same technique and concepts. The three lessons are within the syllabus of the course ME-402 Mechanical Vibration for the mechanical engineering students in the 7th term. They appear in the combo box lessons as Lesson-1, Lesson-2 and Lesson-3. The first two lessons are examples of a vibrating undamped mechanical system with a single degree of freedom whilst lesson-3 demonstrates the free vibration behaviour of undamped two-degrees of freedom system. A hemi-spherical shell rolling on a horizontal surface is presented in the lesson-1 with its radius, gravitational acceleration and initial deflection as its controlling variables.

A mass-spring system is demonstrated by the next example. In the third example, two masses connected to three springs freely vibrate with no damping. The initial data required for this example are the stiffness of each spring and the two masses to have the two modal frequencies and shape factors calculated and typed in the corresponding labels. The velocities and displacements of the masses should be initialized to start the desired interaction. Student can analyze a problem of that type mathematically to have its closed form solution with no visualization of the general vibration behaviour of the two masses and the effects of changing any of the controlling variables. This example, again, shows superiority over its traditional main competitor, the textbook. Even vibration educators cannot successfully explain in classroom the interpretation of the resulting forms.

D. Quick Return Motion

The example shows the kinematics behaviour of the return mechanism commonly used in workshop machinery. That behaviour is mainly dependent on the dimensions of different components of the mechanism. For every location of the mechanism during its operation, the velocity and acceleration diagrams are plotted to scale on the page. Mechanical Engineering students actually solve a maximum of two problems of such type during their 4th term of study in the course ME-209, Mechanics of Machines-I. Traditional education methodologies do allow the student neither to recognize the relation between the geometry of the mechanism and its behaviour nor to feel the events taking place during the operation of that mechanism. Besides overcoming those points of weakness in traditional static education methodologies, the present software helps in building the students abilities in creativity by answering many questions of the type, *what happens if.*

V. Discussion

An advantage of the document based teaching material is that it can be available to all teaching staff to modify as they see fit [5]. Exchanging knowledge and co-operation between the staff of different universities is beneficial. Significant time-savings can be achieved.

Current software has a long way to go before it is as easy to use, accessible, and subject to such quality control as a book. Although sometimes poorly applied, the method of using a book or a lecture is almost universally understood. Published material is relatively easy to referee. Further, both book and lecture can be used almost everywhere. A successful software must do something better than the medium it replaces, be straightforward to use and be right as far as possible. None of these points is trivial.

A lot of effort has been done by numerous educators in universities all over the world searching for innovative teaching in engineering [6]. Good software takes a long time to write. A refereed publication ensures both credit and distribution. Further, software publication requires testing and editing. Each of these phases is more difficult in terms of money and time.

The strength of graphical CALS is that it can be used to make visible processes which cannot be seen, either because they occur on too fine scale or because they are inaccessible. Many topics can be illustrated more clearly in animated software than they can be in a book. Therefore, understanding will be improved. Foremost among the advantages of this medium is that its developer learns a lot. It is impossible to code a simulation correctly without having understood the science behind.

An obvious use of this software is for student self-assessment. The study guide of a course allows its contents divided into separate lessons with a lecturing timetable. The approach is based on a data bank of multiple choice questions carefully prepared for each lesson. The assumption, here, is that the student will have an access to the computer lab at his own convenience. The student has the option either to review a part of a course or to attend a relevant test at a time arbitrarily chosen by him within an interval dictated by the instructor of the course. The material to be tested conforms to what is taught. Obviously, the review phase can continue up to the satisfaction of the student to perform a test. Certification may be automatically issued.

The main objective of the authors, now, is to convince the authorities of UET Taxila Pakistan, to adopt a project in order to produce such software for all possible topics contained within the courses of the undergraduate curriculum. That, of course, needs human resources of different talents, capabilities and knowledge. Software in the form described in this paper necessitates the cooperation of a team consisting of a knowledge expert and at least five persons who should have knowledge in programming, mechanics, drawing, scenario authoring and mathematics. The software of one lesson consumes about one man-week which may cost around US$1000. Financing such a project through the cooperation of different universities in the same country or in different countries will, obviously, prove its economical benefits in implementing such a project.

VI. Conclusions

Microcomputers software offers many advantages and has a place in many areas of teaching. Its rate of introduction has the limitation of academic time, money and manpower for coding. These problems can be alleviated by ensuring cooperation and encouraging the faculty in various universities. This activity generates credit for authors. Such material is very much suitable for education at far distances.

It is recommended to establish multimedia centers or committee in every engineering institution for that purpose. From the economic point of view, collaboration of engineering institution at either national or international levels can prove great benefits in this respect. Planning of available resources are necessary to share effort, time and expenses. In return the output can be shared also.

Acknowledgment

The authors would like to thank Mr. Adnan Shah, the web developer in the network centre of UET Taxila, for his efforts to maintain the courseware online.

References

[1] P. J. Goodhew, "*Computers in the teaching of materials science and engineering, Innovative teaching in engineering*", edited by R.A. Smith, Published by Ellis Horwood, West Sussex, England, pp. 128-33, 1991.

[2] T. W. Clyne and F. J. Humphreys, "*Atomic scale diffusion simulation*", Institute of Metals, Engineering Materials Software series, London, 1988.

[3] T. Muneer and S. H. Hawley, "*Design of computer-based monitoring systems for engineering education*", Innovative teaching in engineering, edited by R.A. Smith, Published by Ellis Horwood, West Sussex, England, pp 145-55, 1991.

[4] *A. Linse, J. Turns, J. Yellin and T. VanDeGrift.*, "*Preparing future engineering faculties: Initial outcomes of an innovative teaching portfolio program*", Proceedings of the 2004 American Society for Education Annual Conference & Expositio, 2004.

[5] C. B. Hammond, "*what good am I, if I know and don't do*?", Innovative teaching in engineering, edited by R.A. Smith, Published by Ellis Horwood, West Sussex, England, pp 134-9, 1991.

[6] H. Kuznetsov, "*Technology-based innovative teaching methods*", Proceedings of the 2002 American Society for Education Annual Conference & Exposition, 2002.

International Conference on Engineering Education, Instructional Technology, Assessment, and E-learning (EIAE 05)

Language Test for Accreditation: the experience of C.L.A.M. (Language University Centre, Messina).

Francesco Stagno d'Alcontres
fstagno@unime.it
Centro Linguistico d'Ateneo Messinese
University of Messina - Italy

Abstract **– Since June 2004, C.L.A.M. has introduced a language test for accreditation for University students wholly created, administered and delivered in electronic format. C.L.A.M.' s target users include mainly University students who need to take a test in a foreign language, such as English, German, French and Spanish to get credits for their linguistic competence in receptive skills (i.e. listening and reading comprehension, including grammar and vocabulary). In the last exam session, almost 300 students per day were evaluated in a multimedia lab with 50 PCs connected by an Intranet. During the four annual exam sessions, students can log into on-line simulations and exercises via C.L.A.M. website (also available via *mobile* version through PocketPC) creating a flow of almost 20.000 visited pages (almost 150 unique accesses per day) in the period considered. This paper illustrates technical-oriented choices underlying the creation of the standard for the tests, specific tools, automation processes – from students' subscription to the exam – and the making of on-line simulations, exercises and tests supported by *QuestionMark* platform software *Perception*. The procedure for the creation of the different parts of the test will be displayed, highlighting pros and cons of *Perception*, integration of multimedia objects and the software created by our Language Centre for automatic publishing of original materials produced by Item Writers directly in .doc and .txt extension. Furthermore, the various phases which each student has to accomplish will be analysed also considering the solutions, related to the automation, security and monitoring of operations.**

I. INTRODUCTION

Since June 2004 C.L.A.M. has introduced a language test for accreditation (TA) for university students. The credits are for linguistic competence in receptive skills: reading and listening comprehension, including grammar and vocabulary. There are different levels of the test according to the requests of the faculties. The format created by C.L.A.M is represented below:

LEVEL A2	LEVEL B1[1]
• Lexicon and grammar functions	• Lexicon and grammar functions
• Communicative structure	• Reading comprehension
• Reading comprehension	• True/false reading comprehension
• Short listening comprehensions	• Sort listening comprehensions
• Long listening comprehension	• Long listening comprehension

The need for competence in the receptive skills is highly felt in the first level degrees at the Faculty of Engineering, more felt than the need to acquire productive skills, which are more practised in second level degrees. Furthermore, receptive skills include grammar exercises and vocabulary, according to topics and functions established by the Common European Framework of Reference[1] (CEF) for each level. As for competence in foreign languages, the University of Messina requires the adherence of those levels proposed by the Council of Europe in CEF, and the Faculty of Engineering is no exception.

Structure of TA.

Perception v3[2] produced by *QuestionMark*, used at C.L.A.M. for the management of the TA allows to create a great range of simple Learning Objects containing questions: their presentation is through a common Web Browser. The didactic format is more complex because it should provide items with multimedia elements/objects. The making of a TA and the sophisticated structure of the format, have imposed a revision to the standard creation flow. This procedure aimed at the simplification in the creation of the items, in order to make them more usable and to overcome the technological barriers imposed by

[1] All the C.L.A.M. services are structured following the Common European Framework of Languages (CEF). The CEF describes the linguistic competences that are necessary to reach a specific level. The framework includes six levels ranging from basic (A1) to proficiency (C2). The categories serve the purpose to standardize linguistic competence for each level.

[2] It is a software platform which allows the creation, reporting and delivery of tests via Internet or a local intranet.

K. Elleithy et al. (eds.), Advances in Computer, Information, and Systems Sciences, and Engineering, 367–371.

Perception. Furthermore we create some library allowing the synchronization between *Perception*, *GestCLAM* (software for the management of student data) and the Centre website. The following table (cfr. Tab. 1) show the details of each section of the TA:

Lexicon and grammar functions - LGF A reading passage where there are some missing words. The student can choose each missing word from a list of options among which only one is the right one.
Reading comprehension A short passage followed by five multiple-choice (MC) questions.
Comunicative structure 5 MC questions.
Short listening comprehensions Five short (30 seconds) listening comprehensions each with a MC question.
Long listening comprehensions - LL One long (120 seconds) listening comprehension followed by 5 MC questions.

Tab. 1 – Description of the sections contained in a TA.

II. CREATION OF THE TEST

The idea to use an electronic format was devised for several reasons: first of all, the huge number of students to be examined required a system which could combine transparency, reliability and a fast processing of data. Moreover, the electronic format was ideal in a context such as the Engineering Faculty, where all students are more than familiar with the devices used in delivering the language test. Before the final implementation, a survey about the new exam was delivered to some pilot classes in order to evaluate first reactions to the test and students' individual outcomes, which were also compared to outcomes of the same students in tests delivered in hard copies of the previous exam format. The latter presented a more complex format in that it featured four abilities (reading, writing, listening and speaking) instead of the two abilities (reading and listening) tested in the new version of the test. The previous exam was also delivered in at least two separate days (with consequent increase in cost of time and money) and the final result was given at least after one-two weeks due to the big number of students and small number of teachers (with a consequent loss of efficiency).

In comparison to other Faculties (i.e. Arts or Law), the electronic test has also shown the positive impact on engineering students, mainly because of the familiarity that students have with PCs.

The well-known criteria observed in creating the format are listed below:

1) validity,
2) reliability,
3) discrimination,
4) motivation,
5) administrability,

where particular relevance is found in criteria 1), with reference to a subcategory called *face validity*, which brings in the idea that students need to immediately recognize types of exercises in the test without being hindered in any way in the completion of the test. The test presented in this paper fully satisfies the above mentioned criteria in that all students, being asked about the clarity and face validity, answered that the instructions, typologies of exercises and layout were fully understandable and clear. The criterion of motivation was nonetheless met, since students show a penchant for anything delivered in electronic format, as said, because they feel it user-friendly, faster and more fair in terms of the final grade.

The general feedback was that the test was easy to cope with, independently from the outcomes. The main aim of the survey was in fact not to assess the student's ability to pass the test but to assess its validity and reliability. The test also proved to be reliable in that an *a posteriori* evaluation of the percentage of passes and fails showed the reliability in measuring the levels of competence in foreign languages as required by the CEF, cornerstone in the assessment of foreign language skills in Italy.

The test lasts max 60 minutes and it is entirely carried out, recorded and corrected on computer. The minimum percentage to pass the test is 60% and it is shown at the end in the following way:

- *Pass*
- *Fail* – in this case the student can take another test in the next session.

The logistic organization for the creation of a TA consists of seven stages:

1. Creation of the items;
2. Correction of the items;
3. Production of audio/video material;
4. Data entry in the software platform;
5. Revision;
6. Publication;
7. Testing.

There are several professional figures such as Item Writer (IW), programmer, graphic designer, network administrator and sound engineer involved in the creation of a test.

A *standard de facto* has been realized in order to create a system that it is easy to work with for IW, who often do not have specific informatics competence. The system is made of a "*protocol of agreement*" between programmers and IW, which is based on some given rules for electronic tests and a software for the conversion of documents in a form which can be imported in *Perception*.

Perception Organization Standard (POS)

The *standard* contains a series of rules and guidelines that an IW has to keep in mind while typing the *items*:

- the file must be Microsoft Word or text format;
- the document must contain only normal letters (NO heavy type, NO italics, etc) and particular Microsoft Word features must not be used (alignment, table, list etc, etc);

- it is not possible to use some reserved letters in the typing of questions and answers;
- every section is identified in the document by "#" followed by the identification code, a space and the title of the section (option used for identification inside the *Perception*) between square brackets;
- in every section the correct answers will be preceded by "*" (asterisk);
- due to some technical limits of the *Perception* system, the questions belonging to the section LGF cannot contain the same answer more than once;
- every question can contain TAG HTML.

In the *standard* there are all the specific instructions and examples in order to type every section of the TA. There are also some conventional names for the items and the assessment.

Software Converter

Document written in Microsoft Word or text in accordance with the standard POS will be changed to be imported on the Question Manager[3] (QM) which works with file written on a meta-language (QML – Question Manager Language) that derives from XML[4] (eXtensible Markup Language).

The Converter software that we have created simplifies the creation of these item files, allowing the IW to create a document (*.doc* o *.txt*) ready for importing into QML, following the instruction of standard POS. The Converter contemplates a series of operations which lead to the creation of a *.qml* file which is ready to be imported in QM. In details the Converter:

1. requires the file's path;
2. checks possible mistakes or contradictions with the standard guidelines ;
3. analyses the exercises one by one requiring possible introduction of specific parameters (file's name of output, inside name of QM, score, etc);
4. causes output and a log of the operations made.

Access to the data is performed by a parser which uses a SAX type (Simple Api for XML) approach described as an *event driver* analysis model (Fig. 1). Parser sends an input when a given events happens: for example, when a new *item* is found (indicated in the standard with the symbol "#") it sends a sign with the same data to the module that required it. At its reception, a call-back function is automatically made. The call-back function is personalized in order to deal transferred data according to given instructions. This system is also known as *signal/call-back framework*.

[3] Authoring applicative for questions belonging to Perception.

[4] XML, as HTML, was born from SGML (Standard Generalized Markup Language), and it is considered an informatics metalanguage that defines other informatics languages.

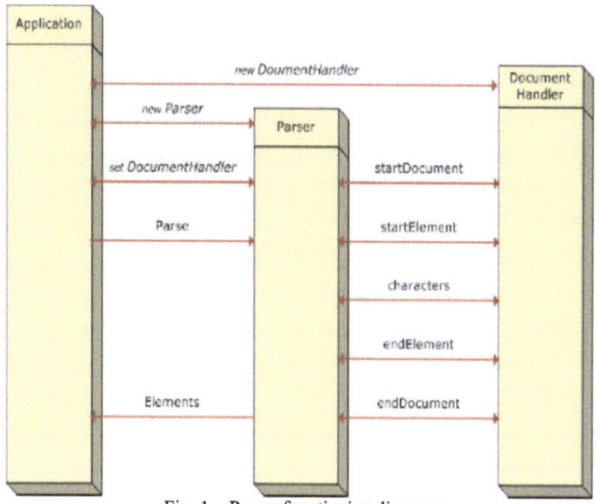

Fig. 1 – Parser functioning diagram

Figure 1 shows:

- Application or *Converter*;
- Document handler showing the code to examine (for example the tag "#") inside the document;
- Parser.

The latter performs a semantic check of the document including mistake, producing a series of interlude data, that needs the action of IW to be completed or expanded (following some parameters of options). These data are different according to the analyzed section of the format. The missing data could be substituted with *default* ones setup by the IW when creating the document; for example, when there is not any *feedback* on a *item*, *Converter* requires whether it is necessary to produce an automatic or manual one or whether it is not necessary.

E.g.: Realization of an item of the section Long Listening

The LL section consists in MC-type questions related to an audio file (i.e. a dialogue of about 180 sec); students can listen to the file twice through a multimedia Macromedia Flash object .

The *Perception* does not allow to automatically run hybrid HTML and Flash questions; either we use the MC-type (that does not allow to play the sound file) or we create each item belonging to this kind of question, using Macromedia Flash. This last solution is not practical, because it would need too many human resources (graphic designer, Action Script programmer, IW). Following the *standard* POS, we create a MC-type item with a link to a Flash object which can be parameter by command line in Converter software. The path relative to the file mp3 connected to the item is inside the link. To simplify the operation of putting the items in the platform, the path is specified according to the *standard* of the document. So it is the Converter's task to create a straightforward link that will permit a correct viewing/fulfilment of the object Flash in the TA.

```
#ASL.MLC [T_MC.S-(Weekend Dialogue)]
|Malcom: Hi Judith, how are you?
Judith: I'm fine Malcom, how was your day?
Malcom: It was pretty boring. Just a couple of families and a
group of 15 school kids but those kids were unbelievable! They
drank almost 50 cups of cola. Each kid manage to drink more cola
than all the other people together. How was your weekend?
Judith: Oh, it was fun! My boyfriend and I rented a small cottage
and we stayed at the beach for two days. The weather was not so
good but we enjoyed the beach.|
|%SERVER.GRAPHICS%/ascoltobreve/inglese/a2/settembre2004/|

|How many cup of coke did the kids drink?|
*50
-5
-15
|1|
|La risposta esatta è: <I>50</I>|
|S|

|What was the weather like last weekend?|
*not very good
-good
-very good
|1|
|La risposta esatta è: <I>not very good</I>|
|S|
```

Tab. 3 – Long Listening document typed according to POS *standard.*

Synchronization

GestCLAM is the software, created by the informatics group, for the management of all the students that come into contact with the centre. In order to use all the services offered by the Centre, the student receives a badge called CLAM Card. Through a form of synchronization, *GestCLAM* manages to update the servers for the access to the online services and automate the registration procedures to get into the right TA language and level. *GestCLAM* contains functions that allow the automatic organization of the exam sessions, create *schedules* and groups for the tests on the *Perception* servers and automatically synchronize the database in order to update the students' records.

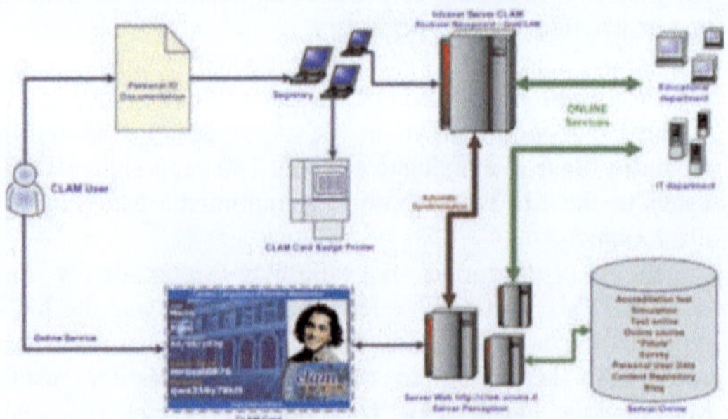

Fig. 2 –C.L.A.M. flow diagram for linguistic TA

III. STUDENTS AND ACCREDITATION TEST

Students who want to take a TA must have a CLAM Card. In order to get it they must hand in the required documentation at the Centre secretariat. The staff will immediately introduce the student in the *GestCLAM* software. After this phase, the student can have access to the Centre website, try the TA simulations, check his/her subscription to a particular session of exams and/or check the results of a test already done. An experimentation on learning services has also started through *mobile* devices. The access to the *mobile* service is extremely simple: you need to type the URL on the display and then the system will automatically redirect your device to a specific version. The service is operating on most devices with *Symbian* and *Windows CE* systems. The access to TA simulations is perfectly operating only from the PocketPC version due to some limits of resources of mobiles with *Symbian* system.

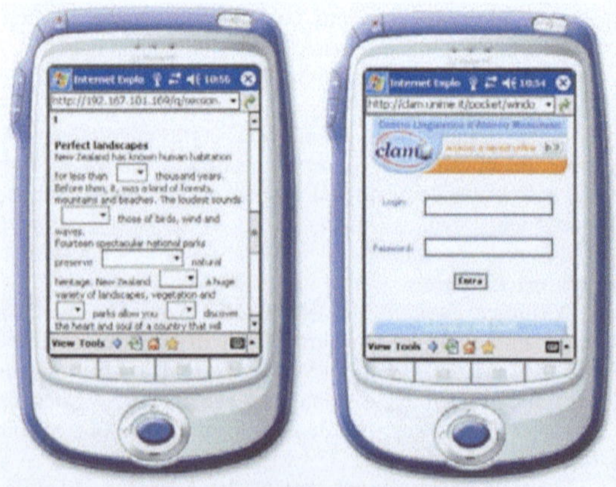

Fig. 3 – Mobile version of TA simulations.

The synchronization between Web and Intranet takes place twice a day to guarantee the integrity of the information published on Web.

Students should address themselves to the office of the Centre secretary, show an identification card and the CLAM Card to enrol for the test. The staff will than register the student and print a registration receipt through the *GestCLAM.*

The TA takes place at CLAM multimedia labs. On the day of the test, the student must show an identification card, the CLAM Card and the registration receipt to the invigilators. The multimedia labs are supplied with intranet and *Perception* server, including tests and data for the access of the registered students. As soon as the test is completed, the *GestCLAM* will take the results from the *Perception* server. CLAM staff will automatically send the results to the secretaries of the different faculties which will formalize the exam.

The synchronism module identifies the various sections forming the TA and assigns the score gotten from the student in every section for each linguistic skill (comprehension, reading and grammar).

The final results will be kept in *GestCLAM* as the total score obtained in the test and divided in the different skills. The division of the score among the different skills allows a student to sit for an exam which will allow him/her to integrate the missing skills (written and oral production) and to obtain the linguistic degree certificate.

The results are delivered to the faculty office, by electronic mail under the supervision of the teaching staff.

CONCLUSION

Along with the didactic and scientific pros above sketched, a significant improvement in terms of saving of money and time is doubtless, if we consider that no teacher is involved any longer in delivering, correcting and evaluating the test. Corrections are in fact completely carried out by the software and the final outcome only depends on the number of correct answers given. Students are aware of the transparency and consider as priorities the opportunity to have immediately after the exam the outcome of the test and the almost non-existent possibility of error-correction. No significant drawback was suggested by students, with the exception of some difficulties in passing the exam that are not relevant in this context.

The result of C.L.A.M. experience is a bug-free template for TA which can be easily integrated in many contexts, in virtually any other fields.

ACKNOWLEDGEMENTS

I would like to thank all the colleagues who contributed to this study, in particular thanks to Mari Grazia Sindoni and Mariavita Cambria for their helpful contributions.

REFERENCES

Fallon, C. *E-Learning Standards*, CRC Press, 2005

Clark, S. et al. *VBScript Programmer's Reference*, Wiley Publishing, Indianapolis, 2003.

Jones, A. et al. *SQL Functions Programmer's Reference*, Wiley Publishing, Indianapolis, 2005.

European language portfolio (elp) - principles and guidelines. Available from: http://culture2.coe.int/portfolio (accessed 15 September 2005).

Kalinin, V. and Rafalovich, V. *Palm & Pocket PC Programming*, A-List publishing, 2003.

Hughes, A. *Testing for Language Teachers*, Cambridge Press, United Kingdom, 2003.

Candlin, C. and Hall, R. *Language Testing and Validation*, Palgrave Macmillan, Great Britain, 2005.

Stephenson J. *Teaching and Learning Online: New Pedagogies for New Technologies (Creating Success)*, Kogan Page, London, 2001.

Perception Server Installation Guide 3.4.3, QuestionMark. Available from: http://www.questionmark.com (accessed 10 September 2005).

Windows™ Deployment Guide for Perception Version 3, QuestionMark.
Available from: http://www.questionmark.com (accessed 28 August 2005).

Use of a Web-based Teaching Collaborative Platform at Third Level: A Qualified Success?

A. Mullally A. Jennings C. O'Connor D. Dolan A. Parkinson J.A Redmond

Department of Computer Science
Trinity College
Dublin 2
Ireland

Contact: redmond@cs.tcd.ie

Abstract- **Web-based collaborative platforms appear to show significant potential for improving teaching and learning productivity and flexibility at Third Level. Two pilot studies were conducted at Trinity College Dublin (TCD), one within TCD and one between TCD and the University of Reading to explore pertinent planning, operational, technological and organizational issues with a view to identifying pointers for the future in relation to planning and implementation from student, lecturer and administrative perspectives.**

I. INTRODUCTION

This research initiative focuses on the impact of online synchronous learning using a web-based collaborative platform with part-time, mature, evening Information Systems university students, in full-time employment, and a cohort of full-time Master's students in Computer Science. However there was some asynchronous learning in that the students could recall the saved lectures and replay them at a later date.

Online Synchronous learning can be defined as: a real-time, instructor-led online learning event in which all participants are logged on at the same time and communicate directly with each other [1], [2]. In this virtual classroom setting, the instructor maintains control of the interaction with the class, with the ability to "call on" participants. On most platforms, students and teachers can use an electronic "whiteboard" to see work in progress and share knowledge. Interaction may also occur via audio- or video-conferencing, Internet telephony, or two-way live broadcasts.

Asynchronous learning can be defined as: learning in which interaction between instructors and students occurs intermittently with a time delay [2]. Examples of available asynchronous learning are self-paced courses taken via the Internet, CD-ROM or DVD [3]. Some research results on Collaborative Learning in Asynchronous mode over a long period of time are given in [4].

There is a dearth of research material in the area of web-based online synchronous delivery of learning in traditional universities *"Considering the massive adoption of e-learning, what is surprising and cause for concern, is that we know so little about the use of this medium to facilitate learning" [5] [6].*

The Trinity College Dublin (TCD) project was funded under the European Union GENIUS (Generic E-Learning Environments for the new Pan-European Information and Communication Technologies Curricula) programme [7][8]. The overall purpose of the project was to explore the real-life practical issues associated with applying a web-based collaborative platform embodying both synchronous and asynchronous dimensions and with particular regard to Planning, Operational, Technological and Organisational issues. One of the goals was to assess the efficacy of the course presentation via web-based collaborative platform versus the traditional lecturing approach. Two pilot studies were carried out using the Web-based Collaboration platform, LearnLinc [9]. It was envisaged that this would provide a basis for more substantial studies with these technologies in the future.

The purpose of these pilot studies was to investigate practical and operational aspects and issues to do with using such a tool-set. On this part of the project, a Trinity College Dublin, (Ireland) staff member (DD) presented a course entitled "IT and the Enterprise" to a group of mature, evening attendance, computer-literate, undergraduate Information Systems students. Of this cohort of students more than 76% worked with computers greater than 30 hours per week. A number of weeks later another member of TCD staff, (AM) presented a course entitled "Managing the IS/IT Infrastructure" from the TCD base to a group of forty M.Sc. Computer Science

K. Elleithy et al. (eds.), Advances in Computer, Information, and Systems Sciences, and Engineering, 373–378.

students at the University of Reading in the UK. This was a trans-national study in pursuit of one of the goals of the Genius project. Andragogical considerations were considered to be outside the scope of the study, the major concerns being student, lecturer, administrative, technical and operational issues at this early stage.

II. MATERIALS AND METHODS

The use of LearnLinc as a collaborative web-based platform was a requirement of the overall GENIUS project. LearnLinc provides two separate environments, the virtual "campus" and the virtual "classroom". The virtual campus is modeled on a physical college campus in that it provides administrative functions with registration of students for courses, course creation, class creation, adding of course materials and assigning lecturers to lectures. The virtual classroom provides an environment with whiteboard area, synchronized web browser, application sharing, text chat, hand raising, questions and answers, feedback, attendance list and an agenda for the class.

A. The Virtual Classroom - Lecturer

Having entered the virtual campus, the Lecturer joins the classroom and is automatically given the floor. His/her photograph is displayed on the students' desktop. The Lecturer can see which students have joined the class and those who are currently joining. Each student's username appears under People along with a count of the number currently logged in. Figure 1 shows the Lecturer view of the virtual classroom.

Figure 1: The Virtual Classroom – Lecturer View

B. The Virtual Classroom - Student

When the student has logged on to the virtual campus he/she then 'joins' the class using a password. On joining the class students are presented with a classroom environment where they can see a still image of the Lecturer who has logged in and has the floor. The students can also see the usernames of the other students who have joined the class, the agenda for the class, the current text chat, the whiteboard or other application share that is being used at the time. They can actively participate in the class by listening to the lecture, asking questions – by "hand-raising" or text chat, using feedback and participating in Questions and Answers. When the students are given the floor, a still image of the student is displayed. (Alternatively, if video is used, a video image is displayed). Figure 2 shows a Student view of the virtual classroom.

Figure 2: The Virtual Classroom - Student View

C. Equipment Requirements

A participating student should be equipped with a computer conforming to at least the minimum specification as set out by LearnLinc and a network connection fast enough to support the LearnLinc server connection [8]. Students were also required to have downloaded the client software. The students in a computer laboratory environment were issued with headsets (microphone and earphones) so that they can listen and speak to the lecturer without sound distortion and acoustic feedback from such a noisy environment. The individual student can communicate with the lecturer through text chat either privately, where only the lecturer sees it, or publicly where everyone in attendance sees the message. The student can also communicate by symbolically 'raising the hand' on the interface. The lecturer sees the indicator for the hand raise and can then give the floor to that student. A photograph of the student appears and he/she can speak to the lecturer and the class.

D. The Virtual Classroom - The Assistant

The Assistant is a member of staff who joins the class in order to support the Lecturer and to ease the information overload problem in a busy virtual classroom. In addition the Assistant may have many other roles and duties. He/she can be an observer monitoring the text chat just in case the Lecturer misses something such as a query.

III. TCD STUDIES

Three questionnaires were used in the TCD studies.

A. Pilot Study One

Some forty nine Information Systems second year undergraduate students (average age was 29 years, about 75% male) of the Trinity College Dublin, Computer Science Department took part in this pilot study. The students had full-time jobs and attended lectures in the evenings from 6 to 9 p.m. They were given a (sub)-course in "IT and the Enterprise" by a TCD lecturer (DD) using LearnLinc. This course, consisting of 4 weekly 2-hour evening slots, was part of a larger 22-week course. The whole 22-week course was examined conventionally, with one question devoted to the aspects of "IT and the Enterprise" covered in the web-based contribution. The students were distributed across various locations, at home, on campus (in computer laboratories) or at places of business.

B. Pilot Study Two

This second Pilot Study involved a TCD staff member (AM) delivering a part of a course ("Managing the IS/IT Infrastructure") to a group of forty M.Sc. Computer Science students at the University of Reading in the UK. This was accomplished over four successive weeks, using one and a half hour time slots, in the afternoon. The Reading students were a multi-cultural, full-time cohort, all on campus. These students later underwent a conventional examination process.

C. Questionnaire One - Pilot Study One

This questionnaire was presented on the first night of term to the students after the Lecturer had advised the students of the forthcoming teaching collaboration project. Questionnaire 1 was used to gather information about the availability of student computers with the required specification to partake in online lectures. Ideally students were to use either a computer in their workplace, in their home, or in a College computer laboratory, whichever location suited best. The main purpose of this questionnaire was to find out technical requirements and support information so that a support team, administrative issues and a computer laboratory could be made available for the students. Forty six students completed the questionnaire.

D. Questionnaires Two and Three - Pilot Study One

Questionnaires 2 and 3 were designed to capture the before and after mindset of the students. The questionnaires were in two parts, Section A and Section B. Section A had twenty seven quantitative questions and Section B had ten qualitative questions. The twenty seven quantitative questions used a nine-point Likert scale varying from 1 (Strongly Disagree) to 9 (Strongly Agree).

Questionnaire 2 was used to assess the students' expectations of the upcoming online eLearning experience before the experiment started. It contained the core twenty seven questions and the ten qualitative questions. Questionnaire 3 was used to assess the students' opinions on the performance of the eLearning experience after the experiment finished. It contained the twenty seven core questions, the ten descriptive questions as in Questionnaire 2 and an additional twenty one questions focusing on the use of the facilities and functions of LearnLinc as used in the online lectures. These two questionnaires were used to assess the effectiveness of the use of Internet technology to create a virtual classroom to support or enhance the learning experience within the course 'IT and the Enterprise'.

E. Other Data Capture facilities

The LearnLinc environment offers other means of capturing data about the project. The LearnLinc server has a log file analyzer, which captures all logins to the Virtual Campus and shows what facility the students used in the Virtual Campus, e.g. join a classroom, play a recording. This was useful for seeing who connected to the system after the last lecture was over, either to listen to recordings or download material. It was also useful to see if students were disconnected during the online lecture session.

In the Virtual Classroom, recordings were made of all the lectures and also the training session. These recordings could be played back later to check on student interaction and participation during the virtual class. Text chat was saved and was also scanned for interactivity and relevance of text chat to the lecture material. Finally attendance lists were checked to establish if people were consistent in their attendance [10].

As part of the support infrastructure for the project a "Listserv" (an email discussion forum) was set up to enable students to collaborate with each other and learn from each other's problems and issues. It was constantly monitored and responded to by the Genius support team.

Analysis of the text chat, the Listserv, email and use of recordings gave an indication of the students' comfort factor in using these facilities. The content helped identify any problems that students were having.

Conversations and informal discussions that took place after the lecture sessions with the lecturer, support people and students were found to be relevant to evaluating the performance and usability of the collaborative platform [10].

An assignment focused on the eLearning experience was set over the Christmas break. The responses to the assignment were analysed for topics and themes to help in evaluating the collaborative experience [10].

F. Studies of University of Reading Data

Similar, but not identical questionnaires to those of Questionnaires 1, 2 and 3 of Pilot Study 1, were used with the University of Reading students. Both TCD and Reading sample groups were similar in age, range, work experience and professional background. The results were similar in scope and range.

IV. RESULTS

The results and conclusions below are a summary of the main quantitative and qualitative findings [10]. The differences between the questionnaire questions for the Expectations (Questionnaire 2) and Post Experience (Questionnaire 3) Questionnaires are given in [10].The findings are discussed under five headings: Planning, Operations, Technology, Organisation and General.

A. Planning

A considerable amount of planning was necessary before the first pilot study, which is an additional burden compared with the traditional lecturing situation. The responsibilities and roles of different personnel had to be made clear. Equipment and software were procured and appropriate training provided. Lecturers were trained in the use of Learnlinc tools and features. Quite a wide variety of skills were needed to cope with the rich functionality of LearnLinc. Technical support and administration staff were also trained. Standby support staff had also to be inducted and scheduled for each session. A preliminary training session evening was also given to the students.

B1. Operations: Classroom Interactions

It was found, not surprisingly, that interactive teaching requires considerably more effort from the lecturer's perspective than the traditional approach. It also would appear that the student needs to concentrate more. The absence of body language from the class was, in hindsight, an obvious difficulty for the lecturer. Constant monitoring of the virtual class by the lecturer was required in addition to delivering the material. This required continuous surveying of the chat room facility, the 'Hands up' facility and the results of online real-time questions/quizzes by the Assistant and/or the Lecturer. The rapid recovery from breakdowns in programs and equipment was also clearly important. A training session for the students to acquaint them with the equipment and packages was held before the pilot studies started. In general the LearnLinc features worked well.

B2. Operations: Equipment and Software

The bandwidth for the network was crucial. Students accessing via 56 kb modems in some cases had difficulties. To ease the network load, it was found that still images, rather than video, of the lecturer and each student were found to be satisfactory. In addition, somewhat ironically, it was found necessary to "impoverish" some Powerpoint slides previously used in a "traditional" presentation to also help reduce the bandwidth problem. Other equipment difficulties included students using older machines with out of date graphics cards or operating systems.

In general, the experiments were expensive in terms of resources, time and personnel. To justify them economically the usage of LearnLinc would need to be substantially scaled up.

Some problem areas highlighted by the pilot studies were:

- the students needed to get used to wearing a headset for the duration of a 50 minute session
- setting the audio level properly on their computers was a problem for many
- setting up LearnLinc
- downloading slides before the lecture
- viewing an active screen for problems
- for support staff, the ability to be able to quickly restart in case of failure was crucial.

C. Technology

The LearnLinc platform used was relatively effective and reliable. However, licensing arrangements and agreements were needed. Some of its competitors were more expensive and appeared to be more geared to the corporate market. To operate the platform effectively, installation and support was needed. The LearnLinc features included:

- *Slideshow Facility enabling slides to be imported and displayed, Annotated online with text, graphic or highlighter*
- *Online Chat room on screen*

- *Salient still image to identify speaker or student availing of 'Hands up' facility*
- *Full Audio – one to many: Lecturer to Student group or Student to Lecturer & Class*
- *Attendees list on screen*
- *Whiteboard facility, For Drawing or typing Text on screen*
- *Interrupt (Hands up) facility under lecturer control*
- *Facility to institute Quizzes or polls online with instant results*
- *Facility to import Web sites for universal display across class.*
- *Ability to send private messages between lecturer and individual class members*

All the above facilities appear on lecturer's screen and on each student's screen.

D. Organisation

Staffing and training were needed in the following areas: Students, Lecturers, Technical Support and Administration. The following training was necessary for the Web-based Collaborative platform: one week for the Learnlinc System Administrator; two days for each of the Research Assistants; the two lecturers trained themselves on LearnLinc and several internal workshops for the TCD team were run by the Research Assistants. All students and TCD project staff were registered and their accounts uploaded to the LearnLinc server. The Listserv email discussion forum was monitored by the Research Assistants. Extra up-to-date equipment was needed including a new server. Sixty LearnLinc licenses were also purchased.

V. GENERAL CONCLUSIONS

Salient still images of lecturer and students worked fine in place of video and helped reduce the bandwidth requirement. In certain situations still images are less distracting and preferable to video. However stills may exacerbate the problem of lack of feel by the lecturer of feedback from the student body (body language in this instance).

- Broadband is desirable if not essential for the remote students.
- Two screens for the lecturer would help a lot. One for the lecture material and one for other items to be monitored.
- This type of interactive course is hard work for the lecturer because of the need to monitor other things while lecturing (e.g. 'hands up' facility).
- One important consequence is that two lecturers rather than one could be perceived to be much better because it removes some of the pressure, helps with the pauses and gives time for one to collect his/her thoughts and navigation while the other takes over. The best analogy is with presenters of news and sports programmes on television where a minimum of two presenters seems to be necessary. During a pause, the network chat amongst those in the virtual class rises dramatically with just one lecturer present. However there are obvious economic ramifications of this policy.
- Two lecturers also help if there is a series of questions or comments to be handled.
- Lecturers are needed who are suited to this type of medium. It is a heavy load on one lecturer to prepare and deliver such a course.
- While the emphasis in these studies was more on technical and operational feasibility than economics, there is no doubt that at present these applications are very expensive to equip, set up and run.

A. Further Work Needed

Each of these web-based sub-courses covered four sessions. Four sessions are probably too few to overcome the novelty effect for the students and also for the lecturers to move sufficiently further along the learning curve for this new medium. Longer courses and more courses are needed. Longer courses would allow more familiarity with the system and equipment for lecturers, support staff and students. More courses would also identify which courses and material were more suitable. The issue of the andragogical implications and their evaluation need to be addressed in the longer term. A full examination of the economic implications of this medium is needed.

E. In The Future

It is obvious that some of the issues which gave rise to problems in this study will diminish or disappear as Broadband and computer equipment continues to speed up.

However many other issues will not go away so easily. In particular, a lot of students preferred the traditional lecture over web-based presentation [8]. The main advantage for students is reduced travelling time to lectures. However most of these students spend quite a large percentage of their time in front of computer screens at their work already and report that they see little advantage, other than travel, in having the course material presented by screen. A blended approach, that is a mix of traditional lecturing with web-based presentation is probably what is needed with the blend, perhaps, being 90:10 in favour of traditional at present.

A quantitative and qualitative evaluation was carried out in these pilot studies. There was broad satisfaction with the process by students, lecturers, technical support and administration. But there was also considerable constructive criticism from all quarters. It appears that the students' rather blue skies expectations were not fulfilled. A rerun of the studies, with one course of this medium already completed, would probably give rise to more realistic expectations of the medium from the students.

A better evaluation template for the process with the use of a control group is needed. A cross-over study with half the students getting traditional teaching for half the course while the other half get the web-based collaborative platform for that half, and vice-versa for the remainder of the course is needed in a more complete study.

The eLearning paradigm provides opportunities for the facilitation of individual differences. In future applications, this issue could also be addressed [11][12].

These pilot studies illustrate the difficulties of exploring virtual student/lecturer interactions in eLearning environments. One unexpected result is a much deeper appreciation of how much is involved in the "traditional" lecturing environment.

ACKNOWLEDGEMENTS

We thank the Technical Support staff for their many strenuous efforts.

REFERENCES

1. http://www.learningcircuits.org/glossary.html - for glossary type definitions.
2. http://www.e-learningsite.com/elearning/character/synchr.htm
3. http://www.careerspace.com/whats_new/news2.php?idn=56
4. S.R. Hiltz "Collaborative Learning in Asynchronous Learning Networks: Building Learning Communities" Invited Address at "WEB98" Orlando Florida November 1998 and also available at http://eies.njit.edu/~hiltz/collaborative_learning_in_asynch.htm
5. S. W. Gilbert, (2000), "A new vision worth working toward – connected education and collaborative change" as referenced in "Beyond Institutional Boundaries: reusable learning objects for multi-professional education" available at http://www.tltgroup.org/gilbert/NewVwwt2000--2-14-00.htm
6. D.R. Garrison, and T. Henderson, *E-Learning in the 21st Century: A Framework for Research and Practice*, Routledge Falmer, 2003.
7. D. Dolan, C. O'Connor, A. Mullally, and A. Jennings "Experience in the use of synchronous eLearning in a traditional university for non-traditional learners" Proceedings of Second International Conference on Multimedia and ICTs in Education (Eds.: Antonio Méndez-Vilas and J.A.Mesa González), m-ICTE Badajoz, Spain Dec 3-6, 2003 and in "Advances in Technology-based Education: Towards a Knowledge-based Society" Edited by A.Méndez-Vilas, J.A.Mesa González, J.Mesa González Volume II, (Pages 659-1335): 84-96212-11-4 ISBN Published by: JUNTA DE EXTREMADURA, Consejería de Educación, Ciencia y Tecnología (Badajoz, Spain), 2003
8. D. Dolan, C. O'Connor, A. Mullally, and A. Jennings, The implementation of on-line synchronous eLearning for non-traditional learners at traditional universities (In press)
9. http://www.parsecinfo.nl/products/learnlinc.htm
10. A. Jennings, *Implementing an Integrated Web-Based Synchronous eLearning Collaboration Platform at Tertiary Level for Part-Time Mature Evening Students* TCD-CS-2005-67 University of Dublin, Nov 2005 (http://www.cs.tcd.ie/publications/tech-reports/reports.05/TCD-CS-2005-67.pdf).
11. A. Parkinson, and J.A. Redmond, The Accommodation of the Field-dependent Learner in Web Design *The Psychology of Education Review* Vol 29 No 1 March 2005 pp 43-53.
12. J.A. Redmond, C. Walsh, and A. Parkinson, "Equilibriating Instructional Media for Cognitive Styles" Inroads - SIGCSE Bulletin Vol 35, No 3 September (2003) pp 55-59 ACM Press New York. also in Proceedings of the 8th Annual Conference on Innovation and Technology in Computer Science Education (ITiCSE 2003) ed. David Finkel June 30 - July 2 Thessaloniki, Greece (2003) and also published in the online ACM Digital Library (http://portal.acm.org)

Multilingual Technology for Teaching Mathematics

Olga Caprotti, Wanjiku Nganga, Mika Seppälä
University of Helsinki

Abstract. This paper describes the experiences acquired and the goals of the European project Web Advanced Learning Technology, WebALT, in developing a multilingual showcase of exercise problems in mathematics to be used by university students.

I. Introduction

Mathematics and e-learning are a winning combination. Not only is mathematics one of the few subject areas that has been given attention to by the World Wide Web Consortium from the beginning, but also, by its own nature, mathematics lends itself naturally to being taught using computer technologies. Students begin using pocket calculators already in elementary school and in this way they become used to getting assistance for solving mathematical problems by a machine. In fact, today's calculators' abilities (just check for example Texas-Instruments products) go way beyond the arithmetic or logarithmic operations one may expect. They are able to plot functions, compute derivatives and offer the possibility to load applications in finance, engineering, statistics and business. For this generation of students, the step to online training and assistance in doing their college and university courses in mathematics is very small.

The effort devoted to the preparing and to the grading of conventional exercises is enormous and often the main reason why professors at US universities are strongly against teaching large classes. Hence mathematics is often taught in small sections. This gives further impetus to develop e-learning materials in mathematics, and, in particular, exercise problems for online assessment of students. With the cost of tutoring rising, and the number of mathematics graduates decreasing, it becomes increasingly difficult to ensure high-quality educational programs in sciences. Proper use of the possibilities of the new media offers a long term solution.

A variety of Learning Environments provide assessment tools, however, from our survey, it turns out that mathematics is still only partially supported in interactive exercises. The obvious reason for this limited support is that the handling of mathematical objects requires sophisticated software, and is often beyond the expertise of the developers of the learning environment. Most systems allow for a teacher to write a multiple choice question in which the pre-defined set of answers may contain a mathematical expression. The author of the exercise then also chooses the correct answer so that the system only checks whether the student has indeed made the correct choice. This kind of exercise problem is technologically rather simple to support since it does not involve any understanding of the underlying mathematics. Specifically designed systems such as MapleTA, by exploiting computer algebra software, offer advanced features like algorithmically generated exercises, partial grading that takes into account students' performance in previous questions, and automated validation of symbolic answers for open question problems. These features, combined with online accessibility, can be used to alleviate the task of preparing, delivering and finally grading mathematics classes.

Resistance to the adoption of this kind of technology can be motivated by many factors. First of all, good exercises are hard to produce: they require the author to know the inner workings not only of the e-learning system, but also of the underlying computer algebra system. Additional special care has to be spent if the exercises are produced in an algorithmic way, namely by using some lines of programming code, often employing random variables and thus generating a new version of the exercise question every time the student reloads the problem. And even if one could have access to a collection of

K. Elleithy et al. (eds.), Advances in Computer, Information, and Systems Sciences, and Engineering, 379–386.

good exercises, chances are that the notation used by the original author is somewhat different from the one used in class, or in a specific country or scientific area, and this would make the exercises hard to use. Finally, for basic courses such as calculus, presentation of the question in the native language of the students is still a major requirement since one aspect of teaching mathematics is also teaching its specific jargon, "Let x be arbitrary but fixed…".

These last two drawbacks of exercises produced using current systems are tackled by the goals of the Web Advanced Learning Technologies, WebALT, project [5] sponsored by the European Community in the eContent framework. One of the results of the project will be a database of multilingual and multicultural exercises for students and teachers of secondary and tertiary level mathematics. This paper presents the project goals and the results obtained so far in developing multilingual technology for application in the automatic generation of natural language verbalization of mathematical content as found in problem exercises. The paper focuses especially on the technology used for the generation of natural language from mathematical content. Other aspects of the project, like architecture and design of the overall system, are discussed elsewhere, see [5].

The structure of the paper is as follows. Section II reports the experience of teaching the same course in class and online. This experience provides evidence that employing e-learning technologies in teaching can be very rewarding. Section III explains what the multilingual issues encountered in teaching mathematics online or in bilingual communities are. The actual outcome of the multilingual technology developed by WebALT is then described in Sections IV-V-VI. The last section is devoted to discussion and conclusions.

II. Learning from Experience

To understand what kind of methods work in on-line learning of calculus, the same basic calculus course was offered, in fall 2004, both as an on-line course and as a traditional lecture course at the University of Helsinki. The on-line materials were Power Point slide shows developed to support traditional instruction. The lectures, i.e. the Power point slide shows, were used as basis for instruction also in the traditional course. The on-line students took proctored examinations together with the traditional students, and the examinations were graded by the same grader.

The result was that the on-line students faired better in both examinations in fall 2004 than the traditional students, and the drop-out rate among the on-line students was lower than in the traditional class. Hence on-line instruction is a serious alternative to traditional instruction.

In this experiment the learning materials ([4]) comprised of a collection of Power Point slide shows. These are well suited for the computer monitor. Majority of students did, however, prefer pdf versions of the slide shows. Most likely because they either did not have proper software in their computers (MS Office, or Power Point viewer) or they simply preferred to print the slide shows and read the materials from the printed pages.

At Florida State University the on-line materials ([4]), the Power Point slide shows, were used to support regular teaching based on the Stewart text (Early Transcendentals). Around 10th week of classes an anonymous survey was conducted among the students. The purpose of the survey was to assess the value of the on-line materials in a traditional setting.

Practically all students had the text book. The following answers were given to the question: **"When studying calculus, what materials you are using?"**

Question: When studying calculus, what materials you are using? Answer Choices	**Percentage of respondents**
Only Stewart's book	9.5
Mainly Stewart's book accompanied by the on-line materials	0
Mainly the on-line lecture notes accompanied by Stewart's book	14.3
On-line materials are my primary source for the course	76.2

The use of MapleTA was incorporated to the calculus course offered in Helsinki (and at Florida State University) in spring 2005. MapleTA makes it possible to design algorithmic problems, i.e., problems which contain random parameters and appear different every time it is invoked.

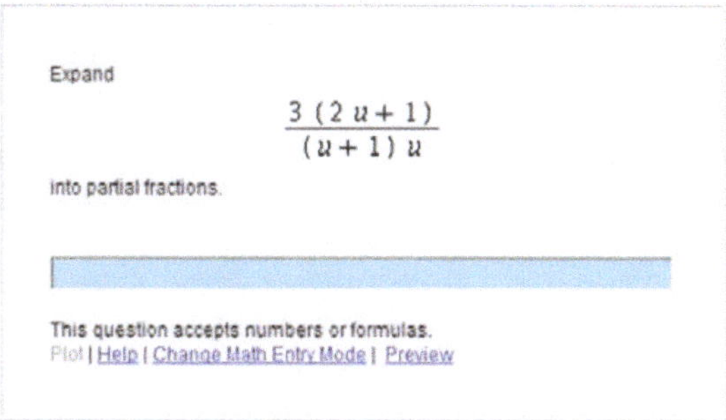

The above screen shot shows a typical algorithmic problem. The problem itself is a simple program which has single digit integer valued parameters. Every such problem leads to a simple answer and all the problems are equally difficult or easy. The program produces over 100 000 such problems. From the students' point of view, this is infinitely many problems.

Students' responses get automatically graded and grades entered into students' grade books. Programs like MapleTA are wonderful tools when teaching mechanical tasks to students. Thanks to the algorithmic problems, the examinations can be published beforehand. Students can practice as much as they want, and when they come to the examination they still get different versions of the problems. Hence memorizing certain solutions by heart is not useful. They must learn the procedure in order to be able to pass the exam.

This way of studying suits most students very well. Monitoring students' behaviours shows that many students use these tools in a serious way, solving hundreds of problems when practicing tasks like computing limits, differentiating or integrating. These tasks form the basics of calculus.

The WebALT project will bring tools like MapleTA to the next level by adding features like multilinguality. WebALT will offer an adaptive database of multilingual exercises that can be used with a variety of platforms including MapleTA. In addition to that WebALT will produce its own exercise player that can be used with Moodle to track students' performance.

The eContent produced by the WebALT project forms a new way to deliver mathematics instruction. The recently introduced programs, like Skype and Festoon, provide important infrastructure making it possible to offer on-line instruction that, in no way, lacks the interactivity of a class room contact lecture. This will result into significant savings in the cost of instruction world wide. Students can study from the comfort of their homes and several of the tasks of the instructor can be automated without reducing quality. Advanced portals will also offer educational institutes the possibility to export education.

III. MULTILINGUAL ISSUES IN MATHEMATICS

Since different types of context may influence the choice of a mathematical notation and language, WebALT has opted to derive localized presentations of the exercises from language-independent semantic representations of the content of its subject material. Among the many factors that influence the way mathematics is written, the following are always taken into account:

1. semantics of the mathematical operation e.g., the kind of multiplication, notation for cross-product differs from that of scalar product of vectors

2. level of mathematical sophistication of the target audience e.g., using "×" notation for simple multiplication in elementary school, but "invisible" multiplication in secondary and higher education or expressing formulas in natural language instead of using a compact symbolic presentation

3. typographic conventions arising from the specific area of study or from the geographic location e.g. using (a,b) or $]a,b[$ for denoting the open interval between two points a and b , or using i or j for the imaginary unit depending whether it occurs in complex analysis or in electrical engineering

4. individual stylistic choices (e.g. the use of a mirrored capital E vs. a "big Or" notation for existential quantification, with

corresponding changes in layout positions for the respective parts of the quantified expression)

5. cultural or linguistic distinctions (e.g. “tan” vs. “tg” in different parts of the world, and the different notations for the greatest common divisor in different languages – “gcd” in English, “ggT” in German, “mcd” in Spanish, “MCD” in Italian, and so on in different languages, all abbreviating the respective languages’ translation)

6. choice between formal or informal rendering (e.g. “f where x is an element of S” vs. “λx:S . f”)

7. choice between different rendering modes (e.g. visual vs. aural)

IV. WebALT Multilingual Technology

The multilingual technology used by WebALT relies on a representation of the exercise problem that is language-independent, yet rich enough to convey: 1) the mathematical content of the problems unambiguously, 2) the natural language constructs and sentence flavour desired by the authors.

For question and interactive testing, the QTI standard language is widely used and supported by e-learning platforms. MathQTI is an extension of QTI that allows the mathematics appearing in the exercise text or feedbacks to be encoded in a semantically safe manner by using OpenMath. WebALT has chosen to adopt MathQTI as the source language-independent format for archiving the exercises. While in a generic MathQTI exercise, natural language text is intermixed with fragments of mathematical markup, WebALT exercises contain only OpenMath objects.

i. *OpenMath*

OpenMath [1] is a language for the electronic representation of mathematical objects that focuses on capturing the content, semantic meaning and aspect of the mathematics. The development of a language to encode the meaning of mathematical expressions was started by a European Union funded project in 1993. By 1995 it became clear that there was a need for a limited version of the language that could serve for the majority of needs and yet be simple to implement. Hence the development of MathML was started by the W3C. The work was carried out in collaboration with the OpenMath group whose aim was to develop a method to manage general mathematical expressions correctly in the internet.

There has been a considerable overlap throughout between the developers of MathML [2] and OpenMath and as a result, the OpenMath markup encoding for mathematical formulas provides an extension mechanism for MathML-Content. Interested readers may find details on how to create and use OpenMath objects from [1].

Mathematical objects in OpenMath are defined abstractly by using symbols which come from the definitions given in "Content Dictionaries." These are collections of names for mathematical concepts, whose definitions are given in "Content Dictionaries". Shown below is the OpenMath XML representation of the mathematical expression “n times x cubed”, $n \cdot x^3$:

```
<OMOBJ xmlns="http://www.openmath.org/OpenMath"
    version="2.0"
    cdbase="http://www.openmath.org/cd">
    <OMA><OMS cd="arith1" name="times"/>
          <OMV name="n"/>
          <OMA><OMS cd="arith1" name="power"/>
                <OMV name="x"/>
                <OMI>3</OMI>
    </OMA></OMA>
</OMOBJ>
```

The operations `times` and `power` are explained in the OpenMath Content Dictionary `arith1`, available from the cdbase http://www.openmath.org/cd,. n and x are the names of variables (OMV), while 3 is an integer (OMI).

OpenMath Content Dictionaries provide a mechanism for adding new symbols to the OpenMath language in order to represent mathematical concepts. If the mathematical notions pertaining to an area of application are not yet covered by the existing Content Dictionaries, a user may introduce new symbols in a custom-made new Content Dictionary. This procedure mimics very closely the way mathematics proceeds by introducing new definitions.

Currently, WebALT has extended OpenMath Content Dictionaries (cdbase http://www.webalt.net/cd) with new symbols for describing

properties of functions: `asymptote`, `inflection_point`, `injective`, `slope`, `surjective`,

geometric properties: `cube`, `curve`, `parabola`, `perimeter`, `rectangle`, `region`, `solid`, `surface`.

Natural language hints for verbalizing the mathematical objects are represented in OpenMath via attribution.

ii. Extending OpenMath with NLG attributes

The NLG content dictionary defines natural language attributes and constants which influence the final form of the generated natural language output. They specify linguistic information that is required for generation to succeed, as well as information that is specific to the generation of mathematical sentences. Some of these attributes are explained below:

`mood` specifies the type of the sentence that is to be generated, and can be one of the following: `declarative` (for normal sentences, where the subject is included), `imperative` (for command sentences) and `interrogative` (for questions), with the latter two being more common in mathematics.

`directive` specifies the main verb or action of the sentence, and corresponds with the main mathematical task that is to be performed e.g. `solve`, `prove`, `determine` etc. By allowing several values, this attribute makes it possible to achieve a varied range of natural language output.

`render` specifies how the sub-expression for which it is defined should be rendered. The `formula` constant is used to indicate that everything within the given sub-expression should be rendered as a formula, as opposed to natural language, which is the default.

`type` specifies the type of the sub-expression that is to be rendered as a formula. If specified, it provides the word that introduces the formula during natural language generation, else language-specific default values are used. These defaults are applicable, for those languages where such an introductory word is always required e.g. Finnish and German. The values for this attribute include `formula`, `expression`, `equation`, `matrix` etc.

OpenMath objects that contain attribution from the NLG Content Dictionary have the characteristic specified in section II: they encode the mathematical meaning in an unambiguous form and additionally contain enough hints for a natural language generation tool to produce a verbalization of the mathematics in a variety of languages.

V. MULTILINGUAL TECHNOLOGY

WebALT's multilingual generator, which produces natural language variants of mathematical content from OpenMath, has been implemented using the Grammatical Framework (GF). GF is a parsing and generation formalism developed by Aarne Ranta [3]. The GF formalism separates general language resource grammars from domain-specific application grammars. The former are used as libraries and are written by resource grammarians, while the latter are application-specific and are necessarily developed by application grammarians who, with their knowledge of the domain application, use the resource libraries to build these special-purpose grammars.

Multilingual generation of mathematical content from OpenMath requires definition of a concept lexicon and specification of generation rules. To this end, WebALT has defined a mathematical concept lexicon which contains terms that define the concepts covered by the OpenMath content dictionaries. Currently, this lexicon lists 442 nouns, 96 adjectives and 74 verbs. In addition, a language-specific lexicon file is defined for each of WebALT's target languages, providing the multilingual knowledge required to render any of these mathematical concepts in natural language. A WebALT generation grammar has also been defined. This grammar comprises of language-specific generation rules which define how a given OpenMath symbol (or any combination of such symbols) is verbalized into natural language, based on specified NLG hints. These two tasks require a close collaboration between mathematicians, with a good command of a given target language, and linguists. While the linguist has the expertise to write the application grammars, the mathematician has the multilingual domain knowledge. Currently, the WebALT generator has a good multilingual capability covering English, Finnish, Swedish, Spanish, French and Italian.

Examples of how the technologies described so far are combined to automatically produce mathematical exercise problems in all the above languages are given in the next section.

VI. MULTILINGUAL TECHNOLOGY AT WORK

The following diagram illustrates the generation pipeline which transforms the input OpenMath object to its natural language equivalent.

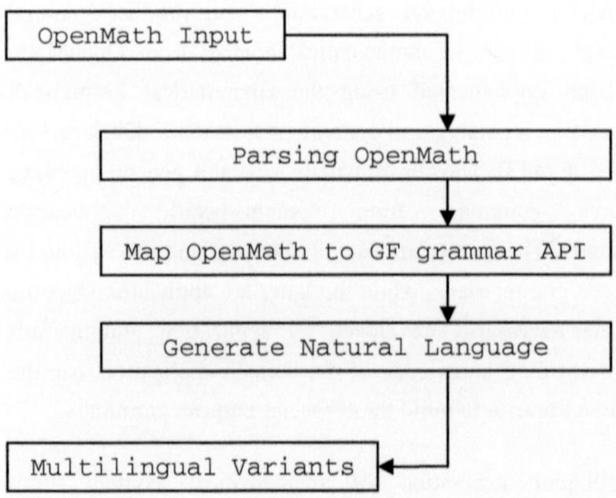

For each of the following examples of OpenMath objects represented in a prefix notation, additional NLG hints have been specified to determine the mood, tense and main directive of the generated sentence. We skip them here to increase readability. Notice that it is possible to obtain for some sub-expressions to be rendered as formulas by appropriately specifying the relevant attributes. For each example, the output is generated in the following order: English, Finnish, Spanish, Swedish, French and Italian.

i. Example.

```
logic1:equivalent(logic1:implies(x,y),
logic1:implies(y,x))
```

Decide if the proposition that x implies y is equivalent to the proposition that y implies x.

Päättele onko propositio, että x implikoi y:n ekvivalentti proposition, että y implikoi x:n kanssa.

Decide si la proposición que x implica y es equivalente a la proposición que y implica x.

Avgöra om påståendet att x implicerar y är ekvivalent med påståendet att y implicerar x.

Décider si la proposition que x insinue y est équivalente à la proposition qu' y insinue x.

Decidi se la proposizione che x implica y è equivalente alla proposizione che y implica x.

ii. Example

```
set1:suchthat(setname1:Z,
fns1:lambda[x].(relation1:eq(
arith1:plus(x,3),5)))
```

Let x be a member of the set of integers such that x + 3 is equal to 5. Find x.

Olkoon x kokonaislukujen joukon alkio siten, että lauseke x + 3 on yhtäsuuri kuin luku 5. Etsi x.

Sea x un miembro del conjunto de enteros tal que x + 3 es igual que 5. Encontra x.

Låt x vara en medlem av mängden av heltal sådan att x + 3 är lika med 5. Finn x.

Soit x un membre de l' ensemble d' entiers relatifs comme tel que x + 3 est égal à 5. Trouver x.

Sia x un membro dell' insieme di numeri interi tale che x + 3 è uguale a 5. Trova x.

iii. Example

```
linalg1:determinant(
      linalg2:matrix(linalg2:matrixrow(a,b),
                linalg2:matrixrow(c,d)))
```

Find the determinant of the matrix $\begin{pmatrix} a & b \\ c & d \end{pmatrix}$.

Etsi matriisin $\begin{pmatrix} a & b \\ c & d \end{pmatrix}$ determinantti.

Encontra el determinante de la matriz $\begin{pmatrix} a & b \\ c & d \end{pmatrix}$.

Finn determinanten av matrisen $\begin{pmatrix} a & b \\ c & d \end{pmatrix}$.

Trouve le déterminant de la matrice $\begin{pmatrix} a & b \\ c & d \end{pmatrix}$.

Trova il determinante della matrice $\begin{pmatrix} a & b \\ c & d \end{pmatrix}$.

iv. *Example*

```
plangeo3:is_midpoint(x,y,z)
```

Determine if x is the midpoint of y and z.

Määritä onko x y:n ja z:n keskipiste.

Determina si x es el punto medio de y y de z.

Bestäm om x är mittpunkten av y och z.

Déterminer si x est le milieu d' y et de z.

Determina se x è il punto medio di y e di z.

v. *Example*

```
(Attrib([mathmltypes:type
        mathmltypes:integer_type],
logic1:and(a,b))),
(logic1:not(quant1:exists[c].
           (logic1:and(relation1:gt(c,0),
            integer1:factorof(a,c),
            integer1:factorof(b,c),
            relation1:lt(c,arith1:lcm(a,b))))))
```

Let a and b be integers. Prove that it isn't true that there exists c such that c is greater than 0, a is a factor of c, b is a factor of c and c is less than the least common multiple of a and b.

Olkoot a ja b kokonaislukuja. Todista, että ei ole tosi, että on olemassa c siten, että c on suurempi kuin luku 0, a on c:n tekijä, b on c:n tekijä ja c on pienempi kuin pienin yhteinen a:n ja b:n monikerta.

Sean a y b enteros. Verifica que no es cierto que hay c tal que c es mayor que 0, a es un factor de c, b es un factor de c y c es menor que el mínimo común múltiplo de a y de b.

Låt a och b vara heltal. Bevisa att det inte är sant att det finns c sådan att c är större än 0, a är en faktor av c, b är en faktor av c och c är mindre än den minsta gemensamma multipeln av a och b.

Soient a et b des entiers relatifs. Prouver qu' il n' est pas vrai qu' il y a c comme tel que c est plus grand que 0, a est un facteur de c, b est un facteur de c et c est plus petit que le plus petit commun multiple d' a et de b.

Siano a e b numeri interi. Prova che non è vero che ci è c tale che c è più grande di 0, a è un fattore di c, b è un fattore di c e c è più piccolo di il minimo comune multiplo di a e di b.

VII. CONCLUSIONS

The experience gained both at the University of Helsinki and at Florida State University during the academic year 2004—2005 shows that an alternative way, based on the web, to deliver instruction of sciences in general and of mathematics in particular, is going to affect a serious change in the academic market place in the coming years. The on-going WebALT project will provide some of the tools that make this revolution possible by offering a language-independent repository of exercise problems in mathematics exploiting existing technologies for the markup of mathematical content and by combining it with natural language generation. In this paper we have shown the current results, even in cases where they are still not optimal, in the generation of mathematical vernacular for several European languages.

The WebALT project will bring tools like MapleTA to the next level by adding features like multilinguality. WebALT will offer an adaptive database of multilingual exercises that can be used with a variety of platforms including MapleTA. In addition to that, WebALT will produce its own exercise editor, exercise player and serve the exercises via a searchable repository, SCORM-compliant and organized by an ad-hoc mathematical taxonomy.

VIII. References

1. Buswell, S., et al. The OpenMath Standard - Version 2.0. http://www.openmath.org/cocoon/openmath/standard/om20-2004-06-30/omstd20html-0.xml
2. Carlisle, D., et al. Mathematical Markup Language (MathML) Version 2.0 (Second Edition). http://www.w3.org/TR/2003/REC-MathML2-20031021
3. Ranta, A., *Grammatical Framework, a type-theoretical grammar formalism.* Journal of Functional Programming, 2004. **14**(2): p. 145-189
4. Seppälä, M. Single Variable Calculus. http://www.webalt.net/Calculus-2004/
5. WebALT. Web Advanced Learning Technologies. http://www.webalt.net

Engineering Education and Errors

Rein Mägi
Ph.D
Tallinn University of Technology, Faculty of Science
Tallinn, ESTONIA 19086

Abstract
Errors in engineering can cause sometimes serious results. But as practice shows there is impossible to avoid them absolutely. The aim of engineering education is to minimize the probability of errors in design and exploitation of engineering equipment.

1. INTRODUCTION

Errare humanum est = erring is human. (Pseudo)logical conclusion might be – the more we err the more human we are? But human errors in engineering can cause **serious results**. For example during the Baltic Sea storm in september 1994 the passenger boat Estonia foundered and 852 passengers wrecked. As the expertise showed one reason for the shipwreck was unequal strength of visor hinges. The diameter of the hinges should have been 2 times larger than it was.

2. ERRORS IN ENGINEERING

Estonian proverb says – nine times measure and then once cut!

Modern exact computer-software turns useless many methods of **error-free calculation** of last century [4]. But computing process, especially creating programs can include special errors – loss of significant digits, iterative instabilities, degenerative inefficiencies in algorithms etc.[5]

J. Talens [6] examined and analysed **the work of designers**. He classified the errors made in the technical documentation (drawings) to constructive errors, calculation errors and dimensional errors. Constructive errors are caused by wrong or ineffective design solution, considering reliability, economic or ergonomic aspect.

H. Petroski [1] introduced and analyzed many case histories of error and judgement in engineering. He concluded that engineering errors are caused mainly by **human factor**. It would be very instructive to learn typical historical errors for avoiding them in the future.

The **reasons** for mistakes should be divided: 1) caused by the lack of knowledge ("know how" or "don't know how") and 2) caused by carelessness (I knew but I did not pay attention). Only learning and training should diminish or avoid making mistakes.

By causes the errors at experimental measurements are divided into **systematic and random errors** [2]. But there is no strict definition of systematic errors, since what is a systematic error for one experiment may not be the same for another. We can diminish measurement errors, however we cannot get absolutely rid of them. [3].

By importance we can differentiate **essential and unessential mistakes**. Essential mistakes can cause serious economic or human defeat. It needs practical experiences to identify the essentiality of mistakes.

Is it an essential error to ignore the **requirements of standards** [7]? For example by the standard ISO 5457 [8] a technical drawing must have borders and a frame, a title block and centring marks. At the same time the distance of the frame-line from the edge of the sheet must be at least 10 mm for sizes A4, A3 and A2, but at least 20 mm for sizes A1 and A0 (Fig. 1).

K. Elleithy et al. (eds.), Advances in Computer, Information, and Systems Sciences, and Engineering, 387–391.

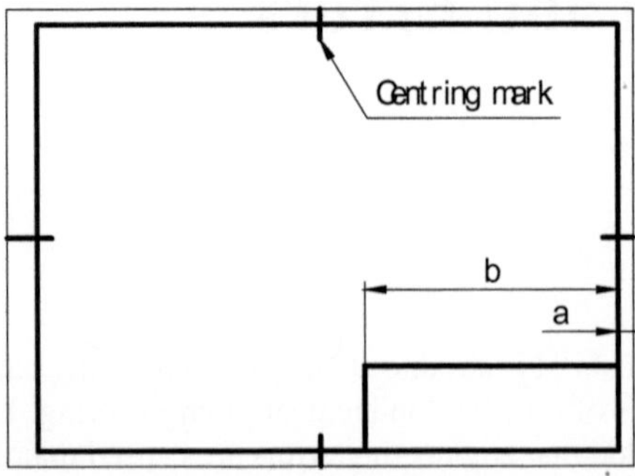

Standard, norm	Format A4, A3, A2		Format A1, A0	
	Dimension, mm			
	a	b	a	b
ANSI (USA)	10-19	159	13-25	193
DIN (Germany)	5	187	5	187
GOST (Russia)	5	185	5	185
JIS (Japan)	10	130	20	130
ISO	10	170	20	170
RT (Finland)	7	178	7	178

Figure 1. Comparison of title block's dimensions by different standards.

Such a large distance is motivated by better fixing the sheet when plotting. In this connexion the maximum horizontal dimension of a title block is 170 mm. Centring marks favour better cadring at microcopying. But nowadays these arguments and therefore also requirements have lost their practical importance. Some comparative dimensions (a, b) of different standards or norms are shown in the figure 1. Some of these variants are offered in CAD template-files. Which of them to use? There is no single answer. In design practice the agreement has to be achieved with the orderer of the project. By my opinion the title block should contain scale and format of the drawing due to their connected character at printing or copying.

International standards have nowadays rather recommending or harmonizing, not commanding character. Using of them is purposeful in case when it concurs with practical effect. Standards should favor but not hinder technical activity. But the problem for lecturers remains– how to explain this to the students?

3. ENGINEERING EDUCATION AND ERRORS

Is it possible to avoid mistakes in **engineering education**? Yes and no! Of course nobody wants to make mistakes in tutorials purposely. But the "discovery" of misprints by students is a good sign of careful reading the book. Even more - some tests are based on differentiating the right and wrong answers, for instance the very popular TV-show "Who Wants to be a Millionaire?".

Very interesting and instructive task for technical gymnasium students was design of spring (Fig. 2). Springs are quite widespread parts in engineering devices. But for working out suitable parameters for a spring, knowledge and skills of various subjects – mathematics, physics and technical drawing have to be used. As experience showed, **nearly 95%(!) of students did mistakes** in their calculations. Every mistake at the beginning would cancel all the following calculation. Therefore the supervisory role of the lecturer was unavoidable. Mistakes were caused by various reasons – inaccurate modification of formulas, incorrect input of values, using wrong units etc. Every mistake was analysed with the aim to understand the reason of errors and to avoid them in the future.

Better err in the engineering studies than in the engineering practice!

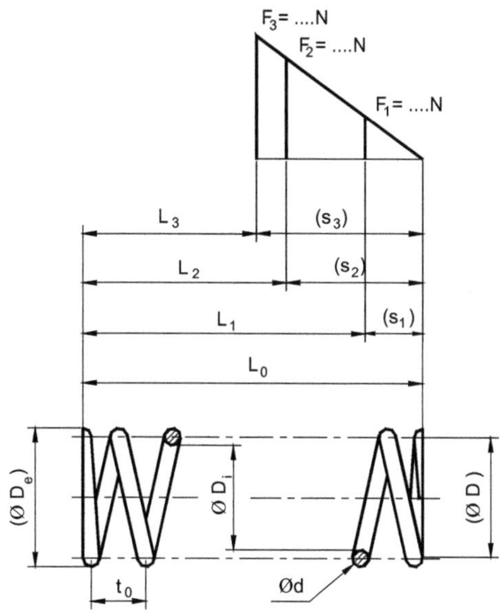

Figure 2. Spring drawing according to standard ISO. In addition to the geometrical dimensions there is a force-deformation diagram.

In engineering graphics **self-testing tasks** (Fig. 3) are created by the principle of selecting the correct variant.

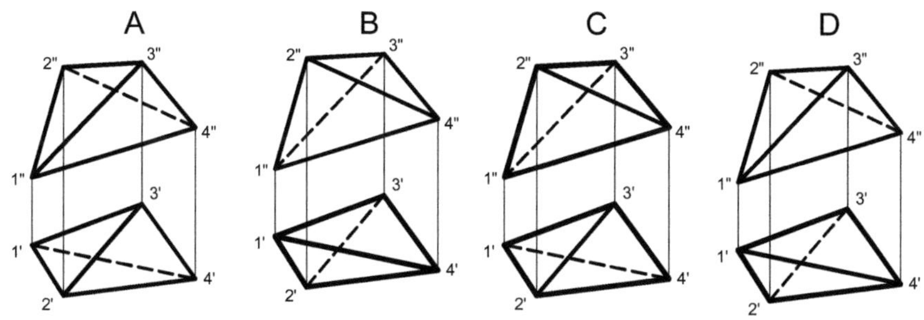

Figure 3. Polyhedron 1234 is represented on two projections – front view and top view. Which variant - A, B, C or D – is correct according to visibility by European system of projection? [9]

The problem is - how to think out wrong answers (mistakes)? But students "help" to solve this problem – they produce the mistakes themselves. The lecturer's role is to select the more typical and didactic mistakes. Of course, every variant should be supplied with explanation – why this answer is wrong or right. It is not correct to say – the teacher corrects the mistakes of a student. The teacher only checks the task and marks out mistakes – the student has to understand and correct his mistakes himself!

How to avoid mistakes in engineering? There are some ways to check correctness of the task – using another solving method, control by units of measure, practical experiment etc. Very sure possibility is graphical control – technical drawing. **Drawing is the language of engineering** – it is understandable for any nationality. Even the standard ISO-128 defines - technical drawings have to be unambiguous, clear and language independent [10]. Drawing as the main visualisation means should help to make solutions more understandable for many subjects –

mathematics, physics, chemistry, economy etc [11].
As our analysis in Estonian companies showed, modern computer-aided drawing (CAD) is used in 100% of the companies examined [12]. CAD is quite inconvenient especially for older designers accustomed to use the large drawing sheet A1 (841x594 mm). As computer display area is nearly 8 times smaller the scale has to be enlarged (*Zoom*) (Fig.4). Too frequent changing the scale disturbs to perceive the reality of the object.

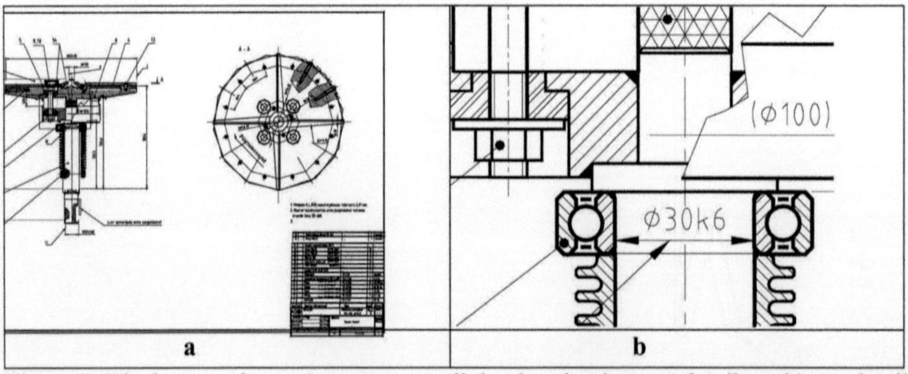

Figure 4. Display paradox – a) we can see all the drawing but not details or b) see details but not all the drawing.

However, computer graphical drawings are mathematically exact; mistakes may appear by human inaccuracy (Fig.5).

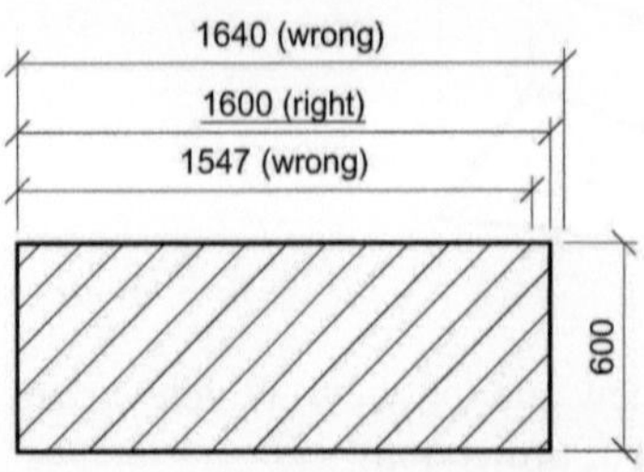

Figure 5. Incautious snapping of lines' endpoints may give wrong dimension value.

In practice only few people (designers, draughtsmen) are directly engaged in making drawings, but nearly every engineer has to handle (read, understand, revise) technical drawings. The handling of CAD-drawings, made by other people, requires additionally specific knowledge, skill and accuracy, otherwise fatal errors could be caused by incompetence [13]. Recognition of invisible elements of drawing (*Lay*er, *Block*, *Attribute*) needs special attention. For example, before *Move*-command, all *Layers* with necessary objects should be switched *On*.

Now about **errors measured in time units** – minutes, hours, days etc. Of course the most important is the final result of engineering work – new project, construction or device, but in the conditions of contemporary competition pressure the duration of the task is equally important. Ignoring prefixed terms could increase logistics costs and lower the prestige of the company. Unfortunately some students (10-30%) cannot complete their tasks or projects in time. The main reason of delay is disability to predict the right time for the task. But the predicting ability may grow only by training.
Famous mathematician G. Polya [14] recommended an universalized order for solving any task: 1) understanding the task; 2) creating the program for solution; 3) realization the program; 4) analysing and checking the solution. The latest phase is very important for solving engineering tasks.
4. CONCLUSIONErring is human. But admitting, understanding and avoiding mistakes is more human! It is indispensable in engineering.

References:

[1] Petroski, Henry, Design Paradigms. Case Histories of Error and Judgement in Engineering. University Press, Cambridge 1995

[2] Barford, N. C. Experimental Measurements: Precision, Error and Truth. (Second Editon), John Wiley & Sons, Chichester 1995.

[3] Taylor, John. R., An Introduction to Error Analysis, The Study of Uncertainties in Physical Measurements. University Science Books Mill Valley, California 1982.

[4] Gregory, R. T., Krishnamurthy, E. V. Methods and Applications of Error-Free Computation. Springler-Verlag, New York 1984.

[5] Acton, Forman S. REAL Computing Made Real. Preventing Errors in Scientific and Engineering Calculations. Princeton University Press, Princeton, New Yersey 1996.

[6] Таленс, Я. Ф. Работа конструктора. Машиностроение, Ленинград 1987. (Talens, J. F. Designer's Work, Leningrad 1987 – *in Russian*)

[7] Mägi, R., Sepsivart, M., Standards – Why Not? or Why? No! //Engineering Graphics BALTGRAF-6. Proceedings of the 6th International Conference. Riga, Latvia, June 13-14, 2002, p.233-236.

[8] ISO 5457:1980 Technical drawing – Sizes and layout of drawing sheet.

[9] http://www.hot.ee/rmagi/Nahtavus-test.html

[10] ISO 128-1:2003. Technical Drawings – General Principles of Presentation.

[11] Meister, K., Mägi, R., Visualisation in Descriptive Geometry //Engineering Graphics BALTGRAF-7. Proceedings of the 7th International Conference. Vilnius, Lithuania, May 27-28, 2004, p.172-175.

[12] Mägi, R., Sepsivart, M. Drawing Management in Estonian Companies //Engineering Graphics BALTGRAF-7. Proceedings of the 7th International Conference. Vilnius, Lithuania, May 27-28, 2004, p.111-115.

[13] Mägi, R., Specific Features of Computer-Aided Draughting //Engineering and Computer Graphics 4, Vilnius, Technika 1998, p. 3-9.

[14] Polya, G. Kuidas seda lahendada. Tallinn, Valgus 2001. (How to Solve It. A New Aspect of Mathematical Method. By G. Polya – *Translation to Estonian*)

Technology Enabled Interdisciplinary Project Based Learning (IPBL)

Osama K. Alshara

Information Technology, Higher Colleges of Technology-ADWC
Abu Dhabi- UAE
osama.al-shara@hct.ac.ae

Fawaz A. Masoud

Department of Computer Information Systems, University of Jordan
Amman - Jordan
fawaz@ju.edu.jo

Abstract

Allocating projects for students to work on is not an easy task let alone allocating real-life projects, which further enhances students learning. Real life projects are multi dimensional; in the sense that they require team based work. Moreover, team members represent different disciplines. Hence, educational institutes Adapting the independent and project based learning require: 1) real life projects for the students to work on. 2) Team building mechanism. 3) And interdisciplinary students and teachers formation. 4) Such projects require strong support and participation from the local community. However, a mechanism has to be proposed and established to bring about an integral relationship between the higher education institutes and the local community.

Any one of the above requirements can be an obstacle of its own. This paper proposes an interdisciplinary project based learning model using the SEE [1] portal. This integral solution is designed to overcome these obstacles.

KEYWORDS

Project based learning, integrated learning, interdisciplinary teams, interdisciplinary based learning and multidisciplinary teams.

1. INTRODUCTION

Due to the dynamic and innovative nature of the work place in today's business environment, educational institutes are required to provide graduates who fit in this dynamic environment. Calls have been initiated by different sources to reengineer the education delivery system to meet such a task [6, 22]. These calls have come at a time when advances in technology and in e-learning in particular will perhaps allow educators to meet this goal.

Many studies [8, 10] show that the use of IT in the classroom improves student learning and the way they work [15, 21]. The implementation of technology in education requires innovation by teachers & the institute, and an organizational culture which supports such an implementation. Furthermore, technology must be used in an appropriate situation, be planned, monitored and evaluated against the goals of the educational initiative.

Advances in technology lend themselves to delivery methods such as independent learning, e-learning and project based learning. The authors' experiences in these methods triggered the desire for a more integrated approach to learning. Projects, especially real life ones, are multi-disciplinary by nature. For example, to work on an e-commerce website, the project team may consist of an interdisciplinary team of IT, Business and Media students in order to fully develop the business case and the website itself. The danger of developing a business application in isolation is that the developer runs the risk of not achieving the sponsors' business goals fully.

Both students and faculty face problems of allocating enough projects for the students to work on. Moreover, there are major differences between the use of hypothetical case studies and projects when compared to real life projects [13, 7]. Local organizations are the best source for real life projects, which triggered the idea of establishing a relationship between the college and local organizations as the best source for these projects.

The SEE project [1] was proposed to serve the students' needs to implement real life projects as assessed course work tasks. This portal brings the students, instructors and local organizations together. Local organizations are required to

K. Elleithy et al. (eds.), Advances in Computer, Information, and Systems Sciences, and Engineering, 393–398.

register as members in the SEE portal before they can load their projects. Each project must include, among other information, description of the project, deadline date and contact person. Furthermore, organizations can browse through the students' innovative ideas/projects and ask to sponsor and implement a project at the organization.

Students would post their ideas/projects, browse posted projects by local organizations to find a suitable project to implement, bid for it and get the instructor's approval on the project. This online process would trigger direct communications between the students, instructor and the organization's representative to implement the project, which will be used by the organization.

Abu Dhabi Women's College http://adw.hct.ac.ae is part of the Higher Colleges of Technology (HCT) http://www.hct.ac.ae in the United Arab Emirates, and the University of Jordan (www.ju.edu.jo) support the independent and project-based learning approaches. Many initiatives have been taken system-wide by the HCT to enhance its educational delivery methods (e.g.. Laptop integration, e-learning strategy, Centers of Specialization). Senior management support for these early initiatives has further encouraged faculty and line management to generate new & innovative ways to utilize technology in the classroom.

2. IMPACT OF TECHNOLOGY ON LEARNING METHODOLOGIES

Organizations are increasingly paying attention to the concept of organizational learning in order to increase competitive advantage, innovation, and effectiveness. "Organizational learning is a competence that all organizations should develop" [3], and is defined as "the acquisition, application, and mastery, of new tools and methods that allow more rapid improvement of processes whose improvement is critical to the success of the organization" [19]. Educational organizations must learn how to employ and implement new technologies to improve the delivery of their learning methodologies.

Most organizations in today's work practices are constantly faced with environmental change to which they must adapt. There is also increased pressure on these organizations to produce employees prepared for the knowledge-driven economy, and to serve as both a source and a producer of knowledge. Educational institutes must always pay attention to such environmental changes since they are the main human resource provider to the market.

Issues about the relationship between information technology and learning methodologies are becoming more critical as IT is becoming a focal point in organizations. The authors assert that in order to understand the role of IT in learning methodologies, educational institutions need to focus on the relationship between learning theories and understanding technology. Technology (especially information systems) can help achieve effective learning, hence improving students' and teachers' overall performance. Information systems can facilitate the learning process by supporting the processes of knowledge management and information distribution, sharing and communication [4]. These processes are very well represented in the educational concepts of:

- *Peer tutoring*: where students learn from each other directly;
- *Learning by example*: students learn from other students' work that was completed previously for the same or a similar course;
- *Independent learning*: this method requires students to access different resources such as peers, libraries, Internet, teachers etc. to acquire information and knowledge. Hence, the educational institute must facilitate access to such resources by its students;
- *E-learning*: promotes independent learning, both synchronous and asynchronous learning. Access to the technology allows for anytime, anywhere learning;
- *Project based learning*: Among other things it promotes team work, information sharing and communication skills [20, 7]
- *Market driven learning*: this is a new concept that we introduce here to emphasis the integral relationship needed between the local organizations (as the potential employer of graduates) and the educational institution that delivers the content [20].
- *Integrated learning*: based on the demand of today's job market, students must have well rounded knowledge about other fields to be able to work together in teams and communicate efficiently and effectively.
- *Interdisciplinary learning teams*: To be able to provide project based learning and integrated learning you need to form

interdisciplinary learning teams of both students and teachers. It is through these teams that members learn the needs of the other disciplines and how to integrate their overall knowledge to form an integrated solution.

3. INTERDISCIPLINARY TEAM BASED LEARNING

Multidisciplinary teams are not new in Education, and in fact have been used widely in the Health industry [5] and within large corporations such as Boeing, DEC and Chrysler [18]. The value of interdisciplinary teamwork is being recognized in many fields. In the educational setting, both students and teachers felt the interdisciplinary approach was a worthwhile and enlightening experience [7]. However, improvement and some fine-tuning of the different models are required.

"Utilizing interdisciplinary teams provides a more realistic development environment, but also comes with many challenges" [18] such as 1) project ownership and 2) team formation - who is the driving force behind this team formation?

3.1 Project ownership

The problem of claimed ownership between participants from different disciplines affects the intra-team communication, consequently affecting the success rate of the overall experience. In [18] they considered the business professor as the real client and students as the developers. In our model, we use the outsourcing method, where a student of any major would contract the project with a real industry-based client. In turn, this student - being the primary contractor - would identify the help/skills that she requires to outsource by teaming up with students from other disciplines. This model ensures project ownership clearly lies with the student.

3.2 Interdisciplinary team formation

Most of the literature showed that teachers were the driving force in building the interdisciplinary teams. There appeared to be two primary methods of building interdisciplinary teams:

1) By synchronizing two interdisciplinary courses with 'obvious' synergies [18]. For example, the Business Dept course 'e-Business' has obvious synergies with the IT Dept. course 'Web Development I'; or

2) By one or more teachers delivering a course that is common across different program disciplines [16]. For example, an 'Introduction to Computing' course that is a core requirement in semester 1 of all programs.

The models that are illustrated by the literature show the formation of predefined interdisciplinary groups [16, 2, 12]. By this we mean that members of the groups were assigned by the teachers, based on their knowledge of the students' academic level, or conveniently grouped by virtue of the fact that the students were in the same class. To the contrary, our model supports non-predefined group formation. Depending on the amount of work needed for each task, student sub-teams can also be formed for each task. For example, if the development is a major task that requires more than one student to execute, an IT sub-team can be formed. Each student or sub-team then assumes a pivotal role of communicating the results from the sub-team back to the primary team. Thus, proper governance and effective communication are critical for project success.

As depicted in Figure 1, our model allows for the formation of the following non-predefined teams:

1- Students' sub-team: students of the same major working on a major specific task
2- Students' team: students from different majors working on the same project
3- Educators' team: includes the teachers (from different disciplines) of the students' who worked on the project.
4- Project team/Community team: includes all parties involved in the development of the project: students, teachers and the project sponsor (usually from industry).

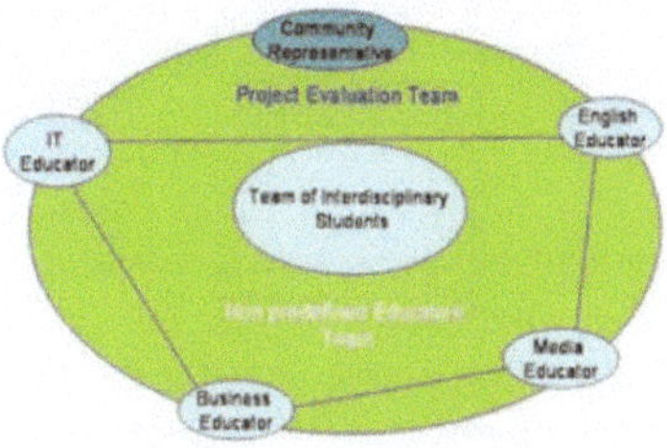

Figure 1: Multi-interdisciplinary teams' formation.

Linder [14] states that "Students in general like to work in teams to the contrary of teachers". Hence our challenge was to develop a model where the interdisciplinary team based learning would not be teacher driven, but rather student driven. Our model focuses on the students' teamwork regardless of the teachers' teamwork. It makes the students the center of the team, which in turn helps promote the formation of the educators' team. This results in a non-predefined team of educators, which fosters an environment where educators can formally or informally meet, leading to increased interdisciplinary cooperation. As a consequence, our model overcomes many of the obstacles faced when forming interdisciplinary educators' teams [14].

Having the presence of an outsider in the team, in the form of a project sponsor, also tends to bring the educators closer to presenting a unified image of their institute.

Students working within the SEE portal form their own groups as described earlier. Figure 2 further illustrates this concept by describing an e-Commerce project which is currently being undertaken. The skill-set needed to successfully implement this project, include business and marketing knowledge to be provided by the Business Department, web development skills from the IT Department, graphic design by the Media department, and content revision by the English department.

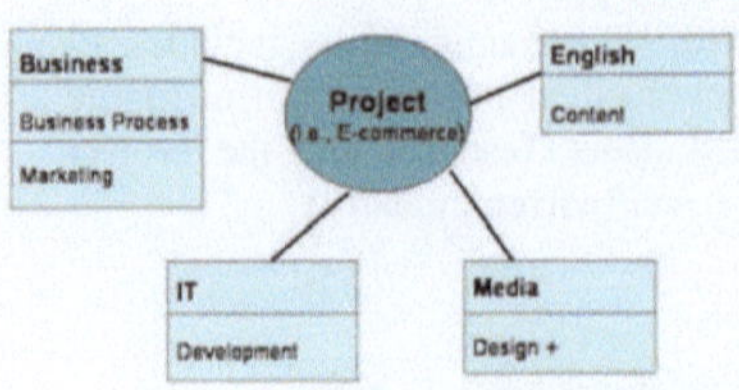

Figure 2: Interdisciplinary project based learning format.

4. PROJECT BASED LEARNING USING SEE

One of the major problems that instructors in our college face is allocating enough suitable projects for students to work on. This is a major requirement when adopting independent and project based learning, which our college does in many advanced courses. So teaming up with the local community, as a rich source for such projects, works as a great idea that serves many objectives. Some of these objectives are real life project based learning, integration between the college and the local community, takes away some of the burden off of the teacher who constantly needs to come up with new projects, encourages students' innovation, etc.

SEE is a multi purpose portal that supports real life project based learning, learning by example, along with other added value services. The OM part of SEE employs the case based method. A literature review shows many OM and OL based systems that use this method such as HICAP [9], SEED [17] and BORE [11].

The SEE portal brings about a strong integration between the institute and its students on one side and the local organizations on the other. This relationship is very crucial in to the success of our proposed model. The organizations in the local community are the best source for real life projects. These organizations would act as sponsors in the project team/ community team. To support the Interdisciplinary based project learning, SEE is divided into three major components:

1- Actors: are the students, instructors and the local organizations that interact and communicate through the portal.

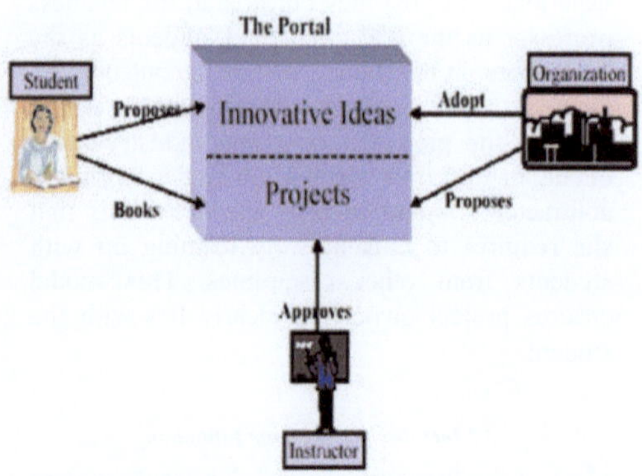

Figure 3.a. Project based learning

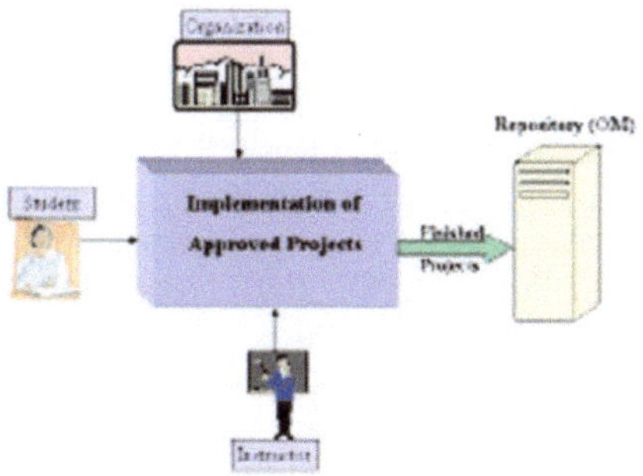

Figure 3.b. Project based learning with the OM

2- The Portal: is the host of all the activities that take place between the actors. The portal houses the innovative ideas proposed by the students to be adopted by organizations, and projects proposed by the organizations to be booked and implemented by the students as shown in Figure 3.a.
3- The OM: it is the repository of all finished projects as shown in figure 3.b.

We find the SEE portal to be a suitable virtual environment to implement the interdisciplinary teams. Most higher education institutes are structured, both physically and logically, in a secular manner. This means that colleges and even departments within the same college are not interweaved, rather are secular. Forming interdisciplinary teams in such environment is very difficult. Changing the structure and attitude of higher education institutions is an immense job if not impossible. Therefore, creating a virtual environment is a much easier approach.

Reviewing figures 1 and 3.a shows the tandem between the SEE design and actors on one hand and the proposed interdisciplinary team formation model on another hand.

5. CONCLUSION AND FUTURE WORK

This paper sets out to espouse the benefits of using technology to enhance education. In particular, it advocates the adoption of integrated project based learning (IPBL) in an interdisciplinary environment. Our model illustrates the benefits of self-defining teams of both students and instructors. Our preliminary conclusions regarding IPBL and interdisciplinary teams is that together these two delivery strategies provide a rich environment for collaborative learning, the application of real-world (as opposed to simulated) activities and a chance for students to experience the positive & negative aspects of teamwork.

Project based and independent learning methods require teachers to provide the students with projects as well as divers examples based on the given projects, which proved to be a very difficult task. The SEE system is built around the fact that local organizations are the suitable source for such projects. Students work on real life projects as part of their assessed course requirements, and interdisciplinary project based learning.

Thus far we present the idea and developed and hosted the tool. Just recently, we proposed the use of the system by Abu Dhabi Women's College. The top management along with the educational technology department has approved our request to pilot the SEE system for a selected set of project based courses in the IT and E-commerce departments. Furthermore, the Imaging Technology Center (ITC) was established which will be using our IPBL proposed method.

References

[1] Alshara O. SEE: A Project Based Learning Initiative To Integrate ADWC With The Local Community. (SETIT 2005) International Conference: Sciences of Electronic, Technologies of Information and Telecommunications March 17 – 21, 2005 TUNISIA

[2] Anewalt, K. Utilizing Interdisciplinary Teams in Teaching E-commerce. Consortium for Computing Sciences in Colleges (CCSC). 2003.

[3] Argyris, C. On Organizational Learning, Blackwell Publishers, 1999. http://www.monitor.com/binary-data/MONITOR_ARTICLES/object/94.pdf

[4] Balasubramanian, V. Organizational Learning and Information Systems, ISWorldNet, May 1995. http://www.e-papyrus.com/personal/orglrn.html

[5] Carlton, B. The Role of the Health Educator in Interdisciplinary Health Team Development: An Organizational Development Strategy. Health Education, 15, 6 (Oct-Nov 1984), 13-15.

[6] Derntl, M., and Renate Motschnig-Pitrik R. Engineering e-learning systems (ELS): Patterns for blended, Person-Centered learning: strategy, concepts, experiences, and evaluation Proceedings of the 2004 ACM symposium on Applied computing (March 2004) 916-923

[7] Fernandez, E. and Williamson, D. M. Software development: Using project-based learning to teach object oriented application development Proceeding of the 4th conference on Information technology education, October 2003.

[8] Fisher, D. & Stolarchuk, E (1998).The effect of using laptop computers on achievement, attitude to science and classroom environment in science. Proceedings Western Australian Institute for Educational Research Forum 1998. Available from: http://www.education.curtin.edu.au/waier/forums /1998/fisher.html.

[9] Flemming, U., Coyne R. & J. Snyder, case-based Design in the SEED System, American Society of Civil Engineers. The Proceedings of First Congress on Computing in Civil Engineering, Washington D.C., 1994. http://seed.edrc.cmu.edu/SD/asce-dist.html

[10] Griffith, R.A., Gu, Y., & Brown, D.G. (1999). Assessment of the Impact of Ubiquitous Computing. Paper presented at the Annual Forum of the Association for Institutional Research. (39th, Seattle, WA, May 30 – June 3, 1999).

[11] Henninger, S. Case-Based Knowledge Management Tools for Software Development, Journal of Automated Software Engineering, 4(1), pp. 319-340. 1992.

[12] Johnson, P. M., Moffett, M. L. and B. T. Pentland, Lessons learned from VCommerce: A virtual environment for interdisciplinary learning about software entrepreneurship. Communications of the ACM, Vol. 46, No. 12, December, 2003.

[13] Levitt, B. and March, J. Organizational Learning, Annual Review of Sociology. 14: 319-340, 1988.

[14] Linder, A. and Ibrahim, A. Issues in technical communication education: Building cross-disciplinary teams in higher education institutions, Proceedings of IEEE professional communication society international professional communication conference and Proceedings of the 18th annual ACM international conference on Computer documentation: technology & teamwork, September 2000.

[15] Lowther, D.L., Ross, S.M., and Morrison, G.R. (2001) Evaluation of a Laptop Program: Successes and Recommendations. In: Building on the Future, NECC 2001: National Educational Computing Conference Proceedings (22nd, Chicago, IL, June 25-27, 2001).

[16] Mosiman, S. and Hiemcke, C. Interdisciplinary Capstone Group Project: Designing Autonomous Race Vehicles. SIGCSE 2000 (Mar. 2000) Austin Tx, USA.

[17] Muñoz-Avila, H. , Aha, D.W., Breslow L.A. & Nau, D. HICAP: An Interactive Case-Based Planning Architecture and its Application to Noncombatant Evacuation Operations, To appear in Proceedings of AAAI/IAAI-99. AAAI Press. 1999.

[18] Rogers, R. L., and Stemkoski, M., J. Reality-Based Learning and Interdisciplinary Teams: an Interactive Approach Integrating Accounting and Engineering Technology. Educational Resources Information Center (ERIC). 1995.

[19] Schneiderman, A. M. Measuring Organizational Learning. 2000. http://www.schneiderman.com/The_Art_of_PM/ measuring_learning/learning.htm

[20] Tan, J. and Phillips, J. Challenges of real-world projects in team-based courses, The Journal of Computing in Small Colleges, Volume 19 Issue 2. December 2003

[21] Tiu, F.S., Guglielmi, J.P., & Walton, W.G. (2002). Assessing a Technology Initiative: Lessons Learned While Integrating Technology into Teaching & Learning. Paper presented at the Annual International Forum of the Association for Institutional Research (42nd , Toronto, Canada, June 2-5, 2002).

[22] Vrasidas, C. Engineering e-learning systems (ELS): Issues of pedagogy and design in e-learning systems. Proceedings of the 2004 ACM symposium on Applied computing March 2004.

Approach to an Adaptive and Intelligent Learning Environment

Dalia Baziukaitė
Department of Computer Science
Klaipėda University
Klaipėda, Lithuania 92294
Email: dalia.baziukaite@ik.ku.lt

***Abstract*—This paper proposes a model of the *component* for a virtual learning environment (VLE), which enriches it with some adaptive and intelligent features. Such a model explains how the VLE as a system should direct its actions towards the maintenance of learner needs in selecting, studying and controlling the study topics and evaluating the obtained knowledge. The adaptivity is supposed to be understood such as the system's ability to form and, during the learning process, uniformly update curriculum that satisfies the needs of the learner, and intelligence is expressed by the ability to initiate actions and, referring to the learning process, perform decision making. The implantation of such a component into a VLE enriches the spectrum of the student's own abilities to purposefully, without the extra help of a tutor, seek some level of acquisition.**

I. Introduction

A. The Scope

The issue of student modeling in web-based learning systems is widely discussed [6], [14], [22], [23], [1], and remains an important point of interest. The research began in the early 70s, it was then that the first computer based tutoring systems were designed. Coming from simple "one scenario" or static student model tutoring systems through more complex dynamic designs, they evolved into web-based learning systems [1], [23], [6]. During this evolution the main issue has not changed much. It remains the same - student modeling, or in other words, finding the solution of forcing the learning environment to adapt in respect to individual student needs. Here we address not only adaptable features of the system, but adaptive ones as well. We can distinguish two main types of courses in online learning systems. First, courses instructed by human beings. This instruction is done using different types of communication (assignments, discussions, email, etc.). The second includes courses that do not have a human as an instructor. The ELM-ART II [24], [25] adaptive tutoring system that supports learning programming in LISP serves as an example of such courses.

From a technical point of view of implementation and application, the second case is more complicated and receives more of our interest. Difficulties appear because it is not clear who then is the instructor or teacher in such courses. The research done in this field usually points out the adaptability of the learning material [6], [22], [23] and the adaptive curriculum sequencing [14], [23]. In both in-class and instructed web-based courses the teacher is responsible that the material given maps student abilities and needs; the teacher as well is able to avoid topics that for various reasons are not suited to be taught in class. In the second type of courses we may think about mechanisms, which are able to achieve intelligent decisions and decide further actions.

In this work we propose an agent-oriented approach for the design and implementation of an adaptive and intelligent component for the VLE. The environment, which is enriched by such a component, we call intelligent Virtual Adaptive Learning Environment (VALE). Results related to this issue are published in [2], [3].

B. Adaptivity and Intelligence in VLE

Adaptivity in general is understood as the ability of a system to adapt to the changing environment in which it is situated. According to the definitions in [26], [27], for a system to be intelligent, it must have several features: it has to be reactive, proactive and adaptive. Intelligence is understood as a computational task of the system.

We next turn to the educational systems, such as Steve agent [17], [16], [20] and Adele agent [18]. Steve agent is adaptive in that it can provide assistance to a learner that is appropriate for the current situation. It can flexibly switch between monitoring, demonstrating and explaining actions. This means that Steve is intelligent as well, because it is able to initiate actions by itself depending on the situation in the environment. Another approach is expressed by a tool called InterBook [8], [9], [13], that is used for developing adaptive textbooks on the web. This tool has only adaptivity-based features. It supports two types of adaptation techniques, that is, adaptive navigation support and prerequisite-based adaptive annotation. A more complete tool for developing adaptive courses on the web is NetCoach[1]. It shares features common to the InterBook and is derived from the ELM-ART adaptive web-based educational system. Finally, Virtual Training Laboratory [19], called Trilogy, maintains both adaptive and intelligent features. It is implemented as a multiagent system. Trilogy uses intelligent agents to present information and provide access to tools in response to user expressed requirements. Agents can proactively suggest other possibilities that a user may not have considered. A similar approach is common to the VALA system[2].

[1] http://www.net-coach.de

K. Elleithy et al. (eds.), Advances in Computer, Information, and Systems Sciences, and Engineering, 399–406.

In this model the adaptivity is defined by a system's ability to form and during the learning process uniformly update curriculum that satisfies the needs of the learner; intelligence is expressed by the ability to initiate actions and, referring to the learning process, perform decision making processes by applying adopted artificial intelligence (i.e. machine learning) techniques.

The following section provides an abstract definition of the problem, introducing it like a game with the name "School". Section III describes main concepts and milestones that are used in construction of the component. Section IV formalizes the idea given in Section III. Section V presents the summary and conclusions.

II. Problem Definition

Imagine a class and two entities in it. One entity knows everything that the second entity might ask and might be interested in. The first, which knows an answer to every question, plays the role of the teacher and the second, who wants to know something, plays the role of the learner. Only these two roles exist in this closed setting: the Teacher and the Learner. The game they are playing we call "School". Assume that the teaching proceeds as follows:

- Learner asks the teacher to train her/him in a certain subject.
- The teacher, before starting to teach Learner a certain subject, asks Learner what topics of the subject are of interest. The reason for asking such a question might be the desire to determine the subject matter that should be more emphasized than others.
- The teacher also tries to identify what knowledge the student already has in the subject area, and assuming that Learner is just a beginner, asks some questions. This allows the teacher to select topics that are unknown for Learner and exclude those that are already well known in the subject area.
- Based on the current knowledge of Learner and taking into account topics of interest and the scope of the whole subject the teacher decides what topics of the subject to include or exclude from the course.
- The teacher asks about preferences Learner may have. For example, this could be done with the intention of choosing teaching materials most applicable to the student's learning style.
- The teacher believes Learner is able to learn and Learner has the same opinion. Also, Learner is motivated to finish the course with success.
- The teacher is responsible for teaching and controlling Learner. At the same time, Learner is responsible for learning well.
- The learning ends if the teacher realizes that Learner has enough knowledge to finish the course.

We might note that our actors are highly motivated. Learner is motivated to learn well and the teacher to teach well. Now, assume that into the room enters a third entity, which also has the role of learner. She/he also wants to be trained in a certain subject. The Teacher has to act in the same way as described for the first learner. That means, if fourth, fifth, sixth and so on entities will enter the room they all will have the ability to be trained in a certain subject in the most suitable way.

In the real world when actors are real teachers and real students it is hard to implement the idealistic situation described above. But, if we relocate the class from the real world to the virtual world of VLE and transform the real actors to virtual ones this might be achievable. The teacher actor might become 100% pure virtual actor, but in the case of the student the situation is a little bit different. A student is a person who actually is trained and she/he cannot become a purely virtual, but some "part" of the student might. The notion describing the concept of virtual actors in the best way is agent in the agent ontology. That is why the modeling of the whole situation in our virtual "School" could be done well using an agent-oriented approach.

We present the conceptual model of a module that implements the issue of a component related with adaptivity and intelligence in VLE. This component can be seen as a plug-in module for the standard distribution of some VLE. In particular we concentrate on MOODLE[3] - Modular Object-Oriented Dynamic Learning Environment; practical implementation (partial)[4] of a developed model is based on this environment.

The model proposed has three main parts that describe the main functionality of the adaptive and intelligent component:

- setting-up the student curriculum;
- teaching a student;
- controlling a student.

Next we describe each part of the component in detail.

III. Description of the Component

A. Setting-Up the Student Curriculum

- This part of the system is responsible for the registration of students in courses. Topics to be studied in the course are selectable. A student chooses courses based on her/his own needs.
- In addition, the student may define the level of excellence she/he wants to achieve, as well as a setting of preferences. Under the preferences we include the ability to specify the method of presentation of the learning material. This can be seen as selection between the number of examples, graphical explanations and the amount of text in one qualified unit of a learning material. The level of desired excellence and preferences are stored into the student model database.

[2] http://www.vala.arizona.edu

[3] http://moodle.org

[4] http://cs.ku.lt/~avma

- The student may select different courses as well as topics or units in particular courses. This setting of courses, units and topics we call curriculum.
- After the curriculum is formed and level of excellence with preferences is set, the essential moment of verification of the given settings has to be performed.
- Verification is performed in order to check if the given settings correspond to the "real" curriculum the student needs. If verification shows positive results the curriculum is stored to the student model database as is. Otherwise, if verification shows that the student already has the defined level of excellence on some selected topics and/or units these are removed from the curriculum; if verification shows that some topics that are not included need to be added in order to achieve the defined level of excellence, the curriculum is updated by adding such topics and/or units. After such modification, the curriculum is stored to the student model database.
- Every registered student has the ability to enroll in self-educated courses. Self-educated means that users have the possibility to form preliminary curriculum for themselves.

Here, the curriculum is implied to be understood as a learning path, consisting of courses and/or units of courses, and/or topics of units. Therefore, Learner forms the curriculum (that is Learner selects what topics to learn) and only after that the system revises the curriculum.

1) Definition of the Curriculum: Every registered student (Learner) is able to construct the curriculum, which expresses specific needs of the user. For that, Learner must be logged in to the learning environment. Courses that may be studied using self-education compliant standards are marked as "Course Name SELFEDU". Learner browses the list of available courses. In case Learner clicks the hyperlink of a course that is not marked "selfedu", she/he is asked to confirm the enrollment to the selected course. In case the hyperlink of a "selfedu" course is clicked, the list of units and topics in each unit within the course is displayed. Now, Learner marks units and/or topics she/he wants to be included as study objects. The whole course also can be marked. If Learner marks the whole course that means all units and topics in each unit are included in the curriculum as study objects. In this case there is no need to mark anything else; everything will be included automatically. If Learner marks the unit, all topics in it are included in the curriculum as well. Lastly, Learner may select a single topic or topics that do not belong to the units already selected. To enroll in the course with such selections Learner has to press "Enroll". Learner may select another "selfedu" course or not. If selected, everything happens as described above. These all form the curriculum of the particular learner, and are seen in the user view in the category "My courses".

2) Definition of the Level of Excellence: For each "selfedu" course, which was included into Learner's curriculum, Learner may define the level of desired excellence she/he wants to achieve in her/his curriculum. There are three possible levels of excellence: excellent (86%-100%), good (70%-85%), satisfactory (50%-69%). Learner defines the level of excellence for every "selfedu" course in a curriculum settings control panel. The default level of desired excellence is "excellent".

3) Definition of Preferences: Learner is able to switch between several types of presentations of learning material. She/he may manage the amount of examples, graphical explanations and text in one qualified unit of the learning material. This is called managing of preferences and is performed through the curriculum settings control panel. For every "selfedu" course different preferences are allowed to be set. Learner may change preferences during the process of learning. The default setting of preferences is - show all available types of presentations in one qualified unit of the learning material.

4) Performance of Verification: Before allowing Learner to deal with the learning material, verification of formed curriculum is performed. Verification starts automatically after enrollment in a "selfedu" course. Learner has the possibility to skip the verification. In this case, all selected qualified units of learning material are included in the "verified" curriculum and Learner is allowed to access the learning material. But she/he still has an opportunity to launch the verification for better satisfaction of her/his needs. If Learner does not skip the verification the system acts towards identification of the current knowledge state of Learner in selected curriculum given the desired level of excellence. If verification shows that the student already has the desired level of excellence on some selected topics and/or units these are removed from the curriculum; if verification shows that some topics that are not included need to be added in order to achieve the desired level of excellence, the curriculum is updated by adding such topics and/or units. If there are no such topics and/or units than nothing is updated.

5) Student Model: The student model consists of information about the courses she/he has selected, the format of these courses; if courses are "selfedu" then such information covers units and topics in them. This is what we call curriculum. Also, to the student model we add information about the desired level of excellence and defined preferences as well as information about the progress of Learner in the selected curriculum. All this data are stored into the student model database. The primary student model is formed at the moment of setting the curriculum and defining the desired level of excellence, and selecting preferences. Later, during the process of learning the student model is designed for updating until the desired level of excellence matches the current knowledge state of Learner in the selected curriculum.

B. Teaching a student

- This part of the system is responsible for the decision making. Decisions made form the activity chain, which builds how a student is trained. By activity, we mean the actions a system could perform, for example show the learning material, give a self-test, give a test for evaluation, etc.

- The activity should be selected according to the belief[5] (student model) the system has about a particular student.
- Decisions should be made in order to guarantee the optimal number of learning units that need to be presented to the student, and are needed for achievement of the predefined excellence level in the preformed curriculum[6].

1) Choosing an Activity: The system decides what kind of activity to choose by applying artificial intelligence techniques, some of which might be closely related with mathematical statistics, for example grouping students according to their knowledge, etc. In particular, Reinforcement Learning (RL) [15], [21] and clustering techniques are the object of the scope. Here, any user side intervention is unacceptable. The decision is made by LE. In this paper we will not go into detail and will not describe the methods.

C. Controlling a Student

- Student's control is defined by deciding when to give an evaluative test or other activity that could assess a particular student's knowledge in a certain part of the curriculum.
- After completion of an evaluative activity the system must update its belief about a particular student in order to change its state.

The decision making process in the student controlling activity is based on the same principals as decision making in the student teaching activity. For details refer to Subsection III-B.

IV. DESIGN OF THE COMPONENT

To fulfill the requirements raised in Section III and strictly define the intelligent and adaptive parts of a learning environment (LE), we create an Agent-Oriented model. This model is expressed in the manner of methodology called MaSE [10], [11], [12]. MaSE is a methodology for developing heterogeneous multiagent systems and is a BDI-compatible multiagent systems modeling technique. It consists of two phases: analysis and design. The created model of the adaptive and intelligent part of LE is described by going through the steps of MaSE and explaining the resulting submodels of the multiagent system. The model is created using *agentTool*[7], which provides implementation of the entire MaSE process.

A. Goal Hierarchy

The overall process of analyzing, designing and building a multiagent system starts from the capture of a set of functional requirements; from that the system goals are derived. First, we start with the functional requirements that are described verbally and also expressed in use cases sequence diagrams. The set of functional requirements is already defined in Section III. Next, we capture goals to make these requirements more precise and structure them into a goal hierarchy diagram (Fig. 1).

The main system goal is to *provide an adaptive and intelligent learning environment.* To achieve this goal the system has to reach the highest-level goals: *set-up student curriculum, teach student* and *control student.* These goals are partitioned, which means that in order to achieve a partitioned goal, the subgoals must be met, and if all subgoals are reached, then the partitioned goal is automatically achieved. The highest-level goal, *set-up student curriculum,* defines the adaptive part and other two, *teach student* and *control student,* the intelligent part of the LE. The first is broken down into three subgoals: *define curriculum, define desired level of excellence* and *define preferences.* The essence of these subgoals meets the requirements expressed in Subsections III-A.1, III-A.2 and III-A.3, respectively.

As we see from Fig. 1, to achieve the subgoal *define curriculum,* the goal *perform verification* has to be reached. The verification process of the curriculum is described in Subsection III-A.4. To achieve subgoals *define desired level of excellence* and *define preferences* the goal *update student model* needs to be realized. To realize the latter partitioned goal the subgoals *perform verification, store curriculum to the student model, store level of excellence to the student model* and *store preferences to the student model* becomes of utmost importance (for details refer to Subsection III-A.5).

The goal *update student model* is the subgoal for the *control student.* This means that for the qualitative control of the student, the data on student curriculum, desired level of excellence and preferences are needed. The goal *control student* has another subgoal *choose activity,* which is also a subgoal for the highest-level goal *teach student.* These latter goals map the intelligent part of modeled LE.

After accomplishing the decomposition of goals, we clearly see what the system needs in order to reach the top-level goal (see Fig. 1).

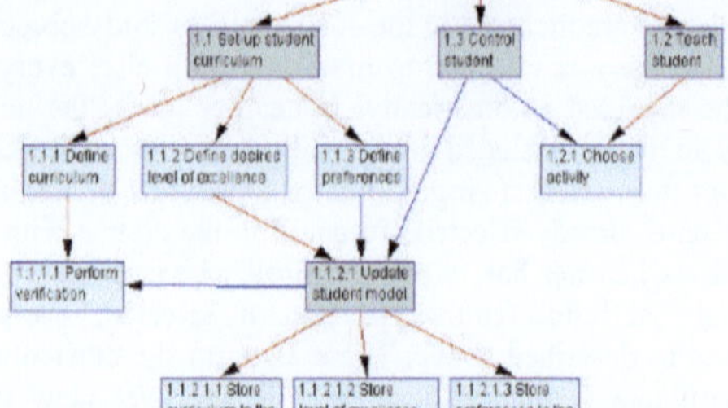

Fig. 1. Goal hierarchy diagram of the Intelligent VALE

[5] Here belief is a synonymous for student model. We use "belief" in order to stress that a system performs actions in an intelligent way.

[6] Note that curriculum is a sequence of course structure elements (such as courses, units, sections).

[7] http://macr.cis.ksu.edu/projects/agentTool/agentool.htm

TABLE I
MAPPING OF SYSTEM GOALS TO THE SET OF ROLES

Role	Goal number in the diagram	Goal
Course Master	1.1.1	Define Curriculum
Excellence Level Definitor	1.1.2	Define Level of Excellence
Preference Definitor	1.1.3	Define Preferences
Curriculum Verifier	1.1.1.1	Perform Verification
Register	1.1.2.1.1; 1.1.2.1.2; 1.1.2.1.3	Store curriculum to the student model database; Store level of excellence and preferences to the student model database
Mentor	-	-
Judge	1.2.1	Choose activity

B. Roles within the System

For easier deriving of the initial set of roles the *applying use cases* step of the MaSE process has to be taken. Use cases help us to validate the system goals and derive an initial set of roles. After writing use cases and creating sequence diagrams, we come to the *refining roles* step. In this step, for our model we get the role diagram depicted in Fig. 2. The role diagram expresses eight roles, each of which maps the concrete goal in the goal hierarchy. The mapping of goals to the set of roles is shown in Table I. The *refining roles* step transforms the structured goals and sequence diagrams into roles and their associated tasks (Fig. 3), which are more suitable for designing a multiagent system. Roles are very important for the agent class definition. They represent system goals during the design phase. From the role diagram (Fig. 3) it becomes clear that each role is expressed by tasks that are actually derived from the goals for which the task is responsible.

The role diagram (Fig. 3) depicts a map of tasks that are interconnected (internal communication protocols) inside a role and outer connected (external communication protocols) among the roles. The role *Course Master* has three tasks that are connected with dashed lines, which denotes the communication protocols that occur between tasks. The *Course Master* role is responsible for displaying the list of courses, detecting the format of the selected course, and saving the enrollment information. The *Excellence Level Definitor* role has two tasks: displaying the level of excellence and saving the newly defined level of excellence. The *Preference Definitor* role also is depicted with two tasks: display preferences and store preferences that by the actions taken are very similar to the earlier two tasks. The *Curriculum Verifier* role responds for revising the curriculum Learner forms and after that allows access for Learner to the learning material. The role *Register* simply accepts information for storing and provides information to the tasks of the other roles if they require it. The role *Mentor* is responsible for displaying learning objects to Learner, collecting required information about Learner and learning objects and asking for the best teaching strategy. The last, *Judge*, has to decide on a strategy and *Mentor* state and update data.

Each task in the role diagram specifies a single thread of control that defines a particular behavior that the role may exhibit [10], [12]. Concurrent tasks are specified graphically using a finite state automaton.

Now the analysis phase of the MaSE process is complete. At this stage we have defined system goals, use cases and their sequence diagrams, and, most importantly, roles defined with their tasks. This ensures that system goals will be implemented in the design phase as well.

C. Agent Classes and Agent Conversations

If in the analysis phase all things dealt with what the system was trying to achieve, now, in the design phase, we focus on how it goes about achieving it [11]. The first step of MaSE in the design phase is *creating agent classes*. We create them from the roles defined in the analysis phase.

1) Agent Classes: From the role diagram (Fig. 3) five main types of agent classes are derived. Within the MaSE technique, agent classes are defined by the roles they play, which differentiate them from the object-oriented model, where classes are defined using attributes and methods. From this, the classes *Definitor, DatabaseInterface, Verifier, UserInterface* and *Teacher* are defined (Fig. 4). Three roles - *ExcLevelDefinitor, PreferenceDefinitor* and *CourseMaster* define the class *Definitor*, because they all have one thing in common in their tasks; they are responsible for defining the learning path (curriculum) or other components of the adaptive and intelligent part of system. The class *Verifier* is defined by only the role of *Curriculum Verifier.* From the role

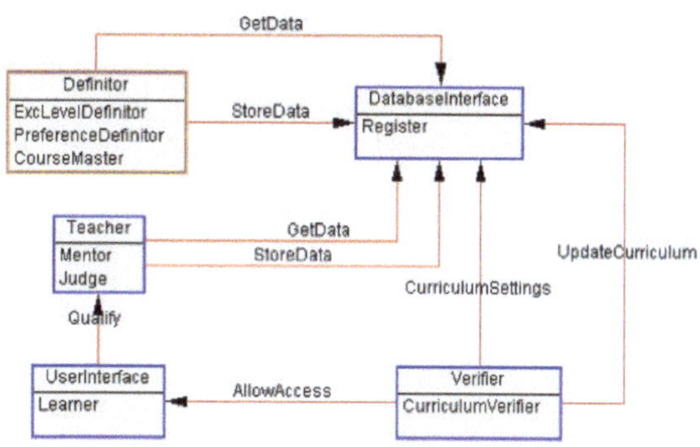

Fig. 4. Agents classes and their conversations

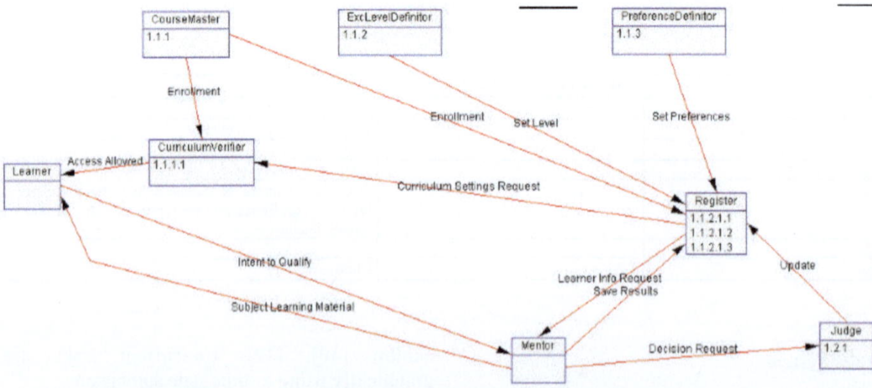

Fig. 2. Role diagram of the Intelligent VALE

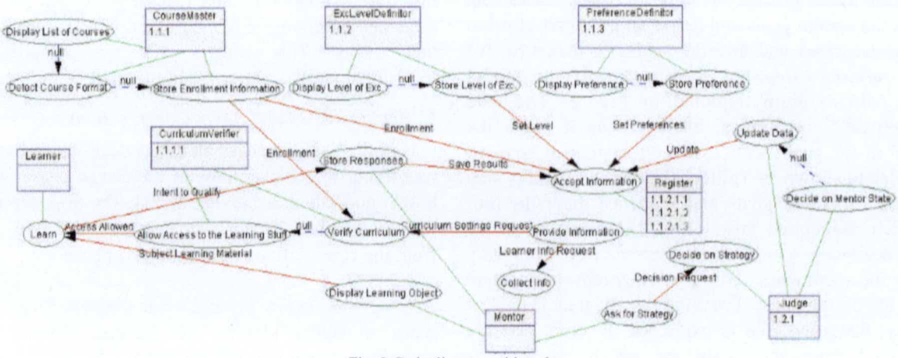

Fig. 3. Role diagram with tasks

diagram *Curriculum Verifier* is responsible for revising the learning path that was selected by Learner and modifying it if necessary. There are no other roles suitable for this class template. Classes *DatabaseInterface* and *UserInterface,* defined by the roles *Register* and *Learner* respectively, actually express the communication of the modeled part of system with the database and user interaction environment. Class *Teacher* is defined by two roles: *Mentor* and *Judge,* which express the tutoring and decision making manner of given roles. This class becomes responsible for the intelligent activities of the modeled system.

2) Agent Conversations: Conversations between agent classes implement the communication described between roles (i.e. in the role diagram and concurrent tasks diagram). The conversations in which different agent classes must participate are denoted with arrows (Fig. 3) pointing to the conversation responder from the conversation initiator. That means the agent *Definitor* participates in two conversations: *GetData* and *StoreData.* Additionally, the agent *Definitor* is the initiator for both conversations. The agent *Verifier* initiates conversations *CurriculumSettings, Update-Curriculum* and *AllowAccess.* The *DatabaseInterface* agent class participates in six conversations as a responder providing necessary information. The agent class *Learner* participates as a responder in the *AllowAccess* conversation and as initiator in the *Qualify* conversation. *Teacher* is a responder in the conversation *Qualify* and is an initiator in the conversations *GetData* and *StoreData.*

The details of each conversation are defined during the *constructing conversations* step of the MaSE process. Each conversation consists of two communication class diagrams: one for the initiator and other for the responder. Communication class diagrams are finite state automatons that define the conversation states of two participating agent classes.

At this point we finish the definition of agent classes and are ready to proceed to assembling agents and system design steps. Those steps finalize the analysis and design process of the overall adaptive and intelligent part of VLE.

3) Agent Architecture: During the *assembling agents* step the internal architecture of every agent identified in the agent template diagram is defined. For each agent we define the components, which build the architecture of the agent class. The architectural component of the agent class is similar to the object in the object diagram and has three parts. The first is the name of the component, the second is the attributes section and the third denotes methods of the agent class [11], [7]. The components of each agent architecture diagram may be

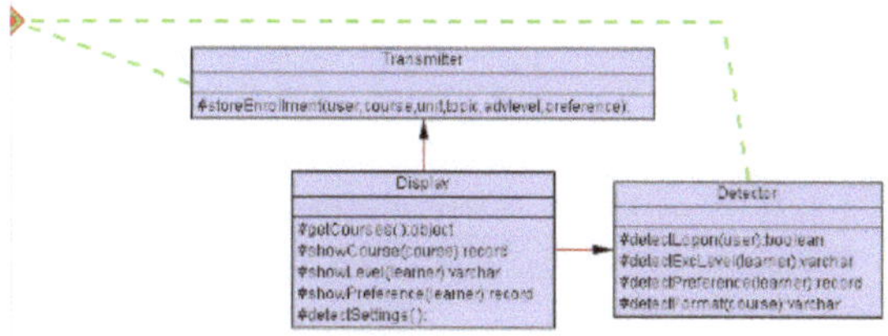

Fig. 5. Definitor agent class architecture

connected using inner-connections or outer-connections. Inner-connections between the agent class components define visibility between components and may call the component to execute an action (method). Outer-connections define the connections with external resources. These are other agents, databases, etc.

Taking the approach that agent architecture is derived directly from the roles and tasks defined in the analysis phase we define the architecture of five agent classes depicted in Fig. 4. From Fig. 3 and concurrent state diagrams that define each task depicted, the architecture of *Definitor* agent class is limited to three components (Fig. 5): *Display*, *Detector* and *Transmitter*. The *Display* component calls the *Detector* in case it needs the level of excellence, preference settings or course format to be detected. As well, *Display* calls *Detector* to verify that Learner is a valid user. In addition, the *Display* calls the *Transmitter* if it needs data to be transmitted to the database. The *Display* has two inner-connections with the *Detector* and *Transmitter* components. Those two have outer-connection with external resources (Fig. 5).

The architecture of *Verifier* and *Teacher* agent classes is shown in Fig. 6 and Fig. 7. The agent class *Verifier* defines two components: *Provider* and *Inspector*. *Inspector* is equipped with methods able to form verification tests and analyze results, as well as if needed, to update the curriculum or simply store it to the database; this is the reason the *Inspector* component has the outer-connection. *Provider* gives access to the learning material after curriculum verification tasks are completed. This component has a relationship with other agents within the system. It also has an outer-connection.

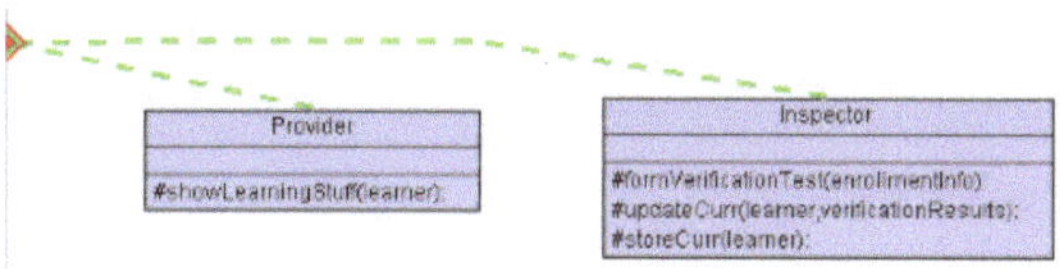

Fig. 6. Verifier agent class architecture

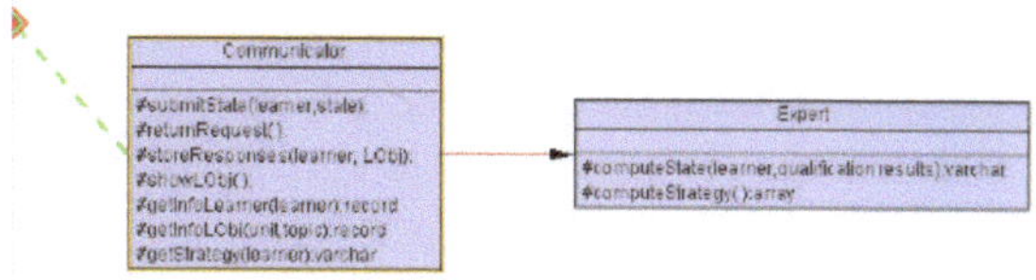

Fig. 7. Teacher agent class architecture

The agent class *Teacher* (Fig. 7) is equipped with methods able to carry out an intelligent decision making process. For that it is divided into two components: *Communicator* and *Expert*. The first one communicates with other agents and calls *Expert* in order to know learning strategy and learner state.

D. System Deployment

To define the final structure of the system the deployment diagram is used. "It may be used to define various configurations of agents and computers to maximize available processing power and network bandwidth" [11].

The deployment we are building proposes that for every *Learner* acting on a different computer there exists a *Teacher* located on another computer, communicating and exchanging information with the Database and other agents located on different computers than where *Teacher* agents are placed.

The reason why this proposition is given is complex computer memory consuming tasks *(computeStrategy, computeState)* that are prescribed for the *Teacher* agent.

V. Conclusion

The proposed model is an attempt to formalize yet another idea of an adaptive and intelligent component for the virtual learning environment. It collects a number of functional requirements that describe the scope of such subsystems; a goal hierarchy leading to the system decomposition to the agents, each of those implements responsibility for the particular goal. The implantation of such a component into a VLE enriches the spectrum of a student's own abilities to purposefully, without the extra help of a tutor, seek some level of acquisition.

This model introduces our individual approach to the adaptive and intelligent component of the virtual learning environment. The model adds some new elements that do not appear in other known systems that address similar issues. This is a feature of the adaptive curriculum formation - a novice feature that allows for each student to independently express a learning goal in terms of "desired" topics. On the other hand, we propose some rarely-used (in this field) AI techniques (that are not the scope of this paper) to be applied for intelligent decision making when the system performs some pedagogical activity.

The main principles of adaptivity in the virtual learning environment from our point of view are discussed in [2], [3], [4], the issues concerning an intelligence is described in [5] and other general information can be referenced in [14], [22], [6], [24]. The proposed model is created using Multiagent system engineering methodology [11]. The creation of appropriate diagrams was made using the agentTool software [10], [12].

REFERENCES

[1] A. Baskas, *E-learning challenges to education,* Proceedings of the Inter national Conference TELDA'03 held at Kaunas University of Technology, Lithuania, Kaunas: Technologija, 2003, pp.129-131. (in lithuanian)

[2] D. Baziukaitė, *Concept of adaptive based virtual learning environment,* Proceedings of the International Conference TELDA'03 held at Kaunas University of Technology, Lithuania, Kaunas: Technologija, 2003, pp.63-66.

[3] D. Baziukaitė, *Adaptive systems in distance studies,* Open and distance education for the knowledge society, Conference proceedings, Vilnius: Vilnius University, 2002. pp.20-24. (in lithuanian)

[4] D. Baziukaitė, A. A. Bielskis, O. Ramašauskas, *Applying adaptive learning principles for the e-studies* Liet.matem.rink, T.42, Spec.nr., 2002, pp.214-218.

[5] D. Baziukaitė, *Making virtual learning environment more intelligent: application of Markov decision process,* Liet.matem.rink, 44, spec. nr., 2004, pp.797-801.

[6] M. A. Benkiran, R. Ajhoun, *An Adaptive and Cooperative Telelearning System: SMART-Learning,* International Journal on E-Learning, April-June, 2002, pp.66-72.

[7] G. Booch, *Object oriented Analysis and Design with applications*, New York: Addison-Wesley Pub. Co., 1994, p.560.

[8] P. Brusilovsky, E. Schwarz, G. Weber, *A Tool for Developing Adaptive Electronic Textbooks on WWW,* WebNet-96 proceedings, 1996.

[9] P. Brusilovsky, J. Eklund, *InterBook: an Adaptive Tutoring System,* UniServe Science News, Volume 12, March, 1999.

[10] S. A. DeLoach, *Analysis and Design using MaSE and agentTool,* Proceedings of the 12th Midwest Artificial Intelligence and Cognitive Science Conference (MAICS 2001), Miami University, Oxford, Ohio, 2001.

[11] S. A. DeLoach, M. F. Wood, C. H. Sparkman, *Multiagent system engineering,* International Journal of Software Engineering and Knowledge Engineering, Vol. 11, No. 3, 2001, pp. 231-258.

[12] S. A. DeLoach, M. F. Wood, *Developing Multiagent Systems with agentTool,* The Seventh International Workshop on Agent Theories, Architectures, and Languages (ATAL-2000), Boston: MA, July 7-9, 2000.

[13] J. Eklund, P. Brusilovsky, E. Schwarz, *Adaptive Textbooks on the World Wide Web,* AusWeb97 Third Australian World Wide Web Conference, Southern Cross University, 1997.

[14] D. L. Hobbs, *A Constructivist Approach to Web Course Design: A Review of the Literature,* International Journal on E-Learning, April-June, 2002, pp.60-65.

[15] A. Gosavi, *Reinforcement Learning for Long-Run Average Cost,* European Journal of Operational Research, 155, 2004, pp.654-674.

[16] W. L. Johnson, *Pedagogical Agents in Virtual Learning Environments,* Proceedings of the International Conference on Computers and Education, 1995.

[17] W. L. Johnson, J. Rickel, R. Stiles, A. Munro, *Integrating Pedagogical Agents into Virtual Environments,* Presence: Teleoperators and Virtual Environments, 7(6), December 1998, pp.523-546.

[18] W. L. Johnson, E. Shaw, R. Ganeshan, *Pedagogical agents on the web,* in ITS'98 workshop on Pedagogical Agents and Workshop on Intelligent Tutoring systems on the Web, 1998.

[19] T. J. Norman, N. R. Jennings, *Constructing a Virtual Training Laboratory using Intelligent Agents,* Int. Journal of Continuous Engineering and Life-Long Learning, 2004.

[20] J. Rickel, W.L. Johnson, *Integrating Pedagogical Capabilities in a Virtual Environment Agent,* in Proceedings of the First International Conference on Autonomous Agents, February 1997, pp. 30-38.

[21] L. P. Kaelbling, M. L. Littman, A. W Moore, *Reinforcement Learning: A Survey,* A Journal of Artificial Intelligence Research, 4, 1996, pp.237-285.

[22] Kinshuk, H. Hong, A. Patel, *Adaptivity through the Use of Mobile Agents in Web-based Student Modelling,* International Journal on E-Learning, July-September, 2002, pp.55-64.

[23] S. Seufert , U. Lechner, K. Stanoevska, *A Reference Model for Online Learning Communities,* International Journal on E-Learning, January-March,2002, pp.43-54.

[24] G. Weber, M. Specht, *User Modeling and Adaptive Navigation Support in WWW-Based Tutoring Systems,* in: Anthony Jameson, Cecile Paris and Carlo Tasso (Eds.), User Modeling: Proceedings of the Sixth International Conference, UM97, Vienna, New York: Springer Wien New York, 1997, pp.289-300.

[25] G. Weber, P. Brusilovsky, *ELM-ART: An adaptive versatile system for Web-based instruction,* International Journal of Artificial Intelligence in Education, Special Issue on Adaptive and Intelligent Web-based Educational Systems, 12(4), 2003, pp.351-384.

[26] M. Wooldridge, N. Jennings, *Intelligent Agents: Theory and practice.* The Knowledge Engineering Review,10(12), 1995, pp.115-152.

[27] D. M. Zhang, L. Alem, K. Yacef, *Using Multi-Agent Approach for the Design of an Intelligent Learning Environment,* in: W Wobcke, M. Pagnucco, C. Zhang (eds.), Agents and Multi-agent Systems, LNAI1441, Springer Verlag, 1998, pp.220-230.

Radio-Chat: Interaction Scenarios for Distance Education in Latin America

Jorge Ramírez; Vladimir Burgos
Graduate Program of Engineering and Technology
Virtual University of Tecnológico de Monterrey
2501 Eugenio Garza Sada Sur
Monterrey, NL 64849 Mexico

Abstract **- The study describes the experience of implementing technologies within academic online courses in a Latin American Virtual University; in particular, the introduction of radio with the option of text-chatting in concurrent time. In distance education, it is essential to encourage effective and qualified interactions between professors and students in order to provide significative learning and valued experiences [5, 7].**

Radio-chat is an integral solution that combines two existing technologies that had been used to offer support to the learning process. Results show that it has been perceived as practical and innovative for the users, in this case for students and professors of the graduate level.

I. Introduction

The internet has the potential to deploy and spread knowledge through distance education in ways hard to be thought before. The key question that we must formulate is what kind of decisions will drive the adoption of emerging technologies to support and leverage the opportunities in distance education when we think about information technologies as a media and not as the driving force behind it.

There is a wide array of technologies to support education and in this particular study we will focus on a distance education model that places learners at the center of the process, supporting their independent learning [7]. Virtual University of Tecnológico de Monterrey (UV-ITESM its acronym in Spanish) has more than 15 years of experience in distance education, has graduated more than four thousand graduate students, has a doctoral and several master degree programs and has contributed to the professional development of more than two hundred thousand Spanish speaking persons all over the world. More than eighty thousand students are enrolled in the learning network each year to attend academic programs, online alphabetization programs for rural and less developed communities or training programs for teachers in the Mexican education system, among others [4, 5, 6].

The education model of the UV-ITESM is centered in the construction of knowledge through self-learning and collaborative learning with the support of technology. It promotes active learning through the development of skills among the students in order to learn, interact and achieve significative learning products (e.g. essays, projects, cases) through electronic media, all of that with the support of advanced didactic techniques and personalized guided processes [7].

These are some of the key success factors that are considered as great educational benefits and advantages for the target audience which is mainly composed of part time students that also have full time professional jobs [2, 3, 4].

Instruction with expert counselors of diverse knowledge areas, approaching the student with the application practices that succeed in the professional field. Professors promote the integration of virtual teams over Internet environments, breaking down the borders of culture, time, space, age and other circumstances –promoting learning without limits.

Flexibility in time and space for the instruction, generating the opportunity of education without limitations of time zones and traditional places for instruction.

Effective use of time and agenda, offering the student to advance progressively according to its individual, personal and professional needs. An asynchronous instruction that facilitates the management of the personal agenda according to the needs of each student, performing and achieving goals in established times in academic calendars for a distance education program.

The development of a social interaction that supports an effective and valuable learning, through the exchange of ideas, points of view and group experiences with classmates, professors and experts. It offers the student the opportunity to adapt theory to the local context, adapting learning to local needs.

Flexibility in the use of educational resources that support the instruction, such as database catalogues with printed materials, services of a digital library, digital video library, multimedia and audiovisual resources, hypermedia documents, etc.

In this study we are going to focus in the Graduate Program of Engineering and Technology (PGIT its acronym in

K. Elleithy et al. (eds.), Advances in Computer, Information, and Systems Sciences, and Engineering, 407–412.

Spanish) which is part of the Virtual University of Tecnológico de Monterrey.

II. Context

The use of several technologies to provide and support distance education has been promoted for a long time. The design of education models to operate through the internet opens the window of opportunities to reach rural and less developed communities in ways never thought before. Also it is possible to get education to people that were difficult to reach in the past, for example, people who are not willing to leave their lands, their work or their family behind in order to travel to a university for education or to obtain a formal degree.

An example is the Community Learning Centers (CCA is the acronym in Spanish) of the UV-ITESM that includes 684 centers, 622 of those in different states in Mexico and 62 in the United States. CCA provides several programs for training and formal education online. The CCA strategy has involved more than thirty thousand teachers; more than thirteen thousand government authorities and more than three thousand civil organizations in more than fifteen Latin American countries. The model implies four key elements: education, innovation, technology and commitment of the parties [8].

Radio broadcast is not a new strategy since it has been used to educate the audience in their socioeconomic and educational development, to provide support and assistance to those living in rural and less developed communities [11]. The particular use and application of broadcast through the internet gives strong individual outreach to the masses, giving distance education institutions and the students an affordable medium of learning and interaction.

The Graduate Program of Engineering and Technology (PGIT) in the UV-ITESM has been promoting the efficient use of technology in order to improve the complex learning process of distance education for graduate students and to facilitate the conditions of such process [9].

In order to improve current students' interactive experiences with professors, an interdisciplinary team was integrated to participate in the design and proposal of a new technological solution. Team members included people from several areas such as telecommunications, informatics, audiovisual production and academics. Several questions arose about the adoption of radio in the graduate program of PGIT:

1) How to make a two way communication channel allowing dynamic interaction in massive groups possible?
2) How to provide a technological solution flexible enough for professors and affordable for students?
3) Which scenario of interaction will be best suited?
4) How many students are expected to participate?
5) How long should each session last?

In the searching process of this technological solution PGIT was conscious of the characteristics needed in the IT application in order to keep consistency with the benefits and advantages of distance education and the education model that UV-ITESM promotes. For example, the attendance to this radio-chat sessions should be optional for the students (in order to keep the principle of flexibility in time and space) but the radio-chat sessions should also be available in an asynchronous way for students not able to attend. Basically the adoption of the technological solution had to be flexible enough so that the radio session could be saved and made available after any transmission, meaning an asynchronous disposition of the audio to generate viable learning objects.

III. Antecedents

The main driver in this approach is the awareness that the interaction process between professors and students may be improved with audio transmission. The inherent benefit is an environment where significant value and experience could be transmitted through professors' knowledge in their professional field and the linking concepts with theory studied in books. This rich interaction process provides a learning motivation with feedback and dialogue to the student taking courses online. The students are expecting from their professors practical counseling to help them deal with the problems they are facing on their own context and regional environment.

In the design of several activities to promote interaction between the professor and his students, PGIT provides several synchronous and asynchronous technological tools that stimulate the learning process. Such activities are moderated with flexibility in time and space. Some of those interaction activities are supported on bulletin boards, electronic mail and conventional chat tools (Inter-Relay Chat, IRC).

In the design of interaction activities through the technological tools mentioned above the professor needs to offer a warm and effective feedback to massive online groups, as well as the need to exhibit live experiences in a practical way with clear intentions and fluid dialogue. Such affective interaction is required because of the distance involved over internet which is typically understood as cold and non-personal.

The need of a higher interaction rises with the constant demand of students to chat directly with their professors in a synchronous way. Students' feedback from PGIT showed the necessity to design new solutions to promote interaction.

The Project "Radio-Chat" and the Solution

PGIT started an evaluation process with audioconferencing to broadcast audio through the internet giving as a result a technological solution called "Radio-Chat". Radio-chat is an integral solution that emerges in the PGIT's academic program to support interaction processes and to promote the creation of learning objects for online courses.

In the design of the solution radio-chat integrates a view as a template. This template integrates the technological tool of a conventional chat-room that supports written messages and a front page identification of the particular online course as an image.

Fig. 1 shows a view of the solution from the student perspective having interaction through the internet. The view can be partitioned in two windows so we have in the superior panel the front page identification of the online course as an image and in the inferior window the conventional chat-room that offers a written messages tool for those participating in the radio-chat session.

This solution offers and promotes synchronous interaction between the student, the professor and other students through written messages. The professor as a moderator has the opportunity to respond with audio plus written messages. A student or international guest can also participate with oral exposition through a telephone call to the moderator station.

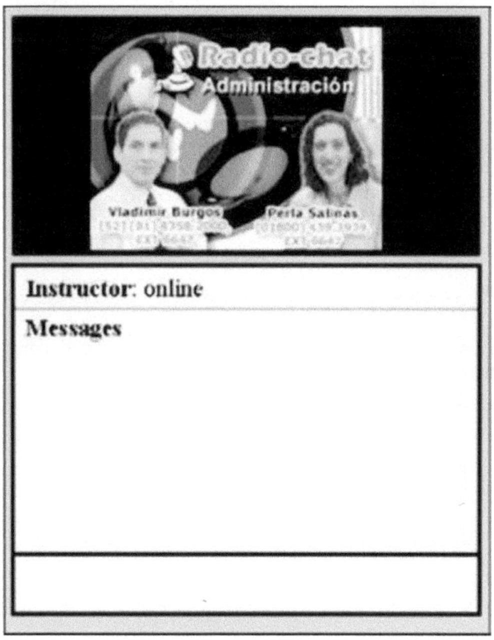

Fig. 1. Radio-chat template

Fig. 2 shows an image that represents the general schema of interaction offered by Radio-Chat solution.

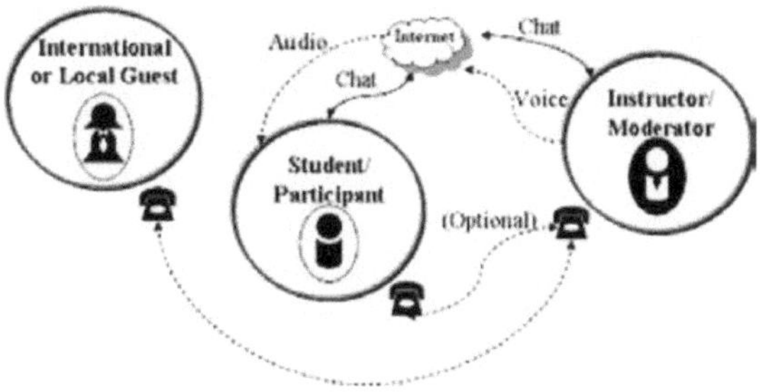

Fig. 2. Radio-chat schema approach

This solution matches the minimal technological requirements for Virtual University' students; for example, internet connection demanded by radio-chat is 28 Kbps versus 56 Kbps requested by the institution, web browser and audio applications, java plugging and others.

One of the essential technological benefits of the broadcast is the possibility to record the radio-chat session for asynchronous disposition later.

Since the instructor is the moderator, he can interact with the students and others guests such as international or local experts. The professor can interact via audio or written messages through the Internet session. The students can interact with classmates and the professor via written messages (chatting), listening at the same time the audio explanation of the professor and if needed they can participate via telephone. In the session other students and colleague professors can participate via audio making a phone call to the moderator station to express an opinion or to introduce a topic or question for discussion.

IV. Research Methodology

Some data is needed to explain the different scenarios designed for the use of radio-chat in distance education. Table I shows a summary of some of the characteristics that collected in the use of this technological solution. A case study research was also developed at PGIT and conclusions are explained with several recommendations in the adoption of radio-chat in some of the online courses offered through internet.

TABLE I
DATA FROM ONLINE COURSES (PGIT RECORDS)

Online Courses	Period of operation	Estimate of		
		Students per course	Sessions	Participants per session
5	Quarterly	120-200	3-4	35 %
4	Semester	150-300	6-8	27 %

Four categories are analyzed to show experiences that resulted in successful practices in the adoption of technologies in distance education [9].

Categories in the study
Use/ Application
Expectation
Impact
Acceptance

The sources of information were:
Direct observation in the operation of radio-chat
Interview with professors
Comments from students (chat sessions and electronic mails)

The case study was based essentially in nine cases (online courses) in the period of January 2003 through May 2004 [9].

V. FINDINGS

A. Use/ Application

One of the main challenges that PGIT faced was how to prepare and design radio-chat sessions where the professors could be able to develop good and warm interactions with high quality with their students. The interaction should not be too formal to inhibit a rich dialogue, but should be under control with a moderation technique to offer a good level of interaction during the session.

Different ideas, warnings and considerations were analyzed and with the experience of several members of the team, some good scenarios of interaction where finally achieved and adapted in the radio-chat sessions.

Each scenario is the result of the intention and purpose of the session during the online course operation, and what this means is that behind each interaction scenario there is a well-defined instructional objective that should be reached.

Six interaction scenarios were designed to promote interactivity with the professor as moderator and the students, but several more could be implemented if combined.

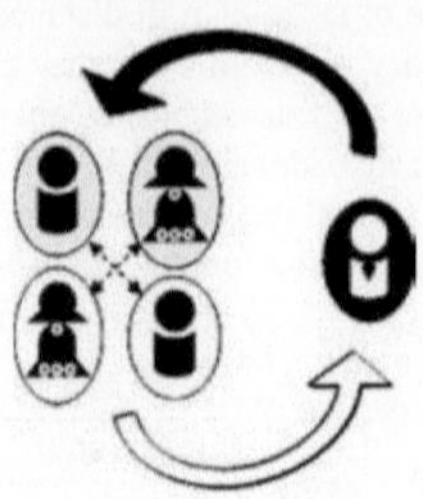

Fig. 3. Interaction process

Feedback support scenario: The professor prepares his class with the objective to give more explicit directions in specific learning activities. For example it can be used to support direct feedback for the consulting activities that students are performing in several companies in their own places; or after group discussions in bulletin boards; or to introduce additional topics to solve particular student's needs or finally to give recommendations and directions in how to focus their efforts in the learning process.

Team consulting scenario: This schema is designed as an interaction environment with small groups (teams) to orient them in specific details for the development of projects or business cases; flexible topics are introduced and discussed to link theory with practice.

Students' presentation scenario: The student does not have a passive role. The objective of this schema is to promote an active participation in the exposition and discussion of topics, results of learning activities or the sharing of professional experiences.

Interviewing scenario: The professor has the opportunity to invite experts in several fields of knowledge and professional practice. This provides a high flexibility in the introduction of relevant news, government announcements or industry agreements that have impact on the topics that have been prepared for the online course.

Experts' panel scenario: The professor has the possibility to have a debate to talk about a social, business, informatics or other dynamic topic. This scenario also offers the possibility to link the discussion activity with an asynchronous forum to get into a deeper discussion.

International guest scenario: With the internet solution provided, the window of opportunities is wide open. The professor has the feasible opportunity to have international experts in several fields of knowledge and industry to give a global opinion in some general ideas or particular application of business strategies, adoption of information technologies, etc.

The most well accepted scenarios in the study cases were the feedback support scenario, team consulting scenario and the interview scenario [9]. Fig. 4 shows an illustrative schema of radio-chat and the interaction scenarios.

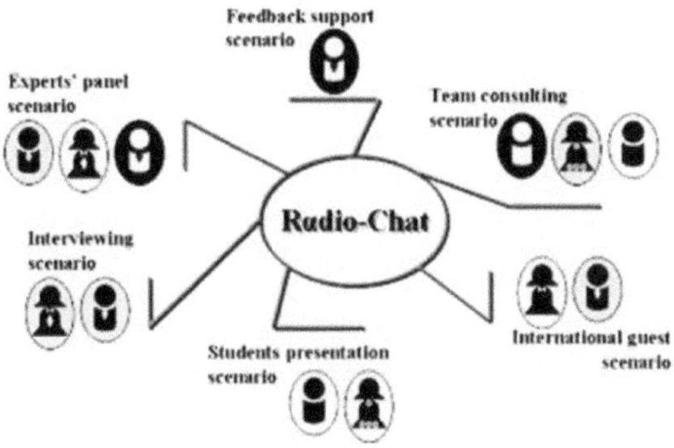

Fig. 4. Interaction scenarios in radio-chat

The professors expressed that the use of radio-chat is a supporting tool in the online courses that promotes an environment that reduces the distance perception and gives a warm feeling that promotes good relationships with their students. The radio-chat makes it easy for moderators to enable a fluid conversation with clear answers and with a direct and simple feedback.

Professors also agree that radio-chat allows them to give an effective feedback to large groups of students, giving them freedom to use their own style in motivating and guiding students through the learning process.

B. Expectations

We can split this category in two parts: in one side there are the professors' goals to provide a good, flexible and efficient chat through the internet considering that the purpose of the activity is to offer an interaction scenario. On the other side, we have the students' needs to have a virtual point of meeting with their professors and classmates. The main expectative of the students is to get additional information of the different topics that are being discussed in the online course with a realistic and practical explanation that helps them link their particular situation with theory, having the opportunity to complement experiences in a collaborative manner [9].

The final results of the case study research in PGIT [9] shows that students and professors expectations were reached in a satisfactory manner, and in some cases even higher than what they had expected initially.

C. Impact

Professors' opinions coincide in that when they were allowed to express their thinking through a microphone instead of a keyboard there was a huge difference in the flexibility to express their ideas and experiences in a fluid dialogue with their students. The online interaction in real time helped them to generate more ideas than if they were forced to think on complex examples planned months before [9].

The restriction of giving feedback in real time to massive groups (large groups) of students online with written messages practically disappears. Before having radio-chat there was a bottleneck on the moderator side when receiving messages concurrently in an exponential way depending on the number of participants in each session.

The students were amazed with the use of information technologies and the solution of radio-chat in their online courses; there was a positive feeling of breaking down the distance perception and isolation [9]. The possibility to consult the recorded session of the radio-chat online and to have the flexibility to hear it again and again in an asynchronous way, gave them support for further reference in their learning activities, including those students that could not be able to attend and participate during the session.

D. Acceptance

Acceptance is divided in two dimensions: the first one is the professors' convenience of using radio-chat as a tool to support an additional activity in their online courses with mayor interaction between their students. On the other side we have students participating online in real time in radio-chat sessions. The radio-chat session is an additional and optional activity in the course that does not affect the final grade and has to be perceived as a tool that is going to give them significant value in their studies.

The implementation of radio-chat in online courses at PGIT during the period of January 2003 tough May 2004 shows an increasing tendency in the number of sessions according to the records and the case study research [9]. In this sense we can see professors' acceptance of radio-chat to provide their students an additional guidance in the learning process.

On the other side, the average percentage of attendance of students in the sessions during the operation of the online courses that were reviewed shows a positive tendency too. The average percentages of attendance are related to the total number of students enrolled in each of the cases [9].

VI. CONCLUSION

The introduction of the concept "Radio-Chat" as a supporting tool in PGIT for online courses has a positive future and good results in its adoption. It has the potential to contribute in the generation of learning objects to increase the heap of the online digital video library of the Virtual University (UV-ITESM). There is a challenge and a tendency for the reuse of materials to help in the design of online courses to give the opportunity to reduce costs and optimize the life cycle preparation.

The design of interaction scenarios has helped in the documentation and classification of the characteristics needed for their reuse in the generation of learning objects as for example, information like subject and content. The designed scenarios offer critical metadata for instructional purposes like the main objective of the audio file to give context for its reuse.

Flexibility and spontaneous chat interaction has helped in the radio-chat sessions. For example, the agenda is designed a few days before the online session giving flexibility to the course for the introduction of recent news and complementary topics. On the other side the spontaneous chat and live interaction of the moderator and participants have helped to reduce the students' fear to participate with the group, reaching at the end a collaborative dialogue.

After the good results given in the implementation of Radio-Chat in the program, there is a positive sense that we could see some other interaction scenarios with the support of technology. We have experienced that technology stimulates and encourages creativity and innovation in distance education.

REFERENCES

[1] Porter, Michael E, "Strategy and the Internet", *Harvard Business Review*, March 2001, Vol. 79 Issue 3, p62, 17p, 1; Retrieved July 2, 2005, from EBSCO Business Source Premier at http://bibliotecauv.tecvirtual.com.mx/

[2] Taylor, James C & Swannell, Peter, "From Outback to Internet: Crackling Radio to Virtual Campus", *University of Southern Queensland*, Australia. 1997; retrieved July 2, 2005, from http://www.usq.edu.au/users/taylorj/readings/geneva.htm

[3] Jerome, Robert W, "The Four T's of Distance Education", *Virtual Forum 2003;* Monterrey NL, México; May 30 2003, Tecnológico de Monterrey' Virtual University; retrieved July 2, 2005, from http://www.ruv.itesm.mx/portal/infouv/eventos/virtualidad/homedoc.htm

[4] López del Puerto, Patricio, "The Virtual University", *Virtual Forum 2003*; Monterrey NL, México; May 29 2003, Tecnológico de Monterrey' Virtual University; retrieved July 2, 2005, from http://www.ruv.itesm.mx/portal/infouv/eventos/virtualidad/homedoc.htm

[5] Virtual University of Tecnológico de Monterrey, "Virtual University: Who we are: corporate information", Monterrey NL, México; retrieved July 2, 2005, from http://www.tecvirtual.itesm.mx

[6] Saucedo, Rocio (2003), "Virtual University Chronicle 1989-2003", *El Tintero*: Year 3, issue 10, October 2003; retrieved July 2, 2005, from http://eltintero.ruv.itesm.mx/

[7] UV-ITESM, "The distance education model of the Virtual University", *El Tintero*: Year 2, issue 8, February 2003; Retrieved July 2, 2005, from http://eltintero.ruv.itesm.mx/

[8] UV-ITESM, "Network of Learning Communal Centers (CCA)", Virtual University of Tecnológico de Monterrey, Retrieved July 2, 2005, from http://www.cca.org.mx/

[9] Burgos, Vladimir, "Radio-Chat: Successful and innovatory practices in the integration of information technologies in distance education, Case study research of the Graduate Program of Engineering and Technology (PGIT)", *El Tintero*: issue 14, February 2004; retrieved July 2, 2005, from http://eltintero.ruv.itesm.mx/

[10] Macmullen, Paul, "Audio/Audio conferencing in Support of Distance Education", *The Commonwealth of Learning*, 2001; retrieved July 2, 2005, from http://www.col.org/

[11] Chaudhary, Sohanvir S & Bansal, Kiron, "Interactive Radio Counseling in Indira Gandhi National Open University: A Study", *Journal of Distance Education/Revue de l'enseignement à distance* (2000). ISSN: 0830-0445. Retrieved July 2, 2005, from http://cade.athabascau.ca/vol15.2/chaudharyetal.html

***Jorge A. Ramirez** received the Ph.D degree in management from the Tecnológico de Monterrey and he is the Director of the Graduate Program of Engineering and Technology (PGIT). jorgeramirez@itesm.mx;*

***J. Vladimir Burgos** received the Master degree in information technologies administration from the Tecnológico de Monterrey and is a consultant professional. vburgos@itesm.mx*

Both authors are full time Professors in Management and Information Systems at the Tecnológico de Monterrey' Virtual University, http://www.tecvirtual.itesm.mx

Assessing senior engineering students with regard to radio communication principles

James Swart, Ruaan Schoeman & Henk De Jager
Vaal University of Technology

Abstract – Universities have long been recognized as institutions organized and incorporated for the purpose of imparting instruction, examining students, and otherwise promoting education in the higher branches such as literature and science, being empowered to confer degrees in the several arts and faculties present in the University [2]. Various theories and ideologies have been put forth over the ages as to how to accomplish this phenomenal task. One of the more recent theories adopted by the Department of Applied Electronics and Electronic Communication at the Vaal University of Technology involves using computer aided frequency spectrum analyzers to teach students radio communication principles. This paper will outline the methods employed by staff to assess senior students understanding of basic radio communication principles by utilizing electronic equipment. It will furthermore highlight the success and efficiency of the co-operative and electronic learning methods.

Index Terms – Facilitators, Instructional facility, Practical instruction, FM, Spectrum analyzer

I. INTRODUCTION

Radio communication was born around 1893 when Nikola Tesla made a demonstration relating to radio communication in St. Louis, Missouri [1]. Shortly afterwards in 1896, Guglielmo Marconi sent radio signals of 300 m on Salisbury Plain in England [1]. From these humble beginnings at the end of the 18th century sprung numerous designs and techniques with regard to radio communications that has led to a proliferation of radio communication devices in the 21st century. Despite the overwhelming abundance of these radio communication devices, the operating principles have remained the same. This paper will address the fundamental operating principles of Frequency Modulation that must be instilled into the minds of senior engineering students. It will outline the design and assessment methods employed to ascertain whether these students have achieved the desired outcomes. Teachers or lecturers will be referred to as facilitators, a classroom or laboratory will be termed an instructional facility and tasks or experiments will be grouped under practical instruction.

II. PRINCIPLES OF FREQUENCY MODULATION

Frequency Modulation, hereafter called FM, is any periodic signal, whose instantaneous frequency is deviated from an average value, called f_c, by an information signal, termed f_m. [4]. Figure 1 highlights an example of a FM signal shown on a spectrum analyzer where the deviation of the carrier is perceived on both sides of the center frequency. Figure 2 illustrates the same FM signal as viewed on an oscilloscope, where the deviation of the carrier is portrayed by the colored band.

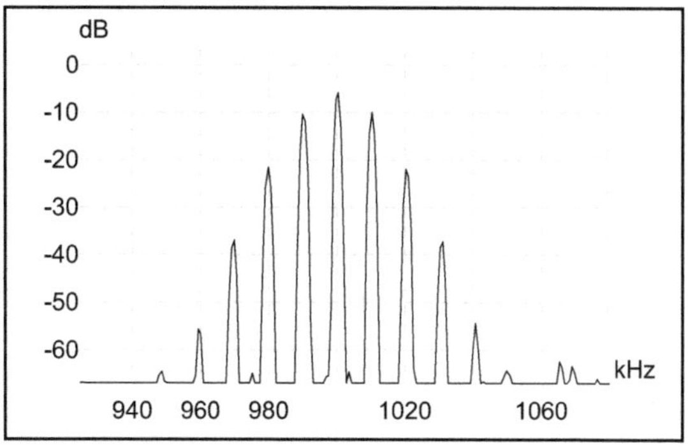

Figure 1. Spectrum analyzer view of a FM signal

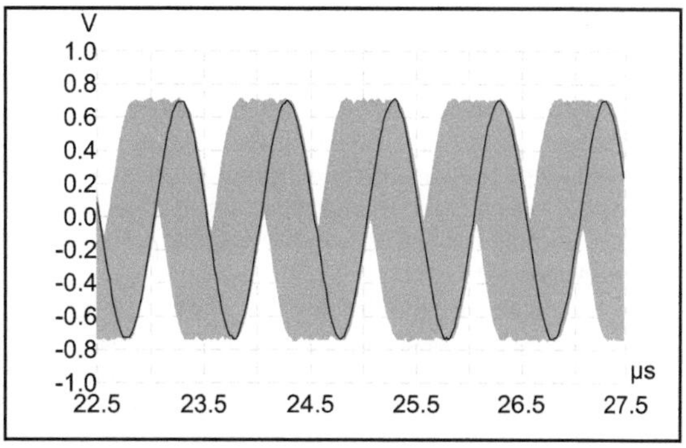

Figure 2. Oscilloscope view of a FM signal

The FM signal illustrated in figure 1 may be expressed as an equation in the form of [3] [4]

$$f_{sig} = f_c + \delta_f \times \sin \omega_m t \qquad \ldots (1)$$

Where f_c ≡ the unmodulated carrier signal in hertz
δ_f ≡ the peak frequency deviation in hertz
ω_m ≡ 2 times π times the modulating signal

The peak frequency deviation may be expressed as [3]

K. Elleithy et al. (eds.), Advances in Computer, Information, and Systems Sciences, and Engineering, 413–417.

$$\delta f = k_f \times E_m \qquad \ldots (2)$$

Where k_f ≡ the modulator sensitivity in hertz per volt
E_m ≡ the peak value of the modulating signal in volts

The foregoing equations suggest the following fundamental principles:

- If f_m increases f_{sig} increases;
- If E_m changes then the peak frequency deviation will also change leading to a new value for m_i.

A third fundamental principle that is unique to FM is that the total power always remains the same, with or without modulation [3]. The last fundamental principle states that the position of the modulating signal exists between the sidebands while the frequency deviation exists between the carrier and the last sideband [4]. How could these four fundamental principles be successfully conveyed to students?

III. INSTRUCTIONAL TECHNIQUES

The Department of Applied Electronics and Electronic Communication at the Vaal University of Technology has experimented with various instructional techniques over the years to try and assist students to understand the principles of radio communication. Teacher-centered instruction such as lecturing and questioning [5] were employed for many years that only succeeded in transferring knowledge from facilitators to students. This transfer or banking of knowledge though did not necessarily mean that students understood what the information meant [6]. Co-operative learning was thus employed where two students form a group assisting each other to complete a given practical instruction as part of independent practice. The approach of co-operative learning is based upon three principles [5]:

- Students construct their own knowledge with regard to the experiment at hand;
- Facilitators serve as a guide and collaborator in the forming of knowledge;
- The instructional facility becomes a learning environment that supports its members.

The approach of independent practice affords the facilitator the opportunity to wander about the instructional facility and monitor the progress of each group of students. This indicates to the students that the facilitator is aware of their behavior and places importance on the practical instruction at hand [5]. Assessment of the practical instruction would occur through either a written report or immediate oral assessment. The students make use of electronic equipment connected to relevant computer aided software to complete the practical instruction. They have to physically turn knobs and change settings and then observe corresponding measurement fluctuations. This utilization of electronic equipment supports the development of high-level thinking in the following two ways [5]:

- It provides students with opportunities to develop their problem-solving skills;
- It may serve as a tool for thinking and problem solving.

A further advantage of electronic equipment and especially computer aided equipment is that a flash disk may be connected to the computer and the relevant information saved to be reviewed at a later stage. The computer may also allow immediate printing of any figures or diagrams for later analysis. What factors though will determine what electronic equipment will be required?

IV. COMPILING THE PRACTICAL INSTRUCTION

The facilitator first determines what outcomes the student must achieve. For example in FM the students will be required to demonstrate an understanding of the four main principles upon which FM is based. The facilitator then selects the desired equipment, as shown in figure 3, which would be able to demonstrate these four principles when used in a specific configuration.

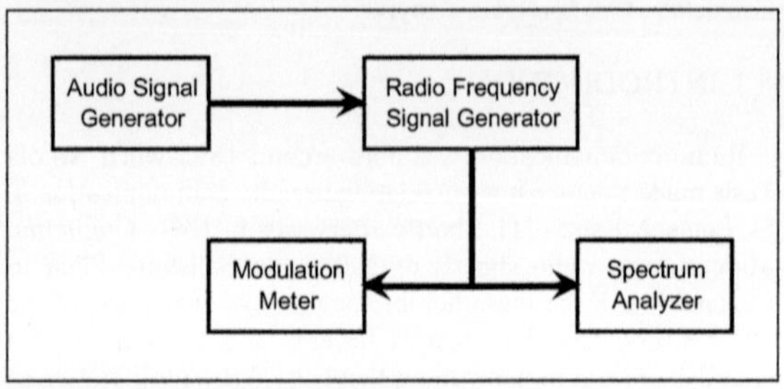

Figure 3. The equipment setup for the FM signal

Reviewing the four main principles of FM then requires electronic equipment capable of generating the carrier and modulating frequencies which are the Audio Signal Generator and Radio Frequency Generator respectively. To measure the frequency deviation and modulation index would require a Spectrum Analyzer and a Modulation Meter. The spectrum analyzer will be computer based due the advantages outlined above. Based upon this configuration, the facilitator then outlines a concise methodology in step form that the student will be required to follow. This is a form of controlled exercise which has the purpose of helping students develop fundamental skills and techniques that have to be mastered [7]. The methodology will primarily involve adjusting equipment settings and observing and measuring corresponding changes. These measurements must then be filled into the appropriate spaces provided on the methodology form. The student will then be required to interpret these changes and thereby prove that an understanding of the four main principles of FM has been achieved. However, before the methodology is given to the student the facilitator will complete the entire practical instruction to see if the results are in line with the desired outcomes. The student will be required to give a written report on the FM signal which will then be assessed.

IV. THE METHODOLOGY

Step 1: Identify the unmodulated carrier signal

An unmodulated carrier signal's frequency component, f_c, is manipulated by a modulating signal, f_m, to provide the FM frequency signal. The student must thus first see this unmodulated carrier signal before any modulating signal is considered. The student is thus required to turn the amplitude dial of the audio generator to 0%, thus eliminating the modulating signal. The radio frequency signal generator now only generates the carrier signal which is presented in figure 4.

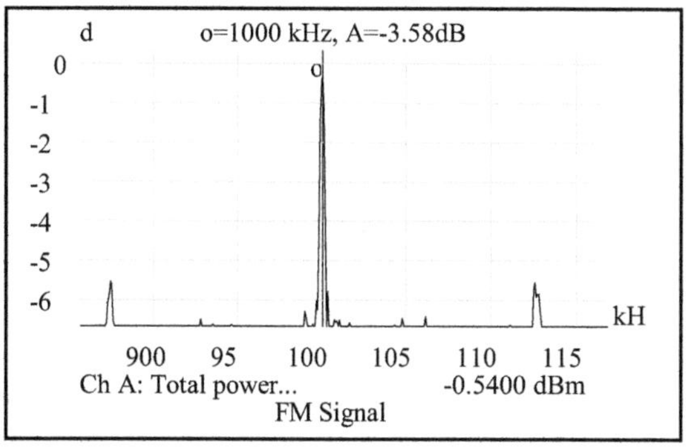

Figure 4. Spectrum analyzer view of an unmodulated carrier signal

The student needs to understand that the large signal in the middle of the screen represents the carrier signal. All other signals in the spectrum represent noise. The first step then in the methodology requires the student to make use of the marker O to determine the value of this carrier signal. If these aspects have been grasped by the student then the correct carrier signal will be found in the written report that is submitted the following week. The student will also include the total power reading in the report, which is -0.54 dBm in this case. dBm is just another way of representing power and is calculated with the following formulae [3] [4]:

$$dBm = 10 \times \log(\frac{P}{1 \times 10^{-3} W}) \quad \dots (3)$$

Step 2: Identify the modulating signal

The second step involves injecting a modulating signal to manipulate the carrier signal's frequency component. For this reason the radio frequency generator must be able to accept an external modulating signal and be capable of producing a frequency modulated signal. The Leader SG 120 has this capability and is utilized in the equipment setup. The modulating signal is supplied by an HC 0204 audio oscillator and has the initial value of 10 kHz. The student now turns the amplitude dial of the audio generator to 50% of its maximum output. The resulting FM signal is illustrated in figure 5.

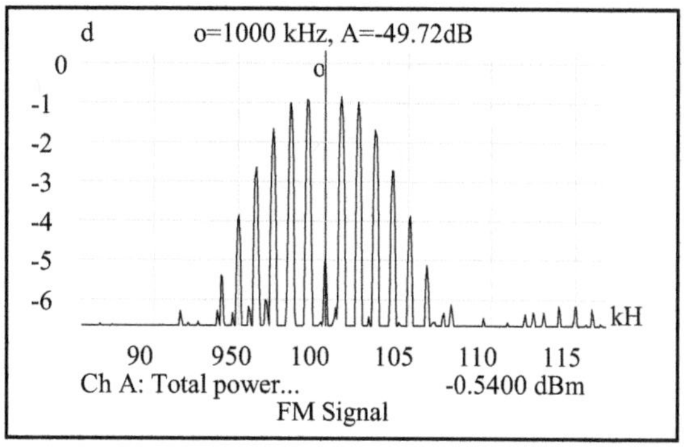

Figure 5. A FM signal with a 50 % amplitude modulating signal

The student needs to take note of the following factors

- Sidebands have been added to the carrier signal indicating modulation has taken place;
- The frequency difference between the sidebands indicates the value of the modulating signal;
- The frequency difference between the carrier signal and the last sideband indicates the value of frequency deviation;
- The total spectrum power at the bottom of the screen.

The student also has to calculate the modulation index of the FM signal at this point. The amplitude dial of the audio signal generator is now turned fully clockwise to 100% amplitude output. An FM signal with a 100% amplitude modulating signal now appears as shown in figure 6. The student must now take note of the same factors as listed above.

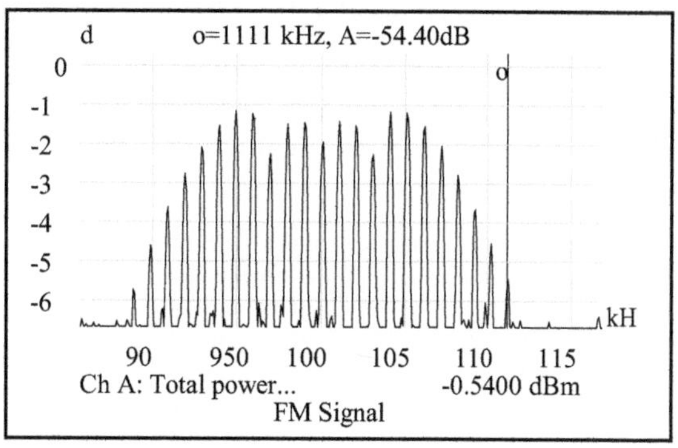

Figure 6. A FM signal with a 100 % amplitude modulating signal

How is the student now assessed, with regard to the factors above? Recall that independent practice is employed in the instructional facility and that the facilitator is free to wander around and monitor the progress of the students. Thus the facilitator glances over the methodology form and especially looks at the obtained measurements in the provided spaces. If these measurements are incorrect, then the facilitator points it out to the students who must then repeat the previous steps. This form of evaluation is termed formative assessment which occurs during the practical instruction, being given formally or informally [8]. A hint or comment is provided informally to assist the student reason upon the faulty measurement and determine where the error has been made. This process continuous until the student has obtained the correct value. The student thus retraces the steps taken and determines himself where the error has been made. This reinforces upon students' minds the basic operating principles of FM. A printout of the spectrum analyzer may be made which is attached to the written report. The student may also save the data to a flash stick to be reviewed at a later stage. This is then where the actual interpretations of the measurements occur, where the important understanding of the FM principles is achieved.

IV. THE FINAL ASSESSMENT

The student must now complete the written report and submit it the following week for final assessment, or for summative assessment. Summative assessment is usually conducted at the end of a lesson or unit and is the final measure of what students have achieved [8]. The final heading in the written report is the findings and conclusions of the practical instruction. Certain facilitators at the Vaal University of Technology have over the past years realized that students struggle to write appropriate conclusions. Students often just repeat the measurements they have taken or repeat what they have done. Few really interpret or understand their measurement results. The facilitator of the radio communication principles thus chose to include some questions in the conclusion that students need to answer based upon their obtained results. These questions serve as hints to help the students interpret or understand their findings. Three questions are posed in the conclusion:

- What happens to the carrier signal at 0, 50 and 100% amplitude?
- What gives rise to a change in modulation index?
- What power in a FM signal remains constant, the carrier or total power?

The student now compares figures 4, 5 and 6 to see the behavior of the carrier signal. The student must discern that the carrier signal's frequency component has not changed while the amplitude has varied. Thus the carrier signal's power component has changed. Next the student compares the number of sidebands in figures 5 and 6 to discern that the higher the output amplitude of the audio signal generator, the higher the modulation index. This modulation index variation has been caused by a variation in the amplitude of the modulation signal, E_m. Finally the student must observe the total power readings of all three figures to understand that the total power in a FM signal remains constant, despite the addition of a modulating signal. The student is now encouraged to make a fourth independent conclusion for which no question is posed. The audio signal generator, representing the modulating signal, has a digital readout giving the value of 10 kHz. The student who correlates this 10 kHz to the frequency difference between the sidebands correctly understands that the modulating signal is found between the sidebands and that it influences to a degree the amount of frequency deviation.

V. RESULTS OF THE CO-OPERATIVE AND ELECTRONIC LEARNING METHODS

What do the results reveal as to the successfulness and of this approach? As mentioned earlier, teacher-centered instruction was employed in the past to try and assist engineering students to better understand the basic principles of radio communication, and in particular FM. FM was primarily discussed in the course Radio Engineering 3 that dealt with various forms of modulation, demodulation and transmission of radio frequency signals. The success of this method of instruction was primarily measured by how many students successfully completed the particular course [9]. Figure 7 illustrates the pass rates of engineering students in Radio Engineering 3 when the teacher-centered method of instruction was utilized.

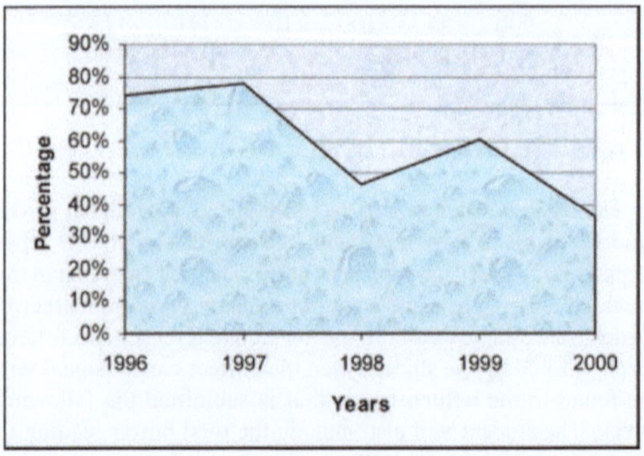

Figure 7. Pass rates of engineering students when the teacher-centered method of instruction was utilized

Concerns that the education system cannot adequately prepare learners for life and work in the 21st century, have prompted people across the world to explore new ways of designing education. Possibly the most significant educational trend operating in the world today is that of Outcomes Based Education [10]. This concern is highlighted in figure 7 where the teacher-centered method of instruction gave rise to a dramatic change in the overall pass rates of Radio Engineering

3 students. The initial success in the years 1996 and 1997 slumped to a pass rate of 46% in 1998 and then further to 36% in the year 2000. Changes had to be made as the engineering students were not benefiting from the teacher-centered method of instruction. Basic radio communication principles in the form of FM were not successfully being conveyed to senior engineering students. A paradigm shift thus occurred where co-operative learning was introduced into the laboratory were students made use of electronic measuring equipment and computer aided software. Figure 8 shows the pass rates of engineering students who were exposed to the co-operative and electronic learning method within the laboratory. Note the lowest pass rate, that of 63% for the year 2004. The highest pass rate, of 84%, has been in 2005. It must be noted that FM comprises only one of the five sections presented in Radio Engineering 3. The other four sections where also converted to the co-operative and electronic learning methods and are thus also represented in figures 7 and 8.

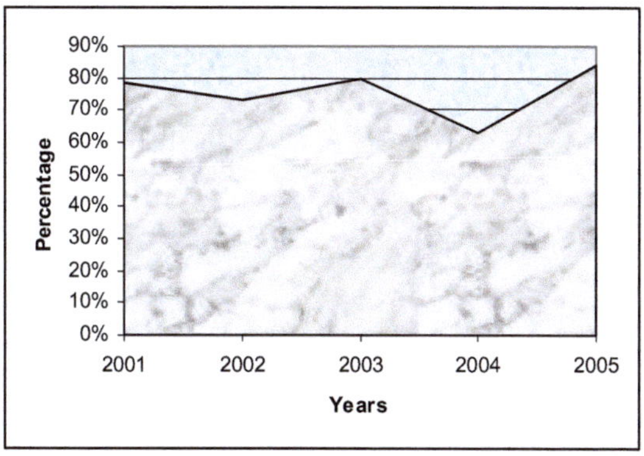

Figure 8. Pass rates of engineering students who were exposed to the co-operative learning method that utilized electronic equipment in the laboratory

VI. CONCLUSIONS

The figures illustrated in this paper may be found in real-life situations where students need to solve radio communication problems based upon relevant data. Authentic assessment occurs when students make use of higher-order thinking skills within real-life contexts to analyze, synthesize and evaluate relevant data [5]. Students analyze or identify the various elements within the FM signal (modulating signal and frequency deviation). Secondly, students synthesize or put some of these elements together to better understand a relevant concept (amplitude of the modulating signal affects the modulation index). Thirdly, students evaluate or judge the efficiency of an FM signal (the total power remains the same with or without modulation). These concepts of analysis, synthesis and evaluation form part of the higher levels of Bloom's taxonomy [5], and are successfully achieved by senior engineering students at the Vaal University of Technology. Criterion-referenced judgments are used in the final assessment to indicate whether the student has achievement the desired outcomes. Both qualitative and quantitative information is presented by the students to show an understanding of FM principles. "Tell Me, I Forget. Show Me, I Remember. Involve Me, I Understand" is a Chinese proverb highlighting the value of involving students in the learning process by means of electronic equipment to better understand basic radio communications principles.

Finally, the results prove beyond doubt that co-operative and electronic learning methods are better suited to the engineering field of radio communications than teacher-centered instruction. The pass rates during the past five years in which this method of instruction was utilized never fell below 63% as compared to the teacher-centered instructional year of 2000 where it dwindled to a mere 36%. Senior engineering students have been able to grasp basic radio communication principles and demonstrate an understanding of them. Higher quality students now exit the university who are better equipped to function within the engineering field of radio communications.

VII. ACHNOWLEDGEMENTS

James Swart would like to thank Prof. H.J. De Jager for his ongoing support and contribution to the research and development of engineering education within the Faculty of Engineering and Technology at the Vaal University of Technology.

VIII. REFERENCES

[1] WIKIPEDIA – THE FREE ENCYCLOPEDIA. 2005. [Online] Available at: <http://en.wikipedia.org/wiki/> Accessed on 5 September 2005.

[2] WEBSTER – DICTIONARY. 2005. *University*. [Online] Available at: <http://www.webster-dictionary.org/definition/university> Accessed on 11 April 2005.

[3] BLAKE, R. 2002. Electronic.Communication Systems. 2nd Edition. New York: Delmar Thomson Learning.

[4] YOUNG, P.H. 1999. Electronic Communication Techniques. 4th Edition. New Jersey: Prentice-Hall.

[5] McCOWN, R, DRISCOLL, M & ROOP, P.G. 1996. Educational Psychology A learning-centered approach to classroom practice. 2nd Edition. Massachusetts: Allyn & Bacon.

[6] LASHLEY, C & WARWICK, B. 2003. 12 Steps to Study Success. London: Continuum.

[7] NEWBLE, D & CANNON, R. 1995. A Handbook for Teachers in Universities and Colleges. A guide to improving teaching methods. 3rd Edition. London: Kogan Page.

[8] VAN DER HORST, H & McDONALD, R. 1997. Outcomes-Based Education. A teacher's manual. Pretoria: Kagiso Publishers.

[9] ELLINGTON, H, PERCIVAL, F & RACE, P. 1995. Handbook of Educational Technology. 3rd Edition. London: Kogan Page Limited.

[10] BOUGHEY, C. 2000. Lecturing. Makoni, S. (ed). 2000, In *Improving teaching and learning in Higher Education: A handbook for South Africa*. Ed. by S Makoni. Johannesburg: Witwatersrand University Press in association with HERDSA.

Technology Student Attitudes Regarding Privacy Scenarios

Jack Freund
Adjunct Faculty
DeVry University
jfreund@devrycols.edu

***Abstract*-This study looks at students' aptitudes for the new role of privacy guardian in the company's IT organizations. The study made use of survey questions designed to gauge how sensitive they were to choosing privacy-neutral questions. Academic tenure, gender, industry experience, and academic major are shown to influence student's privacy choices.**

I. INTRODUCTION

Individuals value their privacy. Consumers are constantly surprised by the extents to which companies will go to in order to extract our personal information. These companies are not unfeeling monoliths—they are comprised of individuals who also are concerned about their own privacy. Their specializations have allowed them to focus with tunnel-vision accuracy on how to achieve the results they need to in order to succeed in their chosen profession. These individuals affect the nature of privacy in their organizations.

For information professionals, their charge is to be the custodians of an organization's data storage and transmission. Their ability to control the technology related to this is paramount in their training, and qualifications for these careers. However, as corporate issues with privacy are becoming increasingly prominent, information technologists are being identified as the professionals responsible for privacy. The ability for individuals to control their privacy in the United States and abroad is based on the capacity of technology professionals to act ethically in dubious scenarios. Privacy rests in the hands of the network administrator, database administrator, network engineer, programmer, and IT manager. Indeed, more responsibility for privacy is being included in IT professionals' job descriptions. Whether voluntary or involuntary, IT professionals will be in control of scenarios where they must consider privacy issues of individuals. This work refers to that role as the privacy guardian.

The problem this research aimed to study was how much aptitude and sensitivity technology students have for this new role of privacy guardian. It is viewed through the lens of ethics, as privacy is a core consideration in many areas of ethical study, and included in the Institute of Electrical and Electronics Engineers (IEEE)/Association of Computing Machinery (ACM) ethical guidelines portion of the Computing Curricula [1]. Specifically, the goals of the present study were to 1) Evaluate students' aptitude and sensitivity for privacy issues they will face in their careers, and 2) Identify any categorical variables that might make a student more or less sensitive to privacy issues such as gender, major, or years of industry and educational experience. This problem is specifically identified by Harris in his 2000 study where he indicates: "It is evident...that there is a continuing need to increase students' awareness to the ethical dilemmas they may face during the tenure of their careers" [2].

Privacy aptitude of students is a relatively new problem. Privacy in general is currently studied as a subset of security, and rarely as a discipline in and of itself. This is further compounded by privacy consideration being subject to ethical considerations and research. As electronic information privacy begins to take its place at the forefront of concern for the general populace, researchers will also begin to take a greater look at the problems that information technology in general is having on privacy considerations. Further, what makes aptitude difficult to ascertain is its nebulous nature. There are many factors that affect the overall aptitude of an entire discipline, not the least of which is the institution one studied at, as well as the professors who gave this instruction. This innate quality of professional aptitude, and privacy aptitude in specific, lends itself to a significant amount of research efforts to identify factors in various environments, with focus on any number of different experimental variables.

Much research has intended to identify shortfalls in student learning and education. For technology professionals, and engineering students in particular, there is always efforts to include classes from different disciplines to offer students a more well-rounded education [1]. This is primarily driven from external forces of the market for graduates of a particular degree programs, and industry, trade, and professional societies. These groups identify that there is a deficiency in graduates and serve as advocates for educational institutions to bridge the gap. Often, there is a push from employers to have universities adjust their programs to alter student's educational outcomes. This study has a similar aim. There is overwhelming evidence that students are being placed in a new role where privacy is becoming their charge—whether explicitly or implicitly. The security industry has advocated that information professionals need to stand up and take charge of data and network security for those areas that do not currently have sufficient coverage and leadership. It is foreseeable that the same call to action will be trumpeted to information professionals about privacy. Understanding the present ability of students to take up the reins of protecting customer privacy, can help us to design programs and instructional techniques to arm the next generation of students

K. Elleithy et al. (eds.), Advances in Computer, Information, and Systems Sciences, and Engineering, 419–426.

with the tools and reasoning ability they need to properly control privacy.

The paper begins with a review of the literature involving educational aims for ethical education. Additionally, this section reviews the methodologies for developing measurement instruments for ethical scenarios. Next, the methodology that was employed for measuring student's attitudes is enumerated. Lastly, there is a review of the results and recommendations for future research.

II. LITERATURE REVIEW

Teaching Privacy and Ethics for Computer Professionals

Computer ethics as a field of educational endeavor has arisen from the need for technologists to consider the effects their work has on the morals of society. It is unreasonable for one to segment their work from its effects. Professional organizations such as the Association of Information Technology Professionals (AITP), ACM, and IEEE have developed codes of ethics as a way for its members to adhere to notions of right and wrong and maintain a moralistic stature for the industry in general. The ACM and the IEEE have developed guidelines for ethics education in the form of their Computing Curricula. This curricula recommendation includes technical competencies as well as professional practice education (such as management, ethics, values, and communicating). The recommendation is to have ethical consideration included as a part of the "computer and ethics" course. Many discussions have ensued regarding which university department is more appropriate to give this instruction: computer science, or ethics [3, 4]. The debate has left universities with pros and cons on both sides, and mostly lament the lack of more courses in the areas of ethics to allow for instruction by professionals from both disciplines. All agree that it is important to take a course in ethics to help develop a moral grounding in the discipline's affect on society. A detailed narrative describing the addition of an ethics course into a business/technology masters program is described in [5]. Further, preparing ethics courses for introduction in an Information Systems/Computer Information Systems curriculum is enumerated by [6]. Several areas identified for instruction include methods of ethical reasoning, intellectual property rights, computer crimes, privacy issues, and professional responsibility. Primary methods for measuring a student's progress and outcomes in ethics are by classroom discussions, case study projects, and a final paper.

Instruction of ethics itself focuses on several different areas for building the foundation of proper ethical behavior. Bardon approached the topic by indicating the different roles that the Internet plays on society, such as social, economic, and cultural [7]. Through these roles, students explore the focus of how this new medium affects notions such as privacy, property, and justice. The primary thrust of Bardon is to focus on the contextual meanings of traditional concepts like privacy as affected by the novel technology that is the Internet.

In Bodi's article, she illustrates that the role of information guardian has changed, and that moral decision making should be guided by a set of principles that uphold Kant's categorical imperative—humans should never be treated merely as a means, but always as an end [8]. Information was at one time held primarily in libraries. A larger library meant that more information was available. It was here that seemed a natural place to put computing technology, as it allowed for the development of information as the domain of libraries. As the Internet evolved, so to did the role of the librarian. Instead of focusing on the role of custodian of information, they have been forced to adapt the role of steward of ethical use of information technology found within the library. In many cases, such use of information technology is the result of abuse by those that mean to violate privacy.

In the classroom, privacy is more than just the ability to control information about oneself. Employers also have an expectation of privacy with regard to their intellectual property. Jennings details a situation where students may inadvertently disclose an employer's information [9]. The primary thrust of this article is that there is an important distinction that students do not check the corporate responsibility and obligations to safeguard employer's privacy during discussions in the educational environment.

Rader talks about the ethical challenges that schools face with increased adoption of Internet usage [10]. Top amongst these issues is the need to preserve the privacy of records on the network. For instance, with students having access to the network, there is an ability for students to access electronic student records or purchasing systems. Anonymity on the Internet is an important factor to consider. The lack of accountability is difficult for students to handle, and may encourage deviant behavior. Rader offers some suggestions for overcoming this potential downfalls. Amongst these is the need for ethical and moral instruction to help extend ethical considerations to Internet usage.

Learning about privacy and ethics in computer science requires a foundation in ethical theory. Staehr articulates this by discussing the ethical instruction in computer science from an Australian perspective [11]. Staehr discusses several methodologies to facilitate instruction. These include using videos that provide hypothetical scenarios. Also recommended is working on ethical case studies, and discussing the issues and cases in class in a formal or informal debate.

References [3, 4] offer a different perspective on the question of who should teach computer ethics. Instead of siding with either social scientists or computer scientists, he offers the perspective that there should be a joint effort in educating students. Using the model developed by [12], Tavani applies the principles to the notion of teaching ethics. Of particular importance in the analysis is that traditional computer ethics focuses on current ethical dilemmas. By narrowing the focus to what is currently an issue, it limits the ability to extrapolate the ethics lessens to other scenarios. Ethical dilemmas which have not yet come to pass may be nonnegotiable by students without a general education in ethics. However, at the same time, it is important for those who give instruction to be familiar with the issues concerning computer technology. Understanding the technologies, sometimes intimately, is just as important as having an intimate knowledge of ethics theory.

Lastly, Harris's presentation of his results after studying ethical attitudes amongst students serves as a foundation for the present work [2]. This study is designed to ascertain student's responses to ethical scenarios. The survey instrument utilized was designed to query several elements. The questions covered several areas of computer science, thus allowing for questions dealing with many areas in which ethical scenarios will arise. The questions are worded in such as way that places the student in the question. This first person wording is important to determine whether it gives the students a more personal approach to the scenario. Harris designed his survey with both personal and impersonal wording with the hope that he may be able to determine if there were any differences between responses to the same question worded in first and third person. The responses are categorized by the respondent's gender, and academic level. The results showed that there is higher sensitivity to these ethical issues amongst students of higher levels of academic exposure. Secondly, females in general have a higher sensitivity across all areas of ethical dilemmas. Finally, the research showed that there was no statistical difference between questions worded in the first or third person.

Harris's work is substantial, in that it represents a broad review of ethical scenarios. Harris indicates some areas for research include sensitivity to ethics in general, and differences between men and women in particular. While Harris's efforts worked to show sensitivity to ethics in general, the present work will focus intently on privacy in information systems.

On Ascertaining Attitudinal Orientations about Privacy

This section describes some of the literature relevant to ascertaining individuals attitudes regarding privacy in varying scenarios. Of particular interest and relevance to the current study, is the ability of researchers to capture attitudes using survey instruments. The current survey uses questions modeled after several of the questionnaires indicated here.

Harris's work has sufficiently built a questionnaire that allows the attitudinal positions of students to be accurately measured. Harris cites [13-19] as examples where survey instruments have been sufficiently utilized to measure respondents attitudes regarding ethical scenarios and situations. Harris's survey was designed to elicit responses in the form of a 5-point Likert scale that measures the respondents response in terms of ethical, acceptable, questionable, or unethical. Harris measures the responses against his hypotheses of sensitivity in ethics due to academic progression, gender, and first or third person wording in the questionnaire.

In Hiltz's study regarding privacy with specific interests in a national ID (NID) card, they used a random dialing function to interview individuals over the phone to ascertain attitudinal positions [20]. Using a semi-structured interview format, they attempted to quantify what was currently known by the respondent, then gave a standard explanation of what the NID card is and then asked open ended questions about how they felt about the prospect of implementing such a measure. They were specifically asked if they would accept such a law requiring such cards, how they felt about the cards overall, and perceived disadvantages/advantages. They were given a list of options and asked to choose what should be on a card like this, and when it should be required. Lastly, they were asked to give their opinion about an implanted chip option.

Reference [21] has conducted an interesting piece on the attitudes of workers related to their expectations of privacy. Specifically, the research questions focus on the opinions they have regarding their own levels of email privacy. Their research questions reflect the nature of their respondents, in that they cover two nationalities, and the research seeks to see if there is any differences between the two in terms of how they view email privacy. Further, they propose that one's perception about email privacy is germane to how they use email in the workplace. The questionnaire used posed many question relevant to this, and is populated with multiple choice and Likert-scale questions.

Several older studies reviewed have attempted to ascertain ethical scenarios in general. In Katz, the researchers look at privacy as a concern amongst their respondents [22]. The results indicate that respondents are quite concerned about the notion of privacy in abstract, although they have limited exposure to problems that relate tot his. The authors show that is empirical data showing an increase in the number of privacy-related incidents. The research is in intended as a follow-up to a 1988 study to see if there has been any changes in attitudes. The questionnaire questions are designed using multiple-choice and Likert scales.

An AT&T Labs study was conducted to measure internet user's attitudes regarding privacy in the online context [23]. The research questions here are designed to garner how users feel about feature of web systems like persistent identifiers, and automatic data transfers. Also measured are opinions of browser features and unsolicited communications. One particularly interesting measure was how users felt about the relative privacy of certain types of data. In short, not all private data was viewed as equally important. Once again, a survey instrument was utilized to research the user's attitudes. Specifically, the questions used multiple choice and Likert-type questions to survey the users.

In reference [24], the author develops a survey methodology for ascertaining attitudes regarding the privacy of several topics such as sexual orientation, drug use and testing, and reproductive issues. Further, the survey looks into opinions of how employers can use such data. The questions are designed to elicit responses to multiple choice styled or Likert-scale questions.

Culnan's study looks to understand attitudes about secondary use of information [25]. The use of information for primary uses (such as eliciting one's name and address to send them a magazine) is fairly commonly understood by many. What is disconcerting for many is the ability for an organization to use this information (name address, and magazine category—fishing for example) and sell it to other organizations. This composition of identifying information, and personal likes can be highly valuable to another company selling products in that area. Sending product offers to people who have already indicating they like this particular are is a leg up in the direct marketing business. Culnan attempts to measure attitudes about such uses in her study. The research question driving the study is whether there is any factors shared by individuals who are information sensitive and those that are not. The survey questions are designed to determine whether the respondent feels the use of personal information is a violation of privacy and therefore viewed negatively, or positively.

Culnan has also prepared a piece for the U.S. Department of Commerce regarding a methodology to assess self-regulation regarding consumer privacy [26]. In this she enumerates the survey work she conducted to ascertain attitudinal positions of individuals. Included is a comparison of survey results from an earlier survey. These question are also multiple choice and Likert-scale questions.

III. APPROACH

In order to ascertain the attitudes of students regarding privacy scenarios, a survey instrument was used. The use of surveys to gauge attitudinal positions is supported in the works review above. The scope of respondents selected for this research is the study body of a technology-oriented University campus in Ohio. The students in this study were attending classes in one of several computing, electronics, and business technology related degree programs. This campus is unique, in that a good deal of the students are non-traditional students. This allowed the author to ask research questions to test the student's aptitude based on industry experience.

In order to gauge the sensitivity to students to these ethical, privacy-laden scenarios, a survey instrument was utilized. The survey was designed to present the student with scenarios that place them in the first person at the helm of the decision making process as in [27]. The scenarios are designed to ask the students what they would do given a particular problem. Unlike the survey presented in the Harris study where the student rated the scenario in terms of how much they felt it violated privacy, the present survey offered multiple choice answers with actual actions the student could take. The question foils are delineated into severe, moderate, and low privacy risk answers. The questions ask what the students should do, given the information presented.

The research questions driving the analysis are as follows:

1. Do the students' aptitude for choosing a privacy-sensitive response change with their academic tenure? The hypothesis tested was:

$H_{0(1)}$: $\mu_{1(1)} = \mu_{2(1)} = \mu_{3(1)} = \mu_{4(1)} = \mu_{5(1)} = \mu_{6(1)}$ vs. $H_{a(1)}$:

The means tested was
$\mu_{1(1)}$ = Not in College
$\mu_{2(1)}$ = Freshman
$\mu_{3(1)}$ = Sophomore
$\mu_{4(1)}$ = Junior
$\mu_{5(1)}$ = Senior
$\mu_{6(1)}$ = Graduate Level

2. Do the student's aptitude for choosing a privacy-sensitive response change based on their gender? The hypothesis tested was:

$H_{0(2)}$: $\mu_{1(2)} = \mu_{2(2)}$ vs. $H_{a(2)}$:

The means tested was
$\mu_{1(2)}$ = Male
$\mu_{2(2)}$ = Female

3. Do the student's attitude for choosing a privacy-sensitive response change based on their years of industry experience? The hypothesis tested was:

$H_{0(3)}$: $\mu_{1(3)} = \mu_{2(3)} = \mu_{3(3)} = \mu_{4(3)} = \mu_{5(3)} = \mu_{6(3)} = \mu_{7(3)}$ vs. $H_{a(3)}$:

The means tested was
$\mu_{1(3)}$ = Less than 1
$\mu_{2(3)}$ = 1
$\mu_{3(3)}$ = 2
$\mu_{4(3)}$ = 3
$\mu_{5(3)}$ = 4
$\mu_{6(3)}$ = 5
$\mu_{7(3)}$ = more than 6

4. Does the student's choice of major affect their aptitude for choosing privacy-sensitive responses? The hypothesis tested was:

$H_{0(4)}$: $\mu_{1(4)} = \mu_{2(4)} = \mu_{3(4)} = \mu_{4(4)} = \mu_{5(4)} = \mu_{6(4)}$ vs. $H_{a(4)}$:

The means tested was
$\mu_{1(4)}$ = Business/Management
$\mu_{2(4)}$ = Biomedical
$\mu_{3(4)}$ = Information Systems Related
$\mu_{4(4)}$ = Electronics and Computer Engineering
$\mu_{5(4)}$ = Graduate Studies
$\mu_{6(4)}$ = Other

The survey instrument was designed with a series of questions that were superfluous to the testing of the hypothesis. This was included to keep the respondent's from trying to deduce which questions were significant. The significant questions fell into two categories. First, were the categorical questions. These categorized the respondents based on age, income level, gender, ethnic background, and course of study in school as well as tenure. The next set of questions are the scenario-based questions that form the basis for this study's analysis. These are based on scenario projects identified by the University of Illinois at Springfield DOLCE (Developing Online Computer Ethics) project [28-31]. These questions are listed here.

LOAN_ZIP
You have advocated and started the use of data mining at a major credit card firm where you work. You soon discover a correlation between customer loan defaults and 25 zip codes. What should you do with this information?

- You should check to see if you can find a correlation between other zip codes and customers who faithfully pay off their loans.
- You should go no further. In fact, you have already gone too far since, through this data mining, you have violated the right to privacy of these individuals.
- You should test these correlations further. Then, if they seem to hold, you should go to your supervisor and recommend not giving loans to people in these zip codes.

EMAIL
You are a supervisor. While reviewing the emails of your employees, you discover that one of them is using the system to operate a weekly football betting pool. What should you do?

- Nothing. You shouldn't have been reviewing emails anyway.
- Fire the employee. He or she is robbing the company of time and money.
- Make it clear to the employee that this is not acceptable. Outline a series of punitive measures that you will take should this activity continue.

MONITOR
You are the supervisor of a medium sized business. You suspect that the employees under your management are goofing off, but you have no proof. Software exists that can monitor employees and measure their productivity. What should you do?

- Purchase the software and comprehensively monitor every action of your employees.
- Send a memo to your employees threatening to monitor their activities if they don't get more productive.
- Purchase the software but only monitor employees whom you suspect are goofing off.
- Implement other measures, short of computer monitoring, to measure the productivity of your employees.

UNI_USE
You are the network administrator for a University. It has been brought to your attention that a student is using the University's Internet access, along with the student's personal computer to post content on a questionable web site she owns, hosted by another Internet Service Provider. All content in question on the student's personal computer is encrypted. What should you do?

- Confiscate the student's computer and retrieve the encrypted files to verify the accusations
- Suspend the student for violating the University's terms of service
- Speak to the student about his activities

CHIP_IMPLANT
A computer chip manufacturer from Florida has announced that a microchip that can be implemented under the skin is ready to be marketed. The company spokesman said that it has conducted extensive tests and has concluded that the chip poses absolutely no threat to human health. The chip is easy to implant and easy to remove. The CEO of the Company, Ms. Doe, told reporters that the company is very proud of this breakthrough achievement and its potential contributions in the areas of security and human health to mention a few. You are the CIO of a major airport. The CEO has asked you to look into implementing these chips for all the operating and security personnel. By implanting these chips, the CEO thinks it will greatly enhance the security in airports. He said, "After 9/11 we all could use more security. This chip will make events such as 9/11 extremely difficult if not impossible in the future." What should you do?

- Prepare a plan to implement the new security chips. The more security the better.
- Tell the CEO that a step such as this should be approached cautiously. Convene a study to look into this more.
- Reject this idea. It is way too invasive.

SPAM
You have invented a really good software product but need a way to market it. Today, you received an e-mail message offering you a CD containing 12 million e-mail addresses plus professional e-mailing software. For the reasonable price of $249 for the CD and for the e-mailing software, you can mass market this product. What should you do?

- Order the mass marketing software. Hopefully someone will order your software.
- Hold off on ordering the software. If you can't get orders any other way, you can use it then.
- Do not ever order the software. Find different ways to market the software

IV. RESULTS

The survey was advertised to students in a campus-wide email to the approximately 2000 registered students; 94 respondents have taken the survey for a response rate of 4.7%. This number is slightly misleading, as it assumes that every student is checking their email and reading the memorandum emails. A cultural issue at the school exists where there is a real problem getting many students to periodically check their email, if at all.

Demographic Data

The dataset collected indicated that the majority of the respondents were Caucasian males, between the ages of 18 and 25 making less than $25,000 a year. Table I shows the distribution of respondent's ages. A majority of respondents were between the ages of 18 and 30, with nearly a quarter of respondents over the age of 30. This is consistent with the view of the University as a non-traditional school, catering to many adults with evening and weekend programs. Table II indicates the income levels of the respondents. The majority of the respondents indicate they had an income level of less than $25,000 a year, with nearly 82% saying they made less than $35,000 a year. Table III shows that nearly 75% of the respondents were male and 25% were female. Lastly, Table IV indicates a 90% response rate by Caucasians, and approximately 10% response rate by minority students.

TABLE I
Respondents' Age

	Frequency	Percent	Cumulative Percent
Under 18	1	1.1	1.1
18 – 25	54	57.4	58.5
26 – 30	16	17.0	75.5
31 – 35	9	9.6	85.1
36 – 40	3	3.2	88.3
Over 40	11	11.7	100.0
Total	94	100.0	

TABLE II
Respondents' Income Level

	Frequency	Percent	Cumulative Percent
Less than $25,000	64	68.1	68.1
$25,001 - $35,000	13	13.8	81.9
$35,001 - $45,000	7	7.4	89.4
$45,001 - $55,000	4	4.3	93.6
Over $55,000	6	6.4	100.0
Total	94	100.0	

Table III
Respondents' Gender

	Frequency	Percent	Cumulative Percent
Male	71	75.5	75.5
Female	23.	24.5	100.0
Total	94	100.0	

TABLE IV
Respondents' Ethnic Background

	Frequency	Percent	Cumulative Percent
African American	6	6.4	6.4
Asian/Pacific Islander	1	1.1	7.4
Caucasian	85	90.4	97.9
Native American	2	2.1	100.0
Total	94	100.0	

Overall Question Analysis

Overall, the responses to the scenario questions supported the observation that students were sensitive to how their actions would affect the information privacy of the subjects in the scenarios. In each of the scenarios, the choices that were considered to be more privacy considerate, were chosen almost overwhelmingly, with half of the questions seeing a choice that was privacy neutral. A privacy neutral response here is considered to be one that has the actor in the scenario stop and consider their actions rather than glibly choosing to either violate privacy, or never implement the feature described. Scenarios where the neutral response was not chosen include the ZIP_LOAN, MONITOR, and SPAM.

For the ZIP_LOAN question, it is interesting to see that 61% of the respondent's indicated that data mining in general seems to violate privacy. This is in stark contrast to the rampant use of data mining today to glean as much competitive advantage as possible from an organizations data. The MONITOR question saw students choose an unnamed methodology for monitoring employee productivity instead of choosing computer-based monitoring 48% of the time. Lastly, for the SPAM question, students chose 71% of the time not to ever order the mass-marketing software. It would seem that spam is an issue that most everybody finds distasteful, and in this scenario, even if it means you may not make sales.

Tenure Sensitivity

The first hypothesis tested was whether there is significance for those based on their progression through school, also called their academic tenure. There is enough evidence to support rejecting the null hypothesis for all but one of the scenario questions (SPAM at the < 0.05 significance level). Two other questions came close (EMAIL, and UNI_USE at 0.062 and 0.083 significance respectively). Taken on a whole, there is sensitivity in this study based on academic progression. Progression through the academic levels cannot be ruled out as a factor contributing to sensitivity to privacy scenarios.

Gender Sensitivity

The second hypotheses tested whether there was any sensitivity based on the respondent's gender. There is enough evidence to support rejecting the null hypothesis for all but one of the questions (ZIP_LOAN at < 0.05 significance level). One other question (UNI_USE) was significant at 0.067. What is interesting about this question is that it is the only one that uses a pronoun in its wording. "She" is the subject of the question, and the respondent is asked to deal with her violations, which are described only as uploading "questionable" content that is encrypted. Female respondents to the question overwhelmingly (95% or 22 out of 23) chose giving the subject a talking to, as opposed to harsher penalties. Male respondents, on the other hand chose harsher penalties 21% of the time. The use of the pronoun "she" may have inadvertently affected the outcome of the responses. Gender cannot be ruled out as a factor effecting the sensitivity of respondent's to the privacy scenarios.

Academic Major Sensitivity

The third hypothesis intended to test whether there is sensitivity for students based on their academic major. There is support to reject the null hypothesis in all but one case (the SPAM question at < 0.05 sensitivity level). It would seem that one cannot rule out major as a factor in the respondent's answers.

Industry Experience and Sensitivity

The final hypothesis was that industry experience would contribute to the respondent's answers. There is support to reject the null hypothesis in all cases, however the UNI_USE question is significant at 0.06. The respondent's experience in the industry also cannot be ruled out as contributing factors to their answers.

V. CONCLUSIONS

This research study attempted to resolve four research questions, namely, whether academic tenure, gender, industry experience, or academic major affected one's sensitivity to privacy issues one may come across in the workplace. Overall, the results show that the respondent's surveyed indicated a high level of sensitivity in general, choosing in most cases the privacy neutral, or the privacy positive response. These choices showed a positive or neutral regard to consumer privacy—the choice of further considering choices before taking action. Further analysis shows that for most questions, there is sensitivity based on the four conditions of the hypotheses.

The results are encouraging, but hardly definitive, that students have the ability to successfully control privacy for organizations. Specifically, the goal is to have student's make choice about information privacy that protect and defend consumer privacy. This was not the case for many of the respondents. This is not to say that the students will not later accumulate the skills necessary to make considerate privacy choices. However, the results have not identified any of the factors tested as having a singular affect on privacy considerations. Indeed, all of them would appear to have an affect.

Further study is needed to fully explore this topic. The sample size in this study is significant, however, still lacking when compared to contemporary studies (such as Harris's). A larger sample size will help to create better data. As identified above, the wording of the UNI_USE question inadvertently used a feminine pronoun. This may have unintentionally affected the choices of some of the respondents. Lastly it may be beneficial to extend the survey to help show a larger delineation between the four hypothesis factors tested. Questions that are more detailed or more questions based on scenarios could help shed light on this.

The future of information privacy rests in the hands of those that are taking up the reins of the information technology profession today. This role will increasingly have privacy for consumer data foisted upon it. It is therefore reasonable to endeavor research that shows educators and employers how prepared new graduates are for this new role. Once practicing technologists, educators, and future employers understand the factors that can make an employee a more responsible social citizen, the foundation can be laid to help ensure they greatest good, for the greatest number of people.

REFERENCES

[1] IEEE Computer Society / Association for Computing Machinery, "Computing Curricula 2001," 2001.

[2] L. Harris, "IS ethical attitudes among college students: a comparative study," *Information Systems Education Conference (ISECON 2000)*, 2000.

[3] H. T. Tavani, "Applying an interdisciplinary approach to teaching computer ethics," *IEEE Technology and Society Magazine*, vol. 21, pp. 32 - 38, 2002.

[4] H. T. Tavani, "Curriculum issues and controversies in computer ethics instruction," *International Symposium on Technology and Society (ISTAS'01)*, 2001.

[5] R. F. Bellaver and J. Gentry, "Teaching business/IT ethics," *International Symposium on Technology and Society (ISTAS'01)*, 2001.

[6] Dudley-Sponaugle and D. Lidtke, "Preparing to teach ethics in a computer science curriculum," *International Symposium on Technology and Society (ISTAS'02)*, 2002.

[7] Bardon, "A conceptual framework for teaching Internet ethics," presented at *International Symposium on Technology and Society (ISTAS'01)*, 2001.

[8] S. Bodi, "Ethics and information technology: some principles to guide students," *The Journal of Academic Leadership*, vol. 24, pp. 459 - 463, 1998.

[9] S. Jennings, "Employed students: ethical and legal issues in the technical communication classroom," *IEEE Transactions on Professional Communication*, vol. 43, pp. 368 - 385, 2000.

[10] M. H. Rader, "Strategies for teaching internet ethics," *Delta Pi Epsilon Journal*, vol. 44, pp. 73 79, 2002.

[11] L. J. Staehr, "Helping computing students develop a personal ethical framework," *IEEE Technology and Society Magazine*, vol. 21, pp. 13 - 20, 2002.

[12] P. Brey, "The politics of computer systems and the ethics of design," *Conference on Computer Ethics: Philosophical Enquiry (CEPE'97)*, 1997.

[13] E. Cohen and L. Cornwell, "A question of ethics: developing information systems ethics," *Journal of Business Ethics*, pp. 431-437, 1989.

[14] J. D. Cougar, "Preparing IS students to deal with ethical issues," *MIS Quarterly*, vol. 13, pp. 210-216, 1989.

[15] K. A. Forcht, *Assessing the Ethical Standards and Policies in Computer-Based Environments*, Boston, MA: Boyd & Fraser Publishing Company, 1992.

[16] K.-A. Kievit, "Information systems majors/non-majors and computer ethics," *Journal of Computer Information Systems*, pp. 43-49, 1991.

[17] D. B. Paradice, "Ethical attitudes of entry-level MIS personnel," *Information and Management*, vol. 18, pp. 143-151, 1990.

[18] D. B. Parker, *Ethical Conflicts in Computer Science Technology*, Montvale, New Jersey: AFIPS Press, 1980.

[19] W. A. Wood, "Computer ethics and years of computer use," *Journal of Computer Information Systems*, pp. 23-27, 1993.

[20] S. R. Hiltz, H.-J. Han, and V. Briller, "Public attitudes towards a national identity "smart card:" privacy and security concerns," *36th Hawaii International Conference on System Sciences (HICSS'03)*, 2003.
[21] R. Agarwal and F. Rodhain, "Mine or ours: email privacy expectations, employee attitudes, and perceived work environment characteristics," *35th Hawaii International Conference on System Sciences (HICSS'02)*, 2002.
[22] J. E. Katz and A. R. Tassone, "A report: public opinion trends: privacy and information technology," *The Public Opinion Quarterly*, vol. 54, pp. 125-143, 1990.
[23] L. F. Cranor, J. Reagle, and M. S. Ackerman, "Beyond concern: understanding net users' attitudes about online privacy," *AT&T Labs - Research Technical Report TR 99.4.3*, 1999.
[24] Cantril, Albert H. and S. D. Cantril, "Live and let live: american public opinion about privacy at home and at work," *American Civil Liberties Union Foundation*, 1994.
[25] M. J. Culnan, " "How did they get my name?" an exploratory investigation of consumer attitudes toward secondary information use," *MIS Quarterly*, vol. 17, pp. 341-363, 1993.
[26] M. J. Culnan, *A Methodology to Assess the Implementation of the Elements of Effective Self-Regulation for Protection of Privacy*, Washington, D.C.: National Telecommunications Information Agency, U.S. Department of Commerce, 1998.
[27] R. O. Mason, "Four ethical issues of the information age," *MIS Quarterly*, vol. 10, pp. 4-12, 1986.
[28] S. Alpert, "Privacy case study question," *University of Illinois at Springfield DOLCE Project*, 2005.
[29] M. Frank, "Marketing e-mail," *University of Illinois at Springfield DOLCE Project*, 2005.
[30] Mahadev, "Innovative new micro chip ready for marketing," *University of Illinois at Springfield DOLCE Project*, 2002.
[31] Frey, "Grey matters in computer ethics," *University of Illinois at Springfield DOLCE Project*, 2005.

The ITESM* Redesigned Model. Outcomes at Campus Estado de Mexico: Engineering and Architecture Division

Aurora González, Blanca Garza, Rubén D. Santiago
*Instituto Tecnológico y de Estudios Superiores de Monterrey
Campus Estado de México

***Abstract*- Every ten years the Instituto Tecnológico y de Estudios Superiores (ITESM) establishes a new mission. The Mission towards 2005 included, among others, the re-engineering of the learning process. To accomplish it, as well as to incorporate some attitudes, abilities and values that the students have to acquire or develop, an Educational Model was declared. This article report is about how this process was implemented at the Engineering and Architecture Division (DIA) of Campus Estado de México (CEM), and the outcomes after creating an Educational Committee in February 2002.**

I. The ITESM Educational Model

A. Theoretical Framework

The design of computer-assisted courses has a long history, since the early attempts in 1950's in Harvard until the use of high technology resources in web-based courses nowadays [1]. During these 50 years, three learning theories have been the theoretical frame of reference for course design. Those theories are Cognitive, Behaviorism and Constructivism. Cognitive Theory has been used to create activities that promote analogical reasoning, solution of algorithmic problems, and schematic organization [2]. Behaviorist Theory has been used to design activities where an "input-output" or "stimulus-response" is expected [3]. Constructivism has been used to create activities where the new knowledge is "built" from the interaction of new situations with the previous knowledge of the student [4]. The course design at Instituto Tecnológico y de Estudios Superiores de Monterrey (ITESM) is a hybrid model. Constructivist activities are the most important, however Cognitive and Behaviorist activities are also used [5].

The design methodology of the ITESM Educational Model (MET) is based in "Theory One" [6], which establishes that to promote learning, a course must include clear information, a final reflection on the process, and strong motivation. The activities are designed to focus on real-life applications. In order to achieve these goals, four teaching techniques are used in ITESM: Collaborative Learning (CL), Problem Based Learning (PBL), Project Oriented Learning (POL) and Case Study Method (CSM)[7]. As pointed out by Vygotsky, learning is a social activity, therefore our courses are designed in order to develop and take advantage of social skills. Nowadays the World Wide Web is a social space where interactions among people are taking place [8]. The courses at ITESM are designed and redesigned to take advantage of this social space and the new possibilities for interaction that it generates [9].

As for the Engineering and Architecture Division (DIA) of Campus Estado de México (CEM), the process caters to the concern of the Mexican engineering schools: to strengthen their programs and curricula considering the advances in sciences, informatics and technology, and to develop students' learning abilities [10].

B. The Training Program

In moving from a traditional model to one in which the students play a more active role [11], the ITESM intends to promote individual and group learning, more efficient use of technology, self-evaluation and the four main teaching techniques mentioned. The courses have been supported by a web-based application –at the beginning it was *Learning Space*, then *Blackboard* and *Webtec*.

This process has been a challenge for the professors. They have had to participate in a training program called *Programa de Desarrollo de Habilidades Docentes* (PDHD- Program for the

[1] Saettler, 1990.
[2] Bednar et al., as cited on Anglin, 1995.
[3] Ertmer and Newby, 1993.
[4] Schwier, 1998 as cited on Mergel, 1998.
[5] Schuman, 1996.
[6] Perkins, 1986; 1992.
[7] ITESM, 1999.
[8] Álvarez, 1999; Crosetti, 2000.
[9] ITESM, 1995; Blázquez,1994.
[10] ANFEI, 2002.
[11] Martín, 2002.

K. Elleithy et al. (eds.), Advances in Computer, Information, and Systems Sciences, and Engineering, 427–431.

Development of Teaching Abilities). This program consists of four stages:

Stage 1. Mobilization towards Change. The objective of this stage is to sensitize and introduce the professor in the Institute's educational model as well as in its Mission.

Stage 2. Teaching Techniques. At this stage professors have to take courses in technology, the structure and design of a course and at least a workshop of Collaborative Learning. Also, each professor can take another teaching technique, especially those who are using it at his/her courses.

Stage 3. Guided Practice. Professors participate in a guided practice during the implementation of the redesigned course. They have to document their results and they are encouraged to continue doing so.

Stage 4. Evaluation and Improvement of the Course. After concluding stage 3, professors use their results and observations to propose improvements to the course they are teaching. In doing this, they participate in the updating of the courses, which are validated every two years.

For the purpose of redesigning a course after the first two stages, we consider these main characteristics: "the behavioral objectives movement; the teaching machine phase; the programmed instruction movement; individualized instructional approaches, computer-assisted learning and the systems approach to instruction" [12].

Fig. 1 presents the relation between professor and student. It includes not only the training program –PDHD- already mentioned, but also other knowledge the professor must have to redesign his/her courses, such as creation of teaching material, evaluation systems, use of the Digital Library, among others, that the Institute provides him/her.

II. THE EDUCATIONAL COMMITTEE

The redesign process began in 1997, two years after the declaration statement of the Institute's Mission. At Campus Estado de México the process was implemented, since the beginning, in a massive way. At that time it was done on paper, and one year later, 1998, the courses were placed in a web-based application, *Learning Space*.

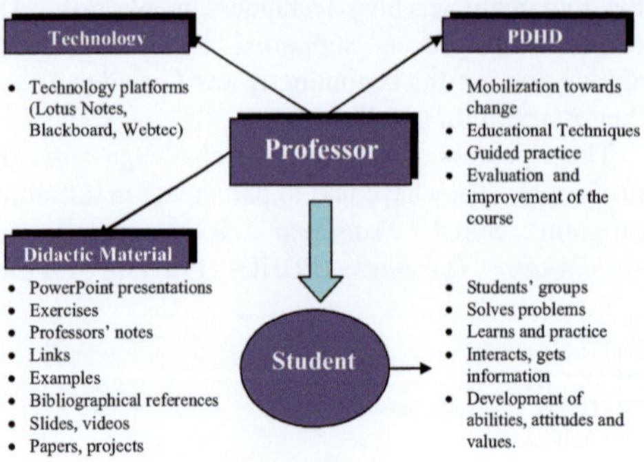

Fig. 1. Professor and student roles

[12] Saettler, 1990 as cited on Mergel, 1998.

At CEM, in February 2002 an Educational Committee was created to assist the faculty staff to develop and implement their courses in a more effective way on a new application, Blackboard, through a workshop, as shown in Fig. 2.

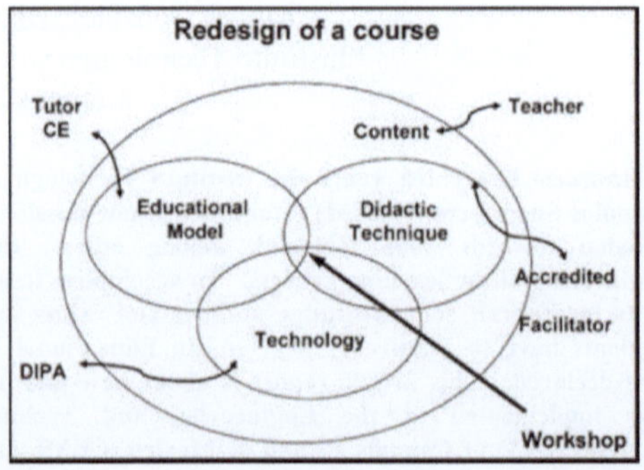

Fig. 2. Elements and relationships of the Educational Committee.

At Fig. 2 we have the different actors: teacher, tutor, teaching technique accredited facilitator and technology facilitator (under the name of the *Dirección de Innovación Para la Academia* –DIPA). The professor is responsible for the content and teaching material of the course; the tutor advises him/her to reach the goals of the MET and provides guidelines him/her about the instructional design; the accredited facilitator is a professor trained on a specific teaching technique at a foreign university and gives support in two ways: s/he teaches the training courses on her/his specific technique and aids in the implementation of the technique in the specific course. Since August 2003 the role of tutor and accredited facilitator has been under by one person.

We have to emphasize two more things: the characteristics of the tutor and the structure of the courses.

A. The Tutor

Each one of the academic staff of the Committee is called a tutor. S/he is a professor that: a) has at least one of his/hers courses redesigned and approved by the Academic Vice-presidency of the Institute; b) has knowledge and/or experience in instructional design; c) has the training in one of the teaching techniques mentioned; d) has completed the PDHD program, and e) collaborates in other academic activities at the Campus in order to contribute in the development of the MET among full and half time professors.

Some of the tutors have a high command of English so they can tutor the process of the courses that are redesigned in English.

B. Structure of a Course

The professor decides how s/he is going to structure his /her course. The DIPA provides the technology and software necessary so the professor can create his/her own web-based course, in which s/he has to incorporate two types of documents: the Information Documents and the Learning Units, whose contents are shown in Table 1.

TABLE 1.
INFORMATION DOCUMENTS AND LEARNING UNITS

Information Documents	Learning Units
Introduction Educative Intentions Objectives Content Teaching Strategy Evaluation System What is Expected of Teacher and Student Information Sources Technology Sources	Title Introduction to the Unit Objectives Unit Activities Teaching Material Related with the Course Content Adopting Teacher's Guide Teaching Documentation

Among the decisions the professor has to make are: define all of the above documents; the number of learning units of the course (from 4 to 16 if s/he decides to structure it by weeks); the activities and experiences in which the student has to participate to obtain the learning goals of the unit/course as well as the requirements; which activities will be held in the classroom and which will be done on-line; number and frequency of collaborative activities and/or using another teaching technique; s/he establishes the evaluation criteria and elaborates the instrument to register them; suggests the information sources as well as the support material, and determines the course policies according to the attitudes and values of the ITESM Mission. All these elements must be congruent with the MET.

An important issue is to create a user friendly course so the student can go through it in an easy way.

III. ACHIEVEMENTS OF THE IMPLEMENTATION OF THE MET AT THE DIA

In this section we present the progress of the implementation of the redesigned courses at the DIA, based on the number of redesigned courses, number of professors that have finished the PDHD, number of groups taught with the MET.

In the last four years, the number of courses approved by de Academic Vice-presidency have increased considerably. The approval is done when the course incorporates all the MET elements.

Table 2 shows the advances, by educational program, of the number of courses and groups that are being taught under the MET, at the DIA, during the semester August-December 2005.

TABLE 2.
COURSES AND GROUPS WITH THE EDUCATION MODEL

Program[a]	Number of courses	Number of courses under the MET	Total groups taught August-December 2005	Total Groups with the MET August-December 2005
Mechanical Electrics Engineering (IME)	58	42	268	236
Architecture (ARQ)	53	35	70	46
Mechanical Business Engineering (IMA)	55	42	261	236
Electronic Systems Engineering (ISE)	55	48	234	212
Computer Sciences Engineering (ISC)	55	44	193	169
Mecatronics Engineering (IMT)	61	47	303	274
Industrial Engineering (IIS)	56	44	226	209
Electronics and Communications Engineering (IEC)	56	52	237	216
B.S. Computer System for Management (LSCA)[b]	55	41	176	158

a The name of the programs is in English but the acronyms are in Spanish.

b Even if this program is a B.S. it belongs to the DIA.

Fig. 3 shows the comparative graphic, by academic program, of the progress of the MET at the DIA.

Fig. 4 shows the percentage of courses redesigned under the different teaching techniques. It is interesting to note that 6% of them have no teaching technique (from the specific four declared by the ITESM), and few of them have incorporated the Case Study Method.

As for the progress in the PDHD, until May 2005, there were 156 professors of the DIA that had concluded it, while 119 were in process of concluding it. Fig. 5 shows the graphic of the percentage of advance of the professors of the DIA by academic department.

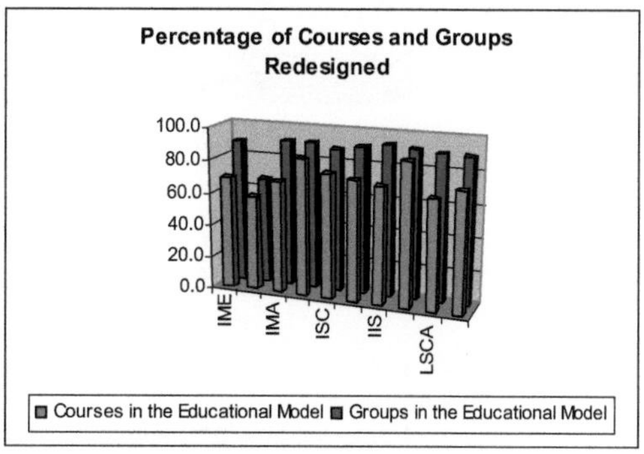

Fig. 3 Redesigned courses and groups at the MET

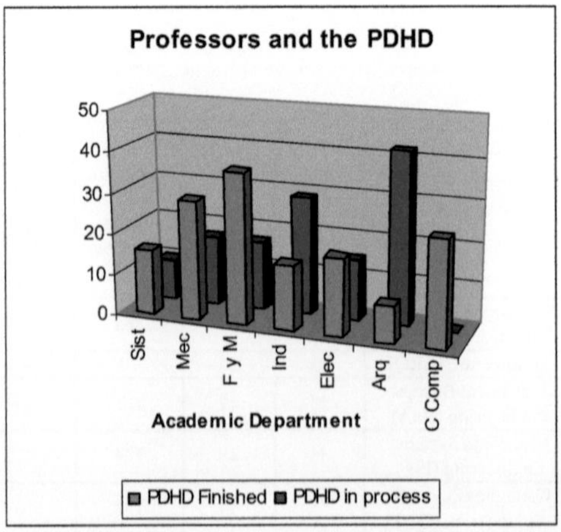

Fig. 4. Percentage of courses with a specific teaching technique.

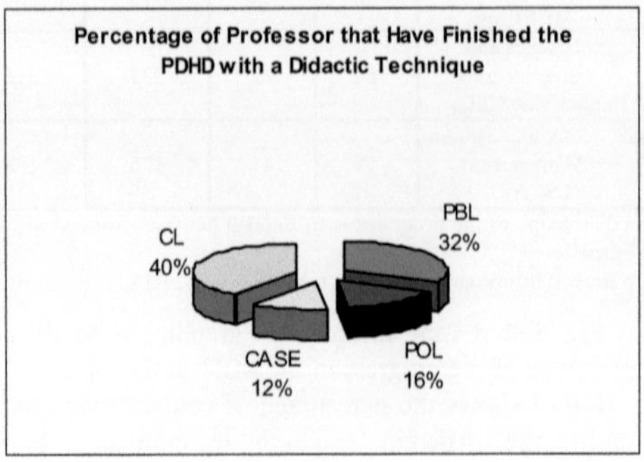

Fig. 5. Professors that have concluded the PDHD at the DIA, by academic department.

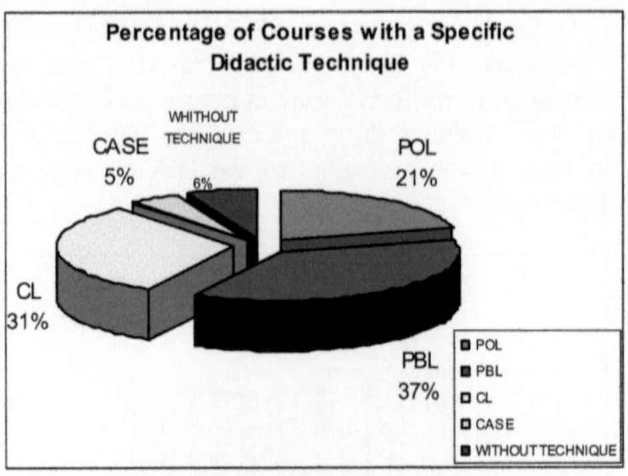

Fig. 6. Percentage of professors that have concluded the PDHD with a teaching technique.

Fig. 6 shows the percentage of trained professors in the different teaching techniques.

The conclusion from the graphics above is that the program has had relative success considering that there is a 5% annual rotation rate; that makes the Institution to permanently reconsider the progress. It is also interesting to notice that a great number of professors get trained in Collaborative Learning and Problem Based Learning.

IV. Outcomes

The conclusion of the process presented can be into divided two aspects: students opinion about their experience with these kinds of courses and progress of the implementation, based on the data of the former section.

A. Students' Opinions

At the end of the semester January-May 2005, 156 students answered a survey about their perception of the implementation of ITESM Education Model in the classroom. The aspects considered were: the Model, teaching techniques, evaluation of the course, professor's performance, technologic application and the course quality.

The results were:

1.ITESM Educational Model. Over 50% of the students considered that the course fulfills the characteristics of the Model. Nevertheless, 97% point out that the course allows the students to develop the ability of identify and solve problems, and 94% thinks that they are developing the ability of self-learning.

2.Teaching Technique. Students recognized that, with the incorporation of a teaching technique in, the course enriches their learning because they develop high level intellectual abilities; even if they say that it still needs to help them develop the efficient use of informatics, creativity and a good oral and written communication.

3.Technologic Application. Their opinion about the technologic application is that they can get the readings and course materials. But they said that is not so easy to participate in debates, discussions and other kinds of interaction with their classmates and teacher.

4.Evaluation. Students agreed that for the course evaluation they were taken into account their class participation, homework, research and the activities from the application as well as those concerned with the teaching technique. They also considered that their professors chose appropriate activities to the course content. Finally, about course quality they said it is good enough but not excellent.

5.Teacher's performance. Over 90% of the students pointed out that the teacher selected the appropriate course material according to the course content, the activities were well planned, the information was new and that s/he was open to student's suggestions.

6.About the Course. The generalized opinion was that the courses had an adequate quantity of work, that is, they to invest from 3 to 5 extra hours per week per course to the classroom, to conclude their assignments. Finally, they established 45% of the time of the course was lecturing, 27% collaborative work and 26% individual work.

In general, we can conclude that the results are good but there is still much more work, not saying that we have already the Mission towards 2015 and we would have to incorporate new elements to our courses related with ethics and citizenship.

B. Conclusions on the Progress of the Process

The creation of an Educational Committee led Campus Estado de México to produce a great number of redesigned courses. The support given by the Committee was both in the tutor's participation as instructors of the PDHD as well as in the process already described (Fig. 2).

The next step will be to incorporate new elements to the courses according to the Mission towards 2015, recently declared, that considers all the aspects of the former.

ACKNOWLEDGMENT

The authors want to acknowledge Noemí Romo and Juan Carlos Barrios for all their support in sharing with us the required data to build the tables and graphics of this document.

REFERENCES

[1] Saettler, P., *The evolution of American educational technology*. Englewood, CO: Libraries Unlimited, Inc., 1990.

[2] Bednar, A.K., Cunningham, D., Duffy, T.M., Perry, J.P., "Theory into practice: How do we link?" in G.J. Anglin (Ed.), *Instructional technology: Past, present and future*. 2nd ed. Englewood, CO: Libraries Unlimited, Inc., 1995, pp. 100-111.

[3] Ertmer, P. A., Newby, T. J., "Behaviorism, cognitivism, constructivism: Comparing critical features from an instructional design perspective," *Performance Improvement Quarterly,* 6 (4), 1993, pp. 50-70.

[4] Schwier, P., cited on Mergel, Brenda. *Instructional Design & Learning Theory*. 1998. [On-line]. Avalilable: http://www.educadis.uson.mx/pagina/ftp/Dise%C3%B1o-Instruc-RPA-B-Mergel-2.doc (September 26, 2005)

[5] Schuman, L., *Perspectives on instruction*. 1996. [On-line]. Available: http://edweb.sdsu.edu/courses/edtec540/Perspectives/Perspectives.html (September 23, 2005)

[6] Perkins, D. N., *Knowledge as design*. Hillsdale, N.J: Lawrence Erlbaum Associates, 1986.
__________ *Smart Schools: Better learning and thinking for every child*. New York: Free Press, 1992.

[7] ITESM, *El modelo educativo del Tec de Monterrey*. 1999. [On-line]. Available: http://www.sistema.itesm.mx/va/dide/modelo/home.htm (September 23, 2005).

[8] Álvarez, A y Del Río, P., *Educación y Desarrollo*. España: Editor Palacios Jesús, 1999.
Crosetti, Bárbara de Benito, *Herramientas para la creación, distribución y gestión de cursos a través de internet*. España: Edutec, 2000.

[9] ITESM, *Tecnología para la educación*. México: ITESM, 1995.
Blázquez, F. et al., *Nuevas tecnologías de la información y comunicación para la educación*. España: Afair, 1994.

[10] ANFEI, *Conclusiones de la XXIX reunión de la ANFEI*. México: ANFEI, 2002

[11] Martín, Marisa, *El modelo educativo del Tecnológico de Monterrey*. Monterrey, México: ITESM, 2002.

[12] Mergel, Brenda. *Instructional Design & Learning Theory*. 1998. [On-line] Available: http://www.educadis.uson.mx/pagina/ftp/Dise%C3%B1o-Instruc-RPA-B-Mergel-2.doc (September 26, 2005)

A Framework for Exploring the Relationships among Pedagogy, Ethics &Technology

Pat Jefferies; Bernd Carsten Stahl; Steve McRobb
Centre for Computing and Social Responsibility
De Montfort University
The Gateway, Leicester, LE1 9BH, UK

Abstract **The 'political push' and technological 'pull' currently prevalent in many higher education (HE) institutions encourages educationalists to experiment with tools that promote e-learning in the belief that this will help in the development of more autonomous, responsible learners. This paper, therefore, explores the relationship between pedagogy, ethics and technology, which are three important constructs for the development of an e-learning strategy. It then proposes a framework that will allow future research to define more clearly the areas where these concepts overlap, and to identify how they affect each other. As a consequence, the framework provides insight into the mutual dependencies of pedagogy, ethics and technology, with the aim of avoiding ethical risks in e-teaching and e-learning.**

I. INTRODUCTION

E-learning, defined as the use of information and communication technology (ICT) for supporting the educational process, is continuously gaining in importance (e.g. [1]). It is now standard in higher education institutions but is also widely used by secondary and even primary schools. It finds application in content delivery, in curriculum design and planning, in examinations, and in communication between students and teachers, between students and between teachers. ICT use is by now so closely interlinked with the educational process that it is hard to imagine a modern educational system without it.

Despite the undisputable importance of ICT in education (see [2] for a review), there clearly remain a number of issues that are not understood sufficiently (e.g. [3]; [4]). These include the relationships between technological tools available for learning delivery and their links with ethics and pedagogy as depicted in Fig. 1.

In order to explore these relationships this paper will first identify and briefly outline some of the relevant issues related to pedagogy, ethics and technology. These particular aspects will then be broken down and discussed with reference to the specific areas of interest for the paper. Next, the overlaps between each of the three circles (pedagogy and ethics, pedagogy and technology, technology and ethics) will be discussed. The three arrows, which represent pressures and constraints on the content of the three circles, raise interesting questions in themselves, but it is not our intention to explore these within this paper. Instead they are left for later research. Finally, the conclusion will explore the centre of the Venn diagram, the overlap between technology, pedagogy, and ethics, in order to prompt the development of an online pedagogy that is cognizant of its ethical implications.

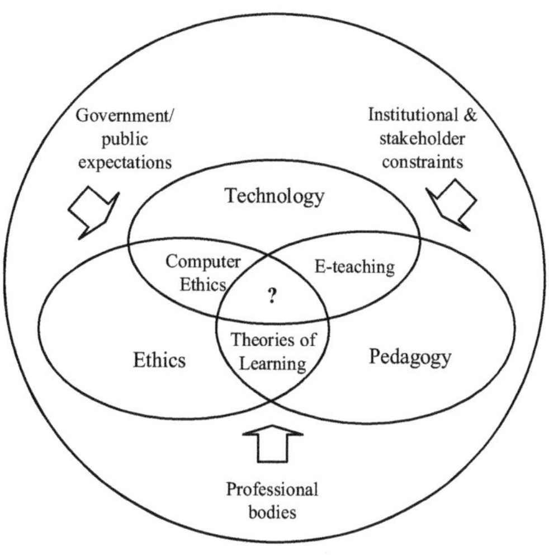

Fig 1. The links between technology, pedagogy and ethics

II. TECHNOLOGY, ETHICS AND PEDAGOGY - THE BUBBLES

In this section we will briefly introduce our understanding of the main constituents of the diagram depicted in Fig. 1 – technology, ethics and pedagogy – in terms of their relevance to the educational context. The first focus of our research will, therefore, be on pedagogy and the role of education in society. The second task will be to examine the ethical issues that may arise in relation to pedagogy in an e-learning context. Next, we will identify particular technological tools currently being used to support the learning and teaching process. This will then allow us to consider the assumptions that are contained within them.

K. Elleithy et al. (eds.), Advances in Computer, Information, and Systems Sciences, and Engineering, 433–439.

Pedagogy: Constructivist versus Positivist Theories

The literature on pedagogy is large. We have, therefore, focused on two groups of pedagogical theories, which can be broadly characterized as positivist and constructivist. E-learning and ICT use in education can, of course, be geared towards either paradigm. Positivist pedagogy starts with a realist ontology, meaning that reality is independent of the observer. Based on this ontological view, truth can be defined as a correspondence relationship between statements and facts. A statement is true if it adequately represents reality. Teaching in this paradigm has the purpose of transmitting objective truths to learners.

On the other hand there are the constructivist pedagogies based on non-positivist ontologies, which assume that reality is constituted by the observer and through the act of observation. Truth then becomes a function of social construction and agreement. As [5] notes, "recent approaches to learning emphasize the active, constructive nature of the knowledge acquisition process wherein the learner is not a passive recipient of information but an active and constructive interpreter of meanings". Thus, the ultimate goal of a constructivist approach is on learning how to construct knowledge appropriate to the situated task – similar to the idea of metacognition, which is the higher order process of reflecting on our own thinking and problem solving processes. This has a powerful problem solving potential. Teaching in this sense cannot be simply the transmission of truth. Rather it is aimed at helping learners to construct their own realities in accordance with social norms.

However, one of the fundamental problems for teachers is to relate the different theories of learning to their teaching strategies as well as to their use of technology. For example, [6] argues that there are two metaphors for learning – the "acquisition metaphor" (that which has been acquired by a learner – the *product* model) and the "participation metaphor" (the learning process itself – the *process* model). In the first metaphor the act of learning may easily be perceived as being actively engaged in the acquisition of knowledge that, in turn, may be packaged and transmitted (the *transmission* model). However, in the second metaphor the implied process is a more collaborative, participatory approach, although this seems to neglect the fact that something (i.e. learning) must be acquired. On the other hand, [7] believes that such a dichotomy is too restrictive, and proposes a "transaction metaphor" that encompasses both acquisition and participation.

A further issue then lies in the fact that "learning" may be seen as either an individual or a social activity. For example, [8] argue against both the domain-centered approach to learning – which they state reflects the *transmission* model of knowledge transfer – as well as learner-centered design – which they suggest is founded on the *information processing* model of cognition. Rather, they propose an "Activity-Centered-Design" model whose central tenets are that: "Activity is mediated by cultural artifacts, that activity must be analyzed at various levels and that internal activity (thinking) first occurs in the social plane (contextualized activity)" [9].

References [9], [10] and many other researchers have similarly argued that learning begins from a social context. For example, Vygotskian social theory proposes that learning occurs as a result of first participating in activities with others who scaffold the process. Learners then internalize and appropriate skills that then allow them to develop from a novice status to more expert [8]; [11]. However, [8] further suggest that mediation changes the nature of the task and that "learning to participate in a cultural practice means moving from partial participation in that practice to full participation". Here again, the actual process of what constitutes "learning" seems to be open to a variety of interpretations although there is clearly a distinction to be made between the conflicting didactic and conversational models of learning as depicted in Fig. 2.

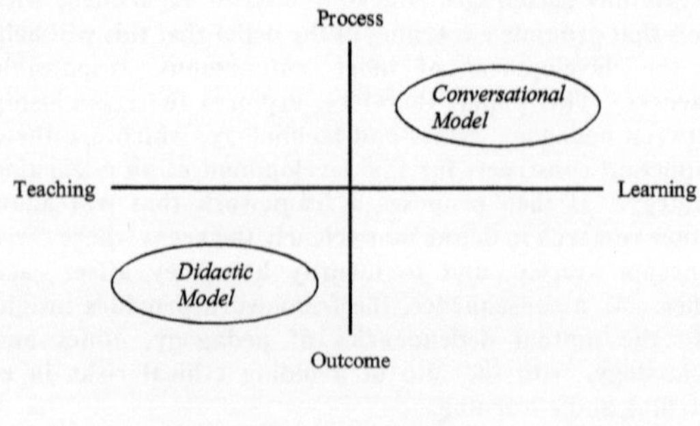

Fig. 2.Conversational versus Didactic model of learning and teaching

Ethics: Computer and Information Ethics

The field of philosophical ethics may be the largest of the three we have chosen to discuss. Rather than even attempt to mention the most important theories and topics, we will concentrate our discussion on the area of computer and information ethics. From its earliest beginnings, arguably in the work of [13], this field has developed to become a well established academic sub-discipline. As is true of most academic disciplines, there is no agreement on appropriate approaches or theories. However, there do seem to be a number of topics that are recurring, which seems to imply an implicit agreement on the subject area of computer ethics. Without claiming completeness, we believe that among the most important issues of computer ethics one can find questions of data quality, access to data and systems, intellectual property, privacy and data protection, change of social structures through technology, and the impact of ICT on our view of human beings [14].

For example if programs are of poor quality and do not fulfill their purpose then this can have repercussion for the working of critical systems, can lead to financial losses, and endanger humans. Similarly, high quality data is necessary for ICT to fulfill its purpose [15].

However, intellectual property and privacy are probably the most visible and salient issues discussed in computer ethics.

While there is no agreement on what exactly the problem is and how it relates to ICT, there is little disagreement that both are important and related to technology. For example, intellectual property increasingly determines the way the economy of the information society works but it is unclear whether traditional concepts can fruitfully be applied to modern technical means [16]. Similarly, there is little agreement on what privacy means or why we value it or how it can be protected. Nevertheless, most scholars agree that privacy is related to the electronic transmission of data and that it can constitute an ethical issue [18].

Technology: E-Learning Tools

Recent surveys have found that a variety of both custom-built and commercially produced virtual learning environments (VLEs), are increasingly being deployed to support education across the HE sector. Each VLE comprises a number of tools that seem to be primarily designed to support content delivery. The tools provided can be used to a) develop repositories that contain a variety of resources (e.g. Powerpoint presentations, Word/PDF documents, Excel/Access files, links to interactive tutorials and other external resources); b) provide assessment through on-line quizzes; and c) provide email communication between tutors and students. There are also tools for supporting discussion forums and synchronous chat; as well as management tools that enable teachers to track student access, to record assignment grades, to manage groups as well as the facility to set up evaluation surveys. It might be perceived that a range of non-routine technology has been developed as a consequence of customized research and development, to support the complex, developing system that is education. (See Fig. 3.)

However, whilst such development is quite likely to have been customized for the educational context, some may view ICT as neutral tools used to achieve the same ends as non-electronic tools previously did. However, such an assumption would patently be false, as there is a large body of literature which states that no technology is value neutral, but rather that all have inbuilt assumptions and ideologies which determine and constrain their possible use (for example, [19]; [20]; [21]). In this, ICT is no different from any other technology.

For example, within the e-learning context, whilst the ability to provide such things as content or online quizzes within a VLE may provide for autonomy and flexibility of access to learning, it also implies that knowledge can be packaged and transmitted. Thus the assumption reified within the use of these tools is that reality can be objectively defined, packaged and transmitted, and the tools actually support a more traditional, objectivist approach to education. For example, [22] report that "Bobbitt argued for a scientific approach to planning school curricula by systematic analysis of those human activities which the curriculum was intended to develop. The activities to be focused on were those which made for efficiency in living as a healthy, gainfully employed citizen. Those activities, analyzed in detail, would be the intended outcomes of the curriculum". This, in simplistic terms, may be seen to reflect a deterministic view of a 'closed' system in which specific outcomes may be defined for appropriately developed processes to achieve. Theories of education such as that proposed by [23] added weight to such approaches through the identification of what were perceived to be desired, observable behaviors. To a great extent, traditional, positivist approaches to education stemmed from particular beliefs of what learners ought to learn, and these, in turn, were based upon the notion that the goal of learning is to understand reality, which is taken to be objective. For theorists such as [20], achievement of learning was also deemed to be measurable through observable behavior. Assessment could therefore be framed in simple stimulus-response terms, reflected in the use of online quizzes where a determined truth is presented to students, and they are graded according to their ability to pick the right answer.

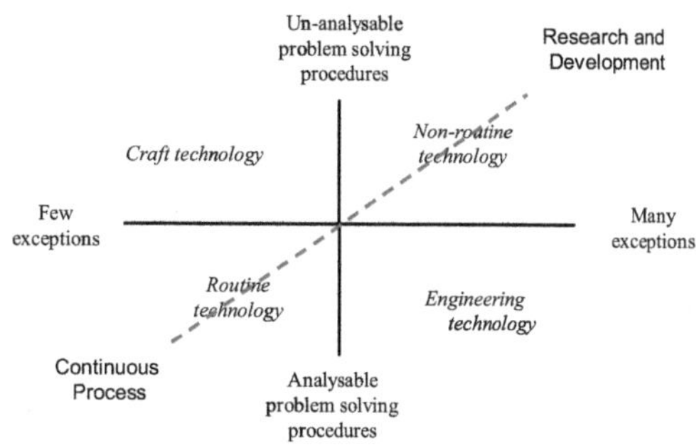

Fig. 3. The Perrow model of technology [5]

However, the use of such tools can also be seen to reinforce a particular relationship between technology and our view of humans [13]; [21]. For example, if we see humans as information processing machines, then failure to process information in the desired way is a failure of the machine, which may require reprogramming or being exchanged. This view may conflict with ethically inspired views of humans.

On the other hand, the use of discussion boards can be seen as supporting a completely different approach, in that the learning to be undertaken is not necessarily gained from acquiring "chunks" of knowledge but instead is constructed through discourse [22]; [23]. This wider scope of discourse in learning is further expressed in the work of academics on reflective learning in professionals [24], action learning sets [25] and cognitive development [10]. Thus use of this type of tool can be seen to relate to an underlying philosophy regarding the role of education in general and the teacher in particular and it is suggested that it is these more complex conversational frameworks that are clearly the type of constructs that [26] identifies as requiring 'Non-routine technology' when seeking a technological solution.

III. INTERSECTIONS

Having established the main content of each of the three aspects of our analysis: pedagogy, ethics and technology, we will now briefly discuss the areas where they overlap.

Ethics and Technology: Computer Ethics and E-teaching Tools

To start this section, we should state that there is a close link between ethics and education in general. Reference [27], for example, identifies several different educational ideologies – Progressivism (meeting individuals needs and aspirations), Instrumentalism (meeting requirements of the socio-economic order), Reconstructionism (moving society in desired ways), Classical humanism (transmission of cultural heritage), and Liberal humanism (creation of a vision of common educational experience). Thus education is usually seen as an ethical good that allows the student to develop independence, autonomy and the skills to provide for his or her livelihood [28]. We have stressed several times already that e-teaching tools are not value neutral but their use has consequences which have ethical importance. To return to the list of issues raised in our earlier discussion of technology, we can look at the impact that e-teaching tools have on privacy, intellectual property, data quality, social structures, and the view of humans.

In all cases, it is relatively simple to observe or imagine relationships between computer ethics and e-teaching tools. If we take the example of VLEs then we can identify aspects of these that are related to ethics. Privacy and data protection, for example may be affected when data on students or staff is collected that previously was not available. As noted earlier, many of the standard commercially available VLEs, such as Blackboard, allow teachers to check the usage of the site by their students. Teachers thus have information as to how often students have accessed which part of the VLE, at what time of the day etc. This can affect the teacher's evaluation of the student and thus have grave consequences for the student. Similarly, a VLE may raise issues of intellectual property. If teaching material is put online and students are able to download it, then they may use and modify it electronically, thus allowing them an easy route to infringement of copyright. Data quality is also important because the quality of the learning experience will be affected by the quality of the data provided. This was certainly true before VLEs but it becomes more salient due to the reliance on quantitative data in an information technology environment.

Another field of ethical issues regarding ICT is the technology's impact on social relations, be they between nations, within societies or organizations. One part of this is the question of digital divides but it also extends to power relationships on the political side and on the organizational level tends to produce social consequences, which have ethical importance. For example, some tools, such as multiple choice tests, are very reliable, easy to use and therefore favored by teachers. What is noticed less often is that they also imply a certain relationship between teachers and learners, namely that of the 'sage on the stage' who tells students the truth rather than 'guide on the side' who encourages students to fulfill their own goals. In effect, this translates into a very strong centralized position of the teacher whereas other teaching tools, such as online chats, voting tools etc, imply different power relationships. Such organizational relationships are then central to the way we perceive humans and the resulting ethical questions. Are students machines of data digestion that need to reproduce the truth as taught and then function in their economic role or are they autonomous persons whose purpose in life is to flourish? Such questions of course go beyond the use of technology in teaching, but they are affected by our use of technology.

Technology and Pedagogy: E-teaching Tools and Pedagogical Theories

As noted earlier the use of a number of VLE tools (email, tutorials, quizzes, Web pages, and Powerpoint) are all clearly related to a positivist pedagogy in that the learners are provided with tools that will scaffold and mediate their learning experience through giving them access to what has been selected and designed by the teacher.

Use of the discussion boards however, has been shown to not only have the potential for supporting the participatory/transaction models of learning but can also facilitate a more androgogic [32], social constructivist pedagogy. There is, for example, a great deal of empirical research that reports the benefits and potential of networked learning (see [2] for a review) and, in particular, the positive effects of social interaction during learning (e.g. [33]; [34]). Additionally, other research has found that collaboration amongst students has been shown to stimulate activity, make learning more realistic and to stimulate motivation (e.g. [35]; [36]). Furthermore, [37] concluded that discussion methods are more effective than didactic methods (e.g. lectures) for stimulating thought, for personal and social adjustment, and for changes of attitude, and were hardly worse than lectures for effectively transmitting information.

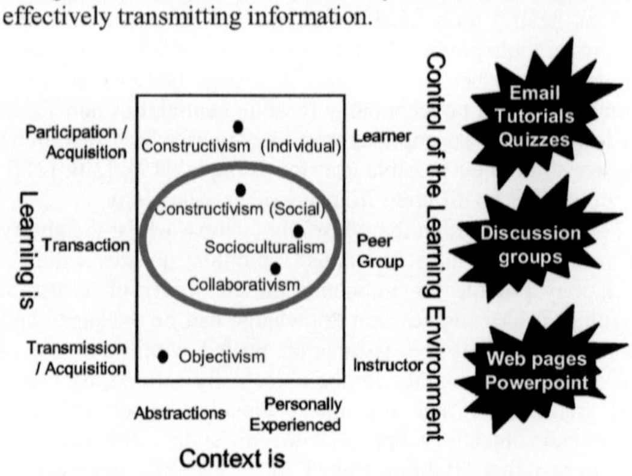

Fig. 4. Dimensions of learning mapped to technological tools

Nevertheless taking the three models of learning that [7] proposes and applying this to a modified version of the Leidner and Jarvenpaa framework [38], the diagram shown in Fig. 4 might more usefully illustrate the dimensions of learning, as well as relating these to the different uses of ICT within a networked learning environment.

Pedagogy and Ethics: Pedagogical Theories and Computer Ethics

In distinguishing earlier between constructivist and positivist pedagogical theories, we have already identified that there may be a link between computer ethics and pedagogy. A closer look at these theories reveals that they are not predominantly of a pedagogical nature but rather result from a general philosophical stance, based on a specific view of ontology. Positivism is based on a realist ontology, which states that the world exists independent of the observer. Constructivism, on the other hand, holds that reality is a (social or individual) construct. If positivist realism is correct, then truth can be described as a correspondence between a statement and reality. True statements can be taught, learned, and reproduced. In the constructivist ontology, truth can only be an agreement by those who are involved in the construction and teaching should concentrate on the process of being involved in creating useful constructions.

If this characterization is correct, then certain pedagogical theories correspond with certain ethical approaches because they are based on similar ontologies and general philosophical worldviews. In the positivist / realist world, one can presume that there are valid ethical theories and that these will lead to determined morally correct actions. The task of computer ethics will thus be to find out what the moral problems are, how they should be evaluated, and to help individuals and collectives do what is ethically desirable. This reasoning is represented by prescriptive approaches to computer ethics, which concentrate on professional duties of the individual. [29]. Reference [30] calls this approach the "engineering model" of ethics, because it assumes that there is one correct solution to an ethical problem.

Ethicists closer to the constructivist understanding of reality would be careful about the possibility of such a determinist approach to ethics. If reality is a social construction, then there can be no objective ways to determine the ethicality of an action and thus no general moral rules that are binding for everybody under all circumstances. For the constructivist, addressing ethical issues of ICT must therefore involve a process of agreement on the definition and understanding of the problem and a collective attempt to define an adequate response. Constructivists will therefore be more participative and discourse oriented, and will try to involve all stakeholders in a solution [31].

While this characterization is more black and white than actual positions tend to be, it does indicate that there is a link and an overlap between certain positions in computer ethics and specific pedagogical theories.

IV. CONCLUSION: THE CENTRE

Having now discussed the relationship between pedagogy, ethics and technology, it is clear that there are problems in designing and developing an e-learning or e-teaching strategy that go deeper than a lack of attention to ethical issues. We have, for example, argued that all three constituents are deeply influenced by more fundamental perceptions of reality. As such, the collection of ontology and epistemology that strongly influences our research methodology also underpins our view of humans and society. This, in turn, has a strong influence on how we perceive and use technology to support pedagogy in an ethical manner. One can argue that these philosophical bases of our world, sometimes called "paradigms" [42], have a pervasive influence on us. It is thus not simply a question of choosing certain pedagogical theories, technical tools, and ethical theories but it is gaining the understanding that such choices will implicitly follow from our view of the world. This raises the fundamental problem as to whether or not there is one right way to do things. If there is not, then we need to determine how we can reconcile different worldviews. We cannot present a simple answer to this question. However, we hope that our paper will help raise awareness to the fact that an ethically relevant overlap of ethics, technology and pedagogy exists. Furthermore, that the framework we have provided will now help us to both understand and address the ethical risks involved in developing our e-learning and e-teaching strategies.

REFERENCES

[1] Kellner, D., "New technologies: technocities and the prospects for democratization" in *Technocities*, Downey, J. and McGuigan J., Eds. Sage Publications, 1999.

[2] Lehtinen, E., Hakkarainen, K., Lipponen, L., Rahikainen, M., and Muukkonen, H., "Computer supported collaborative learning: a review of research and development," *The J.H.G.I. Giesderbs Reports on Education*, 10, Netherlands: University of Nijmegen, Department of Educational Sciences, 1999.

[3] Lipponen, L., "Exploring foundations for computer-supported collaborative learning," in Koschmann, T., Hall R. and Miyake, N., Eds. *Proceedings of the Computer-supported Collaborative Learning Conference (CSCL) 2002: Carrying Forward the Conversation*, Mahwah, NJ: Lawrence Erlbaum Associates, pp. 72 – 81, 2002. . Available at: http://newmedia.colorado.edu/cscl/31.html (accessed 2 May, 2002).

[4] Phipps, R., & Merisotis, J., *What's the Difference?: A Review of Contemporary Research on the Effectiveness of Distance Learning in Higher Education*, Washington, DC: The Institute for Higher Education Policy, 1999. Available at: http://www.ihep.com/Pubs/PDF/Difference.pdf (accessed 28 October, 2003).

[5] Vosniadou, S., "From cognitive theory to educational technology," in Vosniadou, S., De Corte, E. and Mandl, H., *Technology-Based Learning Environments*, Berlin: Springer-Verlag, pp. 11 – 18, 1994.

[6] Sfard, A., "On two metaphors for learning and the dangers of choosing just one," *Educational Researcher*, 27(2), pp. 4 – 13, 1998.

[7] Koschmann, T., "Toward a theory of computer support for collaborative learning," *Journal of the Learning Sciences*, 3, pp. 219 – 225, 1994.

[8] Gifford, B. R. and Enyedy, N. D., "Activity centred design: towards a theoretical framework for CSCL," in Hoadley, C. and Roschelle, J., Eds., *Proceedings of the Computer Support for Collaborative Learning Conference (CSCL) 1999*, Mahwah, NJ: Lawrence Erlbaum Associates, 1999. Available at: http://www.ciltkn.org/cscl99/A22/A22.HTM (accessed 5 March 2001).

[9] Dewey, J., *Psychology and Social Practice*, Chicago, IL: University of Chicago Press, 1901.

[10] Vygotsky, L.S., *Mind in Society: The Development of Higher Psychological Processes*, Cambridge, MA: Harvard University Press, 1978.

[11] Wertsch, J., *Vygotsky and the Social Formation of Mind*, Cambridge, MA: Harvard University Press, 1985.

[12] Taylor, P.H. & Richards, C.M., *An Introduction to Curriculum Studies*, Nfer-Nelson, 1987.

[13] Wiener, N., *The Human Use of Human Beings - Cybernetics and Society*, Garden City, NY: Doubleday Anchor Books, 1954.

[14] Mason, R. O., "Four ethical issues of the information age," in *MIS Quarterly* 10: pp. 5 – 12, 1986.

[15] Littlewood, B. and Stringy, L., "The risks of software," in Johnson, D. G. and Nissenbaum, H., Eds. *Computers, Ethics & Social Values*, Upper Saddle River, NJ: Prentice Hall, pp. 432 – 437, 1995.

[16] Stahl, B. C., "The impact of open source development on the social construction of intellectual property," in Koch, Stefan, Ed. *Free / Open Source Software Development*, Hershey PA: Idea Group Publishing, pp. 259 – 272, 2005.

[17] Stahl, B. C., Prior, M., Wilford, S. and Collins, D., "Electronic monitoring in the workplace: if people don't care, then what is the relevance?" in Weckert, J., Ed. *Electronic Monitoring in the Workplace: Controversies and Solutions*, Hershey PA: Idea-Group Publishing, pp 50 – 78, 2005.

[19] Sachs, P., "Transforming work: collaboration, learning and design," in: *Communications of the ACM* (38:9), pp. 36 – 44, 1995.

[20] Winner, L., "Upon opening the black box and finding it empty: social constructivism and the philosophy of technology," in *Science, Technology and Human Values* (18), pp. 362 – 378, 1993.

[21] Latour, B., "Where are the missing masses? The sociology of a few mundane artefacts," in Bijker, W. and Law, J., Eds., *Shaping Technology / Building Society*, MIT Press, pp. 225 – 228, 1996.

[22] Hoadley, C.M. and Enyedy, N., "Between information and communication: middle spaces in computer media for learning," in Hoadley, C. and Roschelle, J., Eds., *Proceedings of the Computer Support for Collaborative Learning (CSCL) 1999 Conference*, Mahwah, NJ: Lawrence Erlbaum Associates, 1999.

[23] Bernstein, B., *Class, Codes and Control, Vol 3, Towards a Theory of Educational Transmissions*, London: Routledge & Kegan Paul, 1977.

[24] Weizenbaum, J., *Computer Power and Human Reason*, San Francisco: W. H. Freeman and Co., 1976.

[25] Jefferies, P., "ICT in Supporting Collaborative Learning: Pedagogy and Practice," *Journal of Educational Media*, Vol. 28, No. 1, March 2003.

[26] Jefferies, P. & Rogerson, S., "Using asynchronous computer conferencing to support the teaching of computing and ethics: a case study," in *Annals of Cases on Information Technology (ACIT)*, Vol. 5, Hershey PA: Idea Group Publishing, 2003.

[26] Perrow, C., *Complex Organisations*, Scott-Forsman, 1972.

[27] Schon, D., *The Reflective Practitioner*, London: Temple Smith, 1983.

[28] Beaty, L., "Developing your teaching through reflective practice", *SEDA Special Publications*, 1997.

[30] Scrimshaw, P., "Educational ideologies", in *Purpose and Planning in the Curriculum*, Unit 2, E 204, Milton Keynes: Open University Press, 1983.

[31] Stahl, B. C., "E-teaching - the economic threat to the ethical legitimacy of education?" in *Journal of Information Systems Education* (15:2), pp. 155 – 162, 2004.

[32] Gotterbarn, D., "Informatics and professional responsibility," in Bynum, T. W. & Rogerson, S., Eds. *Computer Ethics and Professional Responsibility*, Oxford: Blackwell Publishing, pp.107 – 118, 2004.

[33] van den Hoeven, J., "Computer ethics and moral methodology," in *Metaphilosophy* (28:3), pp. 234 – 248, 1997.

[34] Rogerson, S., "The ethics of software development project management," in Bynum, T. W. and Rogerson, S., Eds., *Computer Ethics and Professional Responsibility*, Oxford: Blackwell Publishing, pp. 119 – 128, 2004.

[35] Knowles, M.S., *The Modern Practice of Adult Education: Androgogy vs. Pedagogy*, New York: Association Press, 1970.

[36] Crook, C., "Computers in the community of classrooms," in Littleton, K. and Light, P. Eds., *Learning with Computers: Analysing Productive Interaction*, London and New York: Routledge, pp. 102 – 117, 1999.

[37] Dillenbourg, P., "Introduction: what do you mean by 'collaborative learning'?" in Dillenbourg, P., Ed., *Collaborative Learning: Cognitive and Computational Approaches*, Amsterdam: Pergamon, Elsevier Science, pp. 1 – 19, 1999.

[38] Harasim, L., "Shift happens: online collaborative learning as a new paradigm in education", keynote speaker at *Fusion 2000*, Glasgow, 2000.

[38] Leidner, E. D. and Jarvenpaa, S. L., "The use of IT to enhance management school education: a theoretical view", *MIS Quarterly*, 19 (3), pp. 265 – 291, 1995.

[39] Veerman, A. and Veldhuis-Diermanse, E., "Collaborative learning through computer-mediated communication in

academic education," in *Proceedings of the Computer Support for Collaborative Learning Conference (CSCL) 2001,* Mahwah, NJ: Lawrence Erlbaum Associates, 2001. Available at: http://www.mmi.unimaas.nl/euro-cscl/Papers/166.doc, (accessed 5 March, 2001).

[40] Bligh, D., *Teach thinking by discussion*, Society for Research into Higher Education, NFER Nelson, 1986.

[41] Orlikowski, W. J. and Baroudi J. J., "Studying information technology in organizations: research approaches and assumptions," in: *Information Systems Research* (2:1), pp. 1 – 28, 1991.

Modern Sensing and Computerized Data Acquisition Technology in High School Physics Labs

Sookram Sobhan, Nerik Yakubov, Vikram Kapila, Magued Iskander, and Noel Kriftcher
Polytechnic University, Brooklyn, NY 11201

***Abstract*—Under a National Science Foundation (NSF) funded GK—12 Fellows project, a solution to invigorate high school students' interest in science, technology, engineering, and mathematics (STEM) careers is being examined and implemented. This paper provides an overview of our strategy and results from the first year of the project.**

I. INTRODUCTION

After World War II, a bitter rivalry ensued between the United States (U.S.) and the Soviet Union developing into the Cold War. In this age, scientific and technological superiority was of paramount concern for both countries as they competed to attain military dominance. The balance of power shifted to the Soviet Union on October 4, 1957 with the successful launch of the Sputnik I. The Soviets continued to flaunt their power with the launch of the Sputnik II on November 3, 1957 [1—3]. The American response to these events came with the creation of the Apollo program in 1961 by President Kennedy, who challenged the U.S.A. to put a man on the moon before the turn of the decade. The resulting decade-long race to the moon led to a surge of student enrollments in STEM programs in American universities [1, 4]. Furthermore, a wide array of theoretical and experimental research conducted in support of the space program benefited the society by leading to the creation of numerous products and processes (e.g., microcomputer, inertial guidance system, etc.). The success of the Apollo program is well documented in scientific publications and mass media [5—9].

The Cold War ended in the late 1980s with the demise of communism in countries that comprised the former Soviet Union. Shortly thereafter, however, new challenges to American leadership of the free world began to arise. First, after a series of strikes on American interests abroad in 1990s, on September 11, 2001, terrorists used passenger airplanes to attack New York City (NYC) and Washington D.C. In the aftermath of the September 11, 2001 attacks, a severe clampdown on immigration to the U.S. ensued. Second, since the late 1990s, the service-sector of the American economy began to shift its operations overseas to low-wage countries, following the model of the manufacturing-sector from an earlier era. Thus, increasing import of services and manufactured goods from abroad has recently led to a growing trade deficits vis-à-vis countries such as China and Japan. Third, in recent years, a growing chorus of civic- and business-leaders has argued [10—13] the paramount importance of protecting American leadership in scientific discovery and technical innovation, to enable the U.S. to bridge the gap with its trading partners and to respond to the asymmetrical terrorist threats through the development of superior technological solutions. Unfortunately, American leadership in science and technology is threatened by a perennial disinterest in STEM disciplines among college-bound, secondary school students. Furthermore, the stringent immigration laws enacted in response to the events of September 11, 2001, have significantly reduced the inflow of foreign-born/trained science, math, and engineering students, who have typically supported the research enterprise at nation's universities.

To sustain the U.S. quest to develop an "innovation economy," our universities must attract, educate, and graduate a large number of qualified scientists and engineers. In responding to this task, the universities face the following challenges: *i)* engineering is held in less esteem than other professions, *ii)* math and science are not perceived as cool by high school students, *iii)* society discourages female high school students from becoming engineers, and *iv)* the typical engineering curriculum is more rigorous than other majors, causing high school graduates to shy away from the engineering discipline. The problem is further exacerbated by negative stereotypes of scientists and engineers held by teenagers.

To develop a creative solution to the problem of attracting more students to STEM disciplines, as engineers, we must begin by analyzing various characteristics of this problem. Thus, we begin by scrutinizing the American K—12 educational environment and discover that the problem is not amenable to any single solution since it spans the entire educational experience of K—12 students. For example, starting with the elementary grades a major preoccupation of the K—12 educational system is literacy (i.e., reading and writing). Unfortunately, this often means math and science are accorded a lower priority, or even are shunned. While in middle school, students are required to take earth and biological sciences, in high school, these science courses are repeated as requirements but physics is left as an elective.

K. Elleithy et al. (eds.), Advances in Computer, Information, and Systems Sciences, and Engineering, 441–447.

Moreover, courses such as marine biology are offered, thus enabling many students to avoid taking physics altogether! Without a rigorous education in math, which forms the foundation of analytical sciences [14], when the relatively few high school students encounter a physics course, their educational understanding and achievement are hampered since abstract concepts and equations lack appeal due to their rigor. To further exacerbate the situation, high schools continue to use lab experiments, which were written decades ago and are accompanied by outdated, turn-of-20th-century equipment. Furthermore, these labs fail to inspire students since they do not connect to practical, real-life applications. The result is, of course, high school graduates with poor appreciation and low enthusiasm for STEM education and careers.

Today's students live in and benefit from a highly technological world, one which includes PlayStation Portable that can play video games and movies, cellular phones with video and MP3 capabilities, and personal digital assistants that are as powerful as laptops, all of which have connectivity to the Internet. Having experienced these technological gadgets of our age, when students enter a lab setting they encounter primitive equipment and perform tasks such as, measure the period of an oscillating body using a manual stroboscope. Not surprisingly, lack of a modern and challenging lab environment further drives these students away from STEM disciplines.

As one remedy to the malaise afflicting the American pre-college educational system, under an NSF GK—12 Fellows Grant, we have formulated and implemented a solution, which is the subject of this paper. Specifically, over a dozen undergraduate and graduate engineering students have partnered with teachers at four NYC public high schools to integrate modern sensing and data collection technologies into the lab section of physics and living environment courses. This paper focuses on project activities related to the physics course. One premise of our approach is that the introduction of tools and techniques that scientists and engineers use on a daily basis will remove the stigma of science being boring and generate enthusiastic response from students. Our approach addresses the following key issues:

- The lab is keeping with the times, that is, students use state-of-the-art equipment in comparison to the equipment used in traditional science labs.
- Sensors and data acquisition tools along with a suitable graphical user interface can provide students with intuitive insights. For example, if a student tosses a ball, it would be rather difficult to record the ball's instantaneous velocity at any point in time during its flight. However, by using a position detector (e.g., ultrasonic sensor), interfaced to a computer running a graphing program, the ball's trajectory and velocity can be recorded and presented graphically [15], which renders an abstract concept concretely.
- Since manual collection and recording of data is not the principal focus of the lab, the students can focus on learning the underlying concepts of the lab and formulating and testing new hypotheses.

Several recent papers [16—18] have reported successful use of robotics technology to attract pre-college students to STEM disciplines. Moreover, [19] suggests using gaming technology to enable engineering education to connect better with today's computer savvy students. In a similar vein, we expect that the modernized physics labs offering interesting exercises will inspire students to learn the fundamental principles of science rather than studying simply to pass a test. We further anticipate that integration of real-world engineering examples as applications of scientific principles and exposure to undergraduate and graduate engineering students will encourage the high school students to consider higher education and professional careers in STEM disciplines.

II. OVERVIEW

The "Revitalizing Achievement by using Instrumentation in Science Education (RAISE)," project is being implemented under an NSF GK—12 Fellows Grant [20]. The project is led by two engineering and a humanities faculty. Project's partner high schools are: George Westinghouse, Marta Valle, Paul Robeson, and Telecommunication Arts and Technology HS. The average SAT score of students from the partner schools is below 850 and fewer than 10 percent of the students score above 1100, the cut-off for admission to the undergraduate science and engineering programs in most universities. Moreover, their passing rates on the required standardized science and math exams and graduation rates are alarmingly low (below 50 percent).

The goals of the project include: *i)* enhance student achievement in standardized exams, *ii)* inspire inner-city students to reach high academic standards while acquiring passion for STEM disciplines and careers, *iii)* provide teachers with technology proficiency, and *iv)* enable GK—12 Fellows (RAISE fellows) to hone their communication skills, leadership skills, and develop a deeper appreciation for STEM disciplines.

The RASIE Fellows develop creative and engaging experiments for physics labs by using modern sensing, instrumentation, and data acquisition tools along with a user-friendly graphical user interface (GUI) for data analysis and plotting. Use of modern data acquisition tools allows high school students to optimize their time in the lab. Furthermore, in collaboration with teachers, RAISE fellows make presentations and conduct demonstrations to introduce appreciation of physics topics from an engineering viewpoint to enhance the educational experience of high school students.

III. TRAINING OF FELLOWS AND TEACHERS

Before being deployed in the four schools, RAISE fellows receive intensive training during the summer months prior to the start of the school year [21, 22]. Fellows are first introduced to modern sensing technology and mechatronics. Topics covered include sensors and signal conditioning, actuators and power electronics, hardware interfacing, and embedded computing. In addition, fellows attend a week-long teaching workshop to develop their pedagogical skills such as lesson planning and effective questioning techniques, student behavior and cognition, learning theory and styles, classroom/group management skills, effective communication and presentation skills, active learning techniques, project-based learning, and evaluation methods.

Teachers from the RAISE supported schools also attend a week-long technical workshop focused on modern sensing technologies. The workshop is conducted by the engineering faculty and the RAISE fellows and it provides the teachers with insight for class activities and prepares them to become technology resources in their schools. A byproduct of this training is that teachers and Fellows have the opportunity to become acquainted with and bond with one another as they begin planning for the upcoming year.

IV. REPRESENTATIVE SENSOR-BASED PHYSICS MODULES

Thirteen sensor based physics experiments have been developed (Table 1). These experiments are intended to support the Regents Physics Lab. In New York State students must pass three science Regents exams with a score of 60% or better to graduate from high school with Regents diploma. Regents Physics is an elective taken by less than 10 percent of students in the partner schools. Students have strong intuitive skills and poor analytical skills. Thus, it is the purpose of these labs to enhance their intuitive skills and build their analytical and math skills. Two sample example labs are described below.

Experiment on Buoyancy and Fluid Density

This experiment is designed to verify the density of a fluid while illustrating Archimedes's principle of buoyancy. Students are first presented with buoyancy effects in design of barges, ships, and submarines. Having seen the applications of this concept, students become motivated in further learning and exploration because they realize the legitimacy of the topic. A conventional way of measuring the density of a fluid involves, finding the mass of a container, then filling it with the test fluid and finding its mass again to determine the mass of the fluid from difference between the two mass measurements. Knowing the volume of the container, the density of the fluid is computed using the mass to volume ratio. Next, Archimedes' principle states the following: "An immersed body is buoyed up by a force equal to the weight of the fluid it displaces," that is,

$$F_b = \rho_f V_d g ,$$

where F_b is the buoyant force, ρ_f is the density of the fluid, V_d is the displaced volume, and g is the freefall acceleration. The buoyant force is countered by the weight of the body mg. The difference between the two produces a net force of

$$F_{net} = mg - \rho_f V_d g .$$

For an immersed body, using a force sensor, one can measure the net force at different levels of fluid displacement (Figure 1). The plot between the net force (i.e., F_{net}) and the product of volume displacement and gravitational acceleration (i.e., $\rho_f g$), should yield a straight line with the slope of the line representing the density of the fluid (Figure 2). This density is then compared to that obtained using the conventional method. By doing this experiment students are presented with an alternative way of measuring fluid density while being familiarized with the concept of buoyancy.

Experiment on Damped Vibrations

The phenomena of vibrations are ubiquitous. Periodic phenomena occur throughout nature. This is evident on a microscopic scale as electrons orbit the nucleus as well as on a colossal scale as the Earth revolves around the Sun. Without the vibration of electromagnetic fields, the world would be dark and cold. With just a few examples, it is apparent that students should be familiar with vibrations and their characteristics. In the examples cited above, as well as many other periodic phenomena, the underlying vibratory systems experience minimal energy loss and thus exhibit properties of simple harmonic motion. When energy loss is significant, the oscillation is no longer simple, but damped in which case the displacement, velocity, and acceleration responses of an oscillating body will resemble an exponentially decaying sinusoid. Although damping is not part of the traditional high school physics curriculum, in the spirit of presenting real-world applications of physics, a lab on damping effects is introduced.

We begin our investigation of damping with a premise in the form of the following question: "Why does a child swinging on a playground swing eventually come to rest?" Equivalence between a pendulum (child on a swing) and a mass-spring system is first described. Students then examine the above question by attaching an accelerometer sensor to the mass of a mass-spring system (Figure 3). The sensor records the acceleration of the mass—a smooth decaying sinusoidal waveform (Figure 4). From this curve, the amount of damping present in the system, the period of vibration, the natural frequency of the system, and the equivalent spring constant can be determined. All of the measured quantities are then reflected back to the playground swing question and conclusions are drawn regarding factors that affect the motion of the body.

V. CLASSROOM IMPLEMENTATION

It is commonly held that high school students who excel in math and science courses succeed in high school and continue their education in college. The difficulty involved in mastering these subjects presents great challenges for students and for the educational system. Moreover, a lack of fully prepared and effective teachers is further limiting the achievement of students in STEM disciplines in U.S. K—12 schools. Teachers lack adequate professional preparation, budgets are limited, and science and math are areas suffering from acute teacher shortages. In addition, the sheer shortage of teachers [23] has led to over-crowded classrooms, further hindering the student achievement in these disciplines.

Motivated by the recognition of these needs, RAISE fellows have been mobilized to *i)* introduce technology to in-service teachers to enhance their technical proficiency, *ii)* serve as an additional resource in the classrooms and labs to provide individual attention to students, *iii)* interact with students as their mentors and coaches to stimulate their interest in math and science, and *iv)* serve as role models to motivate students to pursue careers in STEM disciplines.

Each of our partner schools has been equipped with four computerized lab setups which allow for groups of four to five students per setup. The experimental modules are designed in a way that every member of the group has an active role in the experiment. Furthermore, team members must have constant interaction among each other to complete the lab assignment properly. For example, one student holds the sensor, another operates the computer, the third works with the equipment, and the fourth acts as a manager and monitors that everyone is synchronized. The students have the opportunity to switch roles as most modules have multiple trials. This method of assigning differentiated tasks keeps students engaged and prevents negative behavior before it starts.

Incorporating the Logger Pro Software [24] allows the instructor to convey the material through a wide range of learning styles: *i)* the graphical user interface displays sensor measurements through which visual learners easily pick up the concept, *ii)* the team-based tasks require group effort which benefits auditory/verbal learners, and *iii)* the hands-on lab activities aid the tactile/kinesthetic learners, who grasp the concept by doing the experiment.

VI. OUTREACH EVENTS

Several outreach events, planned and conducted by the RAISE project team are described below.

- On Election Day, November 2, 2004, RAISE fellows conducted a workshop on modern sensing and data acquisition technology and they introduced Vernier sensors to 20 teachers of science and math at George Westinghouse HS.
- On January 22, 2005, RAISE fellows conducted a professional development day for 19 NYC STEM teachers from non-RAISE supported schools. The participating teachers ranged from elementary to high school teachers.
- On April 20, 2005, a career day program was held for students from high schools participating in the RAISE project. The event was attended by over 100 students and teachers. The climax of the event was the showcase of mechatronics-enabled projects, some of which were developed by the RAISE fellows.
- On May 20, 2005, a regional GK—12 grant holder's conference was held at Polytechnic University. At this conference four GK—12 projects at three NYC universities showcased their efforts. This event allowed fellows, teachers, and the project leaders from the four GK—12 projects to exchange ideas. Many GK—12 fellows made presentations to disseminate the results of their efforts. The event was attended by NSF's GK—12 program team and also by personnel from the NYC Department of Education.

VII. EXPECTATIONS AND ASSESSMENT

For high school students, our goal has been to develop and enhance their STEM skills by offering them opportunities to apply STEM knowledge through lab activities. We are also aiming to revitalize student performance on the Regents Exam of Physics [25]. Moreover, we emphasize oral and written communication and opportunities to work in culturally diverse groups. Finally, we aim for students to acquire an appreciation for STEM careers and an opportunity to pursue such a track should they decide to do so.

For the RAISE fellows, our goal is to sharpen their communication, leadership, and STEM skills through curriculum planning, lab development, and instructional delivery. In addition, we expect the fellows to learn to communicate complex engineering concepts to a non technical audience, which will be an essential skill in their future careers as science/technology leaders.

For the high school teachers, we expect them to attain adequate technology proficiency and be able to integrate sensor-based demonstrations in their lesson plans and classroom activities. Moreover, we anticipate that they will improve their pedagogical skills through collaboration and exchange of ideas with engineers.

An independent program evaluator has been retained to assess the degree to which the RAISE project is meeting its stated objectives. It is somewhat early to provide a definitive assessment of the project success in meeting its goals. Nevertheless, the following are the evaluator's results for the first year of the program:

- Teachers at the schools, in general, felt that the presence of the RAISE fellows is helpful in enriching their courses and have rated the program as having a positive effect on their students.
- More students mentioned labs with sensors as compared to labs without sensors as their favorite component of the program.
- Students in most groups generally felt that the RAISE fellows provided substantial educational value.
- Several RAISE fellows have shown instructional skills that suggest that they can be successful teachers with similar populations of students, should they choose to do so.

For many RAISE fellows, teaching science to high school students for the first time was a challenging experience. Dealing with boisterous and sometimes audacious students while teaching science requires a great deal of perseverance and endurance. Nevertheless, the RAISE fellows quickly adapted to the classroom environment and learned the following lessons from their first year experiences:

- While the RAISE fellows are inspired by complex science and math concepts, high school students become easily discouraged. Therefore, the first set of developed experiments had to be revised to accommodate student interest level.
- In order to improve achievement on standardized exams, Regents-type questions have been added at the end of each written lab assignment.
- The attention span of an average high school student is short. As a result the RAISE fellows felt that there was a constant need to inspire their students to do their best.

As William Arthur Ward once said, "When we seek to discover the best in others, we somehow bring out the best in ourselves." This is the kind of outcome we are attempting to achieve for all the participants in the program.

VIII. CONCLUSION

As the program enters its second year, so does our ambitious agenda. Through the use of modern sensing tools in high school Physics classrooms, the RAISE program is undertaking to revitalize science education. The RAISE fellows have designed sensor-based labs that convey physics concepts through the use of modern data acquisition tools. Integrating modern technology into the classroom curriculum will equip students with tools that will benefit them in an increasingly technological society. In addition to student improvement, the fellows develop their leadership and communication skills which are essential for their engineering careers.

ACKNOWLEDGEMENT

This work is supported by the GK—12 Fellows Program of the National Science Foundation under grant DGE–0337668. Polytechnic University provided significant fund for the acquisition of 13 sets each of LabPro Biology Deluxe Package and LabPro Physics Deluxe Package from Vernier. Finally, Vernier Software and Technology provided equipment at a significantly discounted rate to support this project.

REFERENCES

1. P. Dickson, *Sputnik: The Shock of the Century*. Walker and Company, New York, NY, 2001.

2. R. A. Divine, *The Sputnik Challenge*. Oxford University Press, New York, NY, 1993.

3. "Defense: Knowledge is Power." *TIME*, pp. 21—23, 18 November 1957.

4. B. B. Clowse, *Brainpower for the Cold War: The Sputnik Crisis and National Defense Education Act of 1958*. Greenwood Press, Westport, CT, 1981.

5. D. Dooling, "L+25: A Quarter Century after the Apollo Landing." *IEEE Spectrum*, Volume 31, No. 7, pp. 16—29, 1994.

6. R. Launius, R. D. Launius, B. Ulrich, *NASA and the Exploration of Space*. Stewart, Tabori, and Chang, 1998.

7. G. Stix, *et al.*, "Among the Classics Technological Breakthroughs from 1963 to 1988." *IEEE Spectrum*, Volume 25, No. 11, pp. 76—115, 1988.

8. M. Williamson, "Aiming for the Moon: The Engineering Challenge of Apollo." *Engineering Science and Education Journal*, Volume 11, No. 5, pp. 164—172, 2002.

9. M. Williamson, "Man on the Moon: The Technology of Lunar Exploration." *Engineering Science and Education Journal*, Volume 11, No. 6, pp. 217—226, 2002.

10. *Losing the Competitive Advantage? The Challenge for Science and Technology in the United States*. American Electronics Association, February 2005. http://www.aeanet.org/publications/IDJJ_AeA_Competitiveness.asp.

11. *Innovate America: Thriving in a World of Challenge and Change*. The National Innovation Initiative Final Report, December 2004. http://www.publicforuminstitute.org/nde/sources/NII_Final_Report.pdf.

12. T. L. Friedman, The World Is Flat: A Brief History of the Twenty-first Century. Farrar, Straus and Giroux, New York, NY, 2005.

13. B. Gates, "Remarks Delivered at the 2005 National Education Summit on High Schools." http://www.gatesfoundation.org/MediaCenter/Speeches/BillgSpeeches/BGSpeechNGA-050226.htm.

14. G. C. Orsak *et al.*, "High-Tech Engineering for High School: It's Time!" *IEEE Signal Processing Magazine*, pp. 103—108, January 2004.

15. K, Appel, J, Gastineau, C. Bakken, and D. Vernier, "Physics with Computers." *Vernier Software and Technology*, Beaverton, OR, 2003.

16. L. Creighton, "Crash, Bam, Learn." *Prism*, Vol. 12, No. 2, pp. 34—36, 2002.

17. H. Mukai and N. McGregor, "Robot Control Instruction for Eighth Graders." *IEEE Control Systems Magazine*, Vol. 24, No. 5, pp. 20—23, 2005.

18. T. K. Grose, "Jolly Good Fellow." *Prism*, Vol. 15, No. 1, pp. 40—43, 2005.

19. P. Wankat and F. Oreovicz, "Gaming the curriculum." *Prism*, Vol. 15, No. 1, pp. 48, 2005.

20. Online: http://raise.poly.edu, website of the RAISE project.

21. Online: http://mechatronics.poly.edu, website of Mechatronics @ Poly.

22. Online: http://www.parallax.com/html pages/edu/curriculum/sic curriculum.asp, website of Parallax Inc.'s "Stamps in Class" educator's program.

23. W. S. Bacon (Ed.), *Bringing the Excitement of Science to the Classroom: Using Summer Research Programs to Invigorate High School Science*. Research Corporation, Tucson, AZ, 2000.

24. Online: http://www.vernier.com/soft/lp.html, website of Vernier's Logger Pro Software

25. Performance Standards: Science, Board of Education of the City of New York. 1999. ISBN: 1-55839-505-9.

Table 1: Sensor-based physics activities developed under the RAISE project

Experiment	Description
Air Resistance	An ultrasonic sensor is used to measure the velocity of one or more free falling coffee filter(s) and to show that the filter reaches a terminal velocity due to air resistance.
Buoyancy	A force sensor is used to measure the buoyant force of an object immersed in liquid. Knowing the submerged volume of the object the density of the liquid is obtained.
Conservation of Mechanical Energy	An ultrasonic sensor is used to determine the position and velocity of a tossed ball. Using the position and velocity at various locations, principle of conservation of total energy is verified.
Damped Vibration	An accelerometer is used to find the response of an oscillating mass-spring system from which the damping coefficient and natural frequency is determined. The spring constant is also approximated.
Electromagnetism	A magnetic field sensor is used to verify properties of a solenoid such as uniform field strength within the core, negligible field outside the core, direction of poles formed due to current direction, and attenuation of field as measured axially.
Freefall Acceleration	Using a photogate, the acceleration due to gravity of a free falling object is measured.
Heat Transfer	Using a temperature probe, the rate of cooling and heating of water is measured. Insulating properties of different materials are also investigated.
Magnetism	A magnetic field sensor is used to quantify the magnetic properties of different materials as well as to classify the materials as diamagnetic, paramagnetic, or ferromagnetic.
Projectile Motion	Two photogates are used to measure the horizontal component of the initial velocity of a ball being rolled off a table. Using this value, the range of the horizontal landing is calculated.
Simple Harmonic Motion	An ultrasonic sensor is used to measure the amplitude and frequency of a mass-spring oscillator. From this, maximum velocity and acceleration are calculated and the mathematical model of harmonic motion is verified.
Stability	A force sensor is used to pull a block until the block tips or slides. The critical forces are then computed theoretically and compared with the sensor measurements.
Static and Kinetic Friction	A force sensor is used to pull on a wooden block, sliding over a frictional surface, to determine the coefficient of static friction. Kinetic friction is determined using an ultrasonic sensor that measures the deceleration of a sliding block coming to rest.
Vector Addition	Force sensors are used to find the tension in two strings attached to a mass. The resultant force is then computed.

Figure 1—Buoyancy Experiment: A mass, suspended on a force sensor, is placed in a liquid to examine Archimedes Principle

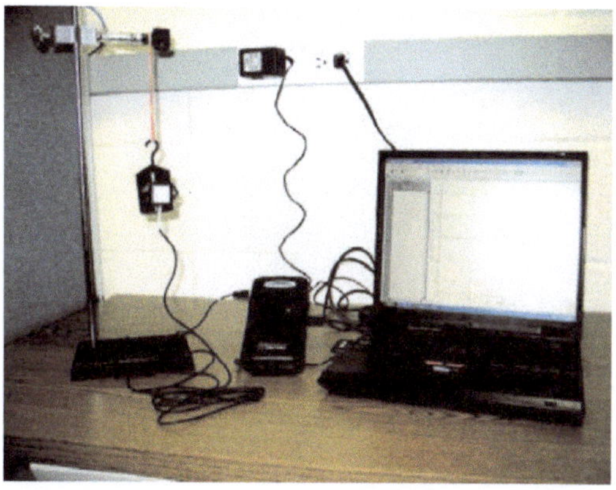

Figure 3—Damped Vibrations Experiment: Mass-spring (rubber band) system with accelerometer

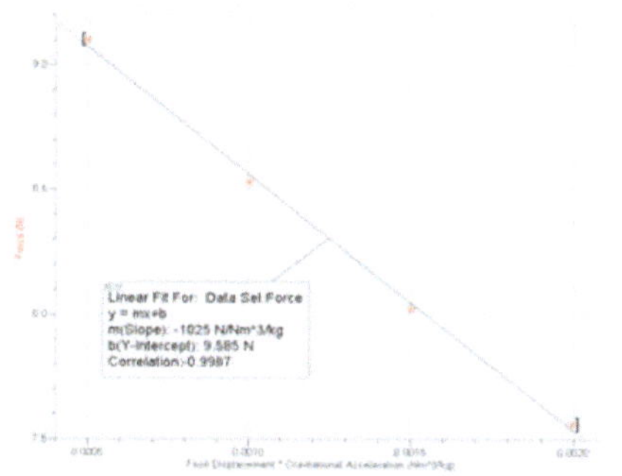

Figure 2—Buoyancy Experiment: Sample result

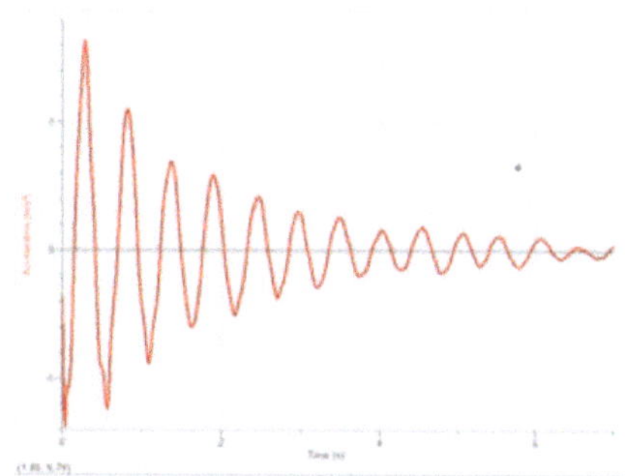

Figure 4—Damped Vibrations Experiment: Sample result

Design & Development Of A Remote Temperature Monitor System Of Web Using Virtual Instruments

Fan yang, Guoping Li, Huipeng Li

Abstract- **The temperature measure and control system is one of the widest application systems in industry fields, and National Instruments' LabVIEW has became a popular programming environment for data acquisition in academia and industry. At Wuhan Institute of Chemical Technology, experimental training for undergraduate students in electric and electronic engineering courses should strengthen students' theoretical knowledge, carry out useful self-training, and acquire basic experience with problems arising in real-world experimental activities. This paper reports the hardware and software of a temperature control system based on a data acquisition card and the experiments of the design are presented.**

Index Terms- **Hardware circuits, data acquisition, LabVIEW, virtual instruments, microprocessor**

I . INTRODUCTION

The Remote Temperature Monitor System Of Web is a system that combines the virtual instruments(VIs) technology, computer communication technology and network technology organically to realize the temperature signal's measure and control, forewarning and remote transmission. As an important unit of the virtual technology, the LabVIEW virtual instrument platform, which is designed by NI company based on Graphics Language, is adopted in this design. LabVIEW provides strong support of DAQ(data acquisition card) [1]-[3]. In this paper, Data acquisition circuit of temperature monitor system is implemented by the microprocessor of AT89C51 and monobus digital temperature sensor DS18B20, and the data of temperature signal is read by PCI-6014 in industrial computer from microprocessor. C51 language program in microprocessor is used to acquire the value of the temperature from DS18B20 and to transmit it to PCI-6014 interface. In industrial Computer, LabVIEW realizes temperature processes, display, forewarning, feedback control and long-distance transmission in web. The experiments of the system simulate successfully in laboratory[4]-[6].

II . PROJECT DESIGN

Considering PCI-6014 is the only available equipments of DAQ card in our laboratorys for student, we decide to use it in the temperature monitor system as the data acquisition hadware. And it is one of the DAQ products of NI company, so LabVIEW programs support it strongly. Temperature sensor DS18B20 runs in a very strict time sequence. It is difficult to Use PCI-6014 to operates DS18B20 directly, and if microprocessor is used, programming to microprocessor will be much easier to operate the signal from DS18B20. So, 8-bit single chip processor AT89C51 of high-quality with 4K byte flash memory and low work voltage which is provided by ATMEL company and single bus digital temperature sensor DS18B20 which is provided by DALLAS company are used to produce the hardware acquisition circuit, while LabVIEW VI programs realize the monitor, control and remote transmission of the system.

A . Whole Design

Connected the data line of DS18B20 with the interface P3.7 of single chip processor, programed to AT89C51 to read or write DS18B20, the temperature data will be acquired. Then, set the data to 4-bit dynamic LED display, which put out bit selection signals from Port 1.0 to Port 1.3 and segment selection signals from Port 0.

The processed data will be send to the digital I/O of PCI-6014 from Port 2, which realizes the data transmission from hardware acquisition circuit to the industrial computer.

The next step is to program in LabVIEW programming environment, which should display the waveforms and data of the signals, forewarn the temperature and send control signals of feedback.

This system has 2 objects of feedback control: the bulb and the fan. The remote transmission of the monitor condition is also realized by LabVIEW programming. The following fig.1 illustrates the whole design of the system.

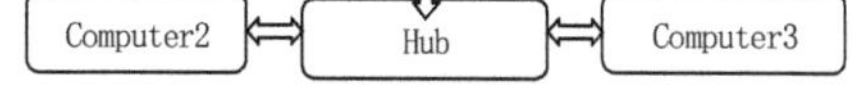

Fig. 1. Whole design of the system

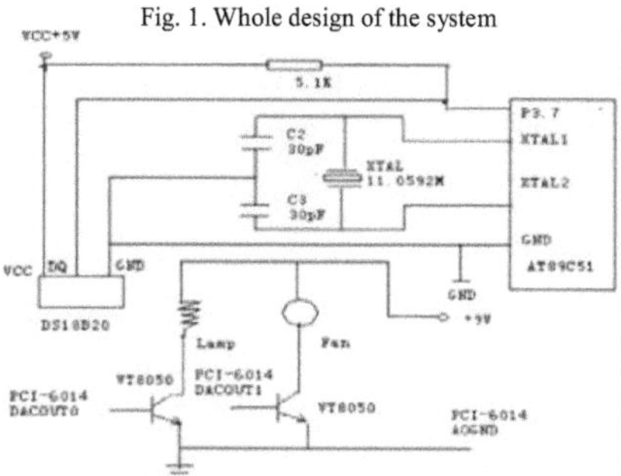

Fig. 2. Hardware circuit of signal measure and control

B . Design of Data Acquisition Circuit

As the important components, AT89C51 and DS18B20 are used to pursue the simpleness in data acquisition hardware

K. Elleithy et al. (eds.), Advances in Computer, Information, and Systems Sciences, and Engineering, 449–452.

circuit design, which main functions are to collect the signal from sensor and transmit the data to LabVIEW's VIs. DS18B20 has two work modes: parasite powered mode and exterior powered mode. The latter one is selected in this system. The pull-up resistor is 5.1 kilohm and power is supplied directly by VDD and the data line of DS18B20 is connected with the interface Port 3.7 of AT89C51. Control circuit is designed as a independent unit, for this part demand a source of high-power. Hardware circuit of signal measure and control is shown in fig.2.

C. Design of Software

The system software contains 2 parts: C51 language program in AT89C51and LabVIEW program.

1) Main Program of C51

The function of this program is to collect sensor data and transmit it to LabVIEW's VIs. The following fig.3 illustrates the flow chart of main program in microprocessor.

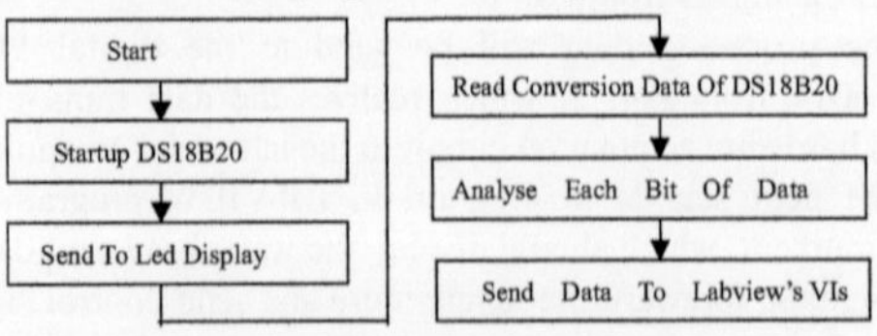

Fig. 3. Flow chart of main program

2)The Operation Sub-program of DS18B20

DS18B20 runs in a very strict time sequence, which can guarantee the write and read on the only I/O bus line. Moreover, the accuracy and integrality of the data can be guaranteed by the strict communication protocols. Several signal sequences are defined in the protocols: initialization sequence, reading sequence and writing sequence. The initialization sequence is illustrated as following fig4.

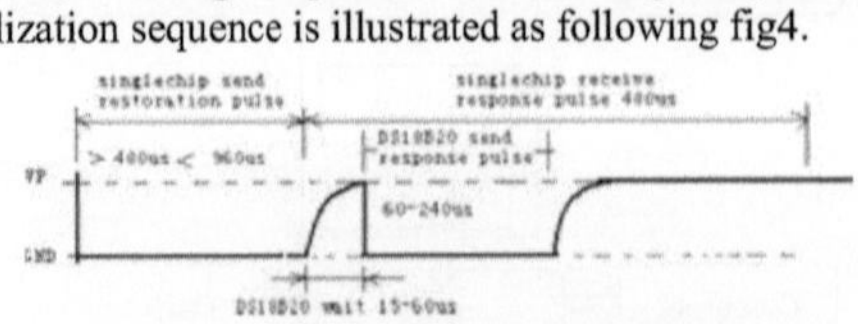

Fig. 4. Initialization procedure of DS18B20

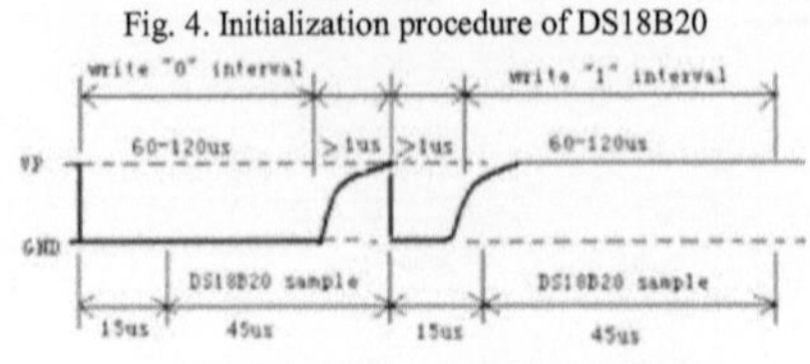

Fig. 5. Write timing diagram of DS18B20

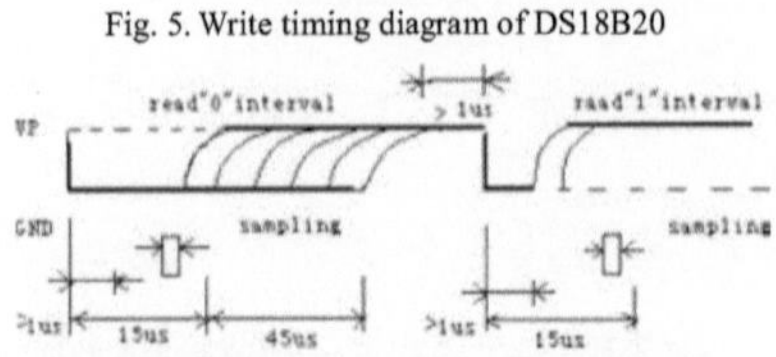

Fig. 6. Read timing diagram of DS18B20

DS18B20 data is read and written through the use of time slots to manipulate bits and a command word to specify the transaction.

A write time slot is initiated when the host pulls the data line from a high logic level to a low logic level. There are two type of write time slots: Write 1 time slots and Write 0 time slots. All write time slots must be minimun of 60 μs in duration with a minimum of a 1-μs recovery time between individual write cycles. The DS18B20 samples the DQ line in a window of 15 μs to 60 μs after the DQ line falls. If the line is high, a Write 1 occurs. If the line is low, a Write 0 occurs (see Fig.5).

The host generates read time slots when data is to be read from the DS18B20. A read time slot is initiated when the host pulls the data line from a logic high level to logic low level. The data line must remain at a low logic level for a minimum of 1 μs; output data from the DS18B20 is valid for 15 μs after the falling edge of the read time slot. The host therefore must stop driving the DQ pin low in order to read its state 15 μs from the start of the read slot (see figure 6). By the end of the read time slot, the DQ pin will pull back high via the external pullup resistor. All read time slots must be a minimum of 60 in duration with a minimum of a 1-μs recovery time between individual read slots.

As described above, the emphases of the c51 are the operation sub-program of DS18B20. The flow of the sub-program of startup of temperature transition and reading of temperature data are described as following fig.7 and fig.8.

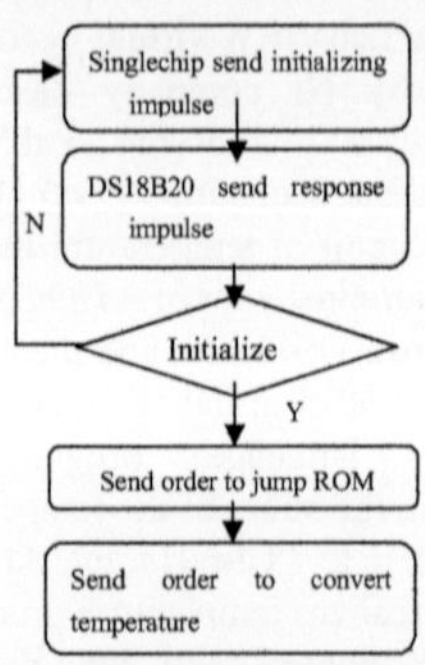

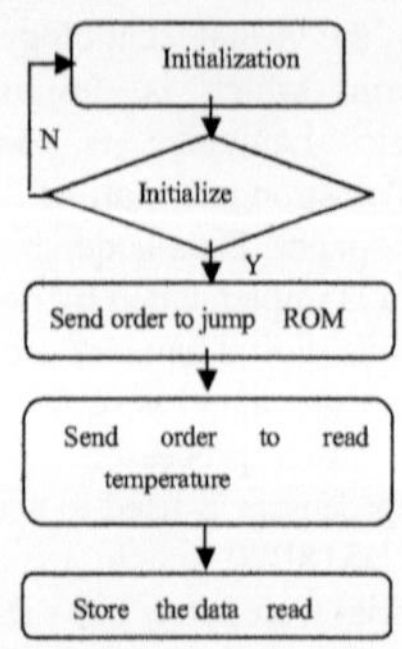

Fig. 7. Flow chart of temperature conversion for start (left)

Fig. 8. Flow chart of reading temperature data (right)

3)Design of VI's Program inLabVIEW

The first step of VI's program is to collect the temperature data from microprocessor. Since it takes about one second for the microprocessor to get the temperature data from DS1820, a 200ms-delay-time sampling interval is set to make VI collect the data from microprocessor for 5 times per second, which will make the wave display of the temperature more consistently. The program of start and stop collecting the data include two cycles--a inner one which collects the data and the outer one which counts the times of sampling. And a boolean switching is used to control the 2 cycles. The enable port of the outer cycle is connected with a not gate, which make the singals of the enable ports of two cycles opposite. So it is realized to collect the signal and count the times of sampling at the same time.

The data store is the part with difficulty in VI program. Specifications are referred in Fig. 9. The design of data processing and exporting control signals are easy parts, which realized mainly by comparison and logical operation.

In this project, DS18B20 is used to Measure the temperature of the surroundings. A alarm for low temperature surroundings and 3 alarms for 3 different high temperature surroundings are used to display the forewarning state. And a fan is used to control the temperature of the surroundings, which is drived by feedback voltage of three levels: 2V, 4V, 6V. Simple fuzzy control method is used in the feedback control. compared the sample value from DS18B20 with the value of 3 diffenrent high temperature condition, the outputs of compare will became “0” or “1”. Then added the outputs up, and amplified the sum by 2 times with multiplying unit, the last result is the feedback export.

The front panel of the VI is shown in Figure 11. A waveform oscilloscope is the core of the front panel, which displays 3 waveforms at the same time: temperature upper limit, temperature lower limit and the value of measure.

Four adjusting knobs are used to set the temperature alarm value of surroundings, and we can adjust the value to making the feedback control more efficient acording to the fact.

The start/stop button is used to start or stop collecting the data, and it also indicates the work state of the VI: when it shows green, the VI stop collecting the data and moving the waveform of the temperature; when it is gray, the VI start to work.

The temperature adjustment button is mainly used to enable or disenable the function of feedback control. When it is enable, the temperature adjusting knob can be used to set the value of the desirable temperature, and the fan and the bulb will work acording to the feedback control signal—when the value of desirable temperature is less than the one of real-time temperature, the bulb will light, and by contraries, the fan will run. When it is disenable, the temperature adjusting knob is invalid, and the program will run in the state of automatic feedback control.

In addition, we set the digital display and the thermometer except for the waveform of the measure temperature. We also set the function of save the data.

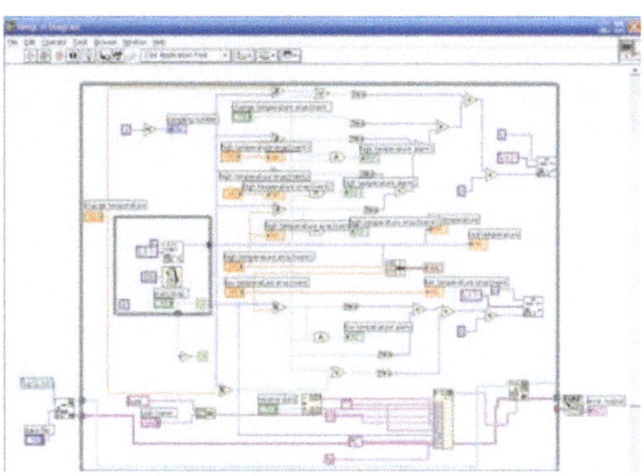

Fig. 9. Flow diagram of LabVIEW

D. The Realization of Remote Transmission

When the IPC is run in the network, we can use it as a server. After configuring LabVIEW panel and opening IE browser on another computer and printing http://the server IP address or the host name/VI program name.htm in address bar, you can activate the front panel of the VI. Fig. 11. shows the front panel seen from web browser.

III. EXPERIMENTS

In this section, we present the Lab simulation experiment of the remote temperature monitor system of web. Fig.10. shows the experimental setup of hardware circuit, and fig.11. shows the VIs Front panel of the system.

Fig. 10. Experimental setup of hardware circuit

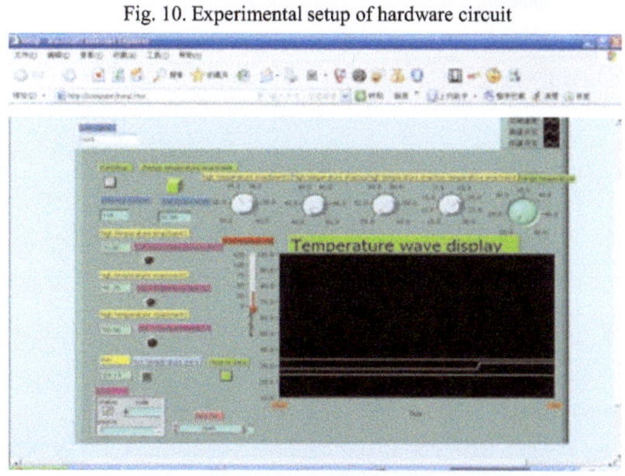

Fig. 11. Front panel seen from web browser

Loaded the system, the value of the led and the VI's temperature wave display are the value of the room temperature.

Then take hold of DS18B20, you will see the curve of temperature ascending gradually.

As soon as it surpasses the value of the first high temperature point, the measure and the control of signal are realized by turning on the No.1 indicator light and driving the fan on the hardware card.

Because the 3 high temperature points are set in certain gradients, with the rising of temperature sequentially, No.2, No.3 indicator lights for warning will light one after the other. At the same time, signal of feedback and control will also change, the velocity of the fan becoming faster.

And only 1 warning light for low temperature is set because it is not convenient to fabricate such a simulant environment. However, the principle of it is the same as in the high temperature state.

Ⅳ. DISCUSSION AND CONCLUSION

The DS18B20 digital thermometer provides 9 to 12-bit (configurable) temperature readings which indicate the temperature of the device. And the measurable range is from -55 $^{\circ}C$ to +125 $^{\circ}C$ and the accuracy is ±0.5 $^{\circ}C$ between -10 $^{\circ}C$ and +85 $^{\circ}C$. Besides, each DS18B20 contains a unique 64 bits silicon serial number, multiple Ds18b20s can exist on the same 1-wire bus. This allows for connecting more sensors to one microprocessor by one cable to realize the real distributed temperature control system.

Certainly, as a remote monitoring and control system, this system has more things to do, for example, the remote access may not be possible if the system is inside a "firewall" and the appropriate port(s) must be opened for remote access.

The system will provide a true environment in which students and faculty members can study engineering, technology, and sciences at a level of detail. And it has proved to be very useful for the experimental training of undergraduates.

REFERENCE

[1] B. Andò, "A new direction for class assistance during laboratory sessions," IEEE Instrumentation & Measurement Magazine, instrumentationnotes, December 2003, pp. 58-63.

[2] Helena Maria G. Ramos, J. M. Dias Pereira, et al., "A virtual instrument to test smart transducer interface modules (STIMs)," IEEE Transactions on instrumentation and measurement, vol. 53, No. 4, August 2004, pp. 1232-1239.

[3] Melanie L. Higa, Dalia M. Tawy, and Susan M. Lord, "An introduction to LabVIEW exercise for an electronics class," the 32nd ASEE/IEEE Forontíes in Education Conference, November 6-9,2002, Boston, MA. session T1D, pp. 13-16.

[4] Dezra Hinkson, Claude Marshall, Sham Robinson, "Design & development of a user interface to remotely control a radio telescope using virtual instruments," in Proc. IEEE Southeastcon 2002, pp. 279-282

[5] Nihnja K. Swain, James A. Anderson et al., "Remote data acquisition, control and analysis using LabVIEW front panel and real time engine," Proceedings IEEE Southeastcon 2003, pp. 1-6.

[6] Piotr Bilski, Wieslaw Winiecki, "Virtual spectrum analyzer based on data acquisition card," IEEE Transactions on instrumentation and measurement, vol. 51, no. 1, february 2002. pp. 82-87.

Fan Yang was bron in Hubei, China, in 1966. She received the M.Sc degree in mechanics and electronic engineering from the Wuhan University of Technology in 2002. In 1988, she joined the department of electric and electronic engineering at At Wuhan Institute of Chemical Technology. Her main research interests are in the area of transducers, Intellectual instrument and measure techniques.

Guoping Li was bron in Hubei, China, in 1978. He received the M.Sc degree in mechanics and electronic engineering from the Wuhan University of Technology in 2005. His main research interests are in the area of instrument and measure techniques.

Huipeng li was bron in Hubei, China, in 1981. A graduate student in the department of electric and electronic engineering at Wuhan Institute of Chemical Technology.

Visual Modeling Using ICT in Science and Mathematics Education

Eugeny I. Smirnov, Vitali Bogun
IEEE Conference Publishing
108 Respublikanskaya street,
Yaroslavl 150000 Russia

Introduction

Utilization of the ICT(Information Communication Technology) gives rise to new opportunities in increase of motivation and efficiency of problem-solving in science as well as mathematical training of the prospective teacher.

One of perspective directions of computerization in a science and mathematical training is utilization of computer-aided mathematical systems (CMS) in scientific research of students and graphic calculators in training of mathematics. CMS are universal mathematical packages of symbolical and numerical calculations (MathCad, Mathematica, Maple, Derive and so on). CMS have joined the category of working instruments for analytical calculations. Utilization of a graphic calculator in teaching of mathematics, being an operative instrument for solving complex computing problems as well as an instrument for recording and visualization of various stages in solving of problems, raises interest to mathematics, makes the spectrum of cogitative operations of students more intense and influences the ways of the training contents presentation.

On the other hand, the prospective teacher should not treat the ICT only as the object of study of their functions, modes, options, communications in order to solve scientific and didactic problems, but as a tool to control cognitive activity of students in their future professional work as well.

The ICT utilization gives a unique opportunity to increase the level of the personal development of a student: growth of computational and algorithmic culture, development of spatial reasoning and graphic culture, expansion of a cognitive circuit spectrum in thinking processes: perception, understanding, representation, etc.

The opportunity of communications as well as utilization of information ideas in the process of exchange of didactic and scientific experience by the Project participants from various countries via Internet for distance training and use of electronic working environments as well as training material is of great importance.

Yet there are a number of contradictions connected with the ICT use in scientific training and mathematical education of the prospective teachers, namely:

• between the rate of development of information technologies and the state of teaching of mathematics in modern teacher's training universities and colleges;

• between opportunities of use the CMS in teaching of mathematics and inadequacy of scientific - methodical development;

• between the necessity of creating in students the skill of construction of algorithmic model, while solving a mathematical problem, and significant volume of the calculations interfering with comprehension of a model structure;

• between the necessity of formation computing skills of students and practical use by students of computer mathematical systems when they solve problems independently.

The problem of the research: define conditions of the ICT integration into the process of becoming proficient in scientific and didactic problems of mathematical training on the basis of visual modelling of objects and processes by students.

The purpose of the research: create an integral system (contents, forms, methods and conditions) of research by prospective teachers in solution of scientific and didactic problems of mathematical education involving of the ICT and utilizing visual modelling of basis and processes.

Application of the CMS for solution of scientific and mathematical problems by students will promote growth of motivation in scientific research as well as in professional development of the prospective teacher on condition that:

1) the practice of visual modelling is included into educational activity during integration of mathematical and information knowledge;

2) students construct projection models while solving scientific and mathematical problems with application of the CMS, which record mathematical optimum procedure mathematical and information actions;

3) students manifest creative activity while learning to use the CMS (a variation of data and analysis of results, formation of hypotheses and their testing, inter-conversion of the sign systems);

4) communicative opportunities for dialogue for groups of students during their educational activity is enlarged by means of information environments (Media, Internet, conferences and so on).

Tasks: (scientific, didactic, information, methodological, professional):

Study functional possibilities and analyze the basic CMS, create the models for modes of work in the information environment;

Reveal didactic conditions and develop a technique of visual modelling utilizing the CMS (the graphic calculator) during teaching of mathematics and solving of scientific problems;

K. Elleithy et al. (eds.), Advances in Computer, Information, and Systems Sciences, and Engineering, 453–457.

Develop a laboratory workshop aimed at solving of physical and mathematical problems utilizing the CMS (the graphic calculator) and the methods of its conducting for students and schoolchildren;

Design interactive volume of information by groups of students on the basis of authoring software products and the results of the CMS (graphic calculator) research;

Visualize the procedure of subject and information actions on the basis of improvement of students' computing and logic culture.

METHODS OF RESEARCH

Visual modelling of objects and procedures

The pedagogical technology of visual-modeling teaching of mathematics plays a fundamental role in the proposed didactic system of mathematical and informatical integration of knowledge. This technology makes it possible to achieve stochastically guaranteed result of teaching of various qualitative levels of learned material as well as integrity of representation of basic mathematical structures.

Visual modeling methods of teaching present:
– "a priori" modeling of the object of perception;
– a process of forming an adequate category of ultimate purpose of the learners' internal actions;
– all teachers' managing actions, modeling of separate pieces of knowledge or an arranged set of knowledge for stabilizing the learners' immediate perception.

Let's turn to the definition analysis (Fig. 1).

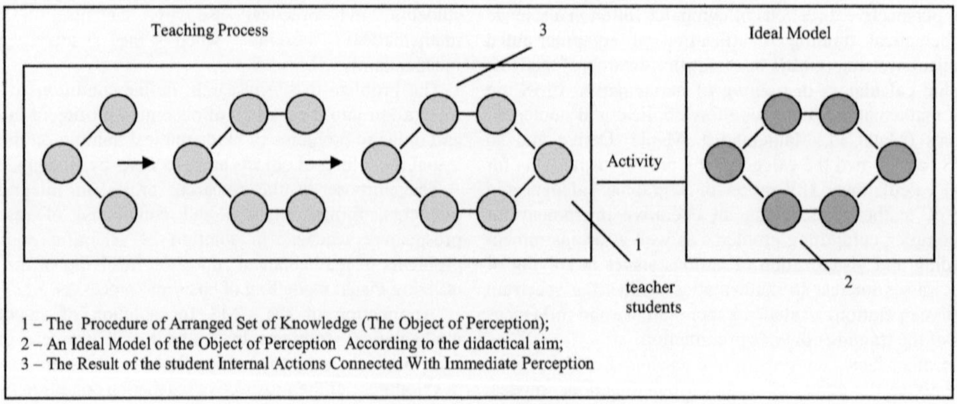

Fig. 1. Visual modeling

The process of perception of the given visual model presupposes all key qualities of the mathematical object. It is especially important when information is of great volume (or contains a mix of mathematical (physical) and informatical knowledge). It is necessary to keep in mind such actions when separate pieces of knowledge or an arranged set of knowledge are given. We can deal with proving theorems, solving problems, teaching some parts of mathematical analysis in its various logical correlations, with a single lesson, a lecture etc.

As has already been mentioned, according to A.N. Leontyev, when visual methods of teaching are used, it is necessary to proceed from the psychological role, which they (methods of teaching) play in the perception of new material. He chooses two functions of visual methods of teaching:
– the first is aimed at extending the sensible experience;
– the second is aimed at developing the essence of the processes or phenomena under study.

In connection with that, external teacher's actions are divided into bearing and structural actions depending on the orientation of the sensible or rational element of perception.

The external bearing actions can be as follows:
– writing down formulas, tables, displaying models, drawing up graphs, formulating theorems, using text-books or manuals.

The structural external actions can be as follows:
– proving theorems, choosing the main theoretical notions and methods, realizing links between different subjects.

According to our concept use of visual methods in teaching of mathematics of a prospective teacher is treated as a special property of psychological images of mathematical objects, the essence of which is considered in an integral paradigm of perception of the basis of the following criterions:
– diagnosable aim-finding of integrity of the mathematical object;
– adequate perception (learner's comprehension of essence of the mathematical object in accordance with aims of teaching);
– stability of perceptive image and presentation under conditions of direct perception;
– cognitive and creating activity on the basis of relaxed and successful teaching.

We should follow of the next structure of visual modeling:

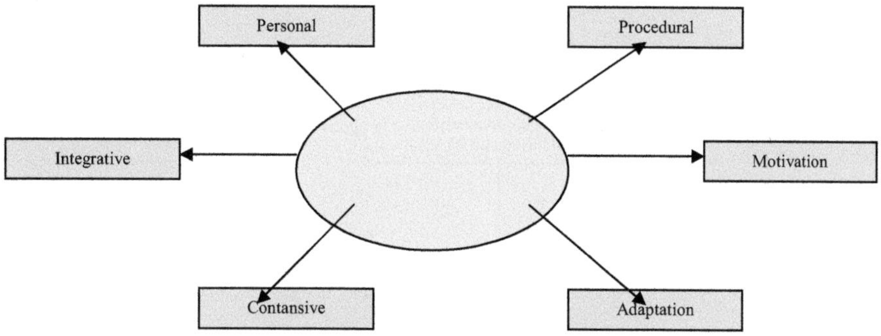

Fig. 2. The structure of visual modeling

Integration of mathematical (natural-science) and information knowledge

Integration of the training contents is most effective on the level of didactic synthesis characterized by integration of organizational forms of training and specificity of didactic problems while dealing with a new material. Integration of the training content (i.e. mathematical (natural science) and information knowledge) is understood as the process and result of interrelation, interaction and synthesis of knowledge as well as ways and kinds of activity with formation of their comprehensive system.

Integration of knowledge assumes possession of the following professional skills:

Skill to solve a problem (formation of a question, finding of scientific information for solution, analysis of a problem situation, setting up of a hypothesis);

Capability for mathematical modelling (definition of the data, conditions and borders of search of the solution, translation of a problem into the language of mathematics, construction of the adequate mathematical device, integration of the solution);

Skill to apply the ICT;

Skill of logical thinking;

Communicative skills.

It is required to teach students to project and investigate mathematical models utilizing the ICT in such components where their application is necessary and justified (complex computing algorithms, visualization and recording of various stages of mathematical actions, construction of complex graphic objects, etc.). Thus construction of integrative information model, which optimizes use of an information resource (functions, commands, modes, algorithms, etc.) is a very important problem.

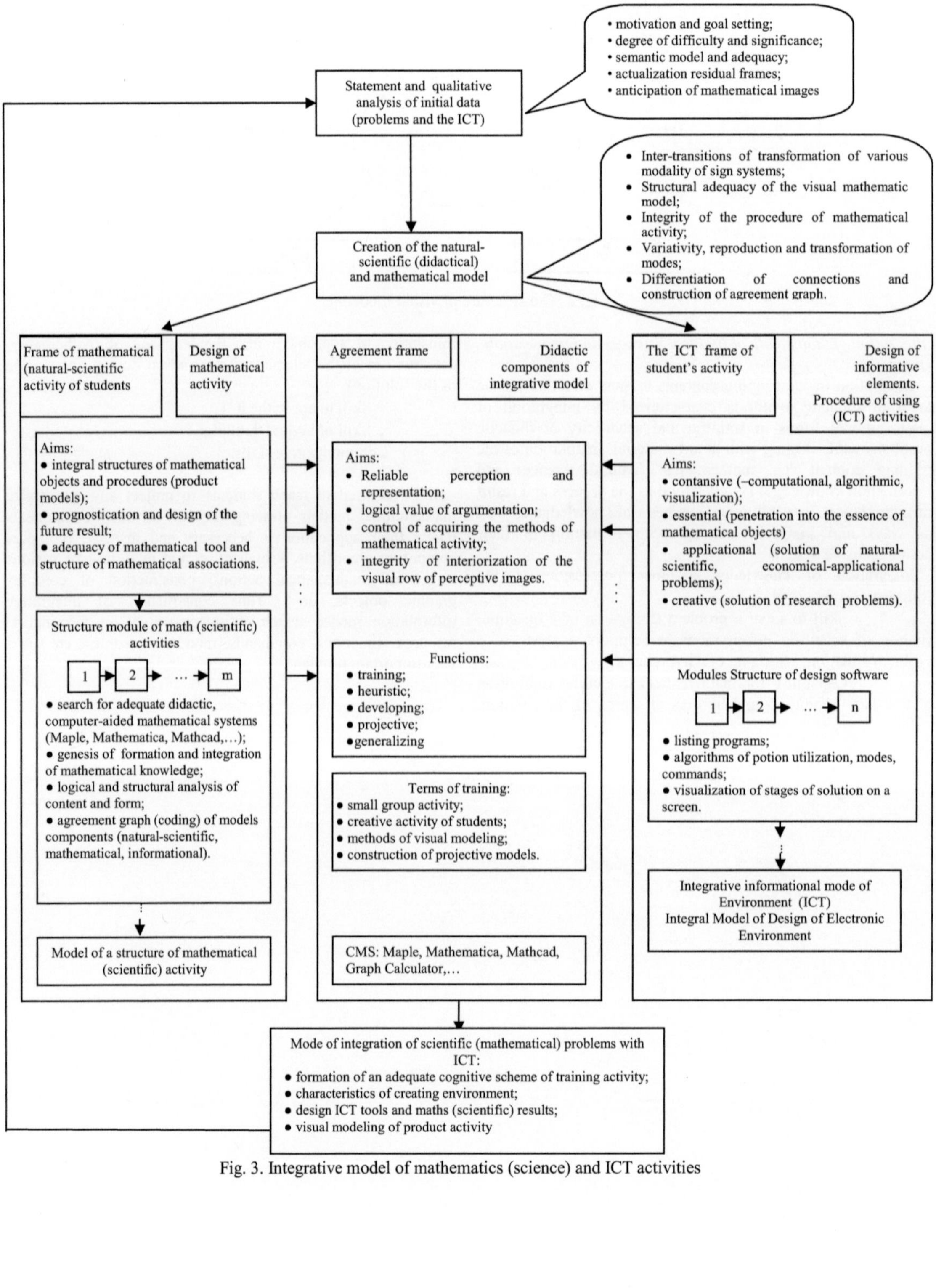

Fig. 3. Integrative model of mathematics (science) and ICT activities

References

[1] V.D. Shadrikov, "Psychology of Activity and Abilities of the Person. Teaching aid," Moscow: Logos, 1996, 318p.

[2] E.I. Smirnov, "Technology of Teaching Mathematics Utilizing Visual Model," Yaroslavl, 1998, 323p.

Software for Self-learning on the Subjects of Cylindrical Involute Gear Meshing

P. Nenov, E. Anguelova, V. Varbanov, S. Ivanov
University of Rousse "Angel Kanchev", Machine and Machine elements
8-th Studentska Str., Rousse, Bulgaria

Abstract - **The topics concerning involute meshing are discussed in numerous disciplines like "Machine Elements", "Theory of mechanism and the machines", "Machine design" and others, which are in the basis of the education of machine designers. Their better understanding is of main importance for wide range of specialists. The theory of the involute meshing includes many abstract terms and definitions. Usually their mastering is assisted with figures, posters, overlaid images, metal and transparent plastic models, etc. The goal of this work is to improve the assimilation of the theoretical matter from the field of involute gearing, with the creation of new and more impressive and memorable multimedia software product.**

I. Introduction

The topics, concerning involute meshing, are discussed in many disciplines from the field of secondary and high education like "Machine Elements", "Theory of the mechanism and machines", "Machine design", specialized courses on "design of gear drives and speed reducers", "design of gear boxes of machine tools", "design of gear boxes for vehicles, tracks" etc. These disciplines serve as a foundation of the education of machine designers, and their better understanding is of great importance for wide range of specialists. The theory of involute meshing [1] along with clear definitions of key areas includes some abstract terms and definitions [2], [3]. Their understanding can be simplified with figures, with poster sets, sets of overlaid images and drawings for projectors, transparent plastic models, metal models, movies etc. The goal of the present work is to aid the assimilation of the theoretical matter from the field of the involute gearing, with the creation of suitable multimedia software product.

II. Method and Basis for Visualization and Animation of Gear Meshing

The "GEARING" software system described below, which is designed for visualization of the problems from the field of the kinematics, geometry and loads of the gear meshing, as well as animations of gears, combines the advantages of the current presentation techniques. Major precondition for its development is the presence of the authors' system GEOMER [5], which facilitates numerous geometrical calculations of cylindrical involute gears, combined with abilities for quick and precise visualization of their teeth, tooth spaces, tooth sectors and whole gear blanks.

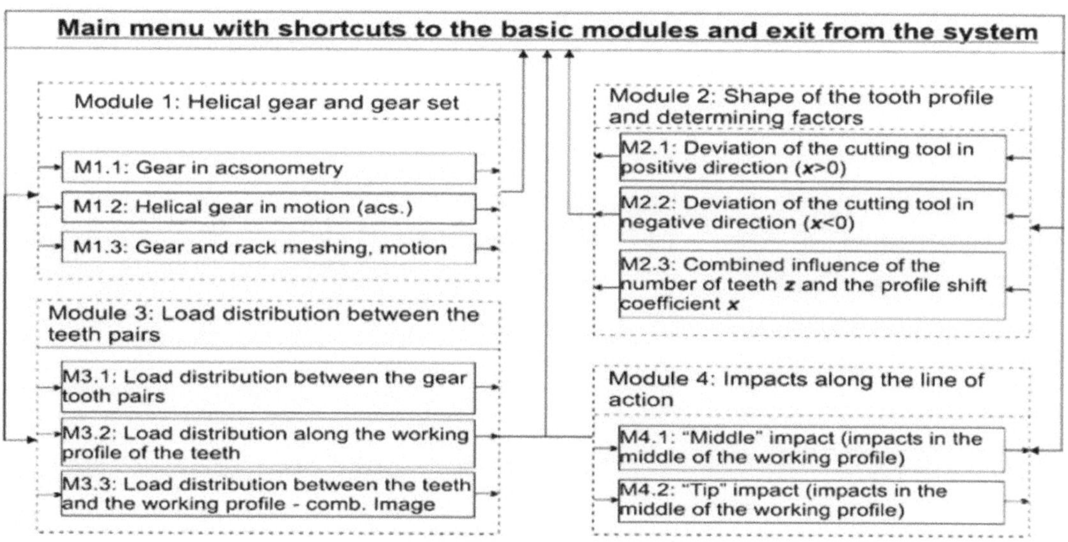

Fig. 1. Structure and scope of the GEARING software

K. Elleithy et al. (eds.), *Advances in Computer, Information, and Systems Sciences, and Engineering*, 459–463.

The philosophy adopted during the development of the software GEARING is to present the basic problems of the theory of meshing not just as a final result. Because it concerns mainly interactions, processes and phenomena, they could be understood better if the mechanism of their appearance and evolution is clarified. Because of that, a key moment is the development of suitable scenarios in which the given problems are to be structured in separate smaller and well-defined basic elements that step by step are overlaid, combined and expanded. In result of that from static picture the visualization is transformed into process, which development follows strict logic. Usually the visualization includes short supplemental texts.

III. SCOPE AND CONTENT OF THE COMPUTER VISUALIZATION AND ANIMATION OF GEAR MESH

The described software system is based on modules. Its structure, scope and capabilities are shown on fig. 1. It includes four basic and a number of program modules, targeted for visualization of:

The gear meshing and its elements (basic module 1);

The shape of the teeth and the basic factors, defining it (basic module 2);

The distribution of the load between the teeth pairs on their working profile (basic module 3);

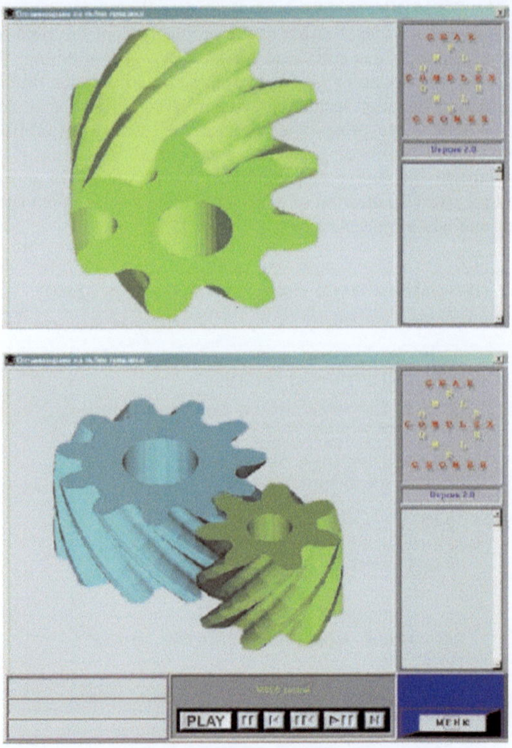

Fig. 2. Photo realistic image of helical gear and gear set

The formation of impacts in the gear mesh (basic module 4).

A. *Visualization and Animation of Gear and Gear Teeth*

The first module visualizes helical gears in motion. The visualization is achieved with 3D photo realistic animation, which could be played with different speeds (fig. 2). The image could be stopped for better viewing of particular details in any moment. The usage of the module helps the development of stable, hard to forget visual images and models. It is obvious form them, that in cylindrical helical gears, the shape of the teeth along their width is helix rather than a straight line, that the inclination of the teeth is different within the different diameters, they enter meshing step-by-step – from the one end to the other, that the meshed gears have opposite angles of their helixes etc.

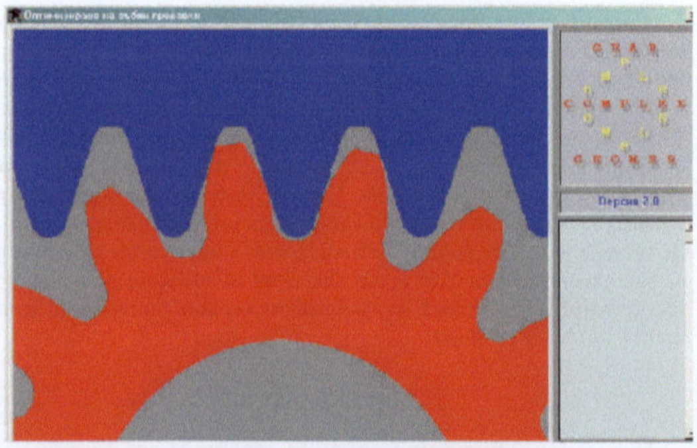

Figure 3. Gear and rack meshing

The program module – M.1.3 "Rack meshing in motion" demonstrates join operation of gear and gear rack, and clarifies in principle one of the most common technique for gear cutting and shaping of the tooth profile – the method of "rolling". The developed visualization proves that the involute is enveloping curve of a family of strait lines. These strait lines are inclined sections from the so-called "basic contour", representing gear rack with trapezoid profile. During "teeth cutting" the same inclined sections are the cutting edges of the rack, the reference line of which rolls without sliding onto the reference circle of the gear and shapes the working profile of the teeth.

B. *Shape of the teeth – working profile and tooth-root fillet*

The second basic module clarifies the influence of the number of the teeth z and the profile shift coefficient x on the shape of the teeth.

It is known from the theory of the gear drives that numerous factors influence the strength and the durability of the gears, most important of which are the mechanical characteristics of the material, the shape of the tooth, the accuracy, the heat treatment, the shaft housing, lubrication regime etc. Practically it is impossible to take into account all of the factors, described above, and to represent them in the gear calculations. In design time, the designer could choose such combination of the number of the teeth z and the profile shift coefficient x, that could make the tooth root thicker, thus to increase its resistance in its root and its bending strength. With the program modules M.2.1 M.2.2 (fig. 1) the influence of the modification of the basic contour in radial directions, on the tooth shape is illustrated. In the example chosen for visualization, the gear is with modulus

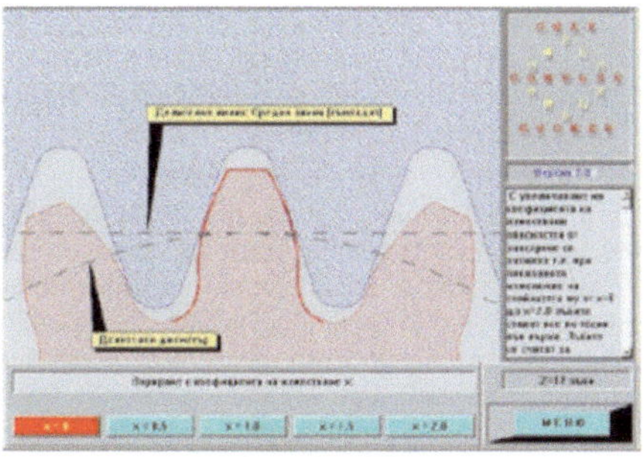

a) without modification

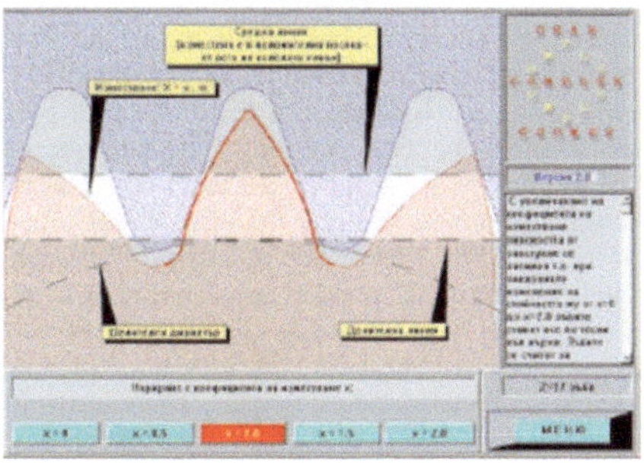

b) with positive modification

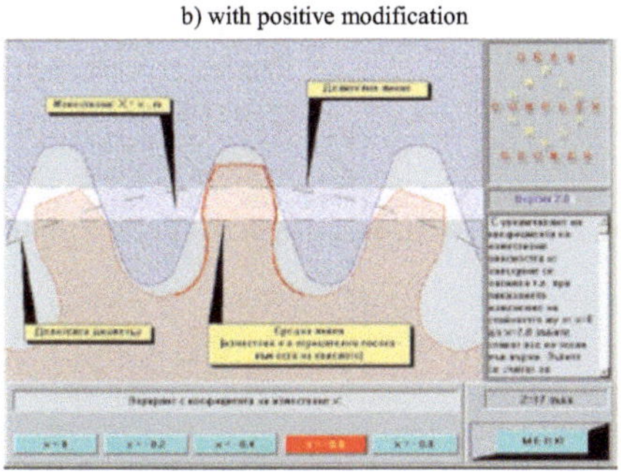

c) with negative modification

Fig. 4. Shape of the teeth at zero a) positive b) and negative c) modification

$m = 4$ mm and number of teeth $z = 17$, and the values of the profile shift coefficient x is from -0.8 to 2.0. The broad scope of $\boldsymbol{x}$ variations creates the possibility for gradient observation of the change of the tooth shape and for clarification of some of the most characteristic distortions in the quality of the meshing – tipping and undercutting (fig. 4). In this module some basic terms like basic contour, reference diameter, reference line, etc are defined.

Good impression for the influence of the number of the teeth z and the profile shift coefficient $\boldsymbol{x}$ on the shape of teeth could be achieved trough visual analysis of the twenty-two precisely generated images, embedded in the software module M.2.3 of the system GEARING. They cover vast varieties of combinations and witch choosing the right window (fig. 5a) the tooth profile of each of them could be shown with its actual shape, including the working profile and the tooth-root fillet (fig. 5b).

In the bottom part of the screen beside the shape of the teeth the basic parameters are also displayed. The chosen examples include gears with small, medium and huge number of teeth, which makes it representative for the practice. The careful study of the shape of the teeth suggests in remembrance manner important conclusions for the designer. For instance, when the gear is with large number of teeth, even high values of the addendum modification coefficient $\boldsymbol{x}$ do not change the shape of the teeth significantly. When the number of teeth is relatively small, with the help of suitably chosen positive corrections, not only can undercut be avoided, but the bending strength of the teeth can be improved etc.

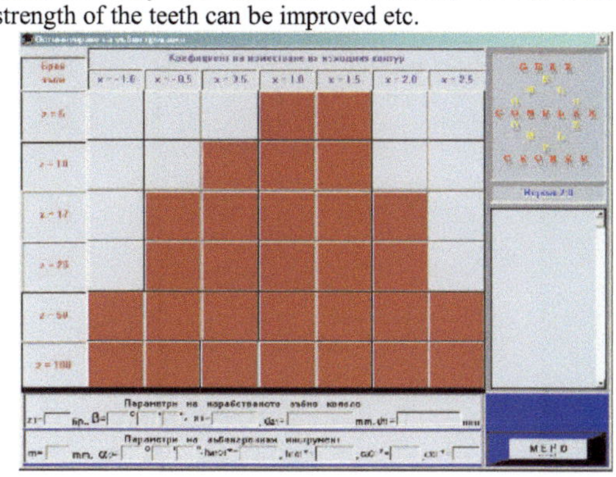

a) combinations of "z-x" with prepared images of the tooth profile

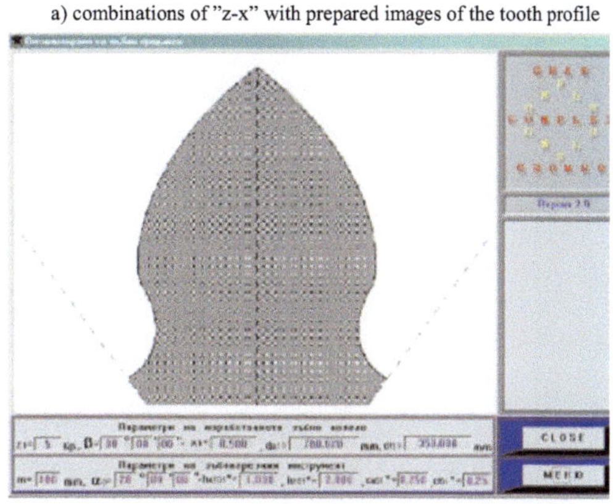

b) tooth contour with precisely generated profile and fillet curve

Fig. 5. Visualisation of the combined influence of z and x on the tooth shape

C. *Load distribution between the gear pairs and on the working profile of the teeth*

The third basic module with its additional modules M.3.1 and M.3.2 shows characteristic stages in the tooth loading (fig.6). These key stages concern the engagement and disengagement of a tooth pair, and are visualizing the sections of single work and the joint work of two gear pairs, as well as the theoretic load distribution between them. The reference between the lengths of the sections is function of the overlap ratio ε_α, which is one of the most important parameters for the gear mesh quality. In high precision gear drives, with the increase of the coefficient ε_α the time during witch the teeth pairs are alone in the line of action is less. This makes the work of the gear quieter, and smoother and increases the load capacity.

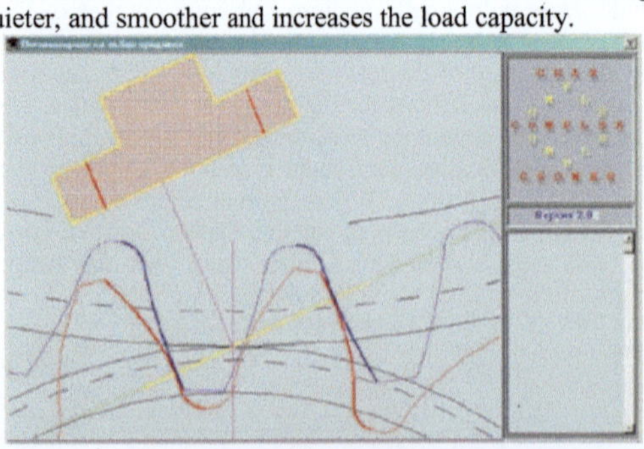

a) overlap ration $\varepsilon_\alpha > 1$ and two tooth pairs

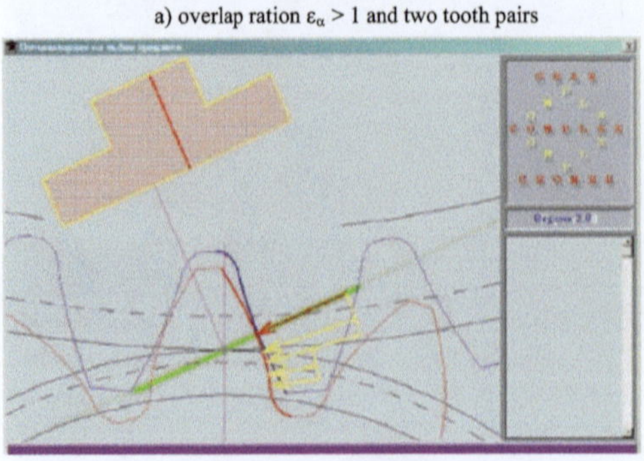

b) single tooth pair in the line of action

Fig. 6. Load distribution in the line of action and on the working profile of the teeth

With the program implementation of the visualization two possibilities are provided – animation of the process of meshing and stage visualization of its key characteristic stages.

D. Middle and tip impact on the line of action

The forth basic module animates the formation of impacts along the line of action, coming from deviations in the basic pitch of the gears. These deviations are influenced mainly by the accuracy, achieved at tooth shaping, which depends on the used techniques, machines, instruments, technologies etc. At the same time even with high accuracy gears, the transmitted torque evokes forces in the meshing, under the influence of which the teeth deform. The arisen deformations (bending) of the teeth decrees the pitch at the base circle p_{bt1} of the pinion and increase the pitch at the base circle p_{bt2} of the wheel. This shows, that when passing trough the line of action there always are differences in the basic pitch of the of the gear pair. These deviations evoke impacts in meshing. The impacts are in the middle of the working profile of the teeth (fig. 7a) when the base pitch of the pinion is larger than that of the wheel, or in the tip of the teeth (fig. 7b) - when $p_{bt1} < p_{bt2}$. That arise the so-called "middle" and "tip" impacts respectively. The additional modules M.4.1 and M.4.2 visualizes the mechanism of damage of the impacts in meshing, which in a lesser or greater extend are inevitable in the work of every gear. The visualization displays the meaning of these gives how to lessen their negative influence – usage of high accuracy gears, initial polish with abrasive pastes, flanking of the tooth profile etc.

IV. Technical Problems of the Visualization and Animation

Basic principle in the development of the software is the achievement of maximal precision of every single picture used.

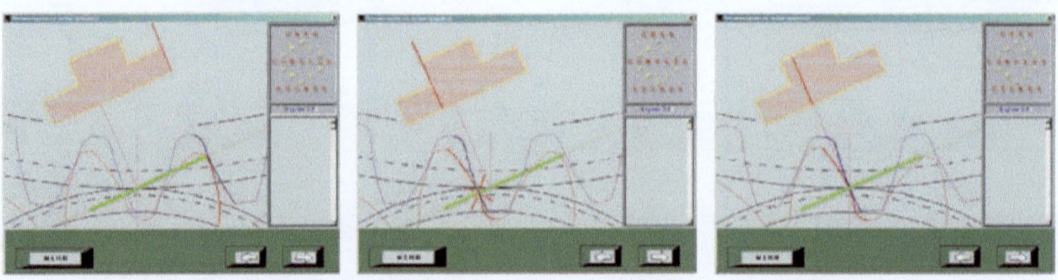

a) impacts in the middle of the working profile – arise when p_{bt1} (pinion) > p_{bt2} (wheel)

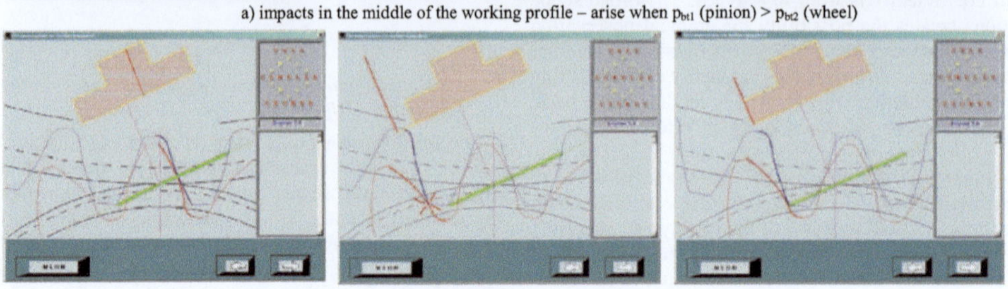

b) impacts in the tips of the teeth – arise when p_{bt1} pinion) < p_{bt2} (wheel)

Fig. 7. Arising of "middle" (a) and "tip" (b) impacts in gears

The visualization is done based on the geometry of real gears and their basic elements. The sequence in transforming the generated gear into graphics is shown on fig 8. The parameters used as input values for the generation of the output contour are: the modulus $\boldsymbol{m}$ of the meshing, the number of teeth z_1 and z_2, the centre distance $\boldsymbol{a}_w$, the tooth width - $\boldsymbol{b}_w$, the helix angle - $\boldsymbol{\beta}$, and the profile shift coefficients $\boldsymbol{x}_1$ and $\boldsymbol{x}_2$. Based on that data with the help of the authors' software GEOMER [4], a text file for the generation of a tooth from given gear is compiled. This file provides information for the coordinates of the points from the tooth profile (of it's involute part and the tooth-root fillet) in Decart coordinate system O-X-Y with centre the gear's axis of rotation.

With the help of an editor a SCRIPT file (text file with .scr extension) understood by AutoCAD is prepared. The generated vector drawing is further edited with AutoCAD and hatches, dimension lines etc. Additional corrections and final editing is done with Corel PhotoPAINT. The final images are exported into raster format, suitable for usage by most presentation software. Thus the database for the final visualization is created. For presentation many technologies could be used, for instance the technologies widely spread for presentation via the Internet. These are the technologies for development of dynamic WEB pages. The integration of the following material as a WEB page gives numerous advantages and simplifies its distribution. The generated images and the supplemental text as well as the animations are embedded in HTML source with the help of HTML editor (for instance MS FrontPage). This allows the source code of the system to be easily modified, without the need of special and expensive tools, and requires only WEB browser to view.

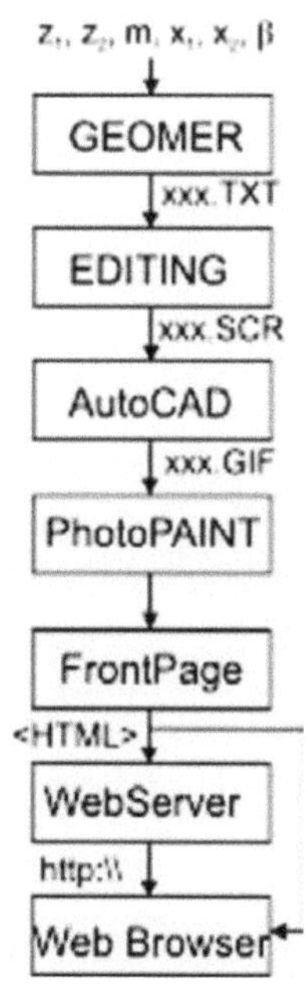

Figure 8. Software GEARING – Scheme for transformation of the data in graphics

Form the animations used in the software GEARING not only constant motion, but also step-by-step presentation is required. This is the reason why they are organized in sets of figures, arranged and shown in sequence, based on the capabilities of HTML and the language JavaScript. The developed product could be uploaded to a web-server and integrated with the existing material, or recorded on media and distributed for local use. This makes the product suitable as a learning tool for e- learning.

V. Conclusion

The software GEARING trough maximum precise images:

- Models basic principles form the field of involute gearing;
- Clarifies visually basic definitions and questions, concerning the influence of different geometrical parameters on the tooth shape.
- Uncovers the nature of the term overlap ratio, simulates the formation of impacts in meshing etc, thus contributing to significantly better and easy learning of key theoretical problem form the scope of involute gearing.

The software is accessible of education on different levels and complies with the modern requirements of CAL-technologies in teaching on the disciplines from the scope of general machine design.

Reference

[1] Гавриленко В.А. Основы теории эволвентной зубчатой передачи. *М. Машиностроение, 1969г.*

[2] Shigley J., Ch. Mischkle. Mechanical Engineering Design. *Fifth edition, McGraw Hill Int. Edition.*

[3] Robert L. Mott. Machine Elements in Mechanical design. *Third edition, Prentice Hall, Ohio.*

[4] Nenov P, A. Anguelova, B. Kaloyanov. "Using of 3-D parametrical blocking contours for optimising external cylindrical gear drives on a "geometrical level" *Int. Design Conf. – Dubrovnik - 2002*

[5] Ненов П. Параметрично оптимизиране на цилиндрични зъбни предавки. *"Техника", С.,2002г. (12 януари, 2004 г. УАПОИ-ОМК)*

The Influence of Cultural Preferences on User Interface Design – Polish Case Study

Adam Wojciechowski, Danuta Zakrzewska
Institute of Computer Science, Technical University of Lodz,
Wolczanska 215, 93-005 Lodz, Poland

Abstract – **Sustainability of international virtual learning communities often appears to be problematic. There are many factors influencing it. The paper deals with the influence of cultural preferences on user interface designing process, which in turn may affect information accessibility and data interaction process – the elements of extended TAM model [1]. We consider Polish users' preferences and needs towards user interface appearance. Paper answers the question if there exist Polish cultural preferences and how they may affect user interface designing. The research is based on an existing Socrates Minerva funded CAB project virtual learning community.**

I. Introduction

Recently, a big amount of international virtual learning communities appeared in the Web. At the beginning they attract the members but sustainability of them is very often problematic. Acceptance of the technology usually plays a big role in it. In [1], there was introduced extended TAM (Technology Acceptance Model)– the model containing all the factors that may influence sustainability of virtual learning communities. Teo, Chan, Wei & Zhang showed, in their paper, that both information accessibility and community adaptivity have significant effects on user perceptions and behavioural intention to use the system. They indicated that information content and amount, access policies, the type of communication channels provided and information organisation, all have an influence on virtual communities. They emphasized the significance of user adaptation ability as one of critical success factors.

Information accessibility is usually connected with functionality and usability of the software and is determined by its ease of use and especially intuitiveness. The key role in co-operation between the human beings and the system plays user interface. Brenda Laurel says that interface is the contact surface of the thing [2]. However in case of computers, user interface can be treated as a place where user has contact with the software. This place should fulfil user's communication and information needs.

People from different cultures (countries) use interface in a different way, anticipate different graphical layouts, have different expectations and patterns of behaviour, use different metaphors. Therefore, the user interface designers should take into consideration the users' needs and anticipations that arise from culture preferences.

In the paper we present the research that was done on the cultural preferences of Polish students belonging into CAB (Collaboration Across Borders) international virtual learning community. After short description of relevant research, we present the research method, based on the specially prepared questionnaire. In the next sections, we analyse the obtained results and finally give some conclusions and advices for user interface designers.

II. Relevant Research

In last years many authors examined the factors that may influence user's perception of the interface. According to Raskin [3] successful user interface design is based on considering both ergonomic and cognetic features. Traditionally, design is subdivided into three macro-categories, based on meaningful characteristics [4]:

- information representation and appearance,
- access-navigation-orientation,
- informative content architecture.

Many authors [2,5,6,7,8] have distinguished five main elements of the interface connected with specific user requirements:

- metaphors,
- mental model,
- navigation,
- interaction,
- appearance.

In the previous research, concerning culture preferences, the focus was made on the role of technical features, in adaptation of user interfaces to specified culture, taking into account such factors as appropriate currencies, measurement systems and the language solutions like translation, adapted terminology and using of idiomatic expressions.

But perfect interface should meet also individual user preferences. Unfortunately, it is very difficult to achieve and such adjustable products are not available so far. Numerous software companies usually provide software with interface affected by programmers bias. The problem, which is not big in case of the local market becomes more significant, while the software is sold globally for users representing different nationalities and different cultures.

Due to users' different localisation, there is a need of interface diversification. To make it thoroughly, all aspects of

The work was supported by Socrates Minerva CAB project 110681-CP-1-2003-1-UK-MINERVA-M.

K. Elleithy et al. (eds.), Advances in Computer, Information, and Systems Sciences, and Engineering, 465–471.

user interface appearance and functionality should be taken into consideration. Our previous work [9] was concentrated on the detailed analysis of cultural aspects affecting user interface design with the focus on cultural dimensions. The last ones were also thoroughly described by Baumgartner [2], Trompenaars [10], Condon [11], Victor [12], Hall [13], Hofstede [14] or Marcus [15]. Each of the authors has concentrated on different cultural aspects but merely the whole set of dimensions is the ideal background for national preferences survey [9]. Since Polish user cultural profile has not been examined yet, this paper aims to retrieve prevailing tendencies in polish people's behaviour and their preferences concerning user interface design.

III. Research Method

While examining the sustainability of CAB project community, several tools connected with different factors influencing it, were created. There were considered groups of methods connected with such aspects as social and pedagogical benefits and acceptance of the technology. The influence of the culture preferences on the last element was one of the goal of the examination.

As the main research tool, we used the questionnaire comprising 42 questions quantifying cultural dimensions' importance and theirs influence on user interface design. Questions were posed in polish language to avoid any misunderstandings. Answering consisted in the choice between strong confirmation or rejection of a certain cultural behaviour, as well as rather confirmation or rather rejection of it or staying neutral. However, at this stage of the research, in the final analysis we didn't differentiate between strong and normal confirmation of the cultural dimension or its strong or normal rejection.

The research was done on the basis of the answers of Computer Science students of the Technical University of Lodz. The amount of 65 students volunteers attending Multimedia Systems classes took part in the survey. All the participants had a basic knowledge about user interface designing.

Questions were divided into four groups, each of which concerning different problems. Some examples of them are presented in the Table I.

The first set of questions was to reveal students attitude towards achievements and their personal activity. Some of the questions concerned time perception, human adaptability and creativity. Some of them were to show how openly feelings should be expressed and how much information people retrieve from the context.

The second group of the questions concerned polish political and economical situation on an international market. Questions were to show authority conception, the culture strength in international comparison and international co-operation. Some questions in this group were to reveal polish economic progress and political decentralisation.

The third group of questions dealt with polish psychological profile. Their goal was to show polish national features like: enthusiasm, hopefulness, pessimism, face-saving ability, perseverance, thrift and spiritual or material goods preferences. Additional goal of the survey was to present polish internal and external relationships. The features comprised: gender roles, subordinate distance, instrumentalism, individualism and collectivism.

Final set of questions were to estimate polish harmonisation with nature, technical development rate, uncertainty accompanying new technologies and intellectual property perception.

The main goal of the answers' analysis was to reveal polish cultural dimensions spectrum [9] and to determine what should be modified in actually used polish user interfaces not appearance deteriorating and intuitiveness declining at the same time. The results of the survey answered the question how different polish user interface might look from actually used patterns and the significance of the investigation of the culture preferences on user interface designing.

TABLE I
EXEMPLARY QUESTIONS

Group	Questions
I -Attitude towards achievements	Do you feel that Polish appreciate achievements more than properties? Do you think that emotions should be hidden while acting? Do you find Polish flexible and creative?
II – Polish culture on international market	Do you find that Polish prefer democratic government rather than authoritative ? Do you feel that Poland have the influence on the international situation? Do you think that it is allowed to enter the public places without asking the permission?
III – Polish psychological profile	Do Polish people like working in teams? Are Polish traditionalists ? Do Polish appreciate money and properties? Do you think that human relations should be stiff and depend on positions?
IV – Polish , technology and intellectual property	Do Polish appreciate the role of the technology? Do you find Polish as open minded? Do you find Polish culture is based on the technology ? Are Polish legalists?

IV. Analysis of the results

A. Attitude towards Achievements

First group of questions was to show how people perceive their achievements and to which extend they ascribe position to titles of nobility [10]. More then a half (54%) of respondents stated that Polish are achievement oriented rather than ascriptive oriented. Almost 40% responded contrarily. Students claim that level of pessimism increases with age of respondents. It means that for young people it is still more important what someone DOES rather than what someone IS. These factors can become important when using titles. For example in ascriptive countries address forms would have the need to provide fields for entering titles of nobility. In case of ascriptive virtual communities it becomes a critical element as the web is a medium that allows for a new status achievement.

High level of activity orientation [11,16] is confirmed by responses to a subsequent question. Nearly 73% of questioned students declared themselves as doing-oriented rather then being-oriented. They claimed to like action, efficiency and to concentrate on things that can be observed or measured. Most of them were unwilling to take time to discuss first and understand well complex things. Activity orientation can play a big role when designing metaphors and interaction process of the interface. Active people prefer working metaphors that show manual work rather than brainwork. Polish activity goes with their polichronic time attention [10,12,13]. People used to do several things at one time and their plans used to be constantly shifted (69%). Persons representing such an attitude prefer browsing through the information space rather than writing precise queries for information retrieval. Yet this factor seems to be age dependant it really suggests how linearly or serendipitously materials should be presented.

Users' poll revealed that over 85% of persons believe that feelings should be expressed openly, reactions might be shown immediately by means of verbal, mimic or body communications (affective culture [10]). Only 12% of students believe that emotions confuse the issues and that is why they should not be expressed and shown to the public (neutral culture). This dimension affect user interface interactive feedback. It might be adjusted to the user profile. Discussion with *neutralists* should focus on the object or proposition while discussion with *affectivists* may be enthusiastic and vehement and need not to concentrate on subject but on persons involved too.

Reasonably great amount of questioned persons (66%) presented themselves as low contextual, what means that they prefer information stated clearly but not derived much from context [12,13]. A low context communication is similar to communication with a computer – not explicit information causes programme distortions. This factor has an impact on all aspects of interface design. Clear understanding of the interface context is critical for having a successful or positive experience. Users may benefit from high-context graphic only in cultures where high-context communication is common and can be understood. High-context orientation lets interface

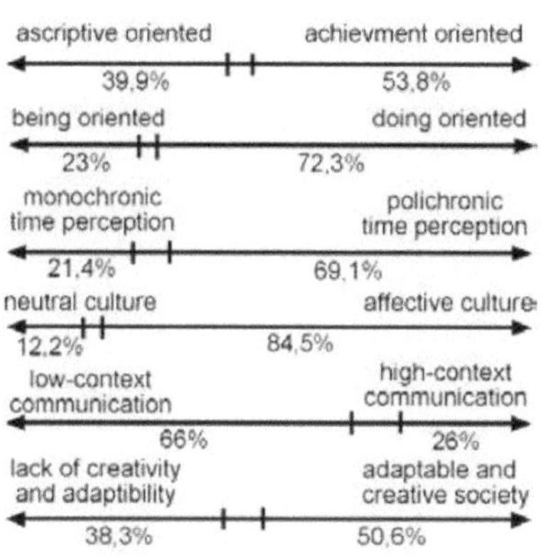

Fig.1. The results of the first set of questions

designers support navigation and interaction process more efficiently. The exact results of the first set questions analysis is presented on Fig.1.

Polish level of creativity and new situations' adaptability was presented ambiguously [17]. Almost 50% of students described themselves as creative and adaptable while above 40% of questioned users expressed their lack of flexibility and adaptability. It should be emphasized that survey was carried out among students and probably adults' opinion would be different (even less of creative and flexible representatives) . The results may affect not only user interface design but economic progress as well. They show that polish prefer already known solutions and are rather unwilling to new concepts. This might be the hangover from the previous communist system. It means that new interface design solutions become more and more popular as the need for experimenting increases.

B. Authority Conception

Authority conception ranges from democratic, through authority-centred to authoritarian [10,12]. Polish respondents opted generally (77%) for democracy. This factor influences users' mental model and interaction methods. Society brought up according to democratic rules claims to have rights to question authority contrarily to authoritarian societies. Such dimensions may influence users' reactions and their involvement in the discussion. Uncertainty of discussion moderator existence or uncertainty of presented materials may determine tone of words, language style and freedom of thoughts expression.

Questions concerning degree of strength of a country/culture in international comparison [17] revealed that polish have the complex of minority. Almost 90% of respondents stated that they have minor position on international market. This feature has shown that Polish are soaked up with western Europe and American patterns. Even people having own point of view different form prevailing solutions hesitate to present it and only accept abroad tendencies. This is mainly the problem of eastern Europe countries and it concerns also Poland. Nevertheless the considerable improvement of Polish international position or increasing of national individualism may cause rapid need for polish interface construction and may speed up process of polish requirements expression.

Political decentralisation dimension [17] pertains believes that decisions made with participation of greater amount of people would be more relevant to diverse interests of society then those made only by political authorities. 49% of respondents present lethargy and indifferent attitude whereas only 40% represent enthusiastic and full of energy point of view. Lack of hopefulness was revealed in answers regarding power of distance perception [14]. Very scarcely less powerful members of institutions and organisations accept that power is distributed equally. Huge unemployment level and intimidation, that take place in Poland, cause that relation between boss and subordinate is strictly ruled and dependant on the decisions of the boss (68%). Only several institutions governed in a modern manner tend to diminish distance between bosses and subordinates, that try to work closely together and consult with each other. These observations are verified by the dimension of space [13] perceiving. It refers to the invisible boundary around the individual that is considered as "personal". 74% claim that they would not accept entering a public space (office) without prior permission. That culture dimension may affect fields of "feedback mechanisms", "open/restricted access", "error messages" and the "use of colours". It may also influence sharing of a workstation (i.e. individual vs. group access), providing personal information (registration dialogues) or interfaces in public space. Overwhelming level of pessimism explains why Polish are not eager to express their own feelings and requirements. The fact that low power distance model become more and more popular is consoling and it means that this research is very important. All the results of the second part of the survey are shown on Fig.2.

C. Psychological Profile

Polish perceive themselves as fast moving market where plans are made in short-term perspective. We stand for fostering of virtues related to the past or present. Even though most of us are traditionalists (65%) we do not bother with fulfilling social obligations and 'face' preservation (only 33% respect it). We are go-getters, yet we do not pay attention to thrift and we are not persevering (71%). Most of Polish strive for accumulation of money and other goods (87%) while spiritual maturity is important only for 13% of respondents. Such an economical and material bias may influence regarding feedback like visualisation of progress bars and progress in general. It influences the amount of efforts user is willing to make in order to get to know the product, and to use it. We seems to be rather impatient and our interests are benefits-oriented.

Polish students have appeared not to put attention to face-saving. Almost 57% of them stated that they would be able to loose their prestige and outward dignity for the sake of

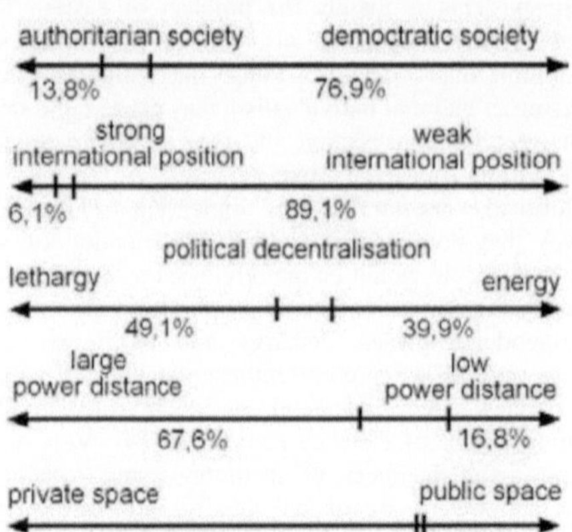

Fig.2. The results of the second part of the survey

negotiation deal. Only 11% definitely stated that face saving is more important then business deals. This society feature affects mental model and interaction process carried out by means of the interface. Respecting considerable differences in this field among certain countries user interface should provide descent business transaction security. This might be critical for interface acceptance and having confidence in it for commerce and social transactions.

When describing gender roles, results were divided into two equal groups [14]. Almost half (47%) of people state that social gender roles are clearly distinct while the second half (48%) claim that gender roles overlap. This reveals considerable change in young people point of view, because traditional polish pattern ascribe to both genders explicit roles (men are supposed to be assertive, tough, and focused on material success, whereas women are supposed to be more modest, tender, and concerned with the quality of life). Unquestionable is the fact that significant gender differences exist, both cultural and biological. Besides public state that default design typically favours the male preferences, both gender's preferences become to be considered. Even though systems have no expectation of masculinity or femininity, calls to action (sale conversation) may require direct appeal to gender role.

Almost each of questionnaire respondents have revealed its nature as basically good (90%). At the same time most of people (70%) declared themselves as being able to change. Essential is the fact that according to human nature orientation, people need to trust the content they read. This dimension shows how a big effort has to be made for gaining users' trust. Polish are generally confidential until the moment when something abuse their trust. On the other hand accomplishing anything should affect in changes. From this point of view Polish pertain to a group of adjustable persons. That is why new, wisely introduced, user interface concepts may meet users acceptance.

Polish have appeared to be individualists (77%) with still growing work-in-group abilities (37%) [10,11,14,16,18]. Whereas about 51% of respondents denied their will for working collectively with other people (see Fig. 3). It means that Polish look after themselves and their family and do not feel integrated into strong cohesive groups. This cultural dimension may affect user interface design when co-operating by means of electronic media. User interface, especially those in groupware category, must be able to deal with varying levels of individuality and anonymity that are required [19]. Culture in which people tend to value relationships for what they can do for them rather than the relationship itself is called instrumental [18]. Polish students revealed such tendency (53%) stating that most relationships have economic bias. Only 38% of respondents showed that pure friendships does really exist. This factor is important especially in establishing a relationship between the interface and the user or group of users, i.e. in a web site striving to attract and retain a certain group of visitors. Society instrumentalism especially may have a great impact on virtual communities

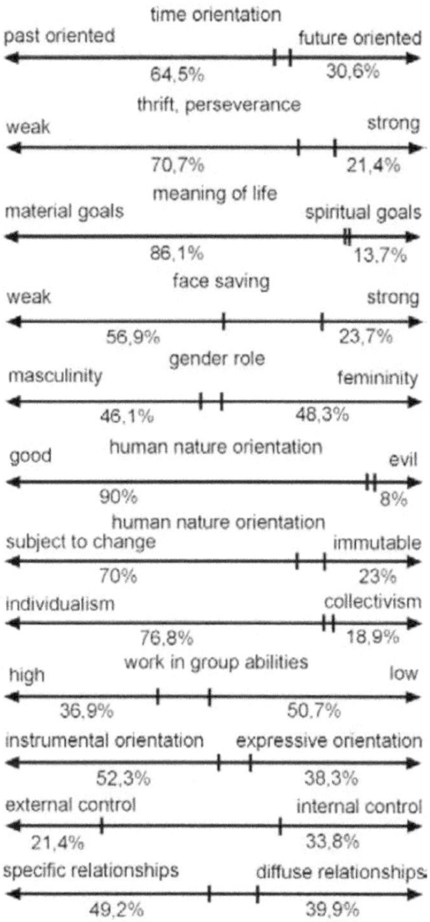

Fig. 3. Results of the third part of the survey.

sustainability. They can mainly survive due to unselfish and honest community participation. People with instrumental attitude perceive virtual co-operation as worthless and lost of time.

Merely 21% of people perceive nature as dominating and more powerful then the individual. Such attitude can be called externalism [2,10,16] and it proclaims harmonisation with nature and its emulation in technique issues. 34% of inquired declared themselves as internalists and see the major force in life, its virtue which resides within the individual. The great amount of people hesitated to confirm their adhesion. This social dimension is crucial for interface design process in order to forecast possible users' reaction and technology acceptance. Internalists have tendencies to blame themselves if they can not use the system, since they are not adapted to the technology yet. Externalists would rather blame the system designer, since they require technical issues to be dominated by people. Individualism predomination cause that ergonomics researchers can not afford ambiguous solutions and declines the margin of freedom and flexibility in a new concepts' introduction.

To find out how much people are willing to commit to communication and collaboration via any interface, the dimension measuring how people get involved with other life space has to be considered. 49% of students opted for (40% against) explicit, regulated, relations with others, what's more 74% of respondents would prefer real and personal contact with the boss and that he took into account their personal life. This social profile do not implies interface design as much as the feature how and why users are introduced and related to each other.

The results of the third part of the survey is presented on Fig. 3.

D. Influence of Technology

Unfortunately we do not have sufficient technological development rate (66%) in Poland. This statement suggests that Polish should be treated as novice users. This is important as developing an interface for advanced users should be considerably different from designing for the beginners. What's more the situation is even worse as great amount of us (39%) are past oriented what means that everything that can occur has took place before and past patterns would be replicated. 40% of respondents perceive themselves as future oriented, implying expectancy of advancement, improvement and progression. Similarity in tendencies of our society shows our stratification and reveals that Polish become still more and more accustomed to technological innovations. On the other hand interface designers should take into account still great amount of people not familiar with new technologies. These social dimensions may influence navigation process and let the users foresee their perspectives – foresee things that may happen within an interface. This dimension becomes also important when designing "business applications" or "providing historical or visional background or motivation".

Low technological development rate mentioned above is strongly connected with the extent to which the members of the culture feel threatened by uncertain or unknown situations [14]. Most of people (75%) perceive new media as uncertain, whereas satisfying the need for certainty will affect the satisfaction the user experiences with the interface. Avoidance of uncertainty cause creation of different values regarding formality, requirements and tolerance for ambiguity. Within interface this factor may highly influence payment possibilities, interaction and navigation design.

Nevertheless answers to subsequent questions revealed that people set their hopes in new technologies (71%). Among control-oriented cultures technology is perceived as a positive good and mastery can be accomplished through technology [12]. Only 19% of respondents definitely stated that human beings should live in a harmony with nature and be subjected to the surrounding environment. These opinions show

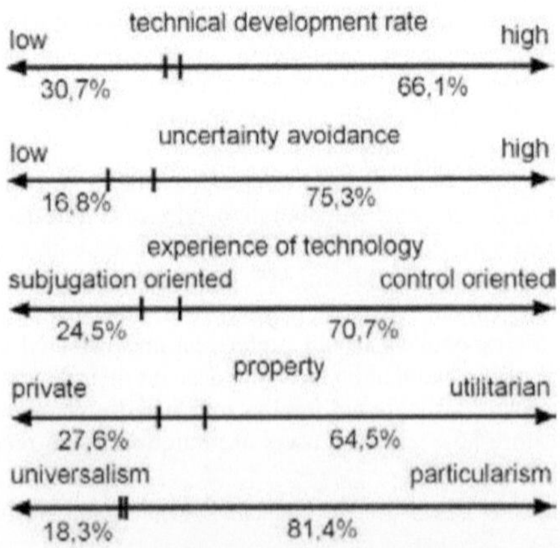

Fig. 4. The results of the last part of the survey.

promising results that Polish are opened for new technologies. This fact is very important as it would be no useful to design the interface if people didn't use it. This factor is critical for the adoption of media as well as the interface. The only problem is suitable interface design, that provides "smart" and "good" technology usage. As far as possible user interface should support user's activities.

The last but disappearing dimension concerns material and intellectual property perception [11]. Almost 65% of Polish represent utilitarian point of view. They treat land and intellectual achievements as common goods and claim that everyone has to take care of them. This situation seems to be a basis of the whole crisis over intellectual property on the web. It corresponds with particular attitude towards the law. Over 81% of respondents are able to fail to obey the rights in case of unique circumstances. The degree of adhering to agreed standards of behaviour is considerably law and reveals our contrariness. From one side Polish would like to be appreciated by foreign countries but from the other they do not appreciate foreign achievements.

The exact results of the last part of the survey can be seen on the Fig. 4.

V. Polish User Interface Profile

The research revealed wide spectrum of Polish users' preferences. These trends and tendencies should be taken into account during the process of user interface designing. The concluding remarks for interface appearance and functionality can be summed up as follows:

- users do not pay attention to titles
- users do not like reading instructions and take up actions without thinking procedure over. Success can be achieved by clear and understandable interface with suggestive (working) metaphors. Lack of successful interface interaction can discourage user from further interaction.
- information should be presented serendipitously with reasonable but descent amount of options rather then linearly with only one discussed topic. It causes that interface is not boring and attractive for enthusiastic and vehement polish users. At the same time polish users are low contextual and that is why information should be expressed clearly.
- Polish hunger for democracy affects discussion process. Users like to have influence on decision making, take part in discussions and forums. On the other hand they do not put attention to face-saving and as a consequence they do not put confidence to on-line arrangements unless properly certificated. Due to such state only signed and certificated transactions make sense.
- Polish are traditionalists however they pin their hopes on new technologies. People are adaptable and creative but due to minority complex they do not trust domestic products. In the most users' opinion Western Europe or American products are better than Polish. Such an attitude can be changed if Polish find domestic solutions cheaper and more helpful than foreign ones. Polish may eventually accept the solution if it brings benefits. They are usually willing for participating in virtual communities as it is new for them, but the economic point of view become more and more predominating and Polish participants will abandon their membership , if the costs of it are too high.
- Polish are individualists and a group work is very difficult to perform on-line. This ability should be learned at school and developed by means of new media.
- Polish do not expect interface gender diversification, however traditional gender roles disappear and femininity becomes noticeable and female users are sensitive over separate treatment. If the interface is to attract women it should address their needs.

VI. Conclusions

In the paper we present some indications concerning the process of building user interface, with Polish cultural preferences. The research was done on the basis of different culture dimensions. The results of investigations will allow to design the interface that will fit Polish users profiles requirements and will cause that the information provided on the portal will be accessible, what is the main key to sustainability of virtual learning communities.

Dimensions overview revealed that the interface for Polish users should differ from the one built for other nationalities. Poland is still the developing country where transformations process has begun quite recently. 15 years of an open market existence changed people minds, but unfortunately post-communism bias left considerable amount of old system remnants, what was easily seen during questionnaire responses analysis. Interface design process for Polish users should be based on Polish cultural dimensions profile, even if people preferences still evolve. However most of older persons cannot catch up with prevailing trends the new generation changes .

Polish are used to take patterns from western countries, have minority complex and claim to have a weak political situation. Additionally Polish have low technological development rate and they are afraid of uncertain and unpredictable situations. On the other hand still increasing creativity, flexibility and adaptability make the research concerning user interface design very fruitful and interesting.

The research revealed that polish user interface designers should put attention to titles and people positions. They should use active metaphors but unfortunately in a low contextual manner. Information within an interface should be presented serendipitously and there is no need to concentrate on one topic. Polish appeared to be impatient, so progress bars are rather required. Subjects can be presented in an open manner as the discussions are not threatened with moderator integration. People present low acceptance level so any programme faults or instabilities evoke uncertainty. On the other hand lack of face saving dimension and lack of haptic issues cause that users treat electronic mass media in a suspicious way. Confidence can be achieved by means of various certificates, published by foreign, high-authority institution. Polish individualism cause that the system should be designed to let people work individually and subsequent coordination process should be carried out at the final stage.

The interest of Polish people in virtual community membership may be also obtained by providing the benefits of participation. In the case of learning communities they are mostly connected with pedagogic and social aspects. The experience and the investigations showed that all the factors should be taken into account while building and developing the community.

References

[1] H.H. Teo, H.C. Chan, K.K. Wei, Z. Zhang , Evaluating information accessibility and community adaptivity features for sustaining virtual learning communities, International Journal of Human-Computer Studies, vol.59, 2003, 671-697

[2] V.J. Baumgartner, A Practical Set of Cultural Dimensions for Global User-Interface Analysis and Design, master thesis, Wien, 2003

[3] J. Raskin, There is no such thing as information design, in: R. Jacobson (Ed.), Information Design, MIT Press, Cambridge, MA, 1999

[4] M. de Marsico, S. Levialdi, „Evaluating web sites: exploiting user's expectations," Int. Journal of Human-Computer Studies, vol.60, 2004, pp. 381-416

[5] P.A. Booth, An Introduction to Human-Computer Interaction, Hove [et. al.]: Erlbaum, 1989

[6] S. McDaniel, What's Your Idea of a Mental Model?, http://www.boxesandarrows.com/archives/print/003253.php

[7] A. Marcus, Dare We Define User-Interface Design?, Interactions vol.9, 2002, pp. 19-24

[8] A. Dix, J. Finlay, G. Abowd, R. Beale, Human-Computer Interaction, New York : Prentice Hall, 1998

[9] A. Wojciechowski, D. Zakrzewska, "Cultural aspects affecting user interface design for virtual learning communities", in: A. Kieltyk (Ed.), Multimedia in Business & Education, Foundation of Modern Management (vol. 2), Bialystok, 2005, pp. 136-142.

[10] F. Trompenaars, C. Hampden-Turner, The seven dimensions of culture, http://www.7d-culture.nl/Content/cont042.htm, last visited 8 May 2005

[11] J.C. Condon, F.S. Yousef, An Introduction to Intercultural Communication, Indianapolis: Bobbs-Meril, 1981

[12] D.A. Victor, Cross-Cultural Aspects of International Business Negotiations, http://www.negotiations.com/Demo/ibndemo.pdf, last visited 8 May 2005

[13] E.T. Hall, Beyond culture, New York: Doubleday, 1989

[14] G. Hofstede, Cultures and Organizations: Software of the Mind, London, McGraw-Hill, 1991

[15] A. Marcus and E. W. Gould, "Cultural Dimensions and Global Web User-Interface Design", ACM Interactions, July/Aug 2000, pp. 32-46.

[16] F.R. Kluckhohn, F.L. Strodtbeck, Variations in value orientations, Evanston: Row, Peterson, 1961

[17] Q. Wright, The Study of International Relations, New York, Appleton-Century-Crofts, 1955

[18] T. Parsons, Talcott Parsons on institutions and social evolutions, selected writings, edited by Leon R. Mayhew, Chicago, University of Chicago Press, 1987

[19] T. Fernandes, Global Interface Design. A Guide to Designing International User-Interfaces, San Diego, Ca. Acad. Press, 1995

Deploy a Successful E-learning Strategy

Joseph Bih
CIS Instructor
Jarvis Christian College
Hawkins, Texas 75765
Email: jbih@jarvis.edu

Abstract: As e-learning is emerging as nontraditional learning approach, it's becoming more acceptable to our society. Although the driving force of evolving Internet and other supporting technologies render the delivery of online courses easily accessible to students, e-learning might create a paradigm in education industry. Evaluation the effectiveness of e-learning is gaining momentum using assessment models. Issues of general concerns are explored in evaluation models, case studies in this paper.

I. INTRODUCTION

E-learning is fast becoming a major learning and skills delivery method within larger companies as a staff development tool. Survey shows that among American colleges and universities in 2002, 11% of students took an online course, 97 % of public institutions offered at least one online or blended course, 49% offered an online degree program, and 67% considering e-learning a critical long-term strategy for their institution [1]. The questions about e-learning have become "how", "why" and "with what outcomes"[2].

E-learning can enhance the competency in new skills and aid knowledge management – thereby boosting productivity, innovation and the spread of best practice. And while the range of courses and materials generally available has been primarily limited to generic and 'soft' skills, there is a body of more product specific e-learning development which can be drawn upon. This raises the question of when e-learning will become an established and integrated part of the total educational process, rather than a fashionable accessory to work-related training in well-resourced businesses.

However there are significant challenges to the successful inculcation of the practice and process of e-learning into the fabric of the educational system. These challenges manifest themselves in many ways – cultural, organizational, financial and curricular.

II. E-LEARNING EVALUATION MODELS

With the daily increase of online course offerings, most universities and corporate training facilities now offer some or all of their courses online. Studies [2] shows, that more than 1,000 corporate universities and online providers offer courses in everything from information technology to cuisine recipe. Although it is clearly advantageous for asynchronous learners to access educational information and content anywhere and anytime, it is difficult to evaluate the quality and effectiveness of online courses and learning modules.

A. Interactive learning Model

As open source learning platforms and public access to online course content are gaining momentum, educational institutes can benefit from joint development efforts and shared resources, resulting in lower cost of online learning. Consortia are sharing volumes of information and courseware based on current technologies. In approach to develop a common, objective scale and summative instrument with which to measure the pedagogical effectiveness of online course offerings, Sonwalker uses the five functional learning styles (as figure 1) - apprenticeship, incidental, inductive, deductive, and discovery (x-axis); the six media elements - text, graphics, audio, video, animation, and simulation (y-axis); and the third axis of the cube (the z-axis), which represents the interactive aspects of learning.

Figure 1: The learning cube [5]
Learning styles: L1 = apprenticeship; L2 = incidental;

K. Elleithy et al. (eds.), Advances in Computer, Information, and Systems Sciences, and Engineering, 473–480.

L3 = inductive; L4 = deductive; L5 = discovery

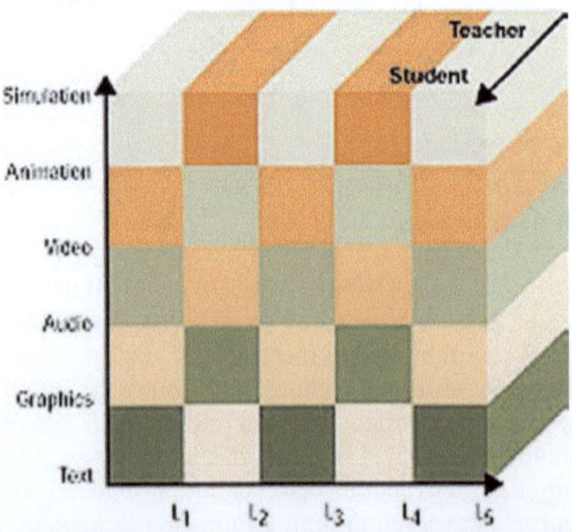

The z-axis indicates the degree to which students are engaged with the learning content, moving from a teacher-oriented to a student-oriented approach. This interactivity axis (z-direction) of the cube can be defined in terms of five elements: system feedback, adaptive remediation and revision, e-mail exchange, discussion groups, and bulletin boards. With this definition of the learning cube, a framework can be constructed to define pedagogy as a 3D space. Pedagogical effectiveness is at the heart of online offerings and defines critical parameters for the evaluation of courses. However, learning management systems provide the essential integrative layer for online courses. For online courses to be delivered in the context of learning management systems, we need to look at several additional factors in any evaluation.

The Pedagogy Effectiveness Index (PEI): [5] The pedagogical effectiveness of an online course can be defined as a summation of learning styles, media elements, and interactivity - equally likely and mutually exclusive, a probability distribution tree diagram (see Figure 4) can be shown to have three branches, with sub-branches represented for each axis of the pedagogical learning cube. A PEI can therefore be determined by a summative rule (see Figure 5). The corresponding probability multipliers can be shown in a simple matrix (see Figure 6).

Figure 4: The probability tree diagram for the pedagogical learning cube

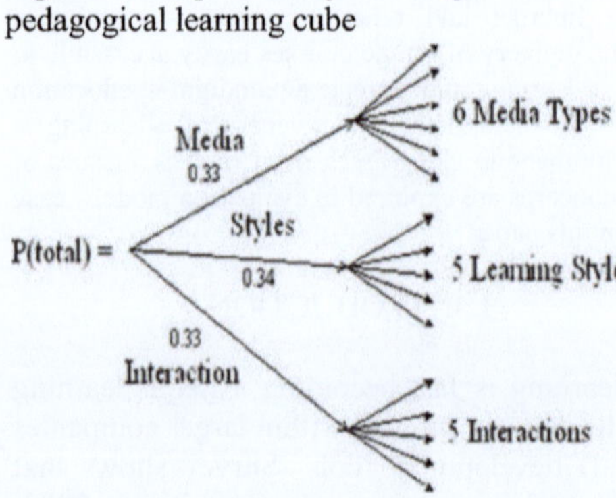

Figure 5: The pedagogy effectiveness index expressed as a summative rule

$$PEI = \sum S_i{}^*p_i + \sum M_j{}^*p_j + \sum I_k{}^*p_k$$

Where S = Style, M = Media, and I = Interaction; the subscripts define the elements' ranges: i = 1 to 5, j = 1 to 6, and k = 1 to 5; and $\sum$ represents summation.

Figure 6: Simple probability distribution matrix

Style	Pi	Media	Pj	Interaction	Pk
Apprenticeship	0.068	Text	0.055	Feedback	0.066
Incidental	0.068	Graphics	0.055	Revision	0.066
Inductive	0.068	Audio	0.055	E-mail	0.066
Deductive	0.068	Video	0.055	Discussion	0.066
Discovery	0.068	Animation	0.055	Bulletin	0.066
		Simulation	0.055		
Total (weighted)	0.34		0.33		0.33

The following are instances of PEI applications for one course offering. [5]

Case 1 - one learning style, one media element, and one interactive element: PEI = 0.068 + 0.055 + 0.066 = 0.189

Case 2 - three learning styles, four media elements, and two interactive elements: PEI = 3*0.068 + 4*0.055 +2*0.066 = 0.556

Case 3 - five learning styles, six media elements, and five interactive elements: PEI = 5*0.068 + 6*0.055 +5*0.066 = 1.0

The above scenario clearly illustrates that the PEI varies from 0 to 1. [5]The probability of the pedagogical effectiveness increases as cognitive opportunity increases with the inclusion of more learning styles, media elements, and interaction. Notice that PEI is based on a simple probability distribution and should be considered an approximate indicator within the bounds of assumptions listed above, specifically relating to the flexible learning approach depicted by the pedagogical learning cube.

Summative rating for online courses: the PEI serves as an indicator of the pedagogical richness of a course. However, successful online course delivery systems are to include content factors; learning factors; delivery support factors; usability factors, and technological factors, with reference to the learning technology standards proposed by IMS, AICC, and SCORM. [5]

Combining PEI with the summative evaluation instrument can be employed as powerful tools (e.g. overall rating = PEI x summative rating score) to evaluate large numbers of online offerings since these criteria often have a clear focus on pedagogically driven design. Use of these tools could guide and motivate online education developers, universities, and training centers toward the successful creation of educational systems.

B. Content Model in the Intelligent Learning Engine [6]

Content delivery is the key of e-learning systems. In his Intelligent Information Delivery System [6], Quinn indicates that the content model, together with other models provide the information to a central learning engine that uses the current information about the situation, and the information from these models, to pull the appropriate content from a content repository to deliver to the learner (see Figure 2).

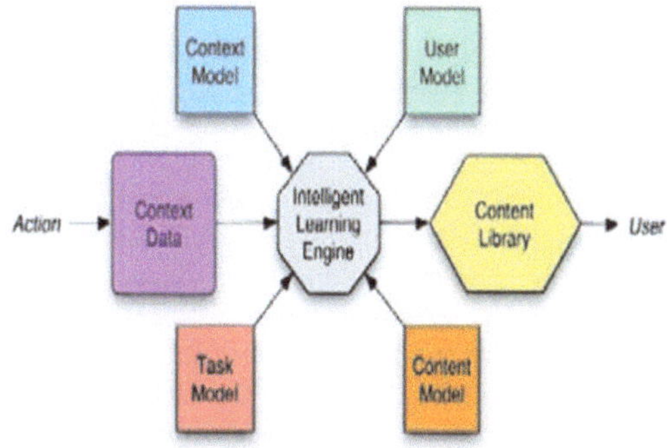

Figure 2. An Intelligent Information Delivery System

The engine uses the models to decide what content would make sense to deliver in this context, and specifies the content to be made available.

Three major categories of information constitute the content model: the different components of information; the metadata used to tag the information with to identify it; the standards that the content conforms to (see Figure 3 [6]).

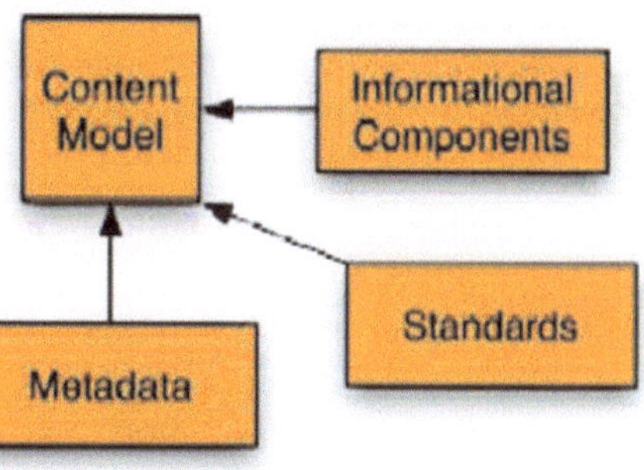

Figure 3. Content Model Components

The Informational Components describe the availability of content types, in terms of their semantic roles. Sets of repair procedures for example, would be of different informational types than customer sales objections job aids. Information concerning the consumers identity and their needs can be employed to build a content model - a structured template detailing what and how to write information that can be transformed through eXtensible Markup Language (XML), and style-sheets into the specific content needed.

Specifications of the standards are important - the Standard Courseware Object Reference Model (SCORM) terms used for learning objects or other standard document formats such as PDF or Flash for instance.

Ideally, [6] the content should be in small granularity, and aggregated into larger chunks but accessible at the smallest level.

C. Content Quality Measures

In order to guarantee better results in e-learning programs, [7] it is necessary to look at content quality measures - the quality of the online education product itself. The National Education Association and Blackboard Inc. examined case studies of six higher education institutions that provide Internet-based degree programs, to ascertain the degree to which various measures of quality identified in previous studies were actually being incorporated into the policies, procedures, and practices of institutions that have distance education learners. A list of twenty-four benchmarks essential to ensuring quality in Internet-based education were grouped under the categories of institutional support, course development, teaching/learning, course structure, student support, faculty support, and evaluation and assessment (The Institute for Higher Education Policy, 2000).

NCREL's (North Central Regional Educational Laboratory) framework builds upon a framework developed by Barbara Means of SRI International. Means identified seven variables that, when present in the classroom, indicate that effective teaching and learning are occurring. These classroom variables are:

- children are engaged in authentic and multidisciplinary tasks
- assessments are based on students' performance of real tasks
- students participate in interactive modes of instruction
- students work collaboratively
- students are grouped heterogeneously
- the teacher is a facilitator in learning
- students learn through exploration

NCREL reorganized them into a set of eight categories of learning and instruction: vision of learning, tasks, assessment, instruction, learning context, grouping, teacher roles, and student roles – all expanded the definitions to include 26 variables or 26 indicators of engaged learning, summarized in Table 1. [7]

Table 1: Indictors of Engaged Learning

Variable	Indicator of Engaged Learning	Indicator Definition
Vision of Learning	Responsible for learning Strategic Energized by learning Collaborative	Learner involved in setting goals, choosing tasks; has big picture of learning and next steps in mind Learner actively develops repertoire of thinking/learning strategies Learner is not dependent on rewards from others; has a passion for learning Learner develops new ideas and understanding in conversations and work with others
Tasks	Authentic Challenging Multidisciplinary	Pertains to real world, may be addressed to personal interest Difficult enough to be interesting but not totally frustrating, usually sustained Involves integrating disciplines to solve problems and address issues
Assessment	Performance-based Generative Seamless and ongoing Equitable	Involving a performance or demonstration, usually for a real audience and useful purpose Assessments having meaning for learner; maybe produce information, product, service

		Assessment is part of instruction and vice versa; students learn during assessment Assessment is culture fair
Instructional Model	Interactive Generative	Teacher or technology program responsive to student needs, requests (e.g., menu driven) Instruction oriented to constructing meaning; providing meaningful activities/experiences
Learning Context	Collaborative Knowledge-building Empathetic	Instruction conceptualizes students as part of learning community; activities are collaborative Learning experiences set up to bring multiple perspectives to solve problems such that each perspective contributes to shared understanding for all; goes beyond brainstorming Learning environment and experiences set up for valuing diversity, multiple perspectives, strengths
Grouping	Heterogeneous Equitable Flexible	Small groups with persons from different ability levels and backgrounds Small groups organized so that over time all students have challenging learning tasks/experiences Different groups organized for different instructional purposes so each person is a member of different groups; works with different people
Teacher Roles	Facilitator Guide Co-learner/co-investigator	Engages in negotiation, stimulates and monitors discussion and project work but does not control Helps students to construct their own meaning by modeling, mediating, explaining when needed, redirecting focus, providing options Teacher considers self as learner; willing to take risks to explore areas outside his or her expertise; collaborates with other teachers and practicing professionals

Student Roles	Explorer Cognitive Apprentice Teacher Producer	Students have opportunities to explore new ideas/tools; push the envelope in ideas and research Learning is situated in relationship with mentor who coaches students to develop ideas and skills that simulate the role of practicing professionals (i.e., engage in real research) Students encouraged to teach others in formal and informal contexts Students develop products of real use to themselves and others

III. CASE STUDIES: SHOESTRING, CDC AND USLEARNING

A. Shoestring Case Analysis Simulation: [8]

In a case analysis simulation, (See Figure 7.) The first screen has three key elements: Simulation assignment, which includes any relevant icons and printable worksheets that guide participants' analysis and solutions (learners must complete all worksheets and present them to facilitators and peers in the next group workshop); Thumbnail photos or illustrations of people with names and titles of staff members, customers, or anyone else involved in the assignment, such as local community leaders, charity fund raisers, and auditors; Menu of knowledge management resources, including corporate policies, recommended procedures and strategies, and guidelines for solving typical challenges.

Developed with Ronin Enterprises as a training program for new managers at a major brokerage firm, the program was blended print self-study resources, an online simulation linked to knowledge management tools, a group workshop, and on-the-job action learning assignments. As a result of each simulation activity, participants improved their ability to: apply concepts and skills to typical job challenges; use online resources in the knowledge management area as guidelines to help them analyze and respond to these challenges; prepare a case analysis for use in group workshop debriefing sessions in which they will share insights with other new managers, receive feedback from facilitators and experienced managers, and improve their ability to resolve the challenge in the simulation; plan activities to use what they learned in the simulation and related workshop sessions in real-life challenges.

The design is cost-effective since the entire simulation is built in basic html. There's no need for costly software, animation, video, or plug-ins that conflict with existing IT systems. [8] Programming costs are relatively low and the thumbnail photos can come from inexpensive stock photo vendors or be created in-house with a digital camera and employees willing to moonlight as actors. The structure of the simulation is also cost effective since a simple linear sequence of a stimulus (the customers comment, question, or objection) with multiple-choice responses. General feedback on each response transitions to the next stimulus. To control costs, the conversation is text only but audio or even video can add to the overall learner experience for additional budget.

Figure 7
This diagram shows the layout of the main screen of the simulation. The top row shows the knowledge management area. The case assignment is in the center. A thumbnail photo of each person involved in the case assignment is shown with name and title.

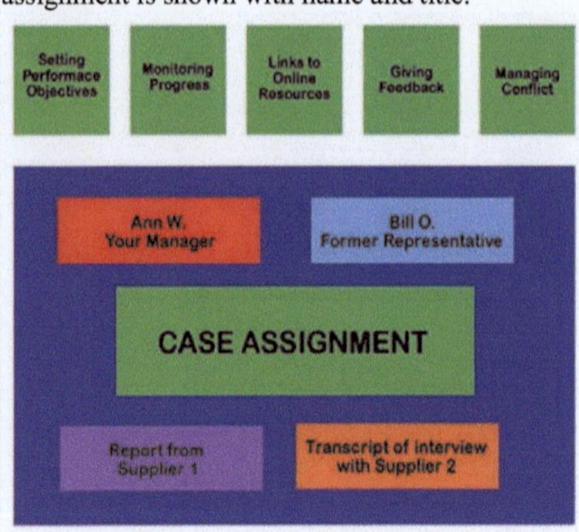

B. Scenario-based e-learning model (SEM) at CDC: [9]

Developers at Centers for Disease Control and Prevention (CDC) used a framework proposed by Clark Aldrich, applied a model that blends characteristics of a simulation with linear e-learning programs. Using a real-world scenario to engage learners, CDC found that developing e-learning programs based on this model required fewer resources than a typical simulation, yet gave the learners a feel as though they were working through a simulation. CDC refers to this model as the scenario-based e-learning model (SEM).

The development process started with identifying a real-life outbreak investigation that would serve as a solid traditional classroom example. The simulation was designed to enable each learner to work through the case study at his or her own pace without the help of an instructor. The outbreak scenario used in the online simulation was more detailed than that provided in the classroom case study--building in characters, places, and specific timelines. Question formats included multiple-choice, yes/no, fill-in-the-blank, and drag-and-drop activities. Learners could access a variety of support tools, such as hints and reference materials to answer questions.

Using the notebook style to present the background information and questions offered the users user-friendly and easy access. [9] To support the notebook metaphor, other interface elements, which aren't included in classroom case studies, were placed on the desktop or clipped to the notebook, including snapshots that depict investigation team activities an epidemic curve that graphically illustrates the outbreak and investigation, which also changes as the scenario progresses an investigation outline that corresponds to the six steps of the outbreak investigation and contains a record of the learners progress related items such as press releases and questionnaires.

To explore more systematically why the linear e-learning approach had the feel of a simulation, CDC examined the program against a framework proposed by Clark Aldrich in Learning Circuits Field Guide to Educational Simulation. The simulation consists of three components: [9] how learners express themselves through input; simulation calculations and branches based on learner input; results and feedback as output that are communicated to learners.

The CDC course expresses itself through input: The questions posed to the learner mirror those one would wrestle with in an actual investigation. This makes the learner feel that his or her response to the questions play into the action of the story and have an effect on the end result; the interface includes tools used when investigating a real-life outbreak. Data from the original investigation is available for analysis. The learners feel at times as though they have the ability to directly manipulate input.

Simulation calculations and branches based on input from learners demonstrated a clear learners' control of available options such as remedial lessons, advanced explorations, data analysis activities, and references. Results and feedback as outputs [9] communicated to learners include customized feedback spontaneously and personally, and snapshots of maps, lists and questionnaires through engaging visual drawings.

C. OPM's USALearning system:

OPM [3] provides training services for various federal agencies. Tremendous cost savings, benefits of convenience and availability save NOAA (National Oceanic and Atmospheric Administration) over 90 percent cost using online solution. Between July 23, 2002, and June 30, 2005, USALearning has registered 261,617 users and 221,491 courses completed. The Office of Personnel Management's e-learning initiative and all three of contracting vehicles – the National Technical Information Services, GoLearn and FasTrac have 1.3 million registered users with 955,000 courses completed. [Source: *Office Personnel Management*]

OPM' USALearning Web portal puts professional development in the hands of federal employees, allows them to take classes at their convenience. The Bush administration expects the program to encourage government-wide adoption of e-learning. For nearly 1 million federal workers have taken e-learning training courses in the past five years, USAlearning's enrollment accounts for only about 15 percent of the federal workforce [3].

The assessment report indicates that students' learning curve, teachers' qualification and productivity outcome are critical factors affecting the success.

Successful learning comes from self-motivation. And often times, a teacher is needed to motivate the learning. Accustomed to classroom environment, the students do not involve the spontaneity necessary for deep learning. However, study indicates that many students find online interaction with instructors stimulating. [4] "Interaction between teachers and students becomes more intense," said, Bill Rust from Gartner, "some students that are too shy to ask a question in a classroom tend to ask more questions online."

E-learning effective depends on the content. How effective the learning will be still depends on the teacher. It is important for the teacher to get training. An instructor who does not understand how e-learning works, sometimes pay more attention to the form than the content. That is a mistake according to experts. "It is all about content," says Jack Kramer at Pathlore Software. "If you don't serve that content up and make it really easy and make people want to come and learn, it's going to fail."[4]

Productivity comes at incentives. Good policies and procedures for earning credits towards promotion upon completing e-learning courses should definitely be in place to insure the productivity outcome.

IV. CONCLUSION

Hall and LeCavalier [10] summarized some firms' economic savings as a result of converting their traditional training delivery methods to e-learning. IBM saved US $200 million in 1999, providing five times the learning at one-third the cost of their previous methods. Using a blend of Web-based (80%) and classroom (20%) instruction, Ernst & Young reduced training costs by 35 percent while improving consistency and scalability. Rockwell Collins reduced training expenditures by 40 percent with only a 25 percent conversion rate to Web-based training. Many other success stories exist. It is important to note that things are not always bright when we look at some firms with a large expenditure on new e-learning efforts have not received the desired economic advantages.

While generally positive economic benefits are so obvious, other advantages such as convenience, standardized delivery, self-paced learning, and variety of available content, have made e-learning a high priority for many corporations. A survey of 500 training directors (*Online Learning News*, [11]) clearly shows the new priorities:

- Sixty percent had an e-learning initiative.
- Eight-six percent had a priority of converting current instructor-led sessions to e-learning.
- Eighty percent will set up or expand knowledge-management programs.
- Seventy-eight percent were developing or enhancing electronic performance support.

ASTD [12], in its *State of the Industry Report*, noted that the year 2000 marked a new era of growth for e-learning. The events of September 11, 2001, have only accelerated this growth as organizations cut back on business travel, improve their security, and increase their e-learning efforts.

V. REFERENCES

[1] Allen, I. E. & Seaman, J. (2003) Sizing the opportunity: The quality and extent of online education in the United States, 2002 and 2003. Needham, MA, The Sloan Consortium

[2] Hitt, J. & Hartman, J. (2002) Distributed learning: New challenges and opportunities for institutional leadership. Washington, American Council on Education.

[3] USALearing: Learn something new – Primer on federal government's e-learning portal, by Megan Lisagor, August 29, 2005, Federal Computer Week.

[4] A e-learning Progress Report, by Judi Hasson, August 29, 2005, Federal Computer Week.

[5] Nishikant Sonwalker: A New Methodology for Evaluation: The Pedagogical Rating of Online Courses (2001) campus Technology from Syllabus Media Group.

[6] Clark N. Quinn: Delivering the Dream – Models for Intelligent Assistance. Learning Circuits, www.astd.org

[7] New Times Demand New Ways of Learning. http://www.ncrel.org/sdrs/edtalk/newtimes.htm

[8] Tita Theodora Beal: Simulations on a Shoestring. ASTD's source for e-learning. www.astd.org

[9] Gathany & Stehr-Green: Scenario-Based E-Learning Model: A CDC Case Study. Learning Circuits, www.astd.org

[10] ASTD. Evaluating the Effectiveness and the Return on Investment of E-learning. What Works Online. 2nd quarter. [Online] Available at: www.astd.org/virtual_community/research

[11] ASTD. E-learning evaluation method gains support in Canada, *ASTD Learning Circuits*, July 2000. [Online] Available at: www.learningcircuits.org

[12] ASTD. (2002). *State of the Industry Report*. [Online] Available at: www.astd.org

Tools for Student Engagement that Facilitate Development of Communication Skills

E. Joseph Derrick
Radford University, Department of Information Technology,
Radford, VA 24142 USA
ejderrick@radford.edu

***Abstract* – Several key elements of Radford University's Information Technology (IT) Senior Seminar course syllabus include (1) training and practice in the research and writing of a technical paper, (2) the delivery of a verbal presentation covering the research topic, and (3) the study of ethics in the IT profession. Traditionally, students are difficult to engage and involve in the seminar classroom since much of their time is spent listening to peers and others delivering verbal presentations. This paper reports on several tools, approaches, and technologies which we have recently tested, adopted and used to improve student engagement during the presentation periods. These provide additional opportunities for speaking and building confidence in communications skills, and aid assessment of the key elements.**

Index Terms – Course assessment, student engagement, verbal communications, and written communications.

BACKGROUND AND MOTIVATION

The Accreditation Board for Engineering and Technology (ABET), the primary accreditor for college and university programs in applied science, computing, engineering, and technology, has recognized the importance of communication skills and professional ethics by mandating their inclusion in the various degree curricula. For example, Reference [1] states in sections IV-15 and 16 that the oral and written communications skills of the student "must be developed and applied in the program." These requirements are mirrored in all program curricula under ABET, further underscoring their importance. [2] These criteria provide necessaray support to the current state of the professional workplace in which (1) employers are demanding that graduates have excellent communication skills and where (2) these skills are essential to career success. [3, 4, 5]

In addition, within the last several years, student engagement has become a pervasive goal throughout higher education. The National Survey of Student Engagement (NSSE) has raised this concept's level of prominence and, as a result, has helped bring focus and insight to the assessment of student learning and academic quality and in the measuring of the effectiveness of colleges and universities. Since it's inception in 1999, the survey has grown with 850 different colleges and universities having participated over the last 5 years. [6] According to the NSSE Institute, "the student engagement concept is very straightforward and almost self-evident; the more students participate in or perform an educationally purposeful activity, the more they generally gain." [7]

Having highlighted the importance of developing these critical skills in graduates and, additionally, the need to engage students more thoroughly in the classroom, significant roadblocks have historically existed for successfully accomplishing these areas. Some of the more common problem areas and questions include [8, 9]:

- Instead of having a dedicated course to teach communications skills, some departments utilize university-wide strategies in which the material is taught in other departments (e.g., in the English or Communications departments. The confronting question here concerns who is most qualified to teach these skills? There are pros and cons no matter how this is viewed.
- Often, instructors find that students nearing graduation are not well-equipped to speak or write within their professional field, yet there is little room within course curricula to devote the time and resources to provide the additional training and practical experience that are necessary to solve these deficiencies.
- If courses which are specifically directed at instruction in communications skills are inadequate in coverage or non-existant, then how can we incorporate speaking and writing within existing technical courses and yet have time to cover the required technical material?
- Even when there is a dedicated course within the department for teaching oral and communication skills, strategies for developing these skills in the classroom suffer from having to use significant amounts of time for the delivery and evaluation of student verbal presentations. This becomes a more difficult issue to solve when class sizes increase, requiring more time for individual presentations and leaving less time for student presentations and other opportunities for student involvement.

These issues continue to challenge us. In our context, we have a semester-long Senior Seminar which is only a one-hour credit course, but which, in part, includes the key elements of (1) training and practice in the research and writing of a technical paper, (2) the delivery of a verbal presentation covering the research topic, and (3) the study of ethics in the IT profession. In seeking an effective solution to give our students more practical experience in their communication

K. Elleithy et al. (eds.), Advances in Computer, Information, and Systems Sciences, and Engineering, 481–483.

skills through improved student engagement, we use a unique combination of tools, approaches, and technologies during the course over a 6 week period that deal with the verbal presentations. This is a difficult task in consideration of the difficulties listed above and the tight constraints of the one-hour format. This paper provides an overview of these techniques and is organized by sections as follows. Course Evolution briefly recaps the hows and whys of course transition to its present form. Course Description discusses the current status and provides details of the tools and the engagement approach (modeled after the training methods of Toastmasters International [10]) that are used. Finally, the Conclusions and Future Plans section introduces (1) the gains and perceived benefits as a result of these changes and (2) the new technologies in use that are planned to further enhance student engagement, student learning, and class management.

COURSE EVOLUTION

We initially lectured on all the ethics topics over a series of weeks and gave an ethics test for assessment. The students did not interact and dialogue with each other when using this approach. Their written papers were due on the presentation date; the presentation was, in reality, the primary and only graded part for assessment of oral skills. The presentations were 15 minutes long, and in order to allow everyone an opportunity to speak, we could only cover only about three students per class. Completing all presentations could take from 5 to 7 weeks, or longer, depending on class size. This is a sizable amount of time. There was no time for other kinds of discussion, and students got bored with nothing to do but listen to their peers during this long stretch.

We added a technical writing text to the required course texts, raised the percentage value worth of the written paper, created a participation component, and improved the presentation component. We hoped to raise the bar and stimulate incentive by placing more emphasis on the oral and writing skills. We also tightened our attendance policy requirements. As far as writing skills, during mid-semester, we began to use one class to cover old student written work from past semesters. We discussed the errors in grammar and style and how to improve these. Furthermore, we spent two additional classes on how to prepare presentations using both lecture and video [11]. The new strategy for teaching oral skills is detailed in the next section.

COURSE DESCRIPTION

During presentation classes, we now more thoroughly engage the students by adopting an approach modeled after Toastmaster International strategies [10]. In addition to performing their primary presentation/report in accordance with a pre-determined schedule, the students volunteer for various roles on a rotating basis, in which they:

- perform and report peer evaluations of fellow student presentations,
- participate in short impromptu discussions over the ethics text topics (called Table Topics in Toastmasters terminology),
- count Ah's or other phrases that are useless fillers,
- record speaking time durations (for Table Topics, Presentations, and Evaluations), report violators, and
- collect ballots on votes for the best Table Topics, Presentations, and Evaluations.

During each 50 minute class, we spend roughly 10 minutes on Table Topics, 20 minutes on Presentations, 10 minutes on Evaluations, and 10 minutes for wrap-up. From previous semesters, we have shortened the presentations to provide more time for the other oral components. The time for each presentation however is dependent upon the number of students and time available. To determine a participation grade, the instructor assigns point values for presenters, Table Topics, roles performed, and winners of the voting. Timing violators and the student with the Most Ahs receive point deductions. In order to ensure that the class flows smoothly and that all elements are covered, the instructor prepares a planning requirements and records sheet before each class/session. Points records are then kept by section using a score sheet. Ballots by section are also tabulated with the voting results. Student evaluators use a Toastmasters evaluation guide [12] to record observations. We require a positive and constructive viewpoint by peers in performing the evaluations. The instructor determines the participation grade by a formula that includes factors for both attendance and participation point totals. The instructor grades and evaluates the oral presentations based on a typical evaluation checklist that covers criteria for the introduction, format and organization, delivery, conclusion and analysis, and length.

CONCLUSIONS AND FUTURE PLANS

Although the above approach requires a lot of detailed record keeping, student evaluations in these courses have improved substantially. Students are more confident in their presentation skills, feel prepared for the workplace, and enjoyed the interactivity in the classroom. Oral presentations are completed in roughly the same time span as before (5-7 weeks), but now the classes have added new opportunities for participation and speaking. Based on the student comments and feedback, we believe the overall effect has been a good one. Students are more engaged and are provided more chances to improve oral communications skills during a presentations class session. As a side benefit, the ethics text topics are covered and incorporated into Table Topics. Additionally, the oral presentations are long enough to make a worthwhile evaluation of student technical presentation skills

To reduce the amount of record keeping, we plan to incorporate the CPS system [13] in the coming semester which will further engage students yet provide a streamlined method for recording and storing class participation data within a class gradebook. In order to provide more meaningful feedback to students on their presentation and delivery, we plan to use Macromedia Breeze Live [14] to create a digital video recording of their presentation for later viewing and comparision with instructor comments. This will reinforce and strengthen the learning process.

ACKNOWLEDGMENT

We thank Toastmasters International [10] for allowing us to use their materials as part of a pilot program.

REFERENCES

[1] ABET Computing Accreditation Commission, "Criteria for accrediting computing programs: effective for evaluations during the 2005-2006 accreditation cycle," ABET, Inc., Baltimore, MD, November, 2004.

[2] ABET Accreditation Criteria, http://www.abet.org/forms.shtml, ABET, Inc. Baltimore MD, accessed October 7, 2005.

[3] K. Alshare and N. Hindi, "The importance of presentation skills in the classroom: students and instructors perspectives," *Journal of Computing Sciences in Colleges*, Vol. 19, No. 4, pp. 6-15, April 2004.

[4] D. Martin, "Evaluating Oral Presentations," *Journal of Computing Sciences in Colleges*, Vol. 20, No. 3, pp. 48-54, February 2005.

[5] P. Gruba and R Al-Mahmood, "Strategies for Communications Skills Development," *Proc. 6th Conf. Australian Comp. Educ. Vol 30*, 2004, ACM Conference Proceedings Series, Vol. 57, pp. 101-107.

[6] National Survey of Student Engagement, http://www.indiana.edu/~nsse/index.htm, accessed October 7, 2005.

[7] National Survey of Student Engagement, http://www.indiana.edu/~nsse/html/info_video.htm, accessed October 7, 2005.

[8] M. Michael, "Fostering and assessing communications skills in the computer science context," *Proc. of 31st SIGCSE*, 2000, pp. 119-123.

[9] L. Pollock, "Integrating an intensive experience with communications skills development into a computer science course," *Proc. of 32nd SIGCSE*, 2001, pp. 287-291.

[10] Toastmasters International, http://www.toastmasters.org, accessed October 7, 2005.

[11] Toastmasters International, "Be prepared to speak," video, Kantola Publications.

[12] Toastmasters International, "The Ice Breaker," http://www.toastmasters.org/pdfs/IceBreaker.pdf.

[13] eInstruction classroom response system (CPS), http://www.einstruction.com, accessed October 7, 2005.

[14] Macromedia Breeze Live, Macromedia, http://www.macromedia.com.

Index

MIX
Papier aus verantwortungsvollen Quellen
Paper from responsible sources
FSC® C105338

If you have any concerns about our products,
you can contact us on
ProductSafety@springernature.com

In case Publisher is established outside the EU,
the EU authorized representative is:
Springer Nature Customer Service Center GmbH
Europaplatz 3, 69115 Heidelberg, Germany

Printed by Libri Plureos GmbH
in Hamburg, Germany